PHOTOMORPHOGENESIS IN PLANTS
AND BACTERIA
3RD EDITION

Photomorphogenesis in Plants and Bacteria
3rd Edition

Function and Signal Transduction Mechanisms

Edited by

EBERHARD SCHÄFER
Albert-Ludwigs-Universität Freiburg, Germany

and

FERENC NAGY
Institute of Plant Biology, Szeged, Hungary

 Springer

A C.I.P. Catalogue record for this book is available from the Library of Congress.

ISBN-10 1-4020-3809-7 (HB)
ISBN-13 978-1-4020-3809-9 (HB)
ISBN-10 1-4020-3811-9 (e-book)
ISBN-13 978-1-4020-3811-2 (e-book)

Published by Springer,
P.O. Box 17, 3300 AA Dordrecht, The Netherlands.

www.springer.com

Printed on acid-free paper

All Rights Reserved
© 2006 Springer
No part of this work may be reproduced, stored in a retrieval system, or transmitted
in any form or by any means, electronic, mechanical, photocopying, microfilming, recording
or otherwise, without written permission from the Publisher, with the exception
of any material supplied specifically for the purpose of being entered
and executed on a computer system, for exclusive use by the purchaser of the work.

Printed in the Netherlands.

This book is dedicated to

Hans Mohr,

a founding member of the AESOP
(Annual European Symposium of Photormorphogenesis),

on the occasion of his 75th anniversary (May 11th 2005).

PREFACE

Plants as sessile organisms have evolved fascinating capacities to adapt to changes in their natural environment. Arguably, light is by far the most important and variable environmental factor. The quality, quantity, direction and duration of light is monitored by a series of photoreceptors covering spectral information from UVB to near infrared. The response of the plants to light is called photomorphogenesis and it is regulated by the concerted action of photoreceptors.

The combined techniques of action spectroscopy and biochemistry allowed one of the important photoreceptors – phytochrome – to be identified in the middle of the last century. An enormous number of physiological studies published in the last century describe the properties of phytochrome and its function and also the physiology of blue and UV-B photoreceptors, unidentified at the time.

This knowledge was summarized in the advanced textbook "Photomorphogenesis in Plants" (Kendrick and Kronenberg, eds., 1986, 1994).

With the advent of molecular biology, genetics and new molecular, cellular techniques, our knowledge in the field of photomorphogenesis has dramatically increased over the last 15 years.

In 2002 the publisher approached us with a suggestion to start a new edition of this advanced textbook. After several discussions we came to the conclusion that a new edition containing only the novel observations would no longer be useful as a textbook. Clearly, all the new molecular information has not erased the validity of the "old" physiological and biochemical data. Even more importantly, it is most unfortunate that in the new generation of researchers the knowledge of the "old" data starts to get lost. Consequently, ample evidence can be found in the literature for over or underinterpretation of results obtained by applying state of art methodologies which can be traced back to lack of in-depth knowledge of classical physiological data.

Therefore, in agreement with the publisher we decided to edit a new textbook focusing on the novel observations and at the same time suggesting the 2nd edition of Photomorphogenesis in Plants (Kendrick and Kronenberg, eds.) to be still available for the interested and motivated reader.

In this new textbook the basis of the physiology and molecular biology of photomorphogenesis is once again summarized in a few intorductory chapters, to support the reading of the new chapters. Nevertheless, reading the 2nd edition is strongly recommended.

The world's leading experts from Europe, Japan, South America and the USA were invited to contribute to this advanced textbook and we are very pleased that almost all of them immediately accepted our invitation.

Despite enormous advances the primary molecular function of photoreceptors is still not known and the UV-B photoreceptor still remains to be identified. Nevertheless, this book attempts to guide the reader through the approaches made with the aim of elucidating how absorption of light by the photoreceptors will be converted into a biochemical signal which then triggers molecular events at cellular level leading to characteristic physiological responses underlying photomorphogenesis of the plant .

Molecular biology, transgenic work, genetics, biochemistry and cell biology techniques have dramatically increased our knowledge in the field of photomorphogenesis. We hope that students, postdocs and academic teachers, like in the past, will again favourably respond to the fascination of photomorphogenesis research and that reading the book in the post-genomic era will stimulate new creative research in this field.

Last but not least we would like to thank the publisher, especially Jacco Flipsen, for his strong support and interest, Prof. Govindjee for invitation and encouragement for this project and Dr. Erzsebet Fejes and Birgit Eiter for excellent assistance in editing.

REFERENCES

Kendrick, R. E. and Kronenberg, G. H. M., *Photomorphogenesis in Plants,* Dordrecht: Martinus Nijhoff Publishers, 1986 (ISBN 90-247-3317-0).

Kendrick, R. E. and Kronenberg, G. H. M., *Photomorphogenesis in Plants, 2nd edition*, Dordrecht: Kluwer Academic Publishers, 1994 (ISBN 0-7923-2551-6).

E. Schäfer[1] and F. Nagy[2]

[1]University of Freiburg
Institute of Biology II/ Botany
Schänzlestr. 1
D-79104 Freiburg
Germany
Eberhard.Schaefer@biologie.uni-freiburg.de

[2]Biological Research Center
Institute of Plant Biology
P. O. Box 521
H-6701 Szeged
Hungary
nagyf@nucleus.szbk.u.szeged.hu

CONTENTS

Preface... VII

Abbreviations.. XXVII

Color plates.. XXXI

PART 1: GENERAL INTRODUCTION AND HISTORICAL OVERVIEW OF PHOTOMORPHOGENESIS

Chapter 1

HISTORICAL OVERVIEW
Eberhard Schäfer and Ferenc Nagy

1.	Introduction..	1
2.	Phytochrome Induction Responses..............................	2
3.	The "High Irradiance Responses"...............................	8
4.	Very Low Fluence Responses.....................................	10
5.	Further reading..	10
6.	References...	10

Chapter 2

PHYSIOLOGICAL BASIS OF PHOTOMORPHOGENESIS
Eberhard Schäfer and Ferenc Nagy

1.	Introduction..	13
2.	Classical action spectroscopy.....................................	13
3.	Mode of function of phytochrome	16
4.	Correlations between in vivo spectroscopical measurements and physiological responses..	18
5.	Phytochrome response types......................................	20
6.	Summary..	21
8.	References..	22

Chapter 3

HISTORICAL OVERVIEW OF MOLECULAR BIOLOGY AND GENETICS IN PHOTOMORPHOGENESIS
Eberhard Schäfer and Ferenc Nagy

	References..	30

Chapter 4

GENETIC BASIS AND MOLECULAR MECHANISMS OF SIGNAL TRANSDUCTION FOR PHOTOMORPHOGENESIS

Eberhard Schäfer and Ferenc Nagy

1. Introduction……………………………………………………….…..	33
2. Phototropism mutants…………………………………………….….	34
3. Photomorphogenic mutants……………………………………….….	34
4. Circadian mutants…………………………………………………....	35
5. Genetic variation, mutants identified by QTL mapping……………...	35
6. Signal transduction mutants……………………………………….…	36
7. Signal transduction at the molecular level…………………………...	37
8. Summary…………………………………………………………….	38
9. References…………………………………………………….……..	39

PART 2: THE PHYTOCHROME

Chapter 5

THE PHYTOCHROME CHROMOPHORE
Seth J. Davis

1. Introduction……………………………………………………….…	41
2. Structure of the phytochrome chromophore………………………....	44
3. Phytochromobilin synthesis…………………………………………	47
3.1 Heme Oxygenases……………………………………….…	50
3.2 Phytochromobilin Synthase…………………………………	53
4. Holo assembly………………………………………………………	55
5. Biophysics of the chromophore…………………………………….	58
6. Personal Perspectives………………………………………………..	59
6.1 Phy chromophore structure…………………………….……	59
6.2 Phy chromophore synthesis…………………………………	59
6.3 Holo-phy assembly and structure……………………………	59
7. References…………………………………………………………....	60

Chapter 6

STRUCTURE, FUNCTION, AND EVOLUTION OF MICROBIAL PHYTOCHROMES
Baruch Karniol and Richard D. Vierstra

1.	Introduction……………………………………………...…..………….	65
2.	Higher plant phys…………………………………………………..…….	66
3.	The Discovery of microbial Phys………………………………....……..	69
4.	Phylogeny of the Phy Superfamily……………………………………….	72
	4.1 Cyanobacterial phy (Cph) family………………………………….	76
	4.2 Bacteriophytochrome (BphP) family………………………..……...	76
	4.3 Fungal phy (Fph) family……………………………………………	83
	4.4 Phy-like sequences…………………………………………………	84
5.	Downstream signal transduction cascades………………………………..	85
6.	Physiological roles of microbial phys………………………………..…..	88
	6.1 Directing phototaxis……………………………………………….	88
	6.2 Enhancement of photosynthetic potential…………………..……...	89
	6.3 Photocontrol of pigmentation………………………………………	91
7.	Evolution of the phy superfamily…………………………………….…..	92
8.	Perspectives……………………………………………………………...	94
9.	References……………………………………………………………….	95

Chapter 7

PHYTOCHROME GENES IN HIGHER PLANTS: STRUCTURE, EXPRESSION, AND EVOLUTION
Robert A. Sharrock and Sarah Mathews

1.	Introduction…………………………………………………………..….	99
2.	Phytochrome gene structures and protein sequences……………………..	100
	2.1 The first phytochrome sequences……………………………..….....	100
	2.2 Phytochrome is a family of related photoreceptors encoded by multiple *PHY* genes in higher plants………………………………	101
	2.3 Phytochrome nomenclature……………………………………..…..	105
	2.4 Heterodimerization of type II phytochromes……………………….	105
3.	Expression patterns of phytochromes in plants…………………………..	106
	3.1 How important are phytochrome expression patterns?……….........	106
	3.2 Assaying phytochromes…………………………………..………...	107
	3.3 Early Expression Studies…………………………………………..	107
	3.4 Patterns of *PHY* gene expression – mRNA levels and promoter fusion experiments……………………………………………...……	108
	3.5 The levels and distributions of phytochromes in plants…………….	112

  3.6 Circadian regulation of *PHY* gene expression............................ 116
4. Evolution of the *PHY* gene family in seed plants................................. 117
  4.1 Phytochrome phylogeny in seed plants.................................... 117
  4.2 Phytochrome functional evolution in seed plants........................ 121
    4.2.1 Angiosperm phyB and Gymnosperm phyP.................. 122
    4.2.2 Angiosperm phyA and Gymnosperm phyN and phyO 124
5. Conclusions.. 125
6. References.. 126

Chapter 8

PHYTOCHROME DEGRADATION AND DARK REVERSION
Lars Hennig

1. Introduction... 131
2. Phytochrome degradation... 132
  2.1 Kinetic properties of phytochrome degradation........................ 132
  2.2 Mechanisms of phytochrome degradation............................... 136
  2.3 Physiological functions of phytochrome degradation.................. 141
3. Dark reversion... 143
  3.1 Kinetic properties of dark reversion....................................... 143
  3.2 Determinants of dark reversion.. 146
  3.3 Functional aspects of dark reversion..................................... 148
4. Concluding remarks... 149
5. Further reading.. 150
6. References... 150

Chapter 9

INTRACELLULAR LOCALIZATION OF PHYTOCHROMES
Eberhard Schäfer, Stefan Kircher and Ferenc Nagy

1. Introduction... 155
2. The classical methods... 155
  2.1 Spectroscopic methods... 155
  2.2 Cell biological methods... 156
  2.3 Immunocytochemical methods.. 156
3. The novel methods.. 157
4. Intracellular localisation of PHYB in dark and light....................... 158
5. Intracellular localisation of PHYA in dark and light....................... 160
6. Intracellular localisation of PHYC, PHYD and PHYE in dark and
  light... 161
7. Intracellular localisation of intragenic mutant phytochromes.......... 162
  7.1 Hyposensitive, loss-of-function mutants................................. 162

7.2	Hypersensitive mutants..	163
8.	Protein composition of nuclear speckles associated with phyB..	163
9.	The function of phytochromes localised in nuclei and cytosol	166
10.	Concluding remarks...	168
11.	References...	168

PART 3: BLUE-LIGHT AND UV-RECEPTORS

Chapter 10

BLUE/UV-A RECEPTORS: HISTORICAL OVERVIEW
Winslow R. Briggs

1.	Introduction...	171
2.	Early history..	172
3.	Phototropism: action spectra can be fickle...................	174
4.	The LIAC: a frustrating digression...............................	179
5.	The cryptochrome story..	180
6.	The phototropin story...	182
7.	Stomatal opening in blue light.....................................	184
8.	Chloroplast movements induced by blue light..............	186
9.	Leaf expansion...	187
10.	The rapid inhibition of growth.....................................	189
11.	Solar tracking...	191
12.	The ZTL/ADO family..	191
13.	Conclusions...	191
14	References	192

Chapter 11

CRYPTOCHROMES
Anthony R. Cashmore

1.	Introduction...	199
2.	Photolyases..	199
3.	The discovery of cryptochrome.....................................	200
	3.1 Cryptochromes of Arabidopsis.............................	201
	3.2 Cryptochromes of algae, mosses and ferns..........	202
	3.3 Drosophila cryptochrome.....................................	202
	3.4 Mammalian cryptochromes..................................	203
	3.5 Bacterial and related cryptochromes...................	203
4.	Cryptochromes and plant photomorphogenesis............	203

5. Cryptochrome and flowering.. 206
6. Plant cryptochromes and circadian rhythms.. 206
7. Arabidopsis cryptochrome and gene expression.................................... 207
8. Cryptochromes and circadian rhythms in animals.................................. 208
   8.1 Drosophila circadian rhythms are entrained through cryptochrome.. 208
   8.2 Mammalian cryptochromes: Negative transcriptional regulators
        and essential components of the circadian oscillator.................... 208
9. The mode of action of cryptochrome.. 210
   9.1 The Arabidopsis cryptochrome C-terminal domain mediates a
        constitutive light response... 210
   9.2 COP1: A signalling partner of Arabidopsis cryptochromes........... 211
   9.3 Intracellular localization of Arabidopsis CRYs............................. 213
   9.4 Phosphorylation of Arabidopsis cryptochromes............................ 214
   9.5 Photochemical properties of Arabidopsis cryptochromes............. 215
   9.6 Drosophila cryptochrome interacts with PER and TIM in a
        light-dependent manner... 216
   9.7 Mouse cryptochromes negatively regulate transcription in a
        light-independent manner.. 217
10. Cryptochrome evolution... 217
11. Conclusions and future studies... 217
12. References.. 218

Chapter 12

PHOTOTROPINS
Winslow R. Briggs, John M. Christie and Trevor E. Swartz

1. Introduction... 223
2. Blue light-activated phosphorylation of a plasma-membrane protein.... 224
   2.1 The protein is likely ubiquitous in higher plants........................... 224
   2.2 Subcellular localization of phot1.. 225
   2.3 Distribution of the phototropins in relation to function................ 226
   2.4 Biochemical properties of the phosphorylation reaction *in vitro*... 227
   2.5 Correlation of phot1 phosphorylation with phototropism............. 228
   2.6 Autophosphorylation occurs on multiple sites.............................. 231
3. Cloning and molecular characterization of phototropin......................... 232
   3.1 The initial discovery of phototropin 1... 232
   3.2 LOV domains function as light sensors.. 234
4. Why two LOV domains?... 234
5. Structural and photochemical properties of the LOV domains.............. 236
   5.1 LOV domain photochemistry.. 236
   5.2 LOV-domain structure... 236
   5.3 The LOV-domain photocycle.. 238
   5.4 Mechanism of FMN-cysteinyl adduct formation.......................... 238

	5.5 The LOV domain back reaction..	240
	5.6 Protein conformational change on photoexcitation...........................	241
6.	The ZTL/ADO family...	242
7.	LOV domains in other systems...	243
8.	A return to physiology: a model for phototropism.............................	244
9.	Future prospects...	245
10.	Note added in proof..	246
11.	References..	247

Chapter 13

BLUE LIGHT PHOTORECEPTORS -BEYOND PHOTOTROPINS AND CRYPTOCHROMES
Jay Dunlap

1.	Introduction..	253
2.	Historical antecedents..	253
3.	The photobiology of Neurospora..	255
4.	Light perception -the nature of the blue light photoreceptor...............	257
	4.1 Flavins as chromophores...	257
	4.2 Genetic dissection of the blue light response...........................	257
5.	Cloning of the white collar genes...	258
6.	WHITE COLLAR-1 is the blue light photoreceptor........................	260
7.	WC-1 and WC-2 -positive elements in the circadian feedback loop...........	263
	7.1 How light resets the clock...	265
8.	VIVID, a second photoreceptor that modulates light responses............	266
9.	Complexities in light regulatory pathways.....................................	268
10.	Other Neurospora photoreceptors..	270
11.	Flavin binding domain proteins as photoreceptors in photosynthetic eukaryotes...	271
12.	Summary and conclusion...	273
13.	References..	274

Chapter 14

UV-B PERCEPTION AND SIGNALLING IN HIGHER PLANTS
Roman Ulm

1.	Introduction..	279
2.	DNA damage and repair..	281
3.	Photomorphogenic responses to UV-B...	284
	3.1 Synthesis of "sunscreen" metabolites.....................................	285
	3.2 Inhibition of hypocotyl growth...	287
	3.3 Cotyledon opening and expansion...	288

4. UV-B perception.. 289
   4.1 Supporting evidence and possible nature of a specific UV-B
       photoreceptor... 289
   4.2 Possible importance of specific UV-B perception?.................. 291
5. UV-B signalling.. 292
   5.1 Reactive oxygen species... 292
   5.2 Plant hormones... 293
   5.3 Calcium... 294
   5.4 Phosphorylation.. 294
   5.5 Nitric oxide... 295
6. Transcriptional response to UV-B radiation...................................... 296
7. Conclusions and perspectives... 298
8. References... 299

Chapter 15

SIGNAL TRANSDUCTION IN BLUE LIGHT-MEDIATED RESPONSES
Vera Quecini and Emmanuel Liscum

1. Introduction.. 305
2. Cryptochrome signalling.. 305
   2.1 Cryptochromes and photomorphogenesis............................... 305
       2.1.1 Cryptochrome signalling and photomorphogenic growth
             responses.. 306
       2.1.2 Cryptochrome signalling and electrophysiological processes 309
       2.1.3 Cryptochrome signalling and the regulation of gene
             expression... 311
3. Phototropin signaling... 311
   3.1 Phototropins and plant movement responses........................... 311
       3.1.1 Phototropins and phototropism.................................... 312
       3.1.2 Phototropins and stomatal aperture control................ 316
       3.1.3 Phototropins and chloroplast movement..................... 318
   3.2 Phototropin signalling and electrophysiological processes.... 320
4. Concluding remarks... 321
5. References.. 321

PART 4: SIGNAL TRANSDUCTION IN PHOTOMORPHOGENESIS

Chapter 16

GENERAL INTRODUCTION
Peter H. Quail

    References 333

Chapter 17

PHYTOCHROME SIGNAL TRANSDUCTION NETWORK
Peter H. Quail

1. Introduction 335
2. Genetically-identified signalling-intermediate candidates 337
3. Phytochrome-Interacting Factors 340
    3.1 PIF3 340
    3.2 PKS1 343
    3.3 NDPK2 344
    3.4 Other phy interactors 344
4. Transcription-factor genes are early targets of PHY signalling 345
5. Biochemical mechanism of signal transfer 353
6. References 354

Chapter 18

THE FUNCTION OF THE COP/DET/FUS PROTEINS IN CONTROLLING PHOTOMORPHOGENESIS: A ROLE FOR REGULATED PROTEOLYSIS
Elizabeth Strickland, Vicente Rubio and Xing Wang Deng

1. Introduction 357
    1.1 Genetic analysis of photomorphogenesis 357
2. A brief summary of the ubiquitin-proteasome system 359
3. Properties and functions of the pleiotropic COP/DET/FUS proteins 359
    3.1 COP1 359
        3.1.1 Nuclear localization of COP1 360
        3.1.2 Light regulation of COP1 360
        3.1.3 Molecular role of COP1 361
        3.1.4 The E3 ubiquitin-protein ligase activity of COP1 363
        3.1.5 COP1 interactors 364

3.2 The COP9 signalosome.. 364
 3.2.1 Interactions and similarities between the CSN and the ubiquitin-proteasome system................................. 365
 3.2.2 Biochemical activities of the CSN..................................... 366
  Derubylation 366
  Deubiquitination 367
  Protein phosphorylation.. 368
  Control of nucleocytoplasmic localization........................... 368
 3.2.3 Independent roles for CSN subunits................................... 369
 3.2.4 Non-photomorphogenic roles of the CSN.......................... 369
3.3 The CDD complex.. 370
 3.3.1 COP 10.. 370
 3.3.2 DET1.. 370
 3.3.3 COP10, DDB1, and DET1 are components of the same CDD complex.. 371
4. Concluding remarks.. 371
5. References.. 373

Chapter 19

BIOCHEMICAL AND MOLECULAR ANALYSIS OF SIGNALLING COMPONENTS
Christian Fankhauser and Chris Bowler

1. Introduction... 379
2. Is phytochrome a light-regulated protein kinase?...................................... 380
3. Phosphorylation in phy mediated signalling... 384
4. G-proteins.. 386
5. Rapid ion fluxes.. 389
6. Cytoplasmic movements... 393
7. Forward and reverse Genetics... 394
8. Interactions with internal cues (growth regulators, circadian clock).......... 399
9. Conclusions... 401
10. References... 401

Chapter 20

THE PHOTORECEPTOR INTERACTION NETWORK
Jorge José Casal

1. Introduction... 407
1.1 Light signals and photoreceptors.. 407
1.2 Shared and specific control of light responses by different photoreceptors.. 408

2.	Photoreceptor interaction during de-etiolation..		409
	2.1	Multiple photoreceptors control de-etiolation..................................	409
	2.2	Redundancy..	410
		2.2.1 The potential action of a photoreceptor can be hidden by the action of others...	410
		2.2.2 Definition of redundancy...	411
		2.2.3 The mechanisms of redundancy..	411
		2.2.4 Redundant photoreceptors are not equally important.............	412
	2.3	Synergism between phytochromes and cryptochromes.....................	412
		2.3.1 Blue light-mediated responsivity amplification towards phytochrome..	412
		2.3.2 cry1 amplifies responsitivity towards phyB............................	413
		2.3.3 The synergism between cry1 and phyB is conditional............	415
		2.3.4 Other manifestations of synergism between phytochromes and cryptochromes..	416
	2.4	Synergistic or antagonistic interaction between phyA and phyB......	418
	2.5	Synergism between phyB and phyC..	420
	2.6	Interactive signalling under sunlight reduces noise/ signal ratio.......	422
3.	Photoreceptor interaction during adult plant body shape formation............		424
	3.1	Redundant control of normal progression of vegetative development by phytochromes and cryptochromes...	424
	3.2	The response to R:FR...	424
4.	Photoreceptor interaction in phototropism..		425
	4.1	Phototropins perceive the unilateral stimulus....................................	425
	4.2	Phytochromes enhance the responses mediated by phototropins......	426
	4.3	The role of cryptochromes..	426
5.	Photoreceptor interaction in clock entrainment..		426
6.	Photoreceptor interaction controlling flowering...		427
	6.1	Different light signals control the transition between vegetative and reproductive growth..	427
	6.2	Roles of cry2, cry1 and phyA in the photoperiodic response............	427
	6.3	Roles of phyB, phyD and phyE in the response to low R:FR...........	428
	6.4	Integration of the responses to photoperiod and R:FR......................	428
7.	Points of convergence in the photoreceptor signalling network..................		428
	7.1	The occurrence of interactions is an emergent property of the signalling network..	428
	7.2	Direct convergence: Physical interaction between photoreceptor pigments..	430
	7.3	Convergence in the control of transcription: HFR1...........................	430
	7.4	Post-transcriptional convergence accounts for the interaction between phyB and phyC..	431
	7.5	Convergence in the control of protein stability: COP1.....................	431
	7.6	Photoreceptor sub-cellular partitioning...	431
	7.7	SUB1...	431
8.	Overview..		432

	8.1	Redundancy...	432
	8.2	Hierarchical action...	432
	8.3	Synergism..	432
	8.4	Sensitivity and homeostasis..	433
	8.5	Connectivity..	433
9.	References..		433

Chapter 21

INTERACTION OF LIGHT AND HORMONE SIGNALLING TO MEDIATE PHOTOMORPHOGENESIS
Michael M. Neff, Ian H. Street, Edward M. Turk and Jason M. Ward

1.	Introduction...		439
2.	Gibberellins...		440
	2.1	Gibberellin biosynthetic genes and seed germination.........................	440
	2.2	Gibberellins and de-etiolation..	442
	2.3	The SPY and PHOR1 genes..	443
	2.4	A possible role for protein degradation..	444
	2.5	Interactions with other hormone signalling pathways........................	445
3.	Auxin..		446
	3.1	Auxin transport...	447
	3.2	Auxin and phototropism...	448
	3.3	Auxin and shade avoidance...	449
	3.4	Auxin responsive genes involved in photomorphogenesis.................	450
	3.5	Auxin and protein degradation..	451
	3.6	Interaction of auxin with other hormone signalling pathways............	452
4.	Brassinosteroids..		452
	4.1	Brassinosteroid-deficient mutants..	454
	4.2	Brassinosteroids and gene expression...	455
	4.3	Further genetic connections between brassinosteroids and light.........	455
	4.4	Brassinosteroids and light signalling: three speculative models.........	456
5.	Ethylene..		460
	5.1	Genetic connections between ethylene and photomorphogenesis.......	461
	5.2	Ethylene mutants and shade-avoidance...	462
	5.3	Ethylene and fruit ripening...	463
6.	Cytokinins...		463
7.	Summary...		465
8.	Further reading..		465
9.	References...		466

PART 5: SELECTED TOPICS

Chapter 22

THE ROLES OF PHYTOCHROMES IN ADULT PLANTS
Keara A. Franklin and Garry C. Whitelam

1. Introduction.. 475
2. The natural light environment... 477
3. R:FR ratio and shade avoidance...................................... 478
4. Roles of different phytochromes in shade avoidance........... 482
   4.1 Roles for phytochrome A in adult plants..................... 486
5. Molecular mechanisms controlling shade avoidance responses.. 489
   5.1 The acceleration of flowering..................................... 489
   5.2 Early events in R:FR ratio signalling........................... 491
6. References... 493

Chapter 23

A ROLE FOR CHLOROPHYLL PRECURSORS IN PLASTID-TO-NUCLEUS SIGNALING
Robert M. Larkin and Joanne Chory

1. Introduction.. 499
2. Chlorophyll biosynthetic mutant, inhibitor, and feeding studies... 500
3. Plastid-to-nucleus signaling mutants inhibit Mg-porphyrin accumulation. 504
4. Mechanism of Mg-Proto/Mg-ProtoMe signaling................. 506
5. Plastid and light signaling pathways appear to interact....... 508
6. Conclusions and perspectives... 509
7. Further Reading.. 510
8. References... 510

Chapter 24

PHOTOMORPHOGENESIS OF FERNS
Takeshi Kanegae and Masamitsu Wada

1. Introduction.. 515
2. Photoreceptors in *Adiantum*... 517
   2.1 Cryptochromes.. 517
   2.2 Phototropins... 518
   2.3 Phytochromes... 519

		2.3.1	Phytochrome 1	520
		2.3.2	Phytochrome 2	521
		2.3.3	Phytochrome 3	522
		2.3.4	Phytochrome 4	523
3.	Mutant analyses			523
	3.1	Methods of mutant selection		523
	3.2	Red light aphototropic mutants		525
	3.3	Mutants deficient in the chloroplast avoidance response		526
	3.4	Dark position-deficient mutants		527
4.	Function of phytochrome3			527
	4.1	Phytochrome3-dependent chloroplast movement		527
	4.2	Phytochrome3-dependent phototropism		528
5.	Function of phototropin2			530
	5.1	Phototropin2-dependent chloroplast movement		530
	5.2	Physiological estimation of the lifetime of phot signals		531
6.	Germination-related genes			532
7.	Concluding remarks			533
8.	References			534

Chapter 25

PHOTOMORPHOGENESIS OF MOSSES
Tilman Lamparter

1.	Introduction			537
2.	Effects of light on moss development			539
	2.1	Spore germination		539
	2.2	Cell differentiation		541
	2.3	Phototropism and polarotropism		541
	2.4	Lights effects on gravitropism		542
	2.5	Chloroplast movement		543
	2.6	Chlorophyll synthesis		544
	2.7	Protoplast regeneration		544
3.	Different photoreceptors in mosses			545
	3.1	Phytochromes		545
		3.1.1	Phytochrome genes and proteins	545
		3.1.2	Mutants	547
			Ceratodon Class 1 mutants	549
			Ceratodon class 2 mutants	550
			Physcomitrella phytochrome knockout mutants	550
		3.1.3	Light direction and polarization	551
	3.2	Cryptochromes and phototropin		556
4.	Signal transduction			557
	4.1	Ca^{2+}		558
	4.2	Cytoskeleton		558

5. Summary 559
6. References 560

Chapter 26

CIRCADIAN REGULATION OF PHOTOMORPHOGENESIS
Paul Devlin

1. Introduction 567
2. The Circadian Clock 568
3. Circadian rhythms 568
4. The circadian clock in plants 569
5. Setting the plant circadian clock 574
6. Driven vs Endogenous Rhythms 578
7. Gating 579
8. Circadian Regulation of Photomorphogenesis 580
   8.1 Circadian regulation of light-induced changes in gene expression 580
   8.2 Circadian regulation of light-mediated inhibition of hypocotyl elongation 581
   8.3 Circadian regulation of light-mediated stimulation of hypocotyl hook opening 581
   8.4 Circadian regulation of light-mediated stimulation of stomatal opening 582
   8.5 Circadian regulation of sensitivity to light allows daylength perception 582
9. Mechanism of circadian regulation of photomorphogenesis 583
10. Mutants affecting circadian regulation of photomorphogenesis 585
    10.1 *early flowering 3 (elf3)* 585
    10.2 *time for coffee (tic)* 588
11. Other possible components involved gating 590
    11.1 Circadian regulation of photoreceptor levels 590
    11.2 Circadian regulation of photoreceptor subcellular localisation ... 591
    11.3 Circadian regulation of photoreceptor signal transduction components 592
        11.3.1 GIGANTEA (GI) 592
        11.3.2 ZEITLUPE (ZTL) 593
        11.3.3 Suppressor of phyA 1 (SPA1) 594
        11.3.4 *early phytochrome responsive 1 (epr1)* 595
12. A twist in the tale: Is there just one circadian clock regulating photomorphogenesis? 596
13. Conclusion: Concerns for photomorphogenic study 598
14. Epilogue 599
15. Further suggested reading 599
16. References 600

Chapter 27

THE MOLECULAR GENETICS OF PHOTO-PERIODIC RESPONSES: COMPARISONS BETWEEN LONG-DAY AND SHORT-DAY SPECIES
George Coupland

1.	Introduction..	605
2.	Genetic model systems...	606
3.	A molecular pathway that controls flowering-time in response to day length in Arabidopsis by generating a long-distance signal from the leaf...	607
4.	An external coincidence model for the day-length response in Arabidopsis	611
5.	Genetic analysis of the photoperiodic control of flowering in rice, a short-day plant...	614
6.	Relationships between photoperiodic control and other environmental cues regulating flowering...	618
7.	Photoperiodic responses other than flowering..	620
8.	Perspectives..	621
9.	References..	621

Chapter 28

COMMERCIAL APPLICATIONS OF PHOTOMORPHOGENESIS RESEARCH
Ganga Rao Davuluri and Chris Bowler

1.	Introduction..	627
2.	Light-mediated responses in the natural environment.................................	628
3.	Manipulation of light responses in agriculture..	629
	3.1 Modulation of day length perception...	629
	3.2 Modulation of shade avoidance responses..	630
	3.3 Modulation of fruit ripening..	632
4.	Light-based biological engineering..	635
5.	Conclusions and perspectives...	636
6.	References..	637

Chapter 29

PHOTOMORPHOGENESIS – WHERE NOW?
Harry Smith

Where are we going, Dad?...	641
Where are we now, Dad?..	641

So what, Dad?..	643
Is that all, Dad?...	646
Why, Dad?..	647
What use is it, Dad?...	649
Are we nearly there yet?..	649
Race you to the beach, Dad!...	650
References...	650
Conclusions...	653
Index...	655

Abbreviations

AFLP	amplified fragment-length polymorphism
APRR	Arabidopsis pseudo response regulator
ATP	adenosine triphosphate
B	blue light
BBP	bilin-binding pocket
Bch	bacteriochlorophyll
BHF	blue light high fluence
BLF	blue light low fluence
BphPs	bacteriophytochrome photoreceptors
BV	biliverdin IXa
CAB	chlorophyll a/b binding proteins
CAT3	catalase 3
CCA	complementary chromatic adaptation
CCA1	circadian clock-associated 1
CCR2	cold circadian clock-regulated
CCT	cryptochrome C-terminal domain
CFB	cytophaga-flexibacter-bacterioides
Chl	chloroplast
CHS	chalcone synthase
CNT	cryptochrome N-terminal domain
CO	constans
COP1	constitutively photomorphogenic 1
CPD	cyclobutane pyrimidine dimmers
Cphs	cyanobacterial Phys
Crt	carotenoids
CRY	cryptochrome
Cry1/ hy4	cryptochrome1/ hypocotyl4
CT	circadian Time"
Cyto	cytoplasm
DBD	DNA-binding domain
DDB1	UV-damaged DNA binding protein
DET1	de-etiolated 1
DET2	de-etiolated 2
DUF	domain of unknown function
ELF3	early flowering 3
ELF4	early-Flowering 4
EPR1	early phytochrome responsive 1
FAD	flavin adenine dinucleotide
FDD	fluorescence differential display
FKF1	flavin-binding kelch repeat F-box 1
FLC	flowering locus C
Fphs	fungal Phys
FR	far-red
FSBA	fluorosulfonylbenzoyladenosine
FT	flowering locus T
G	green light
GA	gibberelin acid
GAF	cGMP phosphodiesterase/adenyl cyclase/FhlA
GAI	GA-insensitive
GFP	green fluorescent protein
GGDEF	Gly/Gly/Asp/Gly/Phe motif
GI	gigantea
GRAS	GAI/RGA and SCARECROW
HAMP	HK/adenyl cyclases/methyl-binding proteins/phophatases domain
Hd	heading date
HIR	high irradiance response
HKD	histidine kinase domain

HKRD	histidine kinase-related domain
HO	heme oxygenase
HPT	histidine phosphotransferase
HWE	His/Try/Asp
HY5	hypocotyl 5
ICGs	interchromatin granular clusters
LFR	low Fluence Response
LHCB	light harvesting chlorophyll a/b-binding protein
LHY	late elongated hypocotyl
LIAC	light-induced absorbance change
LKP2	LOV kelch protein 2
LRE	light-responsive regulatory element
LUC	luciferase
Me-Ac	methyl-accepting chemotaxis protein domain
Mg-ProtoMe	Mg-Protoporphyrin IX monomethyl ester
MS	mass Spectroscopic analysis
MTHF	methenyltetrahydrofolate
NAI2	nitrate reductase
NDPK2	nucleoside diphosphate kinase 2
NLS	nuclear localisation signal
NMR	nuclear magnetic resonance
NO	nictric oxide
NOE	nuclear overhauser effect
NPA	1-naphthylphthalamic acid
NPH	non-phototropic hypocotyl
Nuc	nucleus
ORF	open reading frame
PAC	PAS-like domain C-terminal to PAS
PAS	Per/Arndt/Sim
PCB	3(Z)-phycocyanobilin
Pchlide	protochlorophyllide
PEB	phycoerythrobilin
PER	period
PFT1	phytochrome flowering time 1
Phy	phytochrome
PIF3	phytochrome interacting factor 3
PIL1	PIF3-like 1
PIL2	PIF3- like 2
PIL4	PIF3- like 4
PIL6	PIF3- like 6
PIN1	pinformed 1
PKS1	phytochrome kinase substrate 1
PKS2	phytochrome kinase substrate 2
PLD	PAS-like domain
PM	plasma membrane
PP	pyrimidine-pyrimidinone dimers
PP2C	protein phosphatase-2C
Proto	protoporphyrin IX
PYP	photoactive yellow protein
PΦB	3(Z)-phytochromobilin
QTL	quantitative trait loci
R	red light
RAP2	red light aphototropic 2
RGA	repressor of ga 1-3
RGL	RGA-like
RNAi	RNA interference

ROS	reactive oxygen species
RR	response regulator
Rubisco	ribulose-1,5-bisphosphate carboxylase/oxygenase
SAP	sequestered areas of phytochrome
SCF complex	Skp1 cullin F-box protein
SCN	suprachiasmatic nucleus
SOC1	suppressor of overexpression of co 1
SPA1	suppressor of phyA 1
SPY	spindly
SRD	serine-rich domain
SRR1	sensitivity to red light reduced
TC-HK	two-component histidine kinase
TIC	time for coffee
TIM	timeless
TIR3	toll interleukin resistance domain cotaining protein
toc1	timing of cab expression 1
ULI	UV-B light insensitive
ULI3	UV-B light insensitive 3
UV	ultra violet light
UV-A	320-400 nm UV
UV-B	280-320 nm UV
UV-C	<280 nm UV
VLFR	very low fluence response
ZT	zeitgeber time
ZTL	zeitlupe

Color plate section

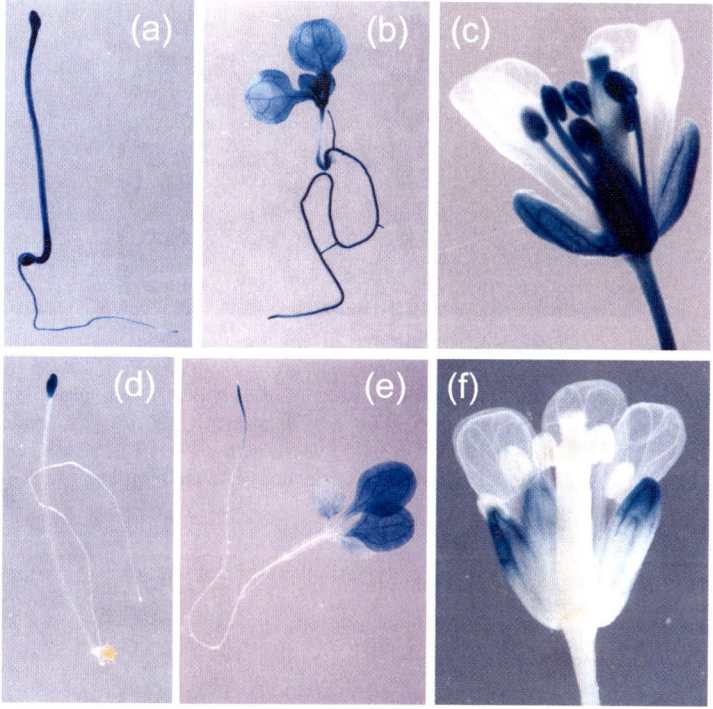

Chapter 7, Figure 5. Histochemical localization of the expression patterns of PHYB::GUS (a-c) and PHYD::GUS (d-f) promoter-reporter fusion genes in Arabidopsis. (a, d) seven day old dark-grown seedlings; (b, e) seven day old light-grown seedlings; (c, f) flowers.

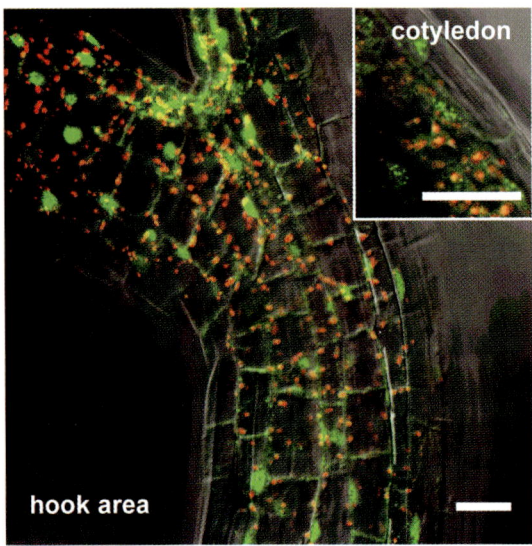

Chapter 9, Figure 1. Localisation of PHYA-GFP fusion proteins in Arabidopsis seedlings. 4d old dark-grown Arabidopsis seedlings expressing fusion proteins of Arabidopsis PhyA and GFP controlled by the Arabidopsis promoter were irradiated briefly with white light. Subsequently bright-field images (greyscale) and confocal images of GFP (green channel) and chlorophyll (red channel) fluorescence have been recorded with a Zeiss LSM510 microscope. The colour-combined images are showing the hook area and an area of the rim of a cotyledon (inlet). Bar= 25 μm.

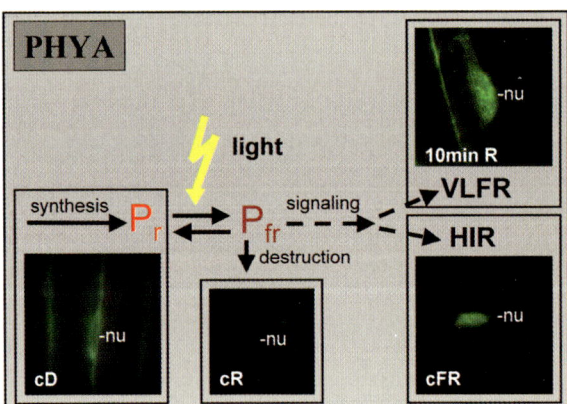

Chapter 9, Figure 2. Model of the light-driven intracellular dynamics of phytochrome A. In dark-grown seedlings phyA is synthesized in its physiological inactive Pr-form (Pr) and stays in the cytosolic compartment. Irradiation establishes a wavelength-dependent equilibrium of the Pr to the active Pfr form. Red light (R) leads to formation of about 80% of Pfr, far-red light (FR) to about 3% Pfr. PhyA Pfr localises to sequestered areas of phytochrome (SAP) in the cytosol and is imported into the nucleus where it forms nuclear speckles. The light-requirements for these intracellular processes overlap with the light requirements for typical physiological responses of phytochrome A. While pulses of light can promote very low fluence response (VLFR, here the effect of a red pulse is shown), continuous irradiation with far-red light (cFR) leads to high irradiance responses (HIR). Due to the instability of the Pfr form of PHYA, continuous red-light (cR) leads to a rapid destruction of the photoreceptor.

xxxiii

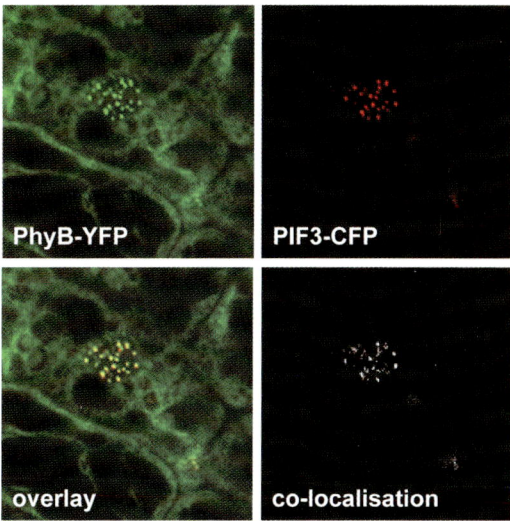

Chapter 9, Figure 3. Co-localisation of Phytochrome B with the bHLH factor PIF3. 4d old dark-grown Arabidopsis seedlings simultaneously expressing fusion proteins of PhyB with YFP and PIF3 with CFP each controlled by the 35S promoter were irradiated briefly with white light. Subsequently, confocal images of YFP (green channel) and CFP (red channel) fluorescence have been recorded with a Zeiss LSM510 microscope. The images are showing epidermal cells of the base of a cotyledon, either representing the PhyB-YFP or PIF3-CFP signals, an overlay of these images resulting in yellow colour for co-localisation of PhyB and PIF3 or an additional co-localisation analysis of both factors using ImageJ software package (NIH).

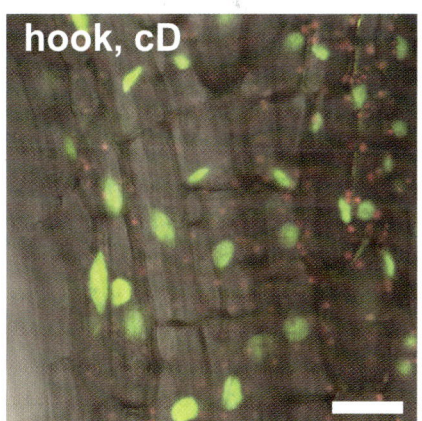 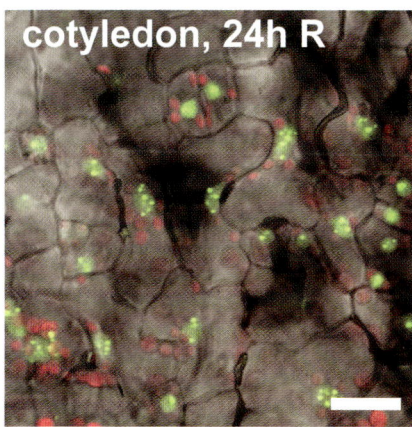

Chapter 9, Figure 4. Localisation of a fusion protein consisting of Arabidopsis PhyB, GFP and a nuclear localisation sequence. 4d old dark-grown Arabidopsis seedlings expressing fusion proteins of Arabidopsis PhyB, GFP and the SV 40 NLS under the control of the Arabidopsis promoter were analysed either after incubation for 24 hours in red light (R) or darkness (cD). Subsequently, bright-field images (greyscale) and confocal images of GFP (green channel) and chlorophyll (red channel) fluorescence have been recorded with a Zeiss LSM510 microscope. The colour-combined images are showing the hook area or an area of a cotyledon. Bar = 25 μm.

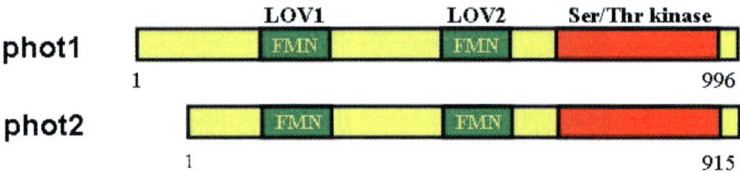

Chapter 12, Figure 1. Domain structures for phototropins 1 and 2.

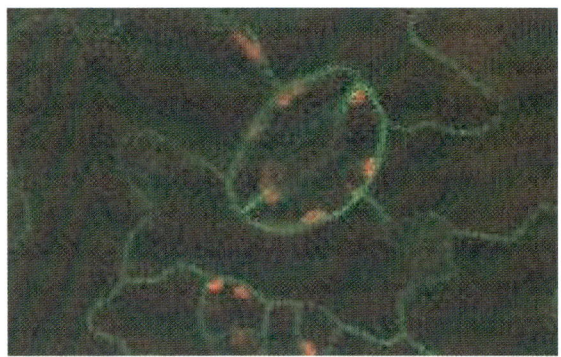

Chapter 12, Figure 2. Localization of phot1-green fluorescent protein (GFP) in guard cells and leaf epidermal cells. Red fluorescence is from chloroplasts. See Sakamoto and Briggs (2002).

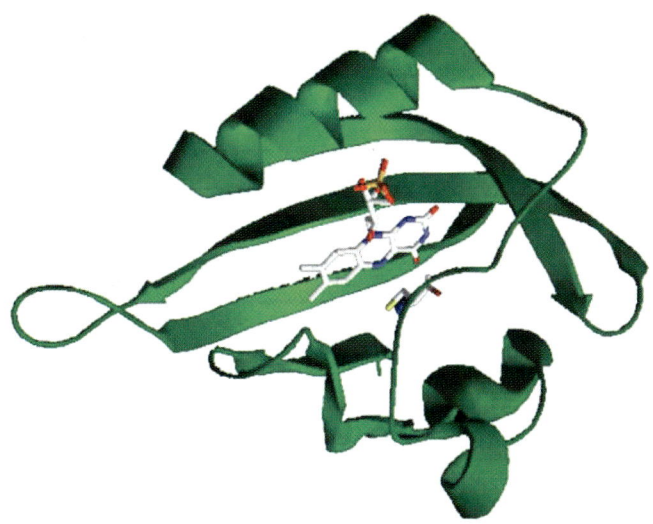

Chapter 12, Figure 6. Structural model of the phytochrome3 LOV domain from Adiantum capillus veneris (after Crosson and Moffat, 2001). FMN and cysteine 39 are highlighted.

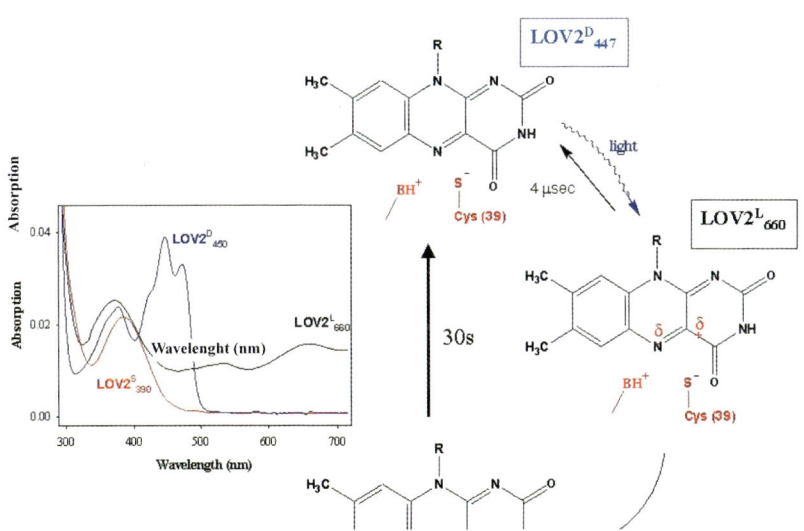

Chapter 12, Figure 7. LOV2-domain absorption of ground state and intermediates and photocycle (see Swartz et al., 2001). Details in text.

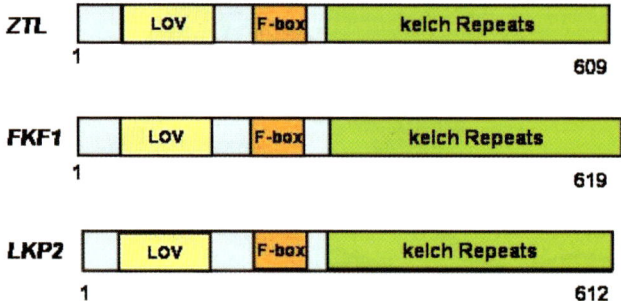

Chapter 12, *Figure 8.* Domain structure of the ZTL/ADO family of putative Arabidopsis blue light receptors.

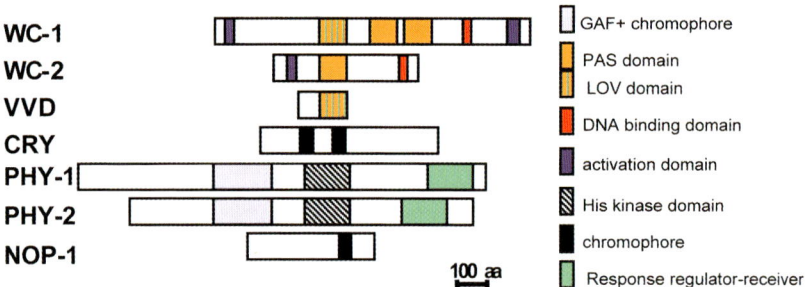

Chapter 13, *Figure 2.* Real and putative photoreceptors in Neurospora. The identity and approximate location of various functional domains in Neurospora proteins are shown. In the list, only WC-1 and VVD are known to bind chromophores and to be true photoresponse mediators, although WC-2 is required for the function of WC-1.

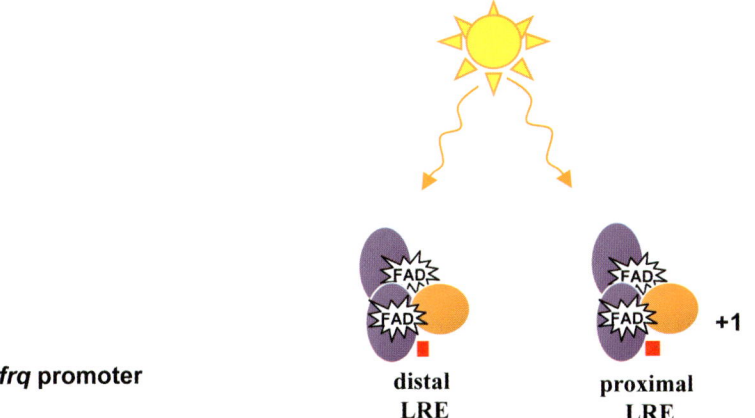

Chapter 13, Figure 3. *Two light regulatory sequences within the frq promoter in Neurospora bind to complexes of WC-1 and WC-2. The large oval represents WC-1 binding to FAD as the chromophore, and the small oval represents WC-2. See text for details (Figure courtesy of A. Froehlich).*

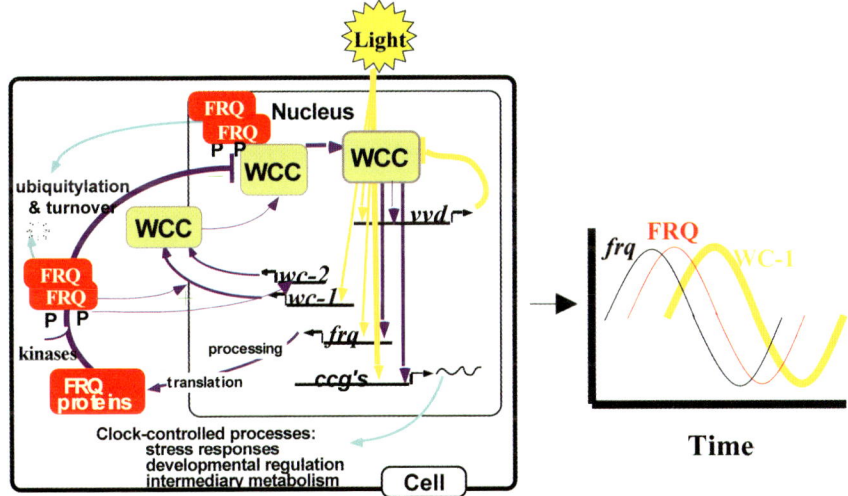

Chapter 13, Figure 6. *Known molecular components in the circadian feedback loop of Neurospora. Light lines represents effects of light, grey or dark lines actions that take place in the dark. Within the cell, the WCC drives expression of FRQ which does three things: (1) Its first and dominant action is to bind to the WCC to block its function, thus constituting a negative feedback loop that is the clock. (2) It promotes the synthesis of WC-1 posttranscriptionally and wc-2 mRNA, actions that promote robustness in the feedback loop; (3) It becomes phosphorylated which eventually leads to its turnover. When this happens, the bolus of WCC whose synthesis was promoted by FRQ is released to start the cycle again (Lee et al., 2000; Loros and Dunlap, 2001). The result of these feedback loops are daily rhythms in expression of frq mRNA, FRQ, and WC-1 whose timing within the day defines biological time; for instance, high frq mRNA doesn't just occur in mid-day, high frq degfines mid-day for the organism (adapted form Dunlap, 2003).*

xxxviii

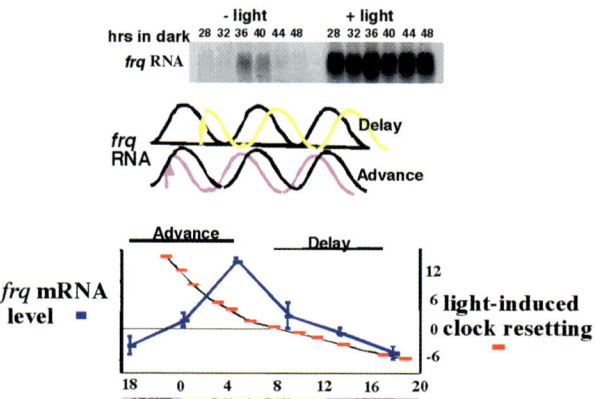

Chapter 13, Figure 7. *Light resetting of the Neurospora clock. At the top is shown a Northern: Absent light on the left, frq expression levels are seen to rise and fall with a circadian rhythm whereas on the right, light exposure at any time of day results in a rapid large induction of frq. In the middle panel is shown a schematic of how this light induction resets the clock: the dark curve represents the control rhythm of frq in the dark. Light exposure when frq is falling (evening) delays the clock whereas light exposure when frq is rising (late night to early morning) advances the clock. The bottom panel shows hour by hour through the day how light induction of frq will reset the clock, by plotting the daily rhythm in frq mRNA on top of the amount of clock resetting elicited by light exposure (adapted from Dunlap, 1999).*

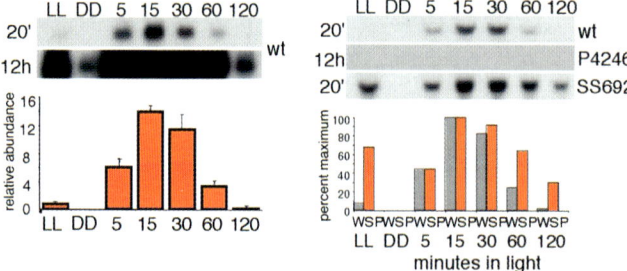

Chapter 13, Figure 8. *vvd is rapidly light induced, and loss of vvd slows the rate of decay of light-induced transcripts. On the right, P and S stand for two alleles, P which makes no transcript and S that makes transcript but no protein; W denotes wild type. (From (Heintzen et al., 2001) with permission).*

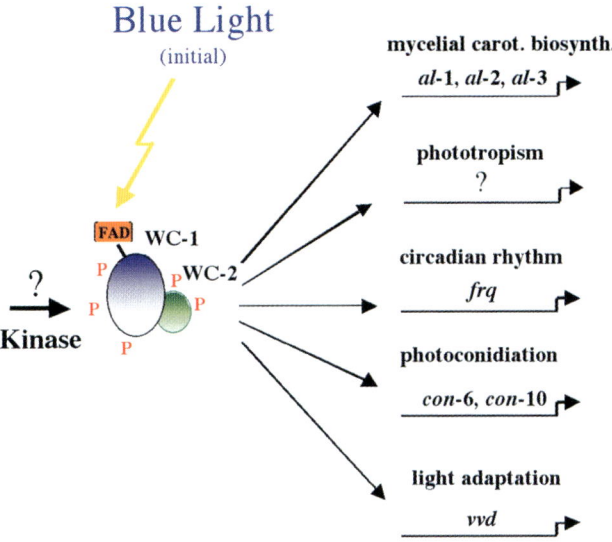

Chapter 13, Figure 9. Summary of light responses in Neurospora (Figure courtesy of C. Schwerdtfeger; all rights reserved).

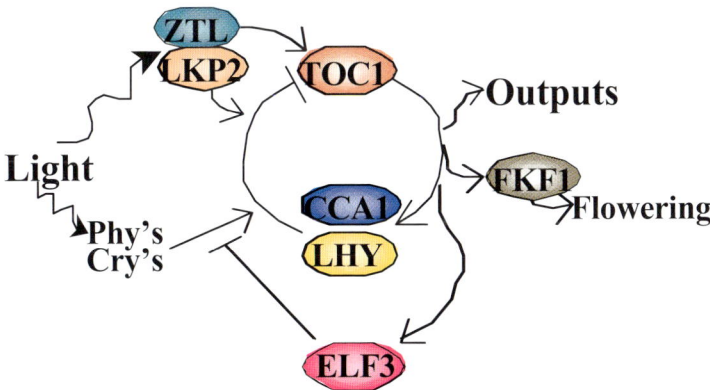

Chapter 13, Figure 10. Placement of LOV domain-associated protein turnover photoreceptors within the circadian system in Arabidopsis. The feedback loop formed by TOC1 and LHY/CCA1 is at the core of the clock. ZTL and LKP2 are associate with light input and FKF1 with output in the regulation of CO in the context of flowering. Clock-regulated ELF3 regulates light signalling pathways. Figure courtesy of Thomas Schulz and Steven Kay; all rights reserved.

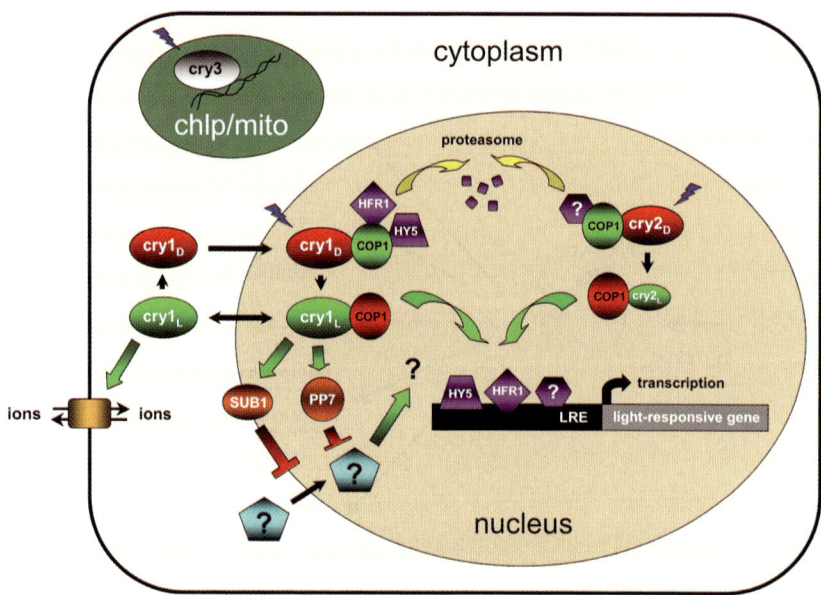

Chapter 15, Figure 1. Depiction of the cryptochrome-associated signalling events discussed in the text. Dark and lit-states of cry2 and cry2 are shown together as they relate to intracellular localization and function. Question marks refer to inferred, but still unidentified, molecules.

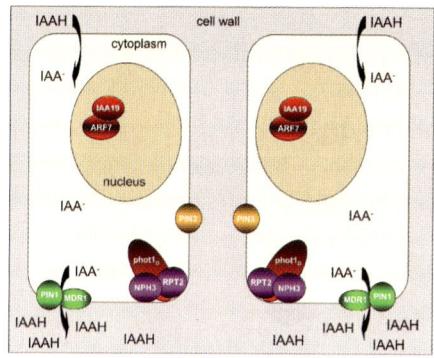

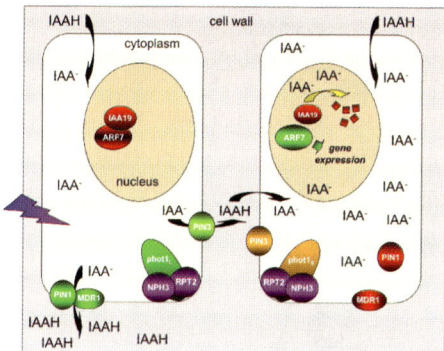

Chapter 15, Figure 2. *Depiction of the phototropin 1 signaling events associated with phototropism. In darkness (top panel) phot1 is in its resting-state, and as such, its signaling is not activated. In the absence of phot1 signaling basipetal polar auxin (IAA) transport predominates, as mediated by passive uptake of IAAH from the cell wall space and active efflux of the IAA anion via action of the PIN1-MDR1 (PGP1, not shown) complex. Also inactive in the dark is the ARF7 transcriptional machinery, as the ARF7 inhibitor, IAA19, persists. Upon unilateral B irradiation phot1 signaling is activated and a dramatic redistribution of auxin occurs. Although it is not clear at present how phot1 activates this process at least two events appear likely: 1) disruption of the the PIN1-MDR1 complex on the shaded side that thus disrupts efflux in that side, and 2) activation of laterally localized PIN3 that results in directional lateral flow of auxin. Both of these events lead to accumulation of auxin on the shaded side of the plant that in turn stimulate the degradation of IAA19 and subsequent activation of ARF7-dependent transcription.*

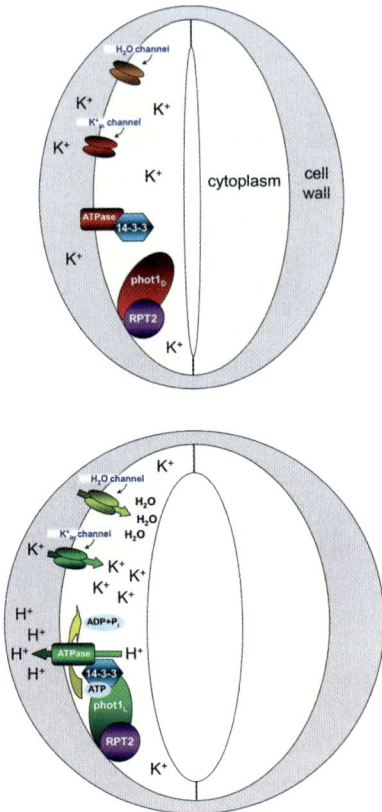

Chapter 15, Figure 3. *Depiction of the phototropin signaling events associated with B-dependent stomatal aperture control. In darkness (or B deficient light) the phot signaling system is in its "off state" and thus the guard cells exhibit no appreciable ion gradient across the plasma membrane (top panel). In the absence of such a gradient no water uptake occurs and a small stomatal pore is maintained. Upon exposure to B (bottom panel), the phot signaling is activated (activation is shown in only one cell for simplicity, but the mirror image processes occur in the second guard cell) whereby the H^+-ATPase is activated leading to a membrane hyperpolarization that activates inward rectifying K^+ channels. The resultant K^+ gradient in turn drives the uptake of H_2O and turgor driven cell expansion, which leads to an opening of the pore.*

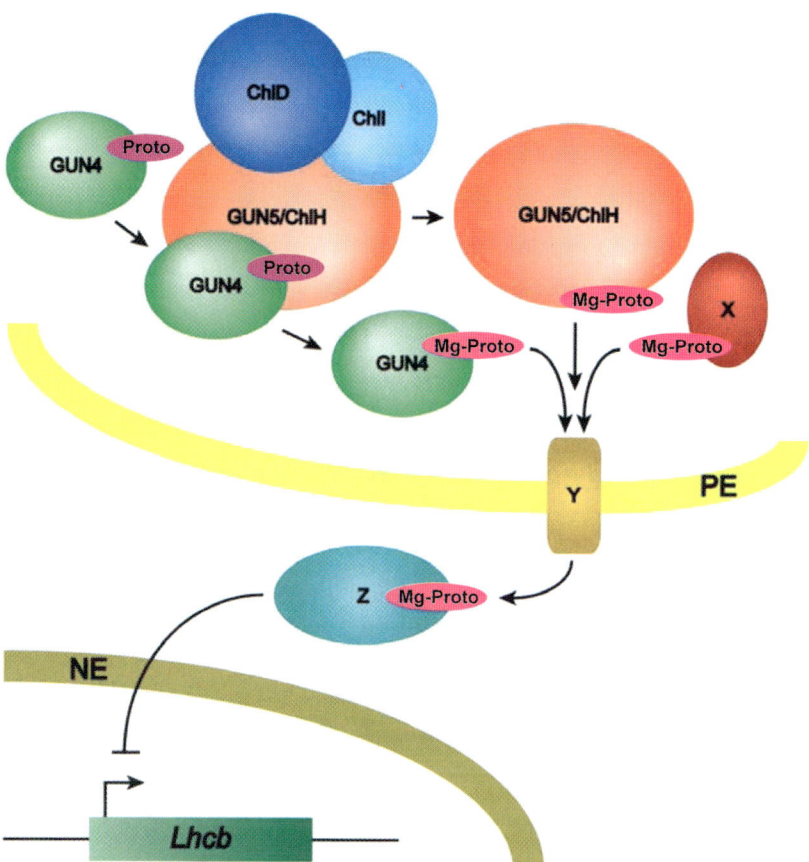

Chapter 23, Figure 2. *Model for Mg-Proto/ Mg-ProtoMe signalling. GUN5, which is also known as ChlH, two other Mg-chelatase subunits named ChlI ChlD insert Mg^{2+} into the Proto ring with assistance from GUN4. GUN4 binds the substrate and product of the Mg-chelatase reaction, Proto and Mg-Proto. Under conditions of Mg-Proto accumulation, Mg-Proto may be guided to a Tetrapyrrole transporter (Y) by GUN4, GUN5 or by another Mg-Proto-binding protein (X). After export from the plastid, a cytoplasmic factor (Z) binds Mg-Proto. Factor Z affects a signalling pathway that represses Lhcb transcription when chloroplast development is blocked; Mg-Proto regulates the activity of factor Z. Mg-ProtoMe signalling may function in a similar manner. This Proto and Mg-Proto trafficking probably takes place in the plastid envelope, but for the sake of clarity, porphyrin trafficking is not shown to be associated with the envelope. The details of the model are explained in the text NE, nuclear envelope; PE plastid envelope.*

A.

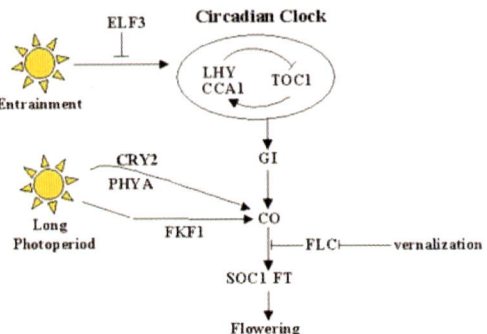

B.

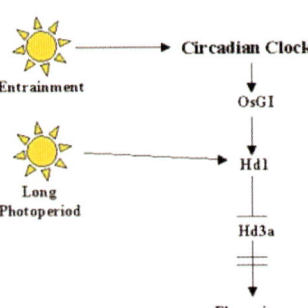

Chapter 27, Figure 1.   *Hierarchy of gene action within the photoperiodic flowering pathways of Arabidopsis and rice. A. The photoperiodic response pathway of Arabidopsis. The circadian clock, which is proposed to comprise of a feed-back loop between the* LHY/CCA1 *and* TOC1 *genes, regulates the expression of an output pathway including the* GI, CO, SOC1 *and* FT *genes. These four genes represent a transcriptional cascade as shown. Light has two roles in the pathway. Exposure to light entrains the circadian clock, and is required for activation of CO function under long photoperiods. The activation of CO involves post-transcriptional regulation by the photoreceptors CRY2 and PHYA, and transcriptional regulation via FKF1. ELF3 modulates entrainment of the clock by light input. B. The photoperiodic response pathway of rice. The logic of the pathway as for A. Hd1 function is activated under long photoperiods, and represses the expression of* Hd3a, *thereby delaying flowering.*

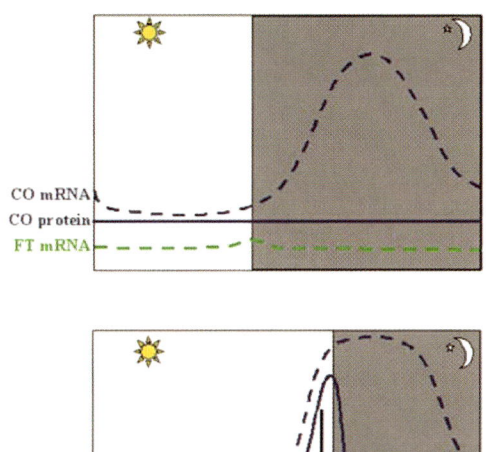
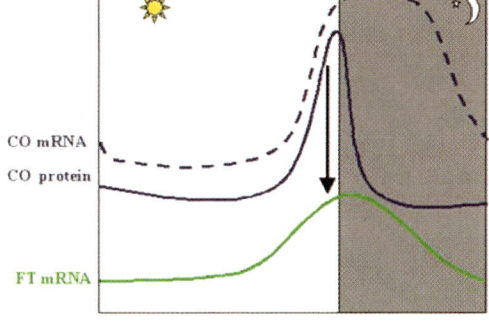

Chapter 27, Figure 2. *Coincidence model for the activation of flowering under long photoperiods in Arabidopsis. Top: Under short days circadian clock regulation of CO causes its mRNA to accumulate in the dark. CO protein does not accumulate, and FT expression is not activates. Bottom: Under long days circadian clock regulation and activation of CO transcription by light via FKF1 causes CO mRNA to accumulate in the light. The CO protein is stabilised through the activity of the PhyA and Cry2 photoreceptors. The presence of the CO protein causes activation of FT transcription, which leads to flowering.*

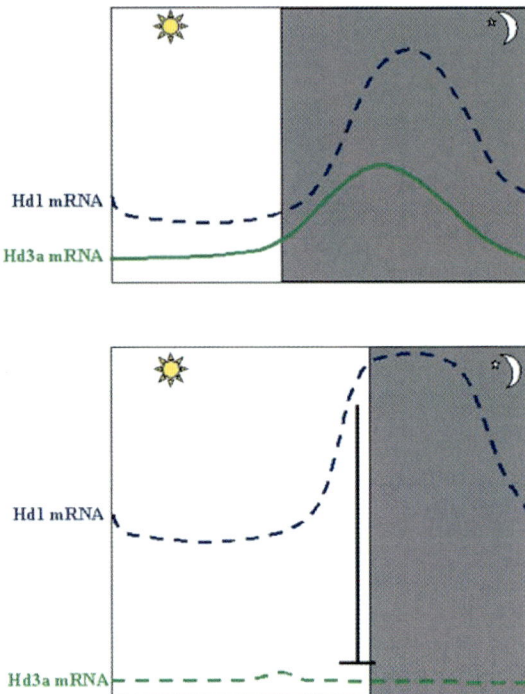

Chapter 27, Figure 3. *Coincidence model for the activation of flowering under short photo-periods in rice. Top: Under short-day conditions, circadian clock regulation of* Hd1 *mRNA causes it to accumulate in the dark.* Hd3a *is expressed and this leads to early flowering. Bottom: Under long-day conditions circadian clock regulation of* Hd1 *mRNA causes it to accumulate in the light. This leads to activation of Hd1 function so that it represses* Hd3a *expression and delays flowering.*

Chapter 28, *Figure 1*. Field trial of transgenic tobacco plants overexpressing the oat PHYA gene. Control plants are shown on the left, PHYA-overexpressing plants are shown on the right. Disabling of the shade avoidance response is clearly apparent in the transgenic PHYA-overexpressing plants. The photograph was taken in 1994 in Rothamstead, UK, and was kindly provided by Harry Smith (University of Nottingham, UK).

Chapter 28, Figure 2. Fruit truss phenotypes from a range of tomato photomorphogenic mutants. From left to right: (1) phyAphyB1, (2) phyAphyB1hp-1, (3) phyAphyB1phyB2, (4) phyAphyB1phyB2hp-1, (5) phyAphyB1cry1, (6) phyAphyB1cry1hp-1. Note that the triple phyAphyB1phyB2 mutant displays elongated truss phenotypes whereas the phyAphyB1cry1 mutant does not. However, both show reduced fruit pigmentation. In each mutant background the hp mutant is epistatic to the photoreceptor mutant phenotypes for fruit pigmentation, but not for the defects in truss architecture. Photograph kindly provided by Ageeth van Tuinen and Dick Kendrick (Wageningen University, NL).

Chapter 28, Figure 3. *Immature fruit phenotype caused by the fruit-specific silencing of the tomoto DET1 gene. The left panel shows fruit from a wild-type plant and the right panel shows fruit from a plant in which DET1 gene expression has been suppressed specifically in the fruits by RNAi.*

PART 1: GENERAL INTRODUCTION AND HISTORICAL OVERVIEW OF PHOTOMORPHOGENESIS

Chapter 1

HISTORICAL OVERVIEW

Eberhard Schäfer[1] and Ferenc Nagy[2]
[1]*Albert-Ludwigs-University Freiburg, Institute of Biology II/ Botany, Schänzlestrasse 1, 79104 Freiburg, Germany (e-mail: Eberhard.Schaefer@biologie.uni-freiburg.de)*
[2]*Biological Research Centre, Institute of Plant Biology, P.O. Box 521, 6701 Szeged, Hungary (e-mail: nagyf@nucleus.szbk.u-szeged.hu)*

1. INTRODUCTION

On germination plants pass through several stages of development culminating in flowering and production of seeds, followed eventually by death. These stages through which the plants pass are not determined by chance but on a combination of fixed genetic factors and specific influence of the environment. One of the most important of these environmental factors for plants is light. The development of the plants is referred as photomorphogenesis.

The effect of light on plants has been divided into various categories. Photosynthesis refers to the effect of light as an energy source. In addition however, there are the effects on movements and bending of plants towards or away from light, the so-called phototropic responses. Finally, there is photomorphogenesis, the control that is exerted by light over the growth, development and differentiation of a plant, independently of photosynthesis (Mohr, 1964).

Photomorphogenesis is, in fact, an ancient study. Theophrastors of Eresos (380 – 287 BC), who is often called the father of botany, refers to the fact that fir trees grown in shady places reach great heights but that the quality of the wood is low and was used primarily for rafters and masts, which must be of great length. Those firs trees grown is sunny places can be used for masts which are "necessarily short but of closer grain and stronger than the others". It seems likely that Theoprastors is describing a photomorphogenic effect of light grown plants, which is now called "shade avoidance effect" (see Chapter 22). Phototropism or heliotropism, the following of the sun by plants, is already mentioned by Marcus Terentius Varro (116– 27 BC) in his "De re rustica".

For the first really scientific observations on photomorphogenesis it was however necessary to wait until the late 17th century when John Ray in his "Historia Plantarum" (1686) commented on the phenomenon known as etiolation. Charles Bonnet described the first experimental observations in 1754 and he introduced the term "etiolement" although observing that the term had long been in use by gardeners. Stephen Hales referred again in 1727 to the effect of shading when he observed that plants growing under trees often had longer stems than those grown in full sunlight. Senebier reported in 1799 that light had considerable effects on plant pigmentation, particularly those known as anthocyanines. This effect of light has since become – besides the greening process – one of the most studied in the attempts to understand the physiological, biochemical and molecular genetic background of photomorphogenesis. Senebier's observations were confirmed and enlarged upon by Julius Sachs (1863). This brings us already to the modern age of plant physiology. In this introductory chapter we will concentrate on the history of phytochrome. W. R. Briggs summarizes that of the blue light responses in Chapter 10.

2. PHYTOCHROME INDUCTION RESPONSES

Caspary observed in a publication of 1860 that seeds of *Bulliarda aquatica* D. C. will not germinate in light and later research showed that many plant species have a light requirement (positive or negative) for seed germination. In the 1930's Flint and McAlister were able to show (1935, 1937) that lettuce (*Lactuca sativa* L.) seed germination was promoted by red light but inhibited (with respect to seeds kept in the dark) by far-red (700-80 nm) light (Figure 1).

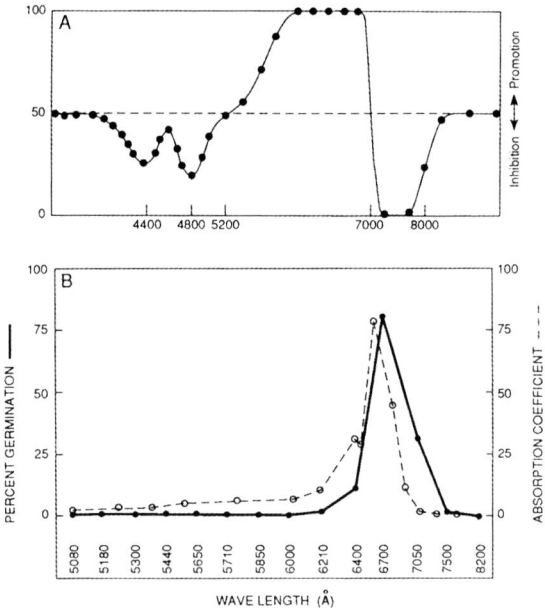

Figure 1. A: Action spectrum for lettuce seed germination. 10 Å = 1nm. (From Flint, 1936.) B: Partial action spectrum for germination of seeds inhibited by blue light (____); absorption of chlorophyll in ether (----). (From Flint and McAlister, 1937. Courtesy of the Smithsonian Institution).

Shortly afterwards the group working under H. Borthwick and S. Hendricks at the U.S.D.A. Laboratory Beltsville, Maryland, U.S.A. began investigating the action spectroscopy of the light requirement for photoperiodism. Photoperiodism refers to the fact that many plants require a specific day length before they will flower. In fact, photoperiodism does not as such refer only to the flowering response although this response was the first studied and was where the principles were first worked out. Nevertheless, many other responses including some in animals are under photoperiodic control (see Vince-Prue, 1975). Photoperiodism had been noticed by Linnaeus (1739) but he had erroneously concluded that it was a response to temperature. It is now known that the length of the dark (night) period is in fact critical. The pioneering work of Tournois (1912, 1914), Klebs (1918) and Garner and Allard (1920) laid the basic principles of the study and showed that even a short (few minutes) light pulse during the dark period could change the response. Important from the photobiological point of view was that light appeared necessary as a trigger and not as an energy source. This led to the problem of photoreceptor identity and it was this problem that Borthwick and co-workers hoped to solve with their action spectroscopy. They first published action spectra for the relative effectiveness of wavelengths in suppressing flowering when given as a short light-break in the dark period of two short day plants, *Glycine max* (L.) Merr. va. Biolxi and *Xanthium strumarium* L. (Parker *et al.*, 1945, 1946). The action spectra were basically the same for both species showing maximum effectiveness between 600 and 680 nm, no action above 700 nm, a minimum at ca. 180 nm with a further increase in effectiveness in the shorter wavelength blue waveband (Figure 2).

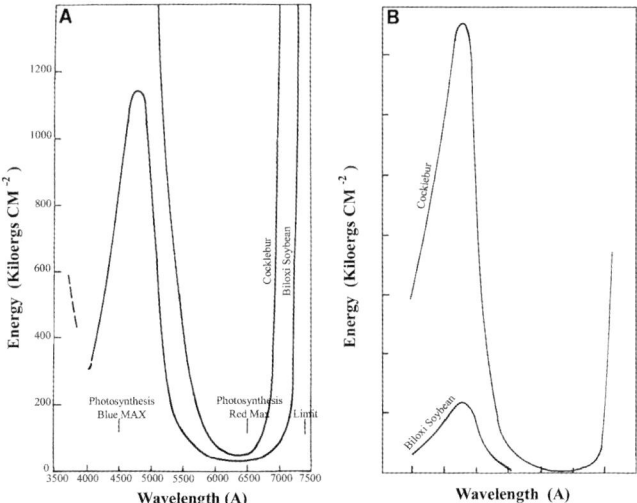

Figure 2. Composite action spectra for suppression of floral initiation in soybean and cocklebur. (A) Points on soybean curve give energy required at middle of 14-h dark period to prevent floral initiation; (B) Points on cocklebur curve give energy required at middle of 12-h dark period to prevent floral initiation (From Parker et al., 1946).

Later experiments showed very similar action spectra for long day plants *Hordeum vulgare* L. cv. Wintex and *Hyoscyamus niger* L. (Borthwick *et al.*,1948, Parker *et al.*,1950). At this stage other light responses were tested, specifically the process known as de-etiolation (i. e. the change from an etiolated to a "normal" light-grown plant). Spectra for inhibition of stem growth and promotion of leaf growth in dark grown *Pisum sativum* L. and *Hordeum vulgare* L. (Parker *et al.*,1949, Borthwick *et al.*,1951) showed again the same basic shape except for a slight shift in the red peak of effectiveness towards longer wavelengths. The experiments on *Hordeum* were of particular interest, having been carried out also using an albino mutant. The action spectra was however the same as that for the non albino (i. e. potentially green) plants. This provided powerful evidence that chlorophyll was not the photoreceptor and photosynthesis was not involved. The next action spectra basically confirmed the work of Flint and McAlister. These (Bortwick *et al.*,1952, 1954) were for germination of *Lactuca sativa*. The action spectra were similar to those for the de-etiolation processes. Important however was the confirmation that far-red (λ c. 720 nm) light inhibited germination (with respect to dark controls). Of tremendous interest was the discovery that they could show that cycles of red and far-red light always gave the response be expected from the last treatment given alone (Figure 3).

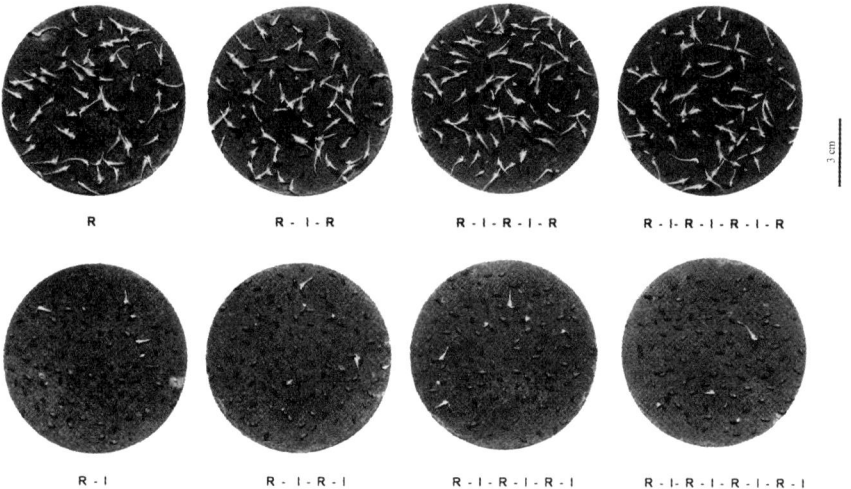

Figure 3. Germination of lettuce seed after alternating exposures to R and FR (I) radiation (From Toole et al., 1953).

This suggested that the photoreceptor existed in two forms one of which could be converted into the other and vice versa. The action spectra suggested that the one had an absorption maximum in the red waveband, the other in the far-red waveband. That this was correct was later shown when the photoreceptor pigment was isolated (Siegelman and Firer 1964) and spectroscopically characterised (Butler *et al.*,1959).

The spectroscopic detectability of phytochromes has only been possible based on the pioneering work of an engineer Karl H. Norris, who developed an electronic egg-grading equipment and then together with Warren L. Butler the dual wavelength ratiospect (Figure 4a, b), which became a very useful instrument to detect phytochrome levels in plant material. This "photochromic" nature of the pigment, which was christened phytochrome by W. L. Butler, also provided a method of assaying the photoreceptor by measuring the absorbance changes caused by alternate irradiation of plant material (or for that matter extracts) with read and far-red light (Butler et al., 1959). It was however necessary to use dark grown (i. e. chlorophyll free) material as chlorophyll not only interferes by its bulk absorbance at 680 nm but also produces red/ far-red reversible absorbance changes not due to phytochrome (Butler 1962).

Figure 4a. Karl H. Norris in 1962 with portable difference meter. Courtesy of USDA.

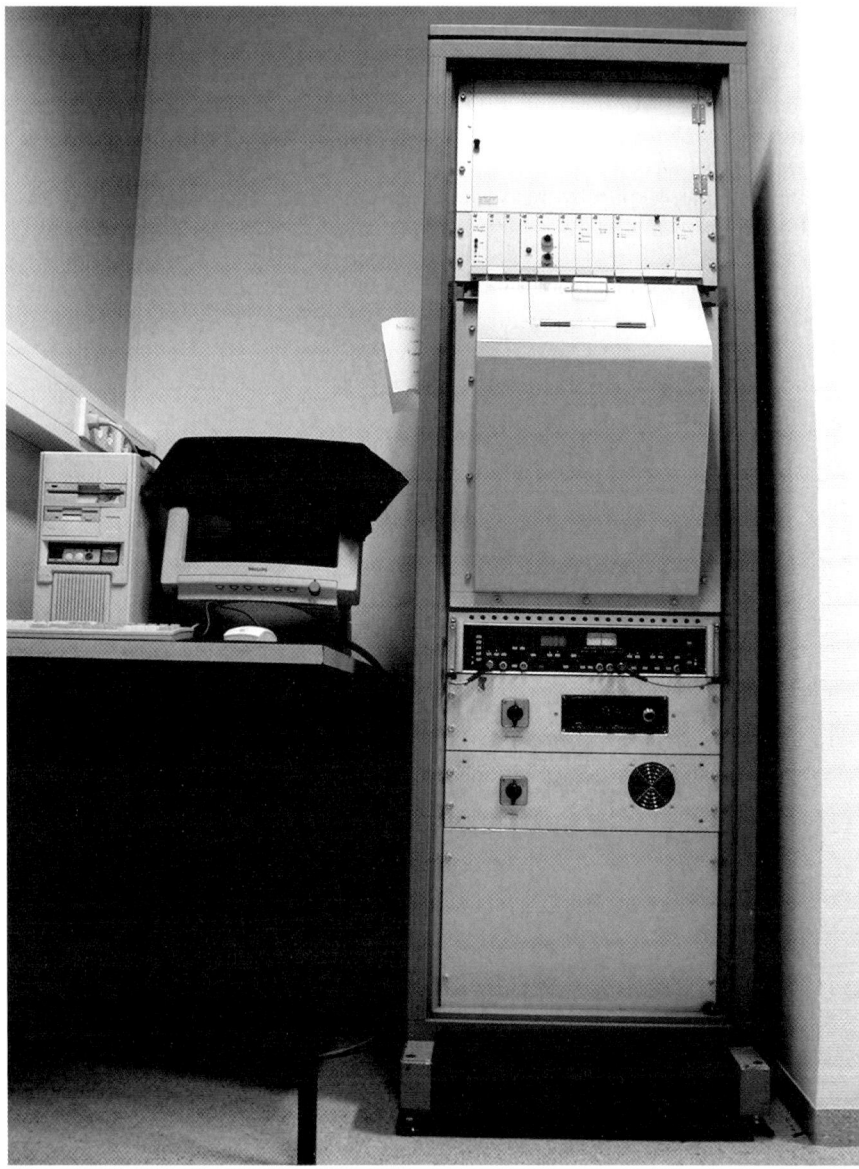

Figure 4b. Costem-built "ratiospect". The construction by the workshop at Freiburg was based on construction plan by Dr. Tilman Lamparter (Berlin). 20 samples can be placed on a rotating wheel and the irradiation programme (controlled time and fluence rate of 660 and 730 nm light) and calibrations and calculations of ΔΔA and Pfr levels is automatically controlled by a computer system.

It was shown that red/ far-red reversible responses were very frequent in plants and occurred at the physiological, biochemical and biophysical level. This property (of red/ far-red reversibility) is now used as one of the operational criteria for the involvement of phytochrome in a response (see Mohr 1972, Smith 1975).

These responses are nowadays often called the phytochrome induction responses and are characterised by their low fluence requirements, that is the response as saturated by a single short pulse of irradiation at low fluence rate, by the above described photo reversibility, their obeying the Bunsen-Roscoe reciprocity law and their having action spectra resembling the absorption spectra of phytochrome *in vitro*, although distortion due to various screening effects may occur. An example of a well-produced induction action spectrum is shown in Figure 5.:

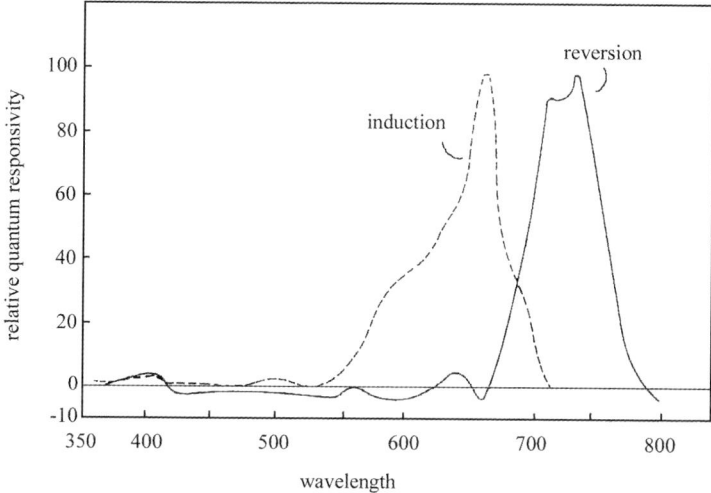

Figure 5. Action spectrum for inductive inhibition of Phaseolus hypocotyl growth and its reversal (from Withrow et al., 1957).

Owing to the fact that the spectrophotometric assay system only functions in chlorophyll free tissue and the necessity to avoid complications due to photosynthesis, a large proportion of the later work on photomorphogenic responses was carried out using etiolated material. This although understandable is regrettable as the etiolated plant is in many respects a rather special case. It is probably however not correct to describe it as an "unnatural" condition as plants do pass through such a condition on germination and the transition from etiolated to de-etiolated must be considered a natural occurrence of great importance to plants. Such experiments of an "induction" or related type carried out for green plants are described in a later section.

Before leaving the red/ far-red reversible induction effects it should be noted that Julius Sachs, the great 19th century plant physiologists who was in many respects

100 years ahead his time, had in 1857 described experiments on leaf movements where he found that *Phaseolus* leaves in the "day position" moved to the "night position" if irradiated with light from a ruby glass. If such a plant was then irradiated with a cobalt glass the leaves moved to the "day position". This cycle could be repeated. Sachs was a careful experimenter and unlike many present day workers described his light sources properly. His ruby source was red and orange, his blue source blue-violet with far-red (many workers to this day are or act as though they are unaware that contamination of blue light sources by far red light is often a serious problem). Sachs was undoubtedly looking at the phytochrome system, which was only properly described 100 years later.

3. THE "HIGH IRRADIANCE RESPONSES"

Since it was apparent that plants living under natural conditions do not inhabit an environment of monochromatic or for that matter white light pulses it was thought to be of importance, after the discovery of the red/ far-red reversible phytochrome system, to study the effects of longer periods of irradiation on various plant responses. Although it was expected that here also the phytochrome system would play an important role, certain previous results, particularly those of Wassink and co-workers using older (more mature) *green* plants suggested that the action spectrum might differ from that for the "classical" phytochrome responses.

The "Beltsville" group accordingly produced action spectra for the effect of longer irradiation on anthocyanin synthesis in *Brassica oleracea* L. and *Brassica rapa* L. and on anthocyanin synthesis and hypocotyl growth inhibition in *Sinapis alba* L. (Siegelman and Hendricks, 1957; Mohr 1957).

The results were surprising, the action spectra differing in all three cases form those for known phytochrome responses. Peaks of effectiveness were found in the far-red (690 – 720 nm) with a secondary peak in the blue waveband. Red was at best a shoulder to the far-red peak. It was thus though necessary to postulate a new photoreceptor for these responses which were called the "High Energy" or later "High Irradiance" responses (HER, HIR) on account of the long irradiation at relatively high fluence rates required for the expression of the responses.

A problem in producing action spectra for this type of response was (and is) that pulse (induction) treatment often also produces the same response due to the "normal" "low energy" phytochrome induction system. It was therefore necessary to separate that portion of the response due to induction from that due to the HIR. This was tried using various techniques. For example Mohr (1959) studying cotyledon growth in *Sinapis alba* determined how much response was obtained using light pulses followed by dark periods. He then assumed this to be the response due to the induction system and corrected the response obtained with continuous irradiation accordingly. The resulting action spectrum did not differ significantly from those already obtained (see above). Another approach was to take a response where the two systems worked antagonistically. Such a response is hypocotyl formation in *Lactuca sativa* L. Here red pulses stimulate hook formation, whereas continuous blue or far-red irradiation reopens the hook. An action spectrum for hook reopening

(Mohr and Noblé, 1960) showed again the main features described above with the exception that there was no effectiveness of the red waveband.

Further action spectra confirmed this general pattern (e. g. Evans et al., 1965). Four further action spectra are however worthy of special mention. One of these is that produced by Hartmann (1967) for the *Lactuca* (*Lactuca sativa* L. var. Grand Rapids tip burn resistant strain) hypocotyl growth inhibition response. This was an extremely detailed action spectrum produced using more than 60 Schott DIL type narrow band interference filters (c. 10 nm half bandwidth) with wavelengths between 320 and 1000 nm. This massive undertaking which required more than 3 years to complete showed an action spectrum with peaks in the near ultraviolet (360 nm), blue and far-red (717 nm) wavebands. The far-red peak was sharp, little of no effectiveness was found in the red waveband and the detailed nature of the action spectrum revealed for the first time a fine structure in the blue with B sub peaks at c. 420, 450 and 480 nm. These blue peaks and the near UV peak are very similar to the fine structure of many "blue light receptor" (BLR) responses and gave rise to speculation that the BLR was responsible for this part of the action spectrum.

Hartmann's action spectrum is reproduced in Figure 6.:

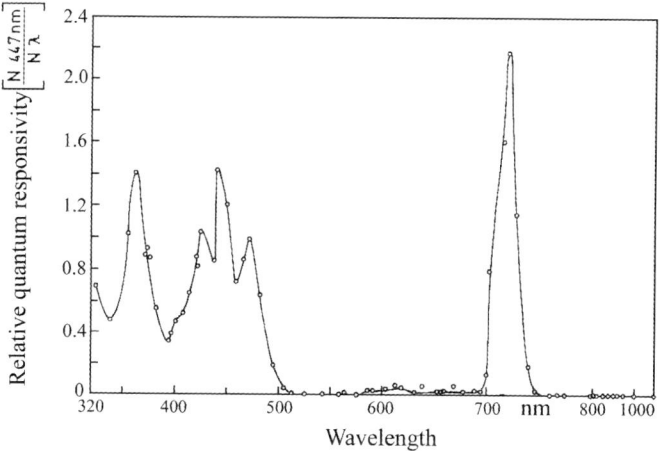

Figure 6. Action spectrum for inhibition of hypocotyl growth in Lactuca (from Hartmann 1967).

Using dichromatic irradiation, i. e. constant fluence rate of long wavelength far-red light with added variable fluence rates of red light Hartmann was able to show that phytochrome is the photoreceptor of the HIR and that a certain photoequillibrin Pfr/ Ptot ratio results in a maximal responsiveness. Various models have been suggested to explain the HIR (peak in far-red) and fluence rate-dependence but the problem seems to be unresolved up to now (see Schaefer et al., 1983 for review, van der Woude, 1987; Hennig et al., 2000).

In 1980 Beggs et al., reported action spectra inhibition of hypocotyl elongation by continuous irradiation in light and dark-grown *Sinapis Alba* L. seedling. These authors observed blue-red and far-red maximum for etiolated seedlings. The later was lost in light-grown seedlings treated or non-treated with the herbicide Norfluorazon. 1981 Holmes and Schaefer reported an even more detailed actions spectrum for the same responses. In this study the authors showed that brief red light pulses given 1–2 h prior to the onset of the continuous light treatment result in a gradual reduction of the far-red peak without a significant wavelength shift of the maximum. It was thus concluded that the light-labile phytochrome (now turned phyA) is degraded after the red light pre-treatment thus resulting in a reduction of the HIR. This type of light regime was almost 15–20 years later used as a strategy to identify hypersensitive mutants for the phyA pathway (Büche et al., 2000; Dieterle et al., 2001).

4. VERY LOW FLUENCE RESPONSES

As a third mode of phytochrome action the Very Low Fluence Responses (VLFR) have been discovered. Mandoli and Briggs (1981) observed the fact that safe light is not safe and that a few photons are sufficient to elucidate inhibition of growth of corn mesocotyles. Schaefer et al. described already 1976 that continuous infrared light (≈ 1000 nm) is sufficient for very strong inhibition of mesocotyl growth of Avena seedlings. Later Shinomura et al. (1996) and Hartmann described the VLFR for seed germination. Hartmann could show that even a few seconds of starlight are sufficient to elucidate a response. This clearly indicates the existence of an extreme amplification mechanism, which is not tried to be identified by genetic methods.

Starting in the early 80ies molecular biology and later also genetical methods were implied to photomorphogenetic research. This will be summarised in chapter 3.

5. FURTHER READING

Kendrick, R. E. and Kronenberg, G. H. M., eds. Photomorphogenesis in Plants. Dordrecht, The Netherlands: Kluwer Academic Publishers, 1994.
Mohr, H. Lectures on Photomorphogenesis. Berlin, Heidelberg, New York: Springer Verlag, 1972.
Sage, L. C. Pigment of the Imagination: A History of Phytochrome Research. San Diego: Academic Press, 1992.
Smith, H. Phytochrome and Photomorphogenesis. London and New York: McGraw-Hill, 1975.

6. REFERENCES

Borthwick, H. A., Hendricks S. B., Parker, M. W. (1948) Action spectrum for photoperiodic control of floral initiation of the long day plant Wintex barley (*Hordeum vulgare*). *Bot Gaz*, *110*, 103–118.
Borthwick, H. A., Hendricks, S. B. Parker, M. W. (1951) Action spectrum for inhibition of stem growth in dark grown seedlings of albino and non-albino barley (*Hordeum vulgare*). *Bot Gaz*, *113*, 95–105.
Borthwick, H. A., Hendricks, S. B. Parker, M. W., Toole, E. H., Toole, V.K. (1952) A reversible photoreaction controlling seed germination. *Proc Natl Acad Sci USA*, *38*, 662–666.

Borthwick, H. A., Hendricks, S. B., Toole, E. H., Toole, V.K. (1954) Action of light on lettuce-seed germination. *Bot Gaz (Chicago), 115,* 205–225.

Büche, C., Poppe, C., Schaefer, E., Kretsch, T. (2000) *Eid1*: A New Arabidopsis Mutant Hypersensitive in Phytochrome A-Dependent High-Irradiance Responses. *Plant Cell, 12,* 547-558.

Butler, W. L (1962) Effects of red and far red light on the fluorescence yield of chlorophyll *in vivo*. Biochim. *Biophys Acta, 64,* 309–317.

Butler, W. L., Norris, K. H., Siegelman, H. W., Hendricks, S. B. (1959) Detection, assay and preliminary purification of the pigment controlling photo responsive development of plants. *Proc Natl Acad Sci USA, 45,* 1703–1708.

Dieterle, M., Zhou, Y.C., Schäfer, E., Funk, M., Kretsch, T. (2001) EID1, an F-box protein involved in phytochrome A-specific light signalling. *Genes Dev, 15,* 939–944.

Evans, L. T., Hendricks, S. B., Borthwick, H. A. (1965) The role of light in suppressing hypocotyl elongation in Lettuce and *Petunia*. *Planta, 64,* 201–218.

Flint, L. H., McAlister, E. D. (1935) Wavelengths of radiation in the visible spectrum inhibiting the germination of light sensitive lettuce seed. *Smithson Misc Collect, 96,* 1–9.

Flint, L. H., McAlister, E. D. (1937) Wavelengths of radiation in the visible spectrum promoting the germination of light-sensitive lettuce seed. *Smithson Misc Collect, 96,.* 1–8.

Garner, W. W., Allard, H. A. (1920) Effect of the relative length of day and night and other factors of the environment on growth and reproduction in plants. *J Agric Res, 18,* 553–606.

Hales, S. (1727) *Statical Essays*, 1, 334

Hartmann, K. M. (1967) Ein Wirkungsspektrum der Photomorphogenese unter Hochenergiebedingungen und seine Interpretation auf der Basis des Phytochroms (Hypokotylwachstumshemmung bei *Lactuca sativa* L.). *Z Naturforsch, 22b,* 1172–1175.

Hennig, L., Büche, C., Schäfer, E., (2000) Degradation of phytochrome A and the high irradiance response in *Arabidopsis*: A kinetic analysis. *Plant Cell Env, 23,* 727–734.

Holmes, M .H., Schäfer, E. (1981) Action spectra for changes in the 'high irradiance reaction' in hypocotyls of *Sinapis alba* L. *Planta, 153,* 267–272.

Klebs, G. (1918) Über die Blütenbildung bei *Sempervivum*. *Flora (Jena), 111,* 128.

Linnaeus, C. (1739) *Rön om växters plantering grundat på naturen*. Kungl. Svenska Vetenskapsakademiens Handlingar 1.

Mandoli, D. F. and Briggs, W. R. (1981) Phytochrome control of two low-irradiance responses in etiolated oat seedlings. *Plant Physiol, 67,* 733–739.

Mohr, H. (1957) Der Einfluss monochromatischer Stahlung auf das Längenwachstum des Hypokotyls und auf die Anthocyanbildung bei Keimlingen von *Sinapis Alba* L. (=*Brassica alba* Boiss.). *Planta, 49,* 380–405.

Mohr, H. (1959) Der Lichteinfluss auf das Wachstum der Keimblätter bei *Sinapis Alba* L. *Planta, 53,* 219-245.

Mohr, H. (1964) The control of plant growth and development by light. *Biol Rev Camb Philos Soc, 39,* 87–112.

Mohr, H. *Lectures on Photomorphogenesis*. Berlin, Heidelberg, New York: Springer-Verlag, 1972.

Mohr, H., Noblé, A. (1960) Die Steuerung der Schliessung und Öffnung des Plumula-Hakens bei Keimlingen von *Lactuca sativa* durch sichtbare Strahlung. *Planta, 55,* 327–342.

Otto, V., Schäfer, E. Nagatani, A., Yamamoto, K. T. Furuya, M. (1985) Phytochrome control of its own synthesis in *Pisum sativum*. *Plant Cell Phyiso,. 25,* 1579–1584.

Parker, M. W., Hendricks, S. B., Borthwick, H. A. (1950) Action spectrum for the photoperiodic control of floral initiation of the long day plant *hyoscyamus Niger*. *Bot Gaz, 111,* 242–252.

Parker, M. W., Hendricks, S. B., Borthwick, H. A., Scully, N. J. (1945) Action spectrum for the photoperiodic control of floral initiation in Biloxi soybean. *Science, 102,* 152–155.

Parker, M. W., Hendricks, S. B., Borthwick, H. A., Scully, N. J. (1946) Action spectrum for the photoperiodic control of floral initiation of short day plants. *Bot Gaz (Chicago), 108,* 1–26.

Parker, M. W., Hendricks, S. B., Borthwick, H. A., Went F. A. (1949) Spectral sensitivities for leaf and stem growth of etiolated pea seedlings and their similarity to action spectra for photoperiodism. *Amer J Bot, 36,* 194–204.

Ray, J. (1686) *Historia Plantarum*, 1, 15.

Sachs, J. (1857) Über das Bewegungsorgan und die periodischen Bewegungen der Blätter von *Phaseolus* und *Oxalis*. *Bot Ztg, 15,* 809–815.

Schäfer, E., Lassig, T.-U., Schopfer, P. (1976) Photo control of phytochrome destruction and binding in dicotyledonous seedlings. The influence of wavelength and irradiance. *Photochem Photobiol, 24*, 267–573.

Schäfer, E., Löser, G., Heim, B. (1983) Formalphysiologische Analyse der Signaltransduktion in der Photomorphogenese. *Ber Dtsch Bot Ges, 96*, 497–509.

Shinomura, T., Nagatani, A., Hanzawa, H., Kubota, M., Watanabe, M., Furuya, M. (1996) Action spectra for phytochrome A- and B- specific photo induction of seed germination in *Arabidopsis*. *Proc Natl Acad Sci USA, 93*, 8129–8133.

Siegelman, H. W., Firer, E. M. (1964) Purification of phytochrome from oat seedlings. *Biochemistry, 3*, 418–423.

Siegelman, H. W., Hendricks, S. B. (1957) Photocontrol of anthocyanin synthesis in turnip and red cabbage seedlings. *Plant Physiol, 32*, 393–398.

Smith, H. *Phytochrome and Photomorphogenesis.* London and New York: McGraw-Hill, 1975.

Tournois, J. (1912) Influence de la lumière sur la floraison du houblon japonais et du chanvre determinées par des semis hatifs. C. R. Hebd. *Sceances Acad. Sci, 155*, 297–300.

Tournois, J. (1914) Études sur la sexualité du houblon. *Ann Sci Naturelles (Botanique), 19*, 49–191.

van der Woude, W. J. (1987) "Application of the dimeric model of phytochrome action to high irradiance responses." In *Phytochrome and Photoregulation in Plants*, Furuya M. (ed) Academic Press, Tokyo, 249–258.

Vince-Prue, D., *Photoperiodism in Plants*. London and New York: McGraw-Hill, 1975.

Withrow, R. B., Klein, W. H., Elstad, V. (1957) Action spectra of photomorphogenic induction and its photo inactivation. *Plant Physiol, 32*, 453–462.

Chapter 2

PHYSIOLOGICAL BASIS OF PHOTOMORPHO-GENESIS

Eberhard Schäfer[1] and Ferenc Nagy[2]

[1]*Albert-Ludwigs-University Freiburg, Institute of Biology II/ Botany, Schänzlestrasse 1, 79104 Freiburg, Germany (e-mail: Eberhard.Schaefer@biologie.uni-freiburg.de)*
[2] *Biological Research Centre, Institute of Plant Biology, P.O. Box 521, 6701 Szeged, Hungary (e-mail: nagyf@nucleus.szbk.u-szeged.hu)*

1. INTRODUCTION

As described in Chapter 1, the phenomenon of photomorphogenesis has been the object of age long observation.

The foremost question arising with respect to any newly observed light-regulated phenomenon is what the photoreceptor(s) controlling the particular response is/are. The first step should always be to investigate the spectral sensitivity.

The correct method to approach this problem is to construct an action spectrum. The problems of action spectroscopy or even analytical action spectroscopy have been reviewed several times (Shropshire, 1972; Hartmann and Unser, 1972; Schäfer *et al.*, 1983), therefore only the basic principles will now be touched upon, with special emphasis on the analysis of the problems involved in the case of photo reversible pigments.

2. CLASSICAL ACTION SPECTROSCOPY

Action spectroscopy is based on Grotthus-Draper's law, which states that only absorbed light can cause photochemical reactions. Of the absorbed photons only a fraction (ϕ_λ the quantum yield of transformation) will cause molecular transformation leading to a certain photo biological response. For monochromatic irradiation with photon fluence rate N_λ the number of molecules transformed per second is:

$$v = P \bullet N_\lambda \bullet E_\lambda \bullet \Phi_\lambda$$

Where P is the concentration of the pigment in the ground state and E_λ is the molar absorption cross section of the pigment.

A major problem arises from the fact that $N\lambda$ is the photon fluence rate at the location of the pigment. This value is normally not directly measurable; it is rather the applied fluence rate that can be easily measured and varied. Due to reflections, scattering and absorption by other pigments, local fluence rate can be variable. This must always be considered when interpreting the fluence rate and wavelength dependence of a physiological response.

The straightforward and easy basis of action spectroscopy is the light-equivalence principle:

The effect of light on the measured response will be equivalent in any two experiments when – and only when – the rates of the primary reactions are equal during the experiments.

Accordingly, in two experiments (a and b), if va equals vb, response a equals response b.

$$P \bullet N_{\lambda a} \bullet E_{\lambda a} \bullet \Phi_{\lambda a} = P \bullet N_{\lambda b} \bullet E_{\lambda b} \bullet \Phi_{\lambda b} \rightarrow R_a = R_b$$

In order to obtain information about the pigment from the measured/measurable responses, the logic of the above equation needs to be reversed, i.e. to conclude from the identity of the photochemical rates. This is strictly possible if the fluence rate response function is a monotonous function. If this does not hold, the analysis has to be restricted to the part that is monotonous.

On this basis one can conclude:

$$R_a = R_b \rightarrow \frac{\sigma_{\lambda a}}{\sigma_{\lambda b}} = \frac{1}{N_{\lambda a}} \div \frac{1}{N_{\lambda b}}$$

Whereby $\sigma\lambda < E_\lambda \Phi_\lambda$ is the photo conversion cross-section.

Thus an action spectrum can be constructed after measuring photon fluence rate response curves and plotting for a specific chosen response level the reciprocal level as a function of wavelength.

This method is still valid nowadays and has been chosen in photomorphogenesis to characterize UVB responses, UVA blue light responses and phytochrome-mediated responses. In UVB to blue light-mediated responses a major problem is that the light gradient in the tissues is very steep due to high light scattering and absorption by many plant components. In phytochrome-mediated responses we are confronted with the problem of dealing with a photo reversible pigment. Clearly, this was the cause of the conflicts in interpreting action spectra under continuous irradiation leading to the high irradiance responses in etiolated seedlings.

Before describing the specific problems of action spectroscopy for photo reversible pigments a few remarks to the interpretation of photoreceptor mutants. In genetic screens very often hyposensitive mutants were obtained and when the cloned gene encodes a chromophore binding protein it is often concluded that the photoreceptor has been identified. This is clearly not <u>necessarily</u> the case. Also a chromophore carrying protein in the transduction cascade or even a screening pigment can result in altered spectral sensitivity. A mutation in the <u>rate</u> limiting photoreceptor should result in the same shift of the threshold fluence rate for any wavelength tested. A mutation in another pigment that modulates the reaction will lead to a change in the slope of the fluence rate response curve. Unfortunately, such tests have not been performed in most cases.

The problem of photo reversibility can be overcome in two ways:

Action spectroscopy after a single light pulse and extrapolation to threshold fluence rates and

By analytical action spectroscopy.

The light-equivalence principle also holds for photo reversible pigments.

In this case, for example for the phytochrome system

$$Pr \underset{k_2}{\overset{k_1}{\rightleftharpoons}} Pfr,$$

we have two light reactions. If – as is the case for dark-grown samples – all the pigments are in the Pr form at very low fluence rate, i. e. in the initial phase – only one photoreaction is dominating

$$Pr \overset{k_1}{\longrightarrow} Pfr$$

and thus an action spectrum can be constructed, which should reflect the Pr absorption spectrum. It should be stated that the analysis of threshold fluence rates is unfortunately forgotten in this age of molecular biology.

The light equivalence principle for photo reversible pigments requires that both light reactions

$$v_1 = Pr \bullet N_\lambda \bullet E_{r\lambda} \bullet \Phi_{r\lambda} \text{ and } v_2 = Pfr \bullet N_\lambda \bullet E_{fr\lambda} \bullet \Phi_{fr\lambda}$$

should be equal in two experiments (a and b), which are compared. In other words, as discussed by Schäfer et al., (1983), the photoequillibrin ($y = k_1 / k_1 + k_2$) and the rate of photo conversion ($k_1 + k_2$) should be identical in the experiments compared. This allows detecting whether, for example, the response in blue light is mediated by phytochrome or an additional blue light receptor is also involved.

This approach has been used most convincingly in the analysis of far-red HIR. Dichromatic irradiation with two wavelengths, which were separately ineffective, resulted in strong inhibition of the hypocotyl growth of *Lactuca* seedlings (Figure 1).

Using this method it was possible to demonstrate that the HIR can be characterized by an optimal responsiveness of the system at a certain phytochrome equilibrium (1 – 10% Pfr in most cases) and fluence rate dependence. The question what are the kinetic and molecular bases of these dependences remains under debate and clearly requires a deeper understanding of the properties of the phytochrome system.

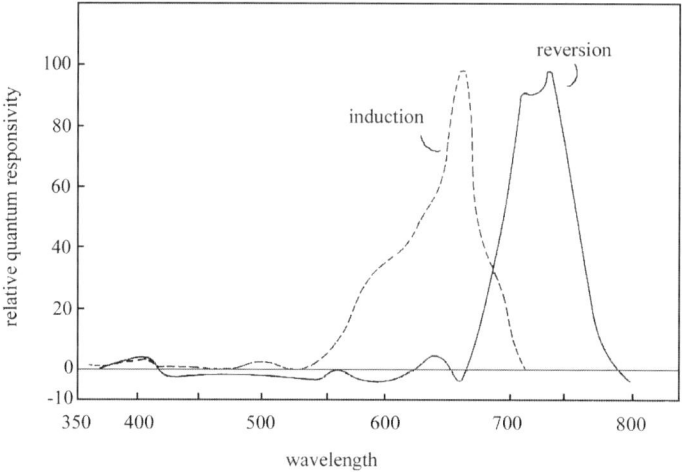

Figure 1. Action spectrum for inhibition of hypocotyl growth in Lactuca (from Hartmann, 1967)

3. MODE OF FUNCTION OF PHYTOCHROME

As soon as the photoreceptor is identified the next burning question is the molecular basis of its function.

The approaches to this question are numerous. In the early days physiologists collected serious responses, which are regulated by phytochromes to get a feeling what is controlled by the photoreceptor and what might possibly be its function. The responses ranged from seed germination, growth responses and enzyme activities to flowering and senescence (Mohr, 1972; Smith, 1974). With the establishment of molecular biology and genetics, it was then later possible to demonstrate rapid control of transcription rates of several genes (Nagy and Schäfer, 2002; Quail, 2002). These data all led – although still indirect – to the support of the hypothesis of gene regulation by phytochromes. On the other side there were several specific responses mediated by phytochrome, i. e. phototropin and polarotropism in ferns and mosses (see Chapters 24 and 25), change of chloroplast orientation in some algae and ferns (Kraml, 1994; Chapter 24), and alterations of membrane potentials (Tanada, 1968; Chapter 19). These observations supported the view of phytochrome being a membrane effector.

Early on in photomorphogenesis research the primary mode of function of phytochrome was subject to intensive debate. Three hypotheses were mainly discussed:

Phytochromes act as light-dependent enzymes
Phytochromes act as membrane effectors
Phytochromes act as regulators of transcription

Although we now know that possibly all three hypotheses are correct, at the time they had only one thing in common, namely that each had only very limited experimental support and no direct proof whatsoever.

Hendricks and Borthwick were convinced that phytochrome was a light-dependent enzyme controlling carbon flow. More than 40 years elapsed before Lagarias' group presented the first data indicating that oat phytochrome A may be a light-dependent kinase (Wong *et al.*, 1986). Only after sequencing several bacterial genomes did it become evident that, firstly, bacteria contain phytochromes and, secondly, these bacterial phytochromes act as light-dependent histidine kinases and are part of a light-dependent two-component phosphorelay system (see Chapters 5 and 6). Thus, bacterial phytochromes are obviously light-dependent enzymes; unfortunately, this is by far not so clear in the case of higher plant phytochromes.

The membrane hypothesis received strong support from analyses of polarotropism and phototropism in ferns and mosses and from the action dichroism of light-dependent chloroplast movement in Mougeotia (see Chapters 23 and 24; and Kraml as well as Wada and Sugai in Kendrick and Kronenberg, 1994). The recent observation that some ferns and also Mougeotia (personal communication, M. Wada) contain a chimeric photoreceptor – first half phy, second half phot, see Chapter 24) – shows that chloroplast orientation is controlled by this chimeric photoreceptors. Despite intensive research mosses do not contain this chimeric photoreceptor, polarotropism and phototropism at least in mosses seem to be controlled by true phytochromes localized in an organized manner that allows action dichroism. Additional support came from observations showing that phytochrome alters membrane permeability. The Tanada effect as well as the observation of phytochrome-induced changes in bioelectrical potential were all used and interpreted as strong evidence for the assumption that phytochrome acts as a membrane effector. However, as aptly pointed out by Quail (1983), all these data are only indirect evidence in favour of this hypothesis. Especially diffusion time is quite sufficient to allow a cytosolic primary function of phytochrome resulting in modulations of membrane properties. Thus, convincing evidence demonstrating that phytochrome is a <u>direct</u> effector of membrane function is still lacking (see Chapter 9). This is especially difficult; because after sequencing phytochrome genes it became obvious that phytochrome cannot be an integral membrane protein. The hypothesis formulated by Mohr (1966), i.e. that phytochrome acts as a regulator of transcription was especially strongly debated. The real question in this case is how direct is the control of transcription. Although the hypothesis was primarily based on measurements of phytochrome-regulated enzyme activities and on inhibitor studies for protein and RNA synthesis, the introduction of molecular biology and transgenic plants clearly showed that phytochrome can turn on and off transcription of target genes (see Chapter 17 and Batschauer *et al.*, in Kendrick and Kronenberg, 1994). The question again is how direct this control is: whether control is exerted via enzyme activity or membrane modulation or whether phytochrome acts via direct transcriptional regulation. Surprisingly, recent data indicate the possibility of direct control of the transcription rate of target genes by direct interaction of phytochrome in its physiologically active Pfr form with a transcription factor, PIF3, bound to promoters (Martinez-Garcia, 2000; Quail, 2002).

Irrespective of these results, the long-time debate whether there is a single primary event leading to the enormous variety of phytochrome-mediated responses observed or whether there are several distinct primary functions is still on.

How has this question been approached?

4. CORRELATIONS BETWEEN IN VIVO SPECTROSCOPICAL MEASUREMENTS AND PHYSIOLOGICAL RESPONSES

Karl Norris made a very important contribution to the discovery of phytochrome by constructing a new type of spectrophotometer, which allowed determining phytochrome levels (measured as Δ (ΔA) in *planta*. Due to the presence of chlorophyll in light-grown seedlings, measurements are only possible in dark-grown, or light grown but chlorophyll-free tissues. This led to the discovery of phytochrome synthesis, degradation in its Pfr form (later also in its Pr form), dark reversion and the establishment of photo equilibria. As we now know, *in vitro* measurements always determined the levels of phyA, the most abundant phytochrome species.

A basic question was what the active form of phytochrome was and whether there was a correlation between the amount of phytochrome in active form and the physiological response.

Although Hendricks had already argued that Pfr is the active form, the issue was only settled when it was discovered that extremely low levels of light could induce several responses. These responses were named Very Low Fluence Response (VLFR) (Mandoli and Briggs, 1981). In these cases at threshold levels sometimes as few as one or only very few Pfr molecules per cell are sufficient to elicit a measurable response. Such responses can even be induced by "safe light" (green light sources commonly used) (Holmes, 1984) or by irradiations in the near infra-red (up to 1093 nm, Schäfer *et al.*, 1982). This allowed to obtain values for the relative photo conversion cross section for Pr from 600 to 1100 nm. This value spans 10^9 orders of magnitude.

Even before *in vitro* spectroscopical measurements were possible, however, Hendricks *et al.*, (1956) were already able to estimate the relative responsiveness of phytochrome by assuming first-order photo conversion kinetics (Figure 2).

By comparing *in vitro* spectroscopical measurements of phytochrome with physiological experiments, good correlations were obtained. Due to the nature of *in vitro* spectroscopy, these correlations were only possible for analysis of dark-grown seedlings after a light pulse treatment.

On the other hand, paradoxes were also reported. The most famous of these are the pisum paradox (Hillman, 1965) and the Zea paradox (Chon and Briggs, 1966). The pisum paradox led to the conclusion that only a small fraction of the spectrophotometrically detectable phytochrome controls pea stem elongation. Now we can probably explain this by the presence not only of phyA, but other phytochromes. The Zea paradox describes that the stimulating effect of very weak red light could be reverted by subsequent far-red light establishing even higher levels of Pfr. A plausible explanation – similar to Hendricks' suggestion – is that weak red light already leads to intracellular translocation (nuclear import see Chapter 9), in the course of which Pfr becomes active. The far red light pulse will then transform Pfr back to Pr at its place of function, thus leading to a reversion of the introduction response, although the second light pulse produces more Pfr.

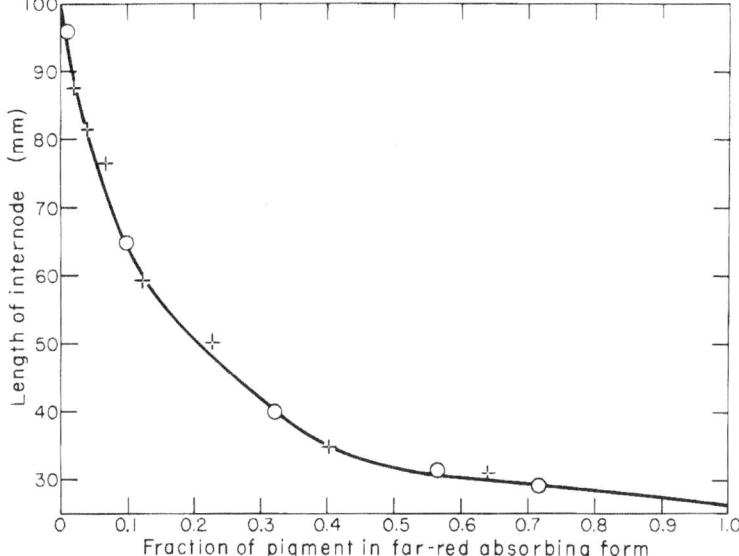

Figure 2. Internode lengths of Pinto beans as a function of pigment conversion: (o) irradiation in R part of spectrum of plants with pigment initially in R-absorbing form; (+) irradiation in FR with pigment initially in FR-absorbing form (from Hendricks et al., 1956).

Two other types of experiments to analyse the fraction of physiologically active phytochrome and its kinetics of signalling were often reported: 1) the null-point-method and 2) kinetics of loss of reversibility. The null-point-method analyses the fraction of phytochrome in Pfr form, which controls physiological responses after a light pulse (normally 5 min red light), followed by a dark period. After a dark period of 3h it is expected that, due to destruction, only 10% Pfr remains, the half-life for destruction of Pfr being determined as 1h. At that time point saturating light pulses establishing different photoequilibria are given and it is determined which newly established photo equilibrium will not alter the response (Hillmann, 1965, Schäfer et al., 1984). This is still considered a highly reliable test of the kinetics of the phytochrome species controlling a certain response. The analysis of the kinetics of loss of reversibility is aimed at acquiring information about the kinetics of signal transduction. In this case, dark control and response after a saturating light pulse are measured. In addition, a far-red light pulse (preferably long wavelength for red light pulse to establish Pfr levels as low as possible) is given, followed by dark periods of various lengths, and the response is analysed after a subsequent prolonged dark period. The differences of the response levels to the red light pulse alone and to the red Δt dark plus far-red light pulse are then plotted as a function of Δt darkness. As discussed by Fukshansky and Schäfer (1984) in detail, these measurements do not analyse the kinetics of signal transduction but rather the kinetics of inactivation of the active form of phytochrome. Thus this method is again still useful, especially to test the dynamics of those phytochromes which are not measurable by *in vitro* spectroscopy at the present time.

In summary one can say that in several cases for phyA-mediated responses a good correlation between the level of active phy and the physiological response has been described. Based on the new information about the complex kinetics of phyA, including cytosolic formation of SAP's (Sequestered Areas of Phytochrome), nuclear import and formation of nuclear complexes, it is questionable which molecules are the functional molecules controlling the physiological response under investigation. Furthermore these correlations were almost exclusively tried for very late responses like growth or anthocyanine formation and therefore should take into account the complex kinetics of phyA.

Also for phyB responses such correlations were obtained. But again as for phyA the new observations of intracellular dynamics of the photoreceptor must be considered. A strong discrepancy between physiological data and spectroscopical data should be mentioned. Using the method of measuring loss of reversibility kinetics decay times for the active phy are in the order of a few hours. This is in contrast to recent measurements of phyB kinetics (Sweere *et al.*, 2001). Although these data nicely show that inactivation of phyB can be mediated by dark reversion, this reaction with half-life of a few minutes is too fast compared with the. This therefore creates a new paradoxon, which has not been solved yet. At a first glance the data indicate that a sub pool of phyB must be very stable in the active state but this pool has not been detected so far! An excuse can be the complex intracellular kinetics and the observation that – at least the PHYB:GFP forms some relative stable nuclear complexes.

5. PHYTOCHROME RESPONSE TYPES

As described in Chapter 1, phytochrome was discovered by the use of action spectroscopy. The surprise was that a single red light pulse could induce several responses and that a subsequent far-red light pulse could reverse this induction. The pressing question whether this phenomenon is due to the function of two antagonistic pigments or one photo reversible pigment was settled by the purification of the photo reversible pigment (Butler *et al.*, 1959). The induction response clearly demonstrates that phytochrome can perform its function in darkness after its light activation. This reaction mode is named Low Fluence Response (LFR. Besides LFR, three other reaction modes have been observed after irradiation of etiolated seedlings: the Very Low Fluence Response (VLFR, the far-red High Irradiance Response (HIR) and the red High Irradiance Response (HIR). After the discovery of two different types of phytochromes, namely type I light-labile phytochromes and type II light-stable phytochromes (Furuya and Schäfer, 1996) and finally the discovery of five phytochrome genes (PHYA – PHYE) in Arabidopsis (Clack and Sharrock, 1994), it was possible to address the question which phytochrome controls what response type. Using phyA-deficient mutants it could be shown unequivocally that both VLFR and far-red HIR are mediated by phyA (Shinomura *et al.*, 1996). These two phenomena are made possible by the high accumulation levels of phyA in etiolated seedlings, allowing the detection of very low levels of light – like rod rhodopsin – and phytochrome degradation in Pfr form (and also in Pr form after light cycling) and its complex kinetic properties. PhyB mutants showed that the LFR and the red HIR are mediated in Arabidopsis dominantly by phyB, but other phytochromes clearly can contribute to these types of response. It is often assumed that phyA does not contribute to these responses, but this problem is not yet decided. Recent analysis of the eid1 mutant resulting in phyA-dependent red light LFR and HIR support the view that phyA can also mediate LFR and red HIR (Dieterle *et al.*, 2001).

The mechanism underlying the red HIR – controlled by light-stable phytochromes – and the inactivation of phyB–E signalling after a light-dark transition is not clear. The recent observations that light-labile phytochromes show a Pfr → Pr dark reversion – after expression and re-constitution of the holoprotein in yeast (Eichenberg *et al.*, 2001) and for over-expressed phyB in *planta* (Sweere *et al.*, 2001) – indicate that rapid dark reversion might be the mechanism for inactivation and fluence rate detection.

Phytochromes are involved in light-controlled development throughout the ontogeny of plants, from germination all the way to flower induction and senescence (Figure 3).

Their major role seems to be the detection of plant neighbours and shading by other plants (see Chapter 22). In this respect, detection of red/ far-red ratios of the spectral distribution of light is dominating. These responses, summarized as shade avoidance syndrome are primarily mediated by light-stable phytochromes (phyB–E) with phyB as the major player. A further mode of action manifests itself in the so-called end-of-day responses, analysed by Downs *et al.*, as early as 1957. This type of response reflects the capacity of the plants to detect, with the help of phytochromes, light quality at the end of the day and to measure Pfr levels after the light-dark transition. How this is achieved – especially in view of the rapidity of dark reversion – and how this contributes to the measurement of day length is unclear. For measuring day length and setting the circadian clock the main light period seems to be of primary importance, because red or far-red light pulses at the end of the day seem not to affect the clock (see Chapter 26).

6. SUMMARY

Physiological experiments gave early demonstration of the multiple functions and multiple modes of action of phytochrome. The discovery of different phytochrome types and five phytochrome genes opened a possibility to address these questions in molecular terms. The use of photoreceptor mutants allowed not only to attribute different response types to different phytochromes but also to solve some of the previously described paradoxes.

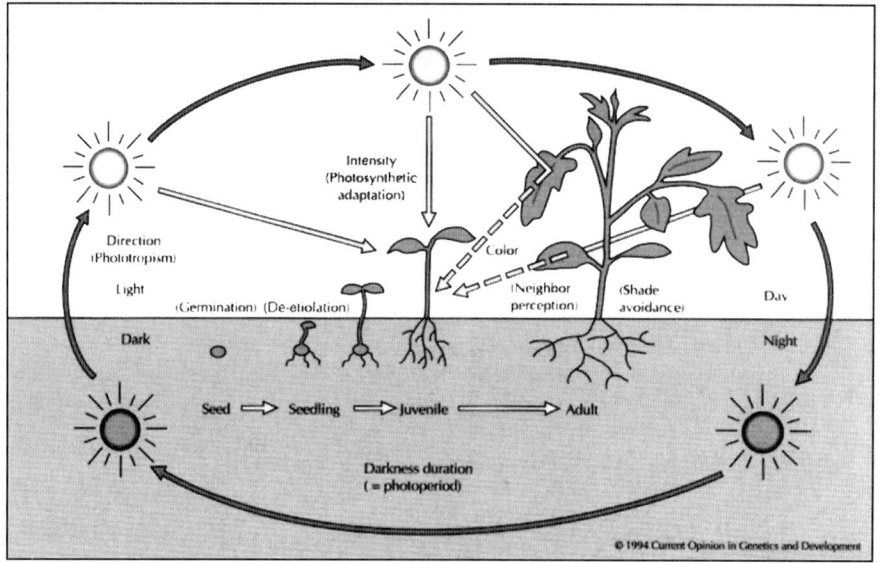

Figure 3. Schematic summary of the major, non-photosynthetical light signals and reactions of plants in response to these signals. The plants receive the changes in this environmental signal throughout their life cycle. The major information is: the depth seeds and young seedlings are buried in the soil, attenuation and light quality changes by shading of reflection from neighbouring plants, duration of the light/ dark period and direction of the light (modified after Quail, 1994).

7. REFERENCES

Butler, W. L., Norris, K. H., Siegelman, H. W., Hendricks, S. B. (1959) Detection, assay and preliminary purification of the pigment controlling photo responsive development of plants. *Proc Natl Acad Sci USA*, 45, 1703–1708.

Chon, H. P. and Briggs, W. R. (1966) Effect of red light on the phototropic sensitivity of corn coleoptiles. *Plant Physiol*, 41, 1715–1724.

Clack, T., Matthews, S., Sharrock, R. A. (1994) The phytochrome apoprotein family in Arabidopsis is encoded by five genes: the sequence and expression of PHYD and PHYE. *Plant Mol Biol*, 25, 413–417.

Dieterle, M., Zhou, Y. C., Schäfer, E., Funk, M., Kretsch, T. (2001) EID1, an F-box protein involved in phytochrome A-specific light signalling. *Genes Dev*, 15, 939 944.

Downs, R. J., Hendricks, S. B., Borthwick, H. A. 1(957) Photoreversible control of elongation of pinto beans and other plants under normal conditions of growth. *Bot Gaz (Chicago)*, 118, 199 208.

Eichenberg, K., Bäurle, I., Paulo, N., Sharrock, R. A., Rüdiger, W., Schäfer, E., (2000) Arabidopsis phytochromes C and E have different spectral characteristics from those of phytochromes A and B. *FEBS Lett*, 470, 107–112.

Fukshansky, L. and Schäfer, E. (1983/1984?) "???" In *Encyclopedia of Plant Physiology New Series, Vol. 16A*, ??? (eds) Verlag, 178–212

Furuya, M. and Schäfer, E. (1996) Photo perception and signalling of 'induction' reactions by different phytochromes. *Trends Plant Sci*, 1, 301–307.

Hartmann, K. M., Unser, I. C. (1972) Analytical action spectroscopy with living systems: Photochemical aspects and attenuance. *Ber Dtsch Bot Ges*, 85, 481–551.

Hendricks, S. B., Borthwick, H. A. (1967) The function of phytochrome in regulation of plant growth. *Proc Natl Acad Sci USA*, *58*, 2125–2130.

Hendricks, S. B., Borthwick, H. A., Downs, R. J. (1956) Pigment conversion in the formative responses of plants to radiation. *Proc Natl Acad Sci USA*, *42*, 19–26.

Hillman, W. S. (1965) Phytochrome conversion by brief illumination and the subsequent elongation of etiolated pisum stem segments. *Physiol Plant*, *18*, 346–358.

Holmes, M. G. (1984) "Light sources". In *Techniques in Photomorphogenesis*, Smith, H. and Holmes, M. G. (eds) Academic Press, London, 43–79.

Kraml, M. (1994) "Light direction and polarisation.". In *Photomorphogenesis in Plants*, Kendrick R.E. and Kronenberg G.H.M. (eds). Kluwer Academic Publishers, Dordrecht, 417–443.

Mandoli, D. F. and Briggs, W. R. (1981) Phytochrome control of two low-irradiance responses in etiolated oat seedlings. *Plant Physiol*, *67*, 733–739.

Martinez-Garcia, J. F., Huq, E., Quail, P. H. (2000) Direct targeting of light signals to a promoter element-bound transcription factor. *Science*, *288*, 859–863.

Mohr, H. (1966) Differential gene activation as a mode of action of phytochrome 730. *Photochem Photobiol*, *5*, 469–483.

Mohr, H., *Lectures on Photomorphogenesis*. Berlin, Heidelberg, New York: Springer-Verlag, 1972.

Nagy, F. and Schäfer, E. (2002) "Phytochromes control photomorphogenesis by differentially regulated, interacting signalling pathways in higher plants." In: *Annu Rev Plant Biology*, Delmer, D., Bohnert, H.J., Merchant, S. (eds), Annual Reviews, Palo Alto, 329–355.

Quail, P. (1983) "Rapid action of phytochrome in photomorphogenesis" In Encyclopedia of Plant Physiology New Series, Vol. 16A, Shropshire, W. Jr. and Mohr, H. (eds) Springer-Verlag Berlin, Heidelberg, 178–212.

Quail, P. (2002) Phytochrome Photosensory Signalling Networks. *Nature Reviews*, *3*, 85-93.

Schäfer, E., Ebert, C., Schweitzer, M. (1984) Control of hypocotyl growth in mustard seedlings after light-dark transitions. *Photochem Photobiol*, *39*, 95–100.

Schäfer, E., Lassig, T.-U., Schopfer, P. (1982) Phytochrome-controlled extension growth of Avena sativa L. seedlings II. Fluence-rate response relationships and action spectra of mesocotyl and coleoptile responses. *Planta*, *154*, 231–240.

Schäfer, E., Löser, G., Heim, B. (1983) Formalphysiologische Analyse der Signaltransduktion in der Photomorphogenese. *Ber Dtsch Bot Ges*, *96*, 497–509.

Shinomura, T., Nagatani, A., Hanzawa, H., Kubota, M., Watanabe, M., Furuya, M. (1996) Action spectra for phytochrome A - and B - specific photo induction of seed germination in Arabidopsis. *Proc Natl Acad Sci USA*, *93*, 8129–8133.

Shropshire, W., Jr. (1972) "Action spectroscopy" In *Phytochrome*, Mitrakos, E. K., Shropshire, W. Jr. (eds) Academic Press, London and New York, 161–181.

Smith, H. *Light and Plant Development*. London, Boston: Butterworths, 1974.

Sweere, U., Eichenberg, K., Lohrmann, J., Mira-Rodado, V., Bäurle, I., Kudla, J. *et al.* (2001) Interaction of the response regulator ARR4 with phytochrome B in modulation red light signalling. *Science*, *294*, 1108–1111.

Tanada, T. (1968) Substances essential for a red, far-red light reversible attachment of mung bean root tips to glass. *Plant Physiol*, *43*, 2070–2071.

Wong, Y. S. Cheng, H. C., Walsh, D. A., Lagarias, J. C. (1986) Phosphorylation of Avena phytochrome *in vitro* as a probe of light-induced conformational changes. *J Biol Chem*, *261*, 12089–12097.

Chapter 3

HISTORICAL OVERVIEW OF MOLECULAR BIOLOGY AND GENETICS IN PHOTOMORPHO-GENESIS

Eberhard Schäfer[1] and Ferenc Nagy[2]

[1] *Albert-Ludwigs-University Freiburg, Institute of Biology II/ Botany, Schänzlestrasse 1, 79104 Freiburg, Germany (e-mail: Eberhard.Schaefer@biologie.uni-freiburg.de)*
[2] *Biological Research Centre, Institute of Plant Biology, P.O. Box 521, 6701 Szeged, Hungary (e-mail: nagyf@nucleus.szbk.u-szeged.hu)*

In the last three decades, the combination of plant molecular biology with genetics and physiology facilitated the isolation of the genes encoding the red/far-red light absorbing phytochromes, the blue light absorbing cryptochromes and phototropins (see Chapters 11 and 12) and has provided a powerful experimental system to tackle some of the central questions in photobiology research, including the exploration of the molecular mechanism by which photoreceptors regulate gene expression and thereby plant growth and development.

In spite of its present dominating role, it is difficult to define, even retrospectively, the exact time when molecular biology became indispensable for cutting edge, innovative photobiology research. At the beginning of the molecular era, in the early 1980's only a few among the many laboratories active in photobiology research embraced this new technology. These studies, by and large, focused on characterising the expression patterns of genes that encoded proteins shown to accumulate in a light-responsive fashion by physiological assays and were amenable to molecular studies at the RNA level. Thus in these years the majority of these laboratories were engaged in defining the expression patterns of genes encoding ribulose-1,5-bisphosphate carboxylase/oxygenase (Rubisco), chlorophyll a/b binding proteins (CAB) and chalcone synthase (CHS) in response to various light stimuli. As the toolbox of plant molecular biology developed and genes encoding the photoreceptor phytochrome itself were isolated, the scope of these studies became broader and much energy was devoted to the elucidation of the molecular mechanism regulating light-responsive transcription of these genes. By the early 1990's molecular biology and genetics dominated research aimed at defining the functional domains of the phytochrome, the cellular events regulating its stability/degradation, the identification of cis- and trans-acting regulatory elements mediating light-dependent transcription etc. These studies in many respects were indeed pioneering and research on light-induced signal transduction fertilised other areas of contemporary plant molecular biology. In this chapter we refrain from giving a detailed account of these studies and only use the data obtained about the expression of genes encoding the Rubisco enzyme and CAB proteins to highlight the most characteristic features and goals of these early studies.

The Rubisco proteins have attracted much interest due to their role in photosynthetic carbon assimilation and their abundance. The large amounts of Rubisco present in leaves facilitated its biochemical characterisation, including the demonstration that the two subunit types, large and small, are encoded within the chloroplast and nuclear genomes, respectively (Ellis 1981). The relatively high abundance of RbcS transcripts in light-grown leaves as opposed to dark-grown leaves made possible the isolation of RbcS cDNA sequences (Bedbrook *et al.*, 1980, Broglie *et al.*, 1981). These cDNA sequences formed the basis of studies into the regulation of RbcS gene expression for the next decade (Tobin and Silverthorne, 1985; Kuhlemeier *et al.*, 1987). Initial studies on RbcS gene regulation relied on the measurement of RbcS steady-state transcript levels, mostly by Northern hybridisation, and demonstrated that it is organ-specific and its abundance is regulated by light (Mansard and Guise, 1988). These early studies used full RbcS cDNA sequences as probes in the assays. Because of the highly conserved nature of the coding sequence of RbcS genes, this type of experiments provided the expression profile of the entire gene family rather than that of individual RbcS genes. The techniques of 5' and 3' S1 nuclease analysis, primer extension with gene-specific oligonucleotide primers, Raze protection experiments etc. enabled the determination of individual RbcS expression levels and revealed striking differences in their expression patterns (Dean *et al.*, 1989).

Nuclear run-on experiments showed that the light-induced increase in RbcS transcript abundance was due primarily to an increase in transcription. In an attempt to unravel the complex pattern of RbcS gene regulation, substantial effort was directed towards the identification and characterisation of cis-regulatory elements that mediate these diverse responses; specifically, attention was focused on the isolation of light-responsive regulatory elements (LREs) (Kuhlemeier *et al.*, 1987). By this time a much awaited and heralded complex molecular technology, i.e. the transfer and incorporation of *in vitro* constructed genes into the genome of host plants and regeneration of fertile plants inheriting the newly integrated gene became a reality (Horsch *et al.*, 1985). This technical breakthrough opened the way for a formerly unimaginable variety of different experimental strategies and the transgenic era of plant molecular biology was launched. The identification of cis-regulatory elements required for light-dependent transcription of RbcS and CAB genes was one of the first problems in photobiology to which this technology was applied (Herrera-Estrella *et al.*, 1984, Morelli *et al.*, 1985). These studies relied on the construction of chimeric genes containing the RbcS promoter fused to various coding sequences and the determination of the expression profile of the chimeric genes in transgenic plants. On the one hand, this approach provided much information about the location and number of cis-acting regulatory elements. On the other hand, the different expression patterns of RbcS genes from various species and the redundancy of regulatory elements found provided the first insight into the organizational complexity of promoters and regulatory circuits mediating transcription in higher plants. From sequence comparisons made possible by the growing body of available sequence information (the beginnings of in silico photobiology) it became apparent that specific sequence motifs are common to many RbcS genes. Several of these conserved DNA motifs were then identified as target sequences for specific DNA-binding proteins by employing a combination of techniques including gel-retardation, DNAse I foot printing and methylation interference studies and *in vitro* analysis of *in vitro* constructed chimeric genes carrying specific mutations in their promoter regions. These experiments led to the identification of the GT-1 binding protein (Green *et al.*, 1987; Gilmartin *et al.*, 1990) and other proteins binding to G-box (Giuliano *et al.*, 1988; Schindler *et al.*, 1992), GATA box (Lam and Chua, 1989), just to name the most important ones. By the middle of the 1990's it became evident that the expression of light regulated genes is mediated through an array of cis-acting regulatory elements and that the trans-acting regulatory proteins interacting with these LREs are themselves typically encoded by small gene families. The unexpected complexity of regulatory circuits mediating the light-controlled gene expression of RbcS genes made this system difficult to use as a model for unravelling the next layer of the regulatory network mediating light-induced signalling.

Parallel with studies aimed at elucidating the regulatory levels for the light-modulated expression of RbcS genes, similar experiments were conducted to identify cis- and trans-regulatory elements required for the photoreceptor-controlled expression of genes encoding the chlorophyll a/b binding proteins (Cab). CAB polypeptides are encoded by nuclear genes and, together with Chl a/b; they form the antenna complexes of PSI and PSII. These protein/chlorophyll complexes are also frequently called light-harvesting complexes, LHCI and LHCII, respectively. CAB proteins are extremely redundant: the eight types of CAB proteins are encoded by fairly divergent members of a multigene family and the expression of these genes was shown to be regulated by light at the level of transcription by nuclear run-off assays already in 1985 (Silverthorne and Tobin, 1985). The light-modulated expression of CAB genes is regulated by a variety of photoreceptors and it was soon recognized that the transcription of CAB genes is regulated in harmony by (i) the LF (low fluence) phytochrome system, (ii) the VLF (very low fluence) phytochrome system, (iii) the HIR (high irradiation), (iv) the BLF (blue light low fluence and (v) the BHF (blue light high fluence) systems. These studies for the first time provided an intellectual frame for the correlation of photoreceptor-controlled physiological responses with molecular events and demonstrated the somewhat unexpected complexity of regulatory networks mediating photomorphogenesis in higher plants. It is worth noting that these results were obtained well before the identification of the blue light receptors or the genes encoding the so-called type I (labile) and type II (stable) phytochromes. In addition, as early as 1985 Kloppstech *et al.*, showed that the steady state level of CAB mRNA oscillates with a 24 h periodicity, thus it is regulated by the plant circadian clock; in 1988 Nagy *et al.*, documented in transgenic plants that the circadian clock controls the expression of wheat Cab genes at the level of transcription and that a small fragment of the promoter is sufficient to confer circadian responsiveness. These results were later used very innovatively by Millar and Kay, who developed the luciferase-based *in vitro* imaging system, thereby laying down the foundation of molecular and genetic studies aimed at dissecting and understanding the molecular mechanism underlying biological time-keeping in higher plants (Millar *et al.*, 1992). By the mid-1990's these studies culminated in the isolation of the first Arabidopsis mutants displaying aberrant circadian phenotypes (Millar *et al.*, 1995), cloning of the first so-called plant clock genes (Schaffer *et al.*, 1998; Wang *et al.*, 1998). Ever since the combination of photo- and chronobiological approaches contributed significantly to research aimed at understanding the molecular events and components involved in regulating flowering time in higher plants and used successfully to unravel the molecular mechanisms regulating many other aspects of plant development.

The molecular cloning of phytochrome genes from a variety of plants including Arabidopsis thaliana (Sharrock *et al.*, 1988) was only achieved by the end of the 1980's. Within a short time, however, studies on these genes using contemporary molecular methods became a rich source of information on the structure, evolution and biological functions of this photoreceptor. Molecular studies on phytochrome genes in the early 1990's, similarly to those on RbcS, CAB and CHS genes were preceded by a multitude of physiological and molecular experiments using less sophisticated methods. As early as 1982, auto regulation of phytochrome expression was established and soon studied in more detail in monocot and dicot species (Otto *et al.*, 1985). Immunocytochemical studies using antibodies against phytochrome helped to gain information about the level and distribution of phytochrome (as it turned out later, mainly about phyA) within the plant (Pratt 1986). These and other experiments provided the first molecular information about the possible degradation mechanism (SAP formation) underlying the physiological function of the photoreceptor (Speth *et al.*, 1987). In the course of these studies type I (labile, phyA) and type II (stable, phyB-E) phytochrome species were detected, providing useful information for the molecular cloning of phy genes from various plant species (Furuya 1989). After having the various phytochrome genes in hand, various laboratories embraced the molecular methodology with a remarkable speed (Dehesh *et al.*, 1990) and applied it to gain information about (i) the physiological role of various phytochrome species *in vitro* (over-expression studies) (Boylan and Quail, 1989) (ii) the cellular events regulating the degradation of photoreceptor (Cherry *et al.*, 1992) (iii) the function of the different domains of the photoreceptor in active signalling (Cherry and Vierstra, 1994) as well as the mechanism mediating the down-regulation of phyA transcription by light (Quail, 1994).

Although these experiments yielded a tremendous amount of information, it also became evident that efforts to decipher light-induced signalling by promoter bashing of the PHYA gene and identification of cis- and trans-acting factors required for its regulated transcription, similarly to the experiments performed on the RbcS, CAB and CHS genes, face conceptual problems because of the complexity of regulation. Molecular biological approaches applied to study photomorphogenesis needed a helping partner. Like in other areas of science, when one methodology proved to be insufficient another was born: the application of genetics and later molecular genetics/cell biology in combination with molecular biology methods revitalised photo biological research from the early 1990's. Pioneering experiments by Koornneef led to the identification of hy mutants in Arabidopsis, a rich source of mutants extensively used ever since (Koornneef *et al.*, 1980). The strategy of

Korneef and his team was to isolate mutants that display aberrant photomorphogenic phenotypes under saturating light conditions. Chory *et al.*,(1989) and later Deng *et al.* (1991) developed a radically new approach by screening for and identifying mutants whose phenotype in dark resembled that of light-grown seedlings. Isolation of the COP (constitutive photomorphogenic) and DET (de-etiolated) mutants opened a Pandora-box for plant molecular biologists and dramatically changed our concepts about signalling cascades initiated by photoreceptors. Despite some resistance from the part of the classical photomorphogenic field, within a few years it became evident that these negative regulators play a pivotal role in controlling not only photoreceptor-mediated signalling but also many other aspects of plant development. Isolation of the COP1 mutant (Deng *et al.*, 1991) was soon followed by the identification of other COP mutants and soon the concept of the COP9 signalosome was born. It also became evident that the overall structure and the components of the COP9 complex show a striking homology to the lid of the 26S proteosome. Because of these developments, in the early 1990's it became obvious that controlled protein degradation plays an important role in regulating photomorphogenesis in particular and plant development in general (for a detailed description of COP1 and COP9 signalosome see Chapter 18). The impact of these results proved to be much more far-reaching than just fertilizing plant molecular biology research. Very recent studies provided compelling evidence that mammalian homologs of the plant COP1 (Dornan *et al.*, 2004) and DET-1 proteins (Wertz *et al.*, 2004) play a critical role in regulating a multitude of cellular responses by controlling the turnover of the p53 or c-JUN proteins, respectively.

In addition to providing the first hy and cop/det mutants, genetics continued to contribute in many ways to photomorphogenic research in the years to come. Mutant screens using specialised saturating light conditions facilitated the isolation of point mutants carrying single amino acid substitutions or lacking functional phyA (Parks and Quail, 1993) and phyB (Reed *et al.*, 1993) photoreceptors (hy8 and hy3, respectively). Thus it is not surprising that today, even by a conservative estimate, more than 100 phytochrome mutants exist and are used extensively by various laboratories. These type of screens gave rise to suppressor screens (spa1, red1) and by employing non-saturating light conditions (fhy1, fhy3) the era of mutant screens aimed at isolating mutants impaired in signal transduction rather than primary light sensing began in about the mid-1990's (for a recent review see Nagy and Schafer 2002). As our knowledge of photoreceptor-controlled signalling continuously improved, this type of experiment gained increasing sophistication and even today is being widely used by laboratories interested in characterising light-induced signal cascades. Similarly to mutants listed above, signalling mutants isolated by this approach such as eid1, pat1, ohr1 etc. (reviewed by Nagy and Schafer, 2002) contributed significantly to our present knowledge about phytochrome-controlled photomorphogenesis.

Whereas early genetic screens facilitated the isolation of specific point mutants generated mainly by chemical mutagenesis, the combination of these screens with other genetic methods, including insertional mutagenesis using Agro bacterium T-DNA or Ac/Ds elements significantly broadened the scope of these experiments. This statement is underlined by the fact that CRY1, the first blue light receptor identified in plants was isolated from a T-DNA insertional mutant collection of this type (Ahmad *et al.*, 1993). The applicability of these methods kept improving during the last decade. By establishing several large insertional mutant collections representing a large percentage of the approximately 25 000 genes expressed in Arabidopsis, this approach has become one of the fastest and most widely used method in photobiology.

The aim of this chapter was to summarize and give a short description of those molecular and genetical methods that contributed most significantly to photo biological research, in a chronological fashion, from the mid-1970's until the mid-1990's. As any such inventory, this one is also inevitably subjective and represents only the authors' view. The scope of this chapter prevented us from giving a detailed summary of all methods arguably belonging to the repertoire of molecular biology and genetics. It was for this reason that we refrained from discussing methods developed in the last ten years, for example the micro-array analysis of gene expression or the application of the green fluorescent protein (GFP) to monitor trafficking and cellular distribution of proteins. Each chapter of this book devoted to a specialised area of photobiology contains an introductory part describing the most important methods and results of the last 10 years of research in the specific area and we hope that these, in combination with this general historical overview will satisfy the needs of most readers interested in the last 30 years of photo biological research.

REFERENCES

Ahmad, M., Cashmore, A. (1993) HY4 gene of Arabidopsis thaliana encodes a protein with characteristics of a blue-light photoreceptor. *Nature, 366,* 162-166.

Bedbrook, J. R., Smith, S. M., Ellis, R. J. (1980) Molecular cloning and sequencing of cDNA encoding the precursor to the small subunit of chloroplast ribulose-1,5-bisphosphate carboxylase. *Nature, 287,* 692-697.

Boylan, M. T., Quail, P. H. (1989) Oat phytochrome is biologically active in transgenic tomatoes. *Plant Cell, 1,* 765-773.

Broglie, R., Bellemare, G., Bartlett, S. G., Chua, N.-H., Cashmore, A. (1981) Cloned DNA sequences complementary to mRNAs encoding precursors to the small subunit of ribulose-1,5-bisphosphate carboxylase and a chlorophyll a/ b-binding polypeptide. *Proc Natl Acad Sci USA, 78,* 7304-7308.

Cherry, J. R, Vierstra, R. D. (1994) "The use of transgenic plants to examine phytochrome structure/function." In *Photomorphogenesis in Plants*, Kendrick R. E. and Kronenberg G. H. M. (eds) Martinus Nijhoff Publishers, Dordrecht, 271-297.

Cherry, J. R., Hondred, D., Walker, J. M., Vierstra, R. D. (1992) Phytochrome requires a 6-kDa N-terminal domain for full biological activity. *Proc Natl Acad Sci USA, 89,* 5039-5043.

Chory, J., Peto, C., Feinbaum, R., Pratt, L., Ausubel, F. (1989) Arabidopsis thaliana mutant that develops as light grown plant in the absence of light. *Cell, 58,* 991-999.

Dean, C., Pichersky, E., Dunsmuir, P. (1989) "Structure, evolution and regulation of RbcS genes in higher plants." In *Annu Rev Plant Physiol, 40,* 415-439.

Dehesh, K., Bruce, W. B., Quail, P. (1990) A trans-acting factor that binds to a GT-motif in the phytochrome gene promoter *Science, 250,* 1397-1399.

Deng, X. W., Caspar, T., Quail, P. (1991) cop1: a regulatory locus involved in light controlled development and gene expression in Arabidopsis. *Genes Dev, 5,* 1172-1182.

Dornan, D., Wertz, I., Shimizu, H., Arnott, D., Frantz, G., Dowd, P., *et al.* (2004) The ubiquitin ligase COP1 is a critical negative regulator of p53. *Nature, 429,* 86-92.

Ellis, R. J. (1981) "Chloroplast proteins - synthesis, transport and assembly." In *Annu Rev Plant Physiol*, 32, 111-137.
Furuya, M. (1989) Molecular properties and biogenesis of phytochrome I and II. *Adv Biophys*, 25, 133-157.
Gilmartin, P. M., Sarokin, L., Memelink J., Chua, N.-H. (1990) Molecular light switches for plant genes. *Plant Cell*, 2, 369-378.
Giuliano, G., Pichersky, E., Malik, V.S., Timko, M.P. Scolnik, P.A., Cashmore, A.R (1988) An evolutionarily conserved protein binding sequence upstream of a light regulated promoter. *Proc Natl Acad Sci USA*, 85, 7089-7093.
Green, P. J. Kay, S. A. Chua, N.-H. (1987) Sequence specific interactions of a pea nuclear factor with light-responsive elements upstream of a plant light regulated gene. *EMBO J*, 6, 2543-2550.
Herrera-Estrella, L., van den Broeck, G., Maenhaut, R., van Montagu, M., Schell, J., Timko, M., Cashmore, A. (1984) Light inducible and chloroplast associated expression of a chimeric gene introduced into Nicotiana tabacum using a Ti-plasmid vector. *Nature*, 310, 115-120.
Horsch, R. B., Fry, J. E., Hoffmann, N. L., Eicholtz, D., Rogers, S., Fraley, R. T. (1985) A simple and general method for transferring genes into plants. *Science*, 227, 1229-1231.
Kloppstech, K. (1985) Diurnal and circadian rhythmicity in the expression of light induced plant nuclear messenger RNAs. *Planta*, 165, 502-506.
Koornneef, M., Rolff, E., Spruit, C. P. J. (1980) Genetic control of light-inhibited hypocotyl elongation in Arabidopsis thaliana (L.) *Heynh Z Pflanzenphysiol*, 100, 147-160.
Kuhlemeier, C., Green, P. J., Chua, N.-H. (1987) "Regulation of gene expression in higher plants." In *Annu Rev Plant Physiol*, 38, 221-257.
Lamm, E., Chua, N.-H. (1989) ASF2: A factor that binds to the cauliflower mosaic virus 35S promoter and a conserved GATA motif in Cab promoters. *Plant Cell*, 1, 1147-1156.
Manzara, T. and Gruissem, W. (1988) Organisation and expression of the genes encoding ribulose -1,5 - bisphosphate carboxylase in higher plants. *Photsynthesis Res*, 16, 117-139.
Millar, A. J., Short, R. S., Chua, N.-H., Kay, A. S. (1992) A novel circadian phenotype based on firefly luciferase expression in transgenic plants. *Plant Cell*, 4, 1075-1087.
Millar, A., Carre, I. A., Strayer, C. A., Chua, N.-H., Kay, S. A. (1995) Circadian clock mutants in Arabidopsis identified by luciferase imaging. *Science*, 267, 1161-1163.
Moesinger, E., Batschauer, A., Schaefer, E., Apel, K. (1985) Phytochrome control of *in vitro* transcription of specific genes in isolated nuclei from barley (Hordeum vulgare). *Eur J Biochem*, 147, 137-142.
Morelli, G., Nagy, F., Fraley, R., Rogers, S., Chua, N.-H. (1985) A short conserved sequence is involved in the light-inducibility of a gene encoding ribulose-1,5-bisphosphate carboxylase small subunit of pea. *Nature*, 315, 200-204.
Nagy, F., Kay, S. A., Chua, N.-H. (1988) A circadian clock regulates transcription of the wheat cab-1 gene. *Genes Dev*, 2, 376-382.
Nagy, F., Schaefer, E. (2002). "Phytochromes control photomorphogenesis by differentially regulated, interacting signalling pathways in higher plants." In *Annu Rev Plant Biol*, 53, 329-355.
Parks, B. M., Quail, P. H. (1993) hy8 a new class of Arabidopsis long hypocotyl mutants deficient in functional phytochrome-A. *Plant Cell*, 3, 1177-1186.
Pratt, L. H. (1986) "Phytochrome: localization within the plant." In: *Photomorphogenesis in Plants*, Kendrick R. E. and Kronenberg G. H. M. (eds) Martinus Nijhoff Publishers, Dordrecht, 61-81.
Quail, P. (1991) "Phytochrome: A light-activated molecular switch that regulates plant gene expression." In *Annu Rev Genet*, 25, 389-409.
Quail, P. (1994) "Phytochrome genes and their expression." In *Photomorphogenesis in Plants*, Kendrick R. E. and Kronenberg G. H .M. (eds) Martinus Nijhoff Publishers, Dordrecht, 71-103.
Reed, J. W., Nagpal, P., Poole, D. S., Furuya, M., Chory, J. (1993) Mutations in the gene for the red/far-red light receptor phytochrome-B alter cell elongation and physiological responses throughout Arabidopsis development. *Plant Cell*, 5, 147-157.
Schaffer, R., Ramsay, N., Samach, A., Corden, S., Putterill, J., Coupland, G. (1998) The late elongated hypocotyl mutation of Arabidopsis disrupts circadian rhythms and the photoperiodic control of flowering. *Cell*, 93, 1219-1229.
Schindler, U., Terzaghi, W., Beckmann, H., Kadesch, T., Cashmore, A. R. (1992) DNA binding sites preferences and transcriptional activation properties of the Arabidopsis transcription factor GBF1. *EMBO J*, 8, 651-656.

Sharrock, R. A. and Quail, P. (1989) Novel phytochrome sequences in Arabidopsis thaliana: structure, evolution and different gene expression of a plant regulatory photoreceptor family. *Genes Dev*, 3, 1745-1757.

Speth, V., Otto, V., Schaefer, E. (1987) Intracellular localization of phytochrome and ubiquitin in red-light-irradiated oat coleoptiles by electron microscopy. *Planta*, *171*, 332-338.

Tobin, E. M. and Silverthorne, J. (1985) "Light regulation of gene expression in higher plants." In *Annu Rev Plant Physiol*, *36*, 569-593.

Wang, Y. Z., Tobin, E. M. (1998) Constitutive expression of the circadian clock associated 1 (CCA1) gene disrupts circadian rhythms and suppresses its own expression. *Cell*, *93*, 1207-1217.

Wertz, I., O'Rourke, K., Zhang, Z., Dornan, D., Arnott, D., Deshaies, J., *et al.* (2004) Human de-etiolated-1 regulates c-JUN by assembling a CUL4A ubiquitin ligase. *Nature*, *303*, 1371-1374.

Chapter 4

GENETIC BASIS AND MOLECULAR MECHANISMS OF SIGNAL TRANSDUCTION FOR PHOTO-MORPHOGENESIS

Eberhard Schäfer[1] and Ferenc Nagy[2]

[1] Albert-Ludwigs-University Freiburg, Institute of Biology II/ Botany, Schänzlestrasse 1, 79104 Freiburg, Germany (e-mail: Eberhard.Schaefer@biologie.uni-freiburg.de)
[2] Biological Research Centre, Institute of Plant Biology, P.O. Box 521, 6701 Szeged, Hungary (e-mail: nagyf@nucleus.szbk.u-szeged.hu)

1. INTRODUCTION

Genetic approaches to unravel the molecular mechanisms of animal development have been widely used in model organisms including *Drosophila*, *C. elegans*, zebrafish and mouse. The first genetic studies aimed at the isolation of mutants specifically impaired in photomorphogenesis were performed by Max Delbrück's group. These researchers used *Phycomyces* as a model organism and phototropism of sporangiophores as a response; their work led to the isolation of several mutants displaying altered phototropic and gravitropic sensitivities. *Phycomyces* lends itself well to genetic but not to biochemical studies and the lack of a reliable transformation technique makes its utilization even more problematic. This might explain that, although physiological characterization of these mutants indicated that genes coding for photoreceptors or antenna pigments were affected (the mutants had different action spectra as compared to wild type), a detailed analysis of these mutants have not yet been accomplished. After these early trials, which were obviously hindered by the shortcomings of supporting molecular methodology, subsequent genetic approaches backed by more sophisticated molecular techniques resulted in the isolation and detailed characterisation of the first mutants in higher plants. We had to learn, however, that completion of the work on these mutants, most of them impaired in phototropic and photomorphogenic responses, did not come fast and easy and in some cases is still continuing. Notwithstanding the emergence of unexpected problems related to the analysis of these mutants, it is clear by now that they heralded the beginning of a new era in photomorphogenetic research.

2. PHOTOTROPISM MUTANTS

Several years after the pioneering effort made by the Delbruck laboratory, Ken Poff's group initiated and later Mannie Liscum and Winslow Briggs performed successful screens to isolate mutants for hypocotyl phototropism of *Arabidopsis* seedlings. They identified several mutants, designated nph1-4, and additional years of systematic work utilising map-based cloning resulted in the isolation of the corresponding genes (see Chapter 12). It became known that *NPH1* encodes a plasma membrane associated serine/threonine kinase with two N-terminal domains (LOV1 and LOV2), which bind FMN. These data and the physiological and biochemical characterisation of the nph1 mutant indicate that *NPH1* – later renamed *PHOT1* – encodes a photoreceptor for phototropism. These results also settled a long-lasting debate based on action spectroscopy, i.e. whether blue light receptors are carotenoids or flavoproteins, in favour of the latter. More recently a second photoreceptor (*NPL1*, renamed *PHOT2*) was identified as the high light intensity photoreceptor. Recently it turned out that *PHOT1* and *PHOT2* control not only phototropism but also chloroplast orientation and light-mediated opening of stomata (see Chapter 12).

3. PHOTOMORPHOGENIC MUTANTS

In photomorphogenesis research, the genetic approach was pioneered by Maarten Koorneef, who introduced *Arabidopsis* as a model organism. In his experiment, seeds mutagenised either by EMS treatment or by fast neutron bombardment were germinated and grown under saturating white light. Seedlings with long hypocotyls were selected and classified as hy mutants (hy 1–5). Several years later cloning of the mutated genes revealed – with the exception of *HY5*, which encodes a bZIP transcription factor – that these genes encode either the apoprotein of various photoreceptors (*HY3, PHYB; HY4, CRY1*) or enzymes of the phytochromobiline (chromophore) biosynthesis (HY1, hemeoxigenase; HY2, phytochromobiline synthase) (see Chapters 5, 7, 11).

The reciprocal screen looking for mutants with short hypocotyls and partial photomorphogenesis development in darkness was pioneered by Joanne Chory and extensively elaborated by Xing-Wang Deng's group (see Chapter 18). These screens led to the identification of many recessive genes, designated DET/ COP/ FUS, all of which, if mutated, result in constitutive photomorphogenesis in darkness. These results are consistent with the view that etiolation is a late discovery of the evolution of plants and it is fully manifested only in angiosperms. This implies that photomorphogenesis is the default pathway and must be actively suppressed by a complex cellular machinery to ensure skotomorphogenesis of seedlings grown in the dark (see Chapter 18).

Parallel with these screens, more targeted and specialised screens using monochromatic light sources were performed by several laboratories. These efforts, within a relatively short time, led to the isolation of specific photoreceptor mutants and revealed that the corresponding genes encode the apoproteins of the five phytochromes, designated phyA-phyE (see Chapter 7) and two blue-light receptors CRY1 and CRY2 (see Chapter 11). The blue light receptors CRY1 and CRY2 are photolyase homologs and thus also flavoproteins. In other systems like Neurospora and Euglena (see Chapter 13) other blue light receptors have been identified. It is still not clear whether PHOT1 and PHOT2 are the only blue light photoreceptors in higher plants other than CRY1 and CRY2. The UVB receptors still await molecular identification (see Chapter 14).

4. CIRCADIAN MUTANTS

In the first part of the last decade of the 20th century it became evident that the expression of the majority of photoreceptor-controlled genes also display a characteristic oscillatory pattern, i.e. they are regulated by the circadian clock. It was also shown that the blue and red/far-red light absorbing photoreceptors play an essential role in the entrainment of the endogenous biological oscillator (see Chapter 26). Recent microarray experiments indicate that at least 6% of Arabidopsis genes are rhythmically expressed (see Chapter 26), thus circadian gene expression produces rhythms that pervade plant physiology including photomorphogenesis and flowering time. It is widely accepted by now that phototransduction to the central circadian clockwork and signal transduction mediating photomorphogenesis are intimately intertwined. Again, genetic screens helped achieve major breakthroughs and provided important details about these signalling pathways and the molecular mechanism essential for a functional biological clock. This approach has been especially successful in *Neurospora* and *Arabidopsis*. The *Neurospora* system is described in a separate chapter (see Chapter 13); here we only mention that one of the major outcomes of these experiments was the identification of a novel photoreceptor as an integral part of the clock. In *Arabidopsis* this approach helped identify some of the central components of the clock (see Chapters 26 and 27), numerous intermediates of phytochrome signalling and a putative blue light receptor (see Chapters 12 and 26). However, it is worth noting that in plants the photoreceptors seem not to be the integral parts of the clock but almost all of the clock genes identified to date encode early components of the phytochrome signalling network (see Chapters 17 and 26). Not too surprisingly, because of the complexity of the system it was shown that in crosses of two Arabidopsis ecotypes having "the same" circadian output pattern – as demonstrated for the circadian clock control of LHCP expression (Swarup *et al.*, 1999; Thain *et al.*, 2000) - the progenies exhibit a variation in the output. This observation indicates that the genetic balance, which is likely perturbed in crosses, is important for the circadian output response.

5. GENETIC VARIATION, MUTANTS IDENTIFIED BY QTL MAPPING

Another approach to identify the genetic basis of photomorphogenesis makes use of the genetic variation of different ecotypes. This has the clear advantage that the ecological significance of the altered genetic programme has been established. The disadvantage is the problem of how to identify the genes contributing to the manifested response variation. In principle, this approach has been used for many years by plant breeders to identify the genetic loci (QTL: quantitative trait loci) important for complex traits like resistance, sugar content or flowering time.

Genetic variation among the different Arabidopsis ecotypes has been observed for photomorphogenetic responses like VLFR (Chapter 20), dark reversion of phyA (Eichenberg *et al.*, 2000) light-dependent inhibition of hypocotyl elongation and shade avoidance reaction (see Chapter 22). In only a few cases this information could be used to clone the corresponding genes, but in many cases photoreceptor mutants (phyA (Lm-2) Maloof *et al.*, 2001; phyD (WS) Aukerman *et al.*, 1997, and cry2 (cvi) El-Assal *et al.*, 2001) were identified.

6. SIGNAL TRANSDUCTION MUTANTS

To identify gene products involved in signal transduction cascades mediated by the individual photoreceptors, more targeted genetic approaches have to be used. Three different strategies have been described:

1) Revertant screens: either for weak photoreceptor alleles or overexpressor lines

2) Screens under non-saturating light conditions

3) Specifically problem-targeted light conditions, like hourly red light pulses followed by far-red light pulses to obtain mutants impaired in loss of reversibility, or alternating 20 min red and far-red light treatments to obtain mutants impaired in phyA degradation or far-red hypersensitivity (see Chapters 17 and 18).

All of these approaches have their specific advantages and a high number of gene products involved in photomorphogenesis have been identified. The number is steadily increasing, yet these approaches so far have not pinpointed the so-called "early" components of signal transduction mechanisms. This means that genetics did not help identify the holy grail of plant photobiology, i. e. the phy, cry, phot interaction partners. The question is: why not? There are possible answers but all are rather speculative. The obvious ones are that the components were not identified either because of redundancy or because a mutation would be lethal. But a more trivial answer may also hold: all the screens employed have been based on the readout of a late event, i. e. changes in growth or seedling habit after 2 – 4 days of irradiation, or expression profiling of a gene positioned as a terminal step of a specific signalling pathway. Especially growth will not only be regulated by light but by numerous other factors, each of which may only have a small contribution, thus the "weak" phenotypes caused by these mutations escape classification as mutants (see Chapter 17).

Despite these apparent problems, various genetic approaches identified several loci involved in phytochrome signalling. Of the genes cloned so far, the majority encode molecules localized in the nucleus, whereas a few present in both cellular compartments or are exclusively cytosolic. The genetic analysis of these mutants clearly supports the view that for phytochrome-mediated responses both cytosolic and nuclear transduction cascades should be involved. Unfortunately, it is at the moment difficult to envisage a plausible functional role for these genes, except for a few, at the molecular level. It is fair to predict that novel screens based on the available genetic, molecular and physiological data and specifically targeted to identify early components should be developed to overcome this apparent deadlock.

7. SIGNAL TRANSDUCTION AT THE MOLECULAR LEVEL

It has to be admitted that, despite decades of intensive research, the primary molecular mechanisms by which the blue and red/far-red light absorbing photoreceptors initiate signal transduction for photomorphogenesis are still unclear. In the case of phototropins, light-dependent phosphorylation is clearly important. How light absorption leads to changes in phosphorylation state and how this leads to coupling signal transduction remains to be elucidated (see Chapters 12 and 15). For cryptochromes a redox reaction chain analogous to photolyases has been postulated, but solid functional proof is still missing. The lack of knowledge about other elements of CRY-specific signal transduction clearly makes further research important (see Chapter 11).

In the case of phytochrome-mediated signal transduction, many possible elements of the signalling cascade are known, but the complexity of the signalling network(s) makes it difficult to come to a general conclusion. The primary function of phytochromes is largely unknown and the discussion about the primary function of phytochromes – (i) are phytochromes light-dependent enzymes, (ii) regulators of membrane properties or (iii) transcriptional regulators – is still far from being finished for the following reasons.

Bacterial phytochromes clearly possess enzyme activity – phytochrome as the input element of a two-component system – and it is generally accepted that bacterial phytochromes are light-regulated histidine kinases (see Chapters 5 and 6). In lower plants, ferns, mosses and some algae, the action dichroism observed for phytochrome-mediated responses supported the membrane hypothesis (see Chapters 23 and 24), but direct proof for membrane function is still missing.

In higher plants a possible membrane function was also concluded based on the results of microinjection experiments. Using the chromophore-deficient aurea mutant from tomato, microinjection of either phyA or phyB could result in light-induced gene expression (see Chapters 19 and 28). Especially the induction of gene expression by activating possibly trimeric G proteins led to the conclusion of cytosolic and probably membrane-associated phytochrome function. Besides trimeric G-proteins, a role for Ca^{2+}, calmoduline, cGMP and kinases was suggested by experiments using the microinjection technique (see Chapter 19). This view was later questioned, when light-dependent nuclear import of phytochromes was observed (see Chapter 9). A way to merge these two sets of data into an integrated picture of phytochrome signal transduction remains to be found.

The present view regarding phytochrome and cryptochrome signal transduction is that major components of both operate in the nuclear compartment. Again, however, our understanding of nuclear events is poor for the following reasons. Both photoreceptors show complex dynamics including formation of photoreceptor-containing light-induced speckles (protein complexes), which then disappear and reappear (see Chapters 9 and 11). In the case of cryptochromes, an interaction with the COP1 complex is suggested as an early step of signal transduction (see Chapters 11 and 18). Whether this is the primary reaction of CRYs or whether COP1-independent pathways are also involved is not clear at the moment. In the case of phytochrome function, novel insights came from the analysis of the phytochrome interacting factors (see Chapters 17, 19 and 28). Especially the observation of Pfr-specific interactions with transcription factors of a subgroup of the bHLH family (see Chapter 17) gave a very attractive hypothesis for the primary function of phytochromes: it was postulated based on data obtained in yeast that after being imported in the nuclei, phytochromes will interact with specific bHLH transcription factors and directly control the transcription of cognate target genes. Unfortunately, this hypothesis was challenged by the most recent observations indicating that PIF3 (phytochrome interacting factor 3) null mutants have only a weak hypersensitive phenotype in red light, thus it is a negative rather than a positive regulator of phytochrome signalling. In addition, it has been shown that the PIF3 protein, even though it co-localises with phytochrome in the nucleus, is degraded very rapidly after short light pulses. The rapid light-induced degradation of PIF3 is controlled by phytochrome. This latter observation indicates a novel function of phytochromes in the nucleus, namely targeting of the pre-existing transcription factors to the degradation machinery.

8. SUMMARY

It is obvious that the use of molecular biology and genetics in photomorphogenic research led to a burst of new results and a much better understanding of the processes involved in photoreceptor-mediated signal transduction. This process is still continuing and will be summarized in several chapters in this book. Nevertheless, even in possession of all these new data, the principal question, i.e. how photoreceptors function, how they initiate the primary signal transduction steps remains elusive. In the case of phytochromes, we do not even know whether a single or several, basically different types of primary events are essential for its functionality. The complexity of phytochrome-mediated events at the cellular/molecular level including its cytosolic retardation, its nuclear transport and its participation in various nuclear complexes points, at the moment, to multiple primary events. In addition, despite the prominent role of nuclear events in phytochrome-mediated signal transduction one should keep in mind that even under saturating light conditions 30 – 50 % of phytochromes still remains cytosolic. Are these cytosolic phytochromes functional? What responses do they control?

Obviously, there are still numerous unanswered questions to be addressed. The editors and contributors of the present book hope that by the time the next edition of this book will be available, novel genetic screens and new molecular, cell biological and biochemical approaches will give us the clues to solve these dilemmas and provide definitive answers.

9. REFERENCES

Aukerman M. J., Hirschfeld, M., Wester, L., Weaver, M, Clack, T., Amasino, R. M. *et al.* (1997) A deletion in the *PHYD* gene of the Arabidopsis Wissilewskija ecotype defines a role for phytochrome D in red/ far-red light sensing. *Plant Cell*, *9*, 1317-1326.

Eichenberg, K., Hennig, L., Martin, A., Schäfer, E. (2000) Variation in dynamics of phytochrome A in *Arabidopsis* ecotypes and mutants. *Plant, Cell and Environment*, *23*, 311-319.

El-Assal, S. E.-D., Alonso-Blanco, C., Peeters, A. J. M., Raz, V., Koorneef, M. (2001) A QTL for flowering tim in *Arabidopsis* reveals a novel allele of *CRY2*. *Nature Genetics*, *29*, 435-440.

Maloof, J. N., Borevitz, J. O., Dabi, T., Lutes, J., Nehring, R. B., Redfern, J. L. *et al.* (2001) Natural variation in light sensitivity of *Arabidopsis. Nature Genetics*, *29*, 441-446.

Swarup, K., Alonso-Blanco, C., Lynn, J. R., Michaels, S. D., Amasino, R. M., Koorneef, M. (1999) Natural allelic variation identifies new genes in the *Arabidopsis* circadian system. *The Plant Journal*, *20*, 67-77.

Thain, S. C., Hall, A., Millar, A. (2000) Functional independence of circadian clocks that regulate plant gene expression. *Current Biology*, *10*, 951-956.

PART 2: THE PHYTOCHROME

Chapter 5

THE PHYTOCHROME CHROMOPHORE

Seth J. Davis
Max Planck Institute for Plant Breeding Research, 50829 Cologne, Germany (e-mail: davis@mpiz-koeln.mpg.de)

1. INTRODUCTION

For vision to occur, light absorption is an absolute physical requirement. The array of photomorphogenic responses all has one thing in common: these processes are initiated only after photo excitation of a holoprotein. These holoproteins generally contain a polypeptide component and a small, light-absorbing ligand termed the chromophore. Plants have evolved an array of photoreceptor systems to detect the colour, intensity, direction, and periodicity of light. For this, a varying collection of polypeptides and associated chromophores are required. Not surprisingly, many of these receptors have been termed "chromes," including the "cryptic" cryptochromes that detect ultra violet-B (UV-B) and blue (B) light (Lin and Shalitin, 2003), and the "phyto" (meaning plant) phytochromes that detect UV-B, B, red (R), and far-red (FR) light (Kevei and Nagy, 2003). Additional chromophore-bearing photoreceptors include the B-light detecting phototropins (Christie *et al.*, 2002), the Zeitlupe family of B-light chromo proteins (Imaizumi *et al.*, 2003), and two physiologically distinct systems that detect UV-B (Ulm *et al.*, 2004). The protein component of these latter detection systems awaits identification (see Chapter 14). It should be noted that photosynthesis itself functions in light detection as a "receptor," as light-induced redox changes in the plastid provide signalling information to the plant cell (Dietz, 2003). And finally, small molecule absorption of light is also "detected" by cells (Vladimirov, 1998), albeit as a photo-damage-sensing mechanism (see Chapter 14). Clearly, plants use an array of photochromic perception systems to optimise growth and development to all aspects of the ambient light environment (Sullivan and Deng, 2003). Of these higher-plant photochromic systems, the most thoroughly studied is the phytochrome family of chromoproteins.

One extensively characterized group of plant photoreceptors are the R-absorbing and FR-absorbing phytochromes (phys) (see Chapter 7). This family of bilin-containing chromoproteins regulates development throughout the life cycle of plants. They play prominent roles in seed germination and de-etiolation, and control cotyledon, leaf, and stem size by regulating cell division and expansion. Phys also help plants perceive and respond to shading and crowding by neighbouring plants, entrain the circadian timer, and influence the seasonal timing of flowering. Ultimately, phys control plant senescence. Clearly phys control a diverse aspect of biology from the birth to the death of a plant. Despite the wealth of knowledge on phy structure and function, many questions remain as to the nature of it assembly. What is currently known is that phys are homodimeric chromoproteins with each subunit consisting of a linear tetrapyrrole chromophore covalently attached to each polypeptide monomer (Figure 1). The molecular mass of each monomer ranges from 118-127 kD, depending on the phy isoform and the plant species (see Chapter 7). The bilin chromophore, called phytochromobilin (PΦB), is attached to a ~300 amino-acid domain within N-terminal half (Bhoo et al., 1997; Lagarias and Rapoport, 1980). Phy apoproteins can autocatalytically bind chromophore *in vitro* to form the spectrally active holoprotein (Li and Lagarias, 1992). Once assemble, the holoprotein dimer appears as a "Y" shaped molecule by electron microscopy (Jones and Erickson, 1989; Nakasako et al., 1990). The N-terminal half of phys is both necessary and sufficient for chromophore binding and the spectral properties. The C-terminal half appears to contain domains necessary for homodimerization (Matsushita et al., 2003).

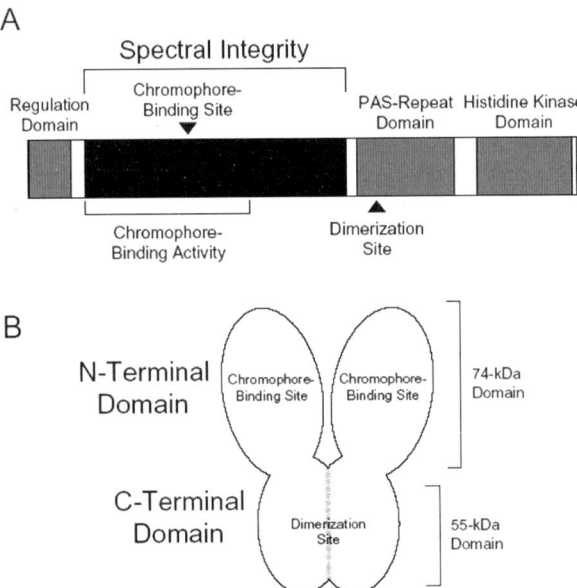

Figure 1. Domains of phy. (A) Functional regions are based on cumulative structure/function studies and genetic analysis of phyA and phyB. The amino-acid boundaries are approximate. See text for a description of each functional domain. (B) Model of the three-dimensional structure of phy based primarily on electron micrographs and X-ray scattering of purified phyA.

Phys have two conformational forms, a R-absorbing form (Pr) and a FR-absorbing form (Pfr) (Figure 2). Pr and Pfr are fully photointerconvertible upon irradiation of R and FR, respectively (Vierstra and Quail, 1983). It is thought that all phys from plants are synthesized in the Pr form, and are converted to Pfr only upon absorption of R. Because physiological responses are initiated as Pfr, this conformation is considered biologically active (see Chapter 17). Whereas FR absorption results in ~100% photoconversion of Pfr to Pr, both Pr and Pfr absorb R (Figure 2). This overlap in R absorption results in the total phy pool reaching a photoequilibrium of 86% Pfr and 14% Pr after a saturating R treatment (Vierstra and Quail, 1983). Phys also absorb UV-B and blue light as both Pr and Pfr. This appears to be physiological relevant, as phys assist phototropin and the cryptochromes in mediating blue-light responses (Franklin et al., 2003a), and in phys assisting the uncharacterised UV-B-perception system (Kim et al., 1998).

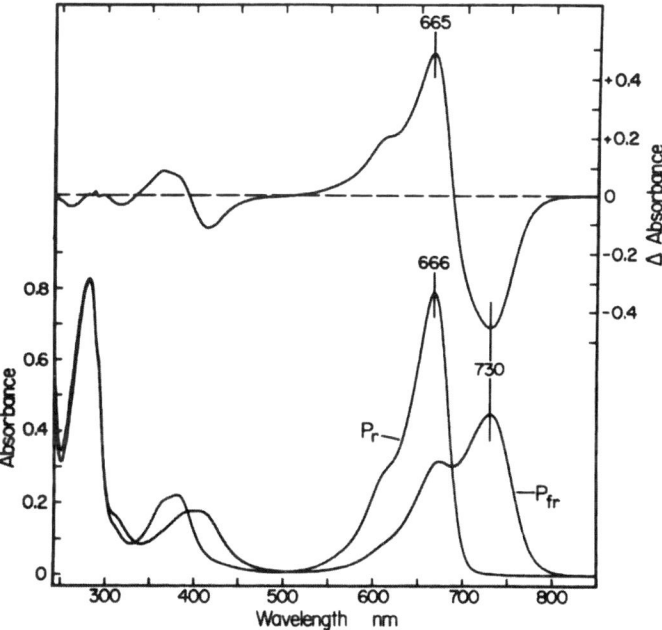

Figure 2. Absorbance spectra of oat phyA as Pr and Pfr. Spectra was generated using phy purified from etiolated oat seedlings (primarily phyA), after saturating R and FR irradiation. The absorbance spectra of phy as Pr (λmax = 666 nm) or Pfr (λmax = 730 nm) are shown at the bottom part of the figure. The top curve shows the FR-minus-R difference spectra, generated by subtracting the spectrum of Pfr from that of Pr [adapted from (Vierstra and Quail, 1983)].

As previously stated, the chromophore of phy is PΦB. This molecule is a bile pigment, and as will be described shortly, it is derived from heme. Once the PΦB chromophore assembles with a phy isoform, the resulting chromoprotein is poised to initiate photomorphogenic processes. The details of signalling and subsequent physiology are dealt with in great depth in the coming chapters. Within this chapter, I will discuss findings and continuing investigations regarding the roles of tetrapyrroles and the derived bilins in light detection and physiology. As many recent findings on this topic revolve around phy photobiology, this will be the near exclusive focus of this chapter.

2. STRUCTURE OF THE PHYTOCHROME CHROMOPHORE

The chromophore of phytochrome A (phyA) has rigorously been proven to be PΦB. This topic was discussed at length in Chapter 4.1 of the previous edition of this book (Kendrick and Kronenberg, 1994), but will be briefly reviewed here. In the 1950's, the absorption spectrum of phyA was found to be similar to that of phycocyanin, a protein used by cyanobacteria in photosynthesis. This was taken as evidence that the chromophores of these two chromoproteins were related. As it was known that the phycocyanin chromophore is the bilin phycocyanobilin (PCB), this was the first suggestion that holo-phyA was a bile-pigmented chromoprotein. The first proof that the phyA chromophore was a bilin was obtained by oxidative degradation of the protein, and subsequent chemical analyses (Rüdiger and Correll, 1969). Additionally, these results indicated for the first time that the chromophore was linked to the phyA by a thioether bond. Further chemical analysis led to the first suggested structure of the phytochromophore as the 3E isomer of PΦB (Figure 3) (Lagarias and Rapoport, 1980).

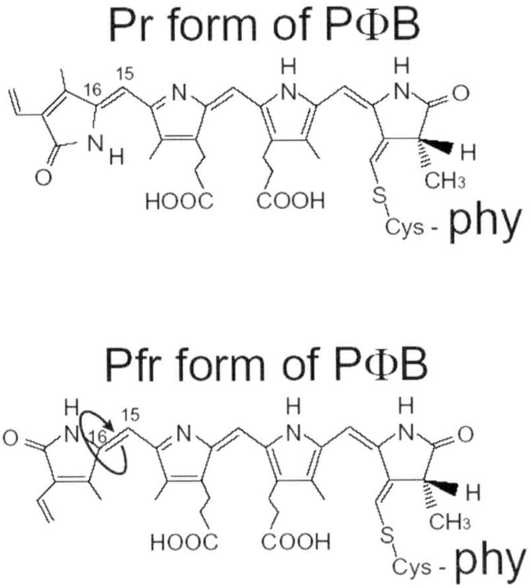

Figure 3. Structure of PΦB as Pr and Pfr. The chromophore is attached to a positionally conserved cysteine in plant phys (Cys322 in oat phyA), and undergoes a cis-to-trans isomerization around the C15=C16 double bond upon photoconversion from Pr to Pfr.

The proposed structure of PΦB was confirmed independently from high-resolution NMR studies of an oat phyA chromopeptide (Lagarias and Rapoport, 1980). What Lagarias and co-workers did was a purification of phyA, followed by proteolysis. A peptide containing the bile chromatic group was purified to homogeneity from this mixture. The chemical structure of the chromopeptide was solved by NMR by using comparisons to PCB. Although it is unlikely the chemical structure of PΦB was changed through the required proteolysis and chemical treatments, absolute proof of the chromophore structure within an intact phyA will only be obtained when native phytochrome structure is fully solved by either x-ray crystallography or protein NMR. This is relevant as will be described later, it is plausible that the alternate isomer 3Z- PΦB could be the chromophore of phyA.

The original chromophore structures were solved from peptides derived from the R-absorbing form of phyA, and thus the resolved structure was the Pr form of PΦB. Later investigations on phyA-chromophore structures focused on the difference between Pr and Pfr forms of this PΦB molecule. The absorption spectrum is clearly different between the Pr and Pfr of holo-phyA (Figure 2) (Vierstra and Quail, 1983). As phy-chromophore interactions could alter the electronic properties of PΦB, early research resolved to identify whether the Pr and Pfr form of PΦB were the same chromophore structure simply in a different protein environment, or were different isomers. The first experiments showing that the Pfr form of PΦB is chemically different from the Pr form were obtained from preparation of Pfr chromopeptides and their comparison to Pr chromopeptides (Rüdiger *et al.*, 1980). The first thing noticed was that the absorption spectrum of these two preparations was different. Thus, the Pr versus Pfr absorbance shift is not strictly protein dependent proving that phototransformation of phyA resulted in a chromophore structural change. Eventually, the structure of the PΦB form of PΦB was solved by NMR (Thümmler *et al.*, 1983) and shown to the 15E isomer PΦB, whereas Pr is the 15Z isomer (Figure 3). All other features of the chemical structure are identical between the Pr and Pfr versions of PΦB, and thus the only photochemical reaction that occurs to the PΦB chromophore upon R-light absorption is a cis to trans isomerization in the D ring. FR-absorption causes the reverse trans to cis reaction. Dark reversion of Pfr to Pr also results in this trans to cis reverse reaction of PΦB.

The experiments that solved the PΦB structure in retrospect simply solved the chromophore of phyA. During those years, there was believed to only be a single phy species within the plant cell. With the current knowledge that plants contain multiple isoforms of phy (see Chapter 7), one can wonder whether each phy species has the same chromophore or if there are structurally different chromophores. To date, two approached has attempted to answer this question: generating transgenic plants ectopically expressing various phytochrome isoforms and subsequent analyses, and *in vitro* tests of bilin binding to recombinant apo-phy isoforms. This is briefly reviewed from Chapter 4.3 of the previous edition of this book (Kendrick and Kronenberg, 1994). Several groups accomplished successful transgenic expression of phyA simultaneously. In all cases, the overexpressed phyA was found to assemble into a photochromic protein spectrally indistinguishable from the native phyA. This result was interpreted as an indication that ectopically expressed phyA assembles into a holoprotein in an identical fashion as native phy apo-proteins. This work was extended when phyB was overexpressed in transgenic plants. The resulting holoprotein had the same absorption spectrum as holo-phyB assembled *in vitro*. This could suggest that endogenous phyB assembles with PΦB *in vivo*. Rigorous proof of this is lacking. Experiments to date to test the chromophore of other type II phys (phyC-phyE) comes from *in vitro* reconstitution, and mutational and cell biological conformation of photobiological action. Work by Schäfer's group revealed that linear bilins efficiently ligate to phyC-E (Eichenberg *et al.*, 2000). Additionally, mutations in phyC, phyD, and phyE all have R and FR physiological phenotypes (Aukerman *et al.*, 1997; Devlin *et al.*, 1998; Franklin *et al.*, 2003a). Finally, the phyC, D, and E proteins exhibit protein translocation throughout the plant cell in response to light (Kircher *et al.*, 2002). This suggests direct action of these proteins in response to light. At a minimum, these collective results implicate that the type II phys all bear an attached bilin. That said, proof of *in vivo* chromophore ligation to the type II phys (phyB-phyE), and resolution of the endogenous chromophore awaits further experimentation.

In higher plants, the question still remains as to whether PΦB is the only chromophore of the phy family of photoreceptors. Not only has there not been rigorous proof that the chromophore for phyB through phyE is PΦB, *in vivo* reconstitution of phyB action hints at multiple chromophores. The first experiments suggesting that alternate chromophores would assemble to phys occurred when oat seedlings were exogenously treated with the bilins biliverdin IXα (BV) and PCB (Elich and Lagarias, 1987b). The endogenous apo-phy pool assembled into a spectrally active, non-natural holo-phy. Thus, alternate chromophores can be ligated onto a phy *in vivo*. This work was confirmed in *Arabidopsis thaliana* seedlings bearing mutations in the PΦB biosynthetic pathway. When such seedlings are fed various bilins, not only does spectrally active phyA result, but phy photobiology is restored (Parks and Quail, 1991). Recent work by Furuya's group extended this via the application of chemically synthesized, synthetic chromophores to PΦB-deficient *Arabidopsis thaliana*. Certain analogue chromophores restored phyA and phyB responses, whereas others restored only phyB responses (Hanzawa *et al.*, 2002). These results can be interpreted in a variety of ways. For one, either alternative chromophores exist in plants, or optionally, the various phys have different structural tolerances to the steric interactions of artificial chromophores.

In lower plants, it has been conclusively demonstrated that phys can harbour a different chromophore than PΦB under natural *in vivo* conditions. Work in Lagarias' lab on the alga *Mesotaenium caldariorum* revealed that this organism uses an alternative chromophore on one of its phys. Specifically, it was found that the dominant phy species purified from this algae had a blue-shifted absorption spectrum relative to that of the higher plant holo-phyA spectrum. The authors thus concluded that the phy analysed in this alga lacked PΦB. Hydrolysis of the chromophore from this phy-related polypeptide, and subsequent structural characterization, revealed it as PCB (Wu *et al.*, 1997). One clear conclusion was that a phy could harbour alternative bilins. In retrospect, these exciting experiment results were amongst the first to reveal that a range of chromophores could be used throughout phy photobiology.

Recent work has revealed the prevalence of phytochromes throughout the prokaryotic kingdom (see Chapter 6). Upon sequence identification of these bacteriophytochrome (*BphP*) encoding genes, it became a major question as to the nature of the *in vivo* chromophores. This currently is a very active field of research by numerous laboratories, and the logic of the chromophore chemistry is extensively explored in the next chapter. Briefly stated here, the BphPs use BV as an efficient chromophore. Whereas CPH1-type phys, which have distinct primary structural differences, appear to use PCB. Interestingly many bacteria contain a suite of linear tetrapyrroles. It remains an intriguing possibility that the cognate phy-like proteins present in diverse eubacteria can switch chromophore use in response to alterations of bilin stoichiometry that occur during photomorphogenic processes, such as chromatic adaptation (Vierstra and Davis, 2000).

3. PHYTOCHROMOBILIN SYNTHESIS

The PΦB chromophore present in higher plants is synthesized entirely in the plastid via a branch of the pathway used to synthesis chlorophyll and heme (Elich and Lagarias, 1987a, 1987b). The first specific precursor leading to the production of the bilins is 5-aminolevolinic acid (ALA). In plants, ALA is produced from the glutamate (C5 pathway). Via a linear pathway shared by both heme and chlorophyll synthesis, ALA is converted ultimately into protoporphrin IX (Cornah *et al.*, 2003). The heme and chlorophyll biosynthetic pathways diverge at this point as ferrochelatase converts protoporphrin IX into heme, whereas Mg^{+2}-chelatase converts protoporphrin IX into Mg-protoporphyrin for chlorophyll production (Beale and Cornejo, 1991a, 1991b, 1991c). Heme is an essential cofactor for a wide range of enzymes, where its primary function is to chelate iron; the bound iron assists in oxidation/reduction reactions and in ligand binding within a very large suite of enzymes and adapter proteins (Platt and Nath, 1998).

The enzymology specific to PΦB biosynthesis was elegantly elucidated by Dr. Beale though metabolic studies using a red alga (Figure 4). The committed step in the synthesis of the phy-chromophore is the oxidative cleavage of heme to BV by a heme oxygenase (HO). In contrast to animal HOs, the HOs partially purified from photosynthetic organisms are soluble enzymes that require NADPH provided by the ferrodoxin-$NADP^+$ reductase (Cornejo and Beale, 1988). Importantly, the first HO from a photosynthetic organism was isolated and the corresponding gene cloned from the cyanobacterium *Synechocystis* sp. PCC6803 (Cornejo *et al.*, 1998). The presence of HOs in cyanobacteria suggests that higher plants use related enzymes in the commitment step of PΦB production. *Synechocystis HO1* provided a crucial molecular probe to address this question, as discussed below.

The final steps in PΦB production involve the reduction and isomerization of BV (Figure 4). The methine bridge of BV is reduced by a phytochromobilin synthase to yield (3Z)-PΦB (Beale and Cornejo, 1991a, 1991c; Parks and Quail, 1991; Terry, 1997). This activity is present in isolated etioplast of higher plants

(McDowell and Lagarias, 2001). Finally, (3Z)-PΦB is isomerised into (3E)-PΦB. An activity that defines this reaction has not been conclusively detected (Terry, 1997). Given that phys can covalently ligate (3Z)-PΦB to generate photochromic holoproteins (Terry et al., 1995), phys themselves could be the phytochromobilin isomerases. This waits to be tested.

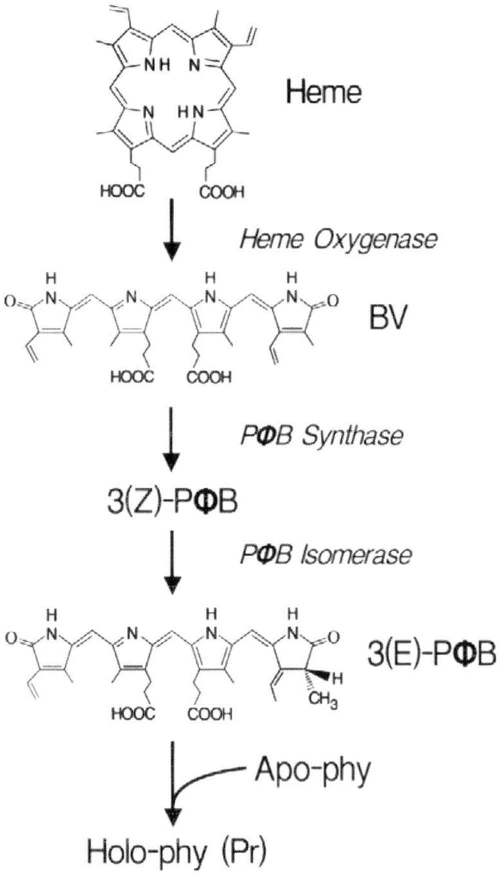

Figure 4. Proposed pathway for phy-chromophore biosynthesis in higher plants.

Mutations that block various steps in the PΦB pathway have been identified in a variety of plants, and as a consequence, phenotypically lack many phy responses. These include *hy1* and *hy2* from *Arabidopsis thaliana* (Chory et al., 1989; Parks and Quail, 1991), *au* and *yg-2* from tomato (Terry, 1996; van Tuinen et al., 1996), *pcd1* and *pcd2* from pea, and *pew1* and *pew2* from tobacco (Kraepiel et al., 1994) (Figure 5). One notable feature of these plants is that they are generally healthy

(Terry, 1997). This observation is inconsistent with the phenotype expected of a heme deficiency, strongly suggesting that the mutations disrupt steps after heme production. Heme levels in several PΦB-deficient mutants have been measured; each contains normal heme levels (Terry, 1996; Weller *et al.*, 1996; Weller *et al.*, 1997). This provided evidence that the genetic lesions affect steps after heme synthesis. The *pcd1* and *yg2* mutants were both shown to lack a plastid-derived HO activity (Terry, 1996; Weller *et al.*, 1996), and the *pcd2* and *au* mutants were found to lack a phytochromobilin-synthase activity (Terry, 1996; Weller *et al.*, 1997). Based on metabolic feeding experiments, Parks *et al.* showed that the *hy1* mutant of *Arabidopsis thaliana* is genetically blocked at the HO step (Parks and Quail, 1991). A phytochromobilin-synthase deficiency was inferred for the *hy2* mutant of *Arabidopsis thaliana* (Parks and Quail, 1991; Terry, 1997).

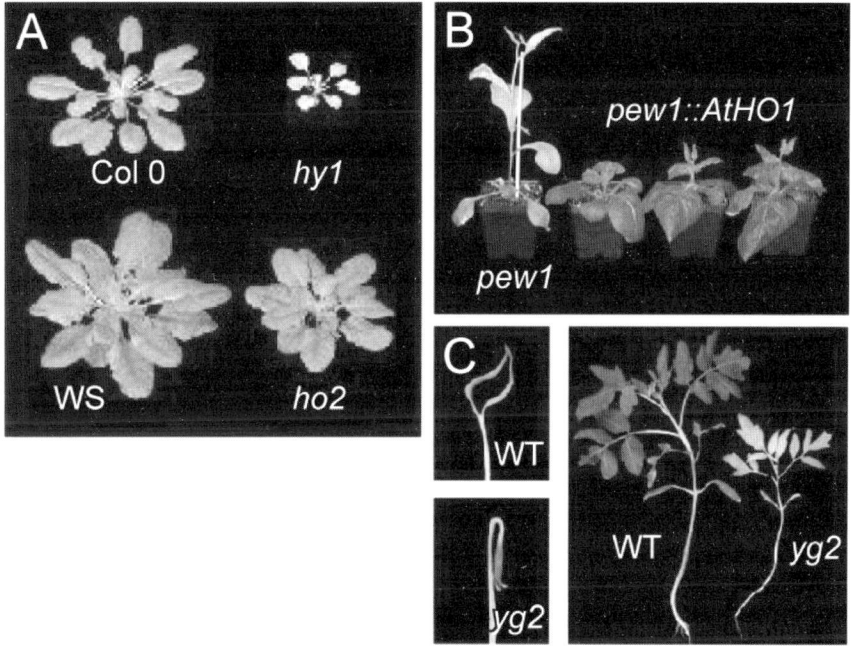

Figure 5. Comparison of heme oxygenase mutants. (A) Phenotype of light-grown Arabidopsis thaliana hy1 and ho2 plants. The respective wild-type parental ecotype of each mutant is included for comparison (Col for hy1 and WS for ho2). Plants were grown for 3 weeks under white light with a short-day photoperiod. (B) Complementation of the pew1 mutation by introducing Arabidopsis thaliana AtHO1. The pew1 mutant and three T1 lines of pew1 independently transformed with a genomic fragment encompassing AtHO1 (labelled pew1::AtHO1) were grown for 2 months under continuous white light. (C) Phenotypic comparison of the tomato yg2 mutant. Wild-type (WT) and yg2 tomato plants grown for 2 weeks under continuous R (left) or for 4 weeks in a greenhouse under natural lighting (right).

3.1 Heme Oxygenases

The first heme oxygenases characterized enzymatically are those from animal systems. These animal counterparts function by binding hemin (iron-bound heme), and use the bound iron and molecular O_2, to cleave heme stereospecifically. BV and carbon monoxide (CO) are released as products (Ishikawa *et al.*, 1998; Matera *et al.*, 1997). BV can further reduced to bilirubin in animals by a biliverdin reductase (Baranano *et al.*, 2002). In many contexts, the CO release after heme cleavage by HO is the desired reaction product for animals (Ryter *et al.*, 2002). Animal HO activity is membrane localized and is primarily located in the endoplasmic reticulum. A membrane anchor present within the primary amino-acid sequence can be removed for recombinant expression of a soluble protein, and this protein uses NADPH provided from a cytochrome P450 (Ishikawa *et al.*, 1998; Matera *et al.*, 1997). NADPH drives the reduction of iron, which is critical to the catalytic mechanism of HO (Ishikawa *et al.*, 1998). Clearly, any work on HOs from photosynthetic organisms can benefit greatly from a comparative analysis to animal HOs.

Protein extract from the red algae *Cyanidium caldarium* proved useful in the biochemical analyses on a HO from a photosynthetic organism. Unlike animal HOs, which are microsomal localized enzymes, and thus insoluble, the algal HO is freely soluble in a lysate. This solubility permitted the partial purification and enzymatic characterization of the enzyme (Cornejo and Beale, 1988). The first notable feature about this HO was that the enzymatic reaction products are identical to that of animal HOs. Via various protein fractionation experiments, it was concluded that three components could be fractionated from this extract to generate HO activity (Cornejo and Beale, 1988). Fraction I is likely a Fe/S cluster protein, and can be replaced by a purified ferrodoxin. Fraction III is a pyridine-binding component and substitution studies suggested that this fraction was a ferrodoxin-NADP+ oxireductase. Fraction II was key, as this contained the heme-binding activity. This most certainly contained the HO enzyme. Collectively what was learnt from these studies was that reducing power from a ferrodoxin (Fraction I) and a cytochrome C-like activity (Fraction III) provide a driving force for the HO (Fraction II) enzyme. The primary differences of the algal system to the animal system were the solubility of HO and the requirement for a different reductant. These clues were useful in the characterization of a higher-plant HO, as described below.

The newly emerged field of genomics, coupled with forward and reverse-genetic approaches aided the identification of a HO from a higher plant. Cornejo *et al.* (Cornejo *et al.*, 1998) identified a HO protein from bacterium *Synechocystis* sp. 6803 that catalyses the enzymatic reaction believed to be absent in *hy1* backgrounds (Terry, 1997). This bacterial enzyme rescues the *hy1* phenotype when added transgenically to plants (Willows *et al.*, 2000). Using this cyanobacterial *HO1* gene as a query, putative *HO* genes from *Arabidopsis thaliana* were identified (Davis *et al.*, 1999a). Two predicted open-reading frames with significant derived amino-acid-sequence homology to *Syn*HO1 were identified in a genetic interval that contained *HY1*. These genes are approximately 40-kbp apart: one was annotated as F18A8.4 and the other as T9J22.22, and are now referred to as *AtHO1* and *AtHO2*, respectively (Figure 6). In addition to candidate HOs that map to the *HY1* locus, two additional HOs were also detected in the *Arabidopsis thaliana* DNA database, referred to as *At*HO3 and *At*HO4 (Davis *et al.*, 2001). It was determined that the multiple alleles of *hy1* all bear lesions in the *AtHO1* gene. The *hy1* alleles currently available likely represent null mutations, and thus the regain of phy responses in this mutant background as plants mature is likely caused by the action of the other HOs. The nature of *HY1* was independently confirmed by a similar positional cloning of the *Arabidopsis thaliana HY1* locus (Muramoto *et al.*, 1999). In this work, it was also shown that, as predicted, *At*HO1 is localizes to the plastid compartment of the plant cell. Both green fluorescent protein (GFP)-reporter studies and western analysis of cellar fractions were used in these localization studies. In this paper, the first reports of a recombinant plant HO activity were also obtained. Briefly, a recombinant *At*HO1 extract was added to a BV-reductase. A time-dependent spectral change in heme resulted, and this was interpreted that *At*HO1 has HO activity. In many ways, the activity of this plant-derived HO was highly similar to the algal purified HO.

Extensive enzymatic analyses have recently been performed on recombinant *At*HO1 (Muramoto *et al.*, 2002). From an *E. coli* expression system, large amounts of soluble *At*HO1 were purified to homogeneity. As is seen with animal HOs, *At*HO1 has tight heme-binding activity. Titration of the heme-*At*HO1 complex revealed a stoicheometry of 1:1. The heme-*At*HO1 complex is enzymatically active after the application of a reduced ferrodoxin. In the presence of iron-chelators, the heme bound to the HO was converted to BV. This rigorously proved that *At*HO1 is a *bona fide* HO. As two protein fractions aided algal HO activity, it was investigated whether multiple reductants aided *At*HO1 activity. This was confirmed as ascorbate assisted ferrodoxin in allowing *At*HO1 to cleave heme. HOs from different organisms can generate different isomers of BV. In PΦB production, BV IXα is the intermediate isomer form. Using an HPLC assay, it was shown that *At*HO1 exclusively generates BV IXα. In animal HO systems, one carbon is lost from heme in its linerization to BV. This leaving product is CO. *At*HO1 was shown to liberate CO from heme. This was assayed spectroscopically with aid from CO binding to myoglobin. As it is now clear that plants have a CO-generating activity, this leads to the intriguing possibility that HOs in plants are components of ethylene signalling. CO can bind to the ethylene receptors, and this results in physiological consequences (Abeles *et al.*, 1992). Further experimentation is required to resolve if *At*HO1 has a role in CO hormone production.

To further define the role of HOs in phy assembly, a genetic analysis of the *AtHO2* locus was performed (Davis *et al.*, 2001). The data presented revealed that *AtHO2* also contributes to PΦB synthesis. A mutation within *AtHO2* was isolated through a reverse-genetic strategy. This mutant showed a phy-deficient phenotype even in the presence of functional *HY1*. Additionally, the phyA polypeptide in etiolated *ho2* seedlings was a mixture of apo- and holoproteins. *AtHO1* and *AtHO2* have overlapping transcriptional expression patterns. Further, *AtHO2* expression was markedly reduced in the *hy1* background, illustrating a possible genetic control of PΦB synthesis. Taken together, these results suggest that *AtHO2* and the *HY1* encoded HO enzymes have distinct, but overlapping roles in regulating photomorphogenesis though holo-phy assembly. That said, it has not been conclusively shown that *At*HO2 functions as a *bona fide* HO. Additionally, while it is known that *AtHO3* and *AtHO4* are expressed in *Arabidopsis thaliana*, there is no data to implicate these genes in PΦB production, nor their encoded proteins bearing HO activity. Further experimentation is required.

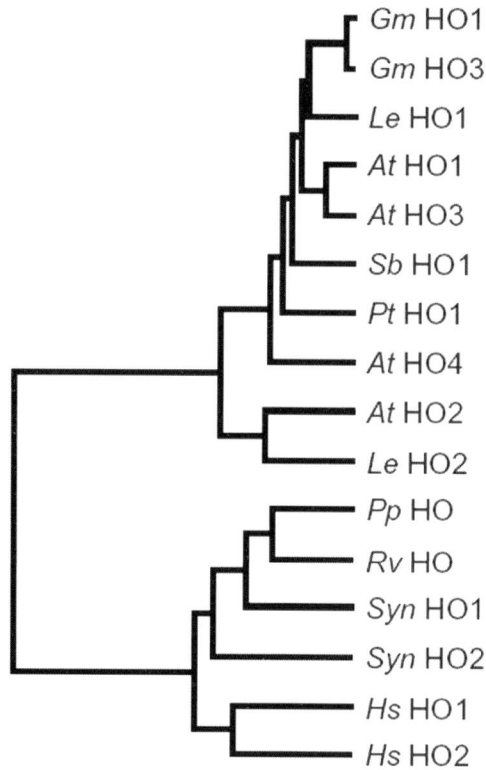

Figure 6. Phylogram showing the amino-acid sequence relationships of the collection of plant HOs with their counterparts from other organisms. The encoded HO proteins described in this chapter were compared by GROWTREE to sequences from Synechocystis (Syn) HO1 (Genbank D90091) and HO2 (genbank D90912), Rhodella violacea (Rv) HO1 (Genbank AF000717), Porphyra purpurea (Pp) HO (genbank P51271), and human (Hs) HO1 (Genbank P09601) and HO2 (Genbank P30519). The distance along the horizontal axis separating two sequences is proportional to their sequence divergence.

Having the *At*HO1 sequence allowed an investigation of holo-phy production in a variety of plants. The *yg2* mutant of tomato has been biochemically described as being deficient at the same PΦB biosynthetic step as *Arabidopsis thaliana hy1* (Figure 5). Using DNA-sequence of *HY1* as a probe, two tomato *HO* genes were identified (Figure 6), one of which bears a mutation the genomic locus of the *yg2* mutant (Davis *et al.*, 2001). Multiple *HO* genes from various other plants were also isolated. Sequence comparisons of the plant HOs to HOs from other organisms illustrated that plant HOs are the most distantly related members of this enzyme class. Figure 6 shows a phylogenetic tree constructed using available higher-plant HOs, two HOs from the cyanobacterium *Synechocystis* sp. PCC6803 (Cornejo *et al.*, 1998), one each from the alga *Porphyra purpurea* and *Rhodella violacea* (Richaud and Zabulon, 1997), and two human HOs (Platt and Nath, 1998). The general features of the tree reveal two main groups of HOs in plants, typified by *At*HO1 and *At*HO2. The HO1-related group included sequences from several dicots, a monocot, and a gymnosperm. The distribution of these proteins is coincides with the expected presence of phys within all land plants. Multiple HO1-like sequences were identified in two dicots (*Arabidopsis thaliana* and soybean), suggesting that all dicots contain a subfamily of HO1 enzymes. Additionally, a HO2 sequence was found in monocots, potentially indicating that all flowering plants contain this second HO subtype. Further investigations are required before speculating on how widespread the HO2-related sequences are in lower plants. The relationships of the plant HOs to the four lower-photoautorophic sequences and the human HOs show that plant HOs are the most evolutionary distant within this group (Figure 6). This is somewhat surprising, as it might be expected that the sequences from the lower photosynthetic organisms would be more related to plant than animal HOs. This divergence may indicate that plants did not obtain their *HO* genes from the plastid genome during the symbiotic evolution of eukaryotic photosynthetic organisms. If this is true, then the plant *HO* genes could have arisen from the progenitor of the nuclear or mitochondrial genome.

3.2 Phytochromobilin Synthase

The synthesis of the phy chromophore requires that BV is converted into (3Z)-PΦB and then into (3E)-PΦB. In plants, BV is directly reduced to (3Z)-PΦB by a ferredoxin-dependant phytochromobilin synthase (McDowell and Lagarias, 2001). The biochemical characterization of this enzyme activity was originally performed in a red alga (Beale and Cornejo, 1991c), and more recently, in oats (McDowell and Lagarias, 2001). The recent purification and characterization of PΦB synthase from plants confirmed that the activity is plastid localized. Using this purified oat protein, the initial biochemical characterization of PΦB synthase revealed several notable features. The PΦB synthase is a low-abundance enzyme, and the turnover rate of PΦB synthase is much higher than that of HO. Collectively, it can be assumed that HO, and not PΦB synthase, is rate-limiting in PΦB production. Additionally, this suggests that relatively low levels of BV exist within the plant. That said, when animal enzymes that catabolize BV are transgenically supplied to plants, a strong phenotypic effect is evident (Franklin *et al.*, 2003b; Lagarias *et al.*, 1997; Montgomery *et al.*, 2001; Montgomery *et al.,* 1999). Thus, BV has at least a limited half-life within the plastid before conversion from BV to 3E-PΦB.

Work by Parks and Quail suggested that the *hy2* mutant of *Arabidopsis thaliana* is deficient in a PΦB synthase (Parks and Quail, 1991). The logic being that *hy2* can be phenotypically rescued by feeding mutant plants PΦB, but not BV. Collaborative work between the labs of Kochi and Lagarias confirmed that the *HY2* locus indeed encodes a PΦB synthase (Kohchi *et al.*, 2001). Using a positional-cloning strategy, the *HY2* locus was fine mapped in *Arabidopsis thaliana*. Towards the end of the walk, a relatively large interval remained with several candidate genes. A candidate gene for *HY2* was isolated based on several criteria. First, the *HY2* gene was expected to encode a novel protein, as the gene sequence was unknown for any ferrodoxin-dependent bilin reductase. Next, it was expected that the conceptually translation of the *HY2* open-reading frame (ORF) would contain an N-terminal plastid transit peptide. This is because the enzymatic activity of PΦB synthase co-localizes to the plastid, and the oat PΦB synthase activity was found to be plastid derived. Finally, it was expected that derived protein-sequence similarity would exist between HY2 and ORFs within cyanobacterial genome sequences, as cyanobacteria are predicted to also have bilin reductase-encoding genes. From this creative logic, a singular *HY2* candidate sequence remained. Multiple *hy2* alleles contained DNA lesions in this candidate confirming the correct identification. Unlike the *HY1* locus, *HY2* encodes an unusual plant protein in that it is not part of a gene family. The N–terminus of this HY2 protein has 45 amino acids with features at a plastid-transit peptide. When these sequences were fused to the coding region of GFP, and introduced into plants, fluorescence localized to the plastid, as expected. They further concluded that, as with the plant-purified oat version of this enzyme, a recombinant form of the *Arabidopsis thaliana* HY2 protein was a ferrodoxin-dependant BV reductase. Using a coupled holo-phy-assembly assay to detect PΦB, HY2 protein was shown to generate PΦB only in the presence of ferrodoxin.

The molecular cloning of *HY2* has provided new insight into the knowledge of bilin biosynthesis in general. Bioinformatic analysis revealed that *HY2* is related to a number of cyanobacterial genes. This lead to the hypothesis that these prokaryotic genes encode proteins involved in PCB and phycoerythrobilin (PEB) synthesis (Frankenberg *et al.*, 2001). This was confirmed through enzymatic assays of recombinant proteins. Proteins converting BV into (3Z)-PCB were identified, and were termed PcyA. The conversion of BV to PEB requires two polypeptides: PebA generated 15, 16-hydrobiliverdin, and this reaction product is a substrate for PebB to generate (3Z)-PEB. The *HY2* super in cyanobacteria thus has a range of roles in bilin metabolism. Each related polypeptide uses similar bilins to generate similar reaction products ensuring that the range of required tetrapyrroles is synthesized.

Final genetic steps in the PΦB-synthesis pathway await isolation. Early work revealed the existence of the *HY6* locus. As this mutant is deficient in both R and FR light perception, this mutant was presumed to be defective in a step of PΦB synthesis (Chory *et al.*, 1989). The only biochemical step where other mutants have not been isolated is in the PΦB isomerase (Figure 4) (Terry, 1997). Unfortunately, the *HY6* mutant has been lost and this mutant cannot be tested if it is blocked in the PΦB isomerase reaction. Further, it remains a question as to whether PΦB isomerases are required at all. The *HY2* super family appears to generate exclusively the 3Z isomers of their respective products (Frankenberg *et al.*, 2001). Numerous studies have established that the 3E isomers are substrates for phycobiliprotein and phy holoproteins (Lagarias and Lagarias, 1989a; Li and Lagarias, 1992; Terry *et al.*, 1995). There appears to be 3 plausible possibilities to resolve this discrepancy. Either 3Z are converted to 3E chemically/non-enzymatically (3E isomer are more thermodynamically stable than the 3Z forms), the bilin isomerases are genetically redundant in a wide variety of plants, and thus have never been isolated mutationally, or, most intriguingly, phycobiliproteins, including the phys, themselves contain the bilin-isomerase activity. This has yet to be tested.

One important point virtually ignored experimentally is that once PΦB is synthesized, it must be excreted from the plastid into the cytosol. The logic being that there is no experimental evidence that apo-phys enter the plastid during holo-phy generation. And if this occurred, a separate translocation out of the plastid would also be required for the holo-chromoprotein to return to the cytoplasm. These two events seem unlikely. Therefore, the PΦB ligand must leave the plastid to attach to apo-phy in the cytosol. It is unclear whether this export occurs though the assistance of chaperons or transporters, or is a simple process of diffusion. BV added to isolated plastids can enter this organelle (Terry and Lagarias, 1991). For this reason, it has been concluded that BV can move freely into and out of plastids *in vivo*. Still, bile pigments are potent inhibitors of cellular metabolism (Vierstra and Sullivan, 1988). It seems plausible that a carrier protein for PΦB exists. This carrier would shuttle PΦB from the plastid to apo-phy. That said neither a biochemical activity nor a candidate gene exists for such an activity. Much work is required to probe this key cell-biological problem.

4. HOLO ASSEMBLY

The primary structure of the phytochrome family reveals two key sequence motifs. The N–terminal domain is the tetrapyrrole binding GAF domain, which is associated with the C–terminal dimerization domain related to histidine kinases (Montgomery and Lagarias, 2002). GAF domains are present in a variety of proteins and are structurally related to PAS domains. The GAF and PAS domain types typically occur multimerized within the phytochrome family. These two domain types compose binding sights for small ligands that regulate the biochemical output of the respective protein. It is within these domains that the binding sites for tetrapyrroles exist.

Biochemical analysis of chromophore ligation to various phy subtypes has advanced through recombinant technologies (reviewed in Chapter 4.3, Kendrick and Kronenberg, 1994). One of the first questions regarding chromophore ligation was whether various phy types can assemble with a chromophore autocatalytically or if associated enzymes are required. When phytochrome cDNAs where first expressed in *E. coli* or yeast expression systems, and the resulting apo-proteins purified, various tetrapyrroles were found to ligate to these proteins. The phys assembled *in vitro* with either PCB or PΦB resulted in a covalent attachment and produced a photoreversible difference spectrum. These results indicate that chromophore ligation is intrinsic to the phy apoprotein. Deletion mutants of these full-length phytochrome cDNAs defined the minimal chromophore lyase domain (Figure 1).

Using *in vitro* assembly systems, the chromophore requirements for assembly where defined. Linear tetrapyrrole chromophore analogs were compared for their ability to bind to phy and shown for reversible absorption changes. Analogs such as 3E-PCB readily form covalent adducts while chromophores lacking a C15 double bond between rings C and D, such as (3E)-PEB, are poor substrates for attachment, and when they attach, these chimeric molecules are photochemically inactive and are locked into a "Pr-like" confirmation (Eichenberg *et al.*, 2000).

PΦB is attached via a thiol-ether linkage to a cysteine residue positionally conserved in all higher-plant phys yet described (Cys321 in oat phyA). Apo-phys purified from plants will bind PΦB autocatalytically, providing *in vivo* evidence that chromophore-binding activity is intrinsic to the phy polypeptide (Elich and Lagarias, 1989). When the linkage cysteine is converted to serine, the resulting mutant protein fails to bind chromophore *in vitro* or when expressed in plants (Boylan and Quail, 1989). This indicated at that time that that the thiol-ether linkage is essential for PΦB ligation, and further, that no other cysteine site in apo-phy can serve as a substitute. This is in contrast to some BphPs that naturally lack this "conserved" cysteine. For example, the BphP from *Deinococcus radiodurans* and the two BphPs from *Agrobacterium tumerfaciens* covalently ligate bile pigments while harbouring non-charged residues at this position (Davis *et al.*, 1999; Karinol and Vierstra, 2003). Further, analyses of holo-BphP purified from *D. radiodurans* indicate that chromophore ligation occurs *in vivo* (Bhoo *et al.*, 2001), proving that the cysteine position absolutely required for thiol-ether ligation of PΦB in plant phy, is not an absolutely necessary chemical reaction for phy-like action.

Within higher plant phys, PΦB is found linked to a conserve cysteine via a thiolether bond. Mutational analysis of the conserve cysteine revealed that this residue is critical for covalent attachment from the chromophore to the protein. It appears that the CPH1-type of bacterial phys also link their cognate tetrapyrrole PCB to this positionally conserved cysteine (Hubschmann *et al.*, 2001). Interestingly, a CPH1-type phytochrome from *Calothrix* PCC7601 maintains photoreversibilty even though it fails in covalent ligation (Jorissen *et al.*, 2002). These were amongst the first report to show that covalent ligation is not required for photoconversion of the chromophore. Additionally, as BphPs lack the ability to generate a thiol-ether linkage at this "cysteine site," alternative-binding mechanisms must exist. Preliminary work showed that molecules such as PCB can bind to histidine residues within the bilin binding pocket of *Dr*BphP perhaps via a Schiff's base linkage (Davis *et al.*, 1999b). But we now know that BV is the true chromophore (Bhoo *et al.*, 2001). Thus, it remains a possibility that these PCB-binding results are not typical of the *in vivo* conditions. In support of this notion, recent studies in with *At*BphP1 indicate that the A-ring of BV ligates via thiol-ether bond, rather than a Schiff's-base linkage to the D-ring (Lamparter *et al.*, 2003). This "new" cysteine is at the more extreme N-terminus, and is also highly conserved amongst the BphPs, CPH1s, and phys. Collectively, one interpretation is that BphPs bind the A–ring of BV by a thiolether linkage, can bind the D-ring of PCB via a Schiff's-base linkage, and that higher plants phytochromes bind the D–ring of PΦB via a thiol-ether linkage. It is intriguing that these could be the steps of the evolution of such phy-chromophore interaction. In both BphPs and higher plant phys, both sides of the bilin must be "held" for a conformational flip of the C15 bond of the respective bilin. Conceptually, it would not seem to matter enzymatically whether the A-ring or the D–ring is covalently attached for this double "holding" to occur during the "twisting" of the chromophore during Pr to Pfr conversion. This is supported by studies where the chromophore isomerises in the absence of covalent ligation (Jorissen *et al.*, 2002). Regardless, it is clear that multiple non-covalent interactions must occur during phototransformation, and that while covalent interactions are prevalent in the synthesis of a holo-phy-type chromoproteins, this covalent ligation is not absolutely required.

Recombinant phy expression systems have provided a tool to study the enzymatic requirements of ligation of linear bilins to apo-phys. This first publication of this was by Lagarias and Lagarias 1989 when they synthesized apo-phyA in a coupled *in vitro* transcription-translation system (Lagarias and Lagarias, 1989b). When PCB was added to this protein preparation, a covalent adduct resulted. This result was important as it revealed that phyA itself is a bilin lyase. The fact that ligation occurs *in vitro* indicates that the enzymatic activity of phy as a bilin lyase is energetically favourable. There is no current evidence to suggest that chaperonins, other adapter proteins, co-factors, or molecules such as ATP or NADH participate in any way in the facilitation of this reaction. As mentioned before, the only potential assistant in the lyase reaction is a putative shuttle protein that could mobilize PΦB from the plastid to the cytoplasm. It is also worth noting that the cell-biological site of PΦB attachment to apo-phy is unclear.

Additional site-directed mutants in plant phy have been studied to identify other amino acids involved in chromophore ligation and the photochromic behaviour of phys. A positionally conserved histidine, adjacent to the cysteine that links PΦB, was shown to be critical for chromophore ligation and the photochromic properties of phyA (Bhoo *et al.*, 1997). Lagarias and Rapoport have proposed that this histidine directly participates in the mechanism of PΦB ligation (Lagarias and Rapoport, 1980). Based on the binding of PCB to this histidine in *Dr*BphP, perhaps plant phys use this histidine as an intermediate step in the thiol-ether linkage that ultimately results. Mutations in a conserved glutamine three residues downstream from the chromophore-attachment site and an isoleucine that defines the boundary of the chromophore-ligation domain of phyA were both shown to be important for PΦB binding (Wu and Lagarias, 2000). It is unclear what the enzymological effects of such mutations caused. Substitution mutants of other highly conserved amino-acid residues in the immediate vicinity of the chromophore-binding pocket generated phys with normal ligation and photochromic activity, showing that not all residues in the chromophore pocket are essential for lyase binding (Bhoo *et al.*, 1997). A mechanistic view regarding the enzymatic ligation of PΦB and the nature of photochromism awaits resolution of the phy three-dimensional structure.

As apo-phy can ligate bilins in the absence of accessories, simple recombinant expression systems have been engineered to facilitate structure/function relationships of recombinant phy. One of the first such systems was identified accidentally when an alga phy was expressed in the yeast *Pycia pastoris*. After a 24-hour period of protein induction, covalently ligate chromophore was evident, and a photoreversible holo-phy was generated (Wu and Lagarias, 1996). More recently, recombinant systems have been engineered to generate holo-phy through the coexpression of bilin synthesis enzymes and a phy. One example is the simple coexpression of *D. radiodurans* HO and BphP (Bhoo et al., 2001). As *Dr*BphP uses the primary linearization product of heme cleavage, the coexpression of these two proteins in an *E. coli* system results in holo-BphP. This also works with plant phys: the coexpression of a HO, a PΦB synthase, and a phy results in a photoreversible chromoprotein (Gambetta and Lagarias, 2001). Such systems have future utility, as bile pigments are relatively difficult to purify or chemically synthesize. Thus, this limitation is removed.

5. BIOPHYSICS OF THE CHROMOPHORE

To a physiologist, it may not be apparent as to why one should care about the intermediate states of the photoconversion of Pr to Pfr. The answer is simple. All physiological processes have a physical basis. To understand the molecular events that are initiated by during phototransformation, one must understand how this is physically and chemically possible. Luckily, the phytochrome field has a rich history of biophysical analyses. Current technology to explore this scientific problem has largely been dominated by spectroscopic techniques. Much of this topic was covered in Chapter 4.3 of the previous edition of this book (Kendrick and Kronenberg, 1994), but will be briefly restated here. The oscillator strength ratio of the visible to the SORT bands is a sensitive assay of the conformation of the tetrapyrrole. A Gaussian fit of the Pr and Pfr spectra followed by determination of the SORT bands revealed that the oscillator strength ratio lies between a fully linear and a cyclic confirmation. From this, it was interpreted that both the Pr and Pfr form of PΦB within the holo-phyA maintain a semi-extended confirmation. It should be noted that these experiments are exclusively with holo-phyA and it remains to be shown that PΦB has a similar semi-extended confirmation that the other phytochrome isoforms. Ultimate proof of the confirmation of Pr and Pfr within the native holo-phy will require detailed structural analysis. Protein crystallography of both the Pr and Pfr forms of phy would reveal two "locked" confirmations. Until then, solid-state NMR with labelled chromophore and unlabelled protein shows promise as a technology to reveal not only the confirmation of the chromophore as Pr and Pfr, but also intermediate transition states, as analogous studies have for rhodopsin (Grobner *et al.*, 2000). Additionally, advances in ultra-low-temperature structural analysis holds promise to resolve the photochemical mechanism of chromophore transformation via the isolation of intermediate structures (Genick *et al.*, 1998).

6. PERSONAL PERSPECTIVES

When I started graduate school working on phys, I thought that tetrapyrroles were a "solved" question and did not merit further study. This changed rapidly when I delved into this field. It is now apparent to me that tetrapyrroles are not a "solved" question, and that a host of important answers need to be found. For this concluding section, I would like to state some that I feel are relevant and tractable. Hopefully these will be tackled by the next generation of bench scientists as they learn of the interests and excitements that come from the little linear molecular that is PΦB.

6.1 Phy chromophore structure

Do type I phys use a different chromophore than type II phys?
Can higher plants use chromophores beyond PΦB?
Can the bacterial phys, including CPH1-type and the BphP-type, use alternate chromophores in response to light?

6.2 Phy chromophore synthesis

What is the nature of the *HY6* locus?
Is there a PΦB isomerase?
Are phycobiliproteins, including the phys themselves, the PΦB isomerases?
What is the fate of PΦB upon proteolysis of a holo-phyA?
Why do plants have multiple HOs?
Is CO production from plant HO used in cross talk with ethylene signalling?
What is the catalytic mechanism of a plant PΦB synthase?
Why is NADPH used as a reducing factor for HO and PΦB synthase?

6.3 Holo-phy assembly and structure

How does a plastidic tetrapyrrole get ligated to a cytoplasmic apo-phy?
Are there PΦB chaperones?
Are two ends of a phy or BphP used in associates with the A and D rings of PΦB/BV?
What is the catalytic enzymology of the phy bilin-lyase domain?
Can a Pfr chromophore "pop off" a non-covalently associated bilin:apoCPH1 complex, and is this free Pfr-PCB re-used?
What is a three dimensional structure of an apo-phy?
What is a three dimensional structure of a Pr PΦB-linked holo-phy?
What is a three dimensional structure of a Pfr PΦB-linked holo-phy?
Can comparative analyses of the above protein structures to the Pr and Pfr structures of BV-linked holo-BphP lead to insights into the chemical and enzymatic activities of these respective proteins?

What steps occur structurally as holo-phy photoconverts, both with regard to the chromophore and with regard to the protein?

7. REFERENCES

Abeles, F. B., Morgan, P. W., Saltveit, M. E.. Ethylene in Plant Biology (2nd ed.). San Diego: Academic Press, Inc., 1992.

Aukerman, M. J., Hirschfeld, M., Wester, L., Weaver, M., Clack, T., Amasino, R. M., Sharrock, R. A. (1997). A deletion in the PHYD gene of the Arabidopsis Wassilewskija ecotype defines a role for phytochrome D in red/far-red light sensing. Plant Cell, 9, 1317-1326.

Baranano, D. E., Rao, M., Ferris, C. D., Snyder, S. H. (2002). Biliverdin reductase: a major physiologic cytoprotectant. Proc Natl Acad Sci USA, 99, 16093-16098.

Beale, S. I. and Cornejo, J. (1991a). Biosynthesis of phycobilins. 3(Z)-phycoerythrobilin and 3(Z)-phycocyanobilin are intermediates in the formation of 3(E)- phycocyanobilin from biliverdin IX alpha. J Biol Chem, 266, 22333-22340.

Beale, S. I. and Cornejo, J. (1991b). Biosynthesis of phycobilins. 15,16-Dihydrobiliverdin IX alpha is a partially reduced intermediate in the formation of phycobilins from biliverdin IX alpha. J Biol Chem, 266, 22341-22345. Beale, S. I. and Cornejo, J. (1991c). Biosynthesis of phycobilins. Ferredoxin-mediated reduction of biliverdin catalysed by extracts of Cyanidium caldarium. J Biol Chem, 266), 22328-22332.

Bhoo, S. H., T., H., Jeong, H.-Y., Lee, J.-G., Furuya, M., Song, P.-S. (1997). Phytochrome photochromism probed by site-directed mutations and chromophore esterification. J Am Chem Soc, 119, 11717-11718.

Bhoo, S. H., Davis, S. J., Walker, J., Karniol, B., Vierstra, R. D. (2001). Bacteriophytochromes are photochromic histidine kinases using a biliverdin chromophore. Nature, 414, 776-779.

Boylan, M. T. and Quail, P. H. (1989). Phytochrome A Overexpression Inhibits Hypocotyl Elongation in Transgenic Arabidopsis. Proc Natl Acad Sci USA, 88, 10806-10810.

Chory, J., Peto, C. A., Ashbaugh, M., Saganich, R., Pratt, L. H., Ausubel, F. M. (1989). Different roles for phytochrome in etiolated and green plants deduced from characterisation of Arabidopsis thaliana mutants. Plant Cell, 1, 867-880.

Christie, J. M., Swartz, T. E., Bogomolni, R. A., Briggs, W. R. (2002). Phototropin LOV domains exhibit distinct roles in regulating photoreceptor function. Plant J, 32, 205-219.

Cornah, J. E., Terry, M. J., Smith, A. G. (2003). Green or red: what stops the traffic in the tetrapyrrole pathway? Trends Plant Sci, 2003(8), 5.

Cornejo, J. and Beale, S. I. (1988). Algal heme oxygenase from Cyanidium caldarium. Partial purification and fractionation into three required protein components. J Biol Chem, 263, 11915-11921.

Cornejo, J., Willows, R. D., Beale, S. I. (1998). Phytobilin biosynthesis: cloning and expression of a gene encoding soluble ferredoxin-dependent heme oxygenase from Synechocystis sp. 6803. Plant J, 15, 99-107.

Davis, S. J., Kurepa, J., Vierstra, R. D. (1999a). The Arabidopsis thaliana HY1 locus, required for phytochrome-chromophore biosynthesis, encodes a protein related to heme oxygenases. Proc Natl Acad Sci USA, 96, 6541-6546.

Davis, S. J., Vener, A. V., Vierstra, R. D. (1999b). Bacteriophytochromes: phytochrome-like photoreceptors from nonphotosynthetic eubacteria. Science, 286, 2517-2520.

Davis, S. J., Bhoo, S. H., Durski, A. M., Walker, J. M., Vierstra, R. D. (2001). The heme-oxygenase family required for phytochrome chromophore biosynthesis is necessary for proper photomorphogenesis in higher plants. Plant Physiol, 126, 656-669.

Devlin, P. F., Patel, S. R., Whitelam, G. C. (1998). Phytochrome E influences internode elongation and flowering time in Arabidopsis. Plant Cell, 10, 1479-1487.

Dietz, K. J. (2003). Redox control, redox signalling, and redox homeostasis in plant cells. Int Rev Cytol, 228, 141-193.

Eichenberg, K., Bäeurle, I., Paulo, N., Sharrock, R. A., Rüdiger, W., Schäfer, E. (2000). Arabidopsis phytochromes C and E have different spectral characteristics from those of phytochromes A and B. FEBS Lett, 470, 107-112.

Elich, T. D. and Lagarias, J. C. (1987a). Phytochrome chromophore biosynthesis: both 5- aminolevulinic acid and biliverdin overcome inhibition by gabaculine in etiolated Avena sativa L. seedlings. Plant Physiol, 84, 304-310.

Elich, T. D. and Lagarias, J. C. (1987b). Phytochrome chromophore biosynthesis: both 5-aminolevulinic acid and biliverdin overcome inhibition by gabaculine in etiolated Avena sativa L. seedlings. Plant Physiol, 84, 304-310.

Elich, T. D. and Lagarias, J. C. (1989). Formation of a photoreversible phycocyanobilin-apophytochrome adduct *in vitro*. J Biol Chem, 264, 12902-11298.

Frankenberg, N., Mukougawa, K., Kohchi, T., Lagarias, J. C. (2001). Functional genomic analysis of the HY2 family of ferredoxin-dependent bilin reductases from oxygenic photosynthetic organisms. Plant Cell, 13, 965-978.

Franklin, K. A., Davis, S. J., Stoddart, W. M., Vierstra, R. D., Whitelam, G. C. (2003a). Abstract Mutant analyses define multiple roles for phytochrome C in Arabidopsis photomorphogenesis. Plant Cell, 15, 1981-1989.

Franklin, K. A., Linley, P. J., Montgomery, B. L., Lagarias, J. C., Thomas, B., Jackson, S. D., Terry, M. J. (2003b). Misregulation of tetrapyrrole biosynthesis in transgenic tobacco seedlings expressing mammalian biliverdin reductase. Plant J, 35, 717-728.

Gambetta, G. A. and Lagarias, J. C. (2001). Genetic engineering of phytochrome biosynthesis in bacteria. Proc Natl Acad Sci USA, 98, 10566-10571.

Genick, U. K., Soltis, S. M., Kuhn, P., Canestrelli, I. L., Getzoff, E. D. (1998). Structure at 0.85 A resolution of an early protein photocycle intermediate. Nature, 392, 206-209.

Grobner, G., Burnett, I. J., Glaubitz, C., Choi, G., Mason, A. J., Watts, A. (2000). Observations of light-induced structural changes of retinal within rhodopsin. Nature, 405, 810-813.

Hanzawa, H., Shinomura, T., Inomata, K., Kakiuchi, T., Kinoshita, H., Wada, K., Furuya, M. (2002). Structural requirement of bilin chromophore for the photosensory specificity of phytochromes A and B. Proc Natl Acad Sci USA, 99, 4725-4729.

Huebschmann, T., Boerner, T., Hartmann, E., Lamparter, T. (2001). Characterization of the Cph1 holo-phytochrome from Synechocystis sp. PCC 6803. Eur J Biochem, 268, 2055-2063.

Imaizumi, T., Tran, H. G., Swartz, T. E., Briggs, W. R., Kay, S. A. (2003). FKF1 is essential for photoperiodic-specific light signalling in Arabidopsis. Nature, 426, 302-306.

Ishikawa, K., Matera, K. M., Zhou, H., Fujii, H., Sato, M., Yoshimura, T., Ikeda-Saito, M., Yoshida, T. (1998). Identification of histidine 45 as the axial heme iron ligand of heme oxygenase-2. J Biol Chem, 273, 4317-4322.

Jones, A. M., and Erickson, H. P. (1989). Domain structure of phytochrome from Avena sativa visualized by electron microscopy. Photochem. Photobiol, 49, 479-483.

Jorissen, H. J., Quest, B., Lindner, I., Tandeau de Marsac, N., Gartner, W. (2002). Phytochromes With Noncovalently Bound Chromophores: The Ability of Apophytochromes to Direct Tetrapyrrole Photoisomerization. Photochem. Photobiol, 75, 554-559.

Karinol, B. and Vierstra, R. D. (2003). The pair of bacteriophytochromes from Agrobacterium tumefaciens are histidine kinases with opposing photobiological properties. Proc Natl Acad Sci USA, 100, 2807-2812.

Kendrick, R. E. and Kronenberg, G. H. M., Photomorphogenesis in Plants (2nd ed.). Dordrecht: Kluwer Academic Publishers., 1994.

Kevei, E. and Nagy, F. (2003). Phytochrome controlled signalling cascades in higher plants. Physiol Plant., 117, 305-313.

Kim, B. C., Tennessen, D. J., Last, R. L. (1998). UV-B-induced photomorphogenesis in Arabidopsis thaliana. Plant J, 15, 667-674.

Kircher, S., Gil, P., Kozma-Bognar, L., Fejes, E., Speth, V., Husselstein-Müller, T., Bauer, D., Adam, E., Schäfer, E., Nagy, F. (2002). Nucleocytoplasmic partitioning of the plant photoreceptors phytochrome A, B, C, D, and E is regulated differentially by light and exhibits a diurnal rhythm. Plant Cell, 14, 1541-1555.

Kohchi, T., Mukougawa, K., Frankenberg, N., Masuda, M., Yokota, A., Lagarias, J. C. (2001). The Arabidopsis HY2 gene encodes phytochromobilin synthase, a ferredoxin-dependent biliverdin reductase. Plant Cell, 13, 425-436.

Kraepiel, Y., Jullien, M., Cordonnier-Pratt, M. M., Pratt, L. (1994). Identification of two loci involved in phytochrome expression in Nicotiana plumbaginifolia and lethality of the corresponding double mutant. Mol Gen Genet, 242, 559-565.

Lagarias, D. M., Crepeau, M. W., Maines, M. D., Lagarias, J. C. (1997). Regulation of photomorphogenesis by expression of mammalian biliverdin reductase in transgenic Arabidopsis plants. Plant Cell, 9, 675-688.

Lagarias, J. C. and Lagarias, D. M. (1989). Self-assembly of synthetic phytochrome holoprotein *in vitro*. Proc Natl Acad Sci USA, 86, 5778-5780.

Lagarias, J. C. and Rapoport, H. (1980). Chromopeptides from phytochrome: The structure and linkage of the Pr form of the phytchrome chromophore. J Am Chem Soc, 102, 4821-4828.

Lamparter, T., Michael, N., Caspani, O., Miyata, T., Shirai, K., Inomata, K. (2003). Biliverdin binds covalently to agrobacterium phytochrome Agp1 via its ring A vinyl side chain. J Biol Chem, 278, 33786-33792.

Li, L. and Lagarias, J. C. (1992). Phytochrome Assembly. J Biol Chem, 267(27), 19204-19210.

Lin, C. and Shalitin, D. (2003). "Cryptochrome structure and signal transduction." In Annual Review of Plant Biology, 54, 469-496.

Matera, K. M., Zhou, H., Migita, C. T., Hobert, S. E., Ishikawa, K., Katakura, K., Maeshima, H., Yoshida, T., Ikeda-Saito, M. (1997). Histidine-132 does not stabilize a distal water ligand and is not an important residue for the enzyme activity in heme oxygenase-1. Biochemistry, 36, 4909-4915.

Matsushita, T., Mochizuki, N., Nagatani, A. (2003). Dimers of the N-terminal domain of phytochrome B are functional in the nucleus. Nature, 424, 571-574.

McDowell, M. T., Lagarias, J. C. (2001). Purification and biochemical properties of phytochromobilin synthase from etiolated oat seedlings. Plant Physiol, 126, 1546-1554.

Montgomery, B. L., Yeh, K. C., Crepeau, M. W., Lagarias, J. C. (1999). Modification of distinct aspects of photomorphogenesis via targeted expression of mammalian biliverdin reductase in transgenic Arabidopsis plants. Plant Physiol, 121, 629-639.

Montgomery, B. L., Franklin, K. A., Terry, M. J., Thomas, B., Jackson, S. D., Crepeau, M. W., Lagarias, J. C. (2001). Biliverdin reductase-induced phytochrome chromophore deficiency in transgenic tobacco. Plant J, 125, 266-277.

Montgomery, B. L. and Lagarias, J. C. (2002). Phytochrome ancestry: sensors of bilins and light. Trends Plant Sci, 7, 357-366.

Muramoto, T., Kohchi, T., Yokota, A., Hwang, I., Goodman, H. M. (1999). The Arabidopsis photomorphogenic mutant hy1 is deficient in phytochrome chromophore biosynthesis as a result of a mutation in a plastid heme oxygenase. Plant Cell, 11, 335-348.

Muramoto, T., Tsurui, N., Terry, M. J., Yokota, A., Kohchi, T. (2002). Expression and biochemical properties of a ferredoxin-dependent heme oxygenase required for phytochrome chromophore synthesis. Plant Physiol, 130, 1958-1966.

Nakasako, M., Wada, M., Tokutomi, S., Yamamoto, K. T., Sakai, J., Kataoka, M., Tokunaga, F., Furuya, M. (1990). Quaternary structure of pea phytochrome I dimer studied with small-angle X-ray scattering and rotary-shadowing electron microscopy. Photochem Photobiol, 52, 3-12.

Parks, B. M. and Quail, P. H. (1991). Phytochrome-deficient hy1 and hy2 long hypocotyl mutants of Arabidopsis are defective in phytochrome chromophore biosynthesis. Plant Cell, 3, 1177-1186.

Platt, J. L. and Nath, K. A. (1998). Heme oxygenase: protective gene or Trojan horse. Nat Med, 4, 1364-1365.

Richaud, C. and Zabulon, G. (1997). The heme oxygenase gene (pbsA) in the red alga Rhodella violacea is discontinuous and transcriptionally activated during iron limitation. Proc Natl Acad Sci USA, 94, 11736-11741.

Rüdiger, W. and Correll, D. L. (1969). Über die Struktur des Phytochrom-Chromophors und seine Protein-Bindung. Liebigs Ann Chem, 723, 208.

Rüdiger, W., Thümmler, F., Cmiel, E., Schneider, S. (1980). Chromophore structure of the physiologically active form (Pfr) of phytochrome. Proc Natl Acad Sci USA, 80, 6244-6248.

Ryter, S. W., Otterbein, L. E., Morse, D., Choi, A. M. (2002). Heme oxygenase/carbon monoxide signalling pathways: regulation and functional significance. Mol Cell Biochem, 234-235(1-2), 249-263.

Sullivan, J. A. and Deng, X. W. (2003). From seed to seed: the role of photoreceptors in Arabidopsis development. Dev Biol, 260, 289-297.

Terry, M. J. and Lagarias, J. C. (1991). Holophytochrome assembly. Coupled assay for phytochromobilin synthase in organello. J Biol Chem, 266, 22215-22221.

Terry, M. J., McDowell, M. D., Lagarias, J. C. (1995). (3Z)- and (3E)-phytochromobilin are intermediates in the biosynthesis of the phytochrome chromophore. J Biol Chem, 270, 11111-11119.

Terry, M. J. (1996). The aurea and yellow-green-2 mutants of tomato are deficient in phytochrome chromophore synthesis. J Biol Chem, 271, 21681-21686.

Terry, M. J. (1997). Phytochrome chromophore-deficient mutants. Plant Cell Environ, 20, 740-745.

Thümmler, F., Rüdiger, W., Cmiel, E., Schneider, S. (1983). Chromopeptides from phytochrome and pycocyanin. NMR studies of the Pfr and Pr chromophores of phytochrome and E,Z-isomeric chromophores of phycocyanin. Z Naturforsch, 38c, 359-368.

Ulm, R., Baumann, A., Oravecz, A., Mate, Z., Adam, E., Oakeley, E. J., Schäfer, E., Nagy, F. (2004). Genome-wide analysis of gene expression reveals function of the bZIP transcription factor HY5 in the UV-B response of Arabidopsis. Proc Natl Acad Sci USA, 101, 1397-1402.

van Tuinen, A., Hanhart, C. J., Kerckhoffs, L. H. J., Nagatani, A., Boylan, M. T., Quail, P. H., Kendrick, R. E., Koornneef, M. (1996). Analysis of phytochrome-deficient yellow-green-2 and aurea mutants of tomato. Plant J, 9, 173-182.

Vierstra, R. D. and Quail, P. H. (1983). Purification and initial characterization of 124-Kilodalton phytochrome from Avena. Biochemistry, 22, 2498-2505.

Vierstra, R. D. and Sullivan, M. L. (1988). Hemin inhibits ubiquitin-dependent proteolysis in both a higher plant and yeast. Biochemistry, 27, 3290-3295.

Vierstra, R. D. and Davis, S. J. (2000). Bacteriophytochromes: new tools for understanding phytochrome signal transduction. Semin Cell Dev Biol, 11, 511-521.

Vladimirov, Y. A. (1998). Free radicals in primary photobiological processes. Membr Cell Biol, 12, 645-663.

Weller, J. L., Terry, M. J., Rameau, C., Reid, J. B., Kendrick, R. E. (1996). The phytochrome-deficient pcd1 mutant of pea is unable to convert heme to biliverdin IXa. Plant Cell, 8, 55-67.

Weller, J. L., Terry, M. J., Reid, J. B., Kendrick, R. E. (1997). The phytochrome-deficient pcd2 mutant of pea is unable to convert biliverdin IXa to 3(Z)-phytochromobilin. Plant J, 11, 1177-1186.

Willows, R. D., Mayer, S. M., Foulk, M. S., DeLong, A., Hanson, K., Chory, J., Beale, S. I. (2000). Phytobilin biosynthesis: the Synechocystis sp. PCC 6803 heme oxygenase-encoding ho1 gene complements a phytochrome-deficient Arabidopsis thaliana hy1 mutant. Plant Mol Biol, 43, 113-120.

Wu, S. H. and Lagarias, J. C. (1996). The methylotrophic yeast Pichia pastoris synthesizes a functionally active chromophore precursor of the plant photoreceptor phytochrome. Proc Natl Acad Sci USA, 93, 8989-8994.

Wu, S. H., McDowell, M. T. Lagarias, J. C. (1997). Phycocyanobilin is the natural precursor of the phytochrome chromophore in the green alga Mesotaenium caldariorum. J Biol Chem, 272, 25700-25705.

Wu, S. H., Lagarias, J. C. (2000). Defining the bilin lyase domain: lessons from the extended phytochrome super family. Biochemistry, 39, 13487-13495.

Chapter 6

STRUCTURE, FUNCTION, AND EVOLUTION OF MICROBIAL PHYTOCHROMES

Baruch Karniol and Richard D. Vierstra
Department of Genetics, University of Wisconsin-Madison, 425-G Henry Mall, Madison, Wisconsin 53706-1574 USA (email: vierstra@wisc.edu)

1. INTRODUCTION

Light is the dominant environmental signal for almost all cellular organisms, providing both the energy (either direct or indirect) necessary for growth and metabolism and the sensory information helpful for adaptation. Like plants, microorganisms are profoundly influenced by their surrounding light environment (Häder, 1987; Armitage, 1997; Loros and Dunlap, 2001; Braatsch and Klug, 2004). For photosynthetic species, various photosensory systems provide positional information to optimise light capture and to help minimize damage inflicted by excess solar radiation. In the short term, these sensory systems help these microbes move/grow toward more favourable light fluences. In the long term, they help the organism adjust to the spectral quality of the light (sun versus shade) by modulating the complement of photosynthetic accessory pigments, and/or entrain their growth and development to the diurnal and possibly seasonal cycles. For non-photosynthetic microorganisms, light is also an important environmental cue. Here, light directs preference/avoidance and adaptive strategies similar to those used by photosynthetic species, which in turn enhances their survival or assists them in locating more favourable ecological niches.

Whereas the architectures of the microbial photoreceptor systems used for energy capture are known in exquisite detail, we have begun to appreciate only recently the repertoire of photoreceptors employed for sensory information. Much of this new understanding has emerged from the exponentially expanding number of completely sequenced genomes that can now be easily searched for signature photoreceptor motifs by the BLAST algorithm. Recent examples include the discovery of retinal-based bacteriorhodopsin, *p*-hydroxycinnamic acid-based xantopsin, and flavin-based LOV-type and cryptochrome-type photoreceptors in a variety of bacterial, fungal and animal species (van der Horst and Hellingwerf, 2004; Venter *et al.*, 2004; see Chapters 11, 13).

Phytochromes (Phys) are one of the best examples where genome analyses have greatly enhanced our understanding of light perception in the microbial world. This class of photoreceptors is defined by the use of a bilin (or linear tetrapyrrole) chromophore (Smith, 2000; Quail, 2002). Once bound to the apoprotein, the bilin enables detection of red (R) and far-red (FR) light by photointerconversion between two relatively stable conformations, a R-absorbing Pr form and a FR-absorbing Pfr form. Through their unique ability to photointerconvert between Pr and Pfr reversibly, Phys act as light-regulated switches by having one form behave as "active" and the other as "inactive". This photochromicity also can provide a crude form of colour vision through measurement of the Pr/Pfr ratio generated by alterations in the relative amounts of R and FR (Smith, 2000).

Phy-type pigments were first discovered over 50 years ago in higher plants based on the ability of R and FR to control many agriculturally important aspects of their life cycle (see Chapters 1, 22; Smith, 2000; Quail, 2002). More recently, genetic analyses and BLAST searches have dramatically expanded their distribution to other kingdoms with the discovery of similar photoreceptors in proteobacteria, cyanobacteria, actinobacteria, filamentous fungi, and possibly slime molds (Wu and Lagarias, 2000; Vierstra, 2002) and B. Karniol and R.D. Vierstra, unpublished). The purpose of this chapter is to review our current understanding of these microbial Phys. As will be seen, they offer simple models to help unravel the biochemical and biophysical events that initiate signal transmission by these novel photochromic pigments. Microbial Phys also provide new clues concerning the evolution of what is now emerging as a superfamily of Phy-type pigments. Their widespread distribution alone implies that light has more important roles in microbial ecology than was previously appreciated, especially for heterotrophic species. Defining these roles will eventually shed new light on the intricate interplay between these species and their surroundings.

2. HIGHER PLANT PHYS

As described in Chapter 7, higher plants contain a small collection of structurally similar phys. In *Arabidopsis thaliana* for example, five plant phy isoforms are present (phyA-E) that have both overlapping and unique roles in light perception (Smith, 2000; Quail, 2002). They all assume a "Y-shaped" structure formed by the dimerization of two identical ~120-kDa polypeptides (Figure 1A). The N-terminal half of each polypeptide contains the bilin-binding pocket (BBP) that functions as the sensory input module. The C-terminal half contains contacts for homodimerization. The BBP binds the bilin 3(Z)-phytochromobilin (PΦB) (Figure 2A), which becomes linked to the apoprotein via a thioether bond to a positionally conserved cysteine in a signature c**G**MP phosphodiesterase/**a**denyl cyclase/**F**hlA (GAF) domain (Wu and Lagarias, 2000). PΦB attachment generates a Pr ground state, the typical absorption spectrum of which is shown in Figure 2B. The Pr molecule exhibits a dramatic red shift and an increase in absorption as compared to free PΦB, presumably caused by a network of chromophore/protein interactions between the BBP and the bilin. Upon excitation of Pr, the bound PΦB undergoes a *cis*-to-*trans* isomerization of the double bond between the C and D pyrrole rings (Figure 2A) and a 31° reorientation of the bilin relative to the polypeptide, and the polypeptide undergoes multiple conformational changes (Quail, 2002). This photoconversion ultimately generates the relatively stable Pfr form with a dramatic

FR shift in absorption (Figure 2B), and presumably with an altered biochemical output.

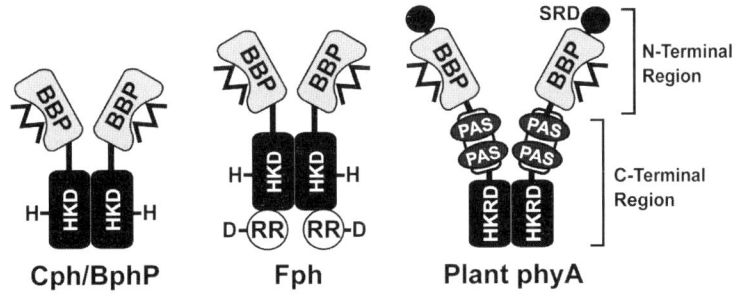

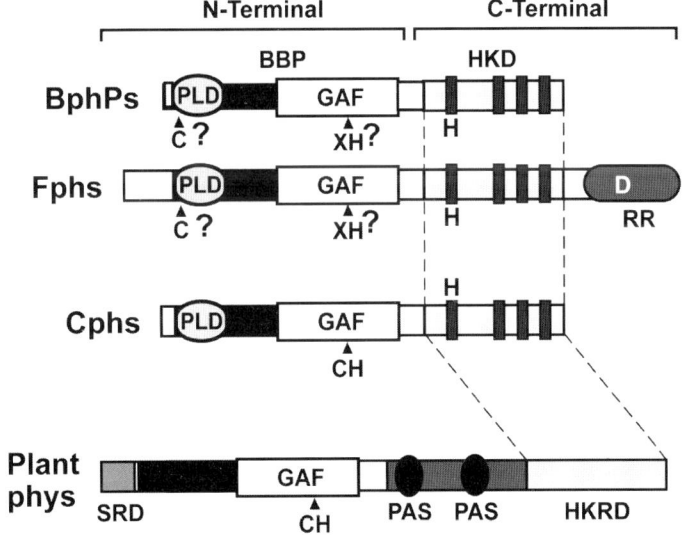

Figure 1. Organization and structural properties of Phy-type photoreceptors. A. Models of a typical Phy from bacteria, fungi and higher plants. B. Linear map of a typical member of the BphP, Fph, and Cph families aligned with a plant Phy. BBP, bilin-binding pocket. PLD, PAS like domain. HKD, histidine kinase domain. HKRD, histidine kinase-related domain. RR, response regulator. PAS, Per-Arndt-Sim domain. SRD, serine-rich domain. The arrowheads identify the bilin attachment site with the amino acid shown below. The question marks identify the two possible sites of chromophore attachment in BphPs and Fphs. The H identifies the histidine that is autophosphorylated by the HKD.

A. Bilin Chromophores

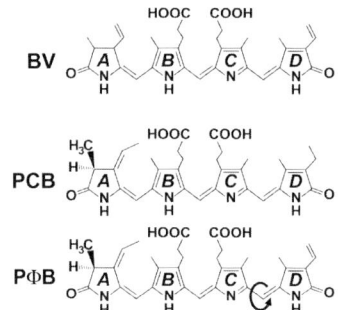

B. Absorbance Spectra

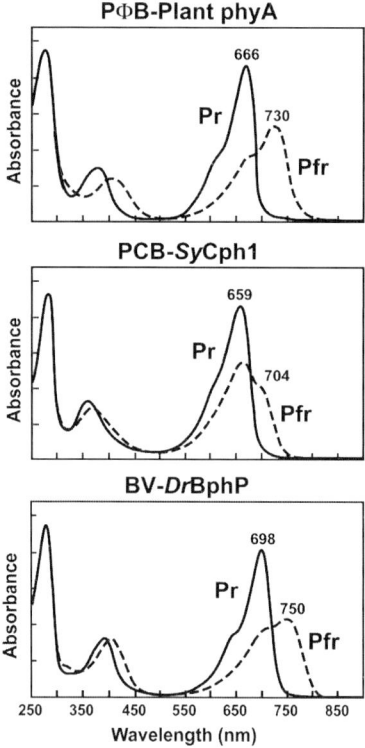

Figure 2. Spectral properties of Phy-type photoreceptors. A. Structure of the BV, PCB and PΦB chromophores. The arrow in PΦB identifies the C15 double bond that undergoes a cis to trans isomerization during Pr to Pfr photoconversion. B. Absorption spectra of the plant phyA following saturating irradiation with FR (Pr) and R (Pfr) as compared to those for a BphP from D. radiodurans and a cyanobacterial Cph from Synechocystis (Cph1). The absorption maxima are indicated.

At present the nature of the output signal from Pfr is unclear, hindered, in part, by the lack of a robust activity and/or obvious sequence motif(s) that would conclusively define an enzymatic function. Previous mutational analyses demonstrated that the two **P**er/**A**rndt/**S**im (PAS) motifs within the C-terminal half are critical for signal transmission (Figure 1 and (Quail, 2002)), with more recent data implicating the BBP as well (Matsushita *et al.*, 2003). One attractive hypothesis is that plant phys act as light-regulated protein kinases. In particular, the realization that a C-terminal region of plant phys is related to the histidine kinase domain (HKD) from bacterial two-component histidine kinases (TC-HKs) led to speculation that plant phys function in similar phosphorelays (Figure 1 and (Schneider-Poetsch, 1992). TC-HKs comprise a family of dimeric protein kinases commonly used by bacteria for environmental adaptation (for reviews see (West and Stock, 2001; Inouye and Dutta, 2003). They work by perceiving a signal through a sensor module, which then promotes an associated HKD to phosphorylate itself and then transfer the phosphate to a cognate response regulator (RR). The phosphorylated RR transmits the signal to appropriate effector pathways.

Based on a similar architecture to TC-HKs, it was speculated that plant phys work in an analogous fashion, using their BBP as a sensor module to activate the histidine kinase-related domain (HKRD) and thus begin a phosphorelay (Schneider-Poetsch, 1992). Despite this homology, it appears unlikely that plant phys are TC-HKs. Their HKRD is missing several important residues that typify a TC-HK, including the histidine residue that serves as the acceptor for the initial phosphorylation step. Furthermore, the assembled chromoprotein acts as a serine/threonine kinase (albeit weakly) and not as a HK, at least *in vitro* (Yeh and Lagarias, 1998). Regardless of its mechanism of action, important subsequent events include a redistribution of Pfr from the cytoplasm to the nucleus and a substantial alteration in gene expression (see Chapters 9 and 17). More specifically, the association of plant phys with DNA-binding proteins imply that they work in close proximity to the nuclear transcriptional machinery (Smith, 2000; Quail, 2002).

3. THE DISCOVERY OF MICROBIAL PHYS

Before 1996, it was assumed that Phy-type pigments are present only in higher and lower plants and a few algae (Mathews and Sharrock, 1997). This perception changed radically with the pioneering work of Kehoe and Grossman working on complementary chromatic adaptation (CCA) in the cyanobacterium *Fremyella diplosiphon* (Kehoe and Grossman, 1996). In a genetic screen for mutants in CCA, they isolated *response to chromatic adaptation E* (*rcaE*), a mutant unable to respond to R or green (G) light. Surprisingly, the affected RcaE protein was discovered to contain a 150-amino-acid, N-terminal region with limited but significant sequence similarity to the signature BBP of plant phys. Attached to this BBP-like domain was a prototypical HKD found in TC-HKs. In contrast to plant phys, all the motifs and residues essential for phosphotransfer were evident in the HKD from RcaE, including the H box that contains the histidine that becomes phosphorylated, and the N, F and G boxes that participate in ATP binding. Both its position within the CCA pathway and its similarity to plant phys suggested that RcaE behaves as a new bacterial Phy-type photoreceptor that initiates a TC-HK phosphorelay (Kehoe and Grossman, 1996, 1997).

Kehoe and Grossman (Kehoe and Grossman, 1996) also noted that other cyanobacteria encode Phy-like sequences in their genomes. In *Synechocystis* sp. PCC6803 for example, five genes with varying degrees of relatedness to *Fremyella diplosiphon RcaE* and higher plant *PHYs* have been detected (Kehoe and Grossman, 1996; Hughes *et al.*, 1997; Wilde *et al.*, 1997; Yeh *et al.*, 1997; Park *et al.*, 2000). Like RcaE, three of the predicted polypeptides also have a C-terminal HKD, implicating them in TC-HK signalling. Lagarias, Hughes, and co-workers then demonstrate that *Synechocystis* Cph1 in particular behaves like a true Phy (Hughes *et al.*, 1997; Yeh *et al.*, 1997). The recombinant apoprotein autocatalytically assembled with bilins like PΦB and 3(Z)-phycocyanobilin (PCB) to generate a dimeric R/FR photochromic chromoprotein *in vitro* (Figure 2B). The resulting holoprotein displayed HK activity with Pr being more active than Pfr.

These observations encouraged a number of investigators to perform similar searches with other bacterial genomes to define the limits of the prokaryotic Phy kingdom. Of the numerous cyanobacterial species surveyed, including *Calothrix* PCC7601, *Oscillatoria* PCC7821, and several *Anabaena*, *Pseudoanabaena*, and *Nostoc* species, all were found to contain one or more Phy-like protein sequences, indicating that these photoreceptors may be common to this phylogenetic group (Herdman *et al.*, 2000; Wu and Lagarias, 2000). The Phy-like protein CikA from *Synechococcus elongatus* PCC7942 was identified genetically as required to reset the circadian clock, thus implicating at least one member of this group in cyanobacterial light perception (Schmitz *et al.*, 2000). Multiple Phy sequences were also detected in the genomes of several photosynthetic α-proteobacteria and purple bacteria, including *Rhodospirillum centenum*, *Bradyrhizobium* ORS278, *Rhodobacter sphaeroides*, and *Rhodopseudomonas palustris*, which further extended the distribution of prokaryotic Phys to photosynthetic eubacteria (Jiang *et al.*, 1999; Wu and Lagarias, 2000; Bhoo *et al.*, 2001; Giraud *et al.*, 2002). As with RcaE and Cph1, these sequences often contained a canonical HKD appended to a BBP, strengthening the view that plant phys evolved from a prokaryotic progenitor TC-HK.

In parallel studies, we extended dramatically the range of Phys beyond photosynthetic organisms with the discovery of Phy-like proteins in numerous non-photosynthetic bacteria, including the α- proteobacteria *Agrobacterium tumefaciens* and *Rhizobium leguminosarium,* the γ-proteobacterium *Pseudomonas syringae*, and *Deinococcus radiodurans* from the Deinococcus/Thermus clade (Davis *et al.*, 1999b; Bhoo *et al.*, 2001; Figure 3). More recently, we found a Phy sequence in *Kineococcus radiotolerans* thus expanding the distribution of Phys even further to actinobacteria (B. Karniol and R.D. Vierstra, unpublished). Collectively, these Phys were important because they provided the first opportunity to study Phy-regulated events in the absence of photosynthesis. Like their photosynthetic brethrens, these Phys contain the signature BBP, which in several cases was demonstrated to attach various bilins autocatalytically *in vitro*. The resulting chromoproteins display R/FR photochromic spectra typical of Phys (Figure 2B and Davis *et al.*, 1999b; Bhoo *et al.*, 2001; Jorissen *et al.*, 2002b; Lamparter *et al.*, 2002; Karniol and Vierstra, 2003). We also isolated native *Deinococcus radiodurans* BphP directly from bacterial cells and confirmed that it assembles with a bilin *in vivo* (Bhoo *et al.*, 2001). Subsequently, we and others found Phy sequences in several filamentous fungi, including *Aspergillus nidulans* and *Neurospora crassa,* indicating for the first time that Phy-type photoreceptors exist in the fungal kingdom (Bhoo *et al.*, 2001; Catlett *et al.*, 2003). Like cyanobacterial Phys, most of these heterotrophic bacterial and fungal sequences contain a C-terminal HKD, and as a result also likely function in TC-HK cascades (Figure 1A, B). However, one important distinction was immediately evident. The GAF domain in these bacterial and fungal Phys is missing the cysteine that was thought to be a prerequisite for bilin attachment (Bhoo *et al.*, 2001). As a consequence these polypeptides must attach their bilin by a different mechanism (see below).

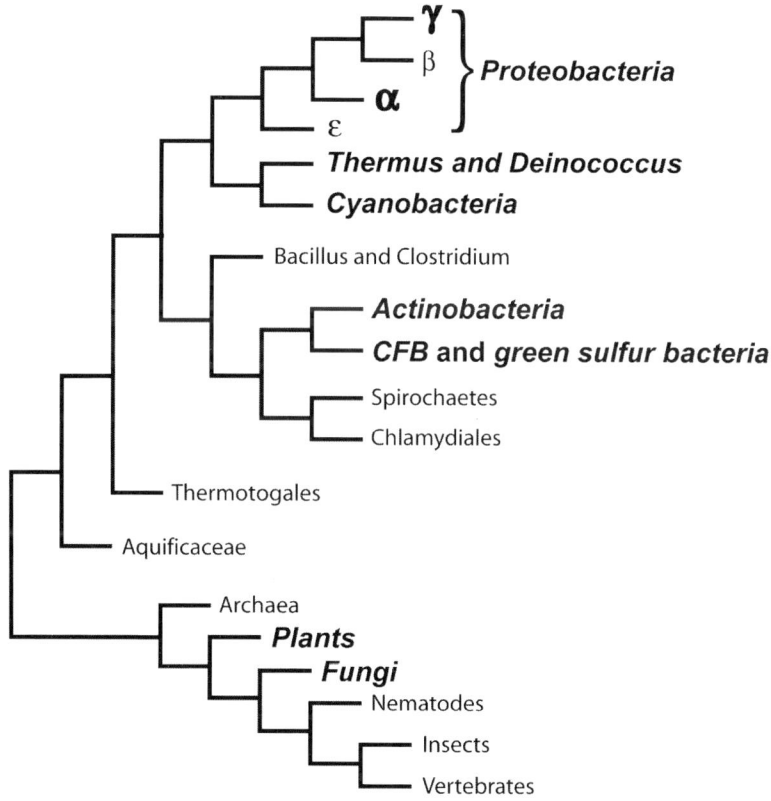

Figure 3. Distribution of Phys among the major divisions of cellular organisms. The divisions/kingdoms in bold identify those where at least one Phy sequence has been identified. CFB, Cytophaga-Flexibacter-Bacterioides.

4. PHYLOGENY OF THE PHY SUPERFAMILY

At present, we have detected more than 50 predicted proteins with the signature GAF domain of Phys in over 30 bacterial and fungal species. As can be seen from an evolutionary tree of cellular organisms in Figure 3, they are present in at least seven major microbial divisions in addition to plants. While defining their physiological functions awaits genetic analyses, it is clear from preliminary biochemical studies that these microbial polypeptides have extensively evolved, with many displaying physico-chemical properties substantially different from their higher plant relatives. For example, chromophore-assembly studies with recombinant polypeptides indicated that some bind bilins other than PΦB and use different attachment sites (*e.g.,* (Bhoo *et al.*, 2001; Jorissen *et al.*, 2002b; Mutsuda *et al.*, 2003; Lamparter *et al.*, 2004), whereas others may not even be photoreceptors (Mutsuda *et al.*, 2003; Terauchi *et al.*, 2004). A striking subset of Phys use the Pfr and not the Pr form as the ground state, indicating that they function backwards, requiring FR and not R to photoconvert the photoreceptor following assembly with the bilin (Giraud *et al.*, 2002; Karniol and Vierstra, 2003). As a consequence, our long-held assumptions regarding characters that define a Phy may need to be relaxed to include these variants. And since most of the predicted microbial Phys await biochemical characterization, more surprises are anticipated.

Unfortunately, the rapid and independent identification of these microbial polypeptides has led to a dizzying menagerie of nomenclatures in the literature. However by combining recent phylogenetic and biochemical characterizations, it appears that this superfamily of Phy polypeptides can be sorted into a few distinct clades. Using the GAF domain alone for phylogenetic comparisons, several families emerge that are clearly distinct from plant phys (Figure 4). Coupled with a grouping based on specific biochemical characteristics (*e.g.,* identity of the bilin and its linkage site), we propose the formation of four major microbial Phy subgroups that reflect their distribution within the bacterial and fungal kingdoms, their photobiological properties, and their possible modes of action.

These divisions encompass the **c**yanobacterial **Ph**ys (Cphs), the **b**acterio**ph**ytochrome **p**hotoreceptors (BphPs), the **f**ungal **Ph**ys (Fphs), and a collection of Phy-like proteins. While we acknowledge that this nomenclature is still preliminary given how little we know about many of these putative Phy sequences, these simple groupings do provide a starting point for discussing critical features of the superfamily. It should be emphasized that this classification does not imply common mechanisms of signal output, as Phys with C-terminal output modules different from the HKD are scattered among the three bacterial families (Figures 4 and 5). Also evident from this classification is that some bacterial species contain members from the different families, indicating that a variety of light signalling systems can co-exist. For example, *Calorthrix* contains both a Cph (CphA) and a BphP type (CphB) (Jorissen *et al.*, 2002b), while *Synechocystis* contains two Cphs (Cph1 and 2) and three Phy-like proteins (TaxD1, PlpA, and an RcaE-like) (Wilde *et al.*, 1997; Yeh *et al.*, 1997; Park *et al.*, 2000; Bhaya *et al.*, 2001; Wilde *et al.*, 2002; B. Karniol and R. D. Vierstra, unpublished).

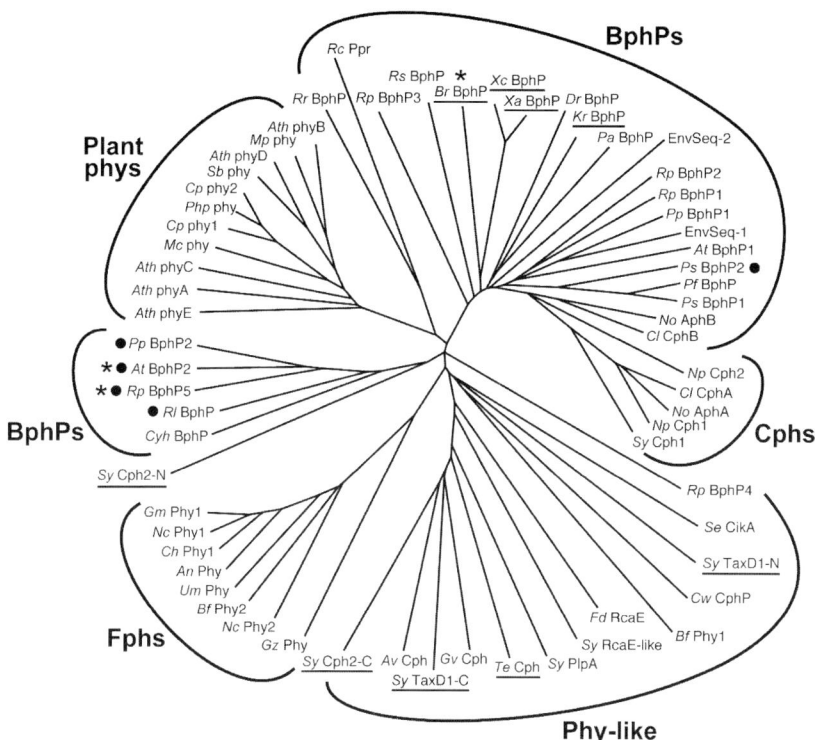

Figure 4. Phylogenetic analysis of the Phy superfamily. Phylogenetic organization of the Phy superfamily based on an alignment of the GAF domain. The major plant Phy, Cph, BphP, Fph and Phy-like subfamilies are indicated. Underline indicates protein sequences without an obvious HKD. Asterisks indicate known members of the bathyBphP subfamily. Closed circles denote HK proteins that belong to the HWE group. Agrobacterium tumefaciens (At), Anabaena variabilis (Av), Arabidopsis thaliana (Ath), Aspergillus nidulans (An), Botryotinia fuckeliana (Bf), Bradyrhizobium ORS278 (Br), Calorthrix PCC7601 (Cl), Ceratodon purpureus (Cp), Cochliobolus heterostrophus (Ch), Crocosphaera watsonii (Cw), Cytophaga hutchinsonii (Cyh), Deinococcus radiodurans (Dr), Fremyella diplosiphon (Fd), Gibberella moniliformis (Gm), Gibberella zeae (Gz), Gleobacter violaceus (Gv), Kineococcus radiotolerans (Kr), Marchantia paleacea (Mp), Mesotaenium caldariorum (Mc), Neurospora crassa (Nc), Nostoc 7120 (No), Nostoc punctiforme (Np), Physcomitrella patens (Php), Pseudomonas aeroginosa (Pa), Pseudomonas putida (Pp), Pseudomonas syringae (Ps), Rhizobium leguminosarium (Rl), Rhodospirillum centenum (Rc), Rhodopseudomonas palustris (Rp), Rhodospirilum rubrum (Rr), Shorgum bicolor (Sb), Synechocystis PCC6803 (Sy), Thermosynechococcus elongatus (Te), Ustilago maydis (Um), Xanthomonas axonopodis (Xa) and Xanthomonas campestris (Xc). EnvSeq-1 (EAI29822.1) and 2 (EAJ02301.1) are environmental sequences described by Venter et al. (2004). Both the N-terminal (N) and C-terminal (C) GAF domains for Synechocystis TaxD1 and Cph2 were included in the comparison.

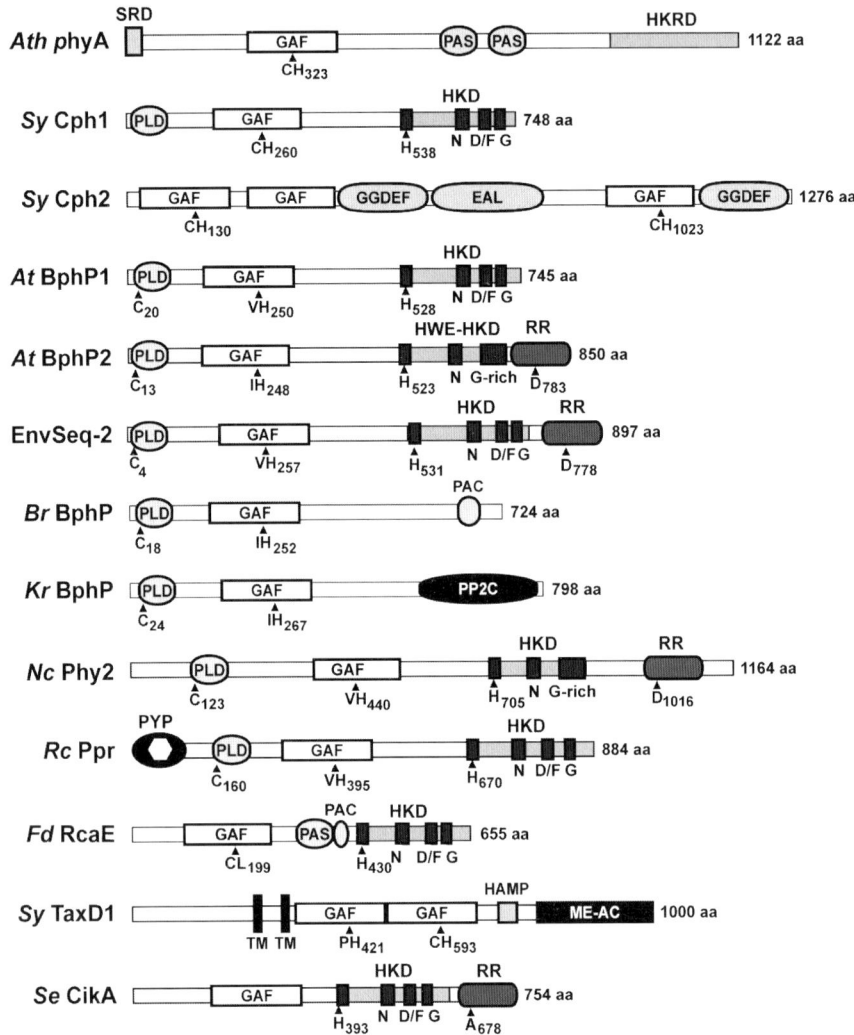

Figure 5. Structure of representative members of the Phy superfamily. Species designations can be found in the legend of Figure 4. EAL, motif bearing a consensus Glu/Ala/Leu sequence. GAF, cGMP phosphodiesterase/adenyl cyclase/FhlA. GGDEF, motif bearing a consensus Gly/Gly/Asp/Glu/Phe sequence. HAMP, HK/adenyl cyclases/methyl-binding proteins/phophatases domain. HKD, histidine kinase domain (the H, N, F and G boxes are indicated). HKRD, histidine kinase-related domain. Me-Ac, methyl-accepting chemotaxis protein domain. RR, response regulator. PAS, Per/Arndt/Sim domain. PYP, photoactive yellow protein. SRD, serine-rich domain. PLD, PAS like domain. TM, trans membrane. PP2C, Protein phosphatase 2C. The position of signature amino acids in the GAF, HKD and RR domains and the N-terminal cysteine that may bind bilins are indicated. EnvSeq-2 is an environmental sequence described by Venter et al. (2004).

The Phy-like family encompasses an expanding collection of unorthodox Phy proteins that have been included in the superfamily mainly based on the presence of a similar GAF domain (Figure 4). Most are found in cyanobacteria, which may be of potential significance to their evolution. Several of these unorthodox proteins have been examined biochemically and found to be missing one or more attributes characteristic of *bona fide* Phys, suggesting that they function in a different manner. It is even possible that these Phy-like species are not R/FR photochromic biliproteins but act as accessory components in Phy-mediated transduction cascades (*e.g., Synechoccocus elongatus* CikA (Mutsuda *et al.*, 2003) and *Synechocystis* RcaE (Terauchi *et al.*, 2004)), or even bind non-bilin chromophores.

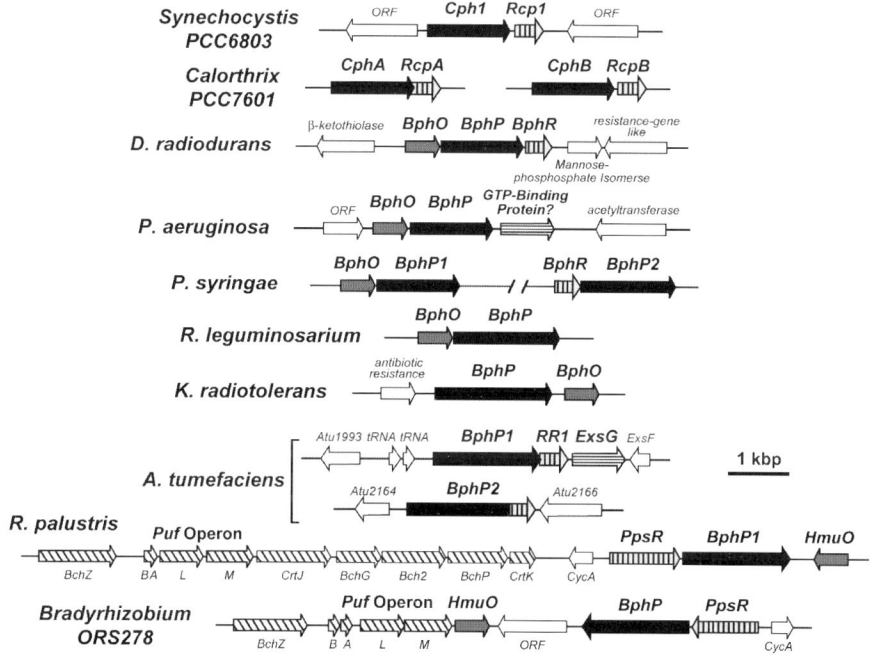

Figure 6. Organization of the genes surrounding the Phy apoprotein gene in various microbial species. Reading frames in black and grey encode the Phy apoprotein and a HO, respectively. Vertically stripped genes encode RRs genetically linked to the BphP. For *Agrobacterium tumefaciens* BphP1, the RR is appended to the Phy apopotein to form a hybrid HK. Horizontal stripped genes indicate other genes within the BphP operon. Diagonally stripped genes in the Bradyrhizobium and Rhodopseudomonas palustris chromosomes identify those linked genes that are predicted to be transcriptionally regulated by the Phy TC-HK cascade. ORF, open reading frame of unknown function. The direction of the arrow indicates the orientation of the reading frame for each gene.

4.1 Cyanobacterial Phy (Cph) family

The Cph family includes a relatively small collection of Phys that is spread among various cyanobacterial species. They appear most related to higher plant phys among the microbial forms but likely use PCB as the chromophore instead of PΦB (Figure 2A) (Hubschmann *et al.*, 2001a; Jorissen *et al.*, 2002a). PCB is synthesized in large quantities as a cyanobacterial photosynthetic accessory pigment and thus is in near unlimited supply for holoprotein assembly. PCB is attached to the same positionally conserved cysteine within the GAF domain as employed by plant phys to bind PΦB (Figure 1B), and likely also becomes attached via a thioether linkage to its A ring ethylidene side chain (Yeh *et al.*, 1997; Park *et al.*, 2000). The founding member of this group, *Synechocystis* Cph1 displays the characteristic R/FR photochromic spectra when assembled with PCB (Figure 2B). The slightly blue-shifted Pr and Pfr absorption maxima for Cph1 holoprotein (654 and 706 nm) as compared to plant phys (666 and 730 nm) are consistent with the loss of one double bond in the π-electron system of PCB versus PΦB. A similar bilin attachment site and R/FR spectra were confirmed for two other members of the Cph family, *Synechocystis* Cph2 (Park *et al.*, 2000; Wu and Lagarias, 2000) and *Calothrix* CphA (Jorissen *et al.*, 2002b)), suggesting that the bilin-BBP interactions among this group have been conserved. *Synechocystis* Cph2 is unusual because it contains three predicted GAF domains (Figure 5). Binding studies with the recombinant polypeptide showed that both the N-terminal and C-terminal GAF domains can bind bilins but only the N-terminal GAF domain generates a R/FR photochromic pigment (Park *et al.*, 2000; Wu and Lagarias, 2000).

Most Cphs have a PAS-like domain (PLD) within the N-terminal region of the BBP that is less obvious in plant phys (*e.g., Synechocystis* Cph1 (Figures 5 and 7 and (Lamparter *et al.*, 2004)). It can be distinguished from a similar PLD in BphPs and Fphs by the absence of a signature cysteine that may help the latter families bind bilins (see below). Most Cphs also contain a C-terminal HKD (*e.g., Synechocystis* Cph1 and *Calothrix* CphA) and thus likely function in TC-HK cascades (Yeh *et al.*, 1997; Jorissen *et al.*, 2002b). One exception is *Synechocystis* Cph2 (Figure 5). Instead of a HKD, Cph2 contains a pair of GGDEF motifs and a single EAL motif (both named for a consensus set of amino acids) that have been proposed to have diguanylate cyclase activity and diguanylate phosphodiesterase activity, respectively (Galperin *et al.*, 2001). (The previous descriptions of these motifs were **d**omain of **u**nknown **f**unction (DUF)-1 and 2, respectively.) Two other microbial Phys have similar GGDEF and EAL motifs at their C-terminal ends instead of the HKD (see below). Cph2 is also missing the N-terminal PLD but still retains normal R/FR photochromic absorption spectra, indicating that at least for Cph2, the PLD is not essential for bilin ligation or the R/FR photochromicity of the holoprotein.

4.2 Bacteriophytochrome (BphP) family

The BphP family currently encompasses the largest collection of Phys in bacteria and appears to be widely dispersed throughout the kingdom. While most known members are from non-photosynthetic α- and γ-proteobacteria and photosynthetic purple bacteria (Davis *et al.*, 1999b; Bhoo *et al.*, 2001; Giraud *et al.*, 2002), potential members are evident in some cyanobacteria (*e.g.*, *Calothrix, Nostoc, Oscilatoria, and Geitlerinema* PCC9228 (Herdman *et al.*, 2000; Bhoo *et al.*, 2001)). These cyanobacterial species appear to contain members of both the Cph and BphP families (Herdman *et al.*, 2000; Jorissen *et al.*, 2002b). The actinobacterial Phy-type sequence discovered in *Kineococcus radiotolerans* genome appears to belong to the BphP family based on its sequence features (see below). In the recently-released shotgun sequence database of microbes present in the Sargasso Sea (Venter *et al.*, 2004), we have identified several new BphPs (Figures 4, 5, and (B. Karniol and R.D. Vierstra, unpublished). In addition to almost 800 new rhodopsin-like photoreceptors in this collection, two full-length BphP sequences and a number of partial sequences were clear. Unfortunately, the host organisms for these BphPs are currently unknown.

Like members of the Cph and plant phy families, tested members of the BphP family can covalently bind various bilins and become R/FR photochromic (Bhoo *et al.*, 2001; Lamparter *et al.*, 2002; Karniol and Vierstra, 2003). Important features that distinguish BphPs from Cphs include the type of bilin used and the nature of its attachment. Assembly studies indicate that BphPs use the PΦB/PCB precursor biliverdin IXa (BV) as the chromophore (Figure 2A). Whereas plant phys and Cphs bind BV poorly or not at all, BphPs easily assemble with this bilin (Bhoo *et al.*, 2001; Karniol and Vierstra, 2003). The slightly red-shifted Pr and Pfr absorption maxima for BV-holoprotein (~698 and ~750 nm) as compared to plant phys assembled with PΦB are consistent with the addition of one double bond in the π-electron system of BV versus PΦB (Figure 2B).

The use of BV as the chromophore of BphPs is also supported by genomic analyses of the bilin biosynthetic pathway (see Chapter 5). In higher plants, PΦB is synthesized by a three-step enzymatic cascade with the first committed step being the oxidative cleavage of heme by a heme oxygenase (HO) to form the linear bilin BV (Davis *et al.*, 1999a; Muramoto *et al.*, 1999). BV is then converted to 3(E)-PΦB by the ferredoxin-dependent reduction of BV by a PΦB synthase (or BV reductase) and then isomerised to 3(Z)-PΦB by a yet to be discovered isomerase (Frankenberg *et al.*, 2001). Bacterial species that contain one or more BphPs invariably express a HO that converts heme to BV. However, none appear to express the BV reductase activit(ies) necessary to convert BV to PCB or PΦB (Bhoo *et al.*, 2001). In several intriguing situations, the bacteria *HO* gene is physically linked to the *Bph* locus (Bhoo *et al.*, 2001; Giraud *et al.*, 2002). The most striking are the *Bph* operons of *Deinococcus radiodurans, Pseudomonas aeruginosa, Pseudomonas syringae,* and *Rhizobium leguminosarium* that include a *HO* gene (*BphO*) within the operon, and the *Bradyrhizobium, Rhodopseudomonas palustris* and *Kinecoccus radiotolerans* genomes where an *HO* gene is nearby (*HmuO* and *BphO* (Figure 6)). Presumably via such a genetic linkage, these bacteria can easily coordinate synthesis of the chromophore with the apoprotein.

Aligning the BBP sequence from members of the BphP family identified two other distinguishing features of this group. One feature is within the GAF domain.

Instead of the conserved cysteine used by plant phys and Cphs to attach bilins, a small hydrophobic residue is often found in this position (Figure 7). Because of this substitution, BphPs must bind BV via a different mechanism and/or a different site than is employed by plant phys and Cphs (Davis *et al.*, 1999b; Bhoo *et al.*, 2001). The second feature is the presence of an N-terminal PLD containing a positionally invariant cysteine (Figures 5 and 7); this cysteine appears to participate either directly or indirectly in chromophore binding (Lamparter *et al.*, 2002; Lamparter *et al.*, 2004) J.R. Wagner and R.D. Vierstra, unpublished).

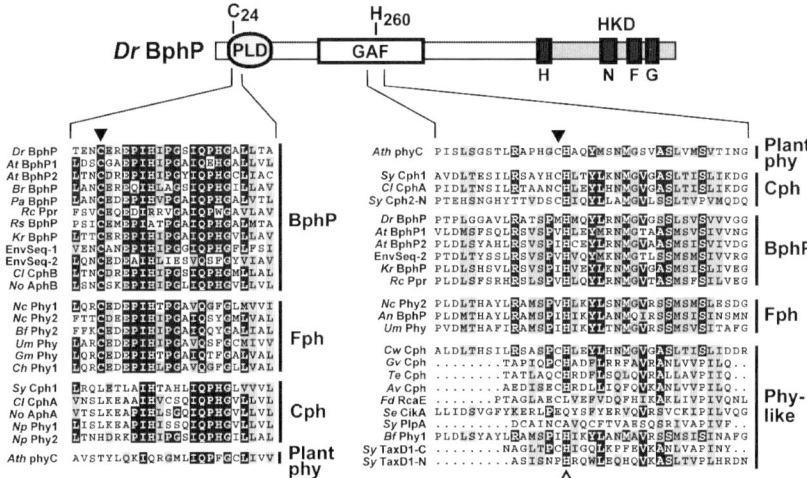

Figure 7. Amino-acid-sequence alignments of the bilin-binding domain within microbial Phy proteins. The position of each domain in the linear Phy sequence is shown using D. radiodurans BphP as the example. Right alignment includes the region of the GAF domain that binds bilins via a cysteine thioether linkage in Cphs and plant Phys or may bind bilins by a histidine Schiff-base linkage in some BphPs. The cysteine and histidine residues are identified by the closed and open arrowheads, respectively. The N-terminal and C-terminal GAF domains of SyTaxD1 are included. Left alignment compares the sequences surrounding the N-terminal cysteine-containing region used by Agrobacterium tumfaciens BphP1 to bind covalently bilins via a cysteine thiolether linkage (Lamparter et al., 2004). Black and grey boxes denote identical and similar residues, respectively. The cysteine is identified by the arrowhead. Species designations can be found in the legend of Figure 4.

The expectation based on sequence similarity within the BBP is that all BphPs bind the BV chromophore by the same method and linkage site. While the situation is far from resolved, preliminary data with several representatives suggest that several different attachment scenarios are possible. For the founding member of this group, BphP from *Deinococcus radiodurans*, mass spectroscopic analysis (MS) implicated a histidine within the GAF domain in bilin binding; it is immediately distil to the site where the cysteine is positioned in Cph1 and plant phys (Davis *et al.*, 1999b). This histidine was previously shown to be important for the bilin lyase activity of plant phys (Bhoo *et al.*, 1997). It has been found since to be conserved in all microbial Phys shown empirically to have bilin-binding activity (Figures 5 and 7). However, a direct role for this histidine in chromophore attachment would infer the use of a Schiff base-type linkage (Bhoo *et al.*, 2001), a bond much less stable than a cysteine-based thioether linkage. The only complication of these studies is that the mass spectrometry studies were performed with PCB since the role of BV as the natural chromophore was not yet known. Hence, it remains possible that BV binds via a different linkage/residue to *D. radiodurans* BphP.

Studies by Lamparter and colleagues using *Agrobacterium tumefaciens* BphP1 (or Agp1) have implicated a different site for BV attachment, at least for this photoreceptor. This alternative site was first suggested by the ability of cysteine modification reagents to block BV binding to the *At*BphP1 apoprotein (Lamparter *et al.*, 2002). Assembly reactions with site-directed mutants subsequently identified a cysteine (Cys20) in the PLD as essential. Notably, this cysteine is conserved in all BphPs identified to date, further supporting its importance (Figure 7). Linkage of BV to the PLD cysteine was recently confirmed by peptide mapping (Lamparter *et al.*, 2004). Following digestion of the BV-*At*BphP1 holoprotein with trypsin, an N-terminal fragment containing the bilin bound to Cys20 was identified by MS. Given the conservation of this cysteine in both BphPs and Fphs, it could be utilized by both of these families (Figure 7). Analysis of various BV derivatives indicated that BV likely binds to *At*BphP1 at the same position in the A ring as PCB and PΦB (vinyl for BV versus ethylidene for PCB/PΦB (Figure 2A)), most likely using a thioether linkage (Lamparter *et al.*, 2003). Such similarity implies that the GAF cysteine and PLD cysteine are close to each other in the three-dimensional structure of the BBP.

Gartner and co-workers have suggested that a third mechanism is possible from recent analysis of *Calothrix* CphB. Recombinant CphB appears to interact non-covalently with bilins even though both the PLD cysteine and the GAF-domain histidine are present (Jorissen *et al.*, 2002b). The holoprotein retains R/FR photochromicity, suggesting that phototransformation between Pr and Pfr does not *a priori* require stable covalent attachment of the chromophore. Strong covalent binding of BV can be induced by introducing a cysteine into the GAF domain, indicating that BV fits into the CphB BBP in a similar arrangement as other BphPs. It has been proposed recently that CphB actually binds BV covalently via the histidine but that this Schiff-base linkage is unstable, thus allowing the chromophore to associate and dissociate readily (Quest and Gartner, 2004).

Spectroscopic studies indicate that most BphPs behave as typical Phys, with the Pr form generated first, following autocatalytic attachment of the bilin (Figure 8A). Pfr is created only upon light absorption. Pfr can either photoconvert back to Pr by FR or revert nonphotochemically from Pfr back to Pr. For *Agrobacterium tumefaciens* BphP1, this dark reversion rate appears to be much faster than for plant phys (Figure 8B), suggesting that the Pfr conformation is less stable for some of these prokaryotic members (Lamparter *et al.*, 2002; Karniol and Vierstra, 2003). Given the overlapping absorption spectra for Pr and Pfr, it is not possible to convert Pr to Pfr completely. Instead, a mixture is generated at photoequilibrium in R that still contains a significant amount of Pr. Surprisingly, analyses of several BphPs by us (Karniol and Vierstra, 2003) and Giraud *et al.* (Giraud *et al.*, 2002) indicate that a novel subfamily of BphPs exists that works in reverse. These backwards (or bathy) BphPs rapidly assemble with BV to generate a Pr-like transient intermediate that quickly converts non-photochemically (t/2 ~5 min at 25°C) to a stable Pfr ground state (Figure 8A). The absorption spectrum of the assembled bathyBphP at completion resembles the predicted spectrum for Pfr without Pr contamination, strongly suggesting that all of the Phy pool eventually becomes Pfr. Pr is generated thereafter only by photoconversion with FR. This Pr is extremely unstable (t/2 <5 min *in vitro*) and rapidly reverts back to Pfr in the dark (Figure 8B).

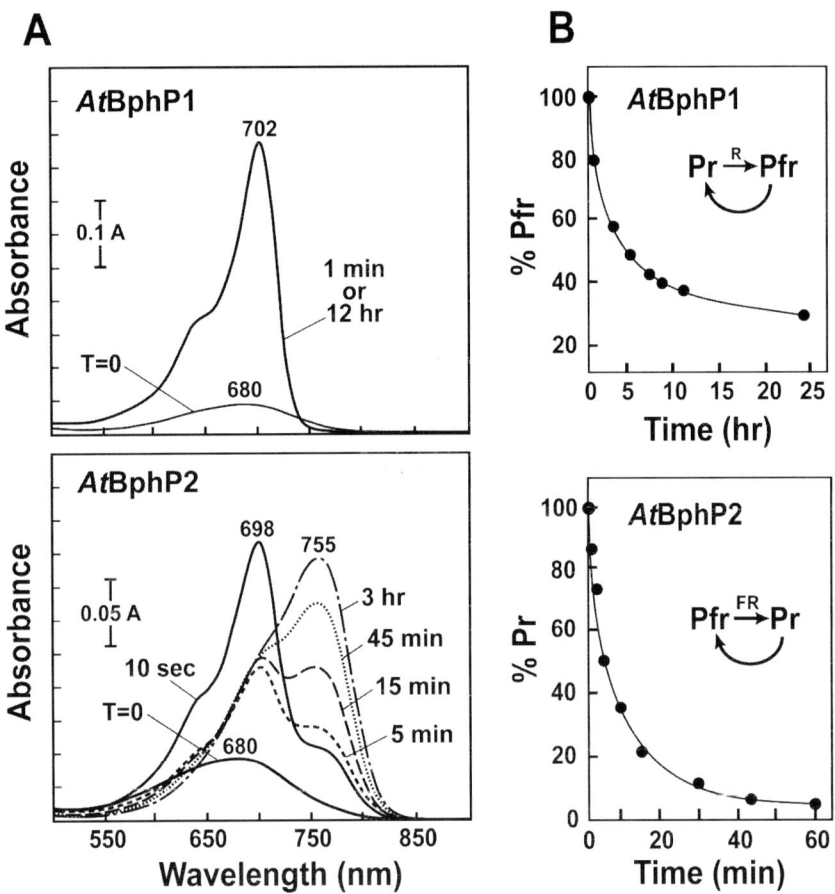

Figure 8. Spectral properties of Agrobacterium tumefaciens BphP1 and BphP2. A. Changes in absorption spectrum of AtBphP1 and AtBphP2 immediately after their assembly with BV. The recombinant appoproteins were added to BV and incubated in the dark. Absorption spectra were recorded at the indicated times. B. Dark reversion of the BphPs. AtBphP1 and AtBphP2 were photoconverted to Pfr and Pr, respectively, and then assayed for their dark reversion back to other spectral form. Adapted from Karniol and Vierstra (2003).

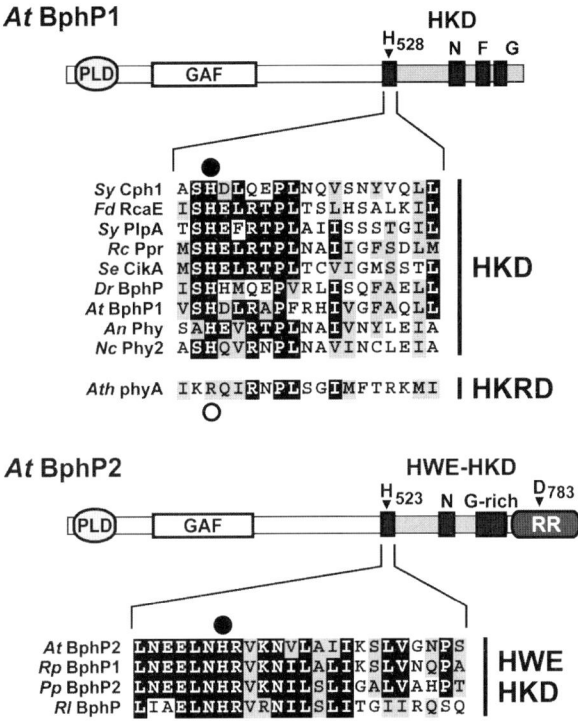

Figure 9. Amino-acid-sequence alignments of the HKD domain within microbial Phy proteins. Upper alignment shows the H-box sequences from typical HKs, using AtBphP1 as an example. Bottom alignment shows the H-box sequences from HWE-HKs, using AtBphP2 as an example. The histidine that serves as the phosphorylation site is identified by the closed circle. For plant PhyA from *Arabidopsis thaliana*, this residue is an arginine (open circle). Species designations can be found in the legend of Figure 4. Black and grey boxes denote identical and similar residues, respectively. Members of each of the HKD types are grouped by the brackets.

To date, members of the bathyBphP subfamily have been confirmed photochemically in three species, *Agrobacterium tumefaciens*, *Bradyrhizobium*, and *Rhodopseudomonas palustris*, with more likely to be found (Giraud et al., 2002; Karniol and Vierstra, 2003). For example, sporulation of the slime mold *Physarum polycephalum* is controlled by a R/FR–absorbing photochromic pigment (Starostzik and Marwan, 1995). In contrast to typical Phy responses, this sporulaton is triggered by FR but not by darkness, R, or FR followed by R, indicating that the photoreceptor is initially synthesized as Pfr and requires FR for phototransformation to Pr. *Pseudomonas putida* BphP2 is another likely candidate given its proximity to other bathy BphPs in the phylogenetic tree of GAF domains (Figure 4). The use of Pfr as the ground state appears to be physiologically relevant for *Bradyrhizobium* and *Rhodopseudomonas palustris* (Giraud et al., 2002). These species use their bathyBphPs to photoregulate expression of the photosynthetic apparatus by detecting FR. Exposure of dark-adapted cells with R has no effect.

Agrobacterium tumefaciens is intriguing because it contains two BphPs; while *At*BphP1 behaves normally (Pr as the ground state), the bathyBphP *At*BphP2 functions photochemically in reverse (Figure 8 and (Karniol and Vierstra, 2003)). The associated kinase activities are also reciprocally modulated with the phosphotransferase activity of *At*BphP1 and *At*BphP2 being maximally active as Pr and Pfr, respectively, and repressed following photoconversion (Karniol and Vierstra, 2003). By simultaneously measuring R and FR, this pair may work either synergistically or antagonistically to detect fluctuating light quality.

Obviously, the photobiological differences between *At*BphP1 and *At*BphP2 reflect the unique way that each of the apoproteins associates with the BV chromophore. This could involve differences in how the bilin is covalently linked and/or differences in the noncovalent interactions that confer the photoreversible spectral properties. *At*BphP2 has a substantially reduced Pr absorbance at its peak at 698 nm versus its Soret peak at 400 nm as compared with *At*BphP1 and other BphPs (Karniol and Vierstra, 2003). Such a reduction is characteristic of bilins that assume a more helical conformation within their binding pockets (Lamparter *et al.*, 1997), implying that the molecular environments of BV when bound to the BBPs of *At*BphP1 and *At*BphP2 are distinct. Further support that the chromophore-binding pockets are different arose from assembly studies with PCB. Whereas PCB-*At*BphP1 adduct is spectrally similar to *At*BphP1 assembled with BV (Lamparter *et al.*, 2002), the spectra of the PCB-*At*BphP2 adduct is aberrant (Karniol and Vierstra, 2003), including a near absence of R/FR photoreversibility, suggesting that the "fit" of the two BBPs is different.

Organization of the BphP proteins indicates that many function as TC-HKs. Often a canonical HKD is found C-terminal to the BBP (Figures 5 and 6). In several species, a RR domain is appended to the C-terminus of the HKD. Such an organization creates a hybrid kinase in which a single polypeptide directs the phosphorylation of the HKD histidine as well as transfer of the phosphate to the asparate residue within the RR (West and Stock, 2001; Inouye and Dutta, 2003). We re-examined several BphPs that at first did not appear to have a HKD and discovered that they bear a new type of HKD, indicating that these BphPs also function in TC-HK relays. This new domain (designated HWE-HKD by the presence of conserved histidine, tryptophan and aspartic acid residues in the HKD) differs from the canonical HKD by substantial sequence alterations within the H, N and G boxes and the absence of an obvious F box (Figure 9 and (Karniol and Vierstra, 2003, 2004)). In addition to BphPs from *Agrobacterium tumefaciens* (*At*BphP2), *Pseudomonas putida*, *Pseudomonas syringae*, *Rhodopseudomonas palustris* and *Rhizobium leguminosarium*, over 80 other potential sensor HKs contain this HWE-HKD, suggesting that it participates in a variety of signalling pathways besides light (Karniol and Vierstra, 2004). Of potential significance is that all but one of the five BphPs with the HWE-HKD cluster together in a phylogenetically distinct subfamily of BphPs (Figure 4). Whether this clustering implies a common mode of action and/or physiological function awaits their analysis.

Not all members of the BphP family bear a recognizable HKD, implying that some do not act as protein kinases. The bathyBphP from *Bradyrhizobium* for example, contains a PAC domain, defined as a PAS-like domain that is often C-terminal to other PAS domains (Figure 5). This PAC domain likely participates in a light-regulated protein/protein interaction (Giraud *et al.*, 2002). Similarly, the BphP from *Kineococcus radiotolerans* has a predicted protein phosphatase-2C (PP2C) domain at the C-terminus instead of a HKD (Figure 5). The presence of this domain suggests that *Kr*BphP acts as a light-regulated phosphatase that dampens or resets a phosphorelay rather than as a HK that initiates a phosphorelay.

Rhodospirillum centenum Ppr also appears to belong within the BphP family. It contains both a GAF domain (with the conserved histidine) and a recognizable cysteine-containing PLD upstream (Figures 5 and 7). It is distinguished from other BphPs by the presence of a **p**hotoactive **y**ellow **p**rotein (PYP) motif appended to the N-terminus of the BBP (Jiang *et al.*, 1999). This 120-amino acid motif binds a single *p*-hydroxycinnamic acid chromophore to generate a photoreceptor capable of detecting blue light (B). Ppr has not yet been reported to bind bilins but given the similarity of its BBP sequence to *bona fide* BphPs, we predict that the apoprotein will have bilin lyase activity. With respect to the known role of *Rhodospirillum centenum* Ppr in light signalling, the PYP motif at a minimum does function in light perception. Bauer and co-workers (Jiang *et al.*, 1999) have shown that a *ppr* mutant displays an altered B-induced expression of the chalcone synthase gene via a response pathway that does not appear to be R regulated.

4.3 Fungal Phy (Fph) family

Our searches of the completed *Neurospora crassa* and *Aspergillus* genomes led to the exciting discovery that various filamentous fungi also have BBP-containing sequences (Bhoo *et al.*, 2001). This revelation confirmed for the first time that Phy-type pigments could be found in a eukaryotic kingdom besides plants (Figure 3). Previous studies by Yager and Mooney predicted a Phy-type pigment in *Aspergillus nidulans* based on their observations that conidiation could be induced by R and reversed by subsequent FR (Mooney and Yager, 1990), but the pigment required had not yet been isolated. The Fph family appears most related in sequence and organization to BphPs. They contain a GAF domain with the conserved histidine preceded by a small hydrophobic residue and an N-terminal PLD with the signature cysteine similar to those in BphPs (Figures 5, 7 and (Bhoo *et al.*, 2001; Catlett *et al.*, 2003)). *Neurospora crassa* Phy1 can covalently assemble with bilins such as BV *in vitro* and become R/FR photochromic, suggesting that the Fphs are indeed Phy-type photoreceptors (B. Noh and R. D. Vierstra, unpublished). All the Fphs identified thus far contain a C-terminal HKD immediately followed by a RR, indicating that they function as hybrid kinases (Figure 1A). Unfortunately phenotypic analysis of several *fphΔ* mutants has failed so far to undercover the photobiological functions of these fungal Phys (Catlett *et al.*, 2003; Froehlich, A. C., Dunlap, J., Vierstra, R. D., unpublished).

4.4 Phy-like sequences

Phylogenetic and biochemical studies have revealed a collection of Phy-like proteins that can be discriminated from the main Phy families by one or more criteria. As can be seen in Figure 4, most fall outside of the main Phy clades when the GAF domain alone is used for sequence comparisons. They are all devoid of an obvious PLD, which is one feature that may ultimately unify this collection (B. Karniol and R. D. Vierstra, unpublished). It is possible that these unorthodox polypeptides either do not bind bilins, bind them in different ways but are still photochromic, or bind bilins but are not photochromic. Many of these GAF domains contain a small deletion upstream of the putative chromophore-binding residue that may be photochemically significant (Figure 7 and (Wu and Lagarias, 2000). Included in the Phy-like group are *Synechocystis* TaxD1, PlpA and RcaE-like, *Fremyella diplosiphon* RcaE, *Synechococcus elongatus* CikA, *Rhodopseudomonas palustris* BphP4, *Crocosphaera watsonii* Cph, *Thermosynechococcus elongates* Cph, *Anabaena variabilis* Cph and a sequence from the fungus *Botryotinia fuckeliana* (Figure 4).

In several cases, biochemical studies have supported the delineation of the Phy-like family from the other main clades. *Fremyella* RcaE for example, has a recognizable GAF domain and can bind bilins both *in vivo* and *in vitro* but the resulting holoprotein is not photochromic (Terauchi *et al.*, 2004). Although RcaE contains a cysteine near the expected site in the GAF domain (Figure 7), this residue is not essential for bilin attachment (at least *in vitro*) leaving the actual binding site unresolved. At present, it is unclear if RcaE is a photoreceptor despite genetic evidence connecting the corresponding locus to photoperception during CCA.

In a similar fashion to RcaE, *Synechococcus elongatus* CikA may not act as a typical Phy even though it appears to participate in photoperception by this cyanobacterium (Schmitz *et al.*, 2000). CikA is missing the positionally conserved cysteine and histidine residues in the GAF domain and does not have a PLD or a similarly positioned cysteine. Despite these deficiencies, recombinant CikA can bind PCB and PΦB (but not BV) *in vitro*, but the resulting holoproteins are not R/FR photochromic (Mutsuda *et al.*, 2003). The CikA protein extracted from *Synechococcus* cells is not a bili-protein, suggesting that the polypeptide does not assemble with these chromophores naturally. The GAF domain of CikA is followed by a typical HKD and a RR motif, implying that CikA functions in a TC-HK cascade (Schmitz *et al.*, 2000). However, the appended RR motif is missing the aspartate residue necessary to receive the phosphate from the HKD. As a consequence, CikA cannot function as a *bona fide* hybrid HK. It has been proposed that CikA is a pseudo-RR that modulates a phosphorelay initiated by another photoreceptor (Mutsuda *et al.*, 2003).

Synechocystis TaxD1 also clusters in the Phy-like group. It has two possible GAF domains with one containing the canonical cysteine-histidine sequence and the other a proline-histidine sequence at the putative bilin-binding site (Figure 7). Although its ability to bind bilins and become R/FR photochromic has not yet been reported, genetic analyses show that TaxD1 is essential for phototaxis toward low-fluence R and FR (Bhaya *et al.*, 2001; Ng *et al.*, 2003). TaxD1 does not contain a C-terminal HKD. Instead, a signalling domain similar to those found in methyl-accepting chemotaxis proteins is evident, suggesting that the phototactic response affected by TaxD1 is regulated by a methylation pathway similar to those utilized during halo bacterial phototaxis and bacterial chemotaxis (Bhaya *et al.*, 2001; Ng *et al.*, 2003). Upstream of the GAF region are two potential transmembrane domains (Figure 5), suggesting that TaxD1 may be part of a membrane-bound signalling cascade. The action spectrum for TaxD1-regulated phototaxis is atypical for a Phy-mediate process, further supporting the potentially unique nature of this unorthodox species (Ng *et al.*, 2003). *Synechocystis* PlpA and *Botryotinia fuckeliana* Phy1p also cluster phylogenetically in the Phy-like group based on the GAF sequence and are missing an obvious PLD. It remains to be determined if their photochemical properties are distinct from the canonical Phys.

5. DOWNSTREAM SIGNAL TRANSDUCTION CASCADES

The structural organization of microbial Phys indicates that many, if not all, function either directly or indirectly in TC-HK phosphorelays. This connection is further supported by the presence of a RR domain, either translationally linked to the HKD thus creating a hybrid HK, or expressed as a separate polypeptide within the operon that encodes the Phy apoprotein (Figures 5 and 6). Even for *Bradyrhizobium* BphP, which does not contain a recognizable HKD, the RR PpsR is transcriptionally linked to the photoreceptor in an operon, suggesting that a TC-HK cascade is employed with PspR serving as the phosphoacceptor for a yet to be discovered HK (Figure 6 and (Giraud *et al.*, 2002)).

The HK activity of several Phys has been confirmed by *in vitro* phosphotransferase assays using recombinant proteins. Importantly, the HK activity (as measured by autophosphorylation of the HKD histidine) is regulated by the spectral form of the assembled holoprotein (Yeh *et al.*, 1997; Bhoo *et al.*, 2001; Karniol and Vierstra, 2003; Mutsuda *et al.*, 2003). In some cases, the holoprotein is more active as Pr and, in other cases, the holoprotein is more active as Pfr. Whether this difference is an artefact of using such purified *in vitro* reactions or reflects intrinsic differences of the photoreceptors is not yet known. For the BphP pair from *Agrobacterium tumefaciens*, the typical Phy *At*BphP1 is a more active kinase as Pr, whereas the bathyBphP *At*BphP2 is a more active kinase as Pfr (Karniol and Vierstra, 2003). As a consequence, both photoreceptors would be maximally active in the dark and be simultaneously repressed by white light. Donation of the bound phosphate to the aspartate in the cognate RR has also been demonstrated for several Cphs and BphPs (Yeh *et al.*, 1997; Bhoo *et al.*, 2001; Hubschmann *et al.*, 2001b; Karniol and Vierstra, 2003). This coupling appears to be specific as little phosphotransfer occurs between the HKD and an unrelated RR.

For most microbial Phys, the transduction cascade that follows RR phosphorylation is not yet clear. Whereas many RRs have an appended output module (*e.g.,* a DNA-binding or protein-interaction motif (West and Stock, 2001; Inouye and Dutta, 2003)) that would be altered by the phosphorylation signal, these modules are often absent in the RRs associated with the microbial Phys. This absence suggests that microbial Phy RRs either interact directly with their targets (in a similar fashion to *Escherichia coli* CheY), or participate in a four-step His→Asp→His→Asp phosphorelay before signal output (*e.g., Escherichia coli* EnvZ). The advantage of such an extended relay may be to provide more points for control and/or for integration with other signals.

A model based on the general rules of TC-HK cascades (West and Stock, 2001; Inouye and Dutta, 2003) is shown in Figure 10A. Signalling begins with light-triggered conformational changes within the BBP altering the autophosphorylation activity of the HKD. This phosphorylation actually occurs in *trans*, using one HKD of the Phy dimer to direct phosphate addition to the HKD histidine of the second HKD. This cross phosphorylation is supported by the fact that all the microbial Phys tested thus far behave as homodimers (at least *in vitro*) (Yeh *et al.*, 1997; Park *et al.*, 2000; Bhoo *et al.*, 2001). The histidine phosphate is then transferred to the aspartate in a cognate RR. The phosphorylated RR could directly bind to an output apparatus (*e.g.*, flagellar motor) to alter its function. Alternatively, the RR could donate the phosphate to the histidine of a histidine phosphotransferase (HPT). This HPT would continue the phosphorelay by transferring the phosphate to the aspartate in a second RR. The second RR could then interact with a separate output factor or contain an appended output domain to relay the signal further. While HPTs and RRs with output domains are evident in the various bacterial and fungal genomes (*e.g.*, (Catlett *et al.*, 2003), those connected to Phy signalling are unknown in most cases. To date, the most complete pictures are from the signalling pathways involving *Fremyella diplosiphon* RcaE that directs CCA (Kehoe and Grossman, 1996, 1997), and *Bradyrhizobium* BphP that controls photosynthetic potential (Giraud *et al.*, 2002). In each case, other members of the sensory cascade, including output modules and targets, can be inferred from genetic and genomic analyses (see Figure 10).

Given the apparent interchangeable nature of TC-HK cascade components, it is possible that other sensory cascades converge at the various phosphotransferase intermediates (*e.g.*, RRs and HPTs) to integrate various environmental signals. Filamentous fungi for example, contain a number of other sensory HKs in addition to Fphs, but appear to contain only a single HPT (Catlett *et al.*, 2003). As result, these HPTs may represent a junction point for a number of TC-HK signalling systems.

Instead of promoting one or more HK cascades, some Phys may dampen such cascades. For example, PP2C domain in *Kinecoccus radiotolerans* BphP implies that it has protein phosphatase activity (Figure 5). This activity could attenuate one or more kinase cascades in a light dependent manner by removing the phosphate signal. The signal transduction chain involving RcaE in *Fremyella diplosiphon* also includes the PP2C-containing CpeR protein, indicating that signal dampening by a phosphatase modulates CCA (Seib and Kehoe, 2002).

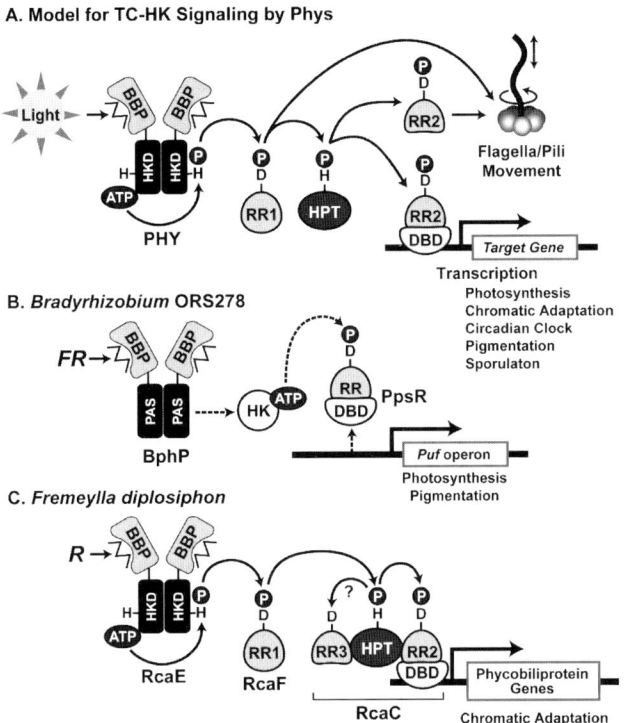

Figure 10. Possible schemes for the function of BphPs in light perception. A. Absorption of light by the BphP homodimer triggers a conformational change in the sensor domain of the photoreceptor that activates (or inactivates) the histidine kinase activity in the HKD. Active BphP cross phosphorylates the conserved histidine (H) in the dimer using adenosine triphosphate (ATP) as the donor, and then transfers this phosphate to an aspartate residue (D) on an associated response regulator (RR1). RR1 can either directly affect an output (flagellar/pilus motor) or transfer the phosphate to a conserved histidine in a histidine phosphotransferase (HPT), which then donates the phosphate to a second RR (RR2). RR2 may interact with output factors or may contain an appended output domain to initiate the response. In this example, RR2 is fused to a DNA-binding domain (DBD) and serves as a transcription factor that affects the expression of genes responsible for various photoresponses. B. Model depicting the action of Bradyrhizobium BphP in controlling photosynthetic gene expression in response to FR (Giraud et al., 2002; Giraud et al., 2004). The dash lines denotes a potentially interaction of BrBphP with a HK, which then triggers a phosphorylation-dependent repression of PpsR. Repression of PpsR allows transcription of various genes necessary for assembly of the photosynthetic apparatus and pigment biosynthesis. C. Model proposed for CCA signal transduction in Fremyella diplosiphon involving the Phy-like protein RcaE. The three components thus far described in this pathway are the putative sensor RcaE and the response regulators RcaF and RcaC (Kehoe and Grossman, 1996, 1997). Phosphorylation is stimulated by R, leading to successive phosphorylation steps from RcaE, to RcaF and then to RcaC that ultimately activates/represses phycobiliprotein gene transcription.

6. PHYSIOLOGICAL ROLES OF MICROBIAL PHYS

Given the varied structural, spectral, and biochemical properties of the microbial Phy families, we predict that they are involved in a diverse array of photosensory processes. Unfortunately, because most of these Phy systems have not yet been dissected at the physiological level, we currently have a rudimentary understanding of just a few. To date, one or more microbial Phys have been connected to the regulation of phototaxis, control of photosynthetic potential, entrainment of circadian rhythms, mitigation from high light fluences and damaging UV irradiation by producing protective pigments, and possibly sporulation/conidiation. In most cases, the transcriptional control of gene expression is the ultimate output, thus implicating the associated TC-HK cascades in the regulation of DNA-binding proteins. Predictably, many of these same responses are also regulated by phys in higher plants (Smith, 2000; Quail, 2002). One potential complication to the study of Phy-mediated responses in microbes is that the responsible bili-proteins may have an unstable and thus transient Pfr (Pr) state (Figure 8B). Consequently, it cannot be presumed that these responses will display the R/FR photoreversibility for induction that is often diagnostic for phy-mediated responses in higher plants (Chapter 8; Smith, 2000; Quail, 2002).

The phy isoforms in plants show both distinct and overlapping functions, indicating that both unique and convergent signal transduction chains are present (see Chapters 17, 20, 22; Quail, 2002). The structure and action of microbial Phys suggest that such redundant and non-redundant sensory chains are present in microbes as well. In *Synechocystis* for example, the protein organization of the five Phys imply distinct output signals. In the phototactic response, TaxD1 helps initiate movement toward R, while Cph2 and PlpA appear to modulate phototaxis toward B (Wilde *et al.*, 1997; Wilde *et al.*, 2002; Ng *et al.*, 2003). *Rhodopseudomonas palustris* expresses five different Phys, of which one of them is a member of the bathyBphP family. Loss of this bathyBphP abolishes light-driven expression of the photosystem apparatus, indicating that this isoform is solely responsible for this phenomenon (Giraud *et al.*, 2002). Likewise, the biochemical properties of normal and bathy-BphPs from *Agrobacterium tumefaciens* suggest distinct functions, but using a similar method of TC-HK signalling (Karniol and Vierstra, 2003).

6.1 Directing phototaxis

As with the role of plant phys in directing growth and morphology, it is likely that microbial Phys also have important roles in regulating phototaxis and phototropism toward more favourable light environments. While few studies have directly implicated Phys in these movement/growth responses, action spectra showing a peak of activity centred around 700-730 nm suggest the involvement of Phy-type pigments. Examples in photosynthetic species include the phototactic behaviours of the cyanobacteria *Anabaena variabilis*, *Phormidium unicinatum*, *Synechococcus elongatus*, and *Synechocystis* (Häder, 1987; Ng *et al.*, 2003). For non-photosynthetic organisms, similar roles are not yet clear. No studies have implicated a R-absorbing pigment in the taxis of non-photosynthetic prokaryotes and the phototropic response of many filamentous fungi is controlled by a B-absorbing pigment.

The best-understood phototactic response in cyanobacteria is that of *Synechocystis*, which displays a complex light response involving at least three photoreceptors that control positive phototaxis toward low-fluence R and negative phototaxis toward high-fluence B. The R phototaxis can be antagonized by simultaneous irradiation with FR (760 nm), supporting the participation of a Phy-type pigment. Genetic analyses have connected several Phys to both the R and B responses. The R response requires TaxD1, a member of the Phy-like clade (Bhaya *et al.*, 2001; Ng *et al.*, 2003). The organization of TaxD1 and its sensory cascade implies that R activates a phosphorelay involving a separate HK TaxAY1, with feedback methylation of TaxD1 possibly regulating the activity of the photoreceptor. Ultimately, the cascade directly impacts the motility motor of Type IV pili and its biogenesis.

Although the B response likely requires another non-Phy receptor in *Synechocystis*, Cph2 appears to modulate this response. Loss of Cph2 increases the sensitivity of *Synechocystis* to B, suggesting that the Cph2 protein functions to repress the B signalling system (Wilde *et al.*, 2002). Loss of the Phy-like protein PlpA derepresses the autotrophic growth of *Synechocystis* in B, inferring a similar connection between a B-absorbing pigment and a Phy in this organism (Wilde *et al.*, 1997). In both these regards, it may be informative that higher plant phys appear to work in concert with the B-absorbing cryptochrome and, in fact, direct binding of the two photoreceptors have been reported (Mas *et al.*, 2000). Consequently, it is tempting to speculate that the modification of B-responses by PlpA and Cph2 reflects a similar association of these Phys with a cryptochrome-type pigment. It may also be relevant for phototactic responses that *Synechocystis* Cph2, and two other bacterial Phys *Rhodobacter sphaeroides* BphP, and *Thermosynechococcus elongatus* Cph, contain a GGDEF motif with potential diguanylate cyclase activity (Figure 5 and B. Karniol and R.D. Vierstra, unpublished). This domain has been found in a several non-Phy proteins involved in motility (*e.g., Pseudomonas aeroginosa* FimX, *Vibrio cholerae* RocS, and *Caulobacter cresentus* PleD (Aldridge *et al.*, 2003; Huang *et al.*, 2003; Rashid *et al.*, 2003)). PleD for example, appears to directly regulate the flagellum motor through a HK cascade (Aldridge *et al.*, 2003).

Little is known about the phototactic response of non-photosynthetic proteobacteria. *Agrobacterium tumefaciens* contains both flagella and Type IV pili and is motile but no photoresponses have been documented. Light does strongly promote the pathogenicity of this soil bacterium toward plant cells (Zambre *et al.*, 2003) but the effective wavelengths have not been established. Since infection and subsequent transformation requires that the bacterium swim to the host, it is possible that a light-regulated motility response directed by *At*BphP1 and/or 2 is involved.

6.2 Enhancement of photosynthetic potential

As in plants (see Chapter 2), Phys from photoautotrophic microbes appear to have important roles in regulating the production, assembly, and modulation of the pigment/protein complexes needed for photosynthetic light harvesting. In both the anoxygenic photosynthetic bacteria *Bradyrhizobium* and *Rhodopseudomonas palustris*, a bathyBphP is essential (Giraud *et al.*, 2002). The corresponding apoprotein genes are close to the cluster of genes responsible for the synthesis of the light-harvesting complex (*Puf* operon), bacteriochlorophylls (*Bch*), and carotenoids (*Crt*) (Figure 6). In the same operon as *BphP* is *PpsR*; it encodes a RR with an appended DNA-binding motif. PpsR represses the expression of photosynthetic genes in other bacteria and likely serves the same function in *Bradyrhizobium* and *Rhodopseudomonas palustris*. Disruption of *Bradyrhizobium BphP* blocks the light-induced activation of the adjacent photosynthetic gene cluster, whereas disruption of the *PpsR* gene constitutively induces their expression (Giraud *et al.*, 2002). The simplest model is that both participate in a TC-HK cascade in which the BphP modulates a separate, yet to be identified, HK that represses transcriptional inhibition by PpsR, thereby stimulating the production of light-harvesting centres (Figure 10B). The photoresponses of wild-type *Bradyrhizobium* and *Rhodopseudomonas palustris* have a maximum at 750 nm, consistent with the Pr form of these bathyBphPs being the active state and the Pfr form being the inactive ground state (Giraud *et al.*, 2002).

In numerous cyanobacteria, Phys are involved not only in producing the light-harvesting centres but also in altering their pigment composition in response to changing light quality. During CCA, reciprocal accumulation of the bili-proteins phycocyanin and phycoerythrin helps certain cyanobacteria optimise light capture in R- and G-rich light environments, respectively. This reversible process is mediated by differential expression of the apoproteins and associated bilins. In *Fremyella diplosiphon*, CCA requires the Phy-like protein, RcaE (Kehoe and Grossman, 1996). Genetic dissection of the response has uncovered other components in the sensory cascade, including RcaC and RcaF (Kehoe and Grossman, 1996, 1997). RcaF is a RR without an obvious output motif. RcaC is a multi-domain protein containing an N-terminal RR followed by a potential DNA-binding motif, a HPT domain, and a second RR. The RRs and the HPT of RcaC contain the catalytic aspartic acid and histidine residues, respectively, indicating that they could be functional in a TC-HK cascade. Such an organization implies that CCA in *Fremyella diplosiphon* is driven by a four-step phosphorelay involving one or more Phys perceiving the light signal, which then initiate a phosphorelay from RcaF to RcaC (Figure 10C). This relay culminates in the activation or repression of the DNA-binding motif in RcaC to effect transcription of appropriate phycobiliprotein and bilin biosynthetic genes. Even though RcaE participates genetically in the sensory cascade for CCA, it is not yet clear whether this Phy is the actual photoreceptor (Terauchi *et al.*, 2004). RcaE can bind bilins but fails to display the characteristic R/FR photochromicity of typical Phys (see above). *Synechocystis* expresses a Phy-like protein related to RcaE but its function(s) are unclear since this cyanobacterium does not display an obvious CCA.

Surprisingly, some unicellular microbes have circadian clocks whose entrainment by day length persists from one generation to another despite doubling times of less than one day. For several cyanobacterial species, this clock helps coordinate the synthesis of photosynthetic genes for maximum light use during the expected day. In the cyanobacterium *Synechococcus elongatus*, entrainment of the clock to the photoperiod requires input from the Phy-like protein, CikA (Schmitz et al., 2000). Both the period and amplitude of the clock as well as its phase entrainment are altered in cikAΔ mutants. How CikA resets the clock is not yet known. It binds bilins poorly and does not become R/FR photochromic (Mutsuda et al., 2003). CikA contains an appended RR but this motif is missing the phosphoacceptor aspartic acid, thus precluding its direct participation in a TC-HK cascade. One possibility is that CikA functions in circadian entrainment but as an accessory factor that modulates the activity of the actual photoreceptor.

6.3 Photocontrol of pigmentation

In several situations, Phys help regulate the synthesis of photoprotective pigments. In *Rhodospirillum centenum*, the BphP Ppr enhances the expression of chalcone synthase, the rate-limiting enzyme in the phenylpropanoid pathway used to synthesize flavonoid pigments (Jiang et al., 1999). In plants, flavonoids production is stimulated by various stresses including high fluence white and UV light, implying that the flavonoids have a photoprotective role. The response in *Rhodospirillum centenum* is toward B and not R, indicating that the *p*-hydroxycinnamic acid chromophore bound to the PYP domain is responsible and not a bilin bound to its predicted BBP (Jiang et al., 1999).

Deinococcus radiodurans BphP appears to have an important role in regulating carotenogenesis. Whereas dark grown cells accumulate carotenoids, the most abundant being a novel form called deinoxanthin, light stimulates this synthesis dramatically (Davis et al., 1999b). In fact, white light-grown *Deinococcus radiodurans* colonies are bright red. This light-induced increase in deinoxanthin is markedly attenuated in a *bphPΔ* mutant. The mutant is also slow growing under the high light conditions, suggesting a photoprotective role for the pigmentation. In higher plants, carotenogenesis is similarly regulated by phys via their ability to transcriptionally up regulate genes for the rate-limiting enzymes, phytoene synthase and phytoene desaturase. Notably, *Deinococcus radiodurans* counterparts to both genes are included within an operon, which would simplify their possible photoregulation by *Dr*BphP (Davis et al., 1999b). In a similar fashion, carotenogenesis appears to be controlled in *Bradyrhizobium* through light-dependent regulation of the carotenoid biosynthetic operon *Crt*. Light signalling through *Br*BphP appears to inactivate the transcriptional repressor PpsR, which in turn allows *Crt* expression (Giraud et al., 2004).

7. EVOLUTION OF THE PHY SUPERFAMILY

The recent discovery of the Phy-type pigments in various proteobacteria, cyanobacteria, and fungi now provides a potential route for the evolution of the Phy superfamily. Given their widespread presence in a number of non-photosynthetic and photosynthetic proteobacteria and cyanobacteria, the BphP family likely represents the progenitor for all of the bilin-containing photochromic pigments. Their use of BV as the chromophore also represents the simplest way to obtain a linear bilin, requiring just one enzymatic step from the cyclic heme precursor (Bhoo *et al.*, 2001).

The precursor of BphPs may have arisen from an ancient protein bearing a GAF domain. This motif, which is related in topology to the PAS domain, is present in a number of proteins implicated in environmental signalling and appears to function as ligand-binding pocket. For example, the *Ectothiorhodospira halophila* PYP motif that binds *p*-hydroxycinnamic acid is an archetypal PAS domain (Genick *et al.*, 1998), whereas the GAF domain in *Escherichia coli* FixL functions as a heme-binding pocket (Gong *et al.*, 1998). To accommodate linear bilins, a PAS domain may have been added to a FixL-type GAF domain and then both evolved to create the extended BBP with its signature Phy-type PLD and GAF domains. The addition of the HKD to the BBP then provided a facile way to connect light to appropriate signalling networks. We presume that the bathyBphPs arose from the BphPs as a way to exploit FR as an environmental signal. However, it should be noted that BphPs are not universally present in bacteria, being absent in many species (*e.g.*, *Escherichia coli* and *Bacillus subtilis*), including those that are closely related to species that do contain a BphP (*e.g.*, other *Pseudomonas* strains and *Thermus aquaticus*). Whether these species lost their BphPs over time or descended from lineages that arose before this photoreceptor is unknown.

Relatives of the BphPs then emerged during the evolution of cyanobacterial, actinobacterial, Cytophaga-Flexibacter-Bacterioides (CFB) and fungal clades. For the Cph family, PCB was adopted as the chromophore, which was readily available given its synthesis as a photosynthetic accessory pigment. Such a switch may have reflected the exploitation of Cphs to enhance light capture under competitive conditions. Whereas the absorption spectrum of the Pr form of BV-BphPs overlaps poorly with chlorophylls, the absorption spectrum of the Pr form of PCB-Cphs overlaps well. As a consequence, the Pr/Pfr ratio of PCB-Cphs at photoequilibrium is much more sensitive to shading by photosynthetic organisms than BV-containing forms. Such shading, which preferentially removes R as compared to FR due to the R-absorption maximum of chlorophyll, can then be monitored by a high Pr/Pfr ratio as opposed to a Pr/Pfr ratio close to one when R and FR are in equal proportions (*e.g.*, full sunlight) (Figure 11). Clearly, the adoption of PCB as the chromophore would require a way to discriminate PCB from its precursor BV. Both problems may have been overcome by the use of the GAF cysteine to attach the chromophore covalently as opposed to the GAF histidine or the PLD cysteine. Various modifications of Cphs then created members of the Phy-like family. This process involved further changes in the BBP, in particular the disappearance of the PLD, and replacement of the HKD with other signalling motifs. The unique structures of these Phy-like sequences certainly suggest a wide diversity of functions. One intriguing scenario is that these unique proteins associate with other non-bilin chromophores, or use the BBP as a bilin sensor and not as a light sensor as proposed by Lagarias and Montgomery (Montgomery and Lagarias, 2002).

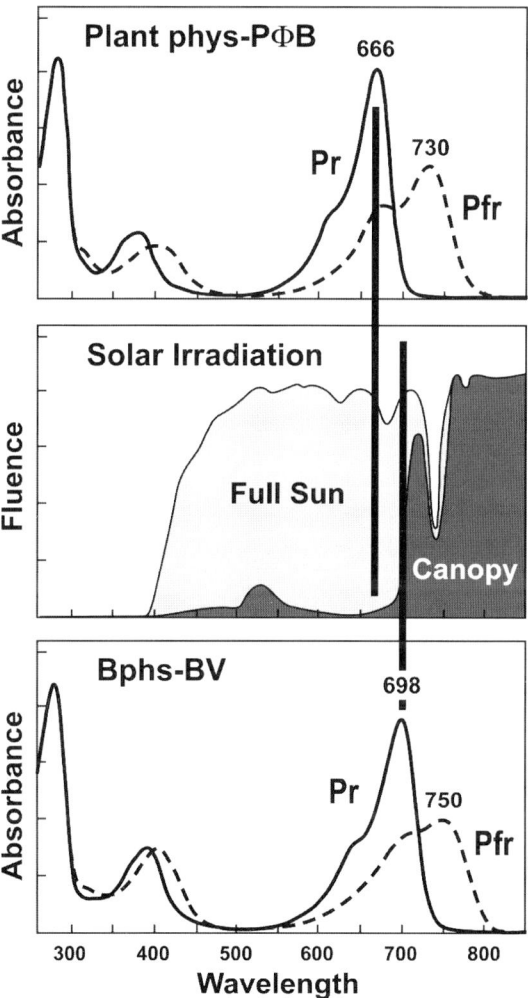

Figure 11. Absorbance spectra comparison of plant phys and BphPs with respect to their suitability for detecting shading by photosynthetic pigments. Top and bottom panels show the Pr and Pfr absorption spectra of a plant phy (oat PhyA) and a BphP (Deinococcus radiodurans BphP). Middle panel shows the fluence distribution versus wavelength of unfiltered sunlight (Full Sun) and sunlight transmitted through a canopy of plants (adapted from Smith, 2000).

Members of the Fph family most likely evolved from BphPs. In addition to a GAF domain bearing a histidine preceded by a small hydrophobic residue, Fphs have the signature N-terminal PLD with the accompanying cysteine. The only exception is *Botryotinia fuckeliana* Phy1, which is missing an obvious PLD and appears by sequence comparisons to be an Phy-like protein (Figure 4). At present, Fphs have been conclusively found only in filamentous fungi with a preliminary action spectrum for light-induced sporulation, suggesting that one may be present in the slime mold *Physarum polycephalum* (Starostzik and Marwan, 1995). Given that Phy-type sequences are not evident in the genomes of single-cell fungi such as *Saccharomyces cerevisiae* and *Schizosaccharomyces pombe* or in another slime mold *Dictyostelium discoidium*, their distribution within the fungal kingdom may be limited (Bhoo *et al.*, 2001; Catlett *et al.*, 2003). Likewise, BLAST searches of various *Archaea* and animal genomes found no evidence that Phys entered these kingdoms.

Plant phys possibly originated from a Cph precursor during the development of the chloroplast from a cyanobacterial endosymbiont. Plant phys became nuclear encoded and evolved further to help control the myriad of growth and developmental responses. This development also included the use of PΦB instead of PCB. However, the green algae *Mesotaenium caldorium* still employs PCB (Wu *et al.*, 1997), suggesting that this substitution occurred later during the evolution of land plants. The major changes in plant phys relative to Cphs were the addition of the two internal PAS domains that help with signal output, and further modifications of the HKD and PLD. In typical TC-HKs, the HKD includes motifs not only for autophosphorylation and subsequent phosphotransfer but also for homodimerization. During the evolution of plant phys, it is possible that the HKRD has retained its dimerization contacts but either lost the HK activity or transformed the kinase domain to one with a serine/threonine-type kinase activity.

8. PERSPECTIVES

The ongoing discovery of Phy-type pigments in microbes has impacted our understanding of the Phy superfamily at many levels. In particular, their participation in TC-HK cascades has helped support the view that plant phys are light-regulated kinases. Whether plant phys use their serine/threonine kinase activity for signal transmission or for autoregulation is not yet clear (Quail, 2002). And like microbial Phys, a main endpoint of plant phy signalling is the alteration of gene expression. Whereas the physiological functions of plant phys are well documented, we know very little about the functions of their microbial counterparts. Preliminary data with photosynthetic bacteria suggest a common theme in optimising light capture. This is accomplished either by directing movement toward more favourable light conditions or by regulating photosynthetic capacity with respect to light fluence, light quality, and/or photoperiod. For some microbes *(e.g., Synechocystis* and *Agrobacterium tumefaciens*), a complex interplay of signals emanating from multiple Phys is likely. In non-photosynthetic species, the roles of Phys are still unclear. Even though red light (R) has important roles in conidiation, circadian rhythms, and sexual development in filamentous fungi, no aberrant phenotypes have been observed yet for *fphΔ* mutants in several species (Catlett *et al.*, 2003; Froehlich, A., Dunlap, J., Vierstra, R. D., unpublished).

With respect to understanding how Phys function mechanistically as R/FR photochromic pigments, it is obvious that these microbial Phys now offer excellent opportunities to study this photoreceptor family. Genomic analyses indicate that several non-photosynthetic bacteria encode only a single BphP and are void of other known photoreceptors, thus providing useful models to study Phy function in the absence of photosynthesis and other light sensing systems (Davis *et al.*, 1999b; Bhoo *et al.*, 2001; Lamparter *et al.*, 2002; Karniol and Vierstra, 2003). These Phys also can be exploited to easily produce homogeneous holoproteins for biochemical and biophysical studies. Several recombinant systems are now available that can assemble unlimited supplies of photoreceptors by co-expressing the apoproteins with appropriate enzymes that synthesize the bilin chromophores (Bhoo *et al.*, 2001; Gambetta and Lagarias, 2001). Furthermore, the availability of bathyBphPs that prefer Pfr as the ground state now affords the first opportunity to study Pfr without significant Pr contamination (Karniol and Vierstra, 2003). Hopefully, biochemical, biophysical, and structural analyses of these pigments will reveal how Phys function as R/FR photochromic regulators of microbial and plant processes. And as we learn more about these microbial Phys, we will hopefully reach the point where biochemical properties and physiological functions can be predicted by placement of the Phy sequence on the phylogenetic tree. The discovery of microbial Phys clearly offers new and intriguing avenues of investigation on Phy-type pigments.

Acknowledgements: We thank various authors for providing information prior to publication. Our work was supported by grants from the U.S. Department of Energy and the National Science Foundation to RDV, and the Binational Research Development Fellowship to BK.

9. REFERENCES

Aldridge, P., Paul, R., Goymer, P., Rainey, P., Jenal, U. (2003) Role of the GGDEF regulator PleD in polar development of *Caulobacter crescentus*. *Molec. Microbiol, 47*, 1695-1708.

Armitage, J. P. (1997) Behavioural responses of bacteria to light and oxygen. *Arch. Microbiol, 168*, 249-261.

Bhaya, D., Takahashi, A., Grossman, A. R. (2001) Light regulation of type IV pilus-dependent motility by chemosensor-like elements in *Synechocystis* PCC6803. *Proc Natl Acad Sci USA, 98*, 7540-7545.

Bhoo, S. H., Davis, S. J., Walker, J., Karniol, B., Vierstra, R. D. (2001) Bacteriophytochromes are photochromic histidine kinases using a biliverdin chromophore. *Nature, 414*, 776-779.

Bhoo, S. H., Hirano, T., Jeong, H. Y., Lee, J. G., Furuya, M., Song, P. S. (1997) Phytochrome photochromism probed by site-directed mutations and chromophore esterification. *J Amer Chem Soc, 119*, 11717.

Braatsch, S. and Klug, G. (2004) Blue light perception in bacteria. *Photosyn Res, 79*, 45-57.

Catlett, N. L., Yoder, O. C., Turgeon, B. G. (2003) Whole-genome analysis of two-component signal transduction genes in fungal pathogens. *Eukaryotic Cell, 2*, 1151-1161.

Davis, S. J., Kurepa, J., Vierstra, R. D. (1999a) The *Arabidopsis thaliana HY1* locus, required for phytochrome-chromophore biosynthesis, encodes a protein related to heme oxygenases. *Proc Natl Acad Sci USA, 96*, 6541-6546.

Davis, S. J., Vener, A. V., Vierstra, R. D. (1999b) Bacteriophytochromes: phytochrome-like photoreceptors from nonphotosynthetic eubacteria. *Science, 286*, 2517-2520.

Frankenberg, N., Mukougawa, K., Kohchi, T., Lagarias, J. C. (2001) Functional genomic analysis of the HY2 family of ferredoxin-dependent bilin reductases from oxygenic photosynthetic organisms. *Plant Cell, 13*, 965-978.

Galperin, M. Y., Nikolskaya, A. N., Koonin, E. V. (2001) Novel domains of the prokaryotic two-component signal transduction systems. *FEMS Microbiol Lett, 203*, 11-21.

Gambetta, G. A., Lagarias, J. C. (2001) Genetic engineering of phytochrome biosynthesis in bacteria. *Proc Natl Acad Sci USA, 98*, 10566-10571.

Genick, U. K., Soltis, S. M., Kuhn, P., Canestrelli, I. L., Getzoff, E. D. (1998) Structure at 0.85 angstrom resolution of an early protein photocycle intermediate. *Nature, 392*, 206-209.

Giraud, E., Fardoux, J., Fourrier, N., Hannibal, L., Genty, B., Bouyer, P., *et al.* (2002) Bacteriophytochrome controls photosystem synthesis in anoxygenic bacteria. *Nature, 417*, 202-205.

Giraud, E., Hannibal, L., Fardoux, J., Jaubert, M., Jourand, P., Dreyfus, B., *et al.* (2004) Two distinct *Crt* gene clusters for two different functional classes of carotenoid in *Bradyrhizobium*. *J Biol Chem, 279*, 15076-15083.

Gong, W. M., Hao, B., Mansy, S. S., Gonzalez, G., Gilles-Gonzalez, M. A., Chan, M. K. (1998) Structure of a biological oxygen sensor: A new mechanism for heme-driven signal transduction. *Proc Natl Acad Sci USA, 95*, 15177-15182.

Häder, D. P. (1987) "Photomovement." In *Cyanobacteria*, Fay, P. and Van Baalen, C. (eds) Elsevier, New York, 325-345.

Herdman, M., Coursin, T., Rippka, R., Houmard, J., Tandeau de Marsac, N. (2000) A new appraisal of the prokaryotic origin of eukaryotic phytochromes. *J Mol Evol, 51*, 205-213.

Huang, B. X., Whitchurch, C. B., Mattick, J. S. (2003) FimX, a multidomain protein connecting environmental signals to twitching motility in *Pseudomonas aeruginosa*. *J Bacteriol, 185*, 7068-7076.

Hübschmann, T., Börner, T., Hartmann, E., Lamparter, T. (2001a) Characterization of the Cph1 holo-phytochrome from *Synechocystis* sp. PCC 6803. *Eur J Biochem, 268*, 2055-2063.

Hübschmann, T., Jorissen, H., Börner, T., Gärtner, W., de Marsac, N. T. (2001b) Phosphorylation of proteins in the light-dependent signalling pathway of a filamentous cyanobacterium. *Eur J Biochem, 268*, 3383-3389.

Hughes, J., Lamparter, T., Mittmann, F., Hartmann, E., Gärtner, W., Wilde, A., *et al.* (1997) A prokaryotic phytochrome. *Nature, 386*, 663.

Inouye, M. and Dutta, R., *Histidine Kinases in Signal Transmission*. New York: Academic Press, 2003.

Jiang, Z., Swem, L. R., Rushing, B. G., Devanathan, S., Tollin, G., Bauer, C. E. (1999) Bacterial photoreceptor with similarity to photoactive yellow protein and plant phytochromes. *Science, 285*, 406-409.

Jorissen, H., Quest, B., Lindner, I., de Marsac, N. T., Gartner, W. (2002a) Phytochromes with noncovalently bound chromophores: The ability of apophytochromes to direct tetrapyrrole photoisomerization. *Photochem Photobiol, 75*, 554-559.

Jorissen, H., Quest, B., Remberg, A., Coursin, T., Braslavsky, S. E., Schaffner, K., *et al.* (2002b). Two independent, light-sensing two-component systems in a filamentous cyanobacterium. *Eur J Biochem, 269*, 2662-2671.

Karniol, B. and Vierstra, R. D. (2003) The pair of bacteriophytochromes from *Agrobacterium tumefaciens* are histidine kinases with opposing photobiological properties. *Proc Natl Acad Sci USA, 100*, 2807-2812.

Karniol, B. and Vierstra, R. D. (2004) The HWE histidine kinases, a new family of bacterial two-component sensor kinases with potentially diverse roles in environmental signaling. *J Bacteriol, 186*, 445-453.

Kehoe, D. M. and Grossman, A. R. (1996) Similarity of a chromatic adaptation sensor to phytochrome and ethylene receptors. *Science, 273*, 1409-1412.

Kehoe, D. M. and Grossman, A. R. (1997) New classes of mutants in complementary chromatic adaptation provide evidence for a novel four-step phosphorelay system. *J Bacteriol, 179*, 3914-3921.

Lamparter, T., Michael, N., Mittmann, F., Esteban, B. (2002) Phytochrome from *Agrobacterium tumefaciens* has unusual spectral properties and reveals an N-terminal chromophore attachment site. *Proc Natl Acad Sci USA, 99*, 11628-11633.

Lamparter, T., Mittmann, F., Gärtner, W., Börner, T., Hartmann, E., Hughes, J. (1997) Characterization of recombinant phytochrome from the cyanobacterium *Synechocystis*. *Proc Natl Acad Sci USA, 94*, 11792-11797.

Lamparter, T., Michael, N., Caspani, O., Miyata, T., Shirai, K., Inomata, K. (2003) Biliverdin binds covalently to *Agrobacterium* phytochrome Agp1 via its ring a vinyl side chain. *J Biol Chem, 278*, 33786-33792.

Lamparter, T., Carrascal, M., Michael, N., Martinez, E., Rottwinkel, G., Abian, J. (2004) The biliverdin chromophore binds covalently to a conserved cysteine residue in the N-terminus of *Agrobacterium* phytochrome Agp1. *Biochemistry, 43*, 3659-3669.

Loros, J. J. and Dunlap, J. C. (2001) Genetic and molecular analysis of circadian rhythms in *Neurospora*. *Ann Rev Physiol, 63*, 757-794.

Mas, P., Devlin, P. F., Panda, S., Kay, S. A. (2000) Functional interaction of phytochrome B and cryptochrome 2. *Nature, 408*, 207-211.

Mathews, S. and Sharrock, R. A. (1997) Phytochrome gene diversity. *Plant Cell Environ, 20*, 666-671.

Matsushita, T., Mochizuki, N., Nagatani, A. (2003) Dimers of the N-terminal domain of phytochrome B are functional in the nucleus. *Nature, 424*, 571-574.

Montgomery, B. L. and Lagarias, J. C. (2002) Phytochrome ancestry: sensors of bilins and light. *Trends Plant Sci, 7*, 357-366.

Mooney, J. L. and Yager, L. N. (1990) Light is required for conidiation in *Aspergillus nidulans. Genes Dev, 4*, 1473-1482.

Muramoto, T., Kohchi, T., Yokota, A., Hwang, I., Goodman, H. M. (1999) The *Arabidopsis* photomorphogenic mutant *hy1* is deficient in phytochrome chromophore biosynthesis as a result of a mutation in a plastid heme oxygenase. *Plant Cell, 11*, 335-348.

Mutsuda, M., Michel, K. P., Zhang, X. F., Montgomery, B. L., Golden, S. S. (2003) Biochemical properties of CikA, an unusual phytochrome-like histidine protein kinase that resets the circadian clock in *Synechococcus elongatus* PCC 7942. *J Biol Chem, 278*, 19102-19110.

Ng, W. O., Grossman, A. R., Bhaya, D. (2003) Multiple light inputs control phototaxis in *Synechocystis* sp strain PCC6803. *J Bacteriol, 185*, 1599-1607.

Park, C. M., Kim, J. I., Yang, S. S., Kang, J. G., Kang, J. H., Shim, J. Y., *et al.* (2000) A second photochromic bacteriophytochrome from *Synechocystis* sp PCC 6803: spectral analysis and down-regulation by light. *Biochemistry , 39*, 10840-10847.

Quail, P. H. (2002) Phytochrome photosensory signalling networks. *Nat Rev Mol Cell Biol, 3*, 85-93.

Quest, B. and Gärtner, W. (2004) Chromophore selectivity in bacterial phytochromes - Dissecting the process of chromophore attachment. *Eur J Biochem, 271*, 1117-1126.

Rashid, M. H., Rajanna, C., All, A., Karaolis, D. K. R. (2003) Identification of genes involved in the switch between the smooth and rugose phenotypes of *Vibrio cholerae*. *FEMS Microbiol Lett, 227*, 113-119.

Schmitz, O., Katayama, M., Williams, S. B., Kondo, T., Golden, S. S. (2000) CikA, a bacteriophytochrome that resets the cyanobacterial circadian clock. *Science, 289*, 765-768.

Schneider-Poetsch, H. A. (1992) Signal transduction by phytochrome: phytochromes have a module related to the transmitter modules of bacterial sensor proteins. *Photochem Photobiol, 56*, 839-846.

Seib, L. O. and Kehoe, D. M. (2002) A turquoise mutant genetically separates expression of genes encoding phycoerythrin and its associated linker peptides. *J Bacteriol, 184*, 962-970.

Smith, H. (2000) Phytochromes and light signal perception by plants - an emerging synthesis. *Nature, 407*, 585-591.

Starostzik, C. and Marwan, W. (1995) A Photoreceptor with characteristics of phytochrome triggers sporulation in the true slime mold *Physarum polycephalum*. *FEBS Lett, 370*, 146-148.

Terauchi, K., Montgomery, B. L., Grossman, A. R., Lagarias, J. C., Kehoe, D. M. (2004) RcaE is a complementary chromatic adaptation photoreceptor required for green and red light responsiveness. *Molec Microbiol, 51*, 567-577.

van der Horst, M. A. and Hellingwerf, K. J. (2004) Photoreceptor proteins, star actors of modern times. *Acc Chem Res, 37*, 13-20.

Venter, J. C., Remington, K., Heidelberg, J. F., Halpern, A. L., Rusch, D., Eisen, J. A., *et al.* (2004) Environmental genome shotgun sequencing of the Sargasso Sea. *Science, 304*, 66-74.

Vierstra, R. D. (2002) "Cyanophytochromes, bacteriophytochromes, and plant phytochromes: light-regulated kinases related to bacterial two-component regulators." In *Histidine Kinases in Signal Transduction* , Inouya, M. and Dutta, R. (eds), 273-295.

West, A. H. and Stock, A. M. (2001) Histidine kinases and response regulator proteins in two-component signaling systems. *Trends Biochem Sci, 26*, 369-376.

Wilde, A., Fiedler, B., Börner, T. (2002) The cyanobacterial phytochrome Cph2 inhibits phototaxis towards blue light. *Molec Microbiol, 44*, 981-988.

Wilde, A., Churin, Y., Schubert, H., Börner, T. (1997) Disruption of a *Synechocystis* sp. PCC 6803 gene with partial similarity to phytochrome genes alters growth under changing light qualities. *FEBS Lett, 406*, 89-92.

Wu, S. H. and Lagarias, J. C. (2000) Defining the bilin lyase domain: lessons from the extended phytochrome superfamily. *Biochemistry, 39*, 13487-13495.

Wu, S. H., McDowell, M. T., Lagarias, J. C. (1997) Phycocyanobilin is the natural precursor of the phytochrome chromophore in the green alga *Mesotaenium caldariorum*. *J Biol Chem, 272*, 25700-25705.

Yeh, K. C. and Lagarias, J. C. (1998) Eukaryotic phytochromes - light-regulated serine/threonine protein kinases with histidine kinase ancestry. *Proc Natl Acad Sci USA, 95*, 13976-13981.

Yeh, K. C., Wu, S. H., Murphy, J. T., Lagarias, J. C. (1997) A cyanobacterial phytochrome two-component light sensory system. *Science, 277*, 1505-1508.

Zambre, M., Terryn, N., de Clercq, J., de Buck, S., Dillen, W., van Montagu, M., *et al.* (2003) Light strongly promotes gene transfer from *Agrobacterium tumefaciens* to plant cells. *Planta, 216*, 580-586.

Chapter 7

PHYTOCHROME GENES IN HIGHER PLANTS: STRUCTURE, EXPRESSION, AND EVOLUTION

Robert A. Sharrock[1] and Sarah Mathews[2]
[1] Montana State University, Bozeman, Montana 59717, USA (e-mail: sharrock@montana.edu)
[2] Arnold Arboretum of Harvard University, Cambridge, Massachusetts 02138, USA (e-mail: smathews@oeb.harvard.edu)

1. INTRODUCTION

Phytochromes play critical roles in monitoring light quantity, quality, and periodicity in plants and they relay this photosensory information to a large number of signaling pathways that regulate plant growth and development. Given these complex functions, it is not surprising that the phytochrome apoproteins are encoded by small multigene families and that different forms of phytochrome regulate different aspects of photomorphogenesis. Over the course of the last decade, progress has been made in defining the number, molecular properties, and biological activities of the photoreceptors that constitute a plant R/FR sensing system. This chapter summarizes our current understanding of the structure of the genes that encode the phytochrome apoproteins (the *PHY* genes), the expression patterns of those genes, the nature of the phytochrome apoprotein family, and *PHY* gene evolution in seed plants.

Phytochrome was discovered and its basic photochemical properties were first described through physiological studies of light-sensitive seed germination and photoperiodic effects on flowering (Borthwick, *et al.*, 1948, Borthwick, *et al.*, 1952). The pigment itself was initially isolated from extracts of dark-grown (etiolated) plant tissue in 1959 (Butler, *et al.*, 1959), but it was not until much later that phytochrome was purified to homogeneity in an undegraded form (Vierstra and Quail, 1983). DNA sequences of gene and cDNA clones for oat etiolated-tissue phytochrome provided the first complete descriptions of the apoprotein (Hershey *et al.*, 1985). Because it accumulates to levels that permit it to be assayed spectroscopically *in planta* and purified in its native form, this dark-tissue phytochrome (now called phyA) remains the most completely biochemically and spectroscopically characterized form of the receptor.

At various times throughout the first 40 years of the study of the abundant etiolated-tissue phytochrome, evidence for the presence and activity of additional forms of phytochrome, often referred to as "green-tissue" or "light-stable" phytochromes, was obtained. Initially, in physiological experiments, it was sometimes not possible to correlate specific *in vivo* phytochrome activities with the known spectroscopic properties of the molecule. Later, direct evidence for multiple species of phytochrome in plants and in plant extracts was obtained using both spectroscopic and immunochemical methods (reviewed in Pratt, 1995). The molecular identities of these additional phytochrome forms were ultimately deduced from cDNA clones that were isolated by nucleic acid similarity to etiolated-tissue phytochrome sequences (Sharrock and Quail, 1989). More recently, analysis of a large number of complete and partial *PHY* gene or cDNA sequences from a broad sampling of plant phylogenetic groups and sequencing of several plant genomes have resulted in a much clearer and more general picture of what constitutes a higher plant R/FR photoreceptor family. It is likely that the major types of long-wavelength photosensing pigments have now been identified and the challenge that lies ahead is to understand how the signalling mechanisms, expression patterns, and interactions of these molecules contribute to plant responses to the R/FR environment. Extending the investigation of phytochrome gene families and their functions to additional angiosperm and gymnosperm genera will be an integral component of this effort and of our ability to utilize this growing understanding of phytochrome function to modify the agricultural properties of plants and to better understand the history of land plants.

2. PHYTOCHROME GENE STRUCTURES AND PROTEIN SEQUENCES

2.1 The first phytochrome sequences

Isolation of cDNA and gene clones encoding phytochrome apoproteins (*PHY* genes) was first accomplished from oat (*Avena sativa*) and depended upon the observation that translatable phytochrome mRNA was readily detectable in dark-grown tissue but strongly reduced in light-grown tissue (Gottman and Schäfer, 1982; Colbert *et al.*, 1983). Differential screening using cDNA libraries constructed from dark and light-grown seedlings led to identification of cDNA clones that encoded an oat phytochrome polypeptide 1128 amino acids in length (Hershey *et al.*, 1984, Hershey *et al.*, 1985). Characterization of oat genomic DNA sequences corresponding to those cDNAs resulted in the first description of the structure of a complete *PHY* gene (Hershey *et al.*, 1987). Figure 1 shows diagrams of the oat *phyA3* gene, the mature *phyA3* mRNA, and the phyA polypeptide product. The gene has the characteristic features of a typical eukaryotic gene including five introns, one located in each of the 5' and 3' untranslated regions, and three within the coding sequence. The *phyA3* transcription start site is ~35 bp downstream of a TATA box region and a AATAAA poly(A)-addition signal is found upstream of the beginning of the poly(A) sequence on the cDNA. In the protein product, cysteine residue 321 has been identified as the attachment site for the chromophore. The description of the oat *phy* mRNA and gene sequences provided the first detailed information on phytochrome apoprotein primary sequence and the genetic elements that control synthesis of those polypeptides. In combination with the biochemical and spectral characterization of the purified pigment, these studies provided the foundations for a molecular understanding of phytochrome structure and function.

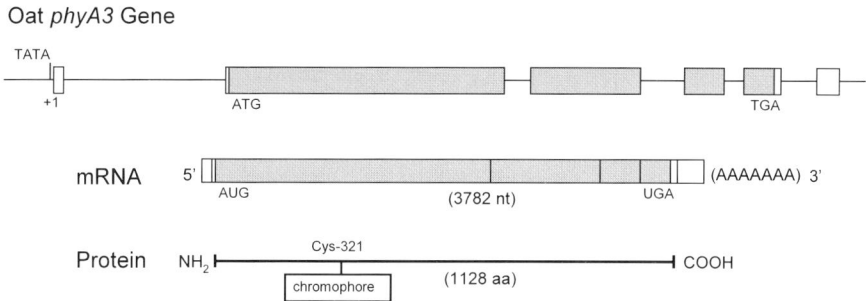

Figure 1. Schematic representation of the oat *phyA3* gene, mature *phyA3* mRNA, and *phyA* protein. Gene: boxes = exons (open = untranslated, filled = translated); lines = introns or flanking DNA sequences. mRNA: boxes = processed RNA sequences.

2.2 Phytochrome is a family of related photoreceptors encoded by multiple *PHY* genes in higher plants

In almost all plants, the relatively abundant phytochrome present in etiolated tissue, such as that encoded by the oat *phyA3* gene (Figure 1), is rapidly degraded upon conversion from Pr to Pfr by red light (see Chapter 8). However, early physiological studies indicated that, while many plant R/FR responses could be explained by the activity of this light-labile phytochrome, other responses appeared to be controlled by a second form of phytochrome that was active for long periods of time as Pfr. For example, in what was referred to as the "Pisum paradox", the effect of a pulse of R on stem elongation in peas could be significantly reversed by a FR pulse given several hours after R, when no spectroscopically detectable Pfr was present or was calculated to be present based on the properties of the known phytochrome (Hillman, 1967). Ultimately, biochemical and spectroscopic studies also suggested the presence of high levels of one phytochrome form in dark-grown plant tissues and much lower levels of immunochemically-distinct phytochrome forms in light-grown tissues. Careful analysis of the degradation kinetics of phytochrome following transfer into R indicated that it was likely that multiple pools of phytochrome with different physical properties existed (Brockmann and Schafer, 1982) and, in 1985, three groups independently reported the presence of secondary, low-abundance phytochrome forms in extracts of light-grown plants (Abe *et al.*, 1985, Shimazaki and Pratt, 1985, Tokuhisa *et al.*, 1985). Hence, there was strong evidence that the phytochrome that predominated in light-grown green plant tissues was a different molecular species than the pigment that had been purified, cloned, and characterized from etiolated tissue.

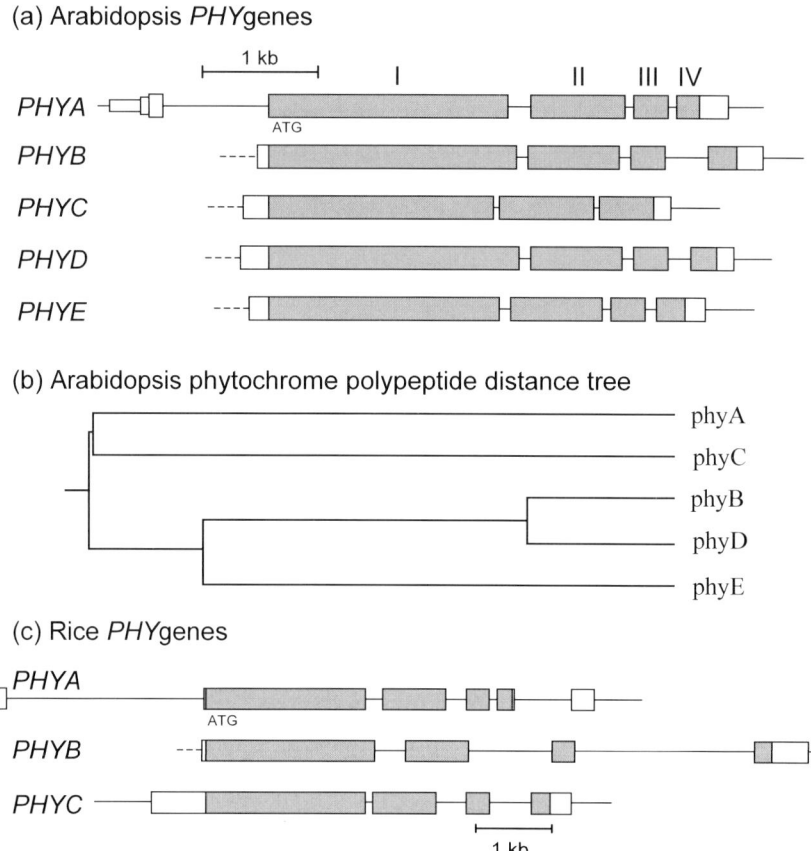

Figure 2. (a) Schematic representations of the five Arabidopsis PHY genes. Boxes = exons (open = untranslated, filled = translated); lines = introns or flanking DNA sequences. The four coding sequence exons are labelled I-IV. The PHYA gene has three transcription start sites and an intron in its 5' UT region. Dotted lines preceding the PHYB-PHYE genes indicate that the transcription start sites for these transcripts have not been mapped and the lengths of the known 5' and 3' UT regions, deduced from cDNA sequences are shown as open boxes. (b) Illustration of the percent amino acid sequence identity of the aligned Arabidopsis phyA-phyE polypeptides in each pair wise combination. (c) Schematic representations of the three rice PHY genes.

Using a nucleic acid hybridisation probe derived from an etiolated-tissue *PHYA* gene under conditions of low hybridisation stringency, Sharrock and Quail (1989) demonstrated the presence of multiple hybridising sequences on Southern blots of Arabidopsis genomic DNA and isolated cDNA clones corresponding to three of those bands. When those cDNAs were sequenced, they revealed clones of a typical etiolated-tissue phytochrome and two additional, related but divergent sequences. The etiolated-tissue phytochrome sequence was assigned the designation *PHYA*, and the two new sequences were called *PHYB* and *PHYC*. Northern blot analysis of the mRNA levels corresponding to these sequences showed that the *PHYA* mRNA was abundant in dark-grown seedlings and was down-regulated in light, as was characteristic of etiolated-tissue *PHY* mRNAs, whereas the *PHYB* and *PHYC* mRNAs were present in lower amounts and were light stable. Hence, the abundant etiolated-tissue phytochrome from Arabidopsis was designated phyA, and the phyB and phyC forms were candidates for "green tissue" phytochromes. Arabidopsis cDNA clones corresponding to the other two hybridising bands were found to encode low-abundance, light-stable mRNAs that encoded apoproteins more closely related to phyB than to phyA or phyC. These were designated phyD and phyE (Clack *et al.*, 1994).

The sequence of the Arabidopsis genome does not contain any additional phytochrome-related genes, so it is clear that the complete *PHY* gene family in this plant consists of these five members. Figure 2a illustrates the structures of the five Arabidopsis *PHY* genes. Four of the five genes have three coding-sequence introns located at sequences that are homologous to those of the oat *phyA3* gene. *PHYC* contains the first two introns but lacks the third. The introns show no significant sequence relatedness when compared between genes. The Arabidopsis *PHYA* gene, like the *PHYA* genes of oat, rice, maize, pea, and tobacco, contains an intron within its 5' untranslated region, whereas the other four *PHY* genes apparently lack such an intron. All five Arabidopsis genes lack introns in their 3' untranslated regions. The polypeptide products of the five *PHY* genes are similar in amino acid sequence along their entire lengths, indicating that their overall structure is conserved and that the phytochrome family contains no intrinsically membrane-bound or plastid-localized forms. These conclusions have been confirmed through *in vivo* localization studies using PHY::GFP fusion proteins in which all five forms are found to be localized to the cytosol as Pr and to be translocated to the nucleus following conversion to Pfr (see Chapter 9). All five proteins have cysteine residues, the predicted chromophore-attachment sites, located within a conserved sequence context at positions similar to that found in oat phyA (Figure 1). When their amino acid sequences are aligned, the Arabidopsis phytochromes show ~50% sequence identity among phyA, phyC, or phyB/D/E, ~65% identity between phyB and phyE, and ~80% identity between phyB and phyD. Figure 2b illustrates these relationships, which indicate a pattern of gene evolution in which ancestral *PHYA*, *PHYC*, and *PHYE* genes, and the common progenitor gene to *PHYB/D*, arose via gene duplication events, with a more recent duplication giving rise to *PHYB* and *PHYD* (see section 4, Evolution of the *PHY* gene family in seed plants, below).

Other seed plant species in which multiple full-length *PHY* genes or cDNAs have been isolated include the dicots tomato, potato, tobacco, pea, soybean, and poplar, and the monocots sorghum and rice. Rice and sorghum each contain a total of three *PHY* genes, which are most closely related to the Arabidopsis *PHYA*, *B*, and *C* genes. The structures of the rice *PHY* genes are shown in Figure 2c (Kay *et al.*, 1989, Dehesh *et al.*, 1991, Basu *et al.*, 2000). The positions of the three coding sequence introns are conserved between Arabidopsis and rice (Figures 2a and 2c), although many of the introns are larger in rice, resulting in larger overall gene size, and the rice *PHYC* gene contains an intron between exons III and IV whereas the

Arabidopsis *PHYC* has apparently lost this intron. Like the oat *phyA3* gene (Figure 1), the rice *PHYA* gene contains an intron in its 3' untranslated region that is not found in rice *PHYB* or *PHYC* or in any of the Arabidopsis genes. The functional or regulatory significance of these differences in intron number among *PHY* genes is not known. Among the complete complements of phytochromes from the dicot Arabidopsis and the monocot rice, the structures of the apoproteins themselves show no major rearrangements or additions or losses of large amino acid sequence domains. The polypeptide products of the three rice *PHY* genes can be aligned with those of the corresponding Arabidopsis genes along almost their entire lengths, and all contain a conserved cysteine residue at their presumptive chromophore-attachment site. Therefore, the overall structure of angiosperm phytochromes appears to be highly conserved.

Comparison of the complete *PHY* gene families from the dicot Arabidopsis and the monocot rice (Figure 2) indicates that these two major groups of angiosperms contain homologs of three genes - *PHYA*, *B*, and *C* - but differ by the presence or absence of a fourth relatively divergent gene, *PHYE*. This observation is supported by an analysis of the *PHY* genes from tomato, a dicot from a different family (Solanaceae) than Arabidopsis (Brassicaceae). Like Arabidopsis, tomato contains genes related to *PHYA*, *PHYC*, and *PHYE*, and a duplicated pair of closely-related *PHYB*-like genes (*PHYB* and *PHYD* in Arabidopsis, *PHYB1* and *PHYB2* in tomato). When it was first described, the sequence of a partial clone of the tomato *PHYC* gene homolog was observed to be more divergent from Arabidopsis *PHYC* than were tomato *PHYA* or *PHYB* from Arabidopsis *PHYA* and *PHYB*, and the gene was initially named *PHYF* (Hauser, *et al.*, 1995). More recent phylogenetic analysis of the full-length tomato *PHYF* gene sequence, however, indicates that this gene is in fact orthologous to Arabidopsis *PHYC* and that its designation as a separate class of phytochromes is not supported (Alba, *et al.*, 2000). The evolution of *PHY* genes in seed plants is discussed further below, however, it is important to note here that, although two divergent dicots, Arabidopsis and tomato, contain very similar five-membered *PHY* gene families and two monocots, rice and sorghum, contain similar three-membered families, it should not be assumed that these gene complements are completely representative for all plants in these major groups of angiosperms. For example, the dicot poplar appears to contain genes homologous to *PHYA* and *PHYB* but no *PHYC* gene (Howe *et al.*, 1998) and this may also be true of the very large legume family (Fabaceae) (Lavin *et al.*, 1998). Nevertheless, it is firmly established that four relatively ancient and widely-distributed *PHY* gene lineages exist in angiosperms and analysis of the functions and regulation of these genes in selected plants, such as Arabidopsis, rice, tomato, and tobacco, is likely to be very relevant to developing an understanding of R/FR sensing and signaling throughout flowering plants.

Additional structural features of some, but not all, *PHY* genes include the presence of multiple transcription start sites, differing in their relative activities, within the promoters of several of the characterized *PHYA* genes (Sato, 1988, Dehesh *et al.*, 1994, Adam *et al.*, 1995) and *PHYB* genes (Heyer and Gatz, 1992, Adam *et al.*, 1996) and the presence of short upstream open reading frames (URF's) in the 5' untranslated regions of many *PHY* genes (Clack *et al.*, 1994, Basu *et al.*,

2000). Specific functions for these features in regulation of the biological activities of the *PHY* genes have not as yet been assigned.

2.3 Phytochrome nomenclature

Several terminologies have arisen to describe the diversity of phytochrome activities and molecular forms observed in plants. The abundant phytochrome found in dark-grown plant tissues, which is rapidly degraded in the light, has been called "light-labile", "etiolated-tissue", "type I", or "phyA" phytochrome. The lower abundance, light-stable forms of phytochrome have been called "light-stable", "green-tissue", "type II", or "phyB-phyE" forms (Furuya, 1993, Pratt, 1995). Although, in a general sense, each of the more descriptive designations has some validity, they can also be misleading. For example, it might be expected that the "light-stable", "green-tissue" phytochromes would be absent from etiolated plants and would show no down-regulation in light. However, these forms of phytochrome are indeed present in dark-grown plants and several of them show a significant degree of down-regulation upon exposure to R (Sharrock and Clack, 2002). The type I/type II terminology is perhaps the most useful way of broadly grouping phytochromes in that it distinguishes phyA, which has a unique role in sensing very low levels of light and continuous FR and is rapidly proteolytically degraded in the light, from phyB-phyE, which function in sensing long term R and the ratio of R:FR and are relatively light-stable (see Chapter 22). Therefore, in this chapter, the precise molecular identity of a phytochrome or its gene, such as phyA (*PHYA*), phyB (*PHYB*), etc. will be given when it is possible and appropriate to do so, whereas the type I/type II designation will be used to distinguish the two broad classes of phytochromes.

2.4 Heterodimerization of type II phytochromes

A number of years ago, analysis of the quaternary structure of phyA phytochrome from dark-grown oat tissue clearly showed that it is a homodimer (Lagarias and Mercurio, 1985, Jones and Quail, 1986). Since that time, it has often been assumed that all phytochromes are homodimers and that the number of different types of phytochrome present in a plant directly correlates with the number of *PHY* genes in that plant's genome. Recently, however, results of co-immunoprecipitation experiments in Arabidopsis suggest that this may not be the case and that the phytochrome array in higher plants likely includes heterodimeric combinations of various type II proteins (Sharrock and Clack, 2004). In these studies, a *myc* epitope-tagged version of one phytochrome was immunoprecipitated from seedling extracts and type-specific anti-phy monoclonal antibodies were used to assay for co-precipitation of other phy proteins. When *myc*-tagged phyB was precipitated, phyC, phyD, and phyE were co-precipitated. This occurred irrespective of whether the seedlings were grown in the light or in the dark. When *myc*-tagged phyD was pulled-down, phyB and phyE were co-precipitated, confirming the interactions of these proteins. However in no case was type I phyA observed to co-precipitate with phyB or phyD. Size-exclusion chromatography demonstrated that the various heteromeric molecules migrate at molecular weights characteristic of dimers. Therefore, it appears that that the type II phytochromes engage in heteromeric binding interactions and that the complexity of the phytochrome array may be significantly higher than previously recognized. The photochemical properties and biological functions of these heterodimers have not yet been assessed. Nevertheless, if they have even subtly different spectral sensitivities or signaling activities from each other, our understanding of the diversity of plant R/FR photoreceptor structure, how that diversity arose during evolution, and how it contributes to plant response to the light environment may need to be reconsidered.

3. EXPRESSION PATTERNS OF PHYTOCHROMES IN PLANTS

3.1 How important are phytochrome expression patterns?

Analysis of light transmittance and reflectance within plants has shown that absorption and scattering by internal pigments and cellular structures results in complex light gradients and wavelength distributions within tissues (Vogelmann, 1994). However, in general, light is quite efficiently propagated within cells and along plant axes, so it is likely that, at least for the above ground parts of plants, the presence or absence of photoreceptors in cells within a tissue will be a major determinant of that tissue's sensitivity to the light environment. Moreover, experiments in which the level of a phytochrome has been changed via altering *PHY* gene dosage or via expression of *PHY* coding sequences from a viral promoter show that, under a given light condition, small changes in the level of expression of a phytochrome can very strongly influence the response of the plant (Cherry *et al.*, 1992, Wester *et al.*, 1994). This indicates that the number of activated photoreceptor molecules, rather than downstream signaling events, is limiting for at least some responses and, therefore, that the expression levels of these receptors are also important determinants of light sensitivity.

On the other hand, how critical the presence or absence of a phytochrome within a particular cell is to the light sensitivity of that cell is not entirely clear. The competence of a cell to respond directly, or in a cell-autonomous fashion, to a stimulus is determined by whether that cell has receptors for the stimulus. However, it has been shown that, within plant tissues, some light responses are not cell autonomous and that cells adjacent to or even distant from a photo-sensitive cell may be activated via intercellular signaling mechanisms. Microbeam irradiation experiments have been performed in which a narrow beam of R or FR (25 μm diameter) is used to illuminate approximately 5 cell diameters in defined regions of etiolated tobacco seedlings and the R/FR response of the activity of a CAB::LUC fusion gene is monitored in surrounding cells (Bischoff *et al.*, 1997, Schütz and Furuya, 2001). These experiments show that propagation of R induction, but not of subsequent FR reversal of induction, of CAB::LUC occurs within minutes over distances as long as the length of the cotyledon. The nature of this intercellular signal is not known. Nevertheless, it must be acknowledged that, although the location of a photoreceptor within a plant tissue is likely to influence the light sensitivity of that group of cells, it cannot be taken to completely define the light responsive region.

3.2 Assaying phytochromes

Phytochromes do not show a readily assayable enzymatic activity. Therefore, detection and quantification of phytochrome in plants or plant extracts has most frequently been performed either spectroscopically or by raising antibodies to the phytochrome apoproteins and measuring their amounts via immunological assays. These methods have individual advantages and disadvantages. Spectroscopic detection involves measuring the R/FR-reversible absorbance change and can be performed in intact tissues or in subcellular fractions. Spectrophotometric assays detect only chromophore-ligated spectrally-active molecules, but are less sensitive than immunological methods and are not able to distinguish among the different forms of phytochrome. Immunological approaches, including immunocytochemistry, ELISA, and immunoblot techniques, which use antibodies raised against the phytochrome apoproteins, can be phytochrome form-specific and are very sensitive. However, they do not distinguish between chromophore-ligated holoproteins and chromophore-less apoprotein molecules.

3.3 Early Expression Studies

Before the multiple phytochrome forms present in plants were described and molecular tools to determine their individual levels became available, studies of phytochrome levels and distribution within plants and plant tissues monitored total phytochrome, which, in dark-grown plants, is principally phyA. Antibodies directed against phytochrome were invariably raised against purified phyA. In the previous edition of this volume, Pratt summarized the results of analyses of both the level and the distribution of phytochrome using these experimental approaches (Pratt, 1994). This body of work suggested that phytochrome(s) are expressed widely throughout roots and shoots of both monocot and dicot plants, in hypocotyls and cotyledons, mature leaves, inflorescences, and fruits. In addition, in immunocytochemical analyses of etiolated grass seedlings, phyA exhibited some degree of cell-type specificity within tissues but was often observed to be heterogeneously distributed, even in groups of cells of the same type. In etiolated seedlings, phyA was found to be most abundant in young, rapidly expanding cells of meristems or of elongation zones recently derived from meristems such as epicotyl and hypocotyl hooks, coleoptilar nodes, and root caps. Importantly, many of the details of phytochrome distribution varied significantly depending upon the plant species chosen for study, underscoring the need for caution when generalizing from data obtained in model systems or in only a limited number of plant species. This caution remains relevant to more recent studies of the expression patterns of the multiple *PHY* genes, described in following sections, which are also restricted to a few selected plant species.

One of the most striking examples of regulation of *PHY* gene expression, and the first to be described, is the auto-regulation of type I phyA. Early in the history of phytochrome research, it was observed that the level of spectrally-detectable phytochrome was markedly lower in light-grown plants than in dark-grown plants (Butler *et al.*, 1963; Gottman and Schäfer, 1983: Otto *et al.*, 1984). It has become clear that this difference is almost entirely due to a difference in the amount of phyA present under etiolated versus photomorphogenic growth conditions in most seed plants and that, with considerable species variation, this reflects both the inhibition of transcription of the *PHYA* mRNA and the conformation-dependent proteolytic degradation of phyA Pfr (see Chapter 8). The degree and the kinetics of down-regulation of transcription of *PHYA* genes in response to R irradiation and the extent to which this is reversible by FR varies among different species, with the strongest effect seen in monocot grasses. Indeed, because it responds very rapidly to a pulse of R, the oat *phyA* gene promoter (see Figure 1) was adopted as a model system for the analysis of direct R/FR transcriptional control and the identification of *cis*-acting sequence elements within the promoter and candidate transcription factors that mediate the response. These experiments were reviewed in the previous edition of this volume (Quail, 1994).

3.4 Patterns of *PHY* gene expression – mRNA levels and promoter fusion experiments

In the initial description of the Arabidopsis *PHY* gene family, transcript-specific hybridisation probes for the five Arabidopsis *PHY* mRNAs were used on Northern blots to investigate their steady-state levels in seedlings grown under varying light conditions, in plants at different stages of growth, and in different plant organs (Clack *et al.*, 1994). Figure 3 shows that, using this method, the phytochrome transcripts are found to be remarkably broad in their expression profiles in dark or light conditions, over the course of rosette development, and in isolated roots, leaves, stems, and flowers. It is important to note that, in Figure 3, blots of different mRNAs cannot be directly compared to each other because they are not normalized for probe labelling or exposure time. Nonetheless, general features of the steady-state patterns of these transcripts are evident. The *PHYA* transcript is unique in showing decreased abundance in light-grown as compared to dark-grown seedlings, although this down-regulation is not nearly as strong as that seen for *PHYA* transcripts in monocots. Aside from reductions in expression as 7 day-old seedlings mature into rosette-stage plants (14-21 days), very little alteration in level of the *PHYB-PHYE* transcripts is seen under any of the conditions tested (Figure 3). Hence, the patterns of transcription of the *PHY* genes in Arabidopsis do not exhibit strong organ-specific or developmental control on a gross level. Results similar to these were observed in Northern blot analysis of potato *PHYA* and *PHYB* mRNA levels, with the exception of a strikingly low expression of *PHYA* in tubers, even under etiolation conditions (Heyer and Gatz, 1992).

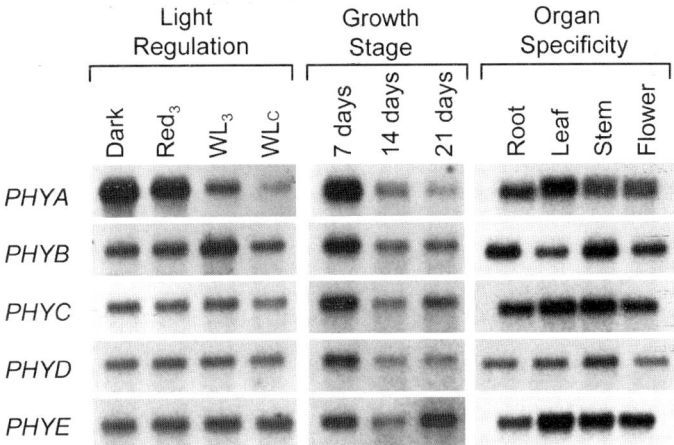

Figure 3. Northern blot analysis of the Arabidopsis PHYA-PHYE transcript levels. Light regulation: seedlings were grown for seven days in darkness or continuous white light (WL$_C$) or in darkness and then placed under R or WL for 3 h prior to harvest. Growth stage: seedlings and rosette-stage plants were grown for 7, 14, or 21 days in white light. Organ specificity: leaves, inflorescence stems, and flowers were harvested from soil-grown plants and roots were harvested from liquid-grown plants. Poly(A)-selected RNA was fractionated and hybridized with transcript-specific probes. Modified from Clack et al. (1994).

A more comprehensive and quantitative analysis of the steady-state levels and distributions of the five tomato *PHY* mRNAs has been performed by Pratt and colleagues (Hauser *et al.*, 1997, 1998). Quantitative hybridisation assays for each of the tomato transcripts were developed, using radioactive RNA hybridisation probes on blots of poly(A)-selected RNA samples and on sense-strand RNA standard curves. The assays were performed on RNA prepared from different tomato organs, over the course of plant development, under various light-dark conditions, and over a diurnal cycle in the greenhouse. With correction for contamination of the samples by residual rRNA, transcript levels were calculated and expressed as a fraction of total mRNA under each condition. An example of this analysis, and of the derived transcript levels in seedlings and seedling organs, is shown in Figure 4. Each of the five *PHY* mRNAs was detected at low levels (1-5 μmol/mol mRNA) in dry seeds. In both seedlings and mature light-grown plants, all five transcripts were detected in all tomato organs but with varying patterns. Surprisingly, the *PHYA* mRNA, which encodes light-labile phyA, was the most abundant *PHY* transcript in most plant organs and throughout plant development (Figure 4). In monitoring the mRNAs from tomato seed imbibition through eight weeks of growth in the greenhouse, a pattern of relatively rapid increase in all of the *PHY* transcripts, beginning at seed germination through the first 6 days, and a subsequent decline over the following week to a plateau level was observed (Hauser, *et al.*, 1998). Diurnal cycling of the *PHYA*, *PHYB1*, and *PHYB2* transcripts in mature plants under greenhouse conditions was also seen. This was detected as two to three-fold changes in the levels of in these mRNAs, with a peak in the dark phase for *PHYA* and *PHYB2* and a peak in the light phase for *PHYB1* (also see section 3.6, Circadian regulation of *PHY* gene expression, below). The results of these analyses indicate that the tomato *PHY* genes are differentially expressed at the mRNA level in ways which may reflect their individual roles in R/FR sensing and signaling. Nevertheless, consistent with the results in Arabidopsis and potato, the overall picture is one in which the distributions of specific *PHY* mRNAs are not restricted to either easily defined stages of plant development or organs within the plant body and variation in expression is quantitative rather than qualitative.

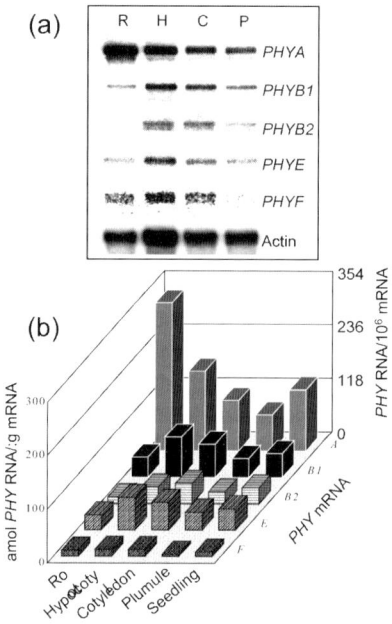

Figure 4. Levels of the five tomato PHY transcripts in the indicated organs of greenhouse-grown 15-21 day-old seedlings. (a) Example of phosphoimager output. (b) Calculated PHY mRNA amounts. From Hauser et al. (1997).

Another approach that has been applied to investigating *PHY* gene expression patterns in whole plants is based upon fusion of *PHY* gene promoter regions to the beta-glucuronidase (GUS) coding region and introduction of these gene fusions into plants as transgenes. In these experiments, the histochemical stain for GUS activity provides visual indication of the pattern of promoter activity in the transgenic lines. In addition, the reporter enzyme activity can be quantified to give an estimate of relative transcriptional activity of the fused gene promoter under different growth conditions or in isolated organs or tissues. Four of the five Arabidopsis *PHY* promoter regions, the exception being *PHYC*, were fused to GUS, introduced back into Arabidopsis, and their temporal and spatial expression patterns characterized (Somers and Quail, 1995a, 1995b, Goosey *et al.*, 1997). In agreement with the mRNA experiments described above, all four promoters drove expression of the reporter sequence at some level throughout the plant, including roots, leaves, stems, and flowers, and over the entire life cycle. However, the *PHYA* and *PHYB* promoters showed stronger and more universal activities than the *PHYD* and *PHYE* promoters. At a tissue level, histochemical staining indicated that many different cell types throughout most of the plant body express both the *PHYA*::GUS and *PHYB*::GUS transgenes (Somers and Quail, 1995b, Goosey *et al.*, 1997). For example, Figure 5 also shows the relatively uniform histochemical staining pattern of whole light and dark-grown seedlings and flowers of a *PHYB*::GUS line. In contrast to this, the staining pattern of a *PHYD*::GUS promoter fusion exhibits clearer tissue and organ specificity, such that, while the *PHYD* transcript is detectable in isolated roots and flowers (Figure 3), this mRNA is likely to be localized primarily in the root tips and in the sepals (Figure 5).

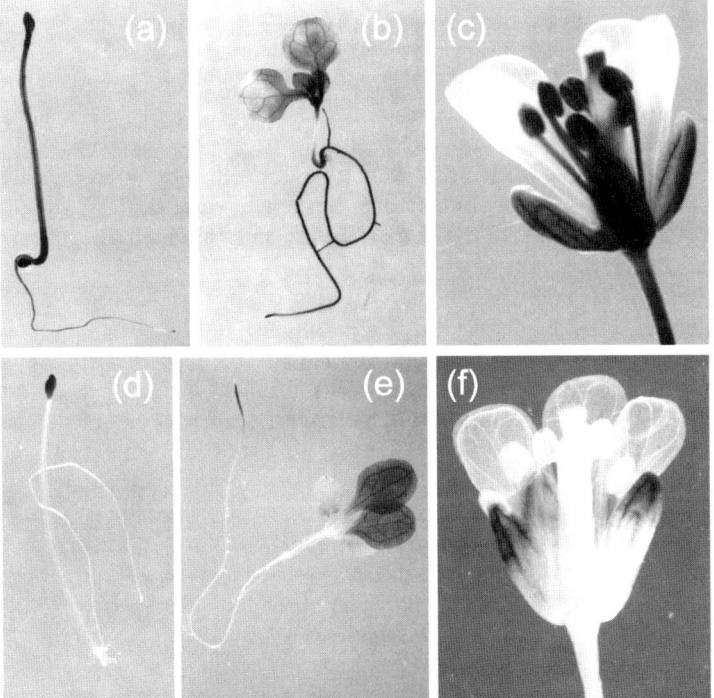

Figure 5. Histochemical localization of the expression patterns of PHYB::GUS (a-c) and PHYD::GUS (d-f) promoter-reporter fusion genes in Arabidopsis. (a, d) seven day old dark-grown seedlings; (b, e) seven day old light-grown seedlings; (c, f) flowers.

The promoter regions of tobacco *PHYA* and *PHYB* genes have also been fused to GUS and their expression patterns analysed in transgenic tobacco plants (Adam *et al.*, 1994, 1996). Compared to the Arabidopsis *PHYA* and *PHYB* promoter fusions, the tobacco constructs showed more highly defined tissue and organ specific expression patterns with, for example, strong histochemical staining of tobacco *PHYA*::GUS activity in hook and root tip regions and absence of expression of *PHYB*::GUS in roots. These studies emphasize again that, while most *PHY* genes appear to be broadly expressed in higher plant tissues and organs, variation in many of the details of expression patterns are likely to be seen between different plant species.

3.5 The levels and distributions of phytochromes in plants

As instructive as it is to know how *PHY* gene transcription and the levels of the *PHY* mRNAs are regulated, ultimately, it is the amounts and locations of the spectrally-active phytochrome receptors themselves that are likely to be important indicators of photosensory potential within the plant. As discussed above, immunological detection of the phytochromes has the sensitivity and specificity to quantify and localize these molecules, although it does not discriminate between chromophore-ligated holophytochrome and chromophore-less apophytochrome. In fact this may not be a major consideration because, when phytochrome apoproteins have been overexpressed by as much as 20-fold from viral promoter regions, co-ordinately elevated levels of spectrally-active phytochrome are produced, indicating that chromophore synthesis and attachment are not limiting in plant cells (Boylan and Quail, 1991, Wagner *et al.*, 1991). Efforts to raise antibodies directed against several or all of the different phytochrome forms in one species have been carried out principally in Arabidopsis and oat. All five of the Arabidopsis *PHYA-PHYE* coding sequences were expressed in *E. coli* and the full-length apoproteins were purified and used as antigens in the generation of type-specific monoclonal antibodies (MAbs) (Hirschfeld *et al.*, 1998). In oat, MAbs that differentially detected each of three different-sized phytochromes present in plant extracts, which likely correspond to the oat phyA, phyB, and phyC-related forms, were identified (Wang *et al.*, 1991).

Sharrock and Clack (2002) used the MAbs specific to each of the Arabidopsis phyA-phyE apoproteins and dilution curves of purified *E. coli*–expressed apoprotein standards to measure phytochrome levels in extracts of seedlings and plants grown under varying light conditions, in plants at different stages of growth, and in different plant organs. Previous determinations of phytochrome levels had been expressed as nanograms of phytochrome per seed or seedling, per gram fresh weight of tissue, or per μg of total protein extracted. However, tissues harvested from seedlings and organs of mature plants grown under various conditions vary in water content, cell size, and protein content, and the question arises of how best to normalize the measured phytochrome levels. One comparison that takes into account several of these variables is an estimate of the average amount of each phytochrome per cell in a given sample. As an approximation of this, the amount of extracted protein loaded on gels was normalized to the DNA content of each extract and levels of the individual phytochromes were expressed as nanograms per "DNA equivalent" (Sharrock and Clack, 2002). Examples of immunoblots from several of these experiments, probed with the five phy-specific MAbs, are shown in Figure 6 and the levels of the phytochrome apoproteins determined by comparison to standard curves of purified antigen are given in Table 1.

Table 1. Levels of the phytochromes in Arabidopsis dark-grown and light-grown seedlings. Phytochrome amounts are expressed as nanograms of apoprotein per DNA equivalent and as % of total phytochrome under each growth condition. A DNA equivalent is defined as 100 μg of total protein from dark-grown and R_{24}-grown seedlings and 130 ug total protein from light-grown seedlings (Sharrock and Clack, 2002).

	Dark		R_{24}		WL	
phyA	355	(85%)	2.8	(10%)	0.7	(4%)
phyB	40	(10%)	12.3	(43%)	7.3	(40%)
phyC	8.1	(2%)	3.1	(11%)	2.4	(13%)
phyD	5.4	(1.5%)	5.5	(19%)	2.9	(16%)
phyE	6.1	(1.5%)	5.0	(17%)	4.8	(27%)
Total	415	(100%)	28.7	(100%)	18.1	(100%)

These experiments confirm that phyA is the only form present in Arabidopsis that has properties corresponding to those expected for a type I phytochrome. Arabidopsis phyA is the most abundant form in the dark and is highly light-labile, declining rapidly in level following transfer into R (Figure 6). The difference in phyA level between dark-grown and light-grown seedlings is over 100-fold (Table 1) and is far greater than the decline in transcriptional activity of the *PHYA* gene indicated by either RNA blots (Figure 4) or promoter-reporter fusion transgenes (Somers and Quail, 1995a). The immunoblot analyses also show that the phyB-phyE forms have properties consistent with their being type II phytochromes. PhyB-phyE are all present at 10- to 60-fold lower levels than phyA in the dark and are much more light-stable. Some degree of differential light-stability is seen among them, with phyB and phyC levels declining several fold after 24 h in R, while phyD and phyE levels are little changed (Table 1). However, it is clear that no example of a true "green-tissue" phytochrome, which would be absent in dark-grown plants and present in the light, is seen. Consequences of the various expression levels and light stabilities of the five Arabidopsis phytochromes include an overall decline of approximately 20-fold in total phytochrome and a strong shift in relative abundance between type I phyA and the type II forms coincident with photomorphogenic development, and the presence of a relatively stable array of type II photoreceptors following this transition (Table 1).

The levels of the four type II phytochromes are maintained in Arabidopsis plants over three weeks of growth in the light, at which point the plants have undergone the transition into reproductive phase and are beginning to bolt. All four are also present in the major plant organs (Figure 6). At the onset of the plant life cycle, the five phytochromes show somewhat greater variation in their levels in seed and in their induction upon seed germination. PhyA is undetectable in imbibed seeds but rises rapidly or very slowly, respectively, during germination in the dark or light. PhyB, D, and E are the most abundant forms in imbibed seeds, whereas phyC is very low, and phyB and phyC both increase over the course of germination while phyD and phyE do not change (Sharrock and Clack, 2002). Hence, even among the relatively constitutively-expressed type II phytochromes, the ratios of photoreceptor forms to each other change over time throughout germination and early seedling growth and development.

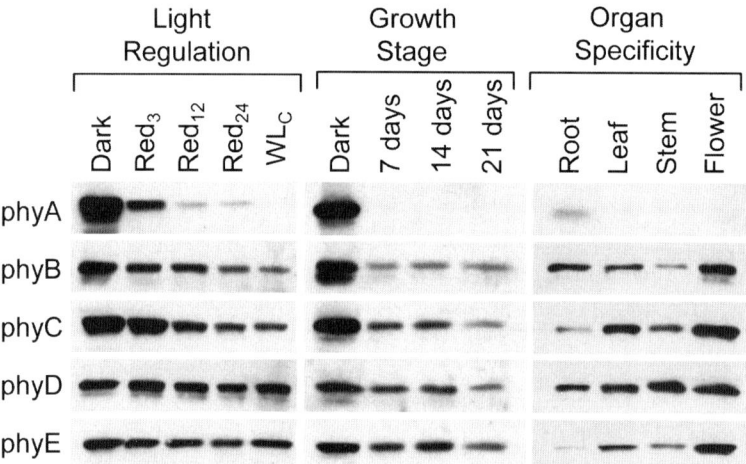

Figure 6. Immunoblot analysis of the five Arabidopsis phytochromes. Light regulation: seedlings were grown for seven days in darkness or continuous white light (WL_C) or in darkness and then placed under R for 3, 12, or 24 h prior to harvest. Growth stage: seedlings were grown for seven days in darkness or white light and rosette-stage plants were grown for 14 or 21 days. Organ specificity: leaves, inflorescence stems, and flowers were harvested from soil-grown plants and roots were harvested from liquid-grown plants. Protein extracts were prepared, fractionated on SDS-PAGE, and immunoblotted with MAbs specific to each of the Arabidopsis phytochromes. Modified from Sharrock and Clack (2002).

If the Arabidopsis *PHY* genes have served as a model for investigation of the expression and function of R/FR receptors in dicots, an important counterpart will be a corresponding analysis in a monocot species. The complete *PHY* gene complement in rice is three genes, which are related in their sequences to the Arabidopsis *PHYA*, *PHYB*, and *PHYC* genes (see Figure 2c). Before this was known, Pratt and his colleagues performed an analysis of three phytochrome proteins that were detected differentially in extracts of oat seedlings that had been grown in the dark versus the light. One of these was purified as the well-characterized and abundant 124 kDa etiolated-tissue phytochrome. The other two were partially purified from green oat leaves, had distinctive apparent monomer sizes of 123 kDa and 125 kDa, and were detected in relatively low amounts (Wang et al., 1991). Relating these to current terminology, the 124 kDa protein is clearly oat phyA and, although it has not been unequivocally established, it is likely that the 125 and 123 kDa proteins are oat phyB-related and phyC-related forms respectively. MAbs that selectively recognized each of these proteins were identified from panels of MAbs raised against the partially-purified antigens and these antibodies were then used to monitor the levels and distributions of these phytochromes in oat seedlings (Wang *et al.*, 1992, 1993a, 1993b). The results of these experiments yield an overall picture of these oat phytochromes that is remarkably similar to that of Arabidopsis phyA, phyB, and phyC.

The three oat phytochromes are all present at low levels in unimbibed seed. The 124 kDa phyA accumulates rapidly to high levels in seedlings grown in darkness but not in the light, whereas the 123 and 125 kDa proteins increase only a few fold over the first several days of growth, irrespective of light conditions. Hence, in 4 day-old dark-grown seedlings, 124 kDa phyA is 40-70 fold higher in abundance than the 123 and 125 kDa forms whereas, in light-grown seedlings, the 123 and 125 kDa forms predominate (Wang *et al.*, 1993b). Oat 124 kDa phyA is highly light-labile and the 123 and 125 kDa forms, like Arabidopsis phyB and phyC (Figure 6), are somewhat light-labile, decreasing a few fold upon transfer from darkness into light. Immunoblot analysis of dissected oat seedlings shows that all three oat phytochromes are present in shoot, scutellum, and root tissues, irrespective of whether the seedlings are grown in the dark or light (Wang *et al.*, 1993a).

For the plant species analysed, results of immunological assays for the phyA-phyE apoproteins, and of the *PHY*::GUS fusion gene experiments (see above), indicate that the phytochromes are a relatively broadly distributed array of photosensory molecules and that many or most plant cells are likely to be individually R/FR photosensitive. Moreover, the levels and the ratios of the five phytochromes change over the course of plant development and in response to ambient light conditions and this regulation of the photoreceptor complement in specific plant cells and tissues likely regulates important aspects of photomorphogenesis. However, the immunoblot analyses performed on the multiple endogenous phytochromes present in extracts of Arabidopsis and oat tissues have only limited organ-specific and no tissue-specific resolution, and although it is possible that immunocytochemical methods might address this in more detail, these experiments have not been attempted with phytochrome type-specific antibodies.

In considering the patterns of expression of the *PHY* genes and the distributions of the phy proteins in various plant organs and tissues, it is important to remember that each of the phy chromoproteins may have distinct spectral characteristics, that they are known to individually regulate different subsets of plant light responses, and that they may form heterodimeric combinations. Clearly, phytochrome expression levels and expression patterns integrate with these differential structural and functional properties. For example, analysis of the spectral properties of *in vitro* adducts of the Arabidopsis phytochromes with phycocyanobilin or phytochromobilin chromophores show that they vary in their light absorption characteristics and dark reversion kinetics (Eichenberg *et al.*, 2000). More importantly, genetic analysis has shown that the different phytochromes have different roles in R/FR light sensing and in interaction with other plant photoreceptor systems. How divergence in phytochrome structure and expression has contributed to the evolution of differential photosensory function remains a vibrant and important research area.

3.6 Circadian regulation of *PHY* gene expression

Phytochromes and the blue light-sensing cryptochromes function as photoreceptors for light cues that entrain the endogenous plant circadian clock to the environmental photoperiod (see Chapter 26). Whether the circadian clock in turn exerts feedback regulation on the expression of *PHY* and *CRY* genes is an important consideration in developing an understanding of how input pathways may interact with the central oscillator. Promoter fusions to the firefly luciferase (LUC) coding sequence have been used to test for circadian regulation of all five of the Arabidosis *PHY* promoters in transgenic Arabidopsis and the regulation of the Arabidopsis and tobacco *PHYB* promoters in transgenic tobacco (Kozma Bognar *et al.*, 1999, Hall *et al.*, 2001, Toth *et al.*, 2001). In these experiments, the *in vivo* bioluminescence assay for LUC activity allows imaging of *PHY*::LUC transgene expression throughout successive 24 hr photoperiods and after transfer of entrained seedlings to continuous light or dark conditions. In transgenic tobacco, both the tobacco and Arabidopsis *PHYB*::LUC fusion genes show clear diurnal cycling and circadian regulation of their expression, with maximal expression during the light phase and amplitude differences ranging from 2-fold to 3-fold in normalized luminescence units. However, the level of endogenous tobacco phyB protein in these lines, as detected on western blots, shows little, if any, cycling following transfer to continuous light (Kozma Bognar *et al.*, 1999).

Transgenes consisting of fusions of the Arabidopsis *PHYA*, *B*, *D*, and *E* promoters to LUC all exhibit diurnal rhythms of activity in transgenic Arabidopsis seedlings, with peaks during the light phase but varying somewhat in their phases, and persistence of the rhythm following transfer to continuous light (Toth *et al.*, 2001). A *PHYC*::LUC fusion shows only low amplitude diurnal rhythm. To confirm these findings, the endogenous *PHY* mRNA levels in wild-type Arabidopsis seedlings, entrained for one week under LD cycles, were followed at intervals through two successive 24 h cycles in constant light. Somewhat surprisingly, all five mRNAs, including the *PHYC* transcript, exhibited clear circadian oscillation (Toth *et al.*, 2001). Hence, the promoter activities and steady-state mRNA levels of all of the Arabidopsis *PHY* genes are regulated by the circadian clock, suggesting the existence of a feedback loop between the receptor components of light input pathways and the clock mechanism. Again, however, the extent to which cycling of *PHY* promoter activities and mRNA levels is converted to cycling of the phytochromes themselves is less clear. Immunoblot analysis, using apoprotein-specific antibodies, indicates that endogenous Arabidopsis phyA and phyC cycle diurnally, with maximal levels during the dark phase, that phyB cycles with a similar phase but only very low amplitude, and that phyE is constant over a light/dark cycle (Sharrock and Clack, 2002). Following a shift to constant light, the levels of phyA and phyC were observed to cycle with very low amplitude whereas phyB and phyE were unchanged over the course of 48 hours. That phyA, phyB, and phyC should oscillate diurnally with maximal abundance in the dark fits with their apparent light stabilities in that all three of these proteins are degraded upon transfer from dark to light (Sharrock and Clack, 2002). However, it is unclear why these cycles in phytochrome protein levels are out of phase with the more evident oscillations in their respective mRNA levels. It is possible that the cycling biosynthetic potential for the phytochromes, in the form of oscillating mRNA levels, compensates for the daily light-induced degradation of the proteins themselves in order to maintain appropriate photoreceptor levels over time.

4. EVOLUTION OF THE *PHY* GENE FAMILY IN SEED PLANTS

4.1 Phytochrome phylogeny in seed plants

The isolation and analysis of *PHY* and *PHY*-related DNA sequences from a broad range of prokaryotes and eukaryotes has resulted in rapid progress in our understanding of the origins and evolution of phytochrome gene families. The structures of the *PHY* genes and gene products in dicot and monocot angiosperms are described in the previous sections of this review. *PHY* sequences and *PHY*-related sequences have been isolated from a number of other groups of seed and nonseed land plants and a picture of the evolution of this receptor family is emerging.

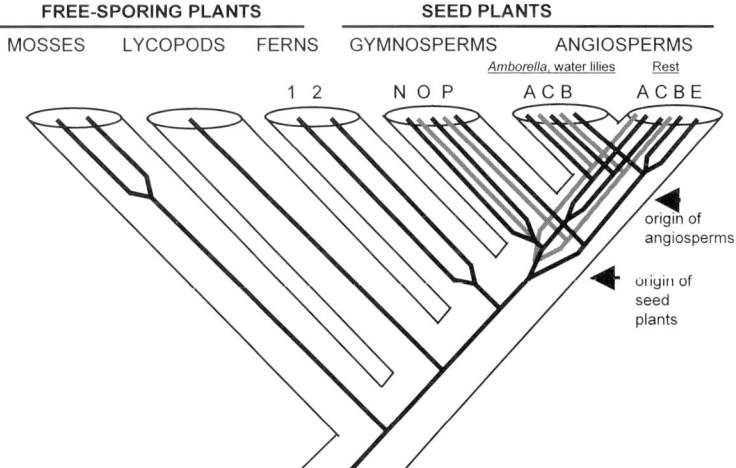

Figure 7. Phytochrome gene phylogeny (solid grey or black lines) within land plant phylogeny (cylinders). Amborella and the water lilies diverge from the rest of the angiosperms before all other groups originate.

All of the expressed phytochrome apoprotein sequences that have been recovered from seed plants and the majority of sequences recovered from eukaryotic organisms share a canonical protein structure of two functional regions: an ~70 kDa N-terminal photosensory region that binds the bilin chromophore and an ~55 kDa C-terminal regulatory region that consists of two PAS domains and a histidine kinase-related domain (Montgomery and Lagarias, 2002). Figure 7 shows a generalized phylogeny of genes that encode these canonical phytochrome proteins within a phylogeny of the major groups of extant land plants. The gene phylogeny suggests that a single *PHY* gene lineage extending from chlorophytes (green algae) persisted through the divergence of the major land plant clades until near the origin of seed plants, when a major split led to two gene lineages that persist in all seed plants. Hence, the phytochrome family diversified early in the history of seed plants, leading to similar levels of *PHY* gene diversity in angiosperms and extant gymnosperms.

Several classes of phytochrome-related proteins occur in prokaryotes, each class having a subset of the domains found in eukaryotic phytochromes (see Chapters 5 and 6). Among these, members of the Cph1 family, collectively known as the bacteriophytochromes, are most similar to eukaryotic phytochromes in sequence and have been shown to bind linear tetrapyrroles to form photochromic holoproteins (Montgomery and Lagarias, 2002). The history of land plant phytochromes can be more recently traced from their chlorophyte progenitors, represented by the *PHY* sequences from *Mesotaenium caldariorum* and *Mougeotia scalaris*. Each of these species contains a single type of canonical eukaryotic phytochrome, encoded by a single gene as in *Mougeotia* (Winands and Wagner, 1996), or by a small family of highly related genes as in *Mesotaenium* (Lagarias *et al.*, 1995). In nonvascular plants, there is limited evidence of gene duplication in the *PHY* family. The moss *Ceratodon purpureus* has two canonical phytochromes (*CpPHY2* and *CpPHY3*), while the moss *Physcomitrella patens* has four canonical phytochromes (*PHY1-PHY4*). *PHY1* and *PHY3* of *Physcomitrella* are sister genes and are most closely related to *CpPHY2*. Similarly, *PHY2* and *PHY4* of *Physcomitrella* are sister genes but are most closely related to *CpPHY3*. *Ceratodon* also has a chimeric sequence (*CpPHY1*) that has a C-terminus related to the catalytic domain of eukaryotic serine/threonine/tyrosine kinases (Pasentsis *et al.*, 1998). *CpPHY1* is of unknown origin, although its phytochrome domain is most closely related to *CpPHY2*. A homolog of *CpPHY1* has not been detected in *Physcomitrella patens*, although it has two homologs of *CpPHY2*. Thus it is likely that the chimeric moss sequence type is absent from other species or has a very restricted distribution. All of the moss sequences form a discrete, well-supported cluster. Together, the data indicate that one gene duplication occurred within mosses before the divergence of *Ceratodon* and *Physcomitrella* and that a second duplication in mosses occurred in *Physcomitrella* or in a near relative. In the early diverging vascular plant, *Selaginella martensii* (a lycopod), a single type of phytochrome has been characterized (Hanelt *et al.*, 1992). However, partial sequences from other lycopods in GenBank provide ambiguous information bearing on the question of gene number in this plant group. Thus, while there currently is no convincing evidence that the earliest diverging vascular plants have more than one type of *PHY*, the question warrants further investigation.

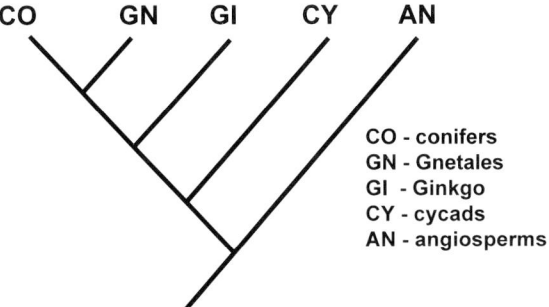

Figure 8. Relationships among major clades of extant seed plants.

In contrast, ferns clearly have two *PHY* genes that encode canonical proteins. These two *PHY* lineages are represented by *Adiantum PHY1* and *PHY2* (Nozue et al., 1998), and by sequence fragments in GenBank that represent homologs of these genes in other ferns. The distinctness of the two clades of fern phytochromes suggests that they result from gene or genome duplication early in the history of ferns and that they might encode divergent functions. *Adiantum* also has two types of chimeric phytochrome-related sequences: *PHY3*, which consists of a phytochrome sensory region fused with a full-length *NPH1* homolog, and *PHY4*, in which a partial exon I sequence is immediately upstream of a *Ty3/gypsy* type retrotransposon (see Chapter 23). The distribution of PHY3 homologs among ferns indicates that this gene originated relatively late in the evolution of ferns (Kawai et al., 2003). It is unclear exactly how these sequences are related to the majority of *PHY* that encode canonical proteins, whose history is summarized in Figure 7. More than two *PHY* also have been detected in *Psilotum nudum*, but there is evidence that some of these may be pseudogenes (Schneider-Poetsch et al., 1994). This would not be surprising given that *Psilotum nudum* also has high chromosome numbers ($2n=104$ or 208). Pseudogenes might be expected in other ferns as well, owing to chromosome numbers that are among the highest in plants. Figure 7 indicates that although *PHY* diversification in ferns may have occurred very early in their history, it likely occurred after they had diverged from the line leading to seed plants, possibly in taxa that are now extinct. This is supported by high levels of support for the monophyly of all seed plant *PHY*, suggesting they are more closely related to each other than to any *PHY* lineage in ferns, which are the sister group of seed plants. Mathews and Sharrock (1997) noted high bootstrap support (92%) for a clade of seed plant *PHY*, and this has been tested further in unpublished analyses of data sets with additional seed plant and fern sequences, resulting in parsimony bootstrap values ranging from 76% to 85% and Bayesian posterior probabilities of 0.8 to 1.0 (S. Mathews unpublished data).

Taken together, the data indicate that phytochrome diversity is limited in nonseed plants and that it increased during the advent and evolution of seed plants. It is important to bear in mind that seed plants have a long history involving the loss of major lines, leaving just five extant lines that are related as depicted in Figure 8 (Donoghue and Doyle, 2000). It is possible that some extinct seed plant groups would attach to the branch leading to angiosperms in Figure 8 while others would attach in the gymnosperm clade. These extinctions limit our ability to reconstruct precisely the history of phytochrome diversification during the early evolution of seed plants. However, phylogenetic analyses strongly indicate that, near the origin of seed plants, a split in the gene lineage extending from free-sporing plants resulted in two distinct lines that persist in all extant seed plants (Figure 7). Each of these lines subsequently diversified in both angiosperms and gymnosperms. Four major forms occur in angiosperms, *PHYA, B, C*, and *E*; *PHYA* and *PHYC* form one related pair, while *PHYB* and *PHYE* form a second related pair (Mathews *et al.*, 1995; Mathews and Sharrock, 1997). Three major forms occur in extant gymnosperms, *PHYN*, *PHYO*, and *PHYP* (Schneider-Poetsch *et al.*, 1998, Schmidt and Schneider-Poetsch, 2002); *PHYN* and *PHYO* form a related pair that diverged from the same line as *PHYA* and *PHYC* of angiosperms, while *PHYP* is a homolog of angiosperm *PHYB/E*. The duplications leading to *PHYN, PHYO*, and *PHYP* are clearly ancient since each gene is present in cycads and *Ginkgo* (Schmidt and Schneider-Poetsch, 2002; S. Mathews and M.J. Donoghue, unpublished), which diverged from other seed plants at least 300 million years ago. It is more difficult to determine if the duplication leading to *PHYA* and *PHYC* is similarly ancient (Figure 7, grey line), or if it occurred after the line leading to angiosperms diverged from other seed plants. Previous phylogenetic analyses have suggested that separate duplications led to the *A/C* and *N/O* pairs (Figure 7, black line), but currently available data from gymnosperms suggest that this is an open question that will require more extensive analysis (S. Mathews, unpublished). The duplication leading to *PHYB* and *PHYE* apparently occurred very early in the history of angiosperms. Both copies are found in *Illicium*, while a single copy has been detected in *Amborella* and the water lilies (S. Mathews, unpublished), the only lines that diverge before *Illicium* and its relatives split from the rest of the angiosperms (Zanis *et al.*, 2002).

Gene loss has altered phytochrome family composition in major clades of both angiosperms and gymnosperms. In the Gnetales, an enigmatic gymnosperm line comprising *Gnetum, Ephedra*, and *Welwitschia*, only two genes have been detected, one related to *PHYP*, and one related to *PHYN* or *PHYO* (Schmidt and Schneider-Poetsch 2002; S. Mathews and M.J. Donoghue, unpublished). It is likely that Gnetales lost one gene copy, since they are nested in the gymnosperms, most closely related to conifers (Donoghue and Doyle, 2000). Similarly in angiosperms, gene number has been reduced in some clades by the loss of *PHYE*, most notably from the monocots (Mathews and Sharrock, 1996), a diverse clade of about 55,000 species, but also from a number of smaller groups, including Piperales (S. Mathews, unpublished data), and members of the willow (Howe *et al.*, 1998) and carnation (S. Mathews, unpublished data) families. *PHYC*, which predates angiosperm origins and has been widely detected in its major clades (Mathews *et al.*, 1995; Mathews and Sharrock, 1997; Mathews and Donoghue, 1999; 2000), also apparently has been lost from some groups, including members of the willow family (Howe, *et al.*, 1998), although it is present in the tropical genus *Dovyalis*, which is the sister genus to the temperate members of the family, *Populus* and *Salix* (C. C. Davis, personal communication). *PHYC* also has not been detected in the legume family despite the use of multilocus primers in a large sample of genera (Lavin *et al.*, 1998).

Conversely, gene duplications have increased phytochrome diversity within some clades of both gymnosperms and angiosperms. The *PHYP* lineage split into two distinct lines within the conifer family Pinaceae, and the *PHYN* lineage split into two distinct lines in the clade comprising all other conifer families (Schmidt and Schneider-Poetsch, 2002; S. Mathews, unpublished data). Multiple copies of *PHYA* occur in the angiosperm families Fabaceae (legumes) and Caryophyllaceae (carnation family), and multiple copies of *PHYB* occur in Brassicaceae (e.g., *PHYD* of *Arabidopsis* and relatives), Solanaceae (tomato family), Apiaceae (carrot family), Salicaceae (willow family), Orobanchaceae (figworts and broomrapes), and in Piperales (Piperaceae and Aristolochiaceae). The duplication leading to *Arabidopsis PHYB* and *PHYD* occurred along the branch leading to Brassicaceae in the strict sense (K.S. McBreen and S. Mathews, unpublished), which is a clade nested within a larger, recently circumscribed Brassicaceae family. Together these observations suggest that the phytochrome gene families of angiosperms and gymnosperms are similarly complex.

4.2 Phytochrome functional evolution in seed plants

The phylogeny of land plant phytochromes suggests that the phytochrome family diversified early in the history of seed plants, leading to similar levels of gene diversity in angiosperms and extant gymnosperms (Figure 7). In angiosperms, functional diversification led to two major pools of phytochrome, a FR-responsive light-labile (type I) pool, and a R-responsive light-stable (type II) pool. As summarized earlier in this chapter, the light-labile pool is encoded by a single *PHYA* gene in the model dicot Arabidopsis and the light-stable pool is encoded by four genes, *PHYB-E*. While there is evidence that gymnosperms also have light-labile and light-stable pools of phytochrome, the genes that encode them remain to be identified. In angiosperms, phyB is the principal mediator of responses under control of the light-stable pool. Gymnosperms have a single gene, *PHYP*, which is related to *PHYB*, and which is the gene most likely to control phyB-like responses in gymnosperms. Conversely, they have two genes that are related to *PHYA*, *PHYN* and *PHYO*, both of which must be considered candidates for the control of phyA-like responses in gymnosperms. Ultimately, the question of whether phytochrome-mediated processes in angiosperms and gymnosperms are indeed homologous, and whether homologous members of the *PHY* gene families in these plants regulate these responses, will require improved resolution of gene history and a more extensive exploration of phytochrome-mediated processes in early-diverging angiosperms and in all lines of extant gymnosperms.

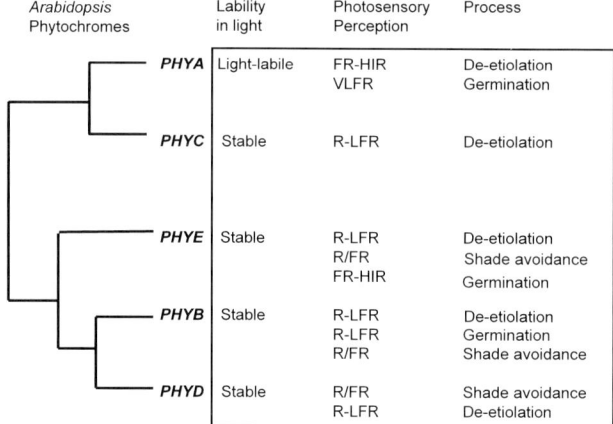

Figure 9. Phylogeny of the Arabidopsis phytochromes. Columns to the right of each gene indicate the stability in light of their product, their photosensory specificity and physiological response mode, and the process that they induce in a particular response mode. FR-HIR, far-red high irradiance response; VLFR, very low fluence response; R-LFR, red low fluence response; R/FR red to far-red ratio.

The availability of molecular tools and genetic resources for analysis of the Arabidopsis phytochromes, including mutants in all five *PHY* genes, has lead to a relatively comprehensive understanding of their functions (see Chapter 23). Figure 9 summarizes basic aspects of these functions relating to seedling responses. In contrast to this, little is known about the functions or expression patterns of phytochrome genes in gymnosperms. The Arabidopsis model and similar work in other angiosperm genera are useful reference points for the formulation of specific hypotheses concerning the functions of gymnosperm *PHY* genes. For fundamental processes that are comparable across seed plants, a reasonable starting point is to infer functional homology based on apparent similarity of the process and the presence of a potential gymnosperm homolog of the gene controlling the process in *Arabidopsis*. It is important to keep in mind that limited insight can be gained in this way and that, to infer functional homology, all lineages of seed plants should be sampled (Figure 8). Evidence that a process is shared by all lineages would suggest that the similarity results from common ancestry, especially in cases where clear homology of angiosperm and gymnosperm *PHY* genes is established. However, evidence that a process is simply shared by two unrelated lineages such as conifers and angiosperms does not provide enough information to discriminate between homology and independent origins.

4.2.1 Angiosperm phyB and Gymnosperm phyP

Gymnosperm *PHYP* is an unambiguous homolog of *PHYB* and *PHYB*-related angiosperm genes (Figure 7). Based on this relationship and on the *Arabidopsis* model (Figure 9), we might expect phyP to mediate responses to R and to changes in the R/FR ratio in gymnosperms, such as induction of seed germination, shade avoidance responses, and de-etiolation. The role of phyB in mediating R induction of seed germination in flowering plants is well established (Casal and Sanchez, 1998). In conifers, R-responsive seed germination has been noted only in two species of shade-intolerant pine (Toole *et al.*, 1961; Fernback and Mohr, 1990), but its distribution in other conifer species, which display a full range of shade tolerance, and in other gymnosperm lines remains undetermined. Furthermore, the large seeds of *Ginkgo biloba* germinate in the dark, and light apparently is not a requirement for germination of cycad seeds. The requirements for germination of seeds in members of the Gnetales remain unknown. Although phyP is a likely candidate gene for the control of R-responsive germination in gymnosperms, the apparent lack of this response in the earliest diverging taxa (cycads and *Ginkgo*; Figure 8) is consistent with possible independent origins of R-responsive germination in conifers and angiosperms, and with the possibility that it is controlled by one of the other gymnosperm phytochromes. Thus the question of whether this response has long been under control of phyP/phyB remains open. In ferns, spores germinate in response to either R or FR, depending on the species, but spore germination is probably best viewed as analogous rather than homologous with seed germination; the former represents an early stage in the development of the gametophyte, the latter an early stage in the development of the sporophyte. Therefore, it is difficult to infer based on its presence in ferns, that R-responsive germination is ancestral in seed plants, predating the major split that led to the two persistent *PHY* gene lineages found in all seed plants (Figure 7).

In Arabidopsis, shade avoidance responses are mediated primarily by phyB, with minor contributions from the other type II phytochromes. Elements of shade avoidance have been observed in conifers, where the magnitude of the response varied as expected with the degree of shade intolerance of the species investigated (Warrington *et al.*, 1988), but it has not been investigated in cycads, *Ginkgo*, or Gnetales, which include both shade-tolerant and shade-intolerant species. Shade avoidance requires the capacity for relatively rapid growth, the retention of meristematic potential through the growing season, and possession of a plant body that has nodes, internodes, and leaves. Growth rates in most gymnosperms are slow relative to angiosperms, and the expression of shade avoidance outside of angiosperms might be further limited by plant form. For example, it is hard to imagine how shade avoidance would be expressed in cycads, which do not have nodes and internodes and which have a limited capacity to respond to conditions that change through a season. Thus, while the capacity for shade avoidance may not be completely limited to angiosperms, its full expression probably relies on traits unique to this group of plants. Gymnosperm phyP may be the mediator of shade avoidance in shade-intolerant conifers, however, as with R-responsive germination, other phytochromes could play a role and the question of whether this response has long been under control of phyP/phyB remains open.

Etiolation can be viewed as a specialized pathway that is used to delay seedling development in dark or dimly lit environments and may have evolved early in the history of vascular plants, soon after plants gained the capacity to drastically modify the light environment through production of a canopy and leaf litter. The capacity for etiolated growth, and thus the need to de-etiolate, characterizes members of all lines of gymnosperms. Ferns are also etiolated in the dark and de-etiolate in response to R or white light. Therefore, the ability of some conifers to de-etiolate in the dark is likely to be a derived condition. In Arabidopsis seedlings, phyB is the principal mediator of R-responsive de-etiolation, with the other type II phytochromes having minor roles. As in angiosperms, it is possible that multiple gymnosperm phytochromes play a role in de-etiolation, and while phyP is again a likely candidate, its role in R induction of de-etiolation remains to be determined.

4.2.2 Angiosperm phyA and Gymnosperm phyN and phyO

Unlike phyP, the relationships of phyN and phyO with angiosperm phytochromes remain ambiguous. As noted above, analytical results suggest that there are two possibilities. Two separate duplications might have led to one pair of paralogs, *PHYN* and *PHYO*, in gymnosperms and to another pair of paralogs, *PHYA* and *PHYC*, in angiosperms (black lines in Figure 7). Conversely, a single duplication early in the history of seed plants might have led to *PHYA* and *PHYN* on one side of the split and *PHYC* and *PHYO* on the other side (grey lines in Figure 7). In this case, *PHYA* and *PHYN* would be orthologs, related by speciation rather than by duplication, as would *PHYC* and *PHYO*. This ambiguity confounds the inference of functional homology. Of particular interest is the identity of the gene controlling the phyA-like processes that have been noted in conifers (Burgin *et al.*, 1999) and *Ginkgo* (Christensen *et al.*, 2002). Based on the *Arabidopsis* model, we would expect a gymnosperm homolog of phyA to control germination in the very low fluence (VLF) response mode, to mediate de-etiolation under continuous FR, and to be labile in light. These processes distinguish the function of phyA from all other angiosperm phytochromes, and may have aided the early establishment of flowering plants (Mathews *et al.*, 2003). Specifically, these functions enhance the capacity of phyA to serve a transient role under conditions where an extremely high sensitivity is required.

Phytochrome A is the primary photoreceptor for VLF responses in angiosperms, allowing dark-imbibed seeds to germinate and de-etiolate in response to millisecond pulses of broad spectrum light. No evidence has been presented that plants other than angiosperms have VLF responses. This might indicate that the capacity to respond to brief pulses of light originated with the origin of phyA (Mathews *et al.*, 2003). However, if *PHYA* is more closely related to *PHYN* or *PHYO* than to *PHYC*, the absence of VLF responses outside angiosperms would suggest that functional innovation in *PHYA* was delayed after its origin by duplication, coinciding instead with the origin of angiosperms. In etiolated angiosperm seedlings, de-etiolation induced by continuous FR (FRc) is also mediated by phyA. Elements of this FR high irradiance response (FR-HIR) have been noted in dark-grown seedlings of two conifers of the pine family, *Pinus elliotii* and *Pseudotsuga menziesii* of intermediate shade tolerance (Burgin *et al.*, 1999). In response to FRc, hypocotyl growth was not inhibited in the seedlings of these species but they showed enhanced growth of the cotyledonary whorl and increased anthocyanin content. FRc also failed to inhibit seedling extension in *Pinus sylvestris* (Fernbach and Mohr, 1990). Similarly, elements of the FR-HIR were noted in *Ginkgo biloba*; in response to FRc, stem growth was not inhibited, but chlorophyll accumulation was enhanced in subsequent white light treatments (Christensen *et al.*, 2002). Thus, both pines and *Gingko* display a subset of the elements that comprise the angiosperm FR-HIR de-etiolation response. If they also were detected in cycads, Gnetales, additional conifers, and basal angiosperms, it would suggest that the FR-HIR has its origins early in the history of seed plants.

A contributing factor to the transient role and the sensitivity of phyA in angiosperms is its accumulation in the dark and rapid, but incomplete, decay in the light, due both to degradation and down-regulation. Burgin *et al.* (1999) detected a light-labile pool of phytochrome in dark-grown seedlings of two conifers (both Pinaceae). However, this pool decayed at a significantly slower rate than did the phyA pool in cucumber and oats. A similar finding has been reported for *Ginkgo* (Christensen *et al.*, 2002). These data suggest that at least some degree of light-lability may be important for phytochrome-mediated responses outside of angiosperms. If a light-labile pool of phytochrome were to be detected in cycads, Gnetales, additional conifers, and basal angiosperms, it would suggest that light lability is ancestral in seed plants and that it is likely to be an important feature of photomorphogenesis outside of angiosperms.

Taken together, these observations suggest that the processes contributing to the unique role of phyA are not fully expressed outside of angiosperms. There is a demonstrated capacity to respond to FRc, but with only a few of the elements that comprise the FR-HIR in angiosperms, and light-lability is less marked in gymnosperms than in angiosperms. Finally, there is no evidence of the VLFR. Since the VLFR operates via a downstream pathway different than that of the FR-HIR (Yanovsky *et al.*, 1997), it is very possible that these two phyA functions might have originated at different times, with elements of the FR-HIR predating the origin of flowering plants. If the *PHYA/C* and *PHYN/O* gene pairs result from separate duplications, both *PHYN* and *PHYO* are candidates for the phyA-like responses observed outside of angiosperms.

5. CONCLUSIONS

In seed plants, the phytochrome apoproteins are encoded by small multigene families. Along with their covalently-attached bilin chromophores, the products of these *PHY* genes constitute a broadly distributed, developmentally regulated array of R/FR-sensing photoreceptors. The phytochromes are highly conserved in overall structure but have varying expression patterns, light stabilities, and photosensory functions. Studies of Arabidopsis and, to a lesser extent, tomato, tobacco, and pea provide a detailed model of the phytochrome family in higher dicots. Two fairly distinct phytochrome types can be defined: type I (phyA), which is very abundant in etiolated tissue, is unstable in light, and functions to regulate responses to very low fluence R and continuous FR, and type II (phyB-phyE), which are more constitutive in expression, are stable in light, and function to regulate responses to low fluence R and the R/FR ratio. Studies in monocots, such as rice and oat, suggest that general features of this model apply to most species of angiosperms. Other seed plants have candidate orthologs of the angiosperm *PHY* genes. Gymnosperm *PHY* genes are structurally similar to those of angiosperms and appear to encode light-labile and light-stable pools of phytochrome in some species. However, based upon the available evidence, aspects of light-regulated development are likely to vary between angiosperms and gymnosperms, and it will be of interest to determine which processes have distinct evolutionary histories in the two groups.

6. REFERENCES

Abe, H., Yamamoto, K. T., Nagatani, A., Furuya, M. (1985) Characterization of green tissue-specific phytochrome isolated immunochemically from pea seedlings. *Plant Cell Physiol*, 26, 1387-1399.

Alba, R., Kelmenson, P. M., Cordonnier-Pratt, M.-M., Pratt, L. H. (2000) The phytochrome gene family in tomato and the rapid differential evolution of this family in angiosperms. *Mol Biol Evol*, 17, 362-373.

Adam, E., Szell, M., Szekeres, M., Schaefer, E., Nagy, F. (1994) The developmental and tissue-specific expression of tobacco phytochrome A genes. *Plant J*, 6, 283-293.

Adam, E., Kozma-Bognar, L., Dallmann, G., Nagy, F. (1995) Transcription of tobacco phytochrome-A genes initiates at multiple start sites and requires multiple cis-acting regulatory elements. *Plant Mol Biol*, 29, 983-993.

Adam, E., Kozma-Bognar, L., Kolar, C., Schäfer, E., Nagy, F. (1996) The tissue-specific expression of a tobacco phytochrome B gene. *Plant Physiol*, 110, 1081-1088.

Basu, D., Dehesh, K., Schneider-Poetsch, H. J., Harrington, S. E., McCouch, S. R., Quail, P. H. (2000) Rice PHYC gene: structure, expression, map position and evolution. *Plant Mol Biol*, 44, 27-42.

Bischoff, F., Millar, A. J., Kay, S. A., Furuya, M. (1997) Phytochrome-induced intercellular signalling activates *cab*::luciferase gene expression. *Plant J*, 12, 839-849.

Borthwick, H. A., Hendricks, S. B., Parker, M. W. (1948) Action spectrum for the photoperiodic control of floral initiation of a long day plant, winter barley (*Hordeum vulgare*). *Bot Gaz*, 110, 103-118.

Borthwick, H. A., Hendricks, S. B., Parker, M. W., Toole, E. M., Toole, V. K. (1952) A reversible photoreaction controlling seed germination. *Proc Natl Acad Sci USA*, 38, 662-666.

Boylan, M. T. and Quail, P. H. (1991) Phytochrome A overexpression inhibits hypocotyl elongation in transgenic Arabidopsis. *Proc Natl Acad Sci USA*, 88, 10806-10810.

Brockmann, J. and Schäfer, E. (1982) Analysis of Pfr destruction in *Amaranthus caudatus* L. Evidence for two pools of phytochrome. *Photochem Photobiol*, 35, 555-558.

Burgin, M. J., Casal, J. J., Whitelam, G. C, Sanchez, R. A. (1999) A light-regulated pool of phytochrome and rudimentary high-irradiance responses under far-red light in *Pinus elliotti* and *Pseudotsuga menziesii*. *J Exp Bot*, 50,: 831-836.

Butler, W. L., Norris, K. H., Siegelman, H. W., Hendricks, S. B. (1959) Detection, assay and preliminary purification of the pigment controlling photoresponsive development in plants. *Proc Natl Acad Sci USA*, 45, 1703-1708.

Butler, W. L., Lane, H. C., Siegelman, H. W. (1963) Nonphotochemical transformations of phytochrome in vivo. *Plant Physiol*, 38, 514-519.

Casal, J. J. and Sanchez R. A. (1998) Phytochromes and seed germination. *Seed Sci Res*, 8, 317-329.

Cherry, J. R., Hondred, D., Walker, J. M., Vierstra, R. D. (1992) Phytochrome requires the 6-kDa N-terminal domain for full biological activity. *Proc Natl Acad Sci USA*, 89, 5039-5043.

Christensen, S., Laverne, E., Boyd, G., Silverthorne, J. (2002) *Ginkgo biloba* retains functions of both type I and type II flowering plant phytochrome. *Plant Cell Physiol*,43, 768-777.

Clack, T., Mathews, S., Sharrock, R. A. (1994) The phytochrome apoprotein family in Arabidopsis is encoded by five genes: the sequences and expression of *PHYD* and *PHYE*. *Plant Mol Biol*, 25, 413-427.

Colbert, J. T., Hershey, H. P., Quail, P. H. (1983) Autoregulatory control of translatable phytochrome mRNA levels. *Proc Natl Acad Sci USA*, 80, 2248-2252.

Dehesh, K., Tepperman, J., Christensen, A. H., Quail, P. H. (1991) phyB is evolutionarily conserved and constitutively expressed in rice seedling shoots. *Mol Gen Genet, 225*, 305-313.

Dehesh, K., Franci, C., Sharrock, R. A., Somers, D. E., Welsch, J. A., Quail, P. H. (1994) The Arabidopsis phytochrome A gene has multiple transcription start sites and a promoter sequence motif homologous to the repressor element of monocot phytochrome A genes. *Photochem Photobiol, 59*, 379-384.

Donoghue, M. J. and Doyle, J. A. (2000) Seed plant phylogeny: demise of the anthophyte hypothesis? *Curr Biol, 10*, R106-R109.

Eichenberg, K., Bäurle, I., Paulo, N., Sharrock, R. A., Rüdiger, W., Schäfer, E. (2000) *Arabidopsis* phytochromes C and E have different spectral characteristics from those of phytochromes A and B. *FEBS Lett, 470*, 107-112.

Fernbach, E. and Mohr, H. (1990) Coaction of blue/ultraviolet-A light and light absorbed by phytochrome in controlling growth of pine (*Pinus sylvestris* L.) seedlings. *Planta, 180,* 212-216.

Furuya, M. (1993) Phytochromes: their molecular species, gene families, and functions. *Annu Rev Plant Physiol, 44*, 617-645.

Goosey, L., Palecanda, L., Sharrock, R. A. (1997) Differential patterns of expression of the Arabidopsis *PHYB*, *PHYD*, and *PHYE* phytochrome genes. *Plant Physiol, 115*, 959-969.

Gottman, K. and Schäfer, E. (1982) *In vitro* synthesis of phytochrome apoprotein directed by mRNA from light and dark grown *Avena* seedlings. *Photochem Photobiol, 35*, 521-525.

Gottman, K. and Schäfer, E. (1983) Analysis of phytochrome kinetics in light-grown *Avena sativa* L. seedlings. *Planta, 157*, 392-400.

Hall, A., Kozma-Bognar, L., Toth, R., Nagy, F., Millar, A. J. (2001) Conditional circadian regulation of *PHYTOCHROME A* gene expression. *Plant Physiol, 127*, 1808-1818.

Hanelt, S., Braun, B., Marx, S., Schneider-Poetsch, H. A. W. (1992) Phytochrome evolution: a phylogenetic tree with the first complete sequence of phytochrome from a cryptogamic plant (Selaginella martensii spring). *Photochem Photobiol, 56*, 751-758.

Hauser, B. A., Cordonnier-Pratt, M.-M., Daniel-Vedele, F. Pratt, L. H. (1995) The phytochrome gene family in tomato includes a novel subfamily. *Plant Mol Biol, 29*, 1143-1155.

Hauser, B. A., Pratt, L. H., Cordonnier-Pratt, M. M. (1997) Absolute quantification of five phytochrome transcripts in seedlings and mature plants of tomato (*Solanum lycopersicum* L.). *Planta, 201*, 379-387.

Hauser, B. A., Cordonnier-Pratt, M. M., Pratt, L. H. (1998) Temporal and photoregulated expression of five tomato phytochrome genes. *Plant J, 14*, 431-439.

Hershey, H. P., Colbert, J. T., Lissemore, J. L., Barker, R. F., Quail, P. H. (1984) Molecular cloning of cDNA for *Avena* phytochrome. *Proc Natl Acad Sci USA, 81*, 2332-2336.

Hershey, H. P., Barker, R. F., Idler, K. B., Lissemore, J. L., Quail, P. H. (1985) Analysis of cloned cDNA and genomic sequences for phytochrome: complete amino acid sequences for two gene products expressed in etiolated Avena. *Nucleic Acids Res, 13*, 8543-8559.

Hershey, H. P., Barker, R. F., Idler, K. B., Murray, M. G., Quail, P. H. (1987) Nucleotide sequence and characterization of a gene encoding the phytochrome polypeptide from Avena. *Gene, 61*, 339-348.

Heyer, A. and Gatz, C. (1992) Isolation and characterization of a cDNA-clone coding for potato type B phytochrome. *Plant Mol Biol, 20*, 589-600.

Hillman, W. S. (1967) The physiology of phytochrome. *Annu Rev Plant Physiol, 18*, 301-324.

Hirschfeld, M., Tepperman, J. M., Clack, T., Quail, P. H., Sharrock, R. A. (1998) Coordination of phytochrome levels in *phyB* mutants of Arabidopsis as revealed by apoprotein-specific monoclonal antibodies. *Genetics, 149*, 523-535.

Howe, G. T., Bucciaglia, P. A., Hackett, W. P., Furnier, G. R., Cordonnier-Pratt, M. M., Gardner, G. (1998) Evidence that the phytochrome gene family in black cottonwood has one *PHYA* locus and two *PHYB* loci but lacks members of the *PHYC/F* and *PHYE* subfamilies. *Mol Biol Evol, 15*, 160-175.

Kawai, H., Kanege, T., Christensen, S., Kiyosue, T., Sato, Y., Imaizumi, T., *et al.* (2003) Responses of ferns to red light by an unconventional photoreceptor. *Nature, 421*, 287-290.

Kay, S. A., Keith, B., Shinozaki, K., Chye, M. L., Chua, N. H. (1989) The rice phytochrome gene: structure, autoregulated expression, and binding of GT-1 to a conserved site in the 5' upstream region. *Plant Cell, 1*, 351-360.

Kozma Bognar, L. K., Hall, A., Adam, E., Thain, S. C., Nagy, F., Millar, A. J. (1999) The circadian clock controls the expression pattern of the circadian input photoreceptor, phytochrome B. *Proc Natl Acad Sci USA, 96*, 14652-14657.

Lagarias, D., Wu, S.-H., Lagarias, J. C. (1995) Atypical gene structure in the green alga *Mesotaenium cladariorum*. *Plant Mol Biol*, *29*, 1127-1142.

Lavin, M., Eshbaugh, E., Hu, J.-M., Mathews, S., Sharrock, R. A. (1998) Monophyletic subgroups of the tribe Millettieae (Leguminosae) as revealed by phytochrome nucleotide sequence data. *Am J Bot*, *85*, 412-433.

Mathews, S., Lavin, M., Sharrock, R. A. (1995) Evolution of the phytochrome gene family and its utility for phylogenetic analyses of angiosperms. *Ann Missouri Bot Gard*, *82*, 296-321.

Mathews, S. and Sharrock, R. A. (1996) The phytochrome gene family in grasses (Poaceae): a phylogeny and evidence that grasses have a subset of the loci found in dicot angiosperms. *Mol Biol Evol*, *13*, 1141-1150.

Mathews, S. and Sharrock, R. A. (1997) Phytochrome gene diversity. *Plant Cell Environ*, *20*, 666-671.

Mathews, S. and Donoghue, M. J. (1999) The root of angiosperm phylogeny inferred from duplicate phytochrome genes. *Science*, *286*, 947-950.

Mathews, S. and Donoghue, M. J. (2000) Basal angiosperm phylogeny inferred from duplicate phytochromes A and C. *Int J Plant Sci*, *161*(6 Suppl.): S41-S55.

Mathews, S., Burleigh, J. G., Donoghue, M. J. (2003) Adaptive evolution in the photosensory domain of phytochrome A in early angiosperms. *Mol Biol Evol*, *20*, 1087-1097.

Montgomery, B. L. and Lagarias, J. C. (2002) Phytochrome ancestry: sensors of bilins and light. *Trends Plant Sci*, *7*, 357-366.

Nozue, K., Kanegae, T., Imaizumi, T., Fukuda, S., Okamoto, H., Yeh, K.-C., et al. (1998) A phytochrome from the fern *Adiantum* with features of the putative photoreceptor NPH1. *Proc Natl Acad Sci USA*, *95*, 15286-15830.

Otto, V., Schäfer, E., Nagatani, A., Yamamoto, K. T., Furuya, M. (1984) Phytochrome control of its own synthesis in *Pisum sativum*. *Plant Cell Physiol*, *25*, 1579-1584.

Pasentsis, K., Paulo, N., Algarra, P., Dittrich, P., Thümmler, F. (1998) Characterization and expression of the phytochrome gene family in the moss *Ceratodon purpureus*. *Plant J*, *13*, 51-61.

Pratt, L. H. (1994) "Distribution and localization of phytochrome within the plant." In *Photomorphogenesis in Plants*, Kendrick R. E. and Kronenberg G. H. M (eds) Kluwer Academic Publishers, Dordrecht, The Netherlands, 163-185.

Pratt, L. H. (1995) Phytochromes: differential properties, expression patterns, and molecular evolution. *Photochem Photobiol*, *61*, 10-21.

Quail, P. H. (1994) "Phytochrome genes and their expression." In *Photomorphogenesis in Plants*, Kendrick R. E and Kronenberg G. H. M (eds) Kluwer Academic Publishers, Dordrecht, The Netherlands, 71-103.

Sato, N. (1988) Nucleotide sequence and expression of the phytochrome gene in *Pisum sativum*: differential regulation by light of multiple transcripts. *Plant Mol Biol*, *11*, 697-710.

Schmidt, M. and Schneider-Poetsch, H. A. W. (2002) The evolution of gymnosperms redrawn by phytochrome genes: The Gnetatae appear at the base of the gymnosperms. *J Mol Evol*, *54*, 715-724.

Schneider-Poetsch, H. A. W., Marx, S., Kolukisaoglu, H. Ü., Hanelt, S., Braun, B. (1994) Phytochrome evolution: Phytochrome genes in ferns and mosses. *Physiol Plant*, *91*, 241-250.

Schneider-Poetsch, H. A. W., Kolukisaoglu, Ü., Clapham, D. H., Hughes, J., Lamparter, T. (1998) Non-angiosperm phytochromes and the evolution of vascular plants. *Physiol Plant*, *102*, 612-622.

Schütz, I. and Furuya, M. (2001) Evidence for type II phytochrome-induced rapid signalling leading to *cab*::*luciferase* gene expression in tobacco cotyledons. *Planta*, *212*, 759-764.

Sharrock, R. A. and Quail, P.. H. (1989) Novel phytochrome sequences in *Arabidopsis thaliana*: structure, evolution, and differential expression of a plant regulatory photoreceptor family. *Genes Dev*, 3, 1745-1757.

Sharrock, R. A. and Clack, T. (2002) Patterns of expression and normalized levels of the five Arabidopsis phytochromes. *Plant Physiol*, *130*, 442-456.

Sharrock, R. A. and Clack, T. (2004) Heterodimerization of type II phytochromes in *Arabidopsis*. *Proc Natl Acad Sci USA*, *101*, 11500-11505.

Shimazaki, Y. and Pratt, L. H. (1985) Immunochemical detection with rabbit polyclonal and mouse monoclonal antibodies of different pools of phytochrome from etiolated and green *Avena* shoots. *Planta*, *164*, 333-344.

Somers, D. E. and Quail, P. H. (1995a) Phytochrome-mediated light regulation of *PHYA*- and *PHYB-GUS* transgenes in *Arabidopsis thaliana* seedlings. *Plant Physiol*, *107*, 523-534.

Somers, D. E. and Quail, P. H. (1995b) Temporal and spatial expression patterns of *PHYA* and *PHYB* genes in *Arabidopsis*. *Plant J, 7*, 413-427.
Tokuhisa, J. G., Daniels, S. M., Quail, P. H. (1985) Phytochrome in green tissue: spectral and immunochemical evidence for two distinct molecular species of phytochrome in light-grown *Avena sativa* L. *Planta, 164*, 321-332.
Toole, V. K., Toole, E. H., Hendricks, S. B., Borthwick, H. A. (1961) Responses of seeds of *Pinus virginiana* to light. *Plant Physiol, 36*, 285-290.
Toth, R., Kevei, E. E., Hall, A., Millar, A. J., Nagy, F., Kozma-Bognar, L. (2001) Circadian clock-regulated expression of phytochrome and cryptochrome genes in Arabidopsis. *Plant Physiol, 127*, 1607-1616.
Vierstra, R. D. and Quail, P. H. (1983) Purification and initial characterization of 124-kilodalton phytochrome from *Avena*. *Biochemistry, 22*, 2498-2505.
Vogelmann, T. C. (1994) "Light within the plant." In *Photomorphogenesis in Plants*, Kendrick R. E. and Kronenberg G. H. M (eds) Kluwer Academic Publishers, Dordrecht, The Netherlands, 491-535.
Wagner, D., Tepperman, J. M., Quail, P. H. (1991) Overexpression of phytochrome B induces a short hypocotyl phenotype in transgenic Arabidopsis. *Plant Cell, 3*, 1275-1288.
Wang, Y.-C., Stewart, S. J., Cordonnier, M. M., Pratt, L. (1991) *Avena sativa* L. contains three phytochromes, only one of which is abundant in etiolated tissue. *Planta, 184*, 96-104.
Wang, Y.-C., Cordonnier-Pratt, M. M., Pratt, L. H. (1992) Detection and quantitation of three phytochromes in umimbibed seeds of *Avena sativa* L. *Photochem Photobiol, 56*, 709-716.
Wang, Y.-C., Cordonnier-Pratt, M. M., Pratt, L. H. (1993a) Spatial distribution of three phytochromes in dark- and light-grown *Avena sativa* L. *Planta, 189*, 391-396.
Wang, Y.-C., Cordonnier-Pratt, M. M., Pratt, L. H. (1993b) Temporal and light regulation of the expression of three phytochromes in germinating seeds and young seedlings of *Avena sativa* L. *Planta, 189*, 384-390.
Warrington, I. J., Rook, D. A., Morgan, D. C., Turnbull, H. L. (1988) The influence of simulated shadelight and daylight on growth, development and photosynthesis of *Pinus radiata, Agathis australis* and *Dacrydium cupressinum*. *Plant Cell Environ, 11*, 343-356.
Wester, L., Somers, D. E., Clack, T., Sharrock, R. A. (1994) Transgenic complementation of the *hy3* phytochrome B mutation and response to *PHYB* gene copy number in *Arabidopsis*. *Plant J, 5*, 261-272.
Winands, A. and Wagner, G. (1996) Phytochrome of the green alga Mougeotia: cDNA sequence, autoregulation and phylogenetic position. *Plant Mol Biol, 32*, 589-597.
Yanovsky, M. J., Casal, J. J., Luppi, J. P. (1997) The *VLF* loci, polymorphic between ecotypes of Landsberg *erecta* and Columbia dissect two branches of phytochrome A signalling pathways that correspond to the very-low fluence and high-irradiance responses of phytochrome. *Plant J, 12*, 659-667.
Zanis, M. J., Soltis, D. E., Soltis, P. S., Mathews, S., Donoghue, M. J. (2002) The root of the angiosperms revisited. *Proc Natl Acad Sci USA, 99*, 6848-6853.

Chapter 8

PHYTOCHROME DEGRADATION AND DARK REVERSION

Lars Hennig
Institute of Plant Sciences, ETH Zürich, Universitätstr. 2, 8092 Zürich, Switzerland (e-mail: Lars.Hennig@ipw.biol.ethz.ch)

1. INTRODUCTION

The complex dynamics of phytochromes have puzzled plant biologists for decades (for a history of phytochrome research see Sage *et al.*, 1992). Although techniques such as *in-vivo* spectroscopy, ELISA detection and life monitoring of GFP fusion proteins have revealed fascinating details of light perception and signal transduction, the understanding of phytochrome dynamics in plants is still far from being complete. From the 1950ies to the 1980ies, several ingenious studies have discovered key properties of phytochromes, e.g. light lability, aggregation and dark reversion (Pratt *et al.*, 1974; MacKenzie *et al.*, 1975; Brockmann *et al.*, 1987). The results, however, often were correlative, inconclusive or even contradictory. The advances of molecular genetics during the last decade have enabled experiments under much better controlled conditions, which improved the understanding of phytochrome dynamics considerably. Three achievements were particularly important: (i) It was discovered that phytochrome is a multi gene family. The subsequent identification of mutants in specific phytochrome genes allowed dissecting the function and dynamics of individual phytochromes. (ii) The heterologous expression of phytochromes in yeast and bacteria and their assembly into photoreversible holoproteins enabled detailed kinetic investigations of phytochrome dynamics. And finally (iii) the efficient generation of transgenic plants expressing mutated or truncated phytochromes paved the road for vigorous tests of functional hypothesis. While early research on phytochrome used various model plants including lettuce, squash, cauliflower, oat, mustard and others, recent work followed the raise of *Arabidopsis* as the dominating model plant. Because this overview will focus on the more recent results it will mainly discuss the *Arabidopsis* phytochromes but frequently relate to findings from other species as well.

Activation of the photoreceptor and the subsequent steps transducing the signal towards the biological response has obviously gained most interest of researchers. An equally important question, however, is what mechanisms allow a signal to be terminated once the inducing stimulus ceases. In a signal transduction chain each molecular step has to be accompanied by a deactivating step, and kinase-phosphatase pairs are maybe the most prominent examples of signal activation and termination. In the case of the activated photoreceptor phytochrome, mainly two inactivating mechanisms have been discussed, namely protein degradation ("phytochrome destruction") and dark reversion. The physiological active, far-red light (FR) absorbing Pfr form of many phytochromes has a much shorter half-life than to the biological inactive, red light (R) absorbing Pr form. In etiolated pumpkin seedlings, for instance, the half-life decreases from more than 100 hrs for Pr to less than 1 h for Pfr (Quail *et al.*, 1973a). The loss of photoreversible phytochrome after conversion into the Pfr form was termed "destruction" and subsequently it was found to involve a *bona fide* protein degradation (see 2.2). However, as not all phytochrome is subject to rapid destruction, the analysis of destruction kinetics lead to the operational distinction of type I phytochrome, which is light-labile, and type II phytochrome, which is more stable in the light. Nonetheless, light stable phytochromes need to be deactivated as well. Spectrometric analysis revealed that most type II Pfr is only meta-stable and undergoes a slow thermal reversion back to inactive Pr. Photoreceptor dynamics, therefore, involve the de-novo synthesis of Pr, which can be approximated by zero order kinetics (Schäfer *et al.*, 1975), photoconversion of Pr and Pfr into each other, destruction for type I phytochrome and dark reversion for probably all type II phytochromes (Figure 1). Importantly, for type I phytochrome of several species dark reversion competes with destruction. In addition to the mentioned reactions, phytochrome dynamics include posttranslational modifications of the protein (e.g. phosphorylation, Hunt and Pratt, 1980) and changes of the intracellular distribution (MacKenzie *et al.*, 1975; Kircher *et al.*, 1999; Yamaguchi *et al.*, 1999).

2. PHYTOCHROME DEGRADATION

2.1 Kinetic properties of phytochrome degradation

Light induced degradation was first discovered in 1959 when H. S Hendricks, W. L. Butler and colleagues planned to demonstrate the newly discovered chromoprotein at the Ninth International Botanical Congress in Montreal and transported etiolated seedlings that contained large amounts of phytochrome in the trunk of their car from Beltsville, Maryland to Canada. To their great disappointment Butler and Hendricks failed to detect any phytochrome during the public demonstration and discovered only later that the occasional openings of the trunk during the transport had induced the destruction of phytochrome (Sage, 1992). The initial observation reported by Hendricks *et al.* (1962) was followed by many experiments targeted to reveal the mechanistic details of phytochrome destruction. It soon became clear that in addition to the pool of rapidly degraded phytochrome a small fraction shows much slower decay (Borthwick, 1964). Therefore, kinetic studies gave the first indications of distinct types of phytochrome.

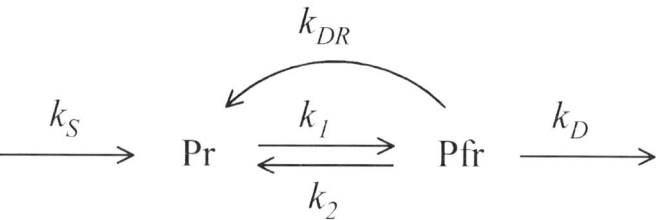

Figure 1. Minimal model of phytochrome dynamics. k_S, synthesis; k_1, k_2, photoconversions; k_{DR}, dark reversion; k_D, degradation. k_1 and k_2 depend on fluence rate and wavelength, for details see Mancinelli (1994).

Partial purification and immunological characterisation of phytochrome from dark or light-grown pea seedlings revealed the existence of two phytochrome species of different primary structure, termed type I and type II phytochrome (Abe *et al.*, 1985). In pea, type I phytochrome proved to make up the phytochrome pool that rapidly disappears after light absorption while type II phytochrome was more stable (Furuya, 1989). Originally the terms type I and type II were defined operationally on strictly immuno-chemical properties of pea phytochrome. Nonetheless, they are now commonly used to differentiate between photolabile (type I) and more stable (type II) phytochrome species.

The cloning of phytochrome genes and the availability of mutants revealed that phytochrome is a small gene family in plants (see Chapter 7) with five members (*PHYA-E*) in *Arabidopsis*, and only phyA-like phytochromes constitute photolabile type I phytochrome (Vierstra, 1994; Clough and Vierstra, 1997). Because transcription of *PHYA* genes is usually much higher in darkness than in the light (see Chapter 7), regulation of the abundance of light-labile phyA occurs not only at the protein level but also at the transcriptional level. This is especially true for monocots while in many dicots including *Arabidopsis* light represses *PHYA* transcription only 2-fold (Quail, 1994). However, because transcriptional regulation is weak in *Arabidopsis*, the dynamics of the photoreceptor *per se* constitute a valid approximation of the dynamic of the complete phytochrome system in this species. As a consequence of transcriptional regulation and phytochrome destruction, type I phytochrome accounts for the majority of phytochrome in etiolated tissue, and etiolated seedlings of various species were used to study phytochrome degradation. In *Arabidopsis* for instance, at least 85% of phytochrome in etiolated seedlings is phyA, and the total level of phyA drops by a factor of 500 in light grown seedlings, in which only 4% of the total phytochrome pool is phyA (Sharrock and Clack, 2002). In *Arabidopsis*, phyB-E together form the pool of type II phytochrome, but also phyB and phyD appear to be degraded upon illumination albeit at a lower rate than phyA (Sharrock and Clack, 2002). The low rate of phyB and phyD loss could also be caused by transcriptional regulation rather than light induced degradation. Regardless of the molecular mechanism involved, many type II phytochromes accumulate to significantly higher protein amounts in the dark than in the light (Stewart et al., 1992; Sharrock and Clack, 2002). Similar observations were made for algal and fern phytochromes that are more similar to the phyB-branch of flowering plants (Morand et al., 1993).

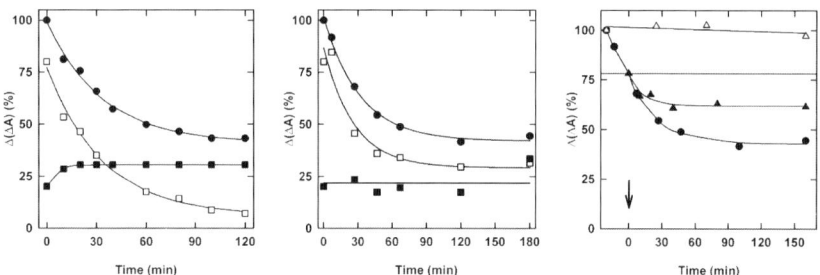

Figure 2. Typical patterns of phytochrome degradation following irradiation with a light pulse. After a pulse of red light, total phytochrome P_{tot} (closed circles) and Pfr (open squares) decrease in mustard (left) and Arabidopsis (middle). Increase of Pr (closed squares) indicates dark reversion and is evident only in mustard. If a pulse of red light is reversed after 15 min (arrow in right panel) by a pulse far-red light (closed triangles), degradation partially continues. A pulse of far-red light alone (open triangles) does not cause destruction. (Redrawn from data in Marmé et al., 1971 and Hennig et al., 1999.)

Destruction kinetic studies showed that degradation of light-labile phytochrome starts immediately after illumination (Figure 2), indicating that the degradation machinery pre-exists in plant cells (Vierstra, 1994). Notably, some studies reported also delays of up to 40 min before the onset of degradation in very young seedlings (Schäfer et al., 1975). In the first years after the discovery of "destruction", several possible mechanisms were taken into consideration, e.g. intra-cellular redistribution. Phytochrome destruction was shown to be temperature-dependent and energy requiring, indicative of an enzymatic process (Butler and Lane, 1964; Schäfer and Schmidt, 1974). A seminal study by Pratt and colleagues (1974), revealed that it is actually the chromoprotein itself that disappears and that phytochrome destruction is based on *bona fide* protein degradation (see section 2.2). This had also been concluded previously from indirect evidence (Quail et al., 1973a). Subsaturating irradiations of tissue revealed a linear dependency of the fraction of phytochrome degraded and the percentage of Pfr established (Butler et al., 1963; Kendrick and Frankland, 1968; Schäfer et al., 1975). Therefore, it was concluded that only phytochrome actually converted to Pfr is subject to degradation. Moreover, phytochrome needs not to be Pfr for destruction because destruction continues even after a reverting FR pulse (Figure 2) (Dooskin and Mancinelli, 1968). Thus, Pr that has been cycled through Pfr is also subjected to destruction (Stone and Pratt, 1979).

Type I phytochrome was observed in monocots and dicots, while phytochromes of lower plants seem to be relatively light stable (Esch and Lamparter, 1998; Morand et al., 1993). Interestingly, the strongly light-labile phyA-like phytochromes evolved only together with the appearance of seed plants (see Chapter 7). In dicots destruction usually shows 1st order kinetics while in monocots zero-order kinetics were measured (Schäfer *et al.*, 1975). From these observations Schäfer and colleagues concluded that in dicots the amount of formed Pfr is limiting the destruction process by a degradation machinery that is present in excess, but that in monocots a limiting amount of degrading activity exists that has a very high affinity for Pfr. Destruction kinetics often depend also on tissue type and organ age (Schäfer, 1978). Destruction can be observed in light grown tissue as well, where it is usually slower than in etiolated plants (Jabben, 1980; Heim *et al.*, 1981).

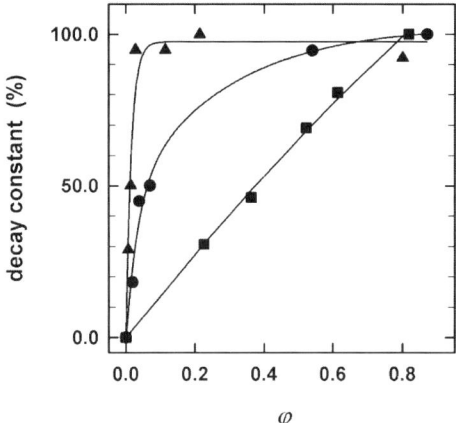

Figure 3. Rate of Ptot destruction as a function of proportion of phytochrome in the Pfr form for Amaranthus (squares), Arabidopsis (circles) and Avena (triangles). (Redrawn from data in Kendrick and Frankland, 1968, Schäfer et al., 1976 and Hennig et al., 1999.)

The minimal kinetic model of phytochrome (Figure 1) predicts a linear relation between the apparent rate of destruction and the photostationary state φ, and the expected linearity was observed in some dicot species (e.g. *Amaranthus*, Kendrick and Frankland, 1968; *Mirabilis*, Kendrick and Hillman, 1971). In contrast, both in oat and *Arabidopsis* a saturation-like dependency was observed (Schäfer *et al.*, 1976; Hennig *et al.*, 1999). Both species are characterised by a much faster degradation in FR than expected (Figure 3). The observed non-linearity implies that other molecular reactions contribute to the observed degradation rate. In particular, it was shown that a simple extension of the minimal model by including a molecular marking step was sufficient to explain the measurements (Figure 4) (e.g. Hennig *et al.*, 1999). Although the specific nature of this postulated marking remains elusive, it could include binding to another protein, posttranslational modification(s) of phytochrome or changing intracellular localisation. Experimental results have shown that phyA is involved in all these processes (see Chapters 4, 9, 19 and section 2.2), but whether they control degradation is largely unknown. The kinetic analysis suggested that it is not Pfr itself that is subject to degradation but Pfr modified by an unknown process (Pfr'). Pfr' appears to be photoreversible and Pr' can be either degraded or transformed back into Pr. This model also predicts that the degradation machinery recognises only the elusive mark but does not differentiate between Pr' and Pfr', thus explaining the observed degradation of 'cycled' Pr (i.e. Pr that was Pfr before). In contrast, the machinery introducing the molecular mark is specific for Pfr and does not recognise Pr. Table 1 lists the rate constants obtained for *Arabidopsis* phyA using the extended model (Figure 4).

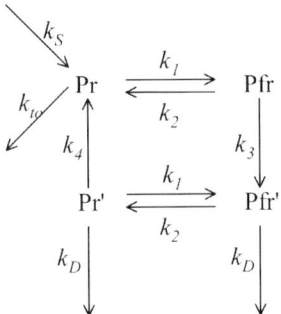

Figure 4. Proposed two-step model of light-induced degradation of phyA. k_S, synthesis; k_1, k_2, photoconversions; k_{to}, turn over of Pr; k_3, molecular tagging; k_4, removal of tag; k_D, degradation.

2.2 Mechanisms of phytochrome degradation

In 1964, Butler and Lane observed with *in vivo* spectroscopy that phytochrome destruction is temperature-dependent and energy requiring (Butler and Lane, 1964). Ten years later, Pratt and colleagues used immunoblots to demonstrate that phytochrome destruction is based on a *bona fide* protein degradation (Pratt *et al.*, 1974). However, the protease specifically degrading Pfr (and cycled Pr) remained elusive. Kinetic studies usually failed to reveal any proteolytic intermediates and the rapid *in vitro* degradation in phytochrome-containing extracts, which accumulate truncated phytochrome polypeptides, was shown to differ from *in vivo* destruction (Furuya and Hillman, 1966). Finally, in 1987 immunoprecipitation with anti-phytochrome antisera followed by probing immunoblots with anti-ubiquitin antisera established that the proteasome-ubiquitin pathway is involved in phytochrome degradation (Shanklin *et al.*, 1987). Specific complexes of ubiquitin-activating enzymes and ubiquitin ligases covalently link the small protein ubiquitin to the ε-amino groups of target proteins. The subsequent formation of poly-ubiquitin chains forms the recognition tag for the 26S proteasome. The proteasome possesses protease activity and degrades the tagged protein completely (see Chapter 18).

Therefore, cleavage intermediates usually cannot be observed. Although polyubiquitylated proteins have a very short half-life, they can be observed after enrichment by immunoprecipitation (Figure 5). Kinetic analysis revealed that the half-life of ubiquitylated phytochrome (ubi-P) is much shorter than that of Pfr indicating that the rate limiting step(s) of phytochrome degradation involve(s) events leading to ubiquitin-conjugation (Jabben et al., 1989a; Jabben et al., 1989b). Importantly, also cycled Pr is ubiquitylated, firmly establishing a close correlation between ubi-P and phytochrome destruction. However, the same authors reported that measured pool sizes and kinetics of ubi-P could only partially account for Pfr destruction. Because there is no evidence for the involvement of another proteolytic system, this kinetic discrepancy is possibly due to the much smaller precision when measuring ubi-P levels compared to Pr and Pfr. Alternatively, additional molecular steps might occur before ubi-P accumulation complicating any kinetic analysis.

Table 1. Rate constants for destruction dynamics of Arabidopsis phyA. Rate constants refer to the model in Figure 4. References are [a] Hennig et al., 2000 and [b] Hennig et al., 1999. n.a., not applicable.

Process	Rate constant	Value	Half-life	Reference
Synthesis	k_S	$5.8 \times 10^{-4}\ \Delta(\Delta A)\ \text{min}^{-1}$	n.a.	[a]
Pr turnover	k_{to}	$5.3 \times 10^{-4}\ \text{min}^{-1}$	22 h	[a]
Degradation in R	n.a.	$0.023\ \text{min}^{-1}$	30 min	[b]
Degradation in FR	n.a.	$0.005\ \text{min}^{-1}$	2 h	[b]
Molecular tagging	k_3	$0.22\ \text{min}^{-1}$	3 min	[a]
Removal of tag	k_4	$0.023\ \text{min}^{-1}$	30 min	[a]
Degradation	k_D	$0.022\ \text{min}^{-1}$	32 min	[a]

Although the involvement of the ubiquitin-system in phytochrome destruction is now well established (Vierstra, 1994; Clough and Vierstra, 1997), most mechanistic aspects remain elusive. In both the Quail and the Vierstra laboratories transgenic plants were constructed expressing reciprocal chimeric phytochromes with the amino terminal and the carboxy terminal domains of phyA and phyB swapped (Wagner et al., 1996; Clough et al., 1999). In both studies phyA/B chimeras were light labile and phyB/A chimeras were light stable, thus demonstrating that the chromophore-bearing amino terminus determines light-stability (Figure 6). Deletion constructs and additional chimeras revealed that a short stretch of 12 amino acids at the carboxy terminus is required for destruction as well. Because the respective stretch can be either a phyA or phyB sequence, this part of the protein does not confer specificity (Clough et al., 1999).

In general, polyubiquitylation of target proteins is achieved by ubiquitin ligases that often are referred to as E2/E3 enzymes. Because no Pfr-specific ubiquitin ligase is known for phyA, another Pfr-specific signal is possibly established first and then read by the unknown E2/E3 complex. Indeed, kinetic simulation of the discussed

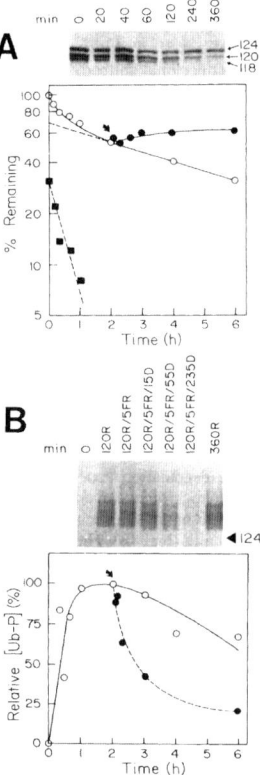

Figure 5. Phytochrome degradation and ubiquitin-phytochrome (Ub-P) accumulation in etiolated tobacco seedlings constitutively expressing oat phyA. Seedlings were irradiated continuously with R (R, open circles) or irradiated with R followed by a FR pulse (FR, closed circles) and further incubation in the dark (D). Top, kinetics of phytochrome degradation. Degradation of tobacco phytochromes is shown for comparison (squares). Bottom, accumulation and loss of oat Ub-P during pyhtochrome degradation. Oat Ub-P were partially purified by immunoprecipitation with an anti-phyA antibody and detected on immunoblots with anti-ubiquitin immunoglobulins. The arrowhead indicates the position of unmodified oat phyA. (From Cherry et al., 1991.)

model of phytochrome dynamics (Figure 4) using experimentally derived rate constants showed that after onset of irradiation a large fraction of phytochrome is expected to be rapidly converted into a tagged state (Hennig et al., 2000). Because ubi-P constitutes only a very small pool of the total light-labile phytochrome at any given time point (Jabben et al., 1989a), it is unlikely that ubiquitin is the tag proposed in Figure 4 that marks phytochrome for degradation Notably, CONSTITUTIVELY PHOTOMORPHOGENIC (COP) 1, a RING motif-containing E3 ligase interacts with the phyA PAS domain, and both the Pr and the Pfr forms of phyA, as well as the PHYA apoprotein, are ubiquitinated by COP1 in vitro (Seo et al., 2004). Moreover, the phyA destruction rate is decreased in cop1 mutants suggesting that COP1 acts as an E3 ligase for phyA degradation. However, light-induced destruction of phyA Pfr occurs even in cop1 mutants, and thus additional E2/E3 ubiquitin ligases for phyA Pfr most likely exist.

A process that kinetically precedes accumulation of ubi-P is the rapid dislocation of dispersed cytosolic phyA into numerous discrete cytosolic areas (sequestered areas of phytochrome, SAP; Mackenzie et al., 1975). SAPs are amorphous structures, about ~1 μm in size not associated with any cellular structures that were observed on electron microscopic pictures and also in vivo using GFP-labelled phytochrome (McCurdy and Pratt, 1986a; Kircher et al., 1999). In Arabidopsis, only phyA Pfr forms SAPs indicating that sequestering is a genuine property of phyA-like phytochromes (Kircher et al., 1999; Kircher et al., 2002; Kim et al., 2000). Sequestering is extremely rapid occurring within seconds after start of irradiation ($t_{1/2}$ < 2min; McCurdy and Pratt, 1986b). Although SAP formation is photoreversible, it takes considerable longer for Pr to return to its uniform distribution ($t_{1/2} \approx 25$ min; Pratt and Marmé, 1976). A sub fraction of phytochrome can be isolated by low centrifugal forces after extraction with buffers containing divalent cations (pelletable phytochrome; Quail et al., 1973b; McCurdy and Pratt, 1986a) but no additional protein compounds could be identified in the preparations (Hofmann et al., 1991). It is commonly assumed that "pelletable phytochrome" is the phytochrome concentrated in SAPs, but this idea has not been firmly established. Furthermore, it is not clear whether "pelletable phytochrome" contains also nuclear phytochrome and therefore consists of diverse sub fractions.

The function of SAPs is unknown, however a possible involvement in phytochrome destruction was suggested (Mackenzie et al., 1975). Indeed, immunolocalization revealed the presence of both phytochrome and ubiquitin in the same electron-dense cytosolic areas, suggesting that SAPs could be sites of ubiquitylation (Speth et al., 1987). To identify molecular determinants of sequestering, a yeast expression system was used to generate photoreversible phytochrome (Kunkel et al., 1993; Kunkel et al., 1995; Eichenberg et al., 1999). Although sequestering was found to be energy-dependent in plants (Quail and Briggs, 1978), it does not require plant-specific factors but occurs spontaneously in yeast (Kunkel et al., 1995). Expression of phyA phyB chimeras (phyA/B) in yeast revealed that both the amino- and the carboxy terminus of phyA are sufficient for sequestering (Eichenberg et al., 1999). While only phyA/B chimeras underwent destruction in plants and phyB/A chimeras were stable (Wagner et al., 1996; Clough *et al.*, 1999), both chromoproteins sequestered in yeast. Because it is possible that SAPs formation is based on different mechanisms in yeast and plants, localisation studies using transgenic plants expressing chimeric phytochromes are needed. Together, the possible role of cytosolic sequestering for phyA destruction still needs rigorous experimental testing.

Because phyA has kinase activity, and phosphorylation is often a signal for ubiquitylation (Vierstra, 1996; Yeh and Lagarias, 1998; Lapko *et al.*, 1999; see also chapter 19 this issue), it was proposed that phosphorylation of phyA actually triggers ubiquitylation and degradation (Otto and Schäfer, 1988; Hennig *et al.*, 2000). Interestingly, methionine 548 of *Arabidopsis* phyA is required both for effective autophosphorylation and destruction (Maloof *et al.*, 2001; see also section 2.3).

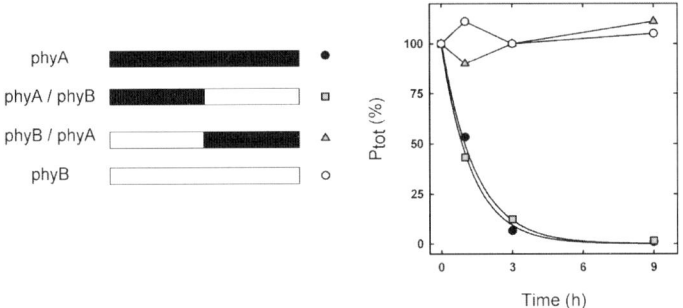

Figure 6. Degradation of wild type and chimeric potato phyA and phyB in tobacco following continuous irradiation. (Redrawn from data in Clough et al., 1999).

Similar to the conclusions drawn from the experiments in yeast, also the use of transgenic plants indicated that degradation of Pfr involves additional plant proteins that are specific for phyA Pfr. In particular, heterologous expression of phyA from various organisms revealed that a phyA from a closely related species is degraded faster than one from a more distantly related species (Cherry *et al.*, 1991). A possible region proposed to be important for Pfr degradation is a PEST domain close to the chromophore-binding domain (residues 323-360 in oat phyA). PEST domains are enriched with the amino acids proline (P), glutamate (E), serine (S), and threonine (T) and were found preferentially in short-lived proteins (Rogers *et al.*, 1986). Consistent with its suggested role in destruction, the PEST domain in phytochrome is more conserved in phyA than in other phytochromes and is more exposed in Pfr than in Pr (Grimm *et al.*, 1988). So far, however, experimental support is lacking for the proposed involvement of the PEST domain in phyA degradation.

Another open issue is the identity of the ubiquitin-attachment site(s). In 1989, Shanklin and colleagues mapped such sites to a conserved lysine rich region between residues 739 and 796 of oat phyA (Shanklin *et al.*, 1989). Subsequent analysis of transgenic plants expressing mutated phyA variants revealed that all lysine residues in this region are dispensable for rapid light-induced phyA degradation (Clough *et al.*, 1999). Chimeric phytochromes showed that the amino terminal half determines protein turnover rates; in contrast, polyubiquitin chains appear to be attached to lysines in the carboxy terminus. Based on these results, the authors question the presence of a few essential and unique ubiquitin attachment sites and suggested that maybe any accessible lysine can function as such a site (Clough *et al.*, 1999). Additional evidence for a role of carboxy terminal sequences in phyA degradation comes from *Arabidopsis* mutants *phyA-106* and *phyA-103*. These plants contain missense mutations in a conserved C-terminal region of phyA (Xu *et al.*, 1995). While phyA-106 is degraded much faster after start of illumination, phyA-103 turns over slower than wild type phyA (Eichenberg *et al.*, 2000b).

2.3 Physiological functions of phytochrome degradation

Phytochromes mediate three classes of responses of plants to light: The very low fluence, low fluence and high irradiance response (VLFR, LFR, HIR; Mancinelli, 1994; Casal *et al.*, 1998). B-type phytochromes mediate the classical R/FR reversible LF responses, and phyA can function both in the VLFR and the HIR mode. Here, I discuss in detail only the classical FR HIR, which is mediated by phyA (see Chapter 2). In addition to its HIR function at high fluence rates, phyA mediates the very sensitive VLFR. The VLFR can be activated by fluences that are in a range of ten orders of magnitude lower than the fluences received during treatments establishing a HIR. The major difference between VLFR and HIR is that the former measures fluences (i.e. total photon numbers) while the latter senses fluence rates (i.e. photons received per time unit). As a prerequisite for this response a differential turnover of Pr and Pfr should serve to maintain differential phyA levels in the dark or in the light. In buried seeds and etiolated seedlings, phyA levels are high, allowing for a very sensitive VLFR. Because phyA levels are much lower in light grown organs, the HIR can function under considerably higher fluences because it is only the fluence rate that is measured in this response. In parallel to destruction, repression of *PHYA* transcription and rapid turnover of *PHYA* mRNA contribute to much lower phyA levels in the light. Thus, the combination of these processes and destruction would serve primarily to extend the dynamic range of the photoreceptor. This tight control of photoreceptor levels is necessary for plants to maintain proper light-responsiveness. Transgenic plants demonstrated that only 2 to 3 fold elevated phyA levels results in a dramatic change in light sensitivity (Cherry *et al.*, 1992). In addition, the signalling from physiological active Pfr needs to be continuously purged to prevent responses in the absence of the inducing light signal and to prepare the system to the next one (Vierstra, 1994). Therefore, destruction also functions in termination of signalling.

If light-induced phyA degradation serves solely to adjust photoreceptor levels to ambient light conditions, destruction could be completely separated from signal transduction. However, several authors proposed that destruction and signal transduction are intimately linked (e.g. Schäfer, 1975), but final experimental proof of this hypothesis is missing so far. Nonetheless, several lines of evidence support a model that destruction is an integral part of phyA function in the HIR. If destruction and HIR could be decoupled, mutants with reduced phyA turnover causing higher steady state levels of the photoreceptor are expected to have an enhanced light sensitivity. In contrast, if destruction and HIR were tightly linked, mutants with enhanced light-stability of phyA are expected to are mainly less sensitive to light. The majority of available evidence supports the second hypothesis. Studies with truncated and chimeric phytochromes or other mutants usually revealed a strict correlation between rapid turnover of a phytochrome protein in the light and its ability to function in the HIR (Wagner *et al.*, 1996; Clough *et al.*, 1999). Furthermore, it is surprising that extensive genetic screens for hypersensitivity did not yield mutants with enhanced protein stability. Obviously, this failure could also be caused by genetic redundancy or the involvement of essential genes causing lethality if defect. In contrast to the prediction that reduced turnover could cause hypersensitivity, destruction of phyA in the *Arabidopsis* accession Lm-2 is greatly impaired and the HIR has a 100fold reduced sensitivity (Maloof *et al.*, 2001). The *PHYA* gene in Lm-2 carries point mutations causing a M548T exchange in the chromo protein. Methionine 548 is located in the hinge region connecting the chromophore bearing amino terminus and the carboxy terminus suggestive for an involvement in interdomain cross talk. When this mutation was introduced into phyB, the mutated protein could complement a *phyB* mutant only at high but not at low fluences of R. This observation indicates similarities in the molecular mechanisms of signal transduction for phyA and phyB. Similar to the natural situation in Lm-2, Casal and colleagues isolated in a genetic screen the *phyA-302* allele, which is more stable under continuous FR but completely lacks the HIR (Yanovsky *et al.*, 2002). Interestingly, the VLFR is normal in *phyA-302* demonstrating that destruction is dispensable for this response. This mutant will probably be a valuable tool to dissect the interaction of destruction and HIR.

Another argument for a direct role of destruction in the HIR came from kinetic studies and modelling: Destruction appeared as an attractive possibility to account for the elusive mechanisms of measuring fluence rates. Analysis of destruction kinetics revealed that the simple model of phytochrome action needs to be extended to account for the experimental data (Hennig *et al.*, 2000). Previously, it was shown that the theoretical properties of such a dynamic and cyclic system resemble several features of the HIR (Schäfer, 1975). In addition to these theoretical considerations, rate constants obtained experimentally in *Arabidopsis* were used to model the dynamics of the extended model and compared to independent data sets (Hennig *et al.*, 2000). Simulated and experimental data for fluence rate response curves, action spectrum and response curves to dichromatic irradiation were very similar for Pfr and cycled Pr. Thus; both molecular species involved in the degradation process could potentially function as mediators of the HIR. Recently, the apparent ability to revert the HIR-inducing effect of FR pulses by R pulses was interpreted to exclude Pfr activity and to support a function of cycled Pr, but the relevance of these findings was questioned because frequent R pulses used to "revert" the HIR reduce phyA levels strongly (Shinomura *et al.*, 2000; Casal *et al.*, 2003).

Because nuclear import of phyA is fluence rate dependent in FR and, therefore, qualifies as an HIR itself, it is likely that cellular transport processes and destruction together contribute to the kinetic properties of the physiological HIR. From an evolutionary point of view, the sole combination of two pre-existing molecular systems, namely the ubiquitin-proteasome machinery and the R-FR reversible phytochromes, which are active in the Pfr form and show light-dependent changes of their sub cellular localization, could have generated a novel light sensing system. This system measures fluence rates rather than fluences, a unique property highly advantageous for plants competing in a natural environment.

3. DARK REVERSION

3.1 Kinetic properties of dark reversion

Degradation of Pfr rapidly decreases the amount of light-labile phytochrome. However, also photo stable phytochrome does not remain in plants in its active Pfr form for very long. It was as early as 1952 when Borthwick and colleagues observed that a single R pulse, which induced germination of lettuce at 20°C, was not sufficient to do so at 30°C (Borthwick *et al.*, 1952). They concluded that Pfr was stable at 20°C but rapidly reverted thermally back to Pr at 30°C and called this process dark reversion. Loss-of-reversibility experiments were very important for the discovery and characterisation of dark reversion. However, such experiments must be interpreted carefully. The escape of a physiological response from photoreversibility is mainly determined by the manifestation of the molecular action of Pfr, thus the kinetic reflects first of all the rate of signalling downstream of Pfr. Loss-of-reversibility curves can give evidently a minimal estimate of the stability of active phytochrome. Active phytochrome as Pfr must be present as long as FR can revert the promoting effect(s) of R.

Direct spectrophotometric measurements of phytochrome content and the Pr : Pfr ratio are obviously much more informative and were possible after the construction of dedicated spectrophotometers, e.g. the RatioSpect (Gross *et al.*, 1984). Dark reversion can most easily be observed for light-stable phytochrome. Because chlorophyll strongly interferes with the detection of phytochrome, measurements were performed in light-grown, naturally chlorophyll-free tissue like cauliflower as well as in plants grown in the presence of antibiotics or the herbicide norflurazon (San9789) to inhibit chlorophyll accumulation (Butler *et al.*, 1959; Butler *et al.*, 1963; Jabben and Deitzer, 1977). For light-labile phytochromes, which show a rapid decrease of Pfr levels after a light pulse, measurements are mostly performed with etiolated seedlings and an increase of Pr rather than the decrease of Pfr is usually considered as the criterion for dark reversion (Figure 2). Recent extensive physiological and spectrophotometric studies revealed that dark reversion is widespread among different species and various phytochromes: It was observed for phytochrome proteins from prokaryotes (Lamparter *et al.*, 2002), mosses (Eichenberg, 1999), ferns (Iino *et al.*, 1997), and spermatophytes (Butler *et al.*, 1963). Comparing results obtained with different species, the following conclusions can be drawn: Light-labile phytochromes of most dicots undergo detectable dark reversion but those from monocots and some dicots (e.g. *Caryophylales*) are thermally stable; nonetheless, they are still subject to destruction (Hillman, 1967). Because plants usually express several *PHY* genes simultaneously, measurements *in planta* are often difficult to interpret. Expression of phytochromes in yeast is one alternative to unequivocally determine the stability of various phytochromes (Kunkel *et al.*, 1995; Remberg *et al.*, 1998; Eichenberg, 1999; Eichenberg *et al.*, 1999; Eichenberg *et al.*, 2000a). In general, light-stable phytochromes have a relatively rapid dark reversion (half life: 6-20 min for phytochromes from *Ceratodon*, tobacco and *Arabidopsis*) while the Pfr form of light-labile phytochromes has a higher thermal stability (half life of 30 min for *Arabidopsis* phyA). This is most obvious in monocots, which usually contain phyA that does not revert in the dark at all (Table 2).

Table 2. *Half lives of Pfr in yeast at 4°C. References are* [a] *Eichenberg et al., 1999,* [b] *Eichenberg et al., 2000 and* [c] *Eichenberg, 1999.*

phytochrome	species	half life	Reference
phyA	rice	> 12 h	(a)
phyA	Arabidopsis	30 min	(b)
phyB	Arabidopsis	< 6 min	(b)
phyC	Arabidopsis	10 min	(b)
phyE	Arabidopsis	10 min	(b)
phyB	tobacco	20 min	(a)
phy2	*Ceratodon*	< 6 min	(c)
phyAB	rice / tobacco	20 min	(a)
phyBA	tobacco / rice	6 min	(a)

In plants, in yeast and *in vitro* dark reversion is strongly temperature dependent, with activation enthalpies of 18-22 kcal/mol for *Arabidopsis* phyA expressed in yeast (Schäfer and Schmidt, 1974; Eichenberg, 1999; Hennig and Schäfer, 2001). Often dark reversion does not affect the entire Pfr pool and the fraction of Pfr that is subject to dark reversion corresponds closely to the predicted amount of PrPfr heterodimers (Brockmann et al, 1987). The hypothesis that dark reversion affects preferentially PfrPr heterodimers was experimentally verified by directly measuring dark reversion of such heterodimers (Hennig and Schäfer, 2001). To this end, a C323S mutant of *Arabidopsis* phyA, which cannot incorporate the chromophore and has a Pr-like tertiary structure, was co-expressed with wild type phyA in yeast. Because only PHYA (C323S) carried an affinity tag, affinity purification yielded spectroscopically inert PHYA (C323S)-PHYA (C323S) homodimers and PHYA (C323S)-phyA heterodimers with a wild type-like difference spectrum. In contrast to the limited dark reversion of phyA, PrPfr heterodimers underwent rapid and complete dark reversion (Hennig and Schäfer, 2001).

Because phytochrome is a dimer *in vivo* and *in vitro* and the phototransformations between Pr and Pfr are not co-operative (i.e. both subunits reach a photoequilibrium independent of each other), the fraction of PfrPfr homodimers can be estimated as $\phi \times \phi = \phi^2$ and the fraction of PrPfr heterodimers as $2 \times (\phi \times (1 - \phi)) = 2\phi - 2\phi^2$, where ϕ is the photostationary state Pfr/(Pr+Pfr) (vanDerWoude, 1985; Jones and Quail, 1986; Brockmann *et al.*, 1987). Thus, a saturating R pulse is expected to establish about 64-68% PfrPfr homodimers and 27-32% PrPfr heterodimers. Indeed, about 30% phytochrome was observed to undergo dark reversion in several species (e.g. Brockmann *et al.*,1987).

$$\text{Pfr Pfr} \xrightarrow{k_{DR,1}} \text{Pfr Pr} \xrightarrow{k_{DR,2}} \text{Pr Pr}$$

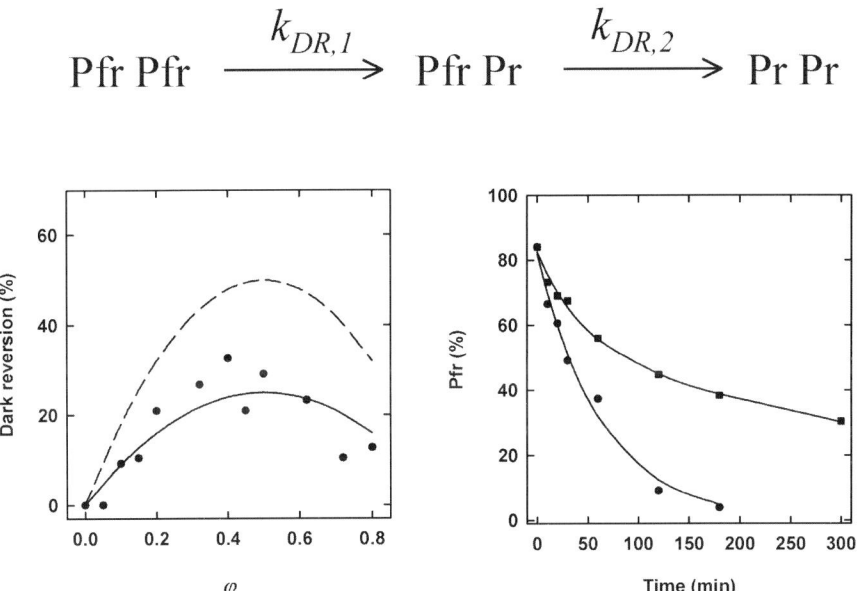

Figure 7. Differential dark reversion of PfrPr heterodimers and PfrPfr homodimers. $k_{DR,1}$, dark reversion rate of PfrPfr; $k_{DR,2}$, dark reversion rate of PfrPr. Left, amount of dark revertible mustard phytochrome in vivo as a function of the proportion of phytochrome in the Pfr form. Expected fractions of PrPfr heterodimers (dashed line) and of Pfr in PrPfr heterodimers (solid line) are shown for comparison. Right, dark reversion of Arabidopsis phyA (squares) and Arabidopsis PHYA(C323S)-phyA heterodimers (circles) in vitro. (Redrawn from data in Brockmann et al., 1987 and Hennig and Schäfer, 2001).

Often more than 30% of Pfr revert back to Pr. Therefore, the initial model of Brockmann was modified to include both a fast dark reversion of PfrPr heterodimers and a slower reversion of PfrPfr homodimers (Figure 7; Hennig and Schäfer, 2001).

Non-linear regression of experimental data showed that the model explains the experimental observations very well. The rate constant k_{DR2} was very similar to the directly measured rate of dark reversion for PfrPr heterodimers confirming the role of dimerization for dark reversion. From the two estimated dark reversion rate constants (k_{DR1} for PfrPfr and k_{DR2} for PfrPr) a ratio $R = k_{DR1} : k_{DR2}$ can be calculated that describes how much the two reaction rates differ, i.e. how much additional thermal stability characterises the PfrPfr homodimer. Table 3 lists these rate constants and their ratio for several phytochromes. In yeast, the ratio $k_{DR1} : k_{DR2}$ was lowest for *Arabidopsis* phyA ($R = 18$) and largest for tobacco phyB ($R > 270$). This suggests that evolution preferentially stabilised PfrPr heterodimers of phyA.

Table 3. Half lives of *PfrPfr* homodimers and *PfrPr* heterodimers generated in yeast. References are [a] Eichenberg et al., 1999, [b] Eichenberg et al., 2000 and [c] Hennig and Schäfer, 2001. n.a., not applicable; n.d. not determined.

Phytochrome	species	T	half life of PfrPfr	half life of PfrPr	Ratio	Reference
phyA	rice	4°C	> 12 h	> 12 h	n.d.	(a)
phyA	Arabidopsis	4°C	5.7 h	19 min	18	(b)
phyB	tobacco	4°C	4.5 h	1 min	270	(a)
phyB	Arabidopsis	4°C	> 12 h	3 min	>240	(b)
phyC	Arabidopsis	4°C	3.6 h	5 min	43	(b)
phyE	Arabidopsis	4°C	>12 h	10 min	>72	(b)
phyA	Arabidopsis	25°C	5.7 h	30 min	11	(c)
Heterodimers	Arabidopsis	25°C	n.a.	44 min	n.a.	(c)

3.2 Determinants of dark reversion

Available data suggest that there are two major determinants of phytochrome dark reversion, namely factors intrinsic to the chromoprotein *per se* and external components. Pfr of oat phyA, which does not revert *in planta*, is also very stable when expressed in yeast (Kunkel *et al.*, 1995). In contrast, *Arabidopsis* phyB shows dark reversion both *in planta* and in yeast (Eichenberg, 1999; Eichenberg *et al.*, 2000a). Therefore, the major determinants for dark reversion are contained in the phytochrome chromo proteins themselves, and plant-specific components are neither required to prevent nor to enable dark reversion. All chimeric phytochromes between non-reverting oat phyA and reverting tobacco phyB expressed in yeast showed dark reversion, demonstrating that determinants throughout the entire primary sequence of oat phyA are required to prevent dark reversion (Figure 8; Eichenberg *et al.*, 1999). So far, it remains elusive which aminoacid motifs confer this additional stability of oat phyA. In the case of phyB, however, two residues are known that strongly control Pfr stability: PhyB-101 (E812K) carries the missense mutation in a part of the carboxy terminus required for signal transduction and is strongly destabilised in its Pfr form (Elich and Chory, 1997). The mutation in phyB-401 (G564E) is located in the amino terminal domain close to the central hinge region and confers extra stability of Pfr suppressing dark reversion (Eichenberg, 1999; Kretsch *et al.*, 2000). These data support the idea that residues in both the amino and carboxy terminus contribute to Pfr stability. Finally, the much faster rate of dark reversion of PfrPr heterodimers compared with PfrPfr homodimers demonstrates that inter-subunit interactions contribute considerably to Pfr stability (Hennig and Schäfer, 2001).

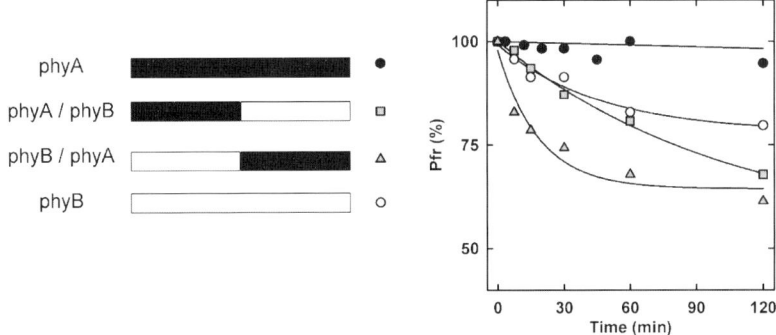

Figure 8. Dark reversion of rice phyA, tobacco phyB and their chimeric fusion products in yeast. (Redrawn from data in Eichenberg et al., 1999).

In vivo, various cellular components appear to modulate dark reversion. Phytochromes from *Cucurbita* and pea, for instance, revert thermally *in vivo* but not *in vitro* suggesting the action of an accelerator of dark reversion in plants (Vierstra and Quail, 1985). Similar to these observations *in planta*, *Arabidopsis* phyA expressed in yeast showed much faster dark reversion in living yeast cells than in extracts (Hennig and Schäfer, 2001). It is, therefore, possible that some general factor not specific to plants accelerates dark reversion *in vivo*. It is tempting to speculate whether a member of the large family of protein chaperones and folding catalysts is involved. In addition to such enhancers, there is evidence for inhibitors of dark reversion in plants as well (Manabe and Furuya, 1971; Shimazaki and Furuya, 1975). In *Arabidopsis*, dark reversion of phyA cannot be detected in most accessions including Columbia, Landsberg *erecta* and Nossen (Hennig *et al.*, 1999; Eichenberg *et al.*, 2000b). Interestingly, phyA has the same aminoacid sequence in the accessions RLD and Col, but dark reversion could be observed directly in RLD but not in Col. Therefore, cellular control of phytochrome dynamics differs between *Arabidopsis* accessions, and the general assumptions that phytochromes from monocots and *Caryophylales* do not show dark reversion while those from the remaining dicots do is probably an over-simplification (Eichenberg *et al.*, 2000b).

3.3. Functional aspects of dark reversion

Dark reversion reduces the level of Pfr, the active form of phytochrome. Therefore, altered dark reversion kinetics are expected to confer hypo- or hypersensitivity to plants. Confirming this hypothesis, phyB-101 (E812K) is a phytochrome with normal chromophore incorporation and unperturbed difference spectrum but strongly destabilised Pfr and reduced light sensitivity (Elich and Chory, 1997). While the control of seedling development in *phyB-101* by phyB is only partially reduced in constant light, end-of-day FR treatments are completely ineffective, supporting the hypothesis that enhanced dark reversion of phyB is the main reason for reduced light sensitivity in *phyB-101*. A *phyB* mutant with an opposite effect to *phyB-101* was identified in another genetic screen (Kretsch *et al.*, 2000). This screen was designed to find mutants in which an inducing R pulse cannot be reverted by subsequent FR. The identified *ohr* mutant (ohne Revertierung = without reversion) was found to be a *PHYB* allele (*phyB-401*, G564E). Like for phyB-101, difference spectra, photoconversion rates and protein stability were similar for phyB-401 and wild type phyB. In contrast to phyB-101, however, Pfr of phyB-401 was strongly stabilised displaying much slower dark reversion. The reduced dark reversion explains reduced petiole length and early flowering of plants grown in short days as well as enhanced light sensitivity of seedlings to very low photon fluences of red light. However, slower dark reversion alone cannot account for loss or reversibility suggesting that in *phyB-401* also interactions with components of the signalling cascade are altered.

Often, protein stability and difference spectra of mutated phytochromes are used to identify residues or sequence motifs that are potentially involved in signal transfer to downstream targets of the signalling cascade. The mutants mentioned above demonstrate that photoreceptor dynamics need to be tested as well, and the same conclusion was derived from a study using a series of *phyA* mutants (Eichenberg *et al.*, 2000b). In this case, mutants had altered destruction or were impaired in dark reversion. However, photoreceptor dynamics and signalling activity are probably tightly linked in plants. Energetic considerations, for instance, suggest that binding of an interaction partner to Pfr can stabilise phytochrome in its active form and reduce dark reversion (Eichenberg, 1999). Indeed, the *Arabidopsis* response regulator ARR4 binds to phyB *in vitro* and *in vivo* and reduces dark reversion rates of phyB in yeast and *in planta* (Sweere *et al.*, 2001). This interaction is specific for phyB and involves the amino terminal 173 aminoacids of the photoreceptor (Figure 9). Importantly, overexpression of ARR4 does not only stabilise phyB Pfr *in planta* but confers also enhanced light sensitivity (Sweere *et al.*, 2001).

PhyB-like phytochromes are characterised by a much higher thermal stability of PfrPfr homodimers than PfrPr heterodimers (i.e. large ratios $k_{DR1} : k_{DR2}$; Hennig and Schäfer, 2001). Interestingly, phyB was suggested to be active mainly as PfrPfr homodimer in the LFR (vanDerWoude, 1985; Furuya and Schäfer, 1996), and light treatments that establish a large fraction of PfrPr heterodimers can revert phyB photoresponses (Poppe *et al.*, 1998). PhyA-like phytochromes are considered evolutionary advanced and act also as PfrPr heterodimers in VLFR and HIR (vanDerWoude, 1985; Furuya and Schäfer, 1996). Thus, it appears that during evolution of phyA the acquisition of novel modes of action involving PfrPr activity was intimately coupled to the thermal stabilisation of such PfrPr heterodimers. This stabilisation was most pronounced in grasses, which completely lack dark reversion of phyA, but can also be observed for *Arabidopsis* phyA.

The ability of plants to both accelerate and reduce dark reversion is probably extremely important in natural environments. Plants as sessile organisms are often subject to a vast range of environmental conditions they cannot simply escape from. In particular the strong temperature dependence of dark reversion rates in yeast suggests that some kind of buffering mechanism(s) exist in plants to prevent strongly different light sensitivities under different temperatures (Hennig and Schäfer, 2001).

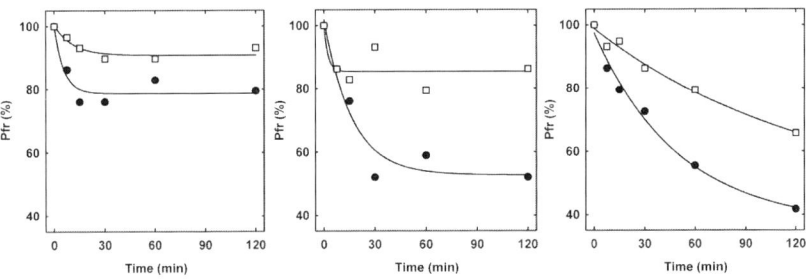

Figure 9. Arabidopsis response regulator ARR4 stabilizes Pfr of wild type phyB (left) and Pfr of phyB-101 (middle) in yeast. To measure phyB dark reversion in planta, the phyB overexpressing line ABO was used (right). Symbols are: circles, phyB alone, squares, phyB and overexpression of ARR4. (Redrawn from data in Sweere et al., 2001).

4. CONCLUDING REMARKS

Together, it becomes evident that photoreceptor dynamics are an intrinsic and essential component of phytochrome signal transduction. Evolution seems to have modified phytochrome dark reversion and destruction to establish novel modes of action including HIR and VLFR. Phytochrome dynamics provide a large evolutionary potential to adapt light sensitivity and signalling to various environmental conditions, and differences can be observed even between different accessions of the same species (Eichenberg *et al.*, 2000b). These observations urge the need for careful characterisation of photoreceptor dynamics in mutants to prevent premature or misleading conclusions about the consequences of the molecular defect(s) introduced. It is expected that further genetic and biochemical studies will soon reveal components of the machinery that recognises phyA Pfr and targets it for destruction. Similarly, the identification of additional modifiers of dark reversion will provide a better understanding of phytochrome signal transduction.

5. FURTHER READING

Clough R. C. and Vierstra R. D. (1997). Phytochrome degradation. *Plant Cell Environ, 20.* 713-721.
Deng, X.-W. (2004) Regulated proteolysis: the role of genes controlling constitutive photomorphogenesis. Chapter 18.
Mancinelli A. L. (1994). "The physiology of phytochrome action." In *Photomorphogenesis in Plants*, Kendrick R. E. and Kronenberg G. H. M. (eds) Kluwer Academic Publishers, Dordrecht, 211-270.
Schäfer, E. and Nagy, F. (2004) Physiological basis of photomorphogenesis, Chapter 2.
Sharrock, R. (2004) Phytochrome genes in higher plants and their expression. Chapter 7.
Vierstra R.D. (1994). "Phytochrome degradation." In *Photomorphogenesis in Plants*, Kendrick R. E. and Kronenberg G. H. M. (eds) Kluwer Academic Publishers, Dordrecht, 141-160.
Vierstra, R. D. (1996) Proteolysis in plants: mechanisms and functions. Plant Mol Biol, *32*, 275-302.

6. REFERENCES

Abe, H., Yamamoto, K. T., Nagatani, A., Furuya, M. (1985) Characterization of green tissue-specific phytochrome isolated immunochemically from pea seedlings. *Plant Cell Physiol, 26,*
Borthwick, H. A., Hendricks, S. B., Parker, M. W., Toole, E. H., Toole, V. K. (1952) A reversible photoreaction controlling seed germination. *Proc Natl Acad Sci USA, 38,* 662-666.
Borthwick, H. A., Toole, E. H., Toole, V. K. (1964) Phytochrome control of *Paulownia* seed germination. *Isr J Bot, 13,* 122-133.
Brockmann, J., Rieble, S., Kazarinova-Fukshansky, N., Seyfried, M., Schäfer, E. (1987) Phytochrome behaves as a dimer *in vivo*. *Plant Cell Environ, 10,* 105-111.
Butler, W. L., Norris, K .H., Siegelmann, H. W., Hendricks, S. B. (1959) Detection, assay and preliminary purification of the pigment controlling photoresponsive development of plants. *Proc Natl Acad Sci USA, 45,* 1703-1708.
Butler, W. L., Lane, H. C., Siegelman, H. W. (1963) Nonphotochemical transformations of phytochrome in vivo.. *Plant Physiol, 38,* 514-519.
Butler, W. L. and Lane, H. C. (1964) Dark transformations of phytochrome *in vivo*. II. *Plant Physiol, 40,* 13-17.
Casal, J. J., Sánchez, R. A., Botto, J. F. (1998) Modes of action of phytochromes. *J Exp Bot, 49,* 127-138.
Casal, J. J., Luccioni, L. G., Oliverio, K. A., Boccalandro, H. E. (2003) Light, phytochrome signalling and photomorphogenesis in *Arabidopsis*. *Photochem Photobiol Sci, 2,* 625-636.
Cherry, J. R., Hershey, H. P., Vierstra, R. D. (1991). Characterization of tobacco expressing functional oat phytochrome. *Plant Physiol, 96,* 775-785.
Cherry, J. R., Hondred, D., Walker, J. M., Vierstra, R. D. (1992). Phytochrome requires the 6-kDa N-terminal domain for full biological activity. *Proc Natl Acad Sci USA, 89,* 5039-5043.
Clough, R. C., Vierstra, R. D. (1997) Phytochrome degradation. *Plant Cell Environ, 20,* 713-721.
Clough, R. C., Jordan-Beebe, E. T., Lohman, K. N., Marita, J. M., Walker, J. M., Gatz, C., Vierstra, R. D. (1999) Sequences within both the N- and C-terminal domains of phytochrome A are required for PFR ubiquitination and degradation. *Plant J, 17,* 155-167.
Dooskin, R. H., Mancinelli, A. L. (1968) Phytochrome decay and coleoptile elongation in *Avena* following various light treatments. *Bull Torrey Bot Club, 95,* 474-487.
Eichenberg, K. *Die Bedeutung der dynamischen Moleküleigenschaften von Phytochrom für die Lichtsignaltransduktion.* Universität Freiburg (ed), PhD thesis (1999).
Eichenberg, K., Kunkel. T., Kretsch, T., Speth, V., Schäfer, E. (1999) *In vivo* characterisation of chimeric phytochromes in yeast. *J Biol Chem, 274,* 354-359.

Eichenberg, K., Bäurle, I., Paulo, N., Sharrock, R. A., Rüdiger, W., Schäfer, E. (2000a) *Arabidopsis* phytochromes C and E have different spectral characteristics from those of phytochromes A and B. *FEBS Lett, 470*, 107-112.
Eichenberg, K., Hennig, L., Martin, A., Schäfer, E. (2000b) Variation in dynamics of phytochrome A in *Arabidopsis* ecotypes and mutants. *Plant Cell Environ, 23*, 311-319.
Elich, T. D. and Chory, J. (1997) Biochemical characterization of *Arabidopsis* wild-type and mutant phytochrome B holoproteins. *Plant Cell, 9*, 2271-2280.
Esch, H. and Lamparter, T. (1998) Light regulation of phytochrome content in wild-type and aphototropic mutants of the moss *Ceratodon purpureus*. *Photochem Photobiol, 67*, 450-455.
Furuya, M. and Hillman, W. S. (1966) Rapid destruction of the Pfr form of phytochrome by a substance in extracts of *Pisum* tissue. *Plant Physiol, 41*, 1242-1244.
Furuya, M. (1989). Molecular properties and biogenesis of phytochrome I and II. *Arch Biophys, 25*, 133-137.
Furuya, M. and Schäfer, E. (1996) Photoperception and signalling of induction reactions by different phytochromes. *Trends Plant Sci, 1*, 301-307.
Grimm, R., Eckerskorn, C., Lottspeich, F., Zenger, C., Rüdiger, W. (1988) Sequence analysis of proteolytic fragments of 124-kilodalton phytochrome from etiolated *Avena sativa* L.: conclusions on the conformation of the native protein. *Planta, 174*, 747-803.
Gross, J., Seyfried, M., Fukshansky, L., Schäfer, E. (1984) "*In vivo* spectrophotometry." In: *Techniques in Photomorphogenesis*, Smith H. and Holmes M. G. (eds): Academic Press, London, 131-157.
Heim, B., Jabben, M.,Schäfer, E. (1981) Phytochrome destruction in dark- and light-grown *Amaranthus caudatus* seedlings. *Photochem Photobiol, 34*, 89-93.
Hendricks, S. B., Butler, W. L., Siegelman, H. W. (1962) A reversible photoreaction regulating plant growth. *J Phys Chem, 66*, 2550-2555.
Hennig, L., Büche, C., Eichenberg, K., Schäfer, E. (1999) Dynamic properties of endogenous phytochrome A in *Arabidopsis* seedlings. *Plant Physiol, 121*, 571-578.
Hennig, L., Büche, C., Schäfer, E. (2000) Degradation of phytochrome A and the high irradiance response in *Arabidopsis*: a kinetic analysis. *Plant Cell Environ, 23*, 727-734.
Hennig, L. and Schäfer, E. (2001) Both subunits of the dimeric plant photoreceptor phytochrome require chromophore for stability of the far-red light-absorbing form. *J Biol Chem, 276*, 7913-7918.
Hillman, W. S. (1967). The physiology of phytochrome. *Ann Rev Plant Physiol, 18*, 301-324.
Hofmann, E., Grimm, R., Harter, K., Speth, V., Schäfer, E. (1991) Partial purification of sequestered particles of phytochrome from oat (*Avena sativa* L.) seedlings. *Planta, 183*, 265-273.
Hunt, R. E., Pratt, L. H. (1980) Partial characterization of undegraded oat phytochrome. *Biochemistry, 19*, 390-394.
Iino, M., Shitanishi, K., Wada, M. (1997) Phytochrome-mediated phototropism in *Adiantum* protonemata: 2. Participation of phytochrome dark reversion. *Photochem Photobiol, 65*, 1032-1038.
Jabben, M., Deitzer, G. (1977) Spectrophotometric measurements of phytochrome in light-grown plants. *Plant Physiol, 59*, suppl. 100.
Jabben, M. (1980) The phytochrome system in light-grown *Zea mays* L.. *Planta, 149*, 91-96.
Jabben, M., Shanklin, J., Vierstra, R. D. (1989a) Red light-induced accumulation of ubiquitin-phytochrome conjugates in both monocots and dicots. *Plant Physiol, 90*, 380-384.
Jabben, M., Shanklin, J., Vierstra, R. D. (1989b) Ubiquitin-phytochrome conjugates. *J Biol Chem, 264*, 4998-5005.
Jones, A. M. and Quail, P. H. (1986) Quaternary structure of 124kDa phytochrome from *Avena sativa* L.. *Biochemistry, 25*, 2987-2995.
Kendrick, R. E. and Frankland, B. (1968) Kinetics of phytochrome decay in *Amaranthus* seedlings. *Planta, 82*, 317-320.
Kendrick, R. E. and Hillman, W. S. (1970) Dark reversion of phytochrome in *Sinapis alba* L.. *Plant Physiol, 46*, 596-598.
Kim, L., Kircher, S., Toth, R., Adam, E., Schäfer, E., Nagy, F. (2000) Light-induced nuclear import of phytochrome-A:GFP fusion proteins is differentially regulated in transgenic tobacco and *Arabidopsis. Plant J, 22*, 125-133.
Kircher, S., Kozma-Bognar, L., Kim, L., Adam, E., Harter, K., Schäfer, E., Nagy, F. (1999) Light quality-dependent nuclear import of the plant photoreceptors phytochrome A and B. *Plant Cell, 11*, 1445-1456.

Kircher, S., Gil, P., Kozma-Bognar, L., Fejes, E., Speth, V., Husselstein-Muller, T. et al. (2002) Nucleocytoplasmic partitioning of the plant photoreceptors phytochrome A, B, C, D, and E is regulated differentially by light and exhibits a diurnal rhythm. *Plant Cell, 14*, 1541-1555.

Kretsch, T., Poppe, C., Schäfer, E. (2000) A new type of mutation in the plant photoreceptor phytochrome B causes loss of photoreversibility and an extremely enhanced light sensitivity. *Plant J, 22*, 177-186.

Kunkel, T., Tomizawa, K.-I., Kern, R., Furuya, M., Chua, N. H., Schäfer, E. (1993) *In vitro* formation of a photoreversible adduct of phycocyanobilin and tobacco apophytochrome B. *Eur J Biochem, 215*, 587-594.

Kunkel, T., Speth, V., Büche, C., Schäfer, E. (1995) *In vivo* characterization of phytochrome-phycocyanobilin adducts in yeast. *J Biol Chem, 270*, 20193-20200.

Lamparter, T., Michael, N., Mittmann, F., Esteban, B. (2002) Phytochrome from *Agrobacterium tumefaciens* has unusual spectral properties and reveals an N-terminal chromophore attachment site. *Proc Natl Acad Sci USA, 99*, 11628-11633.

Lapko, V. N., Jiang, X. Y., Smith, D. L., Song, P. S. (1999) Mass spectrometric characterization of oat phytochrome A: Isoforms and posttranslational modifications. *Protein Sci, 8*, 1032-1044.

MacKenzie, J. M., Coleman, R. A., Briggs, W. R., Pratt, L. H. (1975) Reversible redistribution of phytochrome within the cell upon conversion to its physiological active form. *Proc Natl Acad Sci USA, 72*, 799-803.

Maloof, J. N., Borevitz, J. O., Dabi, T., Lutes, J., Nehring, R. B., Redfern, J. L. et al. (2001) Natural variation in light sensitivity of *Arabidopsis*. *Nat Genet, 29*, 441-446.

Manabe, K. and Furuya, M. (1971) Factors controlling rates of nonphotochemical transformation of *Pisum* phytochrome in vitro. *Plant Cell Physio., 12*, 95-101.

Mancinelli, A. L. (1994) "The physiology of phytochrome action." In: *Photomorphogenesis in Plants*, Kendrick R. E. and Kronenberg G. H. M (eds) Kluwer Academic Publishers, Dordrecht, 211-270.

Marmé, D., Marchal, B., Schäfer, E. (1971) A detailed analysis of phytochrome decay and dark reversion in mustard cotyledons. *Planta, 100*, 331-336.

McCurdy, D. W. and Pratt, L. H. (1986a) Kinetics of intracellular redistribution of phytochrome in *Avena* coleoptiles after its photoconversion to the active far-red-absorbing form. *Planta, 167*, 330-336.

McCurdy, D. W. and Pratt, L. H. (1986b) Immunological electron microscopy of phytochrome in *Avena*: identification of intracellular sites responsible for phytochrome sequestering and pelletability. *J Cell Biol, 103*, 2541-2550.

Morand, L. Z., Kidd, D. G., Lagarias, J. C. (1993) Phytochrome levels in the green alga *Mesotaenium caldariorum* are light regulated. *Plant Physiol, 101*, 97-104.

Otto, V. and Schäfer, E. (1988) Rapid phytochrome-controlled protein phosphorylation and dephosphorylation in *Avena sativa* L.. *Plant Cell Physiol, 29*, 1115-1121.

Poppe, C., Sweere, U., Drumm-Herrel, H., Schäfer, E. (1998) The blue light receptor cryptochrome 1 can act independently of phytochrome A and B in *Arabidopsis thaliana*. *Plant J, 16*, 465-471.

Pratt, L. H., Kidd, G. H., Coleman, R. A. (1974) An immunochemical characterization of the phytochrome destruction reaction. *Biochim Biophys Acta, 365*, 93-107.

Pratt, L. H. and Marmé, D. (1976) Red-light enhanced phytochrome pelletability: re-examination and further characterization. *Plant Physiol, 58*, 686-692.

Quail, P. H., Marmé, D., Schäfer, E. (1973a) Particle-bound phytochrome from maize and pumpkin. *Nat New Biol, 245*, 189-191.

Quail, P. H., Schäfer, E., Marmé, D. (1973b) Turnover of phytochrome in pumpkin cotyledons. *Plant Physiol, 52*, 128-131.

Quail, P. H. and Briggs, W. R. (1978) Irradiation-enhanced phytochrome pelletability: requirement for phosphorylation energy *in vivo*. *Plant Physiol, 62*, 773-778.

Quail, P. H. (1994) "Phytochrome genes and their expression." In: *Photomorphogenesis in Plants*, Kendrick R. E. and Kronenberg G. H. M. (eds), Kluwer Academic Publishers, Dordrecht, 71-104.

Remberg, A., Ruddat, A., Braslavsky, S. E., Gärtner, W., Schaffner, K. (1998) Chromophore incorporation, Pr to Pfr kinetics, and Pfr thermal reversion of recombinant N-terminal fragments of phytochrome A and B chromoproteins. *Biochemistry, 37*, 9983-9990.

Rogers, S., Wells, R., Rechsteiner, M. (1986) Amino acid sequences common to rapidly degraded proteins: the PEST hypothesis. *Science, 234*, 364-368.

Sage, L. C., *Pigment of the Imagination: History of Phytochrome Research*. New York: Academic Press, 1992.

Seo, H. S., Watanabe, E., Tokutomi, S., Nagatani, A., Chua, N. H. (2004) Photoreceptor ubiquitination by COP1 E3 ligase desensitises phytochrome A signalling. *Genes Dev, 18*, 617-622.

Schäfer, E. and Schmidt, W. (1974) Temperature dependence of phytochrome dark reactions. *Planta, 116*, 257-266.

Schäfer, E. (1975) A new approach to explain the "High irradiance responses" of photomorphogenesis on the basis of phytochrome. *J Math Biol, 2*, 41-56.

Schäfer, E., Lassig, T.-U., Schopfer, P. (1975) Photocontrol of phytochrome destruction in grass seedlings. The influence of wavelength and irradiance. *Photochem Photobiol, 22*, 193-202.

Schäfer, E., Lassig, T.-U., Schopfer, P. (1976) Photocontrol of phytochrome destruction and binding in dicotyledonous versus monocotyledonous seedlings: The influence of wavelength and irradiance. *Photochem Photobiol, 24*, 567-572.

Schäfer, E. (1978) Variation in the rates of synthesis and degradation of phytochrome in cotyledons of *Cucurbita pepo* L. during seedling development. *Photochem Photobiol, 27*, 775-780.

Shanklin, J., Jabben, M., Vierstra, R. D. (1987) Red light induced formation of ubiquitin-phytochrome conjugates: identification of possible intermediates of phytochrome degradation. *Proc Natl Acad Sci USA, 84*, 359-363.

Shanklin, J., Jabben, M., Vierstra R. D. (1989) Partial purification and peptide mapping of ubiquitin phytochrome conjugates from oat. *Biochemistry, 28*, 6028-6034.

Sharrock, R. A. and Clack, T. (2002) Patterns of expression and normalized levels of the five *Arabidopsis* phytochromes. *Plant Physiol, 130*, 442-456.

Shimazaki, Y. and Furuya, M. (1975) Isolation of a naturally occurring inhibitor for dark Pfr reversion from etiolated *Pisum* epicotyls. *Plant Cell Physiol, 16*, 623-630.

Shinomura, T., Uchida, K., Furuya, M. (2000) Elementary processes of photoperception by phytochrome A for high-irradiance response of hypocotyl elongation in *Arabidopsis*. *Plant Physiol, 122*, 147-156.

Stewart, S., Pratt, L., Cordonnier-Pratt, M.-M. (1992) Phytochrome levels in light-grown *Avena* change in response to end-of-day irradiations. *Plant Physiol, 99*, 1708-1710.

Stone, H. J., Pratt, L. H. (1979) Characterization of the destruction of phytochrome in the red absorbing form. *Plant Physiol, 63*, 680-682.

Sweere, U., Eichenberg, K., Lohrmann, J., Mira-Rodado, V., Bäurle, I., Kudla, J. *et al*. (2001) Interaction of the response regulator ARR4 with phytochrome B in modulating red light signalling. *Science, 294*, 1108-1111.

van der Woude, W. J. (1985) A dimeric mechanism for the action of phytochrome: evidence from photothermal interactions in lettuce seed germination. *Photochem Photobiol, 42*, 655-661.

Vierstra, R. D. and Quail, P. H. (1985) Spectral characterization and proteolytic mapping of native 120-kilodalton phytochrome from *Cucurbita pepo* L. *Plant Physiol, 77*, 990-998.

Vierstra, R. D. (1994) "Phytochrome degradation." In: *Photomorphogenesis in Plants*, Kendrick R. E. and Kronenberg G. H. M. (eds) Kluwer Academic Publishers, Dordrecht, 141-160.

Vierstra, R. D. (1996) Proteolysis in plants: mechanisms and functions. *Plant Mol Biol, 32*, 275-302.

Wagner, D., Fairchild, C. D., Kuhn, R. M., Quail, P. H. (1996) Chromophore-bearing NH2-terminal domains of phytochromes A and B determine their photosensory specificity and differential light lability. *Proc Natl Acad Sci USA, 93*, 4011-4015.

Xu, Y., Parks, B. M., Short, T. W., Quail, P. H. (1995) Missense mutations define a restricted segment in the C-terminal domain of phytochrome A critical to its regulatory activity. *Plant Cell, 7*, 1433-1443.

Yamaguchi, R., Nakamura, M., Mochizuki, N., Kay, S.A., Nagatani, A. (1999) Light-dependent translocation of a phytochrome B-GFP fusion protein to the nucleus in transgenic *Arabidopsis*. *J Cell Biol, 145*, 437-445.

Yanovsky, M. J., Luppi, J. P., Kirchenbauer, D., Ogorodnikova, O. B., Sineshchekov, V. A., Adam, E. *et al*. (2002) Missense mutation in the PAS2 domain of phytochrome A impairs subnuclear localization and a subset of responses. *Plant Cell, 14*, 1591-1603.

Yeh, K.-C. and Lagarias, J. C. (1998) Eukaryotic phytochromes: Light-regulated serine/threonine protein kinases with histidine kinase ancestry. *Proc Natl Acad Sci USA, 95*, 13976-13981.

Chapter 9

INTRACELLULAR LOCALIZATION OF PHYTO-CHROMES

Eberhard Schäfer[1], Stefan Kircher[1], Ferenc Nagy[2]
[1]Albert-Ludwigs-University Freiburg, Institute of Biology II/ Botany, Schänzlestrasse 1, 79104 Freiburg, Germany (e-mail: Eberhard.Schaefer@biologie.uni-freiburg.de)
[2]Biological Research Centre, Institute of Plant Biology, P.O. Box 521, 6701 Szeged, Hungary (e-mail: nagyf@everx.szbk.u-szeged.hu)

1. INTRODUCTION

Despite intensive efforts and significant progress, the primary processes of phytochrome-mediated responses are not yet fully understood. The most widely accepted three hypotheses describe the molecular functions of phytochrome as an enzyme, membrane effector and transcription regulator, respectively. These molecular models and recent results clearly indicate that detailed knowledge about the intracellular localisation of these photoreceptors is an essential pre-requisite for understanding early events in phytochrome-mediated signalling. Thus, in this chapter we describe data obtained about the distribution and localisation of phytochrome by classical and contemporary methods. Beyond listing these observations, we also evaluate in detail some of the key findings and explain how they helped to develop novel molecular concepts for light-induced signalling.

2. THE CLASSICAL METHODS

The classical studies employed spectroscopic, immunocytochemical and cell biological/ biochemical techniques to characterise the intracellular localization of phytochromes. For more detailed description of these studies and methods see Chapter 4 in Plant Photomorphogenesis (Kendrick and Kronenberg, 1994).

2.1 Spectroscopic methods

Prior to the onset of the molecular era, micro-beam irradiation was a major tool to obtain information about the intracellular localization of phytochromes. In their pioneering experiments Etzold (1965) and Haupt (1970) observed an action dichroism for photo- and polarotropism of the chloronemata of ferns and chloroplast orientation in the green alga *Mougeotia*. These findings suggested that the absorption dipole moment of Pr is parallel and that of Pfr perpendicular to the cell surface. The responses induced by a micro-beam pulse appeared to be local, since they could only be reversed by a subsequent far-red pulse given to the same spot. Thus it was concluded that the intracellular mobility of phytochrome in these cases is very limited.

The group led by M. Wada further refined these experiments and clearly demonstrated that the micro-beam must hit a region including the cell wall, the plasma lemma, and part of the cytosol to initiate the response. It was therefore concluded that phytochromes mediating these responses are not associated with plastids, mitochondria, or nuclei, but localised close to the plasma membrane. We note that although these experiments clearly indicated an ordered localisation of phytochromes, the physical association of the photoreceptor molecules with the membrane could not be proved by this method.

Attempts to use similar techniques in higher plants failed primarily because light is scattered within the tissue and no strictly localised responses mediated by phytochromes are known in higher plants. In contrast, results obtained by Marmé and Schäfer (1972), who used polarised light to induce photoconversion of phytochrome *in vivo* indicated partial action dichroism, i.e. an ordered localisation of the photoreceptor. Interpretation of these experimental data, however, was complicated since the role of differential light attenuation in causing the measured differences could not rigorously be excluded.

2.2 Cell biological methods

In addition to spectroscopic studies, cell fractionation was also considered as an efficient tool to determine whether phytochromes are associated with membranes (Kraml,1994). In summary, these studies indicated that phytochrome could be associated with various organelles and also the plasma membrane. The biological significance of these findings, however, has not yet been demonstrated and there is considerable doubt whether these observations indeed reflect localisation of the phytochrome molecules in vivo. We note, however, that Quail *et al.* (1973) using the same method reported red/ far-red reversible pelletability of phytochrome. This result was later confirmed using immunocytochemical methods when light-dependent formation of SAPs (sequestered areas of phytochrome) was reported (MacKenzie *et al.*, 1975; Speth *et al.*, 1986).

2.3 Immunocytochemical methods

Due to the technical problems inherent in the cell fractionation method, the next approach, pioneered by the Pratt laboratory, was immunocytochemistry. McCurdy and Pratt (1986) showed that the immunodetectable phytochrome (phyA) in dark-grown oat coleoptiles is homogenously distributed throughout the cytoplasm. No association with organelles or membranes was observed and irradiation very rapidly – with a half-life of a few seconds – induced formation of SAPs. In darkness these SAPs disappeared with a half-life of about 30 minutes (Speth *et al.*, 1986). Co-localisation of SAPs and ubiquitin indicated that the SAPs might be the place of phyA degradation (Speth *et al.*, 1987), a process mediated by the 26S proteasome. We should note, however, that this attractive hypothesis is still being debated and yet to be proved experimentally.

Ten years later Mösinger and Schäfer (1984) and Mösinger *et al.* (1985) demonstrated that transcription rates could be regulated by irradiating isolated nuclei. It is worth noting that these observations were ignored and forgotten for the following ten years, even though these data suggested that at least a fraction of phytochromes is localised in the nucleus during signal transduction.

3. THE NOVEL METHODS

After the genes encoding phytochrome were cloned and protein sequences available it was concluded that phytochromes are not integral membrane proteins and do not contain canonical nuclear localisation signals (NLS). Thus, it became generally accepted that phytochromes are soluble cytosolic proteins, which probably associate with membranes only after binding to a molecule that itself is membrane-localised. Pioneering work performed by Sakamoto and Nagatani (1996) seriously challenged this view. These authors reported for the first time the enrichment of phyB in nuclear extracts isolated from light-grown Arabidopsis seedlings. Moreover, the same authors demonstrated that a fusion protein consisting of the C-terminal part of *Arabidopsis thaliana* PHYB fused to the GUS reporter is constitutively localised into the nucleus in transgenic plants. These data obviously contradicted the membrane model but were overlooked for several year.

The situation, however, changed dramatically two years later, when Ni *et al.*, (1998) reported interaction of phyA and phyB with PIF3 (phytochrome interacting factor 3), a transcription factor belonging to the family of bHLH (basic helix-loop-helix) proteins. This finding implied that phyA and phyB have to be localised in the nucleus, in order to interact with this transcription factor at least temporarily, to mediate light-induced signal transduction. In 1999, independent of this hypothesis, Nagatani's group and ourselves demonstrated beyond reasonable doubt by analysing the nucleo/cytoplasmic distribution of the PHYB:GFP (green fluorescent protein) fusion protein that light indeed induces nuclear import of this photoreceptor in transgenic Arabidopsis and tobacco plants (Yamaguchi *et al.*, 1999; Kircher *et al.*, 1999). The appearance of the characteristic PHYB overexpression phenotype of transgenic plants (Kircher *et al.*, 1999) and complementation of an *Arabidopsis thaliana* (Yamaguchi *et al.*, 1999) or a *Nicotiania plumbagenifolia* mutant lacking functional phyB (Gil *et al.*, 2000) by the expression of the PHYB:GFP chimeric protein demonstrated that these fusion proteins represent photobiologically active photoreceptors.

Kircher *et al.* (1999) also studied nucleo/ cytoplasmic distribution of a chromophore-less mutant of phyB (the cysteine- encoding codon of the chromophore attachment site was mutated to code for an alanine) fused to GFP in transgenic plants. These authors found that this fusion protein is constitutively localised in the cytosol. Thus they concluded – based on the hypothesis that this mutant version has a conformation similar to the Pr form - that the Pr conformer of the photoreceptor is not compatible with nuclear import. The same holds for the N-terminal fragment fused to GFP (Matsushita *et al.*, 2003), whereas the C-terminal half of PHYB must contain a functional NLS(s), since chimeric proteins containing the C-terminal part of PHYB fused to the GFP reporter showed constitutive nuclear localisation (Nagy *et al.*, 2000; Matsushita *et al.*, 2003). Having these various transgenic lines in hand that expressed an easily detectable, biologically functional PHYB:GFP photoreceptor, photobiological studies of the molecular mechanism regulating the intracellular localisation of phyB and other phytochromes became feasible.

4. INTRACELLULAR LOCALISATION OF PHYB IN DARK AND LIGHT

In 6-day-old dark-grown seedlings the PHYB:GFP fusion protein is localised mainly in the cytosol. If the expression level of the transgene is high, occasionally a weak diffuse nuclear fluorescence indicating nuclear localisation of the fusion protein is also observable (Kircher *et al.*, 1999; Yamaguchi *et al.*, 1999; Kircher *et al.*, 2002; Matsushita *et al.*, 2003). Results obtained by Kircher *et al.* (2002) suggest that light treatment of imbibed seeds to promote homogenous germination can induce nuclear import of phyB. Thus it is conceivable that the weak nuclear staining detected in 6-day-old etiolated seedlings represents phyB molecules that were imported into the nucleus during this early phase of development. However, independently of the occasional diffuse staining, irradiation with either red or white light induces nuclear import of PHYB:GFP and subsequent accumulation of the photoreceptor in the nucleus. PhyB of nuclear localisation is not distributed homogenously in the nucleoplasm: it preferentially accumulates in characteristic structures termed speckles (Kircher *et al.*, 1999, 2002; Yamaguchi *et al.*, 1999).

Detailed studies showed that the nuclear import of PHYB:GFP as well as the formation of PHYB:GFP containing speckles is a slow process, which saturates in about 4 h and shows a strong fluence-rate dependence (Gil *et al.*, 2000). The wavelength dependence of these processes – tested under 6 h continuous irradiation – paralleled the described wavelength dependence of phyB-mediated seed germination (Shinomura *et al.*, 1996). The almost complete lack of responsiveness to wavelengths longer than 695 nm establishing a Pfr/ Ptot ratio of ca. 40 % was quite astonishing. Tests with light pulses showed that a single light pulse was almost ineffective, but three consecutive 5-minute pulses given at hourly intervals induced import and formation of speckles containing the PHYB:GFP fusion protein. The inductive signal was reversible by a subsequent far-red light pulse, indicating that the nuclear import of phyB has the characteristics of a typical Low Fluence Response (LFR) (Kircher *et al.*, 1999; Nagy *et al.*, 2000). Physiological experiments have shown that responsiveness to an inductive light pulse is often poor in etiolated seedlings, but could be strongly enhanced by pre-irradiation activating either phyB (red light), phyA (far-red light) or cry1, cry2 (blue light). Gil *et al.* (2000) reported that pre-irradiation with red and blue, but not with far-red light enhanced nuclear import and the formation of phyB-containing speckles. Moreover, the same authors also showed that the effectiveness of pre-irradiation with red light slowly decreased, thus the inductive effect of a 5 s red light treatment was completely lost after a 24 h dark period.

The PHYB:GFP fusion protein localised in the nucleus disappears slowly, with a half-life about 6 h, in seedlings transferred back to darkness. First the speckles are dissolved. This stage is transient and the nuclei display diffuse staining. The next stage, i.e. the complete loss of nuclear staining takes about 10 h (Gil et al., 2000). Whether the slow disappearance of nuclear phyB is due to the slow export or slow turnover of the photoreceptor remains to be determined. We note, however, that the disappearance of nuclear staining could be accelerated by about 2 h by irradiating the seedlings with a FR pulse before the transfer to darkness, which is a typical end-of-day response.

Taken together, these data indicate that light-induced nuclear import of phyB exhibits the characteristics of a typical phyB-mediated physiological response. Namely, it displays low responsiveness to single pulses, red/ far-red reversibility of multiple pulses (LFR), sharp decline of responsiveness to wavelengths longer than 695 nm, fluence rate dependence, and responsiveness amplification. Moreover, it became evident that the light-induced import of phyB into the nuclei is followed by the rapid formation of large sub-nuclear complexes, termed speckles, which harbour the bulk of the photoreceptor detectable in the nuclei.

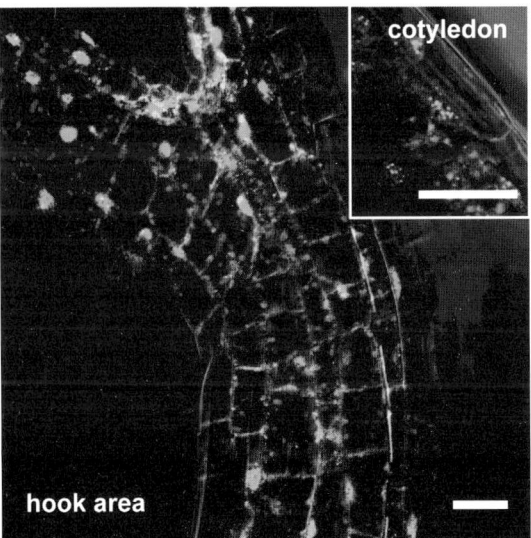

Figure 1. Localisation of PHYA-GFP fusion proteins in Arabidopsis seedlings. 4d old dark-grown Arabidopsis seedlings expressing fusion proteins of Arabidopsis PhyA and GFP controlled by the Arabidopsis promoter were irradiated briefly with white light. Subsequently, bright-field images (greyscale) and confocal images of GFP (green channel) and chlorophyll (red channel) fluorescence have been recorded with a Zeiss LSM510 microscope. The colour-combined images are showing the hook area and an area of the rim of a cotyledon (inlet). Bar = 25 μm.

5. INTRACELLULAR LOCALISATION OF PHYA IN DARK AND LIGHT

The immunocytological experiments performed in the 70's and 80's characterised the localisation of phyA primarily in monocotyledonous plants. They showed that light treatment results in a rapid rearrangement of cytosolic phyA and leads to the formation of phyA-containing cytosolic complexes (SAPs). Using transgenic tobacco and Arabidopsis seedlings expressing the PHYA:GFP fusion protein (Figure 1), the light-dependent intracellular localisation of this photoreceptor was re-investigated (Kircher et al., 1999; Kim et al., 2000; Kircher et al., 2002). As in the case of PHYB:GFP, the functionality of the PHYA:GFP fusion protein was verified by successful complementation of a PHYA null mutant. These studies demonstrated that both, the rice PHYA:GFP and the Arabidopsis PHYA:GFP fusion proteins, were exclusively cytosolic in dark-grown transgenic tobacco and Arabidopsis seedlings, respectively. However, in contrast to PHYB:GFP, the intracellular distribution of the different PHYA:GFP fusion proteins showed a very rapid change after irradiation. Whereas a single far-red light pulse is sufficient to induce nuclear import in all these cases a red light pulse leads to a rapid formation of cytosolic spots, followed by translocation of the PHYA:GFP fusion protein to the nuclei. We note that the cytosolic PHYA:GFP spots are reminiscent of the SAPs previously described in monocotyledonous seedlings (MacKenzie et al., 1975; McCurdy and Pratt, 1986; Speth et al., 1986).

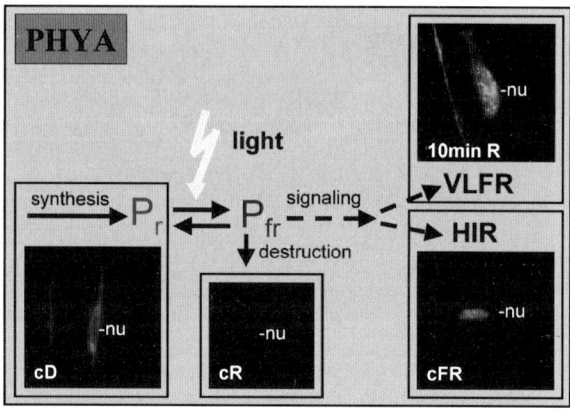

Figure 2. Model of the light-driven intracellular dynamics of phytochrome A. In dark-grown seedlings phyA is synthesized in its physiological inactive Pr-form (Pr) and stays in the cytosolic compartment. Irradiation establishes a wavelength-dependent equilibrium of the Pr to the active Pfr form. Red light (R) leads to formation of about 80% of Pfr, far-red light (FR) to about 3% Pfr. PhyA Pfr localises to sequestered areas of phytochrome (SAP) in the cytosol and is imported into the nucleus where it forms nuclear speckles. The light-requirements for these intracellular processes overlap with the light requirements for typical physiological responses of phytochrome A. While pulses of light can promote very low fluence responses (VLFR, here the effect of a red pulse is shown), continuous irradiation with far-red light (cFR) leads to high irradiance responses (HIR). Due to the instability of the Pfr form of PHYA, continuous red-light (cR) leads to a rapid destruction of the photoreceptor.

Nuclear import of PHYA:GFP was also followed by the formation of nuclear speckles, similarly to PHYB:GFP. However, the PHYA:GFP-containing speckles appeared very rapidly and both, their size and number, were much reduced as compared to those of PHYB:GFP speckles (Kim *et al.*, 2000; 2002). These data demonstrate that light-mediated nuclear import of phyA is a typical phyA-mediated Very Low Fluence Response (VLFR). PhyA can also mediate the far-red High Irradiance Response (HIR). In both, transgenic tobacco and Arabidopsis seedlings, continuous far-red light led to nuclear import of PHYA:GFP (Kim *et al.*, 2000). The import process is fluence-rate and irradiance dependent (Kim, 2002). Thus it reflects a typical far-red HIR. A further characteristic of a far-red HIR is that it is diminished after a pre-treatment with red light (Beggs *et al.*, 1980; Holmes and Schaefer, 1981). The nuclear import of PHYA:GFP could also be almost completely abolished by 24 h pre-treatment with red light (Kim *et al.*, 2000). We note that similar results were obtained by Hisada *et al.* (2000), who analysed continuous far-red light and light pulse-dependent intracellular localisation of phyA in pea seedlings, using cytochemical methods.

In summary it can be concluded that (i) PHYA:GFP is localised exclusively in the cytosol in dark grown seedlings, (ii) irradiation initiates rapid formation of cytosolic SAPs and (iii) nuclear import is followed by formation of nuclear speckles containing the PHYA:GFP fusion protein. These processes display complex dynamics and are mediated by VLFR and HIR (Figure 2).

6. INTRACELLULAR LOCALISATION OF PHYC, PHYD AND PHYE IN DARK AND LIGHT

To complete the characterisation of the nucleo/ cytoplasmic partitioning of all members of the phytochrome gene family, Kircher *et al.* (2002) produced transgenic Arabidopsis lines expressing PHYA–E: GFP fusion proteins under the control of the 35S cauliflower mosaic virus promoter. These authors found that in dark-grown seedlings the PHYD, PHYC and PHYE:GFP fusion proteins are primarily localised in the cytosol, similar to PHYA and PHYB:GFP. Upon irradiation all phytochromes undergo nuclear import and speckle formation; however, both of these processes display phytochrome-specific kinetics and light-dependence. Nuclear transport of PHYD–E is red and white light inducible. Interestingly, although PHYB and PHYD are closely related genes, they showed the largest difference. PHYD:GFP displayed a very slow nuclear import and only one or two larger speckles per nucleus were detectable even after an 8 h irradiation by white light (Kircher *et al.*, 2002). The speckle formation of all PHYs, except that of PHYD, showed a robust diurnal regulation under light/ dark cycles. The start of speckle formation even before the light-on signal indicates a circadian control. This phenomenon could be most clearly shown for PHYB:GFP (Gil, 2001; Kircher *et al.*, 2002).

7. INTRACELLULAR LOCALISATION OF INTRAGENIC MUTANT PHYTOCHROMES

Various approaches aimed at identifying signal transduction components for phytochrome-mediated responses resulted in the isolation of intragenic PHYA and PHYB mutants. These mutants can be classified as loss-of-function (hyposensitive, Yanovsky *et al.*, 2002) and hypersensitive mutants (Kretsch *et al.*, 2000; Casal *et al.*, 2002). Dark and light-dependent intracellular localisation of some of these mutant photoreceptors has been examined by expressing them as PHYA:GFP and PHYB:GFP fusion proteins in transgenic Arabidopsis lines.

7.1 Hyposensitive, loss-of-function mutants

PhyA and phyB was shown to interact with the transcription factor PIF3 in yeast (Ni *et al.*, 1998). In vitro experiments demonstrated that the interaction of the photoreceptors with PIF3 is regulated by light, thus it is mediated by the biologically active conformer, namely the Pfr form of phyA and phyB (Ni *et al.*, 1999). These authors also showed that the interaction of the transcription factor with photoreceptors encoded by a number of mutant alleles of phyA and phyB was significantly weakened. These mutants displayed hyposensitive phenotypes in vivo, thus the perturbed signalling was explained by the lack of interaction between the photoreceptor and PIF3. The majority of phyA and phyB mutants tested in these experiments carried missense point mutations in a specific region, termed the Quail box, of the photoreceptors. Kircher *et al.* (2002) investigated whether these point mutations affected just the interaction of the photoreceptors with PIF3 or also the nucleo/ cytoplasmic distribution of phyA and phyB. To this end they raised transgenic plants expressing the mutant PHYA and PHYB genes fused to GFP under the control of the 35S promoter and characterised the light-induced nuclear import of the fusion proteins in detail. The majority of the mutant photoreceptors were imported into the nuclei in a light-induced fashion, with no significant difference as compared to wild type phyA and phyB. The only exception observed so far was the G767R point mutant of PHYB. In this case the mutant protein accumulated to a significantly lower level in the nuclei of irradiated seedlings. We note that the insertion of an extra NLS into the G767R PHYB:GFP construct increased the accumulation of the fusion protein in the nuclei to levels similar to that of wild type PHYB:GFP and led to full complementation of a phyB deficient mutant (Matsushita *et al.*, 2003). In contrast to the seemingly normal nuclear import of the mutant photoreceptors, formation of speckles of PHYA or PHY:GFP fusion proteins was almost completely absent or much reduced in all mutants, including the G767R point mutant. The Quail box of phyA and phyB had been shown in vitro to be essential for the interaction with PIF3. Thus the loss of nuclear speckles in the mutants was interpreted as an indication that these sub-nuclear complexes could be involved in mediating light-induced signalling and can be considered as molecular markers for physiologically active phyA and phyB (Kircher *et al.*, 2002) (see 7, below for further discussion).

In an independent line of experiments Yanovsky *et al.* (2002) reported the isolation of a phyA mutant, which displayed complete loss of HIR but retained VLFR. The phyA -302 mutant was shown to contain the E777R point mutation in the PAS2 domain of the photoreceptor. The PHYA-302:GFP fusion protein was expressed in wild-type and phyA null-mutant (phyA201) backgrounds and in both cases showed normal translocation from the cytosol to the nucleus under continuous far-red light, but failed to produce nuclear speckles. These data again indicated that these sub-nuclear complexes are required for light-induced signalling. Moreover, these results suggested that they are specifically involved in regulating HIR signalling and/ or degradation of phyA, but not VLFR signalling.

7.2 Hypersensitive mutants

In a screen designed to isolate mutants exhibiting loss of reversibility, Kretsch *et al.* (2000) obtained a hypersensitive mutant (phyB-401), that was identified as a point mutant, carrying a G564E substitution in the conserved hinge region of phyB. Transgenic lines expressing the mutant PHYB:GFP fusion protein under the control of the 35S promoter, expressed in a phyB-minus background, showed extreme hypersensitivity both in wild-type and phyB null (phyB-9) backgrounds. The kinetics of light-induced nuclear translocation of the mutant PHYB:GFP differed from that of the wild-type PHYB:GFP. In contrast to wild-type PHYB:GFP, the nuclear import of the mutant PHYB:GFP was induced by a single light pulse. Moreover, the translocation of the mutant PHYB:GFP to the nuclei was followed by an immediate, rapid formation of speckles containing the fusion protein. These speckles were stable for more than 48 h in darkness and a far-red pulse could still induce the disappearance of the speckles even after incubating the seedlings for 24 h in dark. Thus it can be concluded that the phyB-401 mutant is stable in its Pfr form. We note that in a recent screen, designated specifically to isolate mutants displaying aberrant nuclear import and/ or formation of phyB-containing nuclear speckles, two independent, novel phyB alleles were identified showing the same phenotype (Bauer, Essing, Kircher, Schaefer and Nagy, unpublished). These data also suggest that nuclear speckles contain the phyB photoreceptor in its biologically active Pfr conformation.

8. PROTEIN COMPOSITION OF NUCLEAR SPECKLES ASSOCIATED WITH PHYB

Different laboratories demonstrated that the nuclear import of the PHYB:GFP fusion protein is always followed by the rapid formation of PHYB:GFP-containing speckles. It was found that the size and number of the speckles depend on the quality and quantity of the inductive light treatment (Gil *et al.*, 2000; Kircher *et al.*, 2002). Thus it was suggested that these large nuclear structures, associated with or containing the photoreceptor, might play a role in mediating phyB-dependent, light-induced signal transduction (Nagy and Schaefer, 2002).

More recently, in an attempt to screen for mutants impaired in intracellular localisation of phyB, the patterns of nuclear speckle formation were more precisely categorised (Chen *et al.*, 2003). These authors identified four types of speckles and showed that the number and size of the phyB speckles can be correlated with the ratio of Pfr and Pr conformers of the photoreceptor. The majority of isolated mutants displayed hyposensitivity to red light and aberrant formation of speckles co-localising with phyB. A significant proportion of the mutants was identified as intragenic phyB mutations, whereas some were mapped to chromosome regions that do not contain known genes involved in the regulation of light-induced signalling. A similar approach in our laboratory yielded comparable results (Bauer, Essing, Nagy and Schaefer, unpublished). The identification of these mutant genes is expected to shed light on the organisation and function of these sub-nuclear complexes in phyB mediated signalling.

In an independent line of experiments we could show that changes of intracellular distribution of phyB are extremely dynamic. Earlier studies indicated that the nuclear import of phyB is a slow process and GFP fluorescence in the nuclei could not be detected in a reliable fashion within less than about 45-60 min after the beginning of the irradiation (Gil *et al.*, 2000). Optimisation of microscopic techniques, however, made it possible to obtain a better resolution using GFP fluorescence. Thus we were able to show that within a few seconds after the beginning of irradiation many small nuclear PHYB:GFP speckles, not detected in earlier studies, are formed. These are extremely transient: they already disappear after 10 – 20 min and additional irradiation results in the formation of the more stable speckles reported in the earlier studies (Bauer *et al.*, 2004). Analysis of the localisation of the phytochrome-interacting factor PIF3 (Ni *et al.*, 1998) showed that PIF3 is co-localised with the early phyB speckles but not with the late ones appearing after prolonged irradiation (Figure 3). Moreover, we showed that in a PIF3 null background, phyB forms only the late but not the early speckles on irradiation. Western blot analysis and microscopic studies clearly demonstrated that light induces rapid degradation of PIF3 (the half life of the PIF3 protein is about 10 min) and this process is controlled by the concerted action of phyA phyB and phyD (Bauer *et al.*, 2004). These data suggest that early speckles co-localising with PIF3 and phyB might be the site of PIF3 degradation and/ or represent early events of phyB-mediated signalling.

However, to elucidate the exact function of the various types of phyB-associated speckles in phyB signal transduction, it will be essential to obtain information about their molecular composition. To this end our laboratory undertook the following experimental approach: nuclei from 4-week-old transgenic Arabidopsis plants expressing the PHYB:GFP fusion protein were isolated, disrupted and the intact speckles, representing the stable ones described above, were further purified by differential gradient centrifugation. The purified speckles were then solubilized, their components were separated on SDS PAGE and analysed by MALDI-TOF. It was found that about 80% of the more than 25 proteins identified in phyB-containing speckles display significant homology to proteins shown to be present in the interchromatin granual clusters (ICGs) of animal cells. We note that although phyA, like phyB, forms nuclear speckles on photo transformation, all attempts to obtain information about the composition of this phyA-associated sub-nuclear protein complexes have so far proved unsuccessful.

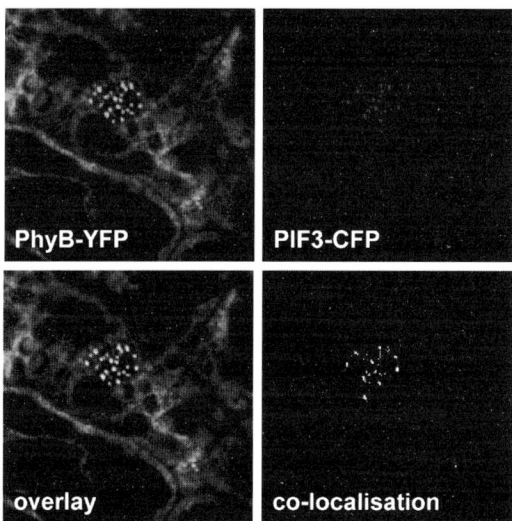

Figure 3. Co-localisation of Phytochrome B with the bHLH factor PIF3. 4d old dark-grown Arabidopsis seedlings simultaneously expressing fusion proteins of PhyB with YFP and PIF3 with CFP each controlled by the 35S promoter were irradiated briefly with white light. Subsequently, confocal images of YFP (green channel) and CFP (red channel) fluorescence have been recorded with a Zeiss LSM510 microscope. The images are showing epidermal cells of the base of a cotyledon, either representing the PhyB-YFP or PIF3-CFP signals, an overlay of these images resulting in yellow colour for co-localisation of PhyB and PIF3 or an additional co-localisation analysis of both factors using ImageJ software package (NIH).

Immunogold co-localization experiments clearly confirmed the co-localisation of phyB with some of the proteins identified by MALDI-TOF. Taken together, these data indicate that the phyB-containing plant nuclear speckles are likely structural homologs of ICGs identified in animal cells. The exact biological role of ICGs is not yet fully understood even in animal cells. ICGs localise close to the actively transcribed regions and contain dozens of proteins involved in mRNA splicing. Thus they are considered to be involved in the storage, modification, and recruitment of factors necessary for transcription and splicing (Bubulya *et al.*, 2002). In plant cells very little is known about these processes, but the association of the photoreceptor phyB with ICG-like complexes indicates that some of these molecular events could be regulated by light (Panihgrahi, Kunkel, Klement, Medzhradszky, Nagy, Schaefer, unpublished results). We note, however, that the N-terminal fragment of PHYB fused to GUS:NLS and the G767R point mutant PHYB fused to GFP:NLS did not show light-induced nuclear speckle formation, yet they successfully complemented the phenotype of phyB deficient mutants (Matsushita *et al.*, 2003; for additional discussion see also section 6.1 and 8). These data indicate that the formation of these nuclear complexes may not be essential for phyB-mediated signalling. Thus we conclude that, given their multiple forms and size, their transient and dynamically changing appearance, to understand the exact molecular function of the light-induced nuclear protein complexes, remains a challenging task.

9. THE FUNCTION OF PHYTOCHROMES LOCALISED IN NUCLEI AND CYTOSOL

Recent results provided compelling evidence that light quality- and quantity-dependent translocation of phytochromes into the nuclei represents a major regulatory step in light-induced signalling. This conclusion was further strengthened by the data recently reported by Huq *et al.* (2003). These authors expressed phyB fused to the glucocorticoid receptor (GR) in a phyB null background and investigated the cellular distribution of the fusion protein and the inhibition of hypocotyl elongation in the transgenic seedlings in red light in the presence or absence of the steroid hormone. They found that the fusion protein remained cytosolic in dark and in red light if no steroid hormone was added. Addition of the hormone allowed light-dependent transport of the fusion protein and complementation of the phyB null mutant phenotype as far as inhibition of hypocotyl elongation is concerned. (Huq *et al.*, 2003).

In an independent line of experiments Matsushita *et al.* (2003) inserted the SV40 NLS in the PHYB:GFP fusion protein and investigated the cellular distribution of PHYB:GFP:NLS in transgenic phyB null mutant Arabidopsis seedlings grown in light and dark. These authors reported that the fusion protein was constitutively localised in the nuclei, irrespective of the light conditions. Moreover, they found that transgenic plants did not exhibit an altered phenotype in dark, but fully complemented the phyB null phenotype when grown in light. Using the same approach (Figure 4), our laboratory obtained somewhat different results. In our hands the expression of a similar fusion protein in Arabidopsis, although in a different genetic background, showed constitutive nuclear localisation but resulted in pronounced hypersensitivity to red light (Kirchenbauer, Kircher, Nagy and Schäfer, unpublished). In contrast, the sub-cellular distribution of the same PHYB:GFP:NLS fusion protein in transgenic tobacco seedlings was not altered; the nuclear import of the chimeric photoreceptor remained light-inducible. Irrespective of the differences, these data clearly show that phytochrome localised in the nucleus is the functional phytochrome, light is still necessary for its activation and there is no obvious major contribution of cytosolic phyB to light-induced signalling underlying early steps of photomorphogenesis. In this context we point out that alteration of the nucleo/ cytoplasmic distribution of PHYA:GFP by insertion of an additional NLS in the fusion protein has not yet been reported.

For phyB, the possibility of driving the photoreceptor constitutively into the nucleus even in darkness by fusing it with an extra NLS also allowed the functional analysis of truncated and mutated phyB molecules that otherwise would have been excluded from the nucleus. In a set of elegant experiments Matsushita *et al.* (2003)

reported that the complete C-terminal domain of the phyB is dispensable for its function as a photoreceptor and it appears to be required to mediate the light-induced nuclear import of the full-length protein. It follows that the N-terminal domain fused to GUS (to facilitate dimerization) and NLS was constitutively imported into the nucleus. More importantly, transgenic seedlings expressing this fusion protein displayed hypersensitivity to red light, but showed no altered phenotype in dark, indicating that N-terminal domain alone can function as a biologically active photoreceptor. The main caveat of these exciting experimental results is that GUS was used as a dimerization domain although it is known to tetramerise. Thus GUS may provide an artificial platform to recruit signalling partners for the otherwise non-functional molecule. Therefore, additional experiments using other dimerization domains could be important to clarify this issue. However, if this is proven not to be the case, the results reported by Matsushita *et al.* (2003) indicate that phyB probably does not function as a kinase or the kinase function of the molecule is not essential for mediating light-induced signalling. This conclusion is based on the fact that the complete histidine kinase-like domain and also parts of the domain believed to be essential for the serine/threonine kinase function are absent from the truncated but biologically active fusion protein. The data described above convincingly prove that phyA and phyB localised in the nucleus are the functional form of these photoreceptors, but light is required to switch on signalling.

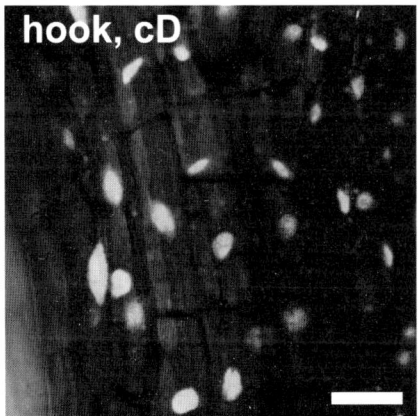
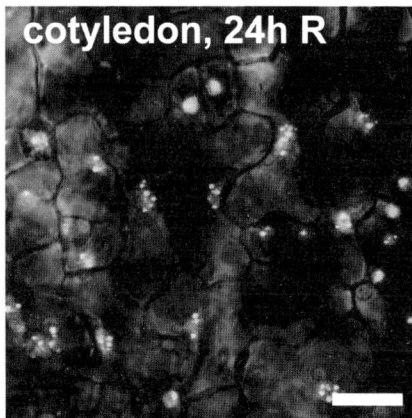

Figure 4. Localisation of a fusion protein consisting of Arabidopsis PhyB, GFP and a nuclear localisation sequence. 4d old dark-grown Arabidopsis seedlings expressing fusion proteins of Arabidopsis PhyB, GFP and the SV 40 NLS under the control of the Arabidopsis promoter were analysed either after incubation for 24 hours in red light (R) or darkness (cD). Subsequently, bright-field images (greyscale) and confocal images of GFP (green channel) and chlorophyll (red channel) fluorescence have been recorded with a Zeiss LSM510 microscope. The colour-combined images are showing the hook area or an area of of a cotyledon. Bar = 25 μm.

In contrast to the wealth of data obtained about the function of nuclear localisation, our knowledge about the function of non-nuclear phytochromes in light-induced signalling is very limited. This is somewhat surprising, since a large fraction of phyA-E remains localised in the cytosol even in plants kept in constant light (Nagy and Schaefer, 2002). Even under saturating light conditions there is still phytochrome remaining in the cytosol. Its localization is preferentially in the periphere of the cytosol as detected by indirect immunocytochemical methods (Kunkel, Panigrahi and Schäfer, unpublished). To define the biological role of this pool of phytochromes, however, radically new methods and approaches are required since in the absence of reliable markers no specific screens aimed at the isolation of such novel mutants can be performed.

10. CONCLUDING REMARKS

Our view about the intracellular localisation of phytochromes has remarkably changed during the last five years. It is obvious that there is a light-dependent nuclear transport of the photoreceptors and that the photoreceptors must be in the activated Pfr form to be functional. But there are still many, many question awaiting answers. What retains phytochromes in the cytosol? Is there an NLS that is masked by folding in Pr and opened after photoconversion to Pfr? Do cytosolic phytochromes have a function not only in ferns, mosses and some algae, but also in flowering plants? Is there a function of the phytochromes associated with the plasma membrane? What are the components of the different types of nuclear complexes and what is their function?

11. REFERENCES

Bauer, D., Viczian, A., Kircher, S., Nobis, T., Nitschke, R., Kunkel, T., et al. (2004) Constitutive photomorphogenesis 1 and multiple photoreceptors control degradation of Phytochrome Interacting Factor 3, a transcription factor required for light signalling in *Arabidopsis*. *Plant Cell*, 16, 1433-1445.

Beggs, C. J., Holmes, M. G., Jabben, M., Schäfer, E. (1980) Action spectra for the inhibition of hypocotyl growth by continuous irradiation in light- and dark-grown *Sinapis alba* L. seedlings. *Plant Physiol*, 66, 615-618.

Bubulya, P.S. and Spector, D. L. (2002) Dasassembly of interchromatin granule clusters alters the co-ordination of transcription and pre-mRNA splicing. *J Cell Biol*, 158, 425-436.

Casal, J. J., Davis, S. J., Kirchenbauer, D., Viczian, A., Yanovsky, M. J., Clough, R. C., et al. (2002) The serine-rich N-terminal Domain of oat phytochrome A helps regulate light responses and subnuclear localization of the photoreceptor. *Plant Physiol*, 129, 1127-1137.

Chen, M, Schwab, R., Chory, J. (2003), Characterization of the requirements for localization of phytochrome B to nuclear bodies. *Proc Natl Acad Sci USA*, 25, 14493-14498.

Etzold, H. (1965) Der Polarotropismus und Phototropismus der Chloronemen von *Dryopteris Filix-Mas* (L.) Schott. *Planta*, 64, 254-280.

Gil, P. (2001) Analysis of the nucleo-cytoplasmic partitioning of phytochrome B and its differential regulation by light and the circadian clock. *Fakultät für Biologie, Universität Freiburg*.

Gil, P., Kircher, S., Adam, E., Bury, E., Kozma-Bognar, L, Schäfer, E., et al. (2000) Photocontrol of subcellular partitioning of phytochromeB:GFP fusion protein in tobacco seedlings. *Plant J*, 22, 135-145.

Haupt, W. (1970) Über den Dichroismus von Phytochrom$_{660}$ und Phytochrom$_{730}$ bei *Mougeotia* Z. *Pflanzenphysiol*, 62, 287-298.

Hisada, A., Hanzawa, H., Weller, J. L., Nagatani, A., Reid, J. B., Furuya, M. (2000) Light-induced nuclear translocation of endogenous pea phytochrome A visualized by immunocytochemical procedures. *Plant Cell, 12*, 1063-1078.

Holmes, M. H. and Schäfer, E. (1981) Action spectra for changes in the 'high irradiance reaction' in hypocotyls of *Sinapis alba* L. *Planta, 153*, 267-272.

Huq, E., Al-Sady, B., Quail, P. H. (2003) Nuclear translocation of the photoreceptor phytochrome B is necessary for its biological function in seedling photomorphogenesis. *Plant J, 35*, 660-670.

Kendrick, R. E. and Kronenberg, G. H. M. (eds) *Photomorphogenesis in Plants*. Dordrecht, The Netherlands: Kluwer Academic Publishers, 1994.

Kim, L., Kircher, S., Toth, R., Adam, E., Schäfer, E., Nagy, F. (2000) Light induced nuclear import of phytochrome-A:GFP fusion proteins is differentially regulated in transgenic tobacco and *Arabidopsis*. *Plant J, 22*, 125-133.

Kim, L. (2002) Analysen zur intrazellulären Lokalisation von Phytochrom A. *Fakultät für Biologie, Universität Freiburg.*

Kircher, S., Kozma-Bognar, L., Kim, L., Adam, E., Harter, K., Schäfer, E., *et al.* (1999) Light quality-dependent nuclear import of the plant photoreceptors phytochrome A and B. *Plant Cell, 11*, 1445-1456.

Kircher, S., Gil, P., Kozma-Bognár, L., Fejes, E., Speth, V., Husselstein, T., *et al.* (2002) Nucleo-cytoplasmic partitioning of the plant photoreceptors phytochrome A, B, C, D and E is differentially regulated by light and exhibits a diurnal rhythm. *Plant Cell, 14*, 1541-1544.

Kraml, M. (1994) "Light direction and polarization." In *Photomorphogenesis in Plants*, Kendrick R. E. and Kronenberg G.H.M (eds), Kluwer Academic Publishers, Dordrecht, 417-443.

Kretsch, T., Poppe, C., Schäfer, E. (2000) A new type of mutation in the plant photoreceptor phytochrome B causes loss of photoreversibility and an extremely enhanced light sensitivity. *Plant J, 22*, 177-186.

MacKenzie, J. M. Jr, Coleman, R. A., Briggs, W. R., Pratt, L. H. (1975) Reversible redistribution of phytochrome within the cell upon conversion to its physiologically active form. *Proc Natl Acad Sci USA, 72*, 799-803.

Marmé, D. and Schäfer, E. (1972) On the localization and orientation of phytochrome molecules in corn coleoptiles (*Zea mays* L.) *Z. Pflanzenphysiol, 67*, 192-194.

Matsushita, T., Mochizuki, N., Nagatani, A. (2003) Dimers of the N-terminal domain of phytochrome B are functional in the nucleus. *Nature, 424*, 571-574.

McCurdy, D. and Pratt, L.H. (1986) Immunogold electron microscopy of phytochrome in *Avena*: identification of intracellular sites responsible for phytochrome sequestering and enhanced pelletability. *J Cell Biol, 103*, 2541-2550.

Mösinger, E. and Schäfer, E. (1984) *In vivo* phytochrome control of *in vitro* transcription rates in isolated nuclei from oat seedlings. *Planta, 161*, 444-450.

Mösinger, E., Batschauer, A., Schäfer, E., Apel, K. (1985) Phytochrome control of in vitro transcirption of specific genes in isolated nuclei from barley (*Hordeum vulgare*) *Eur J Biochem, 147*, 137-142.

Nagy F., Kircher S., Schäfer E. (2000) Nucleo-cytoplasmic partitioning of the plant photoreceptors phytochromes. *Seminars in Cell and Developmental Biology, 11*, 505-510.

Nagy, F. and Schäfer, E. (2002) "Phytochromes control photomorphogenesis by differentially regulated, interacting signalling pathways in higher plants." In: *Annu Rev Plant Biology*, Delmer, D., Bohnert, H. J., Merchant (eds), *53*, 329-355.

Ni, M., Tepperman, J. M., Quail, P. H. (1998) PIF3, a phytochrome interacting factor necessary for normal photoinduced signal transduction, is a novel basic helix-loop-helix protein. *Cell, 95*, 657-667.

Ni, M., Tepperman, J. M., Quail, P. H.(1999) Binding of phytochrome B to its nuclear signalling partner PIF3 is reversibly induced by light. *Nature, 400*, 784-784.

Quail, P. H., Marmé D., Schäfer, E. (1973) Particle-bound phytochrome from maize and pumpkin. *Nature, 245*, 189-191.

Sakamoto, K. and Nagatani, A. (1996) Nuclear localisation activity of phytochrome B. *Plant J, 10*, 859-868.

Shinomura T, Hanzawa H, Schäfer E, Furuya M. (1998) Mode of phytochrome B action in the photoregulation of seed germination in *Arabidopsis thaliana*. *Plant J, 13*, 583-590.

Speth, V., Otto, V., Schäfer, E. (1986) Intracellular localization of phytochrome in oat coleoptiles by electron microscopy. *Planta, 168*, 299-304.

Speth, V., Otto, V., Schäfer, E. (1987) Intracellular localisation of phytochrome and ubiquitin in red-light-irradiated oat coleoptiles by electron microscopy. *Planta, 171*, 332-338.

Yamaguchi, R., Nakamura, M., Mochizuki, N., Kay, S.A., Nagatani, A. (1999) Light-dependent translocation of a phytochrome B-GFP fusion protein to the nucleus in transgenic Arabidopsis. *J Cell Biol, 145*, 437-445.

Yanovsky, J. M., Luppi, P. J., Kirchbauer, D., Ogorodnikova, B. O., Sineshchekov, A. V., Adam, E., *et al.* (2002) Missense mutation in the PAS2 domain of phytochrome A impairs subnuclear localization and a subset of responses. *Plant Cell, 14*, 1591-1603.

PART 3: BLUE-LIGHT AND UV-RECEPTORS

Chapter 10

BLUE/UV-A RECEPTORS: HISTORICAL OVERVIEW

Winslow R. Briggs
Department of Plant Biology, Carnegie Institution of Washington, 260 Panama St., Stanford, CA 94305 U. S. A. (e-mail:briggs@stanford.edu)

1. INTRODUCTION

The past decade has seen phenomenal progress in the identification, characterization, and understanding of blue light receptors not just in higher plants but in a wide range of microorganisms as well. Given this progress, it is easy to forget that interest in what we call blue light receptors (although they may also have action in the UV-A) has a long history extending back almost two centuries and perhaps longer.

It has been known for a century or more that blue light induces such physiological responses as phototropism, stomatal opening, chloroplast movements, leaf expansion, and solar tracking. Here, in the course of presenting an historical background of research on blue light receptors, I have tried to find the earliest reports on light effects on each of these phenomena before bringing the reader up to date on the current status of the field. I will make no attempt to summarize the huge physiological literature on plant responses to blue light that developed over the past century but will rather concentrate on the steps (and missteps) leading to the definitive identification of the photoreceptors involved. As will become clear shortly, conclusive identification was elusive for many years, and in some cases remains controversial. The interested reader should consult the following reviews and review volumes: Briggs, 1963; Briggs and Iino, 1983; Galston, 1959; Presti and Delbrück, 1978; Reinert, 1959; Senger, 1980, 1984, 1987; Senger and Briggs, 1981. Reviews dealing more directly with the blue-light receptor question and signal transduction are: Batschauer, 1998; Briggs and Christie, 2002; Briggs and Huala, 1999; Cashmore, 1997; Cashmore *et al.* 1999; Casal, 2000; Christie and Briggs, 2001; Jenkins *et al.* 1995; Kaufman, 1993; Lin, 2000, 2002; Lin and Cashmore, 1996; Sancar, 2000; Short and Briggs, 1994. As blue-light receptors in unicellular algae and fungi are covered elsewhere in this volume (Chapter 13), they will not be considered more than incidentally here. Reviews dealing with specific responses to blue light will be cited below as appropriate. Although there is a significant body of literature indicating the existence of a UV-B receptor, the responsible photoreceptor or receptors have not to date been identified (Ballaré, 2003).

2. EARLY HISTORY

Almost 200 years ago, scientists first began studying the responses of plants to light of different colours. The earliest report that this author could unearth (thanks to an alert colleague, Prof. Paul Galland at the University of Marburg, Germany) is a paper by Sebastiano Poggioli describing the effects of red and violet light on leaves of the sensitive plant *Mimosa pudica* (Poggioli, 1817). He used a prism and postulated that certain colours of light might be more favourable for the growth of plants. He first noted that the overall growth of the plants was much more vigorous under violet light than under red light. Those in red light were sickly, flaccid, and discoloured, a first indication that red light alone was insufficient for plant growth. He then reported that the leaves changed their position more rapidly in response to violet than to red light, likely a solar tracking response. (None of the early papers distinguish between the differential growth response we call phototropism and solely turgor-driven and completely reversible solar tracking). He also found that violet light was more effective than red in inducing changes in the position of the leaves of *Raphanus rusticanus*. These observations are surely among the earliest studies of the spectral sensitivity of plant growth in general as well as for a specific plant response.

Payer (1842) presented results of a series of simple experiments with seedlings of *Lepidium sativum* (cress). When he grew the seedlings in darkness they grew vertically, but when they received light only from one window in the side of an enclosing box, they curved sharply toward the window. Further, with two windows providing equal light from two different angles, the plants oriented to some resultant angle between the two light sources. If the light sources were unequal, the plants oriented toward the brighter source. If they were equal in intensity and came from opposite sides of the seedlings the latter remained erect. However, if they were unequal, curvature was strongly toward the more intense source. Next he used coloured glass to demonstrate that blue light was the most effective colour in inducing phototropic curvature. In unilateral red, orange, yellow or green light, the plants behaved as though they were in darkness and remained vertical. Finally, he used simultaneous irradiation from two different directions, blue light in one case and violet light in the other, and came to the conclusion that blue light was unequivocally more effective than violet (the first limited action spectrum), something not confirmed until the much more sophisticated action spectroscopy in the middle of the next century (Shropshire and Withrow, 1958; Thimann and Curry, 1960).

One year later Zantedeschi (1843) found that stems of *Oxalis multiflora* and *Impatiens balsamina* remained upright when they were exposed to light passed through red, orange, or yellow glass, but bent sharply toward light of other colours. Though he failed to mention which colours, they were likely (by elimination) blue and violet.

In these early studies on leaf movements and phototropism there was no attempt to determine the relative intensities of the various light sources. However, shortly before, Daubeny (1835), mainly studying photosynthesis, made a heroic effort to do so. Using a range of colours obtained by passing light through coloured glass or coloured solutions, he calibrated the intensities of his light sources in three ways: a) How many layers of wire gauze were required to occlude the light completely? b) How rapidly did the light sources raise the temperature of a blackened thermometer bulb? c) How long did it take each light source to darken a paper soaked in silver nitrate to some "standard point of discoloration"? It is worth noting that one of his sources of red light was light passed through a bottle of port wine (He neglected to report on the quality of the port).

Daubeny (1836) actually made two seminal observations. The first was that light passed through orange glass was very effective in inducing the "secretion of green matter" in bean leaves as they broke through the ground whereas light passed through a solution of copper ammonium sulfate (a deep blue solution) was little better than darkness. This observation is more of relevance to the protochlorophyll and phytochrome fields than that dealing with blue light receptors, but may also represent the first observation of a specific red light response. However, Daubeny's second observation is directly relevant to the present chapter. He used a simple technique of irradiating a plant in an enclosed chamber in the presence of a vial of concentrated sulphuric acid. By weighing the highly hygroscopic sulphuric acid before and after irradiating the plant with particular wavelengths, he could determine the effectiveness of different colours of light in inducing water loss by the relative increase in weight of the sulphuric acid. He was surprised to find that blue light was far more effective in inducing the water loss than other colours. At first he suspected that it was simply a matter of differential heating by the different coloured light sources, but when he tested this hypothesis, it did not hold up and he concluded that "..although heat assists the process, some degree of light is essential to its activity." This early study had clearly uncovered an effect of blue light on stomatal opening, but it was almost a century and a half before this specific effect of blue light was fully confirmed (Hsiao and Allaway, 1973; Zeiger and Field, 1982).

Further semi-quantitative work appears to be that of Julius von Sachs (1864). Sachs used either coloured glass (ruby red glass or dark blue cobalt glass) or a solution of potassium dichromate or of ammoniacal oxide of copper (copper ammonium sulfate or perhaps possibly an ammonium complex of copper, known to have a deep blue colour) as filters to provide red and blue light respectively. He was basically trying to determine whether the effectiveness of the different colours of light on plants was the same as its photochemical effect on silver chloride, another effort to calibrate light sources for a biological experiment. His results confirmed those of the earlier workers: plants did not curve in response to unilateral red light but their response to unilateral blue was as strong as their response to white light (see Sachs, 1887). Darwin (1880) was clearly aware that red light did not induce phototropism. He passed light from a paraffin lamp through a solution of potassium bichromate and used it as a safelight to observe plants that had been or would be stimulated by white light. [Red light continued to be used as a safelight in many laboratories to avoid blue-light responses until the discovery that the far-red-absorbing form of phytochrome, Pfr, was far more labile than the red-absorbing form, Pr, both *in vivo* (Butler *et al.* 1963) and *in vitro* (Butler *et al.* 1964, Briggs *et al.* 1968). Dim green safelights went quickly into vogue thereafter].

Studies on the effect of light on chloroplast position also have a long history. The earliest work that the author could locate was a report by Böhm (1856) that light affected the position of the "chromatophores" in leaves of two species of *Sedum*. When he moved the plants into bright mid-day sunlight the chloroplasts changed their distribution to become clumped on the cell walls. He eliminated heat as responsible and hypothesized that it was the direction of the sun's rays that brought about the change in chloroplast position. In a subsequent publication (Böhm, 1859) he reported that the clumping response occurred rather rapidly when sunlight was filtered through blue glass, but that it required several hours of sunlight passed through red glass to achieve the same chloroplast "clumping," the first indication that the light-activated chloroplast avoidance response was a blue-light response. Subsequently Famitzen (1867) noted that when moss leaves were moved from darkness to light, it was only blue light that induced the accumulation response. By 1908, Senn was able to write an entire book devoted to the effects of light and other factors on chloroplast shape and position in the cell, citing over a hundred references. Senn's book (1908) is a remarkable compendium both of the earlier literature and of his own detailed studies of chloroplast shape and movement under the influence of light and a host of other factors—temperature, water content, gravity, mechanical stress, and chemicals. By that time it was well known that chloroplasts in higher plants moved to the periclinal walls of photosynthetic cells (accumulation response) in dim light and to the anticlinal (avoidance response) in bright light, and these responses were both activated by blue light. In the older literature, these two different orientations are referred to as *epistrophe* and *parastrophe* respectively. Senn confirmed the earlier results and concluded that "only blue-violet (light) according to its intensity called forth epistrophe or parastrophe."

In the following sections, we will trace further development of our understanding of four physiological responses to blue light: phototropism, stomatal opening, chloroplast movement, and leaf expansion with emphasis on the identification of the photoreceptors involved. The article will conclude with brief sections on the rapid inhibition of growth of etiolated seedlings, solar tracking (both responses to blue light), and the ZTL/ADO family of putative blue-light receptors.

3. PHOTOTROPISM: ACTION SPECTRA CAN BE FICKLE

Following the nineteenth century experiments utilizing coloured glass and/or coloured solutions to investigate the spectral sensitivity of the photoreceptor for phototropism there was a long hiatus before the question was re-examined. Among the earliest efforts were those of Blaauw (1909) who reported that the most effective wavelengths for inducing phototropism in oat coleoptiles were between 466 and 478 nm. Blaauw also reported significant activity in the UV-A. Sonne (1929), relying on Blaauw's (1909) observation that the Bunsen-Roscoe reciprocity law (equal response for an equal number of photons independent of the rate of photon delivery, Bunsen and Roscoe, 1862) was valid for first positive curvature, confirmed that this response was activated by blue light. He also found that UV-A (366 nm) light was just over one-fourth as effective as 436 nm. Other spectral studies (Bergann, 1930; Bachmann and Bergann, 1930) confirmed blue light sensitivity and suggested a peak of activity near 445 nm with a shoulder near 465 nm, but failed to report any measurements in the UV-A. Bergann (1930) also demonstrated that unilateral red light did not induce curvature even after an exposure time of 24 hours. [Incidentally, Bergann noted that red light was the most effective, however, in inducing greening in the oat primary leaf, rediscovering the phenomenon first reported by Daubeny (1836) almost a century earlier.] Finally, Johnston (1934), also working with oat coleoptiles, used a monochromator and a balance technique (bilateral illumination with a test wavelength against a standard wavelength) to determine the relative effectiveness of various wavelengths. He confirmed the two maxima just mentioned (440 nm and 475 nm) and the sharp decline near 500 nm but again provided no data at wavelengths shorter than the mercury doublet line at 404-5 nm. Haig (1935) confirmed these results but also failed to make measurements below 400 nm. We will return to action spectroscopy shortly.

The first report of a putative photoreceptor chromophore for phototropism was a four-paragraph paper by Wald and Dubuy (1936) who detected both carotenes and xanthophylls in dark-grown oat coleoptiles, especially concentrated in the coleoptile tips. Since the earlier reports just cited suggested that the phototropic response of oat coleoptiles was a blue light response, Wald and Dubuy concluded that a carotenoid was likely involved as the photoreceptor. This hypothesis remained without contention for more than a decade. Then, Galston and Hand (1949) found that light-induced inhibition of elongation in etiolated pea epicotyls was accompanied by enhanced destruction of the auxin indoleactic acid (IAA) in the medium. In the same year Galston (1949) demonstrated that IAA in solution was rapidly inactivated in the light in the presence of riboflavin. Galston and Baker (1949) subsequently noted that a little riboflavin added to the medium on which pea stem sections were growing resulted in significant growth inhibition, an inhibition that could be overcome by the addition of IAA. Galston and Baker (1949) then compared this *in vitro* reaction with the photodestruction of IAA in the presence of a crude extract from etiolated pea epicotyls clarified by centrifugation. The action spectrum in both cases showed a maximum near 440 nm with a shoulder near 470 nm and a minimum near 420 nm. Although the measurements did not extend far into the UV-A, activity just below 400 nm was in both cases higher than at 440 nm. They found all of these features except the high value below 400 nm in an action spectrum for phototropism in the *Avena* coleoptile and concluded that differential light-activated photodestruction of auxin between illuminated and shaded sides, mediated by riboflavin, led to phototropic curvature (Galston and Baker, 1949, Galston, 1950).

Briggs *et al.* (1957, 1963) challenged the riboflavin hypothesis by demonstrating for maize coleoptiles that phototropically effective unilateral white light did not affect the total amount of diffusible auxin emerging from maize coleoptiles but increased the amount from the shaded side and decreased the amount from the illuminated side in comparison with dark controls. When the coleoptile tips were split at right angles to the light source and a cover slip placed between illuminated and shaded sides, this auxin differential was eliminated (Figure 1) as was curvature. The authors proposed that lateral transport of auxin, an auxin precursor, or some factor limiting auxin synthesis, must be occurring following phototropic induction, in agreement with the pioneering work of Went (1928). Unfortunately (as we shall see shortly) Briggs *et al.* (1957) (and others subsequently) tossed out not just differential auxin inactivation as a mechanism for phototropism but a flavin as chromophore at the same time. Nevertheless, the "flavin/carotenoid controversy," born out of these experiments, had now reached the young-adult stage. Whereas the Briggs laboratory did not actively proselytise for the carotenoid hypothesis, their rejection of flavins at that time left only the carotenoids as viable candidates.

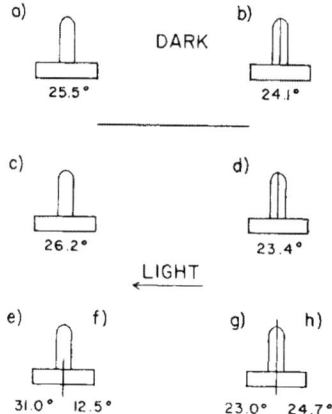

Figure 1. Distribution of auxin between lighted and shaded sides of unilaterally illuminated maize coleoptile tips with or without a vertical barrier extending through the coleoptile tips. Numbers indicate amount of auxin obtained by diffusion into agar blocks expressed as degrees of Avena *curvature (the classic bioassay for auxin). From Briggs et al. 1957.*

More than two decades after the Wald and Dubuy report, Curry (1957, see Thimann and Curry 1960) and Shropshire and Withrow (1958) finally produced high-resolution action spectra for first positive curvature of *Avena* coleoptiles (the most sensitive phototropic response, and the one obeying the Bunsen-Roscoe reciprocity law) extending significantly into the UV-A. Both studies confirmed the two maxima (445 and 475 nm) reported earlier, but showed in addition a shoulder near 420 nm and a broad peak near 365 nm. Eight years later, Everett and Thimann, 1958) obtained a very similar action spectrum for second positive curvature of *Avena* coleoptiles. Finally, Baskin and Iino (1987) obtained a very thorough action spectrum for the phototropic response of alfalfa seedlings. It showed the same features as the two earlier action spectra and showed an additional sharp peak near 280 nm, as illustrated in Figure 2. Since action spectra showing these features are commonly found in organisms reproducing by spores (cryptogams, especially fungi, see Presti and Delbrück, 1980) and remained mysterious and elusive (cryptic), Gressel (1979) proposed the name "cryptochrome" for this putative photoreceptor family. Some of these action spectra are shown in Figure 3, after Presti and Delbrück (1980), their Figure 8.

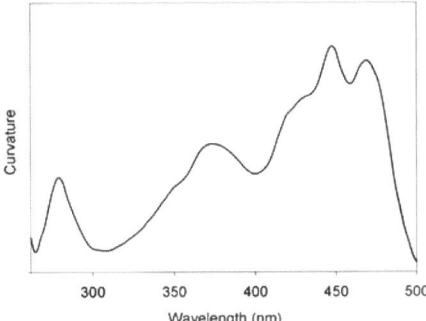

Figure 2. Action spectrum for phototropic curvature of etiolated alfalfa seedlings. After Baskin and Iino 1987.

Unfortunately these action spectra did not conclusively identify any specific chromophore. In fact, they all were inconsistent with either candidate. [Indeed both pterins (Galland and Senger, 1988) and retinal (Lorenzi *et al.* 1994) were subsequently suggested as possible candidate chromophores.] With respect to flavins versus carotenoids, the paradox was simple: the absorption spectra of carotenoids have the appropriate fine structure in the blue but little measurable absorption in the UV-A [with the exception of certain cis-carotenoids: for example, 9,9'-mono-cis-β-carotene has a sharp peak near 345 nm, but a minimum near 370, see Thimann and Curry (1960)]. On the other hand, flavin spectra in general have much more poorly defined fine structure in the blue, but a broad and robust absorption band in the UV-A (Figure 4).

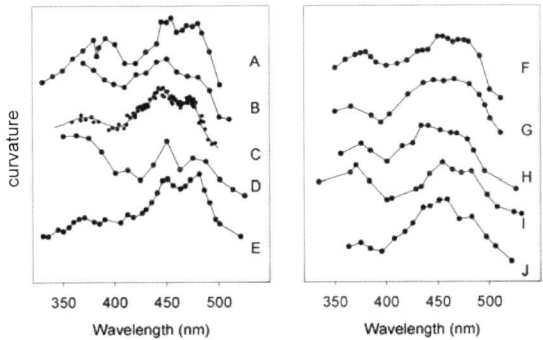

Figure 3. Some "cryptochrome" action spectra. A, phototropism in the fungus Phycomyces; B, Phototropism in the fungus Pilobolus; C, phototropism in Avena *coleoptiles;* D, phototaxis in Euglena; E and F, Stimulation of carotenogenesis in Neurospora *and* Fusarium, respectively; G, stimulation of carotenogenesis in Mycobacterium; H, enhancement of respiration in the unicellular alga Chlorella; I, chloroplast rearrangement in the moss Funaria; J, entrainment in the circadian rhythm pupae emergence in the fruit fly Drosophila. *After Presti and Delbrück 1978.*

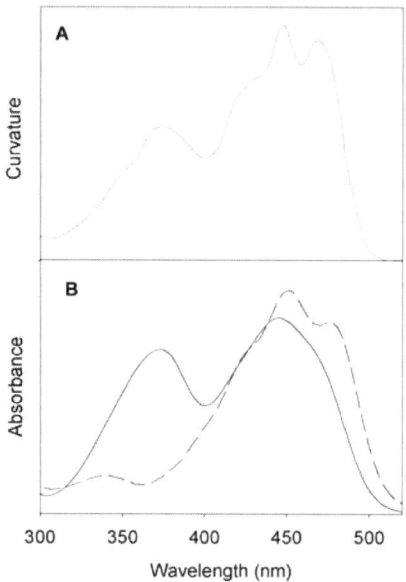

Figure 4. Comparison of action spectrum for phototropism (A) with absorption spectra of a flavin (FMN in water, solid line) and a carotenoid (β-carotene in ethanol, dashed line) (B).

In the proceedings of a meeting held in 1979, Shropshire (1980) elegantly summarized the quandary. In view of the increasing sentiment favouring flavins as chromophores, he was asked to serve as devil's advocate (*Advocatus diaboli*) to present evidence favouring the carotenoid hypothesis. The following is his conclusion: "Finally I wish to close this review with a return to the setting of a moot trial between two protagonists and conclude with a Mulla Nasrudin tale by Idries Shah (1971). These tales have become enormously popular among physicists in the United States. I think you will find this one instructive.

The Mulla was made a magistrate. During his first case the plaintiff argued so persuasively that he exclaimed,

'*I believe you are right!*'

The clerk of the court begged him to restrain himself, for the defendant had not been heard yet. Nasrudin was so carried away by the eloquence of the defendant that he cried out as soon as the man had finished his evidence:

'*I believe you are right!*'

The clerk of the court could not allow this.

'*Your honour, they cannot both be right.*'

'*I believe you are right!' said Nasrudin.*

Therefore if you were to ask me today if the blue-light photoreceptor (cryptochrome) is a flavin or a carotenoid, I would be tempted to answer as Nasrudin: 'Yes, I believe you are right!'"

4. THE LIAC: A FRUSTRATING DIGRESSION

In 1970, Berns and Vaughn made an intriguing discovery: both blue and UV-A light brought about dramatic absorption changes both in intact mycelium and sporangiophores of the fungus *Phycomyces blakesleeanus* and extracts made from them. These light-induced absorbance changes were subsequently designated LIAC's for short. Fluorescence changes paralleled the absorbance changes and in most cases, the changes were reversed in the dark over a period of minutes. Based on these changes the authors hypothesized that a flavoenzyme might be involved. A mutant that lacked the light-growth reaction (transient promotion of sporangiophore growth by light) also lacked the spectral changes indicating that the authors were not simply seeing photo-induced changes in mitochondrial flavoenzymes. Their work was quickly followed by work in Butler's laboratory (Poff *et al.* 1973; Poff and Butler, 1974; and Muñoz and Butler, 1975) describing similar light-induced absorbance changes in the slime mold *Dictyostelium discoideum* and the pink bread mold *Neurospora crassa*. An *in vivo* action spectrum for the *Neurospora* response (Muñoz and Butler, 1975) had all of the earmarks of what Gressel (1979) subsequently called the typical cryptochrome action spectrum and was not dissimilar to the action spectrum for the inhibition of a circadian rhythm of conidiation in *Neurospora* obtained by Sargent and Briggs (1967) and the action spectrum for the light-growth reaction in *Phycomyces* (Delbrück and Shropshire, 1960). Muñoz and Butler concluded that the *Neurospora* LIAC represents the photoreduction of a flavin chromophore that subsequently reduces a *b*-type cytochrome. They hypothesized that this reaction is the primary step in the photocontrol mechanisms for a broad range of blue-light responses.

In a study of both *Neurospora* and maize coleoptiles, Brain *et al.* (1977) first reported a LIAC from a crude membrane preparation from higher plants. In the course of these studies they presented evidence indicating that the LIAC was likely localized to the *Neurospora* plasma membrane but they did not have the marker enzymes to make any conclusive statement about the localization of the maize LIAC. Widell *et al.* (1983) later used a two-phase separation technique to demonstrate that a cauliflower LIAC was also localized to the plasma membrane, and, at least in the blue, had the action spectrum to be expected of a cryptochrome.

During the subsequent decade there appeared a host of LIAC papers (including at least seven more from the author's laboratory). However, repeated attempts to find a correlation—*any* correlation—between LIACs and physiological responses met with universal failure. In a review on LIAC activity in higher plants published in 1991, Asard and Caubergs cautiously stated, "…it has to be mentioned that the physiological relevance of the LIAC measurements is still a matter of debate." A severe blow to the notion of physiological relevance had come from the observation by Schmidt and Butler (1977) that riboflavin in solution mediates the photoreduction of horse-heart cytochrome *c*, hardly a reaction of great physiological significance. Short and Briggs (1994) concluded that "….the LIAC still remains an enigmatic membrane-associated reaction in search of a physiological role." There has been little other mention of LIAC's during the past decade, but the story is an excellent example of how a fascinating phenomenon can capture the imagination of scientists and lead them astray.

5. THE CRYPTOCHROME STORY

With the gradual demise of the LIAC as reflecting the activity of photoreceptor for blue-light responses, the search for a blue-light receptor (stated in many papers as "*the* blue light receptor," *italics mine*) was stalled. As we shall see shortly, however, new biochemical, and molecular genetic approaches applied to the model plant *Arabidopsis thaliana* finally led to the identification and characterization of two different families of blue-light receptors--first the cryptochromes, then the phototropins. They have also led to the identification and preliminary characterization of the putative blue-light receptors in the ZEITLUPE/ADAGIO family.

It was Maarten Koornneef *et al.* (1980) who set the stage for the discovery and characterization of the first plant blue-light receptor. In the course of screening mutated *Arabidopsis* seedlings for altered phenotypes under different light conditions, these authors reported that in contrast to wild type, five different *hy* mutants (designated *hy1* through *hy* 5) failed to respond to light by showing inhibition of hypocotyl elongation. *hy, hy2, and hy3* and *hy5* all showed an impaired response to red light or far-red light but responded normally to blue. By contrast, the *hy4* mutant had a long hypocotyl under continuous blue light and a normal short hypocotyl under red. As the remaining *hy* mutants were deficient in phytochrome responses, *hy4* remained a relatively unexplored curiosity for more than a decade.

The impasse in identifying and characterizing a plant blue-light receptor was finally broken in 1993. Ahmad *et al.* (1993) successfully identified a T-DNA-tagged mutant at the *hy4* locus and used it to clone and characterize the *HY4* gene. The putative HY4 protein showed striking homology to prokaryotic photolyases, well-studied enzymes involved in the photoactivated repair of damaged DNA (Sancar, 1994). Based on this similarity they postulated that the HY4 protein was a blue-light receptor and subsequently designated it cryptochrome 1 (cry1) following Gressell's 1979 suggestion for nomenclature (Lin *et al.* 1995a). However, cry1 possesses a C-terminal extension that is found in none of the photolyases but is essential for cry1 activity (Ahmad *et al.* 1993, 1995; Yang *et al.* 2000; Wang *et al.* 2001; Yang *et al.* 2001).

Photolyases contain two chromophores: a flavin adenine dinucleotide (FAD) plus either a deazaflavin or a pterin as an antenna chromophore. Lin *et al.* (1995b) expressed the *Arabidopsis* gene successfully in Sf9 insect cells and showed that it bound FAD, but could not detect a second chromophore. Malhotra *et al.* (1995) reported that both HY4 from *Arabidopsis* and a similar protein that they had earlier designated SA-PHR1 (and misidentified as a photolyase, Batschauer 1993), expressed in *Escherichia coli*, bound FAD, and in addition, HY4 bound the pterin methenyltetrahydrofolate (MTHF). Neither the HY4 (Lin *et al.* 1995; Malhotra *et al.* 1995) nor SA-PHR1 (Malhotra *et al.* 1995) proteins showed any photolyase activity. (An *Arabidopsis* DNA photolyase gene *PHR1* was subsequently characterized and its encoded protein shown to have authentic DNA photolyase activity, Ahmad *et al.* 1997).

Hoffman *et al.* (1996) and Lin *et al.* (1996) both reported the presence of a second cryptochrome gene in *Arabidopsis*, designated *CRY2*. Lin *et al.* (1998) then demonstrated that the cry2 protein mediated the same inhibition of hypocotyl elongation as cry1, except that it was significantly more sensitive to low light intensities. Searches of the recently completed *Arabidopsis* DNA sequence (Brudler *et al.* 2003) turned up a third member of the cryptochrome family, cryptochrome DASH (so named because closely related cryptochromes are found in **D**rosophila melanogaster, **A**rabidopsis thaliana, **S**ynechocystis spp. 6803, and **H**omo sapiens). These workers also determined the first crystal structure of a cryptochrome, the cry DASH from *Synechocystis*. Finally, some lower vascular plants such as the maidenhair fern, *Adiantum capillus-veneris*, may contain as many as five cryptochromes (Imaizumi *et al.* 2000).

It is worth noting that the cryptochromes appear to have less photosensitivity in the UV-A than in the blue. Indeed, the original mutant used to clone and characterize cry1 showed almost normal hypocotyl inhibition in response to UV-A treatment (Ahmad *et al.* 1993). Thus, despite their designation as cryptochromes, they may fail to show a typical "cryptochrome" action spectrum as envisaged by Gressel. A later action spectrum (Ahmad *et al.* 2002) showed some fine structure in the blue, but did not make measurements into the UV-A below 380 nm.

Thus some one and three-quarter centuries had to pass after Poggioli's initial observations on the blue-light sensitivity of *Mimosa pudica* leaves before the first higher-plant blue-light receptor was characterized. Four cautions need to be expressed regarding the chromophores of the cryptochrome family. First, specific chromophores have only been identified for cry1 from *Arabidopsis*, and not yet from cry2. Second, in both reports cited, the chromophores were identified from *Arabidopsis* proteins expressed in a heterologous system—either insect cells or *E. coli*. Although it is likely that these are the same chromophores as are bound *in planta*, definitive evidence is still lacking. Third, although Lin *et al.* (1996) reported that when expressed in insect cells cry2 is a flavoprotein, they did not report which flavin. And, fourth, there is no information at all about the second chromophore for cry2. However, it seems likely that, at least for cry1, the first chromophore is FAD and the second chromophore is a pterin. If so, the proposal (Galston and Baker, 1949; Galston, 1950) that a flavin is the chromophore for a blue-light receptor and the proposal (Galland and Senger, 1988) that a pterin can serve in that role are both correct.

A very large body of recent genetic and biochemical evidence supports the conclusion that the cryptochromes are genuine plant photoreceptors. Most recently, Shalitin *et al.* (2003) reported that cry1 becomes phosphorylated in a blue-light-dependent manner both *in vivo* and *in vitro*. Although the photochemistry involved is still unknown, these observations provide the first evidence that blue light induces a potentially important biochemical change in a cryptochrome, a key requirement to identifying a molecule unambiguously as a photoreceptor.

6. THE PHOTOTROPIN STORY

Although it was the various action spectra for phototropism that guided the confused but ultimately successful search for a blue light receptor in higher plants, progress in the actual identification of the phototropins lagged behind progress with the cryptochromes. Again it took the extraordinary power of modern molecular genetics to come to a solution. Gallagher *et al.* (1988) first noted that blue light activated the phosphorylation of a plasma-membrane protein in etiolated elongating pea stems. This reaction requires three components: the substrate itself, in this case a protein appearing near 120 kDa, a kinase, and a photoreceptor. Short and Briggs (1990) then published the first characterization of the response, demonstrating that it could occur either *in vivo* or *in vitro*, the latter providing a convenient assay for biochemical characterization of the reaction. The researchers faced two problems. The first was to determine whether the system involved one, two, or three separate components. Obviously the response requires the substrate protein itself, a kinase, and a photoreceptor. The second was to identify and characterize the component that was the true photoreceptor. Whereas a significant body of biochemical information was acquired between 1988 and 1997 and this protein was found in elongating etiolated tissues of all plants tested including grass coleoptiles of several species (see Briggs *et al.* 2001, for a detailed summary), biochemical approaches to purifying and characterizing it ultimately failed. Reymond *et al.* (1992) then reported that an *Arabidopsis* mutant deficient in phototropism, *JK224* (Khurana and Poff, 1989), was also deficient in the light-activated phosphorylation reaction, suggesting for the first time that this phosphorylation might be a part of signal transduction for phototropism. Their suspicion (which later turned out to be the case) was that the JK224 protein was in fact the 120 kDa phosphoprotein.

Meanwhile, the carotenoid vs. flavin controversy re-emerged in slightly different form. First, Quiñones and Zeiger (1994) reported that the phototropic sensitivity of maize coleoptiles was proportional to the content of the carotenoid zeaxanthin and proposed that zeaxanthin might be a blue light receptor. However, Palmer *et al.* (1996) reported that maize coleoptiles without any detectable carotenoid content showed normal phototropism. (Note the dramatic switch of allegiance from carotenoid to flavin in this author's laboratory). In a more detailed phototropism study, Lascève *et al.* (1999) reported both normal first positive and second positive curvature from the *Arabidopsis* mutant *npq1*, a mutant unable to synthesize zeaxanthin. Meanwhile, Ahmad *et al.* (1998) reported an absence both of first positive curvature and of blue light-activated phosphorylation from a mutant deficient in both cryptochromes and proposed that the cryptochromes served redundantly as the photoreceptors for phototropism. However, Lascève *et al.* (1999) tested two different *Arabidopsis cry1 cry2* double mutants and found normal first and second positive curvature in both, as well as normal light-activated phosphorylation. Meanwhile, Liscum and Briggs (1995) had described a series of *Arabidopsis* mutants deficient in hypocotyl phototropism. Complementation studies indicated that they represented mutations in four different loci, designated *nph1-nph4* (for *n*on-*p*hototropic *h*ypocotyl). They also determined that *JK224* was in fact allelic to one of the four classes, *nph1*.

Huala et al. (1997) then used the allele *nph1-5* to track down, clone, and sequence the gene encoding the NPH1 protein. The *NPH1* gene's C-terminal half encoded a classic serine/threonine protein kinase. Thus of the three components, only the photoreceptor itself remained unknown: the protein itself was a kinase and the phosphorylation was a light-activated autophosphorylation. The N-terminal half encoded a couple of PAS domains (see Taylor and Zhulin, 1999) with high homology to domains in a wide range of proteins involving signalling in response to **L**ight, **O**xygen, or **V**oltage, from both prokaryotes and eukaryotes, and the authors therefore designated these domains **LOV** domains. Figure 5 shows autoradiograms of SDS gels of membrane and soluble preparations from wild-type *Arabidopsis* and a null mutant *nph1-5*.

The identity of the third component, the photoreceptor itself, was solved when Christie *et al.* (1998) expressed the *NPH1* gene via a *Baculovirus* construct in insect cells kept in the dark. In the absence of any other plant proteins, light induced autophosphorylation of the nph1 protein. The protein was found to bind FMN as its chromophore. The authors concluded that the molecule itself was the photoreceptor for the phosphorylation reaction and therefore for phototropism. A year later Christie *et al.* (1999) demonstrated that it is the LOV domains that non-covalently but stoichiometrically bind the flavin FMN, and named the photoreceptor phototropin (subsequently phototropin 1, Briggs *et al.* 2001). Figure 6 shows the domain structure of phot1 from *Arabidopsis*.

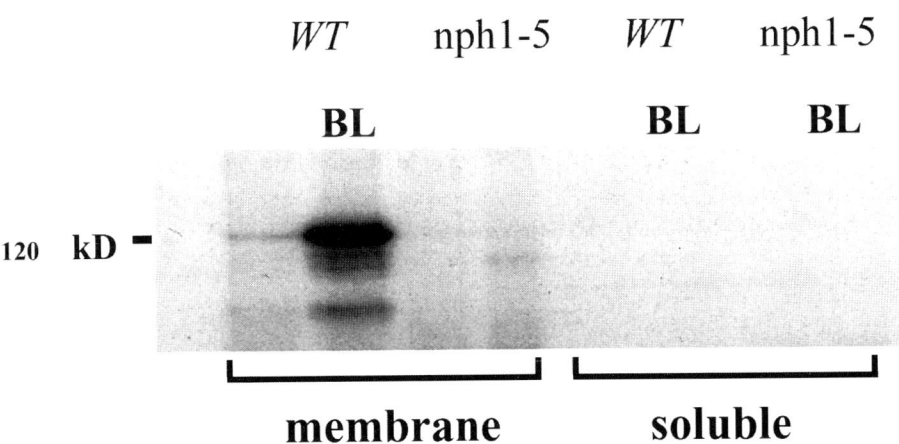

Figure 5. Autoradiograms of SDS gels of membrane and soluble preparations from wild-type Arabidopsis and a null mutant nph1-5 (phot1-5). D, dark control; BL, blue-light treated.

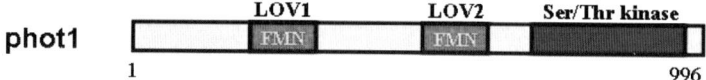

Figure 6. Domain structure of Phot1 from Arabidopsis.

Jarillo *et al.* (1998) reported a second phototropin, now designated phot2, and Sakai *et al.* (2001) subsequently demonstrated that it also mediated light-activated autophosphorylation when expressed in insect cells. Thus a second family of blue light receptors became identified. It is still possible that there are one or more other blue light receptors involved in signalling in phototropism in *Arabidopsis*. Konjević *et al.* (1989, 1992) presented convincing photobiological evidence for two pigments with significantly different absorption spectra mediating phototropism in *Arabidopsis*. Likewise, Sakai *et al.* (2001) reported a small but significant phototropic response from a *phot1 phot2* double mutant at a fluence rate of unilateral light of 10 μmol m^{-2} s^{-1}, a result we have confirmed in our own studies (E. Kaiserli and W. R. Briggs, unpublished results). Hence, although it is clear that the two phototropins are the major players in phototropism in higher plants, the door is not closed for other as yet unidentified photoreceptors, to participate directly in the response. For access to the vast literature on phototropism, the reader should consult the recent reviews by Koller (2000) and Iino (2001).

7. STOMATAL OPENING IN BLUE LIGHT

As was the case with phototropism, the identification of photoreceptors mediating light-activated stomatal opening remained controversial for many decades. Also, as with phototropism, the story is likely not complete. Though we know of two photoreceptors that mediate blue light-activated stomatal opening, physiological studies strongly implicate at least one other player, as we shall see shortly.

Francis Darwin (1898) rediscovered the light effect on stomatal opening over six decades after Daubeny (1836) concluded that there must be some special effect of blue light in inducing an increase in water loss. He then invented an ingenious device called a porometer (Darwin, 1911) that allowed him to measure the rate at which air could be drawn through a leaf by a descending column of water. The device actually allowed him to make the first time-course measurements for stomatal opening in white light and subsequent closing in darkness. Within the resolution of Darwin's measurements, there appeared to be almost no lag following onset of light or darkness and the half-times both for opening in the light and closing in the dark appear to be less than half an hour. It took an additional seven decades, however, to establish that there really was some special effect of blue light beyond its role in photosynthesis in the guard cells. In a review published in 1962, Ketellapper concluded that there was not sufficient evidence to support notion of a specific blue-light receptor for stomatal opening. Two years later Kuiper (1964), working with epidermal strips of the goldenrod *Senecio odora*, came to the same conclusion. He based it on the similarity between his action spectrum for maintaining a fixed stomatal aperture under 20 min continuous light with the action spectrum for photophosphorylation (Black *et al.* 1962). By contrast, an action spectrum by Hsiao and Allaway) 1973) both for stomatal opening and uptake of Rb^+ also strongly implicated a specific blue-light receptor. Using epidermal strips from *Vicia faba* leaves, Lurie (1978) concluded that blue light was more effective than red in maintaining stomata open under steady-state conditions (2.5 h). Citing other studies showing the same thing, she concluded that there was indeed some special effect of blue light over and above its effect on stomatal aperture through guard cell photosynthesis. Ogawa's laboratory (Ogawa *et al.* 1978, Ogawa, 1981) then measured the effects of blue light either alone or in the presence of strong red light (35 W m^{-2}) on transpiration rates of *Vicia faba* and *Allium cepa* leaves. Blue light alone had only a very small effect, but in the presence of red, it induced a large increase in transpiration in both species. Ogawa concluded that blue light did indeed have some special role in inducing opening, likely related to H^+ excretion and accompanying K^+ uptake, but that it required ATP provided by photosynthesis.

A year later Zeiger and Field (1982) combined measurements of stomatal conductance with simultaneous measurements of photosynthesis and demonstrated unequivocally that the effect of weak blue light on the former against a strong red light background could not be accounted for merely by increased photosynthesis driven by the few blue-light photons. Thus after a century and a half, Daubeny's cautious conclusion that there must be something special about blue light in inducing water loss by leaves was finally firmly established.

Identification of a participating blue-light photoreceptor, however, required another two decades. Karlsson (1986) obtained a detailed action spectrum for the effect of blue light on stomatal aperture between 370 and 500 nm for wheat seedlings. It showed the typical features of a "cryptochrome" action spectrum as defined by Gressel (1979): a maximum near 450, a second sharp peak near 470, and a shoulder near 420 nm. Karlsson expressed concern that the values below 400 nm were low by "cryptochrome" standards, but speculated that perhaps there were UV-A-absorbing pigments that served to mask the actinic light from a flavin photoreceptor, or that something about the environment of a putative flavin reduced its UV-A absorption band significantly. Eisinger *et al.* (2000) extended the spectral range tested into the UV-A for stomatal opening in *Vicia faba* leaves and found a broad peak near 350, just where one might have expected a "cryptochrome" peak. Meanwhile, Zeiger and Hepler (1977) demonstrated that blue light could induce the swelling of protoplasts obtained from guard cells. Thus whatever the photoreceptor molecules might be, they were located in the guard cells themselves.

Lascève *et al.* (1999) measured stomatal responses in *Arabidopsis* single mutants deficient in phot1 and single and double mutants deficient in the cryptochromes. As all of the responses were like those of wild type, they concluded that there must be a fourth photoreceptor lurking in the background. At the time, they did not consider the possibility that it might be a second member of the phototropin family.

It was only when investigators turned their attention to phototropin double mutants that a clear picture emerged. The stomata in mutants lacking both phot1 and phot2 were completely unresponsive to blue light (Kinoshita *et al.* 2001). As both single *phot* mutants showed an almost wild-type response, the two phototropins clearly carry out redundant functions in the response. Phot1 localization studies indeed indicate that guard cells contain abundant phot1 in the vicinity of their plasma membranes (Sakamoto and Briggs, 2002). Thus the phototropins are major players in the stomatal response just as in the case of phototropism.

Is the search for the photoreceptors involved in blue light-activated stomatal opening complete? It is highly likely that it is not. Zeiger and Zhu (1998) published the first report of a putative blue-light receptor for stomatal opening. They found that the *Arabidopsis* mutant *npq1* failed to show a stomatal opening response to blue light. Since the mutant fails to synthesize the carotenoid zeaxanthin they hypothesized that zeaxanthin was the photoreceptor chromophore for the response. More detailed studies (Frechilla *et al.* 1999) supported the hypothesis. Subsequently, however, Eckert and Kaldenhoff (2000) obtained a normal stomatal response to blue light from the same *npq1* mutant. Since the experimental conditions were not the same between the two studies, it could well be that one set of conditions favoured a zeaxanthin-based system and another set of conditions favoured a phototropin-based system. It is also possible that the blue-light response reported in the *npq1* mutant was mediated through the blue-absorbing bands of phytochrome.

There is another powerful piece of evidence that the photoreceptor picture for stomata is incomplete. Blue light-stimulated stomatal opening can be reversed by a brief pulse of green light (Frechilla *et al.* 2000). Thus the system can show a phytochrome-like photoreversibility. Since there is no evidence from extensive spectral studies of the LOV domains to indicate that blue-light irradiation causes formation of a long-lived green light-absorbing intermediate, there must be some other pigment doing the work in green light. Thus as with phototropism, the list of photoreceptors involved in the stomatal response to light is still incomplete.

8. CHLOROPLAST MOVEMENTS INDUCED BY BLUE LIGHT

As mentioned above, workers in the nineteenth and early twentieth century established that chloroplast movements in response to light were activated by blue light (Senn, 1908). In a series of detailed quantitative experiments Voerkel (1934) reinvestigated the spectral sensitivity of the accumulation response and confirmed that it was strictly a blue light response. None of his light sources in the visible (relatively broad-band gelatine filters) were sufficiently intense to cause the avoidance response although he did demonstrate an avoidance response with UV-A. Zurzycki (1962) determined the first true action spectra, both for the accumulation and for the avoidance response for chloroplast movements in *Lemna trisulca*. In both cases he obtained a typical "cryptochrome" action spectrum. In a later study of the moss *Funaria hygrometrica* (Zurzycki, 1967) he reported an increase in activity in response to wavelengths below 300 nm. This increase is reminiscent of the UV-B activity subsequently reported by Baskin and Iino (1987) for alfalfa phototropism.

As is the case for phototropism, there is a large literature on the physiology and photophysiology of chloroplast movements (for reviews see Haupt, 1999; Haupt and

Scheuerlein, 1990; Kagawa and Wada, 2002). However, the issue of the photoreceptor remained unresolved for more than three decades following publication of the action spectrum. Then two groups (Jarillo *et al.* 2001; Kagawa *et al.* 2001) reported almost simultaneously that a *phot2* mutant of *Arabidopsis* failed to show any detectable avoidance response. Thus unlike the situation for phototropism and stomatal opening, there is no functional redundancy between phot1 and phot2 for the avoidance reaction. A little later, Sakai *et al.* (2001) reported that at low light intensities, either phot1 or phot2 alone could mediate the accumulation response. Only in the *phot1 phot2* double mutant was the accumulation response completely lacking. Hence, unlike the avoidance response, either phototropin could mediate the accumulation reaction. Thus almost a century and a half was required to identify the photoreceptors involved in the light-activated chloroplast movements first described by Böhm (1856), and it took the tools of molecular genetics to solve the problem. Figure 7 illustrates the respective roles of phot1 and phot2 in chloroplast movement (after Briggs and Christie, 2002).

9. LEAF EXPANSION

Although it was well established during the 19th century that light was required for leaf expansion, it was generally agreed that photosynthesis was the driving force. However, two investigators demonstrated a dramatic effect of short light pulses on leaf expansion. Trumpf (1924, cited in detail by Priestly, 1925) gave brief daily light exposures insufficient to cause "development of chlorophyll" to etiolated seedlings of *Phaseolus multiflorus* and reported formation of well-developed lamina. Using coloured solutions, he made crude attempts to determine the most effective spectral regions and concluded that red light was responsible for the effect. Shortly thereafter, Priestley (1925) induced leaf expansion in etiolated seedlings of *Vicia faba* and *Pisum sativum* with daily pulses of white light as short as ten minutes. These authors both concluded that there must be some specific effect of light, independent of photosynthesis that mobilized reserves to the young leaves.

The next spectral studies were those of Parker *et al.* (1949) working with the Beltsville spectrograph. They measured the action spectrum for the effect of light on stem elongation (negative) and leaf expansion (positive) in *Pisum sativum*. The action spectrum was remarkably similar to action spectra they had obtained for night-interruption effects on flowering and they suggested that the same photopigment might be involved. Red light was by far the most effective, but there was a very small effect near 400 nm as well. This small effect in the blue would later be ascribed simply to the blue-absorbing band of phytochrome. With the discovery that the promotion of lettuce seed germination by red light could be reversed by far red light (Borthwick *et al.* 1952), there was an explosion of papers testing far-red reversibility of other red light-activated processes including leaf expansion. Liverman *et al.* (1955) showed that red light promoted the expansion of leaf disks from etiolated bean seedlings and that far-red light reversed the effect of red. Subsequent reviews of leaf development (Humphries and Wheeler, 1963; Dale, 1988) discussed only the role of phytochrome and made no mention of any blue light effects. Indeed, the proceedings of three conferences on blue light effects (Senger, 1980, 1984, 1987) have not one single article addressing blue-light effects on leaf expansion.

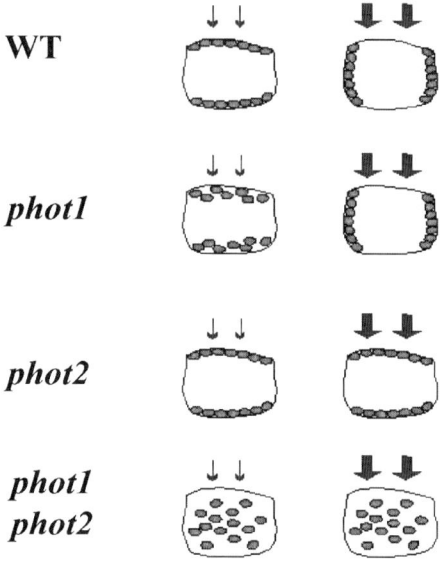

Figure 7. Relative roles of *phot1* and *phot2* in the chloroplast accumulation and avoidance responses. After Briggs and Christie, 2002). Thin arrows are low light, thick arrows are high light.

Bean leaf disks took centre stage again, however, in 1990. Van Volkenburgh and Cleland (1990) demonstrated that the promotion of expansion of the disks by light occurred in the presence of the photosynthesis inhibitor 3-(3,4-dichlorophenyl)-1,1-dimethylurea (DCMU). The action spectrum for the growth promotion showed two equally effective broad peaks, one in the red and one in the blue (Van Volkenburgh *et al.* 1990). The action spectrum was almost the same in the presence of DCMU, eliminating photosynthesis as accounting for either peak. Simultaneous exposure to far red light very much reduced the response to red light, but had no effect on the blue peak. Thus these authors reported for the first time a specific effect of blue light on leaf expansion.

The action spectrum just mentioned was not of sufficient detail to detect fine structure in the blue region of the spectrum, nor did it extend into the UV-A. Thus it was not possible to determine whether it was a typical "cryptochrome" action spectrum or something else. It was only with the availability of photoreceptor mutants that one could address the question of the responsible photoreceptor(s). Sakai *et al.* (2001) mention briefly that the leaves of the *phot1 phot2* double mutant are sharply curled downward, and Sakamoto and Briggs (2002) illustrate the same phenomenon. If either phototropin is present, the leaves appear like those of wild type. Only when both are missing do the leaves show slower growth and abnormal epinasty (Figure 8).

As with phototropism, chloroplast accumulation, and stomatal opening, there is photoreceptor redundancy. At present, however, there are no data available on the relative sensitivity of the two phototropins in mediating the leaf-expansion response. Van Volkenburgh (1999) has reviewed recent progress in research on leaf expansion.

Figure 8. Phot1 and phot2 redundantly mediate leaf expansion in Arabidopsis (Sakamoto and Briggs 2002).

10. THE RAPID INHIBITION OF GROWTH

Compared to the physiological responses to blue light discussed above, study of the rapid inhibition of the growth of etiolated seedlings is a real newcomer. Although effects of light on stem elongation were well known, equipment simply was not available with sufficient resolution to measure minute and rapid changes in growth rate. It was Meijer (1968) who first described the phenomenon. Using a sensitive position transducer he found that blue light induced a rapid and dramatic decrease in the growth rate of etiolated gherkin hypocotyls (*Cucumis sativus*) with a lag period of four minutes or less. Although red light, presumably acting through phytochrome, also induced a decrease in growth rate, the effect was much slower and the lag period was at least half an hour. Gaba and Black (1979) obtained even more dramatic results for partially de-etiolated seedlings of *Cucumis sativus*. The lag period prior to the decrease in growth rate was about 5 minutes for blue light but over five hours for red light. Likewise recovery of the initial dark growth rate was far faster following blue light (about 20 min) than red light (close to 19 hours).

Cosgrove (1981) confirmed the rapid inhibition response in etiolated seedlings of a number of different species. In the case of *Cucumis sativus* the lag period before a precipitous drop in growth rate was between 20 and 30 seconds, far too fast to be accounted for by some hormonal change. He confirmed not only that there were responses to red light as well, but also that they had a far longer lag period (15 minutes for *Pisum sativum*). He also reported a pattern in which following blue light treatment the growth rate recovered fully and remained near the initial value and a pattern in which following recovery, the growth rate subsequently declined again in darkness. Finally, he demonstrated that it was the growing region itself that had to be irradiated to induce the growth inhibition. Figure 9 shows the time course of growth inhibition of an etiolated pea epicotyl by a brief pulse of blue light.

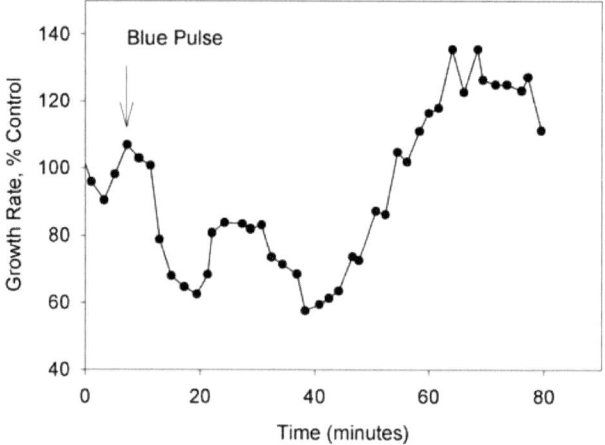

Figure 9. Inhibition of growth of an etiolated pea epicotyl by a brief pulse of blue light (after Laskowski and Briggs, 1989).

Once again it was the availability of *Arabidopsis* photoreceptor mutants that shed light on the identity of the photoreceptors involved. Mutants of *Arabidopsis* lacking either cry1 or cry2 or both showed the normal rapid inhibition of hypocotyl growth brought on by blue light, but subsequently recovered to the dark growth rate over a period of about an hour despite continued irradiation (Folta and Spalding, 2001). By contrast, *phot1* null-mutant seedlings in which both cryptochromes were functional completely lacked the rapid inhibition of growth but showed a gradual decline in growth rate that started within ten minutes or so of the start of irradiation and continued for at least two hours after the onset of illumination. Evidently both cryptochromes are necessary for the sustained inhibition seen in wild-type seedlings. Thus three different blue-light receptors—both cryptochromes and one phototropin—play a role in the suppression of growth that the earlier workers had described. Only phot1 is involved in the very rapid and transient phase whereas both cryptochromes are involved in the slower but persistent phase. As Folta and Spalding used a fluence rate of 80 μmol $m^{-2} s^{-1}$, far above what is needed to activate phot2 in phototropism, and failed to see any rapid response at all, it seems likely that as with the chloroplast avoidance response, only a single phototropin is involved in the rapid inhibition of growth—in this case, phot1 and not phot2. It could simply be that phot2 is present at too low a level to play a role in completely etiolated seedlings.

11. SOLAR TRACKING

Solar tracking, where leaves adjust their position to be at right angles to an incident light source (in nature, the sun) has likely been known for almost two centuries. The changes in leaf orientation that Poggioli (1817) observed for *Mimosa pudica* were almost certainly a solar tracking response. Both Iino (2001) and Koller (2000) have reviewed the extensive literature on this blue light-activated leaf movement (sometimes called diaphototropism). As early as 1936, Yin noted that blue light activated the solar-tracking response in *Malva neglecta* and not red light. He also concluded that in this species it was not a growth response but rather was the completely reversible turgor-related extension and contraction of the two sides of the petiole that accounted for the movement.

The identification of the responsible photoreceptor(s) for this response is still uncertain. *Arabidopsis* leaves are known to change their orientation in response to changes in the direction of incident light and anecdotal evidence indicates that this response is completely lacking in the *phot1 phot2* double mutant. However, it is not resolved whether this response is simply the phototropic response of elongating petioles, the turgor-driven but fully reversible process of solar tracking described above for *Malva*, or some combination of the two. Thus although the phototropins might well be involved (after all they mediate the reversible turgor changes that cause stomatal opening), a strong case has not yet been made that they are the photoreceptor(s) for solar tracking.

12. THE ZTL/ADO FAMILY

Recent studies indicate that LOV domains in higher plants are not confined to the phototropins. Three other *Arabidopsis* proteins contain a single LOV domain. These are FKF1 (Imaizumi *et al.* 2003), a protein required for the flowering response to long days, and ZTL/ADO and LKP2, both with circadian clock function. These three LOV domains all show the spectral change to be expected on light-activated cysteinyl adduct formation and require the conserved cysteine for this change. However, unlike their phototropin counterparts, they show no dark recovery in vitro. Their very recent history and the evidence leading to the conclusion that they are likely blue-light receptors is at the conclusion of chapter 2.12 and will not be considered further here.

13. CONCLUSIONS

The above sections are intended to lay out the lengthy and at times tangled search for blue-light receptors in higher plants. Two families of blue-light receptors are now firmly established: the cryptochromes and the phototropins. Both families are involved in a wide range of physiological and developmental responses, responses that have in some cases been under scientific scrutiny for almost two centuries.

Molecular, genetic, and biophysical evidence implicates the *ZTL/ADO* family of genes as encoding photoreceptors as well. However, it has taken molecular genetic approaches unheard of only a decade ago to come to the tentative conclusion that they are also blue-light sensors. It is fascinating to contemplate the phenomenal acceleration of progress in this area of photomorphogenesis. However, the challenges remain to characterize all of the downstream components involved in signal transduction and elucidate how they lead to the specific responses that are mediated by blue light.

Acknowledgements. The author is extremely grateful to the staff of the Museum of Natural History in Vienna for their valuable assistance in helping him obtain many of the early articles referenced above. He is also grateful to Dr. Trevor E. Swartz for his careful review of the manuscript and his skilful assistance with the figures. Work from the author's laboratory cited in this chapter was supported by NSF Grants 0091384 and 0211605. The author is grateful for this support.

14. REFERENCES

Ahmad, M. and Cashmore, A. R. (1993) *HY4* gene of *A. thaliana* encodes a protein with the characteristics of a blue-light photoreceptor. *Nature, 366*, 162-166.

Ahmad, M., Lin, C., Cashmore, A. R. (1995) Mutations throughout an Arabidopsis blue-light photoreceptor impair blue-light-responsive anthocyanin accumulation and inhibition of hypocotyl elongation. *Plant J, 8*, 653-658.

Ahmad, M., Jarillo, J. A., Klimczak, L. J., Landry, L. G., Peng, T., Last, R. L. *et al.* (1997) An enzyme similar to animal type II photolyases mediates photoreactivation in Arabidopsis. *Plant Cell, 9*, 199-207.

Ahmad, M., Jarillo, J. A., Smirnova, O., Cashmore, A. R. (1998) Cryptochrome blue-light photoreceptors implicated in phototropism. *Nature, 392*, 720-723.

Ahmad, M., Grancher, N., Heil. M., Black, R. C., Giovani, B., Galland, P., Lardemer, D. (2002) Action spectrum for cryptochrome-dependent hypocotyl growth inhibition in Arabidopsis. *Plant Physiol, 129*, 774-785.

Asard, H. and Caubergs, R. (1991) "LIAC activity in higher plants." In: *Biophysics of Photoreceptors and Photomovements in Microorganisms*, Lenci, F., Ghetti, F., Columbetti, G., Häder, D.-P., Song, P.-S. (eds) Plenum, New York, 181-189.

Bachmann F. and Bergann, F. (1930) Über die Wertigkeit von Strahlen verschiedener Wellenlänge für die phototropische Reizung von Avena sativa. *Planta, 10*, 744-755.

Ballaré, C. (2003) Stress under the sun: spotlight on ultraviolet-B responses. *Plant Physiol, 132*, 1725-1727.

Baskin, T. I. and Iino, M. (1987) An action spectrum in the blue and ultraviolet for phototropism in alfalfa. *Photochem Photobiol, 46*, 127-136.

Batschauer, A. (1993) A plant gene for photolyase: an enzyme catalysing the repair of UV-light-induced damage. *Plant J, 4*, 705-709.

Batschauer, A. (1998) Photoreceptors in higher plants. *Planta, 206*, 479-492.

Bergann, F. (1930) Untersuchungen über Lichtwachstum, Lichtkrümmung und Lichtabfall bei Avena sativa mit Hilfe monochromatischen Lichtes. *Planta, 10*, 666-743.

Berns, D. S. and Vaughn, J. R. (1970) Studies on the photopigment system in Physomyces. *Biochem Biophys Res Comm, 39*, 1094-1103.

Blaauw, A. H. (1909) Die Perzeption des Lichtes. *Rec trav bot néerl, 5*, 209-372.

Black, C.C., Turner, J. F., Gibbs, M., Krogmann, D. W., Gordon, S. A. (1962) Studies on photosynthetic processes. II. Action spectra and quantum requirement for triphosphopyridine nucleotide reduction and formation of adenosine triphosphate by spinach chloroplasts. *J Biol Chem, 237*, 580-583.

Böhm, J. A. (1856) Beiträge zur näheren Kentniss des Chlorophylls. *Sitzungsber. Mathem.-Naturwiss Classe kais Akademie Wissenschaften, 22*, 479-512.

Böhm, J. A. (1859) Über den Einfluss der Sonnenstrahlen auf die Chlorophyllbildung und das Wachsthum der Pflanzen überhaupt. *Sitzungsber. Mathem.-Naturwiss Classe kais Akademie Wissenschaften, 37,* 453-476.

Borthwick, H. A., Hendricks, S. B, Parker, M. W., Toole, E. H., Toole, V. K. (1952) A reversible photoreaction controlling seed germination. *Proc Natl Acad Sci USA, 38,* 662-666.

Brain, R. D., Freeberg, J. A., Weiss, C. V., Briggs, W. R. (1977) Blue light-induced absorbance changes in membrane fractions from corn and *Neurospora. Plant Physiol, 59,* 948-952.

Briggs, W. R., Tocher, R. D. Wilson J. F. (1957) Phototropic auxin redistribution in corn coleoptiles. *Science, 126,* 165-167.

Briggs, W. R. (1963) Mediation of phototropic responses of corn coleoptiles by lateral transport of auxin. *Plant Physiol, 38,* 237-247.

Briggs, W. R., Zollinger, W. D. Platz, B. B. (1968) Some properties of phytochrome isolated from dark-grown oat seedlings (Avena sativa L.). *Plant Physiol, 43,* 1239-1243.

Briggs, W. R. (1973) The phototropic responses of higher plants. *Annu Rev Plant Physiol, 14,* 311-352.

Briggs, W. R. and Iino M. (1983) Blue-light-absorbing photoreceptors in plants. *Phil Trans Roy Soc B, 303,* 347-359.

Briggs, W. R. and Huala, E. (1999) Blue-light photoreceptors in higher plants. *Annu Rev Cell Dev Biol, 15,* 33-62.

Briggs, W. R., Christie, J. M., Salomon, M. (2001) Phototropins: A new family of flavin-binding blue light receptors in plants. *Antioxidants & Redox Signalling, 3,* 775-788.

Briggs, W. R. and Christie, J. M. (2002) Phototropins 1 and 2: Versatile plant blue light receptors. *Trends Plant Sci, 7,* 204-210.

Brudler R., Hitomi K., Daiyasu H., Toh H., Kucho K., Ishiura M., *et al.* (2003). Identification of a new cryptochrome class: structure, function, and evolution. *Mol Cell, 11,* 59-67.

Bunsen, R. and Roscoe H. (1862) Photochemische Untersuchungen. *Ann Phys Chem, 117,* 529-562.

Butler, W. L., Lane, H. C. and Siegelman, W. H. (1963) Nonphotochemical transformations of phytochrome *in vivo. Plant Physiol, 38,* 514-519.

Butler, W. L., Siegelman, W. H. and Miller, C. O. (1964) Denaturation of phytochrome. *Biochemistry, 3,* 851-857.

Casal, J. J. (2000) Phytochromes, cryptochromes, phototropin: Photoreceptor interactions in plants. *Photochem Photobiol, 71,* 1-11.

Cashmore, A. R. (1997) The cryptochrome family of photoreceptors. *Plant Cell Environ, 20,* 674-767.

Cashmore, A. R., Jarillo, J. A., Wu, Y.-J., Liu, D. (1999) Cryptochromes: Blue light receptors for plants and animals. *Science, 284,* 760-765.

Christie, J. M., Reymond, P., Powell, G. K., Bernasconi, P., Raibekas, A. A., Liscum, E. *et al.* (1998) *Arabidopsis* NPH1: A flavoprotein with the properties of a photoreceptor for phototropism. *Science, 282,* 1698-1701.

Christie, J. M., Salomon, M., Nozue, K., Wada, M., Briggs, W. R. (1999) LOV (light, oxygen, or voltage) domains of the blue-light photoreceptor phototropin 1 (nph1): Binding sites for the chromophore flavin mononucleotide. *Proc Natl Acad Sci USA, 96,* 8779-8783.

Christie, J. M. and Briggs, W. R. (2001) Blue light sensing in higher plants. *J Biol Chem, 276,* 11457-11460.

Cosgrove, D. J. (1981) Rapid suppression of growth by blue light. Occurrence, time course, and general characteristics. *Plant Physiol, 67,* 584-590.

Crosson, S. and Moffat, K. (2001) Structure of a flavin-binding plant photoreceptor domain: Insights into light-mediated signal transduction. *Proc Natl Acad Sci USA, 98,* 2995-3000.

Curry, G. M. (1957) Studies on the spectral sensitivity of phototropism. Ph. D. Thesis, Harvard University, Cambridge, MA, pp. 1-152.

Dale, J. E. (1988) The control of leaf expansion. *Annu Rev Plant Physiol Plant Mol Biol,* 267-295.

Darwin, C., *The Power of Movement in Plants.* (1966 unabridged republication) New York: De Capo Press, 1881.

Darwin, F. (1898) Observations on stomata. *Phil Trans Roy Soc London Ser B, 190,* 100-143.

Darwin, F. (1911) On a new method of estimating the aperture of stomata. *Proc Roy Soc London B Biol Sci, 84,* 136-154.

Daubeny, C. (1836) On the action of light upon plants, and of plants upon the atmosphere. *Phil Trans Roy Soc, 126,* 149-175.

Delbrück, M. and Shropshire, W. Jr. (1960) Action and transmission spectra of Phycomyces. *Plant Physiol, 35,* 455-458.
Eckert, M. and Kaldenhoff, R. (2000) Light-induced stomatal movement of selected *Arabidopsis thaliana* mutants. *J Exp Botan 51,* 1435-1442.
Everett, M. and Thimann, K. V. (1968) Second positive phototropism in the Avena coleoptile. *Plant Physiol, 43,* 1786-1792.
Famintzen, A. (1867) Die Wirkung des Lichtes und der Dunkelheit auf die Verteilung der Chlorophyllkörner in den Blättern von Mnium sp.? *Jahrb Wiss Botanik, 6,* 49-54.
Folta, K. M. and Spalding, E. P. (2001) Unexpected roles for cryptochrome 2 and phototropin revealed by high-resolution analysis of blue light-mediated hypocotyl growth inhibition. *Plant J, 26,* 471-478.
Frechilla, S., Zhu, J., Talbott, L. D., Zeiger, E. (1999) Stomata from *npq1*, a zeaxanthin-less Arabidopsis mutant, lack a specific response to blue light. *Plant Cell Physiol, 40,* 949-954.
Frechilla, S., Talbott, L. D., Bogomolni, R., Zeiger, E. (2000) Reversal of blue light-stimulated stomatal opening by green light. *Plant Cell Physiol, 41,* 171-176.
Gaba, V. and Black, M. (1979) Two separate photoreceptors control hypocotyl growth in green seedlings. *Nature, 278,* 51-54.
Gallagher, S., Short, T. W., Ray, P. M., Pratt, L. H., Briggs, W. R. (1988) Light-mediated changes in two proteins found associated with plasma membrane fractions from pea stem sections. *Proc Natl Acad Sci USA, 85,* 8003-8007.
Galland, P. and Senger, H. (1988) The role of pterins in the photoreception and metabolism of plants. *Photchem Photobiol,* 48, 811-820.
Galston, A. W. (1949) Riboflavin-sensitised photo oxidation of indoleacetic acid and related compounds. *Proc Natl Acad Sci USA, 35,* 10-17.
Galston, A. W. and Hand, M. E.(1949) Studies on the physiology of light action. I. Auxin and the light inhibition of growth. *Amer J Bot, 36,* 85-94.
Galston, A. W. and Baker, R. S. (1949) Studies on the physiology of light action. II. The photodynamic action of riboflavin. *Amer J Bot, 36,* 773-780.
Galston, A. W. (1950) Riboflavin, light, and the growth of plants. *Science, 111,* 619-624.
Galston, A. W. (1959) "Phototropism of stems, roots and coleoptiles." In: *Handbuch Pflanzenphysiol XVII/1,* Ruhland, W. (ed.) Springer-Verlag, Berlin, 492-529.
Gressel, J. (1979) Blue light photoreception. *Photchem Photobiol, 30,* 749-754.
Haupt, W. and Scheuerlein, R. (1990) Chloroplast Movement. *Plant Cell Environ, 13,* 595-614.
Haupt, W. (1999) "Chloroplast movement: from phenomenology to molecular biology." In: *Progress in Botany,* Esser, K. (ed.) Springer-Verlag, Berlin, 3-36.
Hoffman, P. D., Batschauer, A., Hays, J. B. (1996) PHH1, a novel gene from *Arabidopsis thaliana* that encodes a protein similar to plant blue-light photoreceptors and microbial photolyases. *Mol Gen Genet, 253,* 259-265.
Hsiao, T. C. and Allaway, W. G. (1973). Action spectra for guard cell Rb⁺ uptake and stomatal opening in *Vicia faba. Plant Physiol, 51,* 82-88.
Huala, E., Oeller, P. W., Liscum, E., Han, I.-S., Larsen, E., Briggs, W. R. (1997) *Arabidopsis* NPH1: A protein kinase with a putative redox-sensing domain. *Science, 278,* 2120-2123.
Humphries, E. C. and Wheeler, A. W. (1963) The physiology of leaf growth. *Annu Rev Plant Physiol, 14,* 385-410.
Iino, M. (2001) "Phototropism in higher plants." In: *Photomovement,* Häder, D.-P, and Lebert, M. (eds) Elsevier, Amsterdam, 659-811.
Imaizumi, T., Kanegae, T., Wada, M. (2000) Cryptochrome nucleocytoplasmic distribution and gene expression are regulated by light quality in the fern *Adiantum-capillus veneris. Plant Cell, 12,* 81-95.
Jarillo, J. A., Ahmad, M., Cashmore, A. R. (1998) NPL1 (Accession No. AF053941): A second member of the NPH serine/threonine kinase family of Arabidopsis. *Plant Physiol, 117,* 719.
Jarillo, J. A., Gabrys, H., Capel, J., Alonso, J. M., Ecker, J. R., Cashmore, A. R. (2001) Phototropin-related NPL1 controls chloroplast relocation induced by blue light. *Nature, 410,* 592-594.
Jenkins, G. (1997) UV and blue light signal transduction in Arabidopsis. *Plant Cell Environ, 20,* 773-778.
Johnston, K. S., Phototropic sensitivity in relation to wavelength. *Smithsonian Misc Coll, 92,* 1-17 (Publication 3285).
Kagawa, T., Sakai, T., Suetsugu, N., Oikawa, K., Ishiguro, S., Kato, T., Tabata, S. *et al.* (2001) *Arabidopsis* NPL1: A phototropin homolog controlling the chloroplast high-light avoidance response. *Science, 291,* 2138-2141.

Kagawa, T. and Wada, M. (2002) Blue light-induced chloroplast movement. *Plant Cell Physiol, 43*, 367-371.

Karlsson, P. E. (1986) Blue light regulation of stomata in wheat seedlings. II. Action spectrum and search for action dichroism. *Physiologia Plantarum, 66,* 207-210.

Kaufman, L. S. (1993) Transduction of blue-light signals. *Plant Physiol, 102*, 333-337.

Ketellapper, H. J. (1963) Stomatal physiology. *Annu Rev Plant Physiol, 14*, 249-270.

Khurana, J. P. and Poff, K. L. (1989) Mutants of *Arabidopsis thaliana* with altered phototropism. *Planta, 178*, 400-406.

Kinoshita, T., Doi, M., Suetsugu, N., Kagawa, T., Wada, M., Shimizaki, K. (2001) phot1 and phot2 mediate blue light regulation of stomatal opening. *Nature, 414*, 656-660.

Koller, D. (2000) Plants in search of sunlight. *Adv Botan Res, 33*, 35-131.

Konjević, R., Steinitz, B., Poff, K. L. (1989) Dependence of the phototropic response of *Arabidopsis thaliana* on fluence rate and wavelength. *Proc Natl Acad Sci USA, 86*, 9876-9880.

Konjević, R., Khurana, J. P., Poff, K. L. (1992) Analysis of multiple photoreceptor pigments for phototropism in a mutant of *Arabidopsis thaliana*. *Photochem Photobiol, 55*, 789-792.

Koorneef, M., Rolff, E., Spruitt, C. J. P. (1980) Genetic control of hypocotyl elongation in *Arabidopsis thaliana* (L.) Heynh. *Zeit Pflanzenphysiologie, 100*, 147-160.

Kuiper, P. J. C. (1964) Dependence upon wavelength of stomatal movement in epidermal tissue of *Senecio odoris*. *Plant Physiol, 39*, 952-955.

Lascève, G., Leymarie, J., Olney, M. O., Liscum, E., Christie, J. M., Vavasseur, A. *et al.* (1999) Arabidopsis contains at least four independent blue-light-activated signal transduction pathways. *Plant Physiol, 120*, 605-614.

Laskowski M. J. and Briggs W. R. (1989) Regulation of pea epicotyl elongation by blue light. Fluence-response relationships and growth distribution. *Plant Physiol, 89*, 293-298.

Lin, C., Ahmad, M., Gordon, D., Cashmore, A. R. (1995a) Expression of an *Arabidopsis* cryptochrome gene in transgenic tobacco results in hypersensitivity to blue, UV-A, and green light. *Proc Natl Acad Sci USA, 92*, 8423-8427.

Lin, C., Robertson, D. E., Ahmad, M., Raibekas, A. A., Jorns, M. S., Dutton, L. P. *et al.* (1995b) association of flavin adenine dinucleotide with the *Arabidopsis* blue light receptor CRY1. *Science, 269*, 968-970.

Lin, C., Ahmad, M., Chan, J., Cashmore, A. R. (1996) CRY2: A second member of the Arabidopsis cryptochrome gene family (Accession No. U43397). *Plant Physiol, 110*, 1047.

Lin, C. and Cashmore, A. R. (1996) "Cryptochrome and plant photomorphogenesis." In: *Regulation of Plant Growth and Development by Light*, Briggs, W. R., Tobin, E., Heath, R. L. (eds.) Amer Soc Plant Physiologists, Rockville, MD, 30-41.

Lin, C. (2000) Plant blue-light receptors. *Trends Plant Sci, 5*, 337-342.

Lin, C. (2002) Blue light receptors and signal transduction. *Plant Cell*, Supplement S207-225.

Liscum, E. and Briggs, W. R. (1995) Mutations in the *NPH1* locus of Arabidopsis disrupt the perception of phototropic stimuli. *Plant Cell, 7*, 473-485.

Liverman, J. L., Johnson, M. P., Starr, L. (1955) Reversible photoreaction controlling expansion of etiolated bean-leaf disks. *Science, 121*, 440-441.

Lorenzi, R., Ceccarelli, N. Lercari, B., Gualtieri, P. (1994) Identification of retinal in higher plants: is a rhodopsin-like protein a blue light photoreceptor? *Phytochem.* 36: 599-601.

Lurie, S. (1978) The effect of wavelength of light on stomatal opening. *Planta, 140*, 245-249.

Malhotra, K., Kim, S.-T., Batschauer, A., Dawut, L., Sancar, A. (1995) Putative blue-light photoreceptors from *Arabidopsis thaliana* and *Sinapis alba* with a high degree of sequence homology to DNA photolyase contain two photolyase cofactors but lack DNA repair activity. *Biochemistry, 34*, 6892-6899.

Meijer, G. (1968) Rapid growth inhibition of gherkin hypocotyls in blue light. *Acta Botan Néerl, 17*, 9-14.

Muñoz, V. and Bulter, W. L. 1975. Photoreceptor pigment for blue light in *Neurospora crassa*. *Plant Physiol, 55*, 421-426.

Ogawa, T., Ishikawa, H., Shimada, K. and Shibata, K. (1978) Synergistic action of red and blue light and action spectra for malate formation in guard cells of *Vicia fava* L. *Planta, 142*, 61-65.

Ogawa, T. (1981) Blue light response of stomata with starch-containing (*Vicia faba*) and starch-deficient (*Allium cepa*) guard cells under background illumination with red light. *Plant Sci Lett, 22*, 103-108.

Palmer, J. M., Warpeha, K. M. F., Briggs, W. R. (1996) Evidence that zeaxanthin is not the photoreceptor for phototropism in maize coleoptiles. *Plant Physiol, 110*, 1323-1328.

Parker, M. W., Hendricks, S. B., Borthwick, H. A., Went, F. W. (1949) Spectral sensitivities for leaf and stem growth of etiolated pea seedlings and their similarity to action spectra for photoperiodism. *Am J Botan, 36*, 194-204.

Payer, J. (1842) Mémoire sur la tendance des tiges vers la lumière. *Compte Rendu des Seances de l'Académie des Sciences, July 4, 1842*, 1194-1196.

Poff. K. L., Butler, W. L., Loomis, W. F., Jr. (1973) Light-induced absorbance changes associated with phototaxis in *Dictyostelium*. *Proc Natl Acad Sci USA, 70*, 813-816.

Poff, K. L. and Butler, W. L. (1974) Absorbance changes induced by blue light in *Phycomyces blakesleeanus* and *Dictyostelium discoideum*. *Nature, 248*, 799-801.

Poggioli, S. (1817) Della influenza che ha il raggio magnetico sulla vegetatione delle piante. *Bologna – Coi Tipi di Annesio Nobili Opusc Scientif Fasc I*, 9-23.

Presti, D. and Delbrück, M. (1978) Photoreceptors for biosynthesis, energy storage and vision. *Plant Cell Environ, 1*, 81-100.

Priestley, J. H. (1925) Light and growth I. The effect of brief light exposure upon etiolated plants. *New Phytologist, 24*, 271-283.

Quiñones, M. A. and Zeiger, E. (1996) Close correspondence between the action spectra for the blue-light responses of guard cells and coleoptile chloroplasts, and the spectra for stomatal opening and coleoptile phototropism. *Proc Natl Acad Sci USA, 93*, 2224-2228.

Reinert, J (1959) Phototropism and phototaxis. *Annu Rev Plant Physiol, 10*, 441-458.

Reymond, P., Short, T. W., Briggs, W. R., Poff, K. L. (1992) Light-induced phosphorylation of a membrane protein plays an early role in signal transduction for phototropism in *Arabidopsis thaliana*. *Proc Natl Acad Sci USA, 89*, 4718-4721.

Sachs, J. (1864) Wirkungen farbigen Lichts auf Pflanzen. *Botanische Zeitung, 47*, 353-358.

Sachs, J., *Vorlesungen über Pflanzen-Physiologie* (2nd edition), Leipzig: W. Engelmann, 1887.

Sakai, T., Kagawa, T., Kasahara, M., Swartz, T. E., Christie, J. M., Briggs, W. R. et al. (2001) *Arabidopsis* nph1 and npl1: Blue light receptors that mediate both phototropism and chloroplast relocation. *Proc Natl Acad Sci USA, 98*, 6969-6974.

Sakamoto, K. and Briggs, W. R. (2002) Cellular and subcellular localization of phototropin 1. *Plant Cell, 14*, 1723-1735.

Sancar, A. (1994) Structure and function of DNA photolyase. *Biochemistry, 3*, 2-9.

Sancar, A. (2000) Cryptochrome: The second photoactive pigment in the eye and its role in circadian perception. *Annu Rev Biochemistry, 69*, 31-67.

Sargent, M. L. and Briggs, W. R. (1967) The effects of light on a circadian rhythm of conidiation in *Neurospora*. *Plant Physiol,, 42*, 1504-1510.

Schmidt, W. and Butler, W. L. (1976) Flavin-mediated photoreactions in artificial systems: A possible model for the blue-light photoreceptor pigment in living systems. *Photochem Photobiol, 24*, 71-75.

Senn, G., *Die Gestalts-und Lageveränderung der Pflanzen Chromatophoren*. Leipzig: W. Engelmann, 1908.

Senger, H. (ed.), *The Blue Light Syndrome*. Berlin: Springer-Verlag, 1980.

Senger, H. and Briggs, W. R. (1981) "The blue light receptor(s): Primary reactions and subsequent metabolic changes." In: *Photochem Photobiol Reviews*, Smith, K. C. (ed), Plenum, New York, 1-38,.

Senger, H. (ed.) *Blue Light Effects in Biological Systems*, Berlin: Springer-Verlag, 1984.

Senger, H. (ed.) *Blue Light Responses: Phenomena and Occurrence in Plants and Microorganisms*, Vols. I, II. Boca Raton, Florida: CRC Press, 1987.

Shah, I., *The pleasantries of the incredible Mulla Nasrudin*. New York: E. P. Dutton, 1971.

Shalitin D., Yu X., Maymon M., Mockler T., Lin C. (2003) Blue light-dependent *in vivo* and *in vitro* phosphorylation of Arabidopsis cryptochrome 1. *Plant Cell, 15*, 2421-2429.

Short. T. W. and Briggs, W. R. (1990) Characterization of a rapid, blue light-mediated change in detectable phosphorylation of a plasma membrane protein from etiolated pea (*Pisum sativum* L.) seedlings. *Plant Physiol, 92*, 179-185.

Short, T. W. and Briggs, W. R. (1994) The transduction of blue light signals in higher plants. *Annu Rev Plant Physiol, 45*, 143-171.

Shropshire, W., Jr. and Withrow, R. B. (1958) Action spectrum for phototropic tip-curvature of Avena. *Plant Physiol, 33*, 360-365.

Shropshire, W., Jr. (1980) "Carotenoids as primary photoreceptors in blue-light responses." In: *The Blue Light Syndrome*, Senger, H. (ed.) Springer-Verlag, Berlin, 172-186.
Sonne, C. (1929) Weitere Mitteilungen über die Abhängigkeit der lichtbiologischen Reaktionen von der Wellenlänge des Lichtes. Untersuchungen über Phototropismus. *Strahlentherapie, 31,* 778-785.
Taylor, B. L. and Zhulin, I. (1999) PAS domains: Internal sensors of oxygen, redox potential, and light. *Microbiol Mol Biol Rev, 63,* 479-506.
Thimann, K. V. and Curry, G. M. (1960) Phototropism and phototaxis. *Comparative Biochemistry, 1,* 243-306.
Trumpf, C. (1924) Über den Einfluss intermittierender Belichtung auf das Etiolement der Pflanzen. *Botan Archiv, 5,* 381-410.
Van Volkenburgh, E. and Cleland, R. E. (1990) Light-stimulated cell expansion in bean (*Phaseolus vulgaris* L.) leaves. I. Growth can occur without photosynthesis. *Planta, 182,* 72-76.
Van Volkenburg, E., Cleland, R. E. and Watanabe, M. (1990) Light-stimulated cell expansion in bean (*Phaseolus vulgaris* L.) leaves. II. Quantity and quality of light required. *Planta, 182,* 77-80.
Van Volkenburgh, E. (1999) Leaf Expansion – an integrating plant behaviour. *Plant Cell Environ, 22,* 1463-1473.
Voerkel, S. (1934) Untersuchungen über die Phototaxis der Chloroplasten. *Planta, 21,* 156-205.
Wald, G. and DuBuy, H. G. (1936) Pigments of the oat coleoptile. *Science, 84,* 247.
Wang, H., Ma, L.-G., Ki, J.-M., Zhao, H.-Y., Deng, X.-W. (2001) Direct interaction of Arabidopsis cryptochromes with COP1 in light control development. *Science, 294,* 154-158.
Went, F. W. (1928) Wuchsstoff und Wachstum. *Rec trav Bot Néerl, 25,* 1-116.
Widell, S., Cauberges, R. J., Larsson, C. (1983) Spectral characterization of light-reducible cytochrome in a plasma membrane-enriched fraction and in other membranes from cauliflower inflorescences. *Photochem Photobiol, 38,* 95-98.
Yang, H.-Q., Wu, Y.-Jie, Tang, R.-H., Liu, D., Liu, Y., Cashmore, A. R. (2000) The C termini of *Arabidopsis* cryptochromes mediate constitutive light responses. *Cell, 103,* 815-827.
Yang, H.-Q, Tang, R.-H. and Cashmore, A. R. (2001) The signalling mechanism of Arabidopsis CRY1 involves direct interaction with COP1. *Plant Cell, 13,* 2573-2587.
Yanofsky, M. J. and Kay, S. A. (2002) Molecular basis of seasonal time measurement in *Arabidopsis*. *Nature, 419,* 308-312.
Yin. H. C. (1938) Diaphototropic movement of the leaves of *Malva neglecta*. *Amer J Botan, 25,* 1-6.
Zantedeschi, M. (1843) De l'influence qu'exercent sur la vegetation de plants et la germination des graines les rayons transmis à travers des verres colorés. *Compte Rendu des Seances de l'Académie des Sciences*, January 2, 1843, 747-749.
Zeiger, E. and Hepler, P. H. (1977) Light and stomatal function: Blue light stimulates swelling of guard cell protoplasts. *Science, 196,* 887-889.
Zeiger, E. and Field, C. (1982) Photocontrol of the functional coupling between photosynthesis and stomatal conductance in the intact leaf. *Plant Physiol, 70,* 370-375.
Zeiger, E. and Zhu, J. (1998) Role of zeaxanthin in blue light photoreception and the modulation of light-CO_2 interactions in guard cells. *J Exper Botan,* 433-442.
Zurzycki, J. (1962) The action spectrum for the light-dependent movements of chloroplasts in *Lemna trisulca* L. *Acta Soc Bot Poloniae, 31,* 489-538.
Zurzycki, J. (1967) Properties and localization of the photoreceptor active in displacements of chloroplasts in *Funaria hygrometrica*. I. Action spectrum. *Acta Societatis Botanicorum Poloniae, 36,* 133-142.

Chapter 11

CRYPTOCHROMES

Anthony R. Cashmore
Plant Science Institute, Department of Biology, University of Pennsylvania, Philadelphia PA 19104-6018, USA (e-mail: cashmore@sas.upenn.edu)

1. INTRODUCTION

The isolation of a cryptochrome gene from Arabidopsis led to studies aimed at the characterization of the molecular structure, the physiological function and the mode of action of the corresponding cryptochrome (CRY) photoreceptor (Ahmad and Cashmore 1993). Arabidopsis mutants deficient in this photoreceptor, the *cry1/hy4* mutants, are selectively attenuated in their responsivity to blue and UV-A light, exhibiting a long hypocotyl when grown under these lights conditions, characteristic in this respect of dark grown seedlings. By contrast, the mutants appear like wild type when grown under red or far-red light (Koornneef, Rolff *et al.,* 1980; Ahmad and Cashmore 1993). The initial studies of cryptochromes were first confined to plants, but were later extended to include animals and recently, bacteria. Here I will outline these studies and will include both a description of the discovery of various cryptochromes as well as discussions concerning what is known about the function and the mode of action of these photoreceptors.

2. PHOTOLYASES

The initial argument that cryptochrome functions as a photoreceptor, as well as much of the understanding (or at least, the presumption) concerning the mode of action of cryptochromes, was much influenced by the observation that the sequence of cryptochromes are markedly similar to those of a family of proteins known as photolyases. These latter proteins perform an important function in DNA repair (Sancar 2003). DNA that is damaged through the action of UV-B light contains pyrimidine dimers. The satisfactory repair of this damaged DNA is an essential and characteristic feature of all living organisms and reflecting this all species possess mechanisms to achieve this repair. One mode of repair is brought about through the action of photolyases, found throughout the bacterial kingdom as well as in plants and many animals, although not mammals.

Photolyases are flavoproteins that mediate a repair of DNA in response to the absorption of UV-A or blue light (Figure 1). This light absorption reflects the presence of a non-covalently bound flavin chromophore as well as a second light-harvesting chromophore — this latter is either a pterin (methenyltetrahydrofolate;

MTHF) or a deazaflavin. Excitation energy resulting from light absorption by this second chromophore is transferred to the flavin, the primary catalytic chromophore.

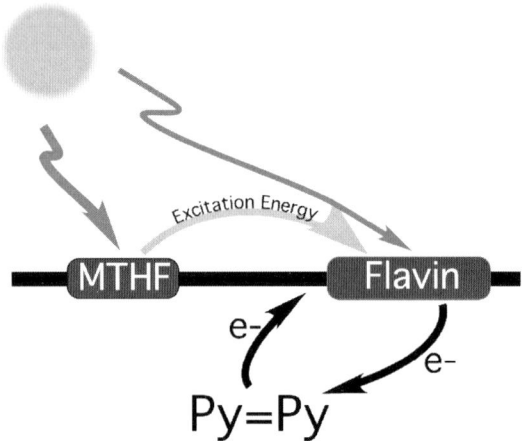

Figure 1. Mechanism of action of photolyase.

Photolyase bind preferentially to DNA sites containing lesions in the form of pyrimidine dimers. This DNA binding activity of photolyases is light independent. In its excited state — generated either through the direct absorption of light, or through transfer of excitation energy from the second light-harvesting chromophore — the flavin transfers an electron to the bound pyrimidine dimer. The pyrimidine dimer, now in its reduced form, isomerises to yield the undamaged monomers, the electron being returned to the flavin. Mechanistically, photolyases function as redox proteins, even though the reaction results in no net change in redox state of either the flavin or the pyrimidine substrate.

The photolyase family of proteins comprises three different sub-families of proteins. These include the type I and II groups, both of which function on cyclobutane pyrimidine dimers, the major products resulting from irradiation of DNA with UV-B light. In addition there is a third class of photolyases, the 6-4 photolyases; these serve to repair 6-4 pyrimidine dimers, a minor product of UV-B irradiation.

3. THE DISCOVERY OF CRYPTOCHROME

Cryptochromes, first identified in Arabidopsis (Ahmad and Cashmore, 1993), have also been found in green algae (Small *et al.*, 1995) and are now known to occur throughout the plant kingdom, including ferns (Kanegae and Wada, 1998) and mosses (Imaizumi *et al.*, 2002; see Chapters 24, 25). Cryptochromes have also been described for animals, including flies (Emery *et al.*, 1998; Stanewsky *et al.*, 1998) and humans (Hsu *et al.*, 1996; Todo *et al.*, 1996) where, as in plants, they play a role in circadian behaviour (see Chapter 26).

3.1 Cryptochromes of Arabidopsis

A description of the molecular properties of a blue light photoreceptor for plants came through the isolation of a T-DNA-tagged Arabidopsis mutant deficient in its response to blue light (Ahmad and Cashmore, 1993). This mutant was shown to be allelic to the previously described *hy4* mutant (Koornneef *et al.*, 1980). The mutant — now commonly referred to as the *cry1* mutant — has a long hypocotyl when grown under blue light, being attenuated in the blue light mediated inhibition of cell expansion that characterizes wild-type seedlings. By contrast, the mutant shows the normal response when grown under red or far-red light and similarly shows no distinct phenotype when grown in darkness; the mutant is selectively impaired in its response to blue light. The sequence of the *CRY1* gene indicated that it was similar to those encoding proteins of the photolyase family (Ahmad and Cashmore, 1993). Furthermore, when expressed in insect cells the Arabidopsis CRY1 protein is observed to bind a flavin, also similar in this respect to photolyases (Lin *et al.*, 1995). However, Arabidopsis CRY1 differs from photolyases in two significant ways: Firstly, and most significantly, CRY1 protein possesses no detectable photolyase activity. Secondly, Arabidopsis CRY1 is larger than *E. coli* photolyase, containing additional distinguishing amino acid residues at its C-terminus (Ahmad and Cashmore, 1993). We refer to this cryptochrome C-terminal domain as CCT and the N-terminal photolyase-like domain we refer to as CNT.

Ectopic overexpression of the *CRY1* gene, in either transgenic tobacco or Arabidopsis plants, confers hypersensitivity to light (Lin *et al.*, 1995, 1996). As with the loss of function *cry1* alleles, this change in light sensitivity of the overexpressing plants was restricted to UV-A and blue light, and to a lesser extent to green light, consistent with the notion that CRY1 is a photoreceptor responding selectively to these wavelengths. The CRY1 overexpressing plants also show a dwarf phenotype when grown under white light, similar in this respect to that observed for plants overexpressing the phytochrome photoreceptors (Keller *et al.*, 1989).

The *cry1* mutants show a semidominant phenotype, reflecting haploid insufficiency (Koornneef *et al.*, 1980; Ahmad and Cashmore, 1993). That is, heterozygous mutants exhibit a phenotype that is intermediate between that of the wild-type plants and the homozygous mutant. This observation simply indicates that both the activity of the photoreceptor and the severity of the associated phenotype are proportional to the amount of the photoreceptor.

As noted, photolyases are characterized by a second light-harvesting chromophore, commonly MTHF (Sancar 2003). Plant cryptochromes also bind MTHF when expressed in *E. coli* (Malhotra *et al.*, 1995), although when recombinant Arabidopsis CRY1 is isolated from insect cells it lacks this second chromophore (Lin *et al.*, 1995). The precise identities of the cryptochrome chromophores in Arabidopsis, and their redox states and absorption properties, are of interest, as in the absence of this information no useful predictions or interpretations of action spectra are possible.

A second Arabidopsis cryptochrome gene, *CRY2*, was identified (Lin et al., 1996; Lin et al., 1998). The encoded CRY2 protein is light-labile, and in keeping with this, *cry2* mutant Arabidopsis seedlings have elongated hypocotyls when grown under low intensity blue light, whereas light inhibition of hypocotyl growth is essentially the same as wild type when the seedlings are grown under blue light of intensity greater than 10 µmoles.meter^{-2}.sec^{-1} (Lin et al., 1998). The most striking phenotypic feature of *cry2* mutants is their altered flowering time (see later).

Cryptochromes have now been characterized for several additional plant species including tomato (Ninu et al., 1999; Weller et al., 2001) and rice (Matsumoto et al., 2003). In both cases, as in Arabidopsis, these cryptochromes apparently play a role in blue light mediated de-etiolation and photomorphogenesis.

3.2 Cryptochromes of algae, mosses and ferns

Cryptochromes have been described for algae (Small et al., 1995), and ferns (Kanegae and Wada, 1998), and recently for mosses (Imaizumi et al., 2002). In the fern *Adiantum capillus-veneris*, spore germination is regulated by blue light, and two of the five cryptochromes described for this fern are thought to be involved in this process. Two *CRY* genes have been described for the moss *Physcomitrella patens*, and disruption of these genes confers an increase in auxin sensitivity in a blue light specific manner (Imaizumi et al., 2002). In the green alga Chlamydomonas, there are blue light specific responses in addition to the phototactic response that is mediated by a rhodopsin-like photoreceptor residing with the eye spot (Deininger et al., 1995). Whether CRY is the photoreceptor mediating any of these other blue light responses, has not been determined.

3.3 Drosophila cryptochrome

A Drosophila cryptochrome was identified and characterized by two independent approaches. A mutant was identified in a screen involving transgenic flies expressing a luciferase reporter driven by the promoter of *PERIOD* (*PER*), a gene that performs an integral role in the central oscillator of the Drosophila circadian clock (Stanewsky et al., 1998). This *cry* mutant exhibited arrhythmic luciferase expression. In a second approach, the Drosophila gene was identified as encoding a protein with sequence similarity to both photolyases and the mammalian and Arabidopsis CRY sequences, prompting the investigators to entertain the possibility that this gene encoded a Drosophila blue light photoreceptor (Emery et al., 1998). In support of this proposal, transcription of the fly *CRY* gene was observed to be under circadian control and influenced by *PER* and *TIM* genes; furthermore, circadian photosensitivity was enhanced in CRY overexpressing strains.

3.4 Mammalian cryptochromes

The earlier-mentioned 6-4 photolyases were first identified in Drosophila (Todo, Takemori *et al.,* 1993). It was of interest to determine if a related sequence existed in mammals, as it had long been believed that mammals lacked any photolyase activity; lesions in mammalian DNA being repaired by an excision repair process. Screening of a human cDNA library led to the identification of a sequence with similarities to the fly 6-4 photolyase gene (Todo *et al.,* 1996). However, expression of this human sequence produced a protein that, although it bound both a flavin and a pterin, possessed no detectable photolyase activity (Hsu *et al.,* 1996). By analogy with the Arabidopsis *CRY* genes it was proposed that this human photolyase-like gene encoded a blue light photoreceptor; furthermore, it was postulated that this photoreceptor might serve to entrain human circadian rhythms (Miyamoto and Sancar 1998).

Cryptochromes have now been identified in a wide variety of animals, including zebrafish (Cermakian *et al.,* 2002; Hirayama *et al.,* 2003). Here, as in mammals (see later), the CRY1 protein is involved in circadian repression of CLOCK:BMAL1-mediated transcription (Hirayama *et al.,* 2003).

3.5 Bacterial and related cryptochromes

Cryptochromes, initially thought to be restricted to eukaryotic organisms, have recently been described for bacteria (Hitomi *et al.,* 2000; Brudler *et al.,* 2003). A photolyase-like sequence from Synechocystis was demonstrated to lack detectable photolyase activity, and hence was designated as a cryptochrome. This sequence was given the name CRY DASH. A crystal structure of the Synechocystis CRY was determined and microarray gene expression studies showed that some genes were upregulated in a mutant lacking the CRY DASH (Brudler *et al.,* 2003). As the protein showed DNA binding properties it was speculated that it might function as a transcriptional repressor.

A related *CRY DASH/CRY3* gene was described for Arabidopsis (Brudler *et al.,* 2003; Kleine *et al.,* 2003). The nuclear encoded CRY3 protein shows the unusual property of being localized to both mitochondria and chloroplasts (Kleine *et al.,* 2003).

4. CRYPTOCHROMES AND PLANT PHOTOMORPHOGENESIS

The phenotype of the *cry1/hy4* mutant seedlings that allowed for the original identification of the Arabidopsis *CRY1* gene was the long hypocotyl observed for seedlings grown under blue or UV-A light (Ahmad and Cashmore, 1993). In these respects the mutant seedlings have a degree of similarity to dark grown seedlings, being attenuated with respect to light induced shortening of their hypocotyl. The *cry1* mutants show no discernable phenotype when grown in darkness, and also appear normal when grown under red or far-red light. Thus the mutants exhibit selectively reduced responsitivity to blue and UV-A light.

Reduced anthocyanin production is an additional readily discernable phenotype associated with the *cry1* mutant, apparently reflecting, at least in part, a reduction in expression of the chalcone synthase gene (Ahmad *et al.*, 1995). Multiple loss of function mutant alleles of the *CRY1* gene have been identified, with several of these mutations affecting the C-terminal domain of the protein, demonstrating the functional importance of this distinguishing sequence. An allelic series of *cry1* mutants were examined in which the severity of both the long hypocotyl phenotype and loss in anthocyanin production were scored (Ahmad *et al.*, 1995). A good correlation was observed for these two phenotypes — mutants severely deficient in one were similarly affected in the other, whereas mutant alleles weakly affecting the long hypocotyl phenotype were also observed to have minimal affects on anthocyanin production (Figure 2).

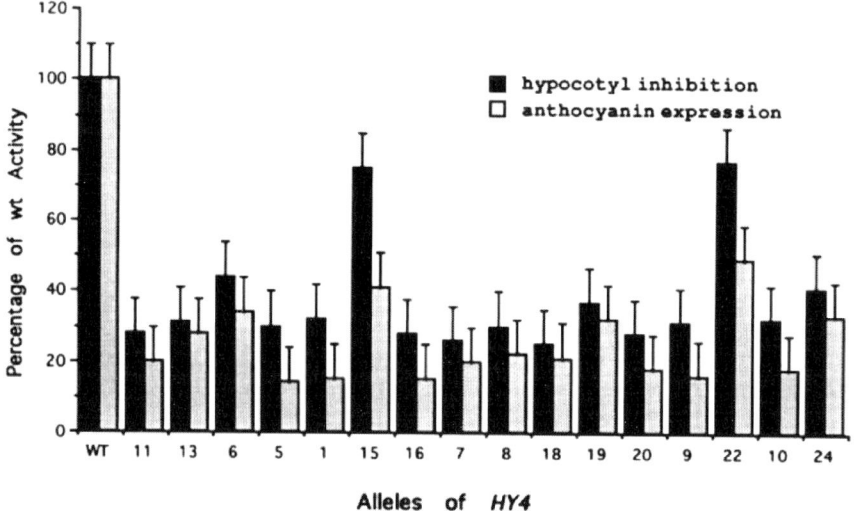

Figure 2. Mutant alleles of Arabidopsis CRY1 (HY4) are attenuated in their responses to blue light for both inhibition of hypocotyl elongation and anthocyanin production. The similarities in the relative severities of the two phenotypes for all of the alleles suggests a corresponding similarity in the cryptochrome signalling mechanism associated with these phenotypes.

Light mediated inhibition of hypocotyl growth can be detected within 30 seconds of stem irradiation with a pulse of blue light (Parks *et al.*, 1998). With continuous irradiation, stem elongation continues at increasingly reduced rates for about 30 minutes after light exposure, whereupon growth continues at a markedly reduced rate for several days. Surprisingly, from mutant studies it appears that this early inhibition observed for the first 30 minutes of exposure to blue light is mediated not by CRY but by PHOT1, the blue light photoreceptor responsible for phototropism

(Folta and Spalding, 2001). However, after 30 minutes the response is largely mediated by CRY1 and CRY2 as, in contrast to the reduced growth rate of wild type plants, normal growth is observed in the *cry1* and *cry2* mutants after about 60 minutes of light exposure (Figure 3). Correlated with this rapid blue light mediated inhibition of growth is an associated depolarisation of the plasma membrane; anion channel blockers have a similar effect to that observed for the *cry1* and *cry2* mutants. From these studies it appears that blue light, acting through CRY1 and CRY2, activates an anion channel resulting in plasma membrane depolarisation, which in turn inhibits cells expansion.

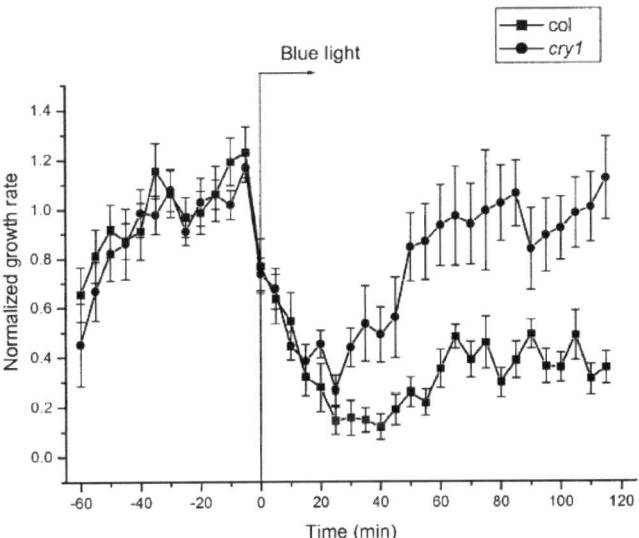

Figure 3. Rapid blue light mediated inhibition of Arabidopsis hypocotyl elongation. The initial inhibition in growth rate is not affected in the cry mutants, however after 30 minutes of exposure to blue light the cry1 (and cry2) mutant recovers the rapid growth rate observed for the dark-grown seedling.

Related observations concerning the role of CRY1 in mediating blue light inhibition of cell expansion have come from studies with Arabidopsis protoplasts. Protoplasts isolated from hypocotyl tissue and kept under continuous red light undergo rapid and transient shrinkage over a period of five minutes subsequent to exposure to a pulse of blue light. This blue light induced protoplast shrinkage does not occur in protoplasts prepared from the *cry1* mutant, demonstrating a role for the CRY1 photoreceptor in this process (Wang and Iino, 1998). The observed responsitivity to blue light requires previous exposure to red light and this response to red light is lost in protoplasts from the *phyA phyB* mutant.

5. CRYPTOCHROME AND FLOWERING

The time at which plants flower, and in some cases whether or not they flower at all, is commonly affected by day length, the so-called photoperiod. Many mutants of Arabidopsis are affected in their flowering time and some of these mutants exhibit an alteration in their sensitivity to photoperiod. Amongst these flowering mutants are those of the phytochrome and cryptochrome photoreceptors.

Mutation of the Arabidopsis *CRY2* gene has a marked effect on flowering, the *cry2* mutant being allelic to the previously described late flowering mutant *fha* (Koornneef et al., 1991; Guo et al., 1998). The late flowering phenotype of *cry2* is observed when plants are grown under white light, or blue plus red light, but not when plants are grown under blue light alone. This red light requirement apparently reflects the fact that the affect of CRY2 on flowering is dependent on PHYB. Mutations in *PHYB* result in early flowering and this early flowering phenotype is also observed in *cry2 phyB* double mutants. These observations lead to a model whereby PHYB negatively regulates flowering and this negative regulation is in turn negatively regulated by CRY2 (Mockler et al., 1999).

Under certain lighting conditions mutation of CRY1 causes a delay in flowering (Bagnall et al., 1996). Furthermore, a *cry1 cry2* double mutant was observed to flower late when grown under blue light (Mockler et al., 1999). As this phenotype is not observed with either of the single mutants under the same blue light conditions, it appears that CRY1 and CRY2 act in a redundant manner to positively regulate flowering. This blue light dependent redundant activity apparently does not involve PHYB, distinct in this respect to the late flowering phenotype associated with the monogenic *cry2* mutant that is observed in white light but not blue light (Mockler et al., 1999; see Chapter 27).

6. PLANT CRYPTOCHROMES AND CIRCADIAN RHYTHMS

Many organisms including plants, as well as some bacteria and some animals, possess biological clocks that enable them to determine circadian time. A distinguishing feature of these clocks is that whereas they are normally entrained by environmental cues such as light, the clock continues to function in the absence of such cues. Such a free running clock will normally exhibit a periodicity of approximately, but not exactly, 24 hours. The precise periodicity of 24 hours that is normally observed commonly reflects daily entrainment by light. In plants, where circadian rhythms are so prevalent, many physiological responses that initially might have been thought of as undergoing daily cyclical oscillations simply in response to light are now known to exhibit circadian responses. Diverse examples of such phenomena include CO_2 fixation, stomatal opening, hypocotyl elongation, and gene expression (Jarillo et al., 2004; see Chapter 26).

Cryptochrome was first shown to participate in Arabidopsis circadian rhythms with the demonstration that extension of rhythmic expression of the catalase *CAT3* gene in constant darkness required the presence of CRY1 as well as PHYA (Zhong et al., 1997). A role for cryptochromes as well as phytochromes in circadian entrainment was demonstrated through a study of the periodicity of the free running clock (Somers et al., 1998). In wild type Arabidopsis, the free running clock in darkness has a periodicity of approximately 27 hours. In constant blue light this free running period decreases with increasing light intensity, reaching a minimum value of approximately 24 hours. In both *cry1* as well as *phyA* mutants, this sensitivity to blue light is markedly reduced, indicating that CRY1 as well as PHYA function as photoreceptors regulating the periodicity of the Arabidopsis circadian clock (Somers et al., 1998). Whereas the periodicity of the free running clock is little affected in the monogenic *cry2* mutant, the double *cry1 cry2* mutant shows substantially greater reduced sensitivity to blue light than that observed for the *cry1* mutant, indicating a degree of redundancy for these two photoreceptors (Devlin and Kay, 2000). Of interest, the double *cry1 cry2* mutant still shows pronounced circadian oscillations, demonstrating that Arabidopsis cryptochromes, in contrast to their mammalian counterparts (see later), are not essential for circadian function (see Chapter 26).

7. ARABIDOPSIS CRYPTOCHROME AND GENE EXPRESSION

An early study of cryptochrome mediated Arabidopsis gene expression was the demonstration that expression of the chalcone synthase gene was down regulated in the Arabidopsis *hy4/cry1* mutant, this observation being in keeping with the pronounced affect that mutations in *CRY1* have on anthocyanin production (Ahmad et al., 1995). Later, more detailed analysis involving microarray hybridisation studies have categorized a large number of Arabidopsis genes, the expression of which is regulated by either CRY1 or CRY2. In one such study it was shown that the majority of blue light regulated gene expression observed in Arabidopsis seedlings was mediated either by CRY1 and/or CRY2, and that this regulated expression was similar to that observed in dark grown seedlings in the *cop1* mutant (Wang et al., 2001). In an examination of seedlings expressing the C-terminal domains of either CRY1 or CRY2 (GUS—CCT1 or GUS—CCT2; see later) it was observed that regulated gene expression for these two transgenic seedlings was very similar to one another and furthermore, the profile was similar, at least qualitatively, to that observed in the study of the *cry* mutants (Wang et al., 2001). In a related study it was shown that many of the genes regulated by cryptochromes under blue light were similarly regulated by PHYA and PHYB during growth under far-red and red light respectively. The global importance of such regulation was exemplified by the estimate that approximately one-third of the Arabidopsis genome was similarly regulated by light (Ma et al., 2001).

In a recent study the requirement for CRY1 was determined for gene expression in Arabidopsis seedlings subsequent to irradiation for 45 minutes with blue light (Folta et al., 2003). At this time point CRY1 is the primary photoreceptor mediating the light-induced inhibition of growth. Even after this relatively short exposure to light, more than 400 genes were differentially expressed in the *cry1* mutant compared with wild-type; these were approximately equally divided between those genes that were upregulated in the mutant and those that were down regulated. The types of CRY1-regulated genes included kinases, transcription factors, and genes involved in cell wall biosynthesis. Of interest, several genes involved in the response to the plant hormones auxin and the synthesis of gibberellic acid were identified. Inhibitor studies supported the notion that the affect of blue light acting through CRY1 is to repress gibberellic acid and auxin levels and/or sensitivity.

8. CRYPTOCHROMES AND CIRCADIAN RHYTHMS IN ANIMALS

8.1 Drosophila circadian rhythms are entrained through cryptochrome

In addition to light, circadian rhythms can also undergo entrainment as a result of daily fluctuations in temperature. In view of this it was of interest that the deficiency in the cyclical *PER* gene expression observed for the *cry* mutant of Drosophila was rescued when mutant flies were subjected to daily temperature changes (Stanewsky *et al.*, 1998). From this observation it was concluded that the *cry* mutant was not deficient in the function of the clock itself, but was defective with respect to light entrainment. In keeping with the similarity of the sequence to photolyases and to Arabidopsis cryptochromes it was proposed that Drosophila CRY was a photoreceptor for entrainment of the circadian clock. Flies, like mammals, undergo daily oscillations in their extent of locomotor activity — in the case of flies they move around more in the daytime than at night. Consistent with this being a circadian response, these daily oscillations in behaviour continue in the absence of any continued entrainment. Somewhat surprisingly the Drosophila *cry* mutant, isolated through its deficiency in circadian gene expression, showed robust circadian oscillations in behaviour. It appears that opsin, required for Drosophila vision but long thought not to play a role in circadian entrainment, is responsible for the circadian behaviour observed for the *cry* mutant. Flies lacking both opsin and cryptochrome are totally deficient in entrainment of circadian behavioural rhythms by light (Helfrich-Forster *et al.*, 2001). Thus, these two quite distinct photoreceptors both play a role in circadian entrainment in Drosophila, similar in this respect to the requirement for both cryptochromes and phytochromes in the entrainment of plant circadian rhythms.

8.2 Mammalian cryptochromes: Negative transcriptional regulators and essential components of the circadian oscillator

The proposal that mammalian cryptochromes might play a role as photoreceptors for circadian entrainment was influenced by several points including not only the similarity to photolyases and Arabidopsis cryptochromes, but furthermore the observation that mouse cryptochromes were noted to reside in the retinal ganglion cells (Miyamoto and Sancar, 1998). A subset of these cells of the inner eye is known to transduce the visual signal to the suprachiasmatic nucleus (SCN), the site of the central clock in the brain. Whereas in mammals, in contrast to flies, the eye is required for circadian photoentrainment, the opsins used for vision are not essential for entrainment (although, see later), similar in this instance to the situation in flies. Also in keeping with the likelihood that cryptochromes play a role in entrainment was the knowledge that blue light was optimum for entrainment (von Schantz et al., 2000).

With this background it was somewhat surprising that mutant mice lacking both CRY1 and CRY2 showed daily oscillations in their behaviour when kept under standard light/dark (LD) conditions (van der Horst et al., 1999). More surprising was the observation that these *cry1 cry2* mutant mice were totally arrhythmic in their behaviour when kept in constant darkness — conditions where normally the free running clock would maintain oscillations in behaviour. From this finding it was concluded that the mice cryptochromes performed an essential role in the functioning of the circadian central oscillator (van der Horst et al., 1999). Subsequent studies have confirmed this conclusion. Both the two CRY genes as well as the PER1 and PER2 genes undergo cyclical transcription through the action of the positive regulatory bHLH proteins CLOCK and BMAL1, as well as the activity of the CRY and PER proteins acting as negative regulators of transcription (Griffin et al., 1999; Kume et al., 1999; Reppert and Weaver, 2002).

Given the essential role that mouse cryptochromes play in the central oscillator it has proven difficult to address the role that these same cryptochromes might play in entrainment. The reason for this simply being, as noted, that mice lacking CRY1 and CRY2 are arrhythmic in their behaviour when kept in constant darkness, and therefore it is not possible to perform the standard experiments to determine if these mutant mice are attenuated in their ability to undergo phase shifting. By contrast, a role for melanopsin in circadian entrainment has recently been clearly demonstrated. Melanopsin is a specialized vertebrate opsin, not used in vision, residing in those retinal ganglion cells that transduce to the SCN and themselves are intrinsically photosensitive (Berson et al., 2002; Hattar et al., 2002). Mice lacking both melanopsin as well as the opsins of the rods and cone cells that are used for vision, are totally deficient in entrainment (Hattar et al., 2003; Panda et al., 2003). Mice lacking just melanopsin show attenuated entrainment in response to pulses of light (Panda et al., 2002) whereas those lacking opsins, undergo essentially normal entrainment. Once again, as in plants and in flies, there is a level of redundancy for the photoreceptors used for circadian entrainment.

In view of the clear demonstration that either melanopsin or the opsins are required for entrainment of circadian locomotor activity, it is now widely believed that mammalian cryptochromes do not function as photoreceptors for this process. Whereas this conclusion may eventually be shown to be correct, the logic of the argument is not correct. The demonstration of an essential requirement for one of the opsins or melanopsin, does not address the requirement for cryptochrome, and it is certainly conceivable that cryptochromes provide input for entrainment in a manner distinct from that provided by the other photoreceptors.

Putting aside the role of cryptochromes in mammalian circadian behaviour, it does appear that these same cryptochromes play a role in both circadian gene expression as well as functioning in the pupillary reflex action in response to light (Selby *et al.*, 2000; Van Gelder *et al.*, 2003). Furthermore, mouse cryptochromes appear to function in the behavioural masking response. This phenomenon of masking is the inhibition of locomotor activity that results in a depression by light in the running activity of mice kept under LD conditions. This is a non-circadian response persisting in mice that are deficient in their central circadian oscillator. As referenced above, mutant *cry1 cry2* mice show apparently normal daily oscillations in behaviour when kept under LD conditions (van der Horst *et al.*, 1999); however this activity is lost in mice lacking both cryptochromes and opsins, indicating a requirement for one or other of these photoreceptors (Selby *et al.*, 2000). Interestingly, the masking response is also absent in mice lacking melanopsin and opsins, showing that cryptochrome itself is not sufficient for this response but serves an overlapping role with the other two classes of photoreceptors (Hattar *et al.*, 2003; Panda *et al.*, 2003). Whether cryptochrome functions as a photoreceptor in these responses, or acts downstream from the actual photoreceptor, is yet to be determined.

9. THE MODE OF ACTION OF CRYPTOCHROME

9.1 The Arabidopsis cryptochrome C-terminal domain mediates a constitutive light response

Considering the sequence relatedness plant cryptochromes and photolyases it would seem reasonable to assume that the modes of action of these two classes of proteins would be similar. The simplest model that one could envisage would be that the distinguishing cryptochrome C-terminal domain (CCT) serves to bind a signalling partner, and that this bound signalling partner is activated through a light-dependent redox reaction. If the mechanistic analogy with photolyase were to hold, this reaction would be an intermolecular redox reaction involving transfer of an electron from the CRY-bound flavin to the signalling partner (Cashmore *et al.*, 1999).

In the simplest form of this model, CCT plays no role in signalling other than binding the signalling partner. According to such a model, CCT does not undergo any change in response to light. We now know that this model for the mode of action of Arabidopsis CRY is incorrect. This conclusion follows from experiments involving transgenic plants over expressing the C-terminal domain of CRY1 (CCT1). These CCT1 plants exhibit a constitutive photomorphogenic (COP) phenotype (Yang *et al.*, 2000). That is, dark-grown CCT1 seedlings exhibit a shortened hypocotyl and expanded cotyledons characteristic of light-grown seedlings. A similar phenotype is observed with transgenic seedlings expressing the C-terminal domain of CRY2 (CCT2); however, this COP phenotype is not observed with mutant CCT1 sequences corresponding to loss-of-function *cry1* mutant alleles. From these experiments it was concluded that the COP phenotype was likely to reflect a true physiological role of CCT.

What does the observation that CCT mediates a COP phenotype tell us about the mode of action of CRY? It indicates, in contrast to the model speculated on above, that CCT plays a role in signalling over and above simply binding the signalling partner. CCT by itself clearly has the potential to signal, and this activity must be repressed in the native CRY molecule in dark grown Arabidopsis seedlings. The action of light is to alleviate repression of the signalling potential of CCT. By analogy with photolyases, this action of light is likely to reflect a light-dependent redox reaction involving the bound flavin. Such a reaction could be intramolecular, giving rise to a conformational change in CRY leading directly to the activation of CCT. An alternative model involves an additional molecule which in the dark serves to negatively regulate CCT. In response to light, the activity of this negative regulator is overcome as the result of an intra- or intermolecular redox reaction (Yang *et al.*, 2000).

The CCTs for CRY1 and CRY2 of Arabidopsis show relatively little sequence conservation. Similarly, orthologous cryptochromes from different species exhibit substantially less sequence conservation for their C-terminal domains than that observed for the corresponding photolyase-like CRY N-terminal domains (CNTs). This relative lack of sequence conservation for CCT clearly does not reflect unimportance in function, as indicated both by mutations within CCT resulting in *cry1* loss of function mutants as well as the above functional studies involving CCTs. In spite of the overall lack of sequence conservation, there are small regions of CCT that are conserved. A serine-rich sequence STAESSSS is conserved between Arabidopsis CRY1 and CRY2 and deletion of this region reduces the level of in vitro phosphorylation of CRY1 by the kinase activity associated with isolated PHYA (Ahmad *et al.*, 1998). Also, the motif DQXVP found in the CCT1 of Arabidopsis CRY1 is also found in the CCTs of the Adiantum CRYs and in Arabidopsis CRY2 (Kanegae and Wada, 1998), as well as the cryptochromes of Physcomitrella (Lin and Shalitin, 2003). The functional significance of this sequence in Arabidopsis CRY1 is supported by the observation that two *cry1* loss of function mutants contain lesions within this DQMVP sequence (Ahmad and Cashmore, 1993; Ahmad *et al.*, 1995).

9.2 COP1: A signalling partner of Arabidopsis cryptochromes

The *cop/det/fus* mutants of Arabidopsis exhibit several developmental features in dark-grown seedlings that are normally restricted to seedlings grown in light (Schwechheimer and Deng, 2000; see Chapter 18). In these respects these mutants are similar to the CCT-over expressing seedlings. One such class of mutants affect COP1, a zinc finger protein containing a WD-40 repeat domain at its N-terminus. COP1 has the properties of an E3 ubiquitin ligase and in dark grown seedlings it facilitates proteasome mediated degradation of proteins such as the bZIP transcription factor HY5 that positively regulate photomorphogenesis (Osterlund *et al.*, 2000; Seo *et al.*, 2003). In this respect COP1, along with components of the COP9 signalosome (CSN), serve to negatively regulate photomorphogenesis in dark grown seedlings. The presence of light serves to suppress this negative regulation (see Chapter 18).

The similarity of the *cop1* mutant and that of CCT expressing plants raised the possibility that the phenotype of the transgenic plants may reflect a direct interaction between CCT and COP1. In order to explore this possibility the yeast two hybrid system and in vitro binding studies were used to demonstrate a selective interaction between COP1 with both CCT1 as well as CRY1. Furthermore, an interaction between CRY1 and recombinant COP1 was observed in plant extracts prepared from transgenic Arabidopsis seedlings expressing a COP1—GFP fusion protein. Neither in yeast nor in Arabidopsis was the interaction between CRY1 and COP1 demonstrably affected by blue light (Yang *et al.,* 2001).

In the above studies no interaction was observed between CCT2 and COP1. This was somewhat surprising as CCT2, like CCT1, confers a COP phenotype on expression in transgenic Arabidopsis seedlings. However in complementary studies an interaction between CCT2 and COP1 was observed in yeast (Wang *et al.,* 2001); the reason for these two contrasting observations in yeast is not clear. Also, in these latter studies COP1 was shown to co-immunopreciptate with CRY2 from Arabidopsis extracts (Wang *et al.,* 2001).

There are several reasons to believe that the above results showing an interaction between COP1 and CRY reflect the fact that COP1 is a major and direct signalling partner of both CRY1 and CRY2. Early genetic studies demonstrated that the *cop1* phenotype was epistatic to *cry1*, indicating that the COP phenotype characteristic of the loss of function *cop1* mutants does not require CRY1 activity. The observation is consistent with the model that COP1 negatively regulates photomorphogenesis in dark grown seedlings, and in response to light this negative regulation is attenuated through a change in the interaction between COP1 and CRY1. The finding that *cop1* loss of function mutants and CCT1 expressing lines display a variety of similar photomorphogenic characteristics supports the notion that a significant component of CRY1 signalling is mediated by an interaction between CCT1 and COP1. In addition to the described effect on hypocotyl development, these phenotypic similarities include anthocyanin production and chloroplast development (Yang *et al.,* 2000), as well as elevated expression of HY5 (Wang *et al.,* 2001). Furthermore, as noted earlier, microarray studies of both CCT1- and CCT2-expressing plants demonstrate changes in gene expression that are similar to those mediated by CRY1 and CRY2 (Wang *et al.,* 2001).

Whereas the interaction of CRY1 and CRY2 with COP1 is likely to be a major signalling mechanism of CRYs, it is unlikely to be the only means of Arabidopsis cryptochrome signalling (Figure 4). The reasons for this conclusion are several. Firstly, the rapidity of the CRY-mediated membrane depolarisation suggests that it occurs via a mechanism that is not dependent on inhibition of proteasome-mediated protein degradation (Parks *et al.,* 1998). Similarly, the blue light and CRY1-dependent changes in Arabidopsis transcripts (Folta *et al.,* 2003) also occurs on a time scale that likely precedes a change, for example, in the amount of chalcone synthase transcripts (Ahmad *et al.,* 1995) — this latter is likely to reflect light and CRY1-dependent inhibition of the degradation of HY5 mediated by COP1. Indeed, in keeping with these conclusions we have recently demonstrated that the N-terminal domain (CNT) of both Arabidopsis CRY1 and CRY2 mediate constitutive light signalling, similar in this respect to the properties of CCT1 and CCT2 (Kang and Cashmore, unpublished).

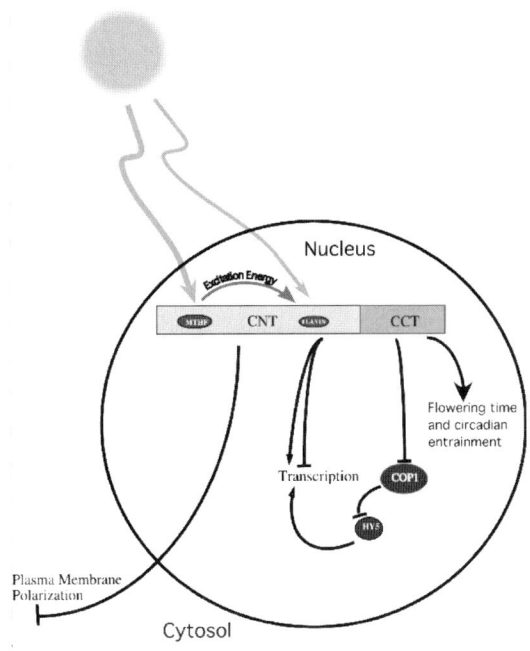

Figure 4. Signalling pathways of Arabidopsis cryptochromes. Both CRY1 and CRY2 are nuclear localized in dark-grown Arabidopsis, as is COP1. In addition to the light dependent negative regulation of COP1 activity, resulting in the accumulation of HY5 (and other targets of COP1 activity), cryptochromes also mediate flowering time and entrainment of the circadian clock, blue-light induced plasma membrane depolarisation, and rapid changes in transcription of a large number of Arabidopsis genes.

9.3 Intracellular localization of Arabidopsis CRYs

Studies involving fusion proteins with GUS have shown that both CRY1 and CRY2 proteins are localized to the nucleus in either onion epidermal cells or in dark-grown Arabidopsis seedlings (Cashmore *et al.*, 1999; Guo *et al.*, 1999; Kleiner *et al.*, 1999) (Figure 5). This nuclear localization is also observed for GUS fusions of either CCT1 or CCT2 in dark-grown seedlings. In light-grown seedlings, CCT2 is similarly localized in the nucleus. In contrast however, CCT1 is primarily localized to the cytosol in light-grown Arabidopsis seedlings (Yang *et al.*, 2000). This light-induced change in distribution of CCT1 is intriguing and it mimics in this respect the properties of COP1 (Osterlund and Deng, 1998). As CCT1 lacks a chromophore, this light-dependent change in intracellular localization must be dependent on one of the Arabidopsis photoreceptors. For Arabidopsis seedlings grown under blue light, the cytosolic relocalisation of COP1 is dependent on CRY1 (Osterlund and Deng, 1998).

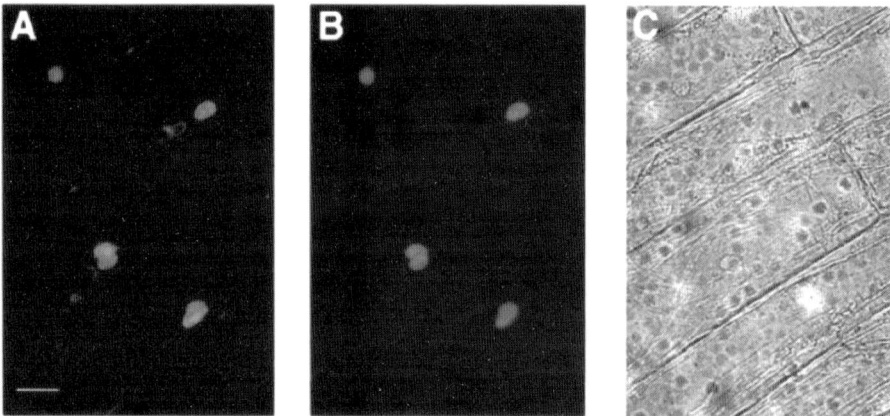

Figure 5. Nuclear localization of Arabidopsis CRY1 protein. A gene encoding the Arabidopsis CRY1 protein fused to GFP was introduced into onion epidermal cells. The fluorescence derived from CRY1:GFP is shown in (A) and this is seen to overlap with the nuclear localized fluorescence of DAPI seen in (B). Cellular structure is shown by the light micrograph in (C).

9.4 Phosphorylation of Arabidopsis cryptochromes

Both CRY1 and CRY2 have been shown to undergo blue light dependent phosphorylation in vivo (Shalitin *et al.*, 2002; Shalitin *et al.*, 2003). In studies involving plants expressing CCT2 it was observed that this C-terminal domain of CRY2 was constitutively phosphorylated and for this reason it was postulated that phosphorylation of CCT2 was causally related to its signalling properties (Shalitin *et al.*, 2002). However, it seems equally plausible that the phosphorylation of CCT2 simply reflects a light-induced change in the properties of CRY2, analogous in this respect to the constitutive signalling properties observed for this C-terminal fragment of CRY (Yang *et al.*, 2000), and that this phosphorylation may simply serve to regulate protein turnover. Indeed, the amount of phosphorylated CRY2 protein (as well as the non-phosphorylated form) decreases on longer exposure to blue light (Shalitin *et al.*, 2002) (Figure 6).

In order to address this question of whether or not phosphorylation of CRY was required for signalling, *cry1* loss-of-function mutants were isolated (Shalitin *et al.*, 2003). Four distinct *cry1* alleles that expressed full-length protein were examined and all showed a corresponding loss of in vivo protein phosphorylation. Whereas these studies certainly infer a close relationship between light-induced activation of CRY1 and phosphorylation, as with the related CRY2 studies, they do not definitively show that phosphorylation is required for signalling.

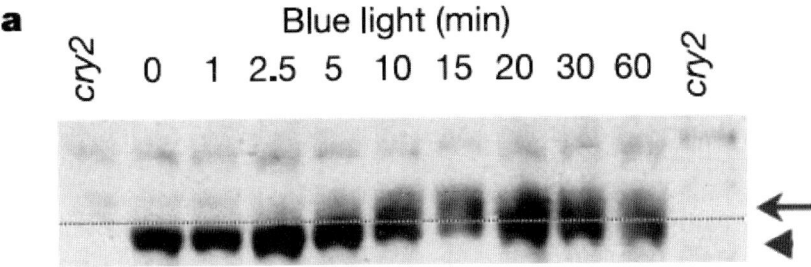

Figure 6. CRY2 is phosphorylated on exposure of Arabidopsis seedlings to blue light. The CRY2 proteins are detected by western blotting and the phosphorylated form is distinguished by its slower migration on SDS/PAGE. The amount of protein decreases with time of exposure to light.

An intriguing observation has been the finding that Arabidopsis CRY1 protein isolated from insect cells is phosphorylated in vitro in a blue light dependent manner (Bouly *et al.*, 2003; Shalitin *et al.*, 2003). Furthermore, similar observations were made for human CRY1 and both the Arabidopsis and human CRYs were shown to bind to an ATP affinity column (Bouly *et al.*, 2003). These observations raise the possibility that the observed phosphorylation of CRY reflects autophosphorylation. As there is no obvious kinase domain within the CRY sequence it was difficult to eliminate the possibility that the observed phosphorylation simply reflected a contaminating kinase. However, a recently reported crystallographic study of the CNT domain of Arabidopsis CRY1 has also demonstrated the binding of ATP (Brautigam *et al.*, 2004). It is of interest that this ATP is bound close to the flavin, in a site analogous to that occupied by the pyrimidine dimer in *E. coli* photolyase.

9.5 Photochemical properties of Arabidopsis cryptochromes

To understand the mode of action of a photoreceptor it is necessary to know the chemical changes of the photoreceptor that occur as a result of the absorption of light. In the case of photolyase, light activates the bound flavin and induces electron transfer. By analogy, it has always been assumed that Arabidopsis CRY would function similarly. In keeping with this line of thinking is the finding that blue light dependent changes in calcium homeostasis and gene expression in Arabidopsis cell suspension cultures involves plasma membrane redox activity (Long and Jenkins, 1998).

Flavins can exist in three different redox states: the fully oxidized FAD, the semiquinone, and the fully reduced $FADH_2$. When isolated from insect cells Arabidopsis CRY1 contains FAD in the fully oxidized form of FAD (Lin *et al.*, 1995). Under anaerobic conditions and on irradiation with blue light this flavin undergoes photoreduction, being converted first to the semiquinone and then to the fully reduced $FADH_2$. These changes can be readily followed by absorption spectrophotometry, as the absorption properties of the different redox state of the flavin are quite distinct. In addition, in this instance, the identity of the neutral flavosemiquinone was independently confirmed by EPR studies (Lin et al., 1995). These findings contrast with comparable studies with *E. coli* photolyase, which is commonly isolated with the flavin in the semiquinone form, although in photoreduction studies the bound FAD is converted to $FADH_2$ without the production of any discernable semiquinone (Jorns et al., 1990).

As is the case with *E. coli* photolyase, the in vitro photoreduction of Arabidopsis CRY1 involves intraprotein electron transfer with one or other tryptophan residues serving as the electron donor (Giovani et al., 2003). From this observation it was concluded that cryptochromes and photolyases share mechanistic similarities in their initial photochemistry. A difficulty with this argument is that the in vitro photochemistry characterized for *E. coli* photolyase is thought not to be relevant to the mechanism of light-induced photoreactivation that occurs in vivo (Li et al., 1991; Sancar 2003). In this latter case, the initial photochemistry does not involve intraprotein electron transfer, but transfer of an electron from the fully reduced FADH2 (in the form of the anion, FADH-) to the bound pyrimidine dimmer.

9.6 Drosophila cryptochrome interacts with PER and TIM in a light-dependent manner

As noted earlier there is very good evidence for Drosophila that CRY acts as a photoreceptor for entrainment of the circadian clock. This is achieved, at least in part, through an interaction with TIMELESS (TIM), which along with PER is an integral component of the Drosophila circadian oscillator, both PER and TIM serving as negative transcriptional regulators. TIM undergoes proteasome-mediated degradation and this activity is both light- and CRY-dependent (Lin et al., 2001). Furthermore, in keeping with earlier discussions about likely modes of action of CRY, this degradation of TIM is sensitive to inhibitors of redox reactions (Lin et al., 2001).

An interaction between Drosophila CRY and TIM has been demonstrated in yeast and remarkably, this interaction is light dependent (Ceriani et al., 1999). However, in Drosophila S2 cells where, as well as in flies and yeast, CRY binds to TIM, this interaction does not appear to be dependent on light. Similarly, Drosophila CRY binds to PER in yeast in a light dependent manner, yet once again in S2 cells this interaction does not require light (Rosato et al., 2001).

These observations are similar in some respects to the interaction described for Arabidopsis CRY and COP1. Here, in spite of the evidence that light affects the properties of COP1 in a CRY1 dependent manner, no effect of light on the binding of CRY1 to COP1 could be demonstrated, neither in yeast nor in Arabidopsis seedlings (Yang et al., 2001).

9.7 Mouse cryptochromes negatively regulate transcription in a light-independent manner

Whereas the initial expectation that mammalian CRYs would function as photoreceptors serving to entrain the circadian clock has not been substantiated (at least not yet), what is clear, and quite unexpected, is that they play an essential role as negative transcriptional regulators of the clock. Indeed the two mouse CRYs, along with PER1 and PER2, form central components of the clock, negatively autoregulating transcription of their own genes (Reppert and Weaver, 2002).

This activity apparently involves interaction between the cryptochromes and PER proteins as well as interaction with CLOCK:BMAL1 transcription factors. Of interest, the ability of the CRY1 and CRY2 mouse proteins to inhibit CLOCK:BMAL-dependent in vitro transcription is independent of light (Griffin *et al.*, 1999). In keeping with this finding, mutants of the mouse CRY1 protein containing lesions within the flavin binding domain are not affected in their transcriptional repressor activities whereas the corresponding mutations do affect the activity of the Drosophila CRY (Froy *et al.*, 2002).

10. CRYPTOCHROME EVOLUTION

Cryptochromes, first identified in plants and then animals, have recently been characterized for bacteria. There are some puzzling and interesting features associated with the evolution of the cryptochrome/photolyase gene family. Firstly, plant and animal CRYs are not orthologous sequences (Cashmore *et al.*, 1999). The Arabidopsis orthologue of mammalian CRYs is a photolyase, specifically the Arabidopsis (6-4) photolyase. This conclusion follows from the fact that the sequence of the human CRYs are markedly more similar to the Arabidopsis (6-4) photolyase sequence than they are to the Arabidopsis CRY1 and CRY2 sequences. Furthermore, there is as yet no described bacterial sequence that is orthologous to the Arabidopsis CRY1 and CRY2 sequences — the Arabidopsis orthologue of the Synechocystis CRY sequence being the recently discovered Arabidopsis CRY DASH/CRY3 sequence.

In early discussions of CRY evolution it had always been assumed that the initial bacterial progenitor sequence was a photolyase — this argument reflecting the belief at the time that no bacterial CRY existed. However, with the description of such a bacterial CRY it seems appropriate to revisit this question. It now seems reasonable to entertain the possibility that the original flavin-based CRY/PHR-like photoreceptor, simply participated in some light dependent redox reaction, not involving DNA repair; such a molecule would be a cryptochrome.

11. CONCLUSIONS AND FUTURE STUDIES

Excitement associated with the original description of the Arabidopsis cryptochrome molecule reflected in part the fact that this was the first identification of a plant blue light photoreceptor. The subsequent study of the animal cryptochromes and their role in circadian rhythms added enormously to the interest in these molecules. Here, the most surprising development was the demonstration that mammalian CRYs, and apparently in some instances those of Drosophila (Krishnan *et al.*, 2001), function in a light-independent manner as essential components of the circadian oscillator. The recent description of bacterial cryptochromes adds even more interest to this family of molecules. Important future developments should include a fuller understanding of the identity of cryptochrome signalling partners as well as the photochemistry associated with this signalling.

12. REFERENCES

Ahmad, M. and A. R. Cashmore (1993) HY4 gene of A. thaliana encodes a protein with characteristics of a blue-light photoreceptor. *Nature, 366,* 162-166.

Ahmad, M., J. A. Jarillo, *et al.*, (1998) The CRY1 blue light photoreceptor of Arabidopsis interacts with phytochrome A in vitro. *Mol Cell, 1,* 939-948.

Ahmad, M., C. Lin, *et al.*, (1995) Mutations throughout an Arabidopsis blue-light photoreceptor impair blue-light-responsive anthocyanin accumulation and inhibition of hypocotyl elongation. *Plant J, 8,* 653-658.

Bagnall, D. J., R. W. King, *et al.*, (1996) Blue-light promotion of flowering is absent in *hy4* mutants of Arabidopsis. *Planta, 200,* 278-280.

Berson, D. M., F. A. Dunn, *et al.*, (2002) Phototransduction by retinal ganglion cells that set the circadian clock. *Science, 295,* 1070-1073.

Bouly, J. P., B. Giovani, *et al.*, (2003) Novel ATP-binding and autophosphorylation activity associated with Arabidopsis and human cryptochrome-1. *Eur J Biochem, 270,* 2921-2928.

Brautigam, C. A., B. S. Smith, *et al.*, (2004) Structure of the photolyase-like domain of cryptochrome 1 from Arabidopsis thaliana. *Proc Natl Acad Sci USA, 101,* 12142-12147.

Brudler, R., K. Hitomi, *et al.*, (2003) Identification of a new cryptochrome class. Structure, function, and evolution. *Mol Cell, 11,* 59-67.

Cashmore, A. R., J. A. Jarillo, *et al.*, (1999) Cryptochromes: blue light receptors for plants and animals. *Science, 284,* 760-765.

Ceriani, M. F., T. K. Darlington, *et al.*, (1999) Light-dependent sequestration of TIMELESS by CRYPTOCHROME. *Science, 285,* 553-556.

Cermakian, N., M. P. Pando, *et al.*, (2002) Light induction of a vertebrate clock gene involves signalling through blue-light receptors and MAP kinases. *Curr Biol, 12,* 844-848.

Deininger, W., P. Kroger, *et al.*, (1995) Chlamyrhodopsin represents a new type of sensory photoreceptor. *Embo J, 14,* 5849-5858.

Devlin, P. F. and S. A. Kay (2000) Cryptochromes are required for phytochrome signalling to the circadian clock but not for rhythmicity. *Plant Cell, 12,* 2499-2510.

Emery, P., W. V. So, *et al.*, (1998) CRY, a *Drosophila* clock and light-regulated cryptochrome, is a major contributor to circadian rhythm resetting and photosensitivity. *Cell, 95,* 669-679.

Folta, K. M., M. A. Pontin, *et al.*, (2003) Genomic and physiological studies of early cryptochrome 1 action demonstrate roles for auxin and gibberellin in the control of hypocotyl growth by blue light. *Plant J, 36,* 203-214.

Folta, K. M. and E. P. Spalding (2001) Unexpected roles for cryptochrome 2 and phototropin revealed by high-resolution analysis of blue light-mediated hypocotyl growth inhibition. *Plant J, 26,* 471-478.

Froy, O., D. C. Chang, *et al.*, (2002) Redox potential: differential roles in dCRY and mCRY1 functions. *Curr Biol, 12,* 147-152.

Giovani, B., M. Byrdin, *et al.*, (2003) Light-induced electron transfer in a cryptochrome blue-light photoreceptor. *Nat Struct Biol, 10,* 489-490.

Griffin, E. A., D. Staknis, *et al.*, (1999) Light-independent role of CRY1 and CRY2 in the mammalian circadian clock. *Science, 286,* 768-771.

Guo, H., H. Duong, et al., (1999) The Arabidopsis blue light receptor cryptochrome 2 is a nuclear protein regulated by a blue light-dependent post-transcriptional mechanism. *Plant J, 19*, 279-287.

Guo, H., H. Yang, et al., (1998) Regulation of flowering time by Arabidopsis photoreceptors. *Science, 279*, 1360-1363.

Hattar, S., H. W. Liao, et al., (2002) Melanopsin-containing retinal ganglion cells: architecture, projections, and intrinsic photosensitivity. *Science, 295*, 1065-1070.

Hattar, S., R. J. Lucas, et al., (2003) Melanopsin and rod-cone photoreceptive systems account for all major accessory visual functions in mice. *Nature, 424*, 75-81.

Helfrich-Forster, C., C. Winter, et al., (2001) The circadian clock of fruit flies is blind after elimination of all known photoreceptors. *Neuron, 30*, 249-261.

Hirayama, J., H. Nakamura, et al., (2003) Functional and structural analyses of cryptochrome. Vertebrate CRY regions responsible for interaction with the CLOCK:BMAL1 heterodimer and its nuclear localization. *J Biol Chem, 278*, 35620-35628.

Hitomi, K., K. Okamoto, et al., (2000) Bacterial cryptochrome and photolyase: characterization of two photolyase-like genes of Synechocystis sp. PCC6803. *Nucleic Acids Res, 28*, 2353-2362.

Hsu, D. S., X. Zhao, et al., (1996) Putative human blue-light photoreceptors *hCRY1* and *hCRY2* are flavoproteins. *Biochemistry, 35*, 13871-13877.

Hsu, D. S., X. Zhao, et al., (1996) Putative human blue-light photoreceptors hCRY1 and hCRY2 are flavoproteins. *Biochemistry, 35*, 13871-13877.

Imaizumi, T., A. Kadota, et al., (2002) Cryptochrome light signals control development to suppress auxin sensitivity in the moss Physcomitrella patens. *Plant Cell, 14*, 373-386.

Jarillo, J. A., J. Capel, et al., (2004) Physiological and molecular characteristics of plant circadian clocks. *Molecular biology of circadian rhythms*. A. Sehgal. Hoboken, John Wiley and Sons.

Jorns, M. S., B. Y. Wang, et al., (1990) Chromophore function and interaction in Escherichia coli DNA photolyase: reconstitution of the apoenzyme with pterin and/or flavin derivatives. *Biochemistry, 29*, 552-561.

Kanegae, T. and M. Wada (1998) Isolation and characterization of homologues of plant blue-light photoreceptor (cryptochrome) genes from the fern *Adiantum capillus-veneris*. *Mol Gen Genet, 259*, 345-353.

Keller, J., J. Shanklin, et al., (1989) Expression of a functional monocotyledonous phytochrome in transgenic tobacco. *EMBO J., 8*, 1005-1012.

Kleine, T., P. Lockhart, et al., (2003) An Arabidopsis protein closely related to Synechocystis cryptochrome is targeted to organelles. *Plant J, 35*, 93-103.

Kleiner, O., S. Kircher, et al., (1999) Nuclear localization of the Arabidopsis blue light receptor cryptochrome 2. *Plant J, 19*, 289-296.

Koornneef, M., C. J. Hanhart, et al., (1991) A genetic and physiological analysis of late flowering mutants in Arabidopsis thaliana. *Mol Gen Genet, 229*, 57-66.

Koornneef, M., E. Rolff, et al., (1980) Genetic control of light-inhibited hypocotyl elongation in *Arabidopsis thaliana* (L.) Heynh. *Z. Pflanzenphysiol. Bd., 100*, 147-160.

Krishnan, B., J. D. Levine, et al., (2001) A new role for cryptochrome in a Drosophila circadian oscillator. *Nature, 411*, 313-317.

Kume, K., M. J. Zylka, et al., (1999) mCRY1 and mCRY2 are essential components of the negative limb of the circadian clock feedback loop. *Cell, 98*, 193-205.

Li, Y. F., P. F. Heelis, et al., (1991) Active site of DNA photolyase: Tryptophan-306 is the intrinsic hydrogen atom donor essential for flavin radical photoreduction and DNA repair *in vitro*. *Biochem, 30*, 6322-6329.

Lin, C., M. Ahmad, et al., (1996) Arabidopsis cryptochrome 1 is a soluble protein mediating blue light-dependent regulation of plant growth and development. *Plant J, 10*, 893-902.

Lin, C., M. Ahmad, et al., (1996) *CRY2*: A second member of the Arabidopsis cryptochrome gene family. *Plant Physiol, 110*, 1047.

Lin, C., M. Ahmad, et al., (1995) Expression of an Arabidopsis cryptochrome gene in transgenic tobacco results in hypersensitivity to blue, UV-A, and green light. *Proc Natl Acad Sci USA, 92*, 8423-8427.

Lin, C., D. E. Robertson, et al., (1995) Association of flavin adenine dinucleotide with the Arabidopsis blue light receptor CRY1. *Science, 269*, 968-970.

Lin, C., D. E. Robertson, et al., (1995) Association of flavin adenine dinucleotide with the Arabidopsis blue light receptor CRY1. *Science, 269*, 968-970.

Lin, C. and D. Shalitin (2003) Cryptochrome structure and signal transduction. *Annu Rev Plant Biol, 54*, 469-496.
Lin, C., H. Yang, *et al.,* (1998) Enhancement of blue-light sensitivity of Arabidopsis seedlings by a blue light receptor cryptochrome 2. *Proc Natl Acad Sci USA, 95*, 2686-2690.
Lin, F. J., W. Song, *et al.,* (2001) Photic signalling by cryptochrome in the Drosophila circadian system. *Mol Cell Biol, 21*, 7287-7294.
Long, J. C. and G. I. Jenkins (1998) Involvement of plasma membrane redox activity and calcium homeostasis in the UV-B and UV-A blue light induction of gene expression in Arabidopsis. *Plant Cell, 10*, 2077-2086.
Ma, L., J. Li, *et al.,* (2001) Light control of Arabidopsis development entails coordinated regulation of genome expression and cellular pathways. *Plant Cell, 13*, 2589-2607.
Malhotra, K., S. T. Kim, *et al.,* (1995) Putative blue-light photoreceptors from Arabidopsis thaliana and Sinapis alba with a high degree of sequence homology to DNA photolyase contain the two photolyase cofactors but lack DNA repair activity. *Biochemistry, 34*, 6892-6899.
Matsumoto, N., T. Hirano, *et al.,* (2003) Functional analysis and intracellular localization of rice cryptochromes. *Plant Physiol, 133*, 1494-1503.
Miyamoto, Y. and A. Sancar (1998) Vitamin B2-based blue-light photoreceptors in the retinohypothalamic tract as the photoactive pigments for setting the circadian clock in mammals. *Proc Natl Acad Sci USA, 95*, 6097-6102.
Mockler, T. C., H. Guo, *et al.,* (1999) Antagonistic actions of Arabidopsis cryptochromes and phytochrome B in the regulation of floral induction. *Development, 126*, 2073-2082.
Ninu, L., M. Ahmad, *et al.,* (1999) Cryptochrome 1 controls tomato development in response to blue light. *Plant J, 18*, 551-556.
Osterlund, M. T. and X. W. Deng (1998) Multiple photoreceptors mediate the light-induced reduction of GUS-COP1 from Arabidopsis hypocotyl nuclei. *Plant J, 16*, 201-208.
Osterlund, M. T., C. S. Hardtke, *et al.,* (2000) Targeted destabilization of HY5 during light-regulated development of Arabidopsis. *Nature, 405*, 462-466.
Panda, S., I. Provencio, *et al.,* (2003) Melanopsin is required for non-image-forming photic responses in blind mice. *Science, 301*, 525-527.
Panda, S., T. K. Sato, *et al.,* (2002) Melanopsin (Opn4) requirement for normal light-induced circadian phase shifting. *Science, 298*, 2213-2216.
Parks, B. M., M. H. Cho, *et al.,* (1998) Two genetically separable phases of growth inhibition induced by blue light in Arabidopsis seedlings. *Plant Physiol, 118*, 609-615.
Reppert, S. M. and D. R. Weaver (2002) Coordination of circadian timing in mammals. *Nature, 418*, 935-941.
Rosato, E., V. Codd, *et al.,* (2001) Light-dependent interaction between Drosophila CRY and the clock protein PER mediated by the carboxy terminus of CRY. *Curr Biol, 11*, 909-917.
Sancar, A. (2003) Structure and function of DNA photolyase and cryptochrome blue-light photoreceptors. *Chem Rev, 103*, 2203-2237.
Schwechheimer, C. and X. W. Deng (2000) The COP/DET/FUS proteins-regulators of eukaryotic growth and development. *Semin Cell Dev Biol, 11*, 495-503.
Selby, C. P., C. Thompson, *et al.,* (2000) Functional redundancy of cryptochromes and classical photoreceptors for nonvisual ocular photoreception in mice. *Proc Natl Acad Sci USA, 97*, 14697-14702.
Seo, H. S., J. Y. Yang, *et al.,* (2003) LAF1 ubiquitination by COP1 controls photomorphogenesis and is stimulated by SPA1. *Nature, 423*, 995-999.
Shalitin, D., H. Yang, *et al.,* (2002) Regulation of Arabidopsis cryptochrome 2 by blue-light-dependent phosphorylation. *Nature, 417*, 763-767.
Shalitin, D., X. Yu, *et al.,* (2003) Blue light-dependent in vivo and in vitro phosphorylation of Arabidopsis cryptochrome 1. *Plant Cell, 15*, 2421-2429.
Small, D. B., B. Min, *et al.,* (1995) Characterization of a *Chlamydomonas reinhardtii* gene encoding a protein of the DNA photolyase/blue light photoreceptor family. *Plant Mol Biol, 28*, 443-454.
Small, G. D., B. Min, *et al.,* (1995) Characterization of a Chlamydomonas reinhardtii gene encoding a protein of the DNA photolyase/blue light photoreceptor family. *Plant Mol Biol, 28*, 443-454.
Somers, D. E., P. F. Devlin, *et al.,* (1998) Phytochromes and cryptochromes in the entrainment of the Arabidopsis circadian clock. *Science, 282*, 1488-1490.

Stanewsky, R., M. Kaneko, *et al.,* (1998) The *cry^b* mutation identifies cryptochrome as a circadian photoreceptor in *Drosophila. Cell, 95,* 681-692.
Todo, T., H. Ryo, *et al.,* (1996) Similarity among the *Drosophila* (6-4) photolyase, a human photolyase homolog, and the DNA photolyase-blue-light receptor family. *Science, 272,* 109-112.
Todo, T., H. Ryo, *et al.,* (1996) Similarity among the Drosophila (6-4)photolyase, a human photolyase homolog, and the DNA photolyase-blue-light photoreceptor family. *Science, 272,* 109-112.
Todo, T., H. Takemori, *et al.,* (1993). A new photoreactivating enzyme that specifically repairs ultraviolet light-induced (6-4) phoptoproducts. *Nature, 361,* 371-374.
van der Horst, G. T., M. Muijtjens, *et al.,* (1999) Mammalian Cry1 and Cry2 are essential for maintenance of circadian rhythms. *Nature, 398,* 627-630.
van Gelder, R. N., R. Wee, *et al.,* (2003) Reduced pupillary light responses in mice lacking cryptochromes. *Science, 299,* 222.
von Schantz, M., I. Provencio, *et al.,* (2000) Recent developments in circadian photoreception: more than meets the eye. *Invest Ophthalmol Vis Sci, 41,* 1605-1607.
Wang, H., L. G. Ma, *et al.,* (2001) Direct interaction of Arabidopsis cryptochromes with COP1 in light control development. *Science, 294,* 154-158.
Wang, X. and M. Iino (1998) Interaction of cryptochrome 1, phytochrome, and ion fluxes in blue-light-induced shrinking of Arabidopsis hypocotyl protoplasts. *Plant Physiol, 117,* 1265-1279.
Weller, J. L., G. Perrotta, *et al.,* (2001) Genetic dissection of blue-light sensing in tomato using mutants deficient in cryptochrome 1 and phytochromes A, B1 and B2. *Plant J, 25,* 427-440.
Yang, H. Q., R. H. Tang, *et al.,* (2001) The signalling mechanism of Arabidopsis CRY1 involves direct interaction with COP1. *Plant Cell, 13,* 2573-2587.
Yang, H. Q., Y. J. Wu, *et al.,* (2000) The C termini of Arabidopsis cryptochromes mediate a constitutive light response. *Cell, 103,* 815-827.
Zhong, H. H., A. S. Resnick, *et al.,* (1997) Effects of synergistic signalling by phytochrome A and cryptochrome1 on circadian clock-regulated catalase expression. *Plant Cell, 9,* 947-955.

Chapter 12
PHOTOTROPINS

Winslow R. Briggs[1], John M. Christie[2] and Trevor E. Swartz[1]
[1]*Department of Plant Biology, Carnegie Institution of Washington, 290 Panama St., Stanford, CA 94305 U. S. A., (e-mail: briggs@stanford.edu)*
[2]*Plant Science Group, University of Glasgow, University Avenue, Glasgow, G12 8QQ Scotland (e-mail: J.Christie@bio.gla.ac.uk)*

1. INTRODUCTION

The history of the lengthy, confused, and often misdirected search for blue-light receptors in higher plants has been summarized in Chapter 10. This earlier chapter covers historical developments starting near the beginning of the 19th century through to the identification of the cryptochromes and the phototropins, with brief mention of the three members of the ZTL/ADO family that are also likely authentic blue-light receptors in plants. The present chapter will focus on the phototropins (phot1 and phot2) and will concentrate on developments related to the phototropins that occurred over the past decade and a half. Related aspects that have already been covered in Chapter 10 will be mentioned only briefly.

Among the many physiological responses of plants to light, the phototropins are known to mediate phototropism, chloroplast rearrangement, stomatal opening, leaf expansion, the rapid inhibition of growth of etiolated seedlings, and possibly solar tracking (see Briggs and Christie 2002 and Chapter 10). The phototropins are blue-light receptor kinases that undergo a unique photochemistry upon absorption of blue light (Salomon *et al.*, 2000). This unique photochemical reaction leads to a protein conformational change (Crosson and Moffat, 2002; Harper *et al.*, 2003; Corchnoy *et al.*, 2003) that in turn results in autophosphorylation of the photoreceptor protein (Christie *et al.*, 1998). The details of these reaction steps with respect to the molecular structure of the phototropins, the putative roles of their chromophore-binding domains, and the biophysical properties of these domains are discussed below, in combination with the cellular and subcellular distribution of these photoreceptors in correlation with the different physiological responses that they mediate. Recent reviews dealing specifically with the phototropins are: Briggs *et al.*, (2001a, 2001b), Briggs and Christie (2002), Briggs and Huala (1999), Christie and Briggs (2001, 2005), Crosson (2005), Lin (2002a, 2002b), Swartz and Bogomolni (2005).

2. BLUE LIGHT-ACTIVATED PHOSPHORYLATION OF A PLASMA-MEMBRANE PROTEIN

2.1 The protein is likely ubiquitous in higher plants

We begin with the discovery and biochemical characterization of the light-activated phosphorylation of a plasma membrane protein originally found in etiolated pea seedlings. Although discovered in 1988 (Gallagher *et al.*, 1988), this protein was not definitively identified as a photoreceptor until a decade later (Christie *et al.*, 1998) and was not designated phototropin (eventually phototropin 1) until 1999 (Christie *et al.*, 1999). We will use the name phototropin (or phot1 or phot2, respectively) throughout discussion of the earlier biochemical and physiological studies for the sake of clarity. The initial study by Gallagher *et al.*, (1988) began a series of biochemical and photophysiological studies aimed at characterizing the reaction. The reaction was found in membrane preparations from etiolated seedlings of all species tested, including several cereal grasses: maize (*Zea mays*) (Hager and Brich 1993; Palmer *et al.*, 1993b); oat (*Avena sativa*) (Salomon *et al.*, 1996, 1997a); wheat (*Triticum aestivum*) (Reymond *et al.*, 1992a, Sharma *et al.*, 1997); barley (*Hordeum vulgare*) and sorghum (*Sorghum bicolor*) (Reymond *et al.*, 1992a) and several dicots: pea (*Pisum sativum*) (Gallagher *et al.*, 1988); sunflower (*Helianthus annuus*), tomato (*Lycopersicum esculentum*), and zucchini (*Cucurbito pepo*) (Reymond *et al.*, 1992a).

As mentioned above, this membrane-associated protein itself turned out to be the photoreceptor for its own phosphorylation and was subsequently given the name phototropin 1 (phot1) (Christie *et al.*, 1999; Briggs *et al.*, 2001a). As will be discussed later, phot1 is a protein that contains a serine threonine kinase domain and two binding sites for the chromophore flavin mononucleotide (FMN), called LOV1 and LOV2 (Huala *et al.*, 1997). A second phototropin was identified in *Arabidopsis* (Jarillo *et al.*, 1998), and designated phototropin 2 (phot2) (Briggs *et al.*, 2001a). Like phot1, phot2 was subsequently also shown to carry out light-activated autophosphorylation (Sakai *et al.*, 2001; Christie *et al.*, 2002). The domain structures of the two phototropins are shown in Figure 1.

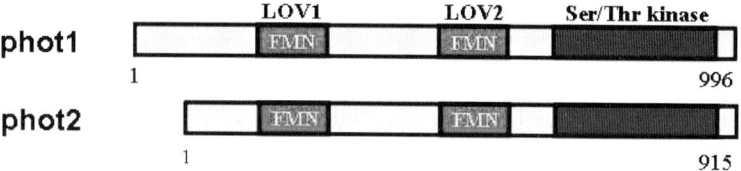

Figure 1. Domain structures for phototropins 1 and 2.

Phototropins have also been reported in the fern *Adiantum capillus-veneris* (Nozue et al., 2000) and, in addition, this same fern contains a unique chimeric photoreceptor that encodes the N-terminal half of a phytochrome followed by a full phototropin sequence including both LOV domains (LOV1 and LOV2) and the kinase domain (Nozue et al., 1998). Thus the phototropins are likely ubiquitous in higher plants. More recently, a phototropin has been identified in the unicellular alga *Chlamydomonas reinhardtii* (Huang et al., 2002; Kasahara et al., 2002) where it serves as the photoreceptor mediating various steps in the cycle for sexual reproduction (Huang and Beck, 2003).

2.2 Subcellular localization of phot1

All membrane-fractionation studies: pea (Gallagher et al., 1988; Short et al., 1993); maize (Hager and Brich, 1993; Palmer et al., 1993b); wheat (Sharma et al., 1997); oat (Salomon et al., 1996) indicate that phot1 in etiolated seedlings is localized to the plasma membrane. Solubilization of right-side-out plasma-membrane vesicles from pea (Short et al., 1993) or maize (Hager and Brich, 1993) with triton greatly increases the level of phosphorylation as does repetitive freezing and thawing of these vesicles (Short et al., 1993), a procedure that produces a mixture of right-side-out and inside-out vesicles (Palmgren et al., 1990). Thus association with the plasma membrane is at the inner surface. However, phot1 is a highly hydrophilic protein with no obvious transmembrane domains (Huala et al., 1997) and to date the nature of its association with the plasma membrane is unknown.

Sakamoto and Briggs (2002) examined the subcellular distribution of phot1 in *Arabidopsis* by fusing phot1 to green-fluorescent protein (phot1-GFP). The construct, driven by the native *PHOT1* promoter, was transformed into a *phot1* null mutant of *Arabidopsis*. In all cells of etiolated seedlings showing green fluorescence, phot1-GFP was located in close proximity to the plasma membrane. However, in both epidermal and cortical cells of the etiolated hypocotyl, a small fraction of phot1-GFP was released to the cytoplasm on illumination with blue light. The green fluorescence in the cytoplasmic fraction disappeared after about an hour, but it is not clear whether the protein became re-associated with the plasma membrane or was actually degraded. Sakamoto and Briggs also found that the phot1-GFP signal was strongest in cortical cells both in the hypocotyl and root, the most intense fluorescence arising from the apical and basal ends of the cells. By contrast, the epidermal signal was much lighter in both hypocotyl and root (barely detectable in the root) and uniformly distributed all around the cell. Stomatal guard cells showed strong green fluorescence adjacent to their walls including that portion of the wall that forms the stomatal pore (Figure 2).

Recent studies using phot1 tagged with hemagglutinin (HA) have shown that phot1 is also associated with the plasma membrane in light-grown *Arabidopsis* (Harada et al., 2003). These findings are consistent with the subcellular distribution of phot1-GFP in light-grown tissues (Sakamoto and Briggs, 2002) and its presence in a membrane fraction. In all of the light-grown tissues examined, phot1-GFP was found in close proximity to the plasma membrane. Included are epidermal cells, mesophyll cells, guard cells, xylem parenchyma, and phloem parenchyma. The green fluorescence in the xylem (and likely the phloem) parenchyma was also concentrated at the apical and basal ends of the cells whereas in other cell types it was more or less uniformly distributed.

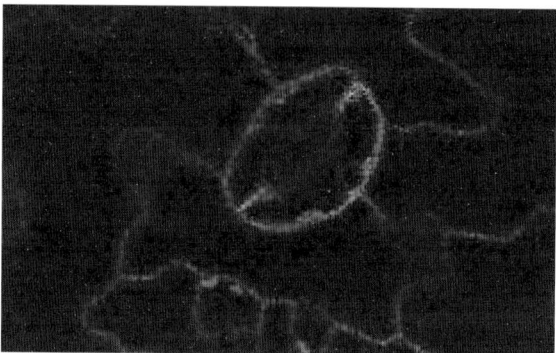

Figure 2. Localization of phot1-green fluorescent protein (GFP) in guard cells and leaf epidermal cells. Red fluorescence is from chloroplasts. See Sakamoto and Briggs (2002).

Recent investigations have also shown that phot2, like phot1, is associated with the plasma membrane. Harada et al., (2003) have demonstrated that phot2, tagged with HA, is localized to the plasma membrane in *Arabidopsis* leaf tissue. Similarly phot2 is associated with a crude membrane fraction obtained from etiolated *Arabidopsis* seedlings (Christie et al., 2002). The similarity in localization of phot1 and phot2 is perhaps not surprising given the known functional redundancy that exists between these photoreceptors (Briggs and Christie, 2002). There are, however, two exceptions to this functional redundancy. First, phot2 evidently plays no role in the rapid inhibition of growth as *phot1* null mutants completely lack this response despite having wild-type phot2 (Folta and Spalding, 2001). The likely explanation is trivial: *PHOT2* gene expression is extremely low in etiolated seedlings. It requires phyA phototransformation by red light and only increases several hours after red light treatment (Tepperman et al., 2001). Second, phot2 alone is the photoreceptor for the chloroplast avoidance response (Kagawa et al., 2001; Jarillo et al., 2001b) although mesophyll cells contain both phototropins (Sakamoto and Briggs 2002, C. Tissier and W. R. Briggs, unpublished). The chloroplast avoidance response is subcellular autonomous in that it only occurs in the irradiated region of a partially shaded cell (Tlalka et al., 1999). Hence phot2 must be in a fixed location with respect to the cell surface and must not migrate during a 40 min irradiation. This behaviour is consistent with the plasma-membrane localization of phot2.

2.3 Distribution of the phototropins in relation to function

Given the role of auxin in phototropism (Koller 2000; Iino 2001), it is perhaps not surprising that phot1 is found localized to those cells thought to mediate polar auxin transport. In the elongate cortical parenchyma cells of the etiolated hypocotyl, phot1 is strongly localized adjacent to the apical and basal walls rather than the side walls (Sakamoto and Briggs, 2002). PIN1, a protein thought to be in some way involved in auxin tranport, has recently been shown to be located at the basal end of these same cells (Gälweiler et al., 1998). In the elongating flowering stem, phot1 is found in both xylem and phloem parenchyma, again localized to the apical and basal ends of the cells (Sakamoto and Briggs, 2002). PIN1, a protein associated in some way with auxin transport, is found at the basal ends of the xylem parenchyma (Gälweiler et al., 1998). PIN3, another protein thought to be involved in auxin transport, is localized to the phloem parenchyma (Friml et al., 2002).

Several studies have demonstrated a close correlation between the tissue distribution of phototropic sensitivity and the level of detectable light-activated phosphorylation in etiolated seedlings: pea (Short and Briggs, 1990); maize (Hager and Brich, 1993; Palmer et al., 1993a); wheat (Sharma et al., 1997); oat (Salomon et al., 1997a). Indeed Salomon et al., (1997b, 1997c) showed that unilateral light induced a gradient in phosphorylation of phot1 across oat coleoptiles, with a higher level of phosphorylation on the irradiated side than on the shaded side. These latter experiments represent the first direct demonstration of a light-inducible biochemical gradient across a phototropically sensitive organ.

The subcellular distribution of phot1 is also consistent with the several physiological responses that it mediates. Its location in both leaf and hypocotyl epidermis is consistent with the epidermal role in regulating growth in these organs—regulation of leaf expansion in developing leaves (van Volkenburgh 1999) and regulation of stem elongation (see Kutschera and Briggs, 1987). Its location in stomatal guard cells is consistent with its role in blue light-stimulated stomatal opening (Kinoshita et al., 2001; Zeiger and Hepler, 1977; Zeiger and Field 1982). Indeed this response shows a typical "cryptochrome" action spectrum (Karlsson 1986; Eisinger et al., 2000). Furthermore, the location of phot1 in the plasma membrane of mesophyll cells is consistent with its role in mediating blue light-activated chloroplast movements (Kagawa et al., 2001; Kagawa and Wada, 2002; Wada et al., 2003). These responses also show typical "cryptochrome" action spectra (Zurzycki, 1962; see Haupt and Scheuerlein, 1990). Although a role for the phototropins in solar tracking has not been definitively established, its location at the end walls of elongate subepidermal cells in the leaf veins (Sakamoto and Briggs, 2002) fits with its putative role in solar tracking. At least in *Lavatera cretica* (in the Malvaceae) it is the veins that perceive the light signal bringing about solar tracking (Schwartz and Koller 1978) and one model, based on results of experiments with polarized light (Koller et al., 1990) hypothesizes this to be the location for the photoreceptor. While there is currently no good action spectrum for solar tracking, Yin (1938) demonstrated many years ago that it was a response to blue light.

2.4 Biochemical properties of the phosphorylation reaction *in vitro*

When elongating etiolated pea stem sections (Short et al., 1992) or growing isolated maize coleoptiles (Hager and Brich 1993; Palmer et al., 1993b) are incubated in ^{32}P inorganic phosphate, to allow them to synthesize radiolabelled ATP, subsequent blue-light irradiation of the treated tissue prior to membrane extraction results in light-activated phosphorylation of phot1 *in vivo* as determined by subsequent gel electrophoresis of the membrane proteins and autoradiography. Likewise, when membranes isolated from etiolated seedlings are irradiated prior to adding (or in the presence of) γ^{32}P-labeled ATP, light-activated phosphorylation can be observed *in vitro*. Hence light-activated phosphorylation can be demonstrated both *in vivo* and *in vitro*, the latter reaction providing a powerful tool to study the biochemical properties of phot1.

In all cases examined: pea (Short *et al.*, 1992; Short *et al.*, 1994); *Arabidopsis* (Reymond *et al.*, 1992b; Christie *et al.*, 1998); maize (Hager and Brich, 1993; Palmer *et al.*, 1993b), light-activated phosphorylation begins without any perceptible lag after the addition of ATP and reaches a maximum within about two minutes at room temperature. In some studies, the level of phosphorylation was observed to decline over the next half hour or so: *Arabidopsis* (Reymond *et al.*, 1992b); pea (Short *et al.*, 1992) whereas in other studies it remained high. The reason for this discrepancy is not clear but it could be related to differences in buffer used, presence or absence of detergent, etc. The reason for the decline itself is also not clear as neither protease nor phosphatase inhibitors prevented it. Addition of cold ATP failed to cause any measurable phosphate turnover of phosphorylated pea phot1(Hager and Brich, 1993; Short *et al.*, 1992). The earlier biochemical studies established several general properties of the phosphorylation system. Both in pea (Short *et al.*, 1994) and maize (Hager, 1996) the pH optimum for phot1 autophosphorylation was near 7.5, with activity declining steeply at lower pH values. The pH optimum is close to the isoelectric point for pea phot1 (Warpeha *et al.*, 1993). Phosphorylation of both pea (Short *et al.*, 1994) and maize (Hager, 1996) phot1 showed high specificity for ATP. GTP was a very poor substrate and both UTP and CTP were completely inactive. Although we now know that the phototropins are flavoproteins, only high concentrations of the flavin quenchers iodide, azide, or phenylacetic acid inhibited the reaction in pea (Short *et al.*, 1992). Likewise, the hydrophobic thiol reagent N-phenylmaleimide was far more effective than its hydrophilic counterparts, iodoacetate or N-ethylmaleimide, in inhibiting the phot1 photoreaction in maize membranes (Rüdiger and Briggs, 1995). Photochemical and structural analyses of the chromophore-binding domain to be described later (section 3.2; Salomon *et al.*, 2000; Crosson and Moffat, 2001) provide a partial explanation: the flavin moieties and a thiol (or thiolate: see section 3.4) with which they react are deeply embedded within a pocket formed by the LOV domains. However, these agents are equally effective if added before irradiation or subsequently but prior to addition of ATP. Other -SH groups required for phosphorylation must be involved.

2.5 Correlation of phot1 phosphorylation with phototropism

The three action spectra for phototropism (Shropshire and Withrow 1958, Thimann and Curry, 1960; Baskin and Iino, 1987) all show a broad action maximum in the UV-A part of the spectrum and an action maximum in the blue with a major peak near 450 nm, a second lesser peak near 470 nm, and a shoulder near 420 nm, as illustrated in Figure 3. As mentioned in Chapter 10, Gressel (1979) at the time assigned this and similar action spectra as indicative an unknown photoreceptor he called "cryptochrome." Given that phot1 is a photoreceptor for phototropism, it is not surprising that the action spectra for light-activated phosphorylation of phot1 either *in vivo* (Hager and Brich, 1993) or *in vitro* (Palmer *et al.*, 1993a) showed the same features. As discussed below, these spectra all match closely the absorption spectrum of phot1 and its chromophore-binding domains.

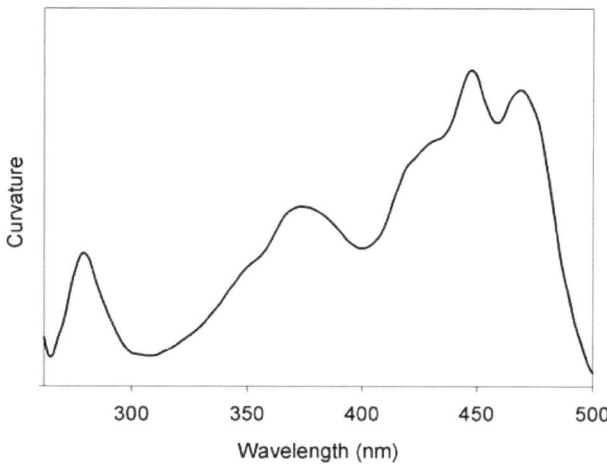

Figure 3. Action spectrum for the phototropic response of etiolated alfalfa seedlings. After Baskin and Iino (1987).

It was Briggs (1960) who first showed that a dose of blue light sufficient to saturate the first positive curvature response but too short to induce second positive curvature rendered coleoptiles completely insensitive to a subsequent pulse of unilateral blue light that would normally induce maximum first positive curvature. However, interposing increasingly long dark periods between the saturating pulse and the second light pulse yielded a gradual recovery of phototropic sensitivity over a period of many minutes. Forty-three years later, Palmer *et al.*, (1992a) and Hager and Brich (1993) demonstrated the same phenomenon with light-activated phosphorylation *in vivo*. These dark-recovery curves are shown in Figure 4. Maize coleoptiles that had been exposed to light sufficient to saturate *in vivo* phosphorylation failed to show *in vitro* phosphorylation of phot1 in membrane preparations isolated immediately after the light exposure. Once again, interposing increasingly long dark periods prior to membrane isolation yielded a gradual recovery of the capacity of the membranes to respond to light with phosphorylation *in vitro*. The time course for this recovery was identical to that originally observed for phototropism (Palmer *et al.*, 1992a). Similar recovery curves were reported for the phosphorylation responses of both oat coleoptiles (Salomon *et al.*, 1997a) and etiolated pea stem sections (Short and Briggs, 1990). At the time it was not clear whether the time course for recovery was owing to dephosphorylation or merely a combination of degradation of phosphorylated phot1 accompanied by its *de novo* synthesis. We will return to this question later.

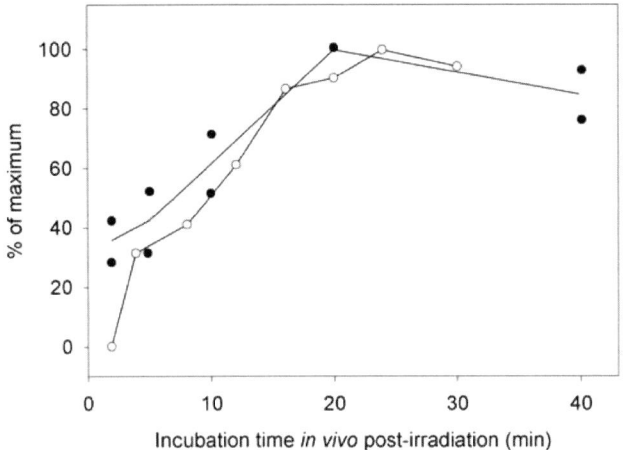

Figure 4. Recovery of phototropic sensitivity (open circles, Briggs 1960) and capacity for light-activated phosphorylation (closed circles, Palmer et al., 1993a) following saturating light treatments.

A number of studies (Short *et al.*, 1992; Hager and Brich, 1993; Palmer *et al.*, 1993b; Salomon *et al.*, 1996) have reported exactly the same dark-recovery phenomenon *in vitro*. Membranes irradiated with blue light and returned to darkness prior to the addition of ATP retained their capacity to respond to light signals for many minutes in the dark. Clearly, the photochemical event was forming a relatively stable species that gradually returns to its dark state. It is now known that this biochemical memory both *in vivo* and *in vitro*, which was initially quite puzzling, almost certainly lies with the unique photochemistry and subsequent dark reactions associated with phototropin photoexcitation. At least in some cases, the phosphorylation response could be reactivated by a second pulse of blue light (Hager *et al.*, 1993; Salomon *et al.*, 1996; J. M. Christie and W. R. Briggs, unpublished), consistent with the photochemical properties now known for the phototropins.

Several workers have reported fluence-response curves for light-activated phot1 phosphorylation both *in vivo* [pea (Short and Briggs, 1990; Short *et al.*, 1992); maize (Palmer *et al.*, 1993a); oat (Salomon *et al.*, 1997a)] and *in vitro* [pea (Short *et al.*, 1992; Short *et al.*, 1993); maize (Hager *et al.*, 1993; Palmer *et al.*, 1993a); oat (Salomon *et al.*, 1996); *Arabidopsis* (Reymond *et al.*, 1992a)]. In contrast to the correlations described above, these curves all show threshold values at minimum ten-fold higher than the threshold required to bring about first positive phototropic curvature. As a result, Palmer *et al.*, (1993a) suggested among other hypotheses that phosphorylation might not be the primary event activating the curvature mechanism, but might possibly serve to desensitise the system by returning it to its ground state. Briggs (1996) hypothesized that the autophosphorylation might be hierarchical, with the most sensitive phosphorylations involved in signal transduction and less sensitive phosphorylations involved in desensitisation. This apparent discrepancy will be addressed in more detail in the following section.

2.6 Autophosphorylation occurs on multiple sites

There was evidence early on that the light-activated phosphorylation of phot1 might occur on multiple sites. Both Short *et al.* (1993) and Liscum and Briggs (1995) found that the phosphorylated phot1 protein showed significantly lower mobility on SDS-PAGE than its unphosphorylated form, consistent with multiple phosphorylations (Beebe and Corbin, 1986). Proteolysis of phosphorylated phot1 (Short *et al.*, 1992; Short *et al.*, 1994; Salomon *et al.*, 1996) yielded multiple stable phosphorylated bands that did not break down further on longer incubation. Phospho-amino acid analysis of phot1 from maize (Palmer *et al.*, 1993b) and pea (Short *et al.*, 1994) yielded a major fraction of phophoserine, at most a trace of phosphothreonine, and no phosphotyrosine.

Although there was early concern whether the phosphorylation reaction involved one, two, or three separate components (the phosphorylation substrate, the kinase, and the photoreceptor; see Briggs *et al.*, 2001b for details), it is now clear that a single protein serves all three functions (Christie *et al.*, 1998). The best biochemical evidence that phot1 is a kinase came from studies with the kinase inhibitor 5'-*p*-fluorosulfonylbenzoyladenosine (FSBA), a reagent that forms a covalent bond in ATP binding sites (Wyatt and Coleman, 1977). Using an antibody against this inhibitor, Short *et al.*, (1993) found a mobility shift for an ATP-binding site identical to the phot1 mobility shift that occurred on phosphorylation. In a related study, Palmer *et al.*, (1993b) demonstrated that FSBA strongly inhibited the phosphorylation reaction *in vitro*, an inhibition that could be relieved with excess ATP. Additional evidence also suggested that phosphorylation may be intermolecular between members of a phototropin dimer. Reymond *et al.*, (1992a) observed that irradiated pea phototropin could cross-phosphorylate non-irradiated maize phototropin, indicating that at least some of the phosphorylation was between phototropin members. It should be mentioned, however, that Hager (1996) was unable to confirm this observation.

Recently, a careful study by Salomon *et al.*, (2003) unambiguously identified eight serines in oat phot1 that became phosphorylated either on illumination of oat phot1 *in vitro* or through phosphorylation mediated by protein kinase A. Two of these sites (S27, S30) occur upstream of LOV1, near the N-terminus of the protein. The remaining six sites (S274, S300, S317, S325, S332, and S349) are located in the hinge region between the LOV1 and LOV2 domains of phot1 (see below). Of special interest, Salomon *et al.*, (2003) found that there was indeed a hierarchy for phosphorylation: the two N-terminal serine residues (S27, S30) were phosphorylated at very low fluences, whereas other sites required either intermediate or high fluences to become phosphorylated. Curiously, this hierarchical pattern was only observed when the phosphorylations were carried out *in vivo*. No such fluence discrimination was detected when the photoreactions were carried out *in vitro*. Thus, at least *in vivo*, the biochemical consequences of low fluences with respect to phosphorylation could be quite different from those of high fluences. Indeed, phosphorylation of specific serine residues in response to different light fluences may be influenced by some interacting factor or factors within the cell. It is therefore possible that sites phosphorylated at the lowest fluences could mediate signal transduction and those phosphorylated at the higher fluences could play some other role, for example photoreceptor desensitisation.

Salomon *et al.*, (2003) also showed that the recovery of unphosphorylated phot1 *in vivo* occurred in a fluence-dependent manner: those sites phosphorylated only by the highest fluences became dephosphorylated first followed by those phosphorylated at intermediate and low fluences respectively (Salomon *et al.*, 2003). Thus the decline in phosphorylation *in vivo* over time in the dark reflects dephosphorylation rather than degradation and resynthesis of phot1.

To date, there is no evidence indicating that phototropin phosphorylation initiates a phosphorylation cascade. Although a number of workers have reported the appearance of bands lower than 120 kDa on SDS-PAGE suggesting light-activated phosphorylation (Palmer *et al.*, 1993b; Reymond *et al.*, 1992b; Salomon *et al.*, 1996; Sharma *et al.*, 1997; Short and Briggs, 1990), these are often found only in older tissue (Salomon *et al.*, 1996; Short and Briggs, 1990) and do not occur with any regularity. It is highly likely that they simply represent phosphorylated phot1 that has been cleaved during sample preparation.

3. CLONING AND MOLECULAR CHARACTERIZATION OF PHOTOTROPIN

3.1 The initial discovery of phototropin 1

Since the historical events underlying the discovery of the phototropins have already been presented in Chapter 10, this topic will be only briefly summarized here. An *Arabidopsis* mutant, JK224 that showed impaired phototropism (Khurana and Poff, 1989) was also deficient in light-activated phosphorylation of the plasma membrane-associated protein now designated phot1. These results suggested that the protein might be involved in phototropism (Reymond *et al.*, 1992b). Liscum and Briggs (1995, 1996) then isolated and characterized additional mutants of *Arabidopsis* deficient in their phototropic responses (designated *nph* for *n*on-*p*hototropic *h*ypocotyl). Among these was the mutant allele *nph1-5* that failed to show any phototropic response to low fluences of unilateral light. Using amplified fragment-length polymorphisms (AFLP), Huala *et al.*, (1997) located, cloned, and sequenced the *NPH1* gene. The deduced protein sequence contains a classic serine/threonine kinase domain at its C-terminal end (Hanks and Hunter, 1995) and two PAS domains (Taylor and Zhulin, 1999). These domains, approximately 110 amino acids in length, are highly similar to domains in other proteins from both prokaryotes and eukaryotes that respond to environmental signals such as light, oxygen, or voltage. Consequently Huala *et al.,* (1997) designated these LOV1 and LOV2 respectively.

Christie *et al.,* (1998) then demonstrated that when the *NPH1* gene was expressed in insect cells kept in the dark, the encoded protein bound the flavin FMN, and could undergo light-activated autophosphorylation. The authors concluded that it was the photoreceptor for its own phosphorylation and was therefore the photoreceptor for phototropism. Further expression studies showed that each LOV domain, when expressed in *Escherichia coli*, bound one molecule of FMN non-covalently (Christie *et al.,* 1999). Moreover, purified LOV domains with bound FMN displayed spectral properties similar to those of the entire photoreceptor protein expressed in insect cells (Christie *et al.,* 1998), showing absorption in the blue/UV-A regions of the spectrum very similar to the bands in the action spectrum for phototropism described earlier. Given the known role for this light-activated phosphoprotein in phototropism, Christie *et al.,* (1999) gave the protein the name phototropin (later phototropin 1; Briggs *et al.,* 2001a). And from these studies it was clear that phot1 was a photoreceptor that stoichiometrically bound not one but two chromophores, both of which were FMN. The close similarity of the absorption spectrum of a LOV domain to the action spectrum for phototropism is shown in Figure 5.

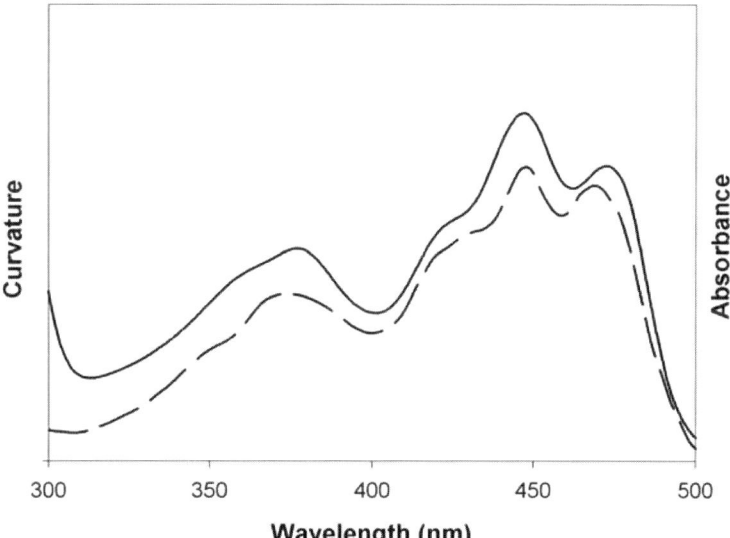

Figure 5. Comparison of action spectrum for phototropism of etiolated alfalfa seedlings (dashed line, after Baskin and Iino, 1987) and the absorption spectrum of the Arabidopsis phot1 LOV2 domain (solid line).

3.2 LOV domains function as light sensors

Salomon *et al.,* (2000) were the first to report that both oat phot1 LOV domains purified from *E. coli* underwent fully reversible photobleaching. However, the light-induced absorption changes detected for the LOV domains were not those typical of a flavin photoreduction (e. g. Massey and Hemmerich, 1978), which typically results in the appearance of a single isosbestic point between 330 and 350 nm, where the absorption of the initial ground state of the flavoprotein and its photoproduct are exactly the same. Instead the light-induced absorption changes associated with the LOV domains had three isosbestic points, at 331, 385, and 407 nm for both LOV domains, reflecting photobleaching in the blue region of the spectrum and formation of a single broad absorption band near 390 nm. This type of absorption change is known to accompany the formation of a covalent linkage between the C(4a) carbon of the flavin isoalloxazine ring and a cysteine residue within the flavoprotein (Miller *et al.*, 1990).

Both LOV1 and LOV2 contain a cysteine situated within a highly conserved subdomain, GRNCRFLQ. This cysteine corresponds to residue 39 within the 110 amino acid stretch of LOV1 and LOV2 and has been designated C39 for simplicity. In the case of LOV2, C39 is the sole cysteine within the domain and hence was likely to be the one involved in the blue light-induced formation of a flavin-cysteinyl adduct. Molecular modelling placed this cysteine very close to the C(4a) carbon of the FMN (Salomon *et al.*, 2000). To test this hypothesis, Salomon *et al.*, (2000) mutated this cysteine to an alanine or a serine. Although flavin binding was unaffected in either LOV domain, the spectral changes observed for the wild-type domains on irradiation were completely eliminated in the C39A and C39S mutated proteins. These findings were consistent with the hypothesis that the primary events following phototropin photoexcitation involved the formation of an FMN-cysteinyl adduct within each LOV domain (see Figure 7).

4. WHY TWO LOV DOMAINS?

Photolyases are photoreceptors with two chromophores, flavin adenine dinucleotide (FAD) and either a deazaflavin or a pterin (Sancar, 1994). In this case, the deazaflavin or pterin serve as antenna pigments passing their absorbed energy to FAD, which then initiates a DNA-repair process. This scenario seems highly unlikely for the phototropins. First, the FMN is tightly bound within the protein pockets of the two LOV domains (see below). Thus, the two chromophores must be a considerable distance apart, making energy transfer between them unlikely. Second, their absorption maxima are sufficiently close (449 nm for LOV1 and 447 nm for LOV2; Salomon *et al.*, 2000), an arrangement that would not favour unidirectional energy transfer. Third, both carry out photocycles that are identical except for their kinetic parameters (see below). It seems more likely that the LOV domains actually serve different functions.

Christie *et al.,* (2002) used the C39A mutation described above to investigate the role of LOV1 and LOV2 in contributing to photochemistry for both phot1 and phot2. They first inactivated one or the other of the LOV domains in a LOV1 + LOV2 fusion protein expressed and purified from *E. coli*. The resulting effects were somewhat different between the phot1 and phot2 fusion proteins. For phot1, the fusion protein with LOV2 inactivated gave a much higher fluorescence yield than when LOV1 was inactivated. Thus LOV2 plays the major role in the initial photochemistry. When full-length phot1 was expressed in insect cells with either LOV1 or LOV2 inactivated, phosphorylation was not perceptibly affected in the absence of photochemically active LOV1, but almost eliminated in the absence of LOV2. In fact, both the time course and the fluence-response curve for phosphorylation without LOV1 were identical to those when both LOV domains were present. Finally, when the various full-length constructs for phot1 were transformed into a null mutant of phot1, photochemically active LOV2 alone (LOV1 inactivated) supported a phototropic response whereas LOV1 alone (LOV2 inactivated) did not. When membrane preparations were obtained from etiolated seedlings of these latter transformants, the same result was obtained: LOV2 alone could support phosphorylation whereas LOV1 alone could not.

For phot2, the situation was somewhat different. Although LOV2 still dominates the photochemistry, inactivation of LOV1 in the LOV1 + LOV2 fusion peptide gave a significant increase in fluorescence, indicating a more prominent role for LOV1 in phot2 photochemistry than in phot1. Likewise, inactivation of LOV2 in full-length protein expressed in insect cells did not completely eliminate light-activated autophosphorylation as appeared to be the case with phot1. Recent unpublished experiments from our laboratory (E. Kaiserli and W. R. Briggs, unpublished) indicate that in the full-length phot2 protein, expressed in a *phot1 phot2* double null mutant, LOV2 alone can support phototropism under light-saturating conditions. As yet, however, comparative quantitative information on fluence-rate versus response between these various transformants is not available. Clearly, the individual roles of the LOV domains in controlling the activity of these two photoreceptors requires more extensive investigation. Most recently, Kagawa *et al.*, (2004) demonstrated that phot2 from the fern *Adiantum capillus-veneris* lacking almost all of the amino acids N-terminal from LOV2 could still fully complement mutants lacking the blue light-induced chloroplast avoidance response. Again, LOV1 was dispensable. These workers also showed that a domain between amino acids 979 and 999 was essential for obtaining complementation.

Since the seminal work of Salomon *et al.*, (2000) there has been an explosion of interest in the photochemical and biochemical properties of the LOV domains. Because these domains offer a completely new type of photochemical reaction and photocycle on activation by light, they have enticed an entirely different group of researchers into probing their properties. Their studies are discussed in the following section.

5. STRUCTURAL AND PHOTOCHEMICAL PROPERTIES OF THE LOV DOMAINS

5.1 LOV domain photochemistry

As mentioned above, the LOV domains undergo a self-contained photocycle following absorption of light. The photocycle is characterized by a series of transient intermediates each involving a protein/chromophore structural perturbation and ending with the system returning to the ground state. As described already, the initial characterization of the LOV domain photocycle identified only one photo-intermediate, which was postulated to be a cysteinyl-adduct species that absorbed maximally at 390 nm (Salomon et al., 2000). This intermediate was designated LOV2$^S_{390}$ (Swartz et al., 2001). In the dark the LOV2$^S_{390}$ species decayed to the ground-state species labelled LOV2$^G_{450}$. All LOV domains characterized to date have similar ground-state absorption and form the LOV2$^S_{390}$ intermediate on blue-light irradiation (Kasahara et al., 2002, Imaizumi et al., 2003, Schwerdtfeger and Linden, 2003).

The reported extinction coefficients for the two domains of oat phot1 (LOV 1, ε_{449} = 12,200 mol^{-1} cm^{-1} and ε_{370} = 10,000 mol^{-1} cm^{-1}; for LOV2, ε_{447} = 13,800 mol^{-1} cm^{-1} and ε_{378} = 8,700 mol^{-1} cm^{-1}; Salomon et al., 2000) are not all that different. However, there are significant differences in the photocycle properties between the two domains (Salomon et al., 2000). The approximate quantum efficiencies for adduct formation are near 0.4 for LOV2, but only 0.045 for LOV1, a ratio close to 10:1. Likewise the dark-recovery rates differ: the half life for recovery of the ground state from the flavin-cysteinyl adduct at room temperature is about 11 s for LOV1 whereas it is near 27 s for LOV2. Differences in approximate quantum yield and kinetics between LOV domains were also noted in *Chlamydomonas* phot, *Arabidopsis* phot1 and phot2, and rice phot1 and phot2 (Kasahara et al., 2002; Kottke et al., 2003; Holzer et al., 2002). The absolute quantum yield of cysteinyl-adduct formation for various LOV domains is disputed. However, the relative quantum yields for the various LOV domains measured within the same study can be compared.

Better understanding the LOV domain photochemistry has involved both structural determination of the protein/chromophore relationships and investigation of the photocycle on a faster time scale than that used by previous workers. Such studies have been successful in identifying the photo-intermediates that appear prior to formation of LOV2$^S_{390}$. These studies are reviewed in the following sections.

5.2 LOV-domain structure

Soon after these initial measurements of LOV domain photochemistry cited above, the crystal structure of a closely related phototropin LOV domain (phy3 LOV2 from *Adiantum capillus-veneris*) was solved to 2.7Å resolution (Crosson and Moffat, 2001). This structure clearly demonstrated the very specific and tight binding of

FMN within the LOV-domain fold. The LOV domain tertiary structure closely resembles that of some other PAS domains, including PYP, HERG and FixL (Crosson and Moffat, 2001). Its protein fold includes 5 β-sheets and 4 α-helices that form a very tight pocket to accommodate the chromophore (Figure 6). The FMN is held in place non-covalently but through a very specific hydrogen-bonding network on the pyrimidime side of the isolloxazine ring and with hydrophobic residues on the dimethylbenzene moiety (Crosson and Moffat, 2001). Specifically, the FMN N3, O2, and O4 all hydrogen bond to protein side chains; in addition, the phosphate group on the FMN ribityl chain interacts with the guanadinum groups of two arginines to form salt bridges. Finally, the hydroxy groups of the FMN ribityl side form hydrogen bonds to the LOV-domain protein. There are two water molecules in close proximity to the chromophore. Both are within hydrogen bonding distance of the hydroxy groups of the FMN ribityl chain.

The most striking feature of the structure is that the sulphur group of cysteine 39, which interacts with the FMN, is located only 4.2Å from the C(4a) carbon of FMN. Rotation around the C_α-C_β of C39 places the sulphur within 2.6Å of the C(4a) of the FMN chromophore. This protein crystal structure was confirmed by the structure of LOV1 of *Chlamydomonas* phot (Federov *et al.*, 2003). The main difference in the structure of *Chlamydomonas* phot LOV1 from that of *Adiantum* phy3 LOV2 is that the sulphur of the active cysteine was found in two conformations instead of only one. It should be noted that the *Chlamydomonas* phot LOV1 structure was solved at 77K, whereas the *Adiantum* phy3-LOV2 structure was solved at room temperature. The structure of phy3 LOV2 at low temperature also showed the cysteine in two conformations (Sean Crosson, personal communication).

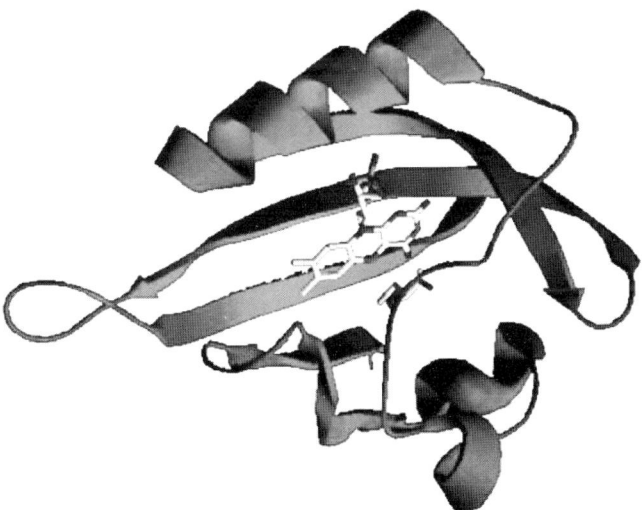

Figure 6. Structural model of the phytochrome3 LOV domain from Adiantum capillus veneris (after Crosson and Moffat, 2001). FMN and cysteine 39 are highlighted.

5.3 The LOV-domain photocycle

The evidence that the long-lived photo-intermediate LOV2$^S_{390}$ involved formation of a flavin-cysteinyl adduct (Salomon *et al.*, 2000) was confirmed from a number of complementary experiments. The first was publication of the NMR chemical shifts that occur during the oat LOV2 photocycle (Salomon *et al.*, 2001). Because the LOV2$^S_{390}$ intermediate decays very slowly back to the ground state, continuous illumination of a LOV2 sample converts almost all of LOV2 into the LOV2$^S_{390}$ species. NMR chemical shifts measured before and after continuous irradiation showed many chemical shifts; in particular, the NMR signal of FMN C(4a) experiences an upfield shift; this shift in association with reduced coupling between the FMN C4 and C(4a) is consistent with sp3 hybridisation of C(4a), which should occur upon flavin-cysteinyl adduct formation (Salomon *et al.*, 2001). Further evidence of a sulphur/FMN bond formation came when Crosson and Moffat (2002) solved the crystal structure of the long-lived photo-intermediate. This structure clearly demonstrated that upon light illumination, the sulphur of cysteine 39 moved to within 1.8 Å of the C(4a) of FMN; in addition, the two atoms shared electron density, further indication of bond formation.

Light-induced formation of this bond was also inferred from difference infrared absorbance spectroscopy (FTIR). Three vibrational bands occurring at 1580 cm^{-1}, 1550 cm^{-1}, and 1350 cm^{-1}, that are strongly coupled to the stretching of the C(4a)-N5 double bond, disappear upon formation of LOV2$^S_{390}$. This outcome is consistent with perturbations expected with formation of a bond at the C(4a) position (Swartz *et al.*, 2002, Ataka *et al.*, 2003).

5.4 Mechanism of FMN-cysteinyl adduct formation

The mechanism of adduct formation was partially elucidated with publication of the photocycle of oat phot1 LOV2. Following light absorption, the LOV domain undergoes a self-contained photocycle characterized by a series of photo-intermediates (Swartz *et al.*, 2001, Kottke *et al.*, 2003). The photo-intermediates are transient species that exist within the photocycle. Upon absorption of a photon, the FMN excited state decays into a red-absorbing species in nanoseconds (Kennis *et al.*, 2003). This first photo-intermediate, a species with a broad absorption band in the red, resembles the triplet state of FMN spectroscopically and is designated LOV2$^T_{650}$ (Swartz *et al.*, 2001). Further refinement of studies on the transient intermediates suggests that the triplet state exists as a mixture of species having the N5 both protonated and unprotonated (Kennis *et al.*, 2003). This triplet state decays in microseconds to the cysteinyl adduct species LOV2$^S_{390}$ (Swartz *et al.*, 2001). There is evidence for a light-activated back reaction from the cysteinyl adduct to the ground state but it is not known whether this reaction is physiologically relevant (Kennis *et al.*, 2004; Kottke *et al.*, 2003). The putative photocycle is shown in Figure 7.

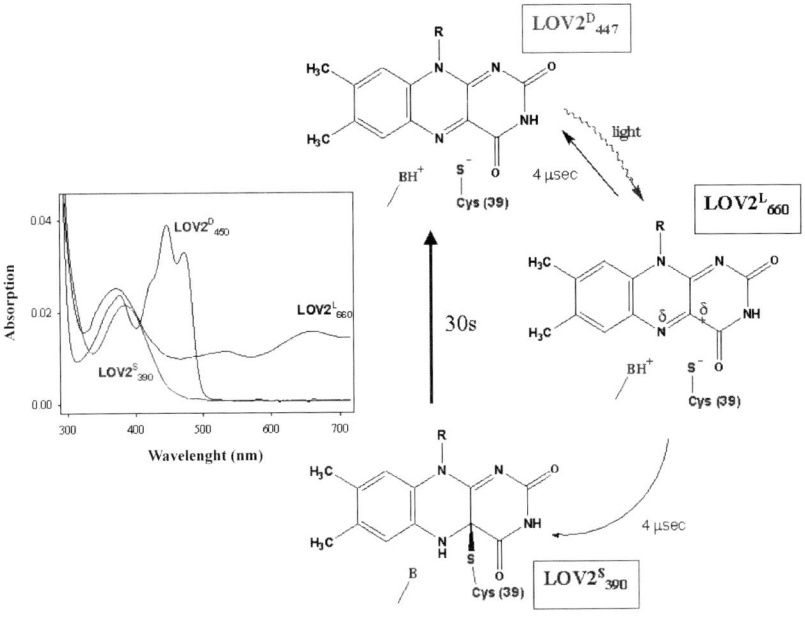

Figure 7. LOV2-domain absorption of ground state and intermediates and photocycle (see Swartz et al., 2001). Details in text.

The FMN triplet state involves a redistribution of charge around the FMN N5-C(4a) double bond, resulting in an increase of basicity of N5. We postulate here that N5 then becomes protonated, thus making N5-C(4a) a single bond and leaving C(4a) as a reactive carbo-cation that is attacked by the sulphur of cysteine 39, forming a flavin-cysteinyl adduct. The rate of adduct formation is 5 times slower in D_2O as compared to H_2O, confirming that the rate-limiting step between triplet state and adduct formation is a proton-transfer reaction (Corchnoy et al., 2003). Consistent with this conclusion, Kottke et al., (2003) have shown that the photocycle is both pH and salt dependent.

It was first suggested by fluorescence pH titrations that the sulphur of cysteine 39 existed as a thiolate (S^-) in the ground state and that the proton came from an unknown proton donor group (Swartz et al., 2001). This result was disputed by difference infrared-absorption measurements showing the disappearance of a vibrational band associated with the SH vibration upon illumination (Iwata et al., 2002; Ataka et al., 2003). This latter observation suggests that the proton that protonates N5 comes from the cysteine 39 sulphur. Following protonation of N5, the S^- attacks the C4a carbon. It is still possible that the ionisation state of the sulphur exists as mixed population with both thiol and thiolate in equilibrium. Shifts in this equilibrium in different LOV domains could explain the varying quantum yields that have been reported. Perhaps a change in pH could affect this equilibrium. The two cysteine conformations measured in the LOV1 domain of *Chlamydomonas* phot (Federov *et al.*, 2003) could possibly be a reflection of this mixed population. Indeed, the *Chlamydomonas* phot-LOV1 photocycle shows two rate constants from the triplet state to adduct formation, suggesting a more complicated scheme. In addition it has also been suggested that an electron transfer occurs during the photocycle (Bittl *et al.*, 2003; Kay *et al.*, 2003; Kottke *et al.*, 2003; Neiss and Sallfrank, 2003). The exact mechanism of the photocycle remains to be clarified, and although the fundamental mechanism for all the LOV domains is probably similar, subtle differences may be found to explain some of the present discrepancies.

5.5 The LOV domain back reaction

The exact mechanism of dark recovery of the ground state $LOV2^D_{450}$ from $LOV2^S_{390}$, which requires breakage of a carbon sulphur-bond, remains a mystery. The dark recovery of oat LOV2 is 3 times slower in D_2O than in H_2O, again suggesting that a proton-transfer reaction is the rate-limiting step (Swartz *et al.*, 2001). Circular dichroism (CD) measurements have also indicated that the protein returns to the ground state with the same kinetics as the chromophore, suggesting concerted events (Corchnoy *et al.*, 2003). The photocycles of at least a dozen different LOV domains have been measured; each has a different rate of dark recovery varying from a few seconds to many minutes (Kasahara *et al.*, 2002; Losi *et al.* 2002; Schwerdtfeger and Linden, 2003) to no dark recovery at all (Imaizumi *et al.*, 2003). The sequences of these various LOV domains present no clear insight into the varying relaxation rates. Abstraction of the N5 proton could initiate the back reaction; however, the structure fails to suggest a basic group in the vicinity of N5. The difference FTIR spectra show hydrogen-bonding perturbations associated with the two buried waters that are in close proximity to the chromophore (Iwata *et al.*, 2003), suggesting that these waters are possibly involved in the photochemistry and may form a proton-conducting channel to a basic group not in the immediate vicinity of the chromophore, a mechanism that has been found in other proteins. It has been suggested that mediation of the structure of a conserved salt bridge occurs during the photocycle and results in a transmitted signal (Crosson *et al.*, 2003).

It is tempting to conclude that the gradual recovery both of phototropic sensitivity and capacity for light-activated *in vivo* phosphorylation (section 2.12.4) reflect the gradual return of the $LOV2^S_{390}$ long-lived intermediate to the $LOV2^D_{450}$ dark state as measured for LOV domains synthesized and purified from *E. coli*. However, the situation cannot possibly be that simple. First, the dark-decay kinetics for single LOV domains differ in a major way from those for heterologously produced peptides containing both LOV domains (Kasahara *et al.*, 2002; Christie *et al.*, 2002). In fact, the presence of the kinase domain in the full-length photoreceptor may well further alter the dark-recovery kinetics. Second, recovery of the capacity for light-activated autophosphorylation *in vivo* is likely accompanied by dephosphorylation, and the rate of dephosphorylation could well be the limiting step in dark recovery at the physiological level. Resolving the relationships between the recovery of physiological sensitivity, the capacity for light-activated phosphorylation *in vivo*, the decay of the long-lived intermediate in purified LOV domains *in vitro*, and the dephosphorylation of activated phototropin *in vivo*, remains a major challenge.

5.6 Protein conformational change on photoexcitation

The long-lived photo-intermediate was initially suggested to be the signalling state of the photocycle, in parallel to other photoreceptors in which the longest lived intermediate was the signalling state (Swartz *et al.*, 2001). This signalling state in some way serves to activate the kinase domain. Presumably, the cell signal generated to activate the kinase requires a protein conformational change. The crystal structure of the LOV2$^S_{390}$ intermediate, compared to that of the ground state, showed only minor, light-induced protein changes, all within the vicinity of the chromophore (Crosson and Moffat 2002). This is not surprising, as the structure was of a relatively small peptide held within a crystal lattice. In contrast, difference FTIR spectra revealed many bands not associated with the chromophore that were perturbed during the photocycle, suggesting that considerable protein conformational changes occur during the photocycle (Swartz *et al.*, 2002; Ataka *et al.*, 2003; Iwata *et al.*, 2003). Circular dichroism (CD) studies provided further evidence that substantial protein structure changes accompanied the chromophore structural alterations (Corchnoy *et al.*, 2003). Specifically, a loss of α-helicity follows formation of the cysteinyl adduct.

Recent NMR experiments have provided the greatest detail on the protein-structural alterations accompanying the LOV domain photocycle. NMR measurements on a LOV2 domain that has an extra 20 residues clearly demonstrated protein structural alterations during the photocycle (Harper *et al.*, 2003). A low-resolution NMR model places 20 amino acids downstream from the C-terminal end of the LOV domain in an amphipathic helix (Jα) that lies almost immediately under the FMN cofactor. This helix, established by both chemical shifts and short range NOEs (*N*uclear *O*verhauser *E*ffect, a measurement of the environment surrounding protons), is moderately flexible, as shown by lack of measurable protection from 2H exchange. Significant light-induced changes in the NMR signature of this helix suggest protein perturbations of the helix initiated by the formation of the cysteinyl adduct and propagated through the LOV domain to this helix, which is over 15Å from the FMN chromophore. Enhancement of proteolysis digestion following illuminating of the LOV domain confirmed unfolding of the Jα-helix during the photocycle.

Measurements of the chemical shifts of the residues in the core of the protein show clear differences between LOV-domain peptides with and without the Jα-helix segment. These differences do not exist in the adduct form of these peptides. The measured quantum yields and back-reaction rates of single phototropin LOV domains have been shown to be altered when part of a peptide contains both LOV domains (Kasahara *et al.*, 2002; see above).

As mentioned above, there are major kinetic differences between the photochemical properties of LOV1 and LOV2, and also between the kinetic properties of single LOV domains and peptides containing both LOV domains. Examination of the protein sequences of all full-length phototropins reported to date indicates that protein sequences that could form the amphipathic Jα helix are only found downstream from LOV2 and not from LOV1. Thus, in addition to the differences in biochemical function of the two LOV domains (see section 4), there may well be differences in the structural consequences of their photoexcitation. At minimum, photoactivation of LOV2 leads sequentially to flavin-cysteinyl adduct formation, protein conformational change, kinase activation, and ultimately to autophosphorylation. A specific role for LOV1 photoactivation remains to be elucidated.

6. THE ZTL/ADO FAMILY

Three other *Arabidopsis* genes encode proteins that have a single LOV domain toward their N-terminus. These are *ZTL/ADO* (Somers *et al.,* 2000; Jarillo *et al.,* 2001a), *FKF1* (Nelson *et al.,* 2000), and *LKP2* (Schultz *et al.,* 2001). Their domain structure is shown in Figure 8. Otherwise, these proteins share no structural homology with the phototropins. Rather they have an F-box (related to targeting proteins for degradation) and six C-terminal kelch-domain repeats that form a propeller-like structure thought to be involved in protein-protein interactions. However, the LOV domains all have the conserved motif GXNCRFLQ containing the cysteine involved in phototropin photochemistry and ten other conserved amino acids found in all phototropin LOV domains sequenced to date. The crystal structure of phy3 LOV2 from the fern *Adiantum capillus-veneris* indicates that these amino acids are all involved either in hydrogen bonding with the FMN ribityl side chain or the flavin isolloxazine ring or van der Waals contacts with the ribityl side chain (Crosson and Moffat, 2001).

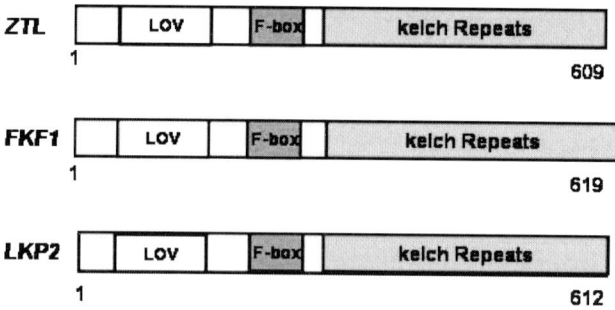

Figure 8. Domain structure of the ZTL/ADO family of putative Arabidopsis blue light receptors.

The above analysis strongly suggests that these three proteins might be blue-light receptors and there is also provocative physiological evidence as well to support this hypothesis.

Mutants at the *ZTL/ADO* locus show a much-lengthened period for their circadian rhythms (Jarillo *et al.*, 2001a; Somers *et al.*, 2000) and altered fluence-rate dependence for the light effect on the circadian period (Somers *et al.*, 2000). Jarillo *et al.*, (2001a) have shown both by yeast two-hybrid analysis and direct *in vitro* studies a physical interaction between the encoded ZTL/ADO protein with both cry1 and phyB. Overexpression of *LKP2* causes arrhythmic phenotypes for several circadian responses and causes a loss of the control of flowering by daylength (Schultz *et al.*, 2001). The deletion of *FKF1* causes a late-flowering phenotype and alters the waveform of the circadian expression of clock-regulated genes (Nelson *et al.*, 2000). Although none of these results establish the encoded proteins as photoreceptors, the presence in all three of a classic LOV domain and their close interaction with both circadian responses and photoperiodism support the possibility (see Chapter 27).

Recently Imaizumi *et al.*, (2003) have investigated the role of FKF1 in greater detail. They confirmed that *fkf1* mutants have lost photoperiodic control of flowering, and demonstrate that a role of FKF1 is to generate a high level of constans (CO) late in the long day in a blue light-dependent fashion. CO protein is known to be required for flowering under long-day conditions in *Arabidopsis* (Yanofsky and Kay, 2002). Imaizumi *et al.* further showed that the LOV domains of all three proteins bind FMN and present spectral evidence that they undergo blue light-activated formation of a cysteinyl adduct as is the case for the phototropin LOV domains (Salomon *et al.*, 2000). Curiously all three LOV domains fail to revert back to the dark state, in dramatic contrast to the phototropin LOV domains. Whether this failure is an artefact of the preparations or is physiologically meaningful remains to be determined, but demonstration of a known photoreaction strengthens the case that they are indeed blue-light receptors.

7. LOV DOMAINS IN OTHER SYSTEMS

There are two proteins in the fungus *Neurospora crassa* that have recently been demonstrated to be photoreceptors in that organism: White collar-1 (Wc-1) and VIVID (see Chapter 13). Both contain a single LOV domain containing the conserved sequence GRNCRFLQ (see Crosson *et al.*, 2003). Other than that domain, neither shares any sequence homology either with the phototropins or with each other. Wc-1 is a zinc-finger protein (Ballario *et al.*, 1996) whereas VIVID is a cytoplasmic protein without homology to other proteins of known function. Froelich *et al.*, (2002) recently demonstrated that Wc-1 protein contains FAD as a cofactor and binds the *frequency* promoter *in vitro* in a light-dependent fashion. Since Wc-1 is well known to be required for most (but not all) of the responses of *Neurospora* to blue light, its long-postulated role as a photoreceptor is confirmed by this study. Efforts to date have failed to demonstrate that the Wc-1 LOV domain undergoes the photocycle described above for the phototropin LOV domains. However, given its remarkable sequence conservation with the LOV domains of the phototropins, it seems highly likely that it does so.

The second *Neurospora* protein, VIVID, has recently been demonstrated to be the photoreceptor that allows the fungus to adapt to changes in light intensity (Schwerdtfeger and Linden 2003). In this case, the full-length protein has been expressed in *E. coli* and shown to undergo the same spectral changes as the phototropin LOV domains on photoexcitation and dark recovery, consistent with the formation of a cysteinyl-C(4a) adduct. As in the case of the phototropin domains, the conserved cysteine is absolutely required for these spectral changes and is also required to complement a *vivid* null mutant *in vivo*.

Finally, extensive sequencing efforts have uncovered single LOV domains in a range of proteins from prokaryotes. Some of them are phosphodiesterases, some are histidine kinases, and some are STAS proteins [found in *Sulfate Transporters and Anti-Sigma* factor proteins (Crosson *et al.*, 2003)]. One of these proteins, YtvA from *Bacillus subtilis*, has been expressed full-length in *E. coli* and shown to undergo exactly the same spectral changes on illumination with blue light as the phototropin LOV domains (Losi *et al.*, 2002). Full-length YtvA undergoes dark recovery but requires over an hour for the process to go to completion.

Although the presence of a LOV domain with its characteristic photocycle strongly suggests photoreceptor function, there is no evidence to date that any of these prokaryotic proteins serve as photoreceptors. Future research is required to elucidate the role of the proteins containing this photochemically active module and to determine whether they play any genuine photoreceptor role.

8. A RETURN TO PHYSIOLOGY: A MODEL FOR PHOTOTROPISM

The fluence-response curve for phototropism consists of two phases: A bell-shaped segment designated first positive curvature and a second rising phase at higher fluences entitled second positive curvature (Figure 9, adapted from Zimmerman and Briggs, 1963). First positive curvature appears to be mediated by first order photochemistry: activation of a phototropin molecule is by a single photon, induction is therefore temperature independent, and the Bunsen-Roscoe reciprocity law is valid (equal numbers of photons induce equal responses independent of delivery time). Second positive curvature, however, is time-dependent and its induction is temperature sensitive, indicating limitation by a chemical reaction (see Briggs 1963). A feasible model is the following: at low fluences, more phototropin molecules are activated on the illuminated side than the shaded side of a responsive organ because of the light gradient generated by unilateral irradiation. The resulting light gradient across the organ leads to a gradient in photoexcitation, indicated by the gradient in light-activated phosphorylation shown by Salomon *et al.* (1997b, 1997c). This biochemical gradient somehow leads to alterations in auxin transport such that auxin is diverted to the shaded side, with the auxin differential thereby created causing the differential growth that leads to curvature. At still higher fluences, however, sufficient light reaches the shaded side that all photoreceptor molecules there are excited and a phosphorylation gradient no longer exists. If the irradiation is short compared to the dark recovery time of the photoreceptor system (see Palmer *et al.*, 1993a and Figure 4) every photoreceptor molecule will have been activated only once. As a consequence there is no longer an auxin gradient and therefore no longer a bending response. However, with longer irradiations, even at very high fluence rates, the rate of re-excitation of phototropin molecules on the shaded side of the organ will be less than the rate of re-excitation on the lighted side. Dark relaxation rates (return to a photoreceptor pool in the unphosphorylated ground state by whatever means) determine the pool of photoreceptors available for re-excitation and involve a chemical reaction(s) that is temperature sensitive. These rates will be similar on the two sides of the organ, but the rates of re-excitation will be different, leading to the maintenance of a phosphorylation gradient (see Salomon *et al.*, 1997c). The coupling of differential phototropin photoexcitation to alterations in auxin transport remains to date a mystery.

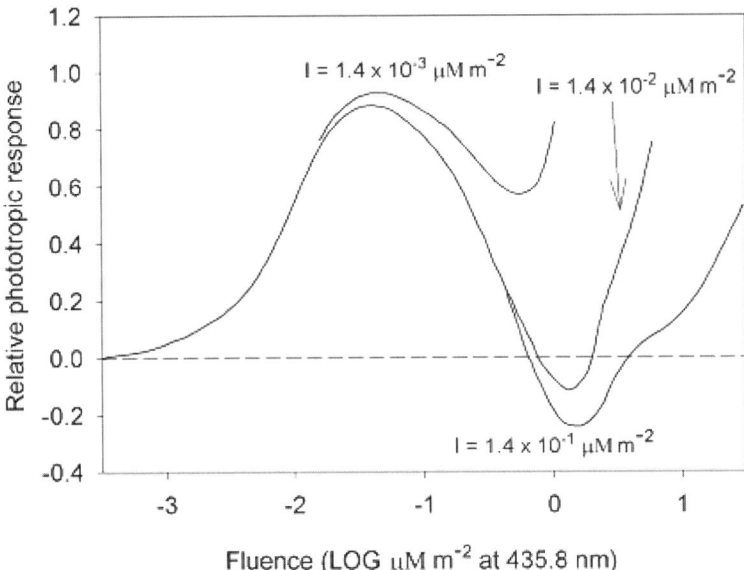

Figure 9. Fluence-response curves for the phototropic response of oat coleoptiles obtained at three different light intensities (I). Adapted from Zimmerman and Briggs (1963)

9. FUTURE PROSPECTS

Although exciting progress has already been made in the few short years since the discovery of the phototropins, much remains to be accomplished. Most of the progress, however, has concerned the photochemical properties of the LOV domains. We still do not know why there are two LOV domains with different properties, and how these differences are reflected in the physiological responses mediated by the phototropins. To date, only a single protein, NPH3, has been identified as interacting directly with a phototropin (phot1) (Motchoulski and Liscum, 1999) and it likely serves a scaffolding function rather than a signalling function. Hence there is still much to learn about events downstream from the phototropins.

It is intriguing to speculate that, despite the major differences between the several physiological responses, a possible common thread is the activation of proton ATPases. Auxin transport, clearly modified following phototropic stimulation, is driven by a proton gradient (see Lomax *et al.*, 1985). Leaf expansion in response to blue light involves light-activated extrusion of protons from epidermal cells (see van Volkenburgh 1999). Stomatal opening in response to blue light is accompanied by proton ATPase activation (see Kinoshita and Shimizaki, 1999). And at least in the alga *Vaucheria sessilis*, chloroplast accumulation in blue light is likely also accompanied by proton extrusion (Blatt *et al.*, 1981). It seems a reasonable hypothesis that phototropin activation is coupled to proton ATPase activation and the specificity of the responses is related to the different cell types involved and the proton ATPases resident in their plasma membranes. Testing this hypothesis will require future investigations for all of the responses mediated by the phototropins.

Acknowledgements. Work from the senior author's laboratory cited in this chapter was supported by NSF grants 0091384 and 0211605.

10. NOTE ADDED IN PROOF

Since the completion of this article, several pertinent studies have come to the authors' attention. Kinoshita *et al.* (2003) clearly demonstrated that both phot1a and phot1b from broad bean (*Vicia faba*) interact with a 14-3-3 protein upon phosphorylation. Moreover, they identified the phosphoserines to which the 14-3-3 protein binds as phosphoserine 358 for Vfphot1a and 344 for Vfphot1b. Both residues reside between LOV1 and LOV2. They suggest that this binding may be a key step in the response of stomata to blue light. Since they detected similar binding in other tissues, they suggest that this reaction may be a common step in other phototropin-mediated responses.

Harper *et al.* (2004) investigated the effects of substituting hydrophilic residues on the hydrophobic face of the Jα-helix just downstream of *Avena sativa* phot1 LOV2 on the LOV-domain structure as determined by NMR. Unlike the wild-type peptide in which the Jα-helix is tightly appressed to the LOV domain in the dark state and is released and loses structure on light-activated formation of the cysteinyl adduct, the mutated peptides failed to show any structure for this region in the dark state. In fact, the mutated peptides resembled the adduct state in darkness and were unchanged by light activation. These workers also showed that when comparable substitutions were made in full-length *Arabidopsis* phot expressed in insect cells, the dark state could be fully phosphorylated. Thus the unfolding of the Jα-helix appears to represent a key step in kinase activation.

Onodera et al. (2005) transformed *Chlamydomonas reinhardtii* phototropin into a *phot1 phot2* double mutant of *Arabidopsis* to see whether the algal photoreceptor (Crphot), mediating developmental changes at the gamete and zygote level in *C. reinhardtii*, could complement the physiological responses mediated by the *Arabidopsis* phototropins. Although leaf and petiole growth were somewhat reduced, leaf development was otherwise normal, indicating that the Crphot complemented the curled- leaf phenotype characteristic of the *phot1 phot2* double mutant. Likewise, strong phototropic curvature was recovered from the transgenic plants, showing photosensitivity somewhere between that of *Arabidopsis* phot1 and phot2. In addition, both chloroplast accumulation in low light and chloroplast avoidance in high light were restored as was blue light-activated stomatal opening. Since Crphot has almost no amino acid residues upstream from LOV1, in particular lacking the serines shown by Salomon et al. (2003) to be phosphorylated, one must conclude that phosphorylation of these serines is not essential for signal transduction for any of the four phototropin-mediated responses.

Salomon et al. (2004) purified the N-terminal half of *Avena sativa* phot1, coupled to a calmodulin-binding protein (CBP) in *E. coli* and showed that it behaved as a dimer on gel exclusion chromatography. As the peptide contained both LOV domains they then investigated which, if either, of the LOV domains served as a dimerization domain. A LOV2-CPB behaved exclusively as a monomer whereas a similar LOV1 peptide occurred as both monomer and dimer, with the proportion as a dimer increasing with incubation time. Thus one possible role for LOV1 is to serve as a dimerization domain. However, still unanswered is the question as to why it should maintain the characteristic photochemistry that is used by LOV2 to activate the kinase domain.

Using phototropin mutants, Kasahara et al. (2002) clearly demonstrated that the chloroplast avoidance response was essential to prevent photodamage in *Arabidopsis* under high light conditions. A *phot2* mutant showed severe chlorophyll bleaching in white light at 1,400 μmol m^{-2} s^{-1}. A phot1 mutant, however, showed a normal avoidance response and was undamaged by the high light treatment.

11. REFERENCES

Ataka, K., Hegemann, P., Heberle, J. (2003) Vibrational spectroscopy of an algal phot-LOV1 domain probes the molecular changes associated with blue-light receptor. *Biophy J, 84*, 466-474.

Ballario, P., Vittorioso, P., Magrelli, A., Talora, C., Cabibbo, A., Macino, G. (1996) White collar-1, a central regulator of blue light responses in Neurospora is a zinc finger protein. *EMBO J, 15*, 1650-1657.

Baskin, T. I. and Iino, M. (1987) An action spectrum in the blue and ultraviolet for phototropism in alfalfa. *Photochem Photobiol, 46*, 127-136.

Beebe, S. J. and Corbin, J. D. (1986) "Cyclic nucleotide-dependent protein kinases." In *The Enzymes, Ed. 3, Vol. XVII, Part A,* Boyer, P. D. and Krebs, E. G. (eds) Academic Press, New York, 44-100.

Bittl, R., Kay, C. W. M., Weber, S., Hegemann, P. (2003) Characterization of a flavin radical product in a C57M mutant of a LOV1 domain by electron paramagnetic resonance. *Biochemistry, 42*, 8506-8512.

Blatt, M. R., Weisenseel, M. H., Haupt W. (1981) A light-dependent current associated with chloroplast aggregation in the alga *Vaucheria sessilis*. *Planta, 152,*: 513-526.

Briggs, W. R. (1960) Light dosage and the phototropic responses of corn and oat coleoptiles. *Plant Physiol, 55*, 951-962.

Briggs, W. R. (1996) "Signal transduction in phototropism." In: *UV/Blue Light: Perception and Responses in Plant*, meeting held in Marburg, Germany, August, 1996, Abstracts p. 49.

Briggs, W. R., Beck, C. F., Cashmore, A. R., Christie, J. M., Hughes, J., Jarillo, J., et al. (2001a) The phototropin family of photoreceptors. *Plant Cell*, *13*, 993-997.

Briggs, W. R. and Christie, J. M. (2002) Phototropins 1 and 2: versatile plant blue-light receptors. *Trends Plant Sci, 7*, 204-210.

Briggs, W. R., Christie, J. M., Salomon, M. (2001b) Phototropins: A new family of flavin-binding blue light receptors in plants. *Antiox Redox Signaling*, *3*, 775-788.

Briggs, W. R. and Huala, E. (1999) Blue-light photoreceptors in higher plants. *Annu Rev Cell Dev Biol*, *15*, 33-62.

Christie, J. M. and Briggs, W. R. (2001) Blue light sensing in higher plants. *J Biol Chem*, *276*, 11457-11460.

Christie, J. M. and Briggs, W. R. (2005) Blue light sensing and signaling by the phototropins. In *Handbook of Photosensory Receptors*, Briggs, W. R. and Spudich, J. L. (eds) Wiley-VCH, Weinheim, 277-303.

Christie, J. M., Reymond, P., Powell, G., Bernasconi, P., Reibekas, A. A., Liscum, E. et al. (1998) *Arabidopsis* NPH1: a flavoprotein with the properties of a photoreceptor for phototropism. *Science*, *282*, 1698-1701.

Christie, J. M., Salomon, M., Nozue, K., Wada, M., Briggs, W. R. (1999) LOV (light, oxygen, or voltage) domains of the blue-light photoreceptor phototropin 1 (nph1): Binding sites for the chromophore flavin mononucleotide. *Proc Natl Acad Sci USA*, *96*, 8779-8783.

Christie, J. M., Swartz, T. E., Bogomolni, R., Briggs, W. R. (2002) Phototropin LOV domains exhibit distinct roles in regulating photoreceptor function. *Plant J*, *32*, 205-219.

Corchnoy, S. B., Swartz, T. E., Lewis, J. W., Szundi, I., Briggs, W. R., Bogomolni, R. A. (2003) Intramolecular proton transfers and structural changes during the photocycle of the LOV2 domain of phototropin 1. *J Biol Chem*, *278*, 724-731.

Crosson, S. (2005) LOV-domain structure, dynamics, and diversity. In *Handbook of Photosensory Receptors*, Briggs, W. R. and Spudich, J. L. (eds) Wiley-VCH, Weinheim, 323-336.

Crosson, S. and Moffat, K. (2001) Structure of a flavin-binding plant photoreceptor domain: Insights into light-mediated signal transduction. *Proc Natl Acad Sci USA*, *98*, 2995-3000.

Crosson, S. and Moffat, K. (2002) Photoexcited structure of a plant photoreceptor domain reveals a light-driven molecular switch. *Plant Cell*, *14*, 1067-1075.

Crosson, S., Rajagopal, S., Moffat, K. (2003) The LOV domain family: photosensitive signaling modules coupled to diverse output domains. *Biochemistry*, *42*, 2010.

Eisinger, W., Swartz, T. E., Bogomolni, R. A., Taiz, L. (2000) The ultraviolet action spectrum for stomatal opening in broad bean. *Plant Physiol*, *122*, 99-105.

Federov, R., Schlichting, I., Hartmann, E., Domratcheva, T., Fuhrmann, M., Hegemann, P. (2003) Crystal structures and molecular mechanism of a light-induced signalling switch: the phot1 LOV domain from *Chlamydomonas reinhardtii*. *Biophys J*, *84*, 2474-2482.

Folta, K. M. and Spalding, E. P. (2001) Unexpected roles for cryptochrome 2 and phototropin revealed by high-resolution analysis of blue light-mediated hypocotyl growth inhibition. *Plant J*, *26*,: 471-478.

Friml, J., Wisniewska, J., Benkova, E., Mendgen, K., Palme, K. (2002) Lateral relocation of auxin efflux regulator PIN3 mediates tropism in *Arabidopsis*. *Nature*, *415*, 806-809.

Froehlich, A., Liu, Y., Loros, J. J., Dunlap, J. C. (2002) White collar-1, a circadian blue light photoreceptor, binding to the *frequency* promoter. *Scienc* , *297*, 815-819.

Gallagher, S., Short, T. W., Ray, P. M., Pratt, L. H., Briggs, W. R. (1988) Light-mediated changes in two proteins found associated with plasma membrane fractions from pea stem sections. *Proc Natl Acad Sci USA*, *85*, 8003-8007.

Gälweiler, L., Guan, C., Müller, A., Wisman, E., Mendgen, K., Yephremov, A. et al. (1998) Regulation of polar auxin transport by AtPIN1 in *Arabidopsis* vascular tissue. *Science*, *282*, 2226-2230.

Gressel, J. (1979) Blue light photoreception. *Photochem Photobiol*, *30*, 749-754.

Hager, A. (1996) Properties of a blue-light-absorbing kinase localized in the plasma membrane of the coleoptile tip region. *Planta*, *198*, 294-299.

Hager, A. and Brich, M. (1993) Blue light-induced phosphorylation of a plasma-membrane protein from phototropically sensitive tips of maize coleoptiles. *Planta*, *189*, 567-576.

Hager, A., Brich, M., Balzen, I. (1993) Redox dependence of the blue light-induced phosphorylation of a 100-kDa protein on isolated plasma membranes from tips of coleoptiles. *Planta*, *190*, 120-126.

Hanks, S. K. and Hunter, T. (1995) The eukaryotic protein kinase superfamily: kinase (catalytic) domain structure and classification. *FASEB J*, *9*, 576-604.

Harada, A., Sakai, T., Okada, K. (2003) phot1 and phot2 mediate blue light-induced transient increases in cytosolic Ca^{2+} differently in Arabidopsis leaves. *Proc Natl Acad Sci USA*, *100*, 8583-8588.

Harper, S. M., Christie, J. M., Gardner, K. H. (2005) Disruption of the LOV/Jα helix interaction activates phototropin kinase activity. *Biochemistry*, *43*, 16184-16192.

Harper, S. M., Neil, L. C., Gardner, K. H. (2003) Structural basis of a phototropin light switch. *Science*, *301*, 1541-1544.

Haupt, W. and Scheuerlein, R. (1990) Chloroplast movement. *Plant Cell Environ*, *13*, 595-614.

Holzer, W., Penzkofer, A., Fuhrmann, M., Hegemann, P. (2002) Spectroscopic characterization of flavin mononucleotide bound to the LOV1 domain of phot1 from *Chlamydomonas reinhardtii*. *Photochem Photobiol*, *75*, 479-487.

Huala, E., Oeller, P. W., Liscum, E., Han, I.-S., Larsen, E., Briggs, W. R. (1997) *Arabidopsis* NPH1: a protein kinase with a putative redox-sensing domain. *Science*, *278*, 2121-2123.

Huang, K. and Beck, C. F. (2003) Phototropin is the blue-light receptor that controls multiple steps in the sexual life cycle of the green alga *Chlamydomonas reinhardtii*. *Proc Natl Acad Sci USA*, *100*, 6269-6274.

Huang, K., Merkle, T., Beck, C. F. (2002) Isolation and characterization of a *Chlamydomonas* gene that encodes a putative blue-light photoreceptor of the phototropin family. *Physiologia Plantarum*, *114*, 613-622.

Iino, M. (2001) "Phototropism in higher plants." In: *Photomovement*, Häder, D.-P, and Lebert, M. (eds), Elsevier, Amsterdam, 659-811.

Imaizumi, T, Tran, H. G., Swartz, T. E., Briggs, W. R., Kay, S. A. (2003) FKF1 is essential for photoperiodic-specific light signaling in *Arabidopsis*. *Nature*, *426*, 302-306.

Iwata, T., Nozaki, D., Tokutomi, S., Kagawa, T, Wada M., Kandori H. (2003) Light-induced structural changes in the LOV2 domain of *Adiantum* phytochrome3 studied by low-temperature FTIR and UV-visible spectroscopy. *Biochemistry*, *42*, 8183-8191.

Iwata, T., Tokutomi, S., Kandori, H. (2002) Photoreaction of the cysteine S-H group in the LOV2 domain of *Adiantum* phytochrome3. *J Am Chem Soc*, *124*, 11840-11841.

Jarillo, J. A., Ahmad, M., Cashmore, A. R. (1998) NPL1 (Accession No. AF053941): A second member of the NPH serine/threonine kinase family of Arabidopsis. *Plant Physiol.*, *117*, 719.

Jarillo, J. A., Capel, J., Tang, R.-H., Yang, H.-Q., Alonso, J. M., Ecker, J. R., *et al.* (2001a) An *Arabidopsis* circadian clock component interacts with both CRY1 and phyB. *Nature*, *410*, 487-490.

Jarillo, J. A., Gabrys, H., Capel, J., Alonso, J. M., Ecker, J. R., Cashmore, A. R. (2001b) Phototropin-related NPL1 controls chloroplast relocation induced by blue light. *Nature*, *410*, 592-594.

Kagawa, T., Kasahara, M., Abe, T., Yoshida, S., Wada, M. (2004) Function analysis of phototropin2 using fern mutants deficient in blue light-induced chloroplast avoidance movement. *Plant Cell Physiol*, *45*, 416-426.

Kagawa, T., Sakai, T., Suetsugu, N., Oikawa, K., Ishiguro, S., Kato, T., *et al.* (2001) *Arabidopsis* NPL1: A phototropin homolog controlling the chloroplast high-light avoidance response. *Science*, *291*, 2138-2141.

Kagawa, T. and Wada, M. (2002) Blue light-induced chloroplast movement. *Plant Cell Physiol*, *43*, 367-371.

Karlsson, P. E. (1986) Blue light regulation of stomata in wheat seedlings. II. Action spectrum and search for action dichroism. *Physiologia Plantarum*, *66*, 207-210.

Kasahara, M., Kagawa, T., Oikawa, K., Suetsugu, N., Miyao, M., Wada, M. (2002) Chloroplast avoidance movement reduces photodamage in plants. *Nature*, *420*, 829-832.

Kasahara, M., Swartz, T. E., Olney, M. O., Onodera, A., Mochizuki, N., Fukuzawa, H., *et al.* (2002) Photochemical properties of the flavin mononucleotide-binding domains from Arabidopsis, rice, and *Chlamydomonas reinhardtii*. *Plant Physiol*, *129*, 762-773.

Kay, C. W. M., Schleicher, E., Kuppig, A., Hofner, H., Rüdiger, W., Schleicher, M., *et al.* (2003) Blue light perception in plants – detection and characterization of a light-induced neutral flavin radical in a C450A mutant of phototropin. *J Biol Chem*, *278*, 10973-10982.

Kennis, J. T. M., Crosson, S., Gauden, M., van Stokkum, I. H. M., Moffat, K., van Grondelle, R. (2003) Primary reactions of the LOV2 domain of phototropin, a plant blue-light photoreceptor. *Biochemistry*, *42*, 3385-3392.

Kennis, J. T. M., van Stokkum, I. H. M., Crosson, S., Gauden, M., Moiffat, K., van Grondelle, R. (2004) The LOV2 domain of phototropin: a reversible photochromic switch. *J Am Chem Soc, 126*, 1412-1413.
Khurana, J. P. and Poff, K. L. (1989) Mutants of *Arabidopsis thaliana* with altered phototropism. *Planta, 178*, 400-406.
Kinoshita, T. and Shimizaki, K. (1999) Blue light activates the plasma membrane H^+-ATPase by phosphorylation of the C-terminus in stomatal guard cells. *EMBO J, 18*, 55548-55558.
Kinoshita, T., Doi, M., Suetsugu, N., Kagawa, T., Wada, M., Shimizaki, K.-I. (2001) phot1 and phot2 mediate blue light regulation of stomatal opening. *Nature, 414*, 656-660.
Kinoshita, T., Takashi, E., Tominaga, M., Sakamoto, K., Shigenaga, A., Doi, M., et al. (2003) Blue-light- and phosphorylation-dependent binding of a 14-3-3 protein to phototropins in stomatal guard cells of broads bean. *Plant Physiol, 133*, 1453-1463.
Koller, D. (2000) Plants in search of sunlight. *Advances Bot Res, 33*, 35-131.
Koller, D., Ritter, S., Briggs, W. R., Schäfer, E. (1990) Action dichroism in perception of vectorial photo- excitation in the solar-tracking leaf of *Lavatera cretica* L. *Planta, 181*, 184-190.
Kottke, T., Dick, B., Federov, R., Deutzmann, I., Hegemann, P. (2002) Irreversible photoreduction of a flavin in a mutated phot-LOV1 domain. *Biochemistry, 42*, 9854-9862.
Kottke, T., Heberle, J., Hehn, D., Dick, B., Hegemann, P. (2003) Phot-LOV1: photocycle of a blue-light receptor domain from the green alga *Chlamydomonas reinhardtii*. *Biophys J, 84*, 1192-2001.
Kutschera, U. and Briggs, W. R. (1987) Differential effect of auxin on *in vivo* extensibility of cortical cylinder and epidermis in pea internodes. *Plant Physiol, 84*, 1361-1366.
Lin, C. (2002a) Blue light receptors and signal transduction. *Plant Cell Supplement*, S207-S225.
Lin, C. (2002b) Phototropin blue light receptors and light-induced movement responses in plants. Science's stke www.stke.org/cgi/content/full/OC_sigtrans;2002/118/pe5: 1-5.
Liscum, E. and Briggs, W. R. (1995) Mutations in the *NPH1* locus of Arabidopsis disrupt the perception of phototropic stimuli. *Plant Cell, 7*, 473-485.
Liscum, E. and Briggs, W. R. (1996) Mutations of Arabidopsis in potential transduction and response components of the phototropic signalling pathway. *Plant Physiol, 112*, 291-296.
Lomax, T. L., Mehlhorn, R. J., Briggs, W. R. (1985) Active auxin uptake by zucchini membrane vesicles: quantitation using ESR volume and ΔpH determinations. *Proc Natl Acad Sci USA, 82*, 6541-6545.
Losi, A., Polverini, E., Quest, B., Gärtner, W. (2002) First evidence for phototropin-related photoreceptors in prokaryotes. *Biophys J, 82*, 2627-2634.
Massey, V. and Hemmerich, P. (1978) Photoreduction of flavoproteins and other biological compounds catalysed by de-aza-flavin. *Biochemistry, 17*, 9-16.
Miller, S. M., Massey, V., Ballou, D., Williams, C. H. Jr., Distefano, M. D., Moore, M. J., et al. (1990) Use of a site-directed triple mutant to trap intermediates: demonstration that a flavin C(4a)-thiol adduct and reduced flavin are kinetically competent intermediates in mercuric ion reductase. *Biochemistry, 29*, 2831-2841.
Motchoulski, A and Liscum, E. (1999) Arabidopsis NPH3: A NPH1 photoreceptor-interacting protein essential for phototropism. *Science, 286*, 961-964.
Neiss, C. and Saalfrank, P. (2003) Ab initio quantum chemical investigation of the first steps of the photocycle of phototropin: a model study. *Photochem Photobiol, 77*, 101-109.
Nelson, D. C., Lasswell, J., Rogg, L. E., Cohen, M. A., Bartel, B. (2000) *FKF1*, a clock-controlled gene that regulates the transition to flowering in *Arabidopsis*. *Cell, 101*, 331-340.
Nozue, K., Christie J. M., Kiyosue, T., Briggs, W. R., Wada M. (2000) Isolation and characterization of a fern phototropin (Accession No. AB037188), a putative blue-light photoreceptor for phototropism. *Plant Physiol, 122*, 1457.
Nozue, K., Kanegae, T., Imaizumi, T., Fukada, S., Okamoto, H., Yeh, K. C., et al. (1998) A phytochrome from the fern *Adiantum* with the properties of the putative photoreceptor NPH1. *Proc Natl Acad Sci USA, 95*, 15826-15830.
Onodera, A., Kong, S-G., Doi, M., Shimizaki, K., Mochizuki, N., Nagatani, A. (2002) Phototropin from *Chlamydomonas reinhardtii* is functional in *Arabidopsis thaliana*. *Plant Cell Physiol*. In press.
Palmer, J. M., Short, T. W., Briggs, W. R. (1993a) Correlation of blue light-induced phosphorylation to phototropism in *Zea mays* L. *Plant Physiol, 102,* 1219-1225.
Palmer, J. M., Short, T. W., Gallagher, S., Briggs W. R. (1993b) Blue light-induced phosphorylation of a plasma membrane-associated protein in *Zea mays* L. *Plant Physiol, 102*, 1211-1218.

Palmgren, M. G., Askerlund, P., Fredrikson, K., Widell, S., Sommarin, M., Larssen, C. (1990) Sealed inside-out and right-side-out plasma membrane vesicles. Optimal conditions for formation and separation. *Plant Physiol, 92*, 871-880.

Reymond, P., Short, T. W., Briggs, W. R. (1992a) Blue light activates a specific kinase in higher plants. *Plant Physiol, 100*, 655-661.

Reymond, P., Short, T. W., Briggs, W. R., Poff, K. L. (1992b) Light-induced phosphorylation of a membrane protein plays an early role in signal transduction for phototropism in *Arabidopsis thaliana*. *Proc Natl Acad Sci USA, 89*, 4718-4721.

Rüdiger, W. and Briggs, W. R. (1995) Involvement of thiol groups in blue-light-induced phosphorylation of a plasma membrane-associated protein from coleoptile tips of *Zea mays* L. *Z Naturforsch, 50*, 231-234.

Sakai, T, Kagawa, T., Kasahara, M., Swartz, T. E., Christie, J. M., Briggs, W. R., *et al.* (2001) Arabidopsis nph1 and npl1: Blue light receptors that mediate both phototropism and chloroplast relocation. *Proc Natl Acad Sci USA, 98*, 6969-6974.

Sakamoto, K. and Briggs, W. R. (2002) Cellular and subcellular localization of phototropin 1. *Plant Cell, 14*, 1723-1735.

Salomon, M., Christie, J. M., Knieb, E., Lempert, U., Briggs, W. R. (2000) Photochemical and mutational analysis of the FMN-binding domains of the plant blue light receptor, phototropin. *Biochemistry, 39*, 9401-9410.

Salomon, M., Eisenreich, W., Dürr, H., Schleicher, E., Knieb, E., Massey, V., *et al.* (2001) An optomechanical transducer in the blue light receptor phototropin from *Avena sativa*. *Proc Natl Acad Sci USA, 98*, 12357-12361.

Salomon, M., Knieb, E., von Zeppelin, T., Rüdiger, W. (2003) Mapping of low- and high-fluence autophosphorylation sites in phototropin 1. *Biochemistry, 42*, 4217-4225.

Salomon, M., Lempert, U., Rüdiger, W. (2004) Dimerization of the plant photoreceptor phototropin is probably mediated by the LOV1 domain. *FEBS Lett, 572*, 8-10.

Salomon, M., Zacherl, M., Rüdiger, W. (1996) Changes in blue-light-dependent protein phosphorylation during early development of etiolated oat seedlings. *Planta, 199*, 336-342.

Salomon, M., Zacherl, M., Luff, L., Rüdiger, W. (1997a) Exposure of oat seedlings to blue light results in amplified phosphorylation of a putative photoreceptor for phototropism and higher sensitivity of the plants to phototropic stimulation. *Plant Physiol, 115*, 493-500.

Salomon, M., Zacherl, M., Rüdiger, W. (1997b) Phototropism and phosphorylation in higher plants: unilateral blue light irradiation generates a directional gradient of protein phosphorylation across the oat coleoptile. *Bot Acta, 110*, 214-216.

Salomon, M., Zacherl, M., Rüdiger, W. (1997c) Asymmetric blue light-dependent phosphorylation of a 116-kilodalton plasma membrane protein can be correlated with first- and second-positive curvature of oat coleoptiles. *Plant Physiol, 115*, 485-491.

Sancar, A. (1994) Structure and function of DNA photolyase. *Biochemistry, 33*, 2-9.

Schwartz, A. and Koller, D. (1978) Phototropic response to vectorial light in leaves of *Lavatera cretica* L. *Plant Physiol, 61*, 924-928.

Schultz, T. F., Kiyosue, T. Yanofsky, M., Wada, M., Kay, S. A. (2001) A role for LKP2 in the circadian clock of Arabidopsis. *Plant Cell, 13*, 2659-2670.

Schwerdtfeger, C. and Linden, H. (2003) VIVID is a flavoprotein and serves as a fungal blue light photoreceptor for photoadaptation. *EMBO J, 22*, 4846-4855.

Sharma, V. K., Jain, P. K., Maheshwari, S. C., Khurana, J. P. (1997) Rapid blue-light-induced phosphorylation of plasma-membrane-associated proteins in wheat. *Phytochemistry, 44*, 775-780.

Short, T. W. and Briggs, W. R. (1990) Characterization of a rapid, blue light-mediated change in detectable phosphorylation of a plasma membrane protein from etiolated pea (*Pisum sativum* L.) seedlings. *Plant Physiol, 92*, 179-185.

Short, T. W., Porst, M., Briggs, W. R. (1992) A photosystem regulating *in vivo* and *in vitro* phosphorylation of a pea plasma membrane protein. *Photochem Photobiol, 55*, 773-781.

Short, T. W., Porst, M., Palmer, J. M., Fernbach, E., Briggs, W. R. (1994) Blue light induces phosphorylation at seryl residues on a pea (*Pisum sativum* L.) plasma membrane protein. *Plant Physiol, 104*, 1317-1324.

Short, T. W., Reymond, P., Briggs, W. R. (1993) A pea plasma membrane protein exhibiting blue light-induced phosphorylation retains photosensitivity following Triton solubilization. *Plant Physiol, 101*, 647-655.

Shropshire, W. Jr., and Withrow, R. B. (1958) Action spectrum for phototropic tip-curvature of Avena. *Plant Physio*, *33*, 360-365.

Somers, D. E., Schultz, T. F., Milnamow, M., Kay, S. (2000) ZEITLUPE encodes a novel clock-associated PAS protein from *Arabidopsis*. *Cell*, *101*, 319-329.

Swartz, T. E. and Bogomolni, R. (2005) LOV-domain photochemistry. In *Handbook of Photosensory Receptors*, Briggs, W. R. and Spudich, J. L. (eds) Wiley-VCH, Weinheim, 305-321.

Swartz, T. E., Corchnoy, S. B., Christie, J. M., Lewis, J. W., Szundi, I., Briggs, W. R. et al. (2001) The photocycle of a flavin-binding domain of the blue light photoreceptor phototropin. *J Biol Chem*, *276*, 36493-36500.

Swartz, T. E., Wenzel, P. J., Corchnoy, S. B., Briggs, W. R., Bogomolni, R. A. (2002) Vibration spectroscopy reveals light-induced structural changes in the LOV2 domain of the plant blue-light receptor phototropin 1. *Biochemistry*, *41*, 7182-7189.

Taylor, B. L. and Zhulin, I. B. (1999) PAS domains: Internal sensors of oxygen, redox potential, and light. *Microbiol Mol Bio Rev*, *63*, 479-506.

Tepperman, J. M., Zhu, T., Chang, H. S., Wang, X., Quail, P. H. (2001) Multiple transcription-factor genes are early targets of phytochrome A signalling. *Proc Natl Acad Sci USA*, *98*, 9437-9442.

Thimann, K. V., and Curry, G. M. (1960) Phototropism and phototaxis. *Comparative Biochemistry*, *1*, 243-306.

Tlalka, M., Runquist, M., Fricker, M. (1999) Light perception and the role of the xanthophyll cycle in blue-light dependent chloroplast movements in *Lemna trisulca* L. *Plant J*, *20*, 447-459.

van Volkenburgh, E. (1999) Leaf expansion: an integrating plant behavior. *Plant Cell Environ*, *22*, 1463-1473.

Wada, M., Kagawa, T., Sato, Y. (2003) Chloroplast movement. *Annu Rev Plant Biol*, *54*, 455-468.

Warpeha, K. M. F. and Briggs, W. R. (1993) Blue light-induced phosphorylation of a plasma membrane protein in pea: a step in the signal transduction chain for phototropism. *Aust J Plant Physiol*, *20*, 393-403.

Wyatt, J. L. and Coleman, R. F. (1977) Affinity labeling of rabbit muscle pyruvate kinase by 5'-p- fluorosulfonylbenzoyladenosine. *Biochemistry*, *16*, 1333-1342.

Yin, H. C. (1938) Diaphototropic movement of the leaves of *Malva neglecta*. *Amer J Botan*, *25*, 1-6.

Zeiger, E. and Field, C. (1982) Photocontrol of the functional coupling between photosynthesis and stomatal conductance in the intact leaf. *Plant Physiol*, *70*, 370-375.

Zeiger, E. and Hepler, P. H. (1977) Light and stomatal function: Blue light stimulates swelling of guard cell protoplasts. *Science*, *196*, 887-889.

Zimmerman, B. K. and Briggs, W. R. (1963) A kinetic model for phototropic responses of oat coleoptiles. *Plant Physiol*, *38*, 253-261.

Zurzycki, J. (1962) The action spectrum for light dependent movements of chloroplasts in *Lemna trisulca* L. *Acta Societatis Botanicorum Poloniae*, *31*, 489-538.

Chapter 13

BLUE LIGHT PHOTORECEPTORS - BEYOND PHOTOTROPINS AND CRYPTOCHROMES

JAY DUNLAP
Department of Genetics, Dartmouth Medical School Hanover, New Hampshire, USA (e-mail: Jay.C.Dunlap@dartmouth.edu)

1. INTRODUCTION

Filamentous fungi are well known to display a variety of responses to blue light, and these have led to the discovery of photoreceptors distinct from the more thoroughly studied phototropins and cryptochromes. Similarly, the pursuit of the biology of photoresponses in Euglena and circadian regulation of flowering time in *Arabidopsis* has led to the discovery of photoreceptor proteins. This chapter will focus on the identification of a novel class of LOV-domain associated photoreceptors first identified in Neurospora in which the light response directly acts to modulate the transcriptional activation potential of the WC-1 protein. This principal blue light photoreceptor was identified in 2002 (Froehlich *et al.*, 2002; He *et al.*, 2002), and the identity of this protein highlighted the LOV protein domain already associated with blue light photoreception in some plants. In addition, the simultaneous emergence of the complete genomic sequence of Neurospora has allowed the identification of additional putative photoreceptors (Galagan *et al.*, 2003; Borkovich *et al.*, 2004). This chapter will begin with a brief history of the discovery of blue light photoreception in fungi and then go on to describe the spectrum of light regulated processes and finally on to the nature of the known and suspected photoreceptors. It will close with a brief overview of photoreceptors from photosynthetic eukaryotes that also use the LOV domain for perception of blue light.

2. HISTORICAL ANTECEDENTS

The origins of photobiology in Neurospora go back over 150 years to the first description of this filamentous fungus. In the summer of 1842, bakeries in Paris were afflicted with widespread colonization by a fungus that grew well on moist bread. Inevitably a commission was set up to study the problem and the organism, and among its findings was the observation that the organism remained white when grown in the dark but turned orange upon exposure to light (Payen, 1843). The organism was *Neurospora crassa*, although prior to 1927 the vegetative phase sometimes went by *Oidium aurantiacum* or *Monila sitophila* (Perkins, 1992).

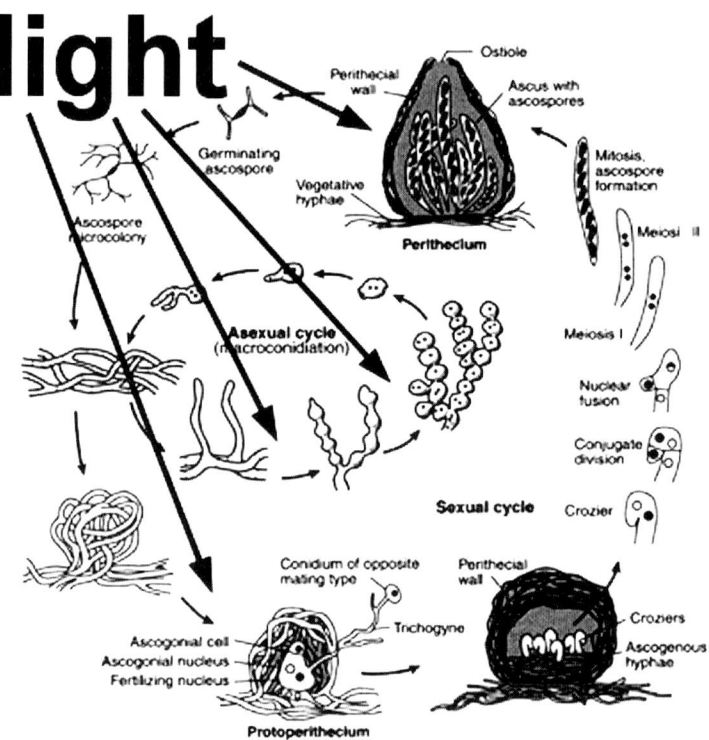

Figure 1. The life cycle of Neurospora. Light influences many aspects of both, the sexual and asexual cycles. See text for detail. Adapted from Davis (2000,) with permission.

The first clues as to the nature of the photoreceptor arose about a century ago in work by the Dutch plant physiologist F.A.F.C. Went who was working at the Bogor Botanic Gardens in Java (Lauter, 1996). Went used a double-walled bell jar whose interstices could be filled with different coloured salt solutions. The result was that different colours of light were transmitted to the fungus (probably the closely related *Neurospora sitophila*) and he deduced, correctly, that the orange colour was due to the production of carotenoids, which were induced by exposure of the organism to blue light (Went, 1904). A similar observation followed again two decades later in the other hemisphere when, in September 1923, an earthquake and subsequent firestorm struck Tokyo. A few days after the fire an orange fungus was seen at many sites burned over by the conflagration. The fungus was identified again as *Neurospora crassa*, the pigment as carotenoids, and the production of the pigment tied to exposure to light (Tokugawa and Emoto, 1924).

3. THE PHOTOBIOLOGY OF NEUROSPORA

Light signals influence many facets of both the sexual and asexual (vegetative) stages of life in this organism as seen in the figure. A brief description of the life and times of this organism will serve to set the stage. Neurospora comes in two mating types, **A** and **a**. In the wild, sexual spores are activated by fires, the heat from which allows them to germinate. Cultures thus emerge after fires (Neurospora is classified as a Pyrenomycete) and the organism spends most of its life growing vegetatively on the burned over substrate (Figure 1). Neurospora grows vegetatively as a syncytium with incomplete cell walls separating cellular compartments. Strains of the same mating type that are otherwise genetically different can often fuse and intermingle their nuclei to form heterokaryons, but strains with different mating types never fuse. As a vegetative culture, Neurospora can exist as surface mycelia or can elaborate aerial hyphae. The tips of aerial hyphae form morphological distinct structures that act as asexual spores (called conidia) that can easily be dispersed by wind. When nutrients get scarce, vegetative Neurospora of either mating type is able to induce sexuality by forming a fruiting body (a protoperithecium) which can be fertilized by a nucleus from a piece of mycelium or an asexual spore of the opposite mating type to form a productive fruiting body, the perithecium. Nuclei of each parent replicate in tandem and eventually fuse to make a transient diploid that immediately undergoes meiosis to produce an 8-spored ascus, each ascus containing the products of meiosis from a single diploid nucleus.

Acute effects of light during the asexual phase of the life cycle include the induction of conidiation, the developmental process leading to the production of the asexual spores. More conidia are produced in light and they are produced faster (Klemm and Ninneman, 1978; Lauter, 1996). Although pigmentation of the conidia is constitutive, as noted above in the historical descriptions of Neurospora, carotenogenesis in mycelia is light-induced (Harding and Shropshire, 1980) and this response is quite rapid, being observable within the first 30 minutes after exposure to light. In a strain defective for light perception or transduction of the light signal, this of course leads to the production of white mycelia underlying yellow/orange conidia; this screen proved to be of central importance in the genetic identification of the photoreceptors as described below. Developmental responses such as light-induced conidiation are slower. Aerial hyphae are reported to display a phototropism (e.g. Siegel *et al.*, 1968) in that they preferentially form on the side of a dish near to light, but it is not clear to what extent this response is distinct from the overall light induction of conidiation. Also somewhat controversial are reports of changes in membrane conductivity (hyperpolarization and an increase in input resistance) in response to light (Potapova *et al.*, 1984).

During the asexual cycle, shown in the centre of the figure, the most global effect of light is to set the phase of the endogenous biological clock which acts to regulate a variety of aspects of the life cycle of the organism (Sargent and Briggs, 1967; reviewed in Loros and Dunlap, 2001). The clock controls the daily timing of a developmental switch that can initiate the morphological changes that lead to conidiation. Light is used to set the phase of the clock in that a light to dark transfer is interpreted as dusk and a dark to light transfer as dawn; the molecular basis of this response will become clear at a later point. Continued light also acts to suppress the expression of the clock. Conidiation (differentiation leading to asexual spore production) can be triggered by environment signals including blue light, desiccation, and nutrient starvation as well as by the endogenous circadian clock in otherwise constant conditions. This asexual development involves a major morphological change that requires many novel gene products. Although the production of asexual spores is the best characterized light-phased circadian rhythm in Neurospora, other persisting rhythms at the physiological level have been described, which include the production of CO_2, lipid and diacylglycerol metabolism (e.g. Lakin-Thomas and Brody, 2000; Ramsdale and Lakin-Thomas, 2000; Roeder *et al.*, 1982), a number of enzymatic activities (e.g. Hochberg and Sargent, 1974; Martens and Sargent, 1974), heat shock proteins (Rensing *et al.*, 1987), and even growth rate (Sargent *et al.*, 1966).

During the sexual phase of the life cycle, light again induces many and varied effects. The overall initiation of the sexual process is enhanced by light (Degli Innocenti and Russo, 1983); this is particularly interesting since once initiated, the process proceeds better in the dark. Not surprisingly, carotenogenesis of perithecial walls is light induced (Perkins, 1988). Perhaps the most interesting photobiology in the sexual phase involves the behaviour of mature perithecia once they are formed. The ejection of spores is induced by light, and the direction in which they are ejected is light regulated - that is, the tips ("beaks') of the perithecia display a distinct phototropism (Harding and Melles, 1983). Finally, even in the dark, the number of spores shot from perithecia is regulated by the light-phased circadian clock.

The processes described above are all light regulated at face value, but Neurospora is also capable of a more subtle responses to light characteristic of added regulatory sophistication: Neurospora can respond to changes in the level of ambient light, a process known as photoadaptation. This response is manifested in two ways, both tied to the response to light at the molecular level. In the first, when the organism initially sees light a response is triggered that peaks within 15 to 30 minutes, but this response generally decays away within two hours or so. If the organism is exposed to light within this two hour period, no additional response is seen. A related phenomenon is observed if the lights remain on; the response decays away, but interestingly, after the two hour latency period, the organism can respond again if the ambient level of light is increased.

Although most research on photobiology in the fungi has been driven by Neurospora, it should be noted in passing that a great deal of excellent research has used *Phycomyces blakesleeanus* which also elaborates carotenoids in response to blue light as well as displaying well described developmental and phototropic responses (Cerda-Olmedo, 2001). Additionally, asexual spore production is light-induced in *P. blakesleeanus*, *Trichoderma harzianum*, and *Aspergillus nidulans* as well as in *N. crassa*, and both cell wall branching and fruiting body production is induced by light in *Schizophyllum commune* as well as in *N. crassa* (reviewed in (Lauter, 1996).

4. LIGHT PERCEPTION - THE NATURE OF THE BLUE LIGHT PHOTORECEPTOR

4.1 Flavins as chromophores

As implied above, all known photoresponses in Neurospora are specific to light in the blue-green region of the spectrum, and no red or far red responses have yet been described. Early efforts at identifying the photoreceptor(s) relied on action spectra (Sargent, 1985) which were consistent with the involvement of either carotenoids or flavins. Since mutation in the carotenogenesis genes *al-1*, *al-2*, and *al-3* eliminated nearly all of these pigments from cells but had no effect on the light responses (Russo, 1986), interest focused on flavins as chromophores, and genetic studies were consistent with this. Sargent and Paietta used flavin biosynthesis mutants *rib-1* and *rib-2* to show flavin-dependence of several light responses including induction of carotenogenesis and phase shifting and photosuppression of circadian rhythmicity (Paietta and Sargent, 1981). In a satisfying test of this hypothesis, they later supplemented the mutants with the flavin analogs 1-deazariboflavin and roseoflavin and showed that the circadian rhythm-specific responses were rescued and had an appropriately altered action spectrum (Paietta and Sargent, 1983). At the time, the weak to nonexistent rescue of other light responses was a concern, but the difference can now be understood in terms of the relative requirement of the photoreceptors for distinct responses as described below. Based on the requirement for flavins, a number of flavoproteins have been suggested and suspected of being the blue light photoreceptor, including nitrate reductase (Klemm and Ninneman, 1978) and cryptochromes. In the end, however, genetic analysis of the phenomenon pointed the way.

4.2 Genetic dissection of the blue light response

The obvious light induction of carotenogenesis leading to orange coloration has proven to be the most durable and successful phenotype for identification of photoreceptor genes. As outlined above, the screens were based on the observation that blind strains are completely developmentally normal except for the fact that mycelia elaborate no coloration, so colonies on plates are white if examined early (before conidia form) and slants kept in the light on the top of a lab bench will have yellow/orange conidia on top of white mycelia on the surface of the agar. When examined from the side, the collar of agar and mycelia at the top of the slant will be white, hence the name for the genes identified in these screens, *white collar*. Extensive screening in the Russo and Macino laboratories identified a number of blind *white collar* strains, but all of the mutants have mapped to two loci, *white collar-1* and *white collar-2* (*wc-1* and *wc-2*). True loss of function *wc* mutants appear to be blind to all known light responses, including the light induction of carotenogenesis (Harding and Shropshire, 1980), light-induction of conidiation (Ninneman, 1991), induction of protoperithecia, and pointing of perithecial beaks

(Harding and Melles, 1983; Degli Innocenti and Russo, 1984b). Photoadaptation is also lost because there is no primary light response to be adapted. Although there are reports of responses that remain in *wc* null strains (e.g. Dragovic *et al.*, 2002), none have been reproduced despite extended efforts to do so and there is good reason to believe that the strains used were not true nulls (for more discussion see Lee *et al.*, 2003 and Cheng *et al.*, 2003).

In addition to *wc-1* and *wc-2*, a number of genes affecting photoresponses have been identified. Historically, the first of these was *poky* which encodes a mitochondrial 19S RNA and appears to cause a b-type cytochrome deficiency that in some way results in reduced sensitivity for photosuppression of clock-regulated conidiation (Brain *et al.*, 1977); the molecular basis of this effect is still not understood. Using a clever colony based assay for the circadian rhythm, Paietta and Sargent identified several *light-insensitive* (*lis*) mutants that fail to show any photosuppression of the circadian rhythm (Paietta and Sargent, 1983); these genes have yet to be cloned and their products are not presently known. Lastly, the *vivid* gene was described based upon a spontaneous mutation that resulted in a much more intense orange colour (Hall *et al.*, 1993); VVD is now known to mediate light adaptation (Schwerdtfeger and Linden, 2001) and to encode a PAS domain-containing protein (Heintzen *et al.*, 2001) (see below). Several other *blr* (*blue light regulator*) genes described as being deficient in light-induction have not held up to further analysis (Carattoli *et al.*, 1995).

5. CLONING OF THE WHITE COLLAR GENES

The molecular dissection of photoresponses in Neurospora began with the cloning of the *wc* genes by Macino and colleagues (Ballario *et al.*, 1996; Linden and Macino, 1997). WC-1 is a 117 kD protein (Ballario *et al.*, 1996) that contains a Zn finger DNA binding domain, two acidic domains and a poly-glutamine stretch of the type typical for transcriptional activators, and three PAS domains (Lee *et al.*, 2000). Vertebrate GATA proteins typically contain paired Zn finger domains with 17 amino acid loops to complex the Zn whereas WC-1 has but a single 18 amino acid finger. In general, PAS domains are believed to mediate protein:protein interactions as has been shown to be the case with the WC proteins. The second and third PAS domains of WC-1 are of this type, but the first is a LOV domain, a subclass of PAS domains associated with proteins that sense light, oxygen, and voltage (Taylor and Zhulin, 1999). In a BLAST search, WC-1 shows strongest similarity via these PAS domains to other proteins, especially to the LOV domain, which identified a number of real and potential plant photoreceptors including phototropins. Next in order of magnitude to the similarities with photoreceptors, but still of unambiguous statistical significance, WC-1 shows strong similarity to a number of animal circadian clock proteins; subsequent work has shown that WC-1 is a sequence and functional homolog of the mammalian clock protein BMAL1 with which it is 48% similar or identical over the entire length of BMAL1(Lee *et al.*, 2000; see below). WC-1 is largely a nuclear protein and its intracellular location is not regulated by light. Its amount, however, is regulated post-transcriptionally at the level of synthesis (Lee *et al.*, 2000) and also through both phosphorylation (Arpaia *et al.*, 1999; Lee *et al.*, 2000) and protein:protein interactions (Cheng *et al.*, 2002; Denault *et al.*, 2001; Talora *et al.*, 1999). (see below). Under constant conditions in the dark, *wc-1* RNA levels do not vary substantially over time, but there is a clear rhythm in WC-1 protein content (Cheng *et al.*, 2001; Lee *et al.*, 2000; Merrow *et al.*, 2001), consistent with post-transcriptional control of WC-1 rhythmicity. In contrast to WC-1, WC-2 is constitutively expressed, giving rise to a 57 kDa (530 amino acid) protein having a single activation domain, PAS domain and Zn-finger. WC-2 is always in the nucleus (Denault *et al.*, 2001) and the protein does not appear to be highly regulated (Denault *et al.*, 2001; Schwerdtfeger and Linden, 2000).

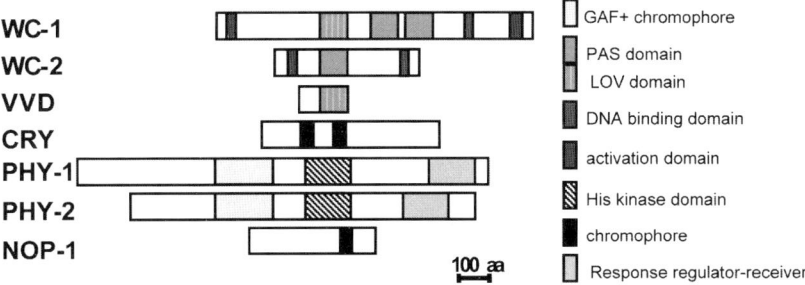

Figure 2. Real and putative photoreceptors in Neurospora. The identity and approximate location of various functional domains in Neurospora proteins are shown. In the list, only WC-1 and VVD are known to bind chromophores and to be true photoresponse mediators, although WC-2 is required for the function of WC-1.

WC-1 and WC-2 interact in the nucleus via their PAS domains to form the White Collar Complex (WCC) (Talora *et al.*, 1999; Schwerdtfeger and Linden, 2000; Ballario *et al.*, 1998; Denault *et al.*, 2001) and with FRQ (Denault *et al.*, 2001; Cheng *et al.*, 2001) in the feedback loop comprising the circadian oscillator (see below). WC-1 is the limiting factor in the complex, since WC-2 is always present in excess (Denault *et al.*, 2001; Cheng *et al.*, 2001). WC-2 is a key element in the WCC, however, in that it mediates the interactions between the regulated components. GST-pull-down assays, sucrose gradient sedimentation, and immunoprecipitations show that WC-2 forms complexes with WC-1 and with FRQ, consistent with the model in which FRQ acts to depress the level of it's own transcript by physically interfering with the activation of the *frq* gene by the WCC (Collett *et al.*, 2001; Denault *et al.*, 2001). WC-2 is the most abundant component in this feedback loop, and, significantly, WC-1 and FRQ do not interact in the absence of WC-2, nor is DNA required for the interactions.

6. WHITE COLLAR-1 IS THE BLUE LIGHT PHOTORECEPTOR

Soon after cloning *wc-1* and *wc-2,* Macino and colleagues proposed that the genes encoded the transcription factors mediating light-induced gene expression (Ballario *et al.*, 1996; Ballario *et al.*, 1998; Linden and Macino, 1997), and that WC-1 might be the photoreceptor itself. This bold guess was later verified (Froehlich *et al.*, 2002; He *et al.*, 2002). The idea had been around for some time based on the fact that it was impossible to identify any genes other than the *white collars* that were essential for photoresponses. However, these data were equally consistent with a model that placed WC-1 and WC-2 as essential elements in the signal transduction cascade leading from a photoreceptor to downstream responses, and the identification in 1997 of an exclusively dark function for the WC proteins in the circadian clock (Crosthwaite *et al.*, 1997) appeared to lend credence to this model. In order to prove that a protein is a photoreceptor, of course, it is necessary not only to prove that mutants lacking it have no photoresponses, and to show that it binds a chromophore, but also that the protein itself undergoes photochemistry with a fluence and wavelength response consistent with the know biology. These were the tests used to establish WC-1 as the blue light photoreceptor in Neurospora (Froehlich *et al.*, 2002).

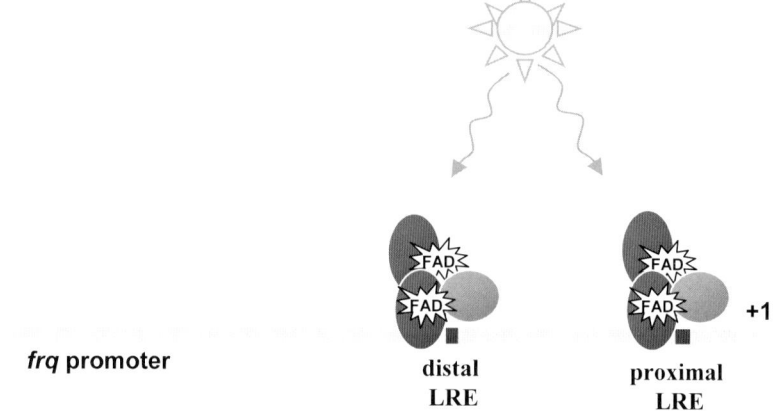

Figure 3. Two light regulatory sequences within the *frq* promoter in Neurospora bind to complexes of WC-1 and WC-2. The large oval represents WC-1 binding to FAD as the chromophore, and the small oval represents WC-2. See text for details (Figure courtesy of A. Froehlich).

WC-1 binds a chromophore, flavin adenine dinucleotide (FAD) as shown by purification of the protein followed by chemical extraction (He *et al.*, 2002) and also by the fact that FAD is essential for function of the in vitro transcribed protein (Froehlich *et al.*, 2002). A molecular genetic route was used to establish photoreceptor function; Froehlich *et al.* first dissected the light regulatory regions of a known highly light-inducible gene, the circadian clock gene *frq*, which turned out to have two separate light regulatory elements (LREs). Light induction of *frq* was known to require both WC-1 and WC-2, and using the LREs as targets in electrophoretic gel mobility shift assays (EMSAs), they showed that the WCC bound to the LREs. Mutants lacking either WC had no binding capacity, and FRQ was not found in the complex, although it is known to bind to the WCC in solution both in vivo and in vitro (Froehlich *et al.*, 2002). The key observation to establishing WC-1 as the photoreceptor was the observation that the mobility of the complex was different depending on whether the cell extracts were isolated, and the EMSA assay was run, in the light or under red light (to which Neurospora is blind). The observation was that extracts from light grown cultures had a slower mobility and, importantly, if cultures were extracted in the dark, frozen, and later exposed to light, they retained the ability to respond to the light. This meant that cell extracts contained a functional photoreceptor that could be purified and identified. Important controls (Figure 4) established that the action spectrum for the conversion of dark complex to light complex had the same wavelength response (action spectrum) as had previously been shown for photosuppression of circadian rhythmicity (Sargent and Briggs, 1967) and also that it has the same fluence response as had previously been shown for the blue light-induced phase shifting of the circadian clock (Crosthwaite *et al.*, 1995). These data established the photoresponse in the test tube as indistinguishable from the biological responses. From this point it was logically straightforward to express the proteins in vitro in a coupled transcription/translation system, so that no other Neurospora proteins would be present, and to show that the light response (the change in mobility on the gel shift) could be achieved. Interestingly, in the absence of any added chromophore, both light and dark mobility complexes were seen, but the addition of FAD (but not FMN) promoted the formation of the dark complex in the dark and the light complex in the light. These data were consistent with the observation of He *et al.* that FAD was present in the complex in vivo, and also clearly consistent with the genetics, and established WC-1, working as a part of the WCC, as the blue light photoreceptor (Froehlich *et al.*, 2002). Because the apparent mobility of the complex increases on exposure to light even when the complex is composed of proteins made in vitro in a test tube, the simplest interpretation is that shown in Figure 3 where light results in a multimerization of the components WC-1 and WC-2. This is consistent with the observation that self association of WC-1 in complexes is seen but self association of WC-2 is not seen (Cheng *et al.*, 2003).

The present model, then, is that the complex of WC-1 and WC-2 binds to DNA at a specific consensus sequence, the light responsive element or LRE. The proteins both contain GATA-type Zn fingers and the WCC binds to LREs having imperfect repeats of GATN (not GATA) (Froehlich *et al.*, 2002), a sequence similar to one previously identified for the Neurospora *al-3* gene (Carattoli *et al.*, 1994). Since the photoactive part of WC-1 is the LOV domain, everything supposed about the photochemistry of the events is based on better studied LOV domains in the phototropins (see Chapter 12). Thus, it is expected that when the FAD chromophore absorbs blue light, the C4a position of the FAD undergoes a transient covalent interaction with the cysteine in the NCRFLQ sequence of WC-1, thereby inducing a conformational change in the protein. Consistent with this, a point mutation changing this Cys to Ser yields a blind WC-1 (Cheng *et al.*, 2003). Since Froehlich *et al.* showed that the response occurs in a test tube containing only in vitro translated proteins and FAD, the WCC by itself must be sufficient with FAD to function as the photoreceptor. This conformational change in turns appears to bring about the quaternary interaction between WCCs, apparently resulting in formation of a multimer and in enhanced transcriptional activation of WCC-bound light responsive promoters.

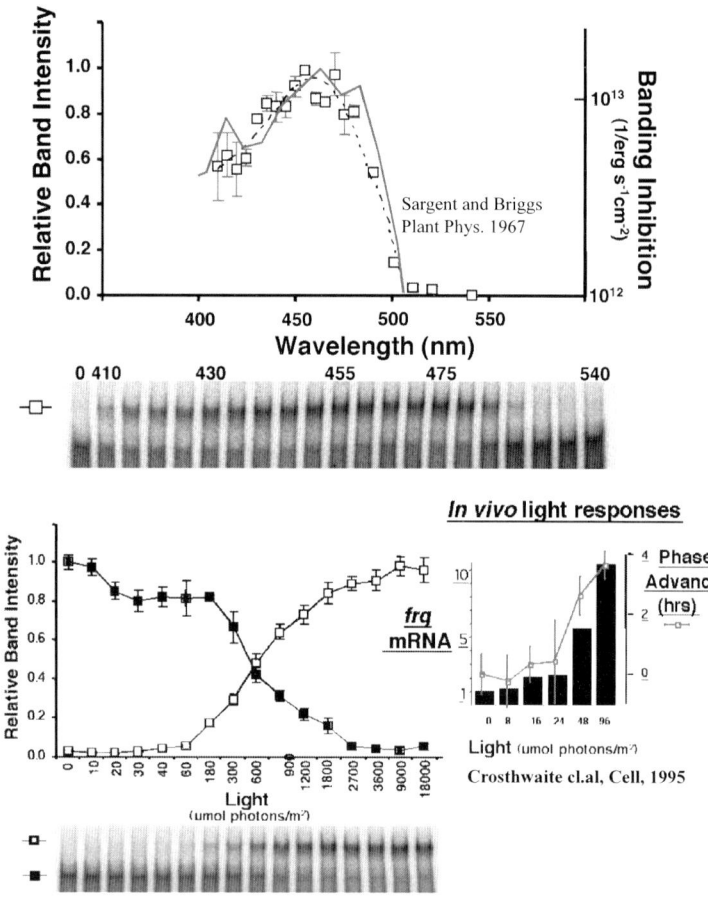

Figure 4. The in vitro light induced WCC/LRE mobility change occurs at biologically relevant wavelengths (top) and light intensities (bottom).

As a result of the exposure to light, WC-1 undergoes quantitative and qualitative changes that appear to be important for regulating its activity. Based on inhibitor studies, Macino and colleagues have advanced a model in which WC-1 is turned over very quickly after lights-on, there is a rapid increase in WC-1 phosphorylation along with a transient decrease in total WC-1 amounts. This is followed by an increase in nonphosphorylated WC-1 (Talora et al., 1999). In the dark, blocking of protein synthesis with cycloheximide has little immediate effect; light, however, causes WC-1 and phosphorylated WC-1 to decrease in the absence of protein synthesis (Figure 5). These data are interpreted as phosphorylation-induced turnover followed by resynthesis of unphosphorylated WC-1. Further, the kinetics of light induction of most genes closely parallels this, and the decrease in phosphorylated WC-1 is accompanied by a refractory period to light stimulation. In vvd^{null} strains (see below) total WC-1 decreases after light exposure but phosphorylated WC-1 levels remain high instead of decreasing, and the refractory period is missing; this is consistent with a model in which phosphorylated WC-1 is the active form of the protein (Heintzen et al., 2001; Schwerdtfeger and Linden, 2001; Talora et al., 1999).

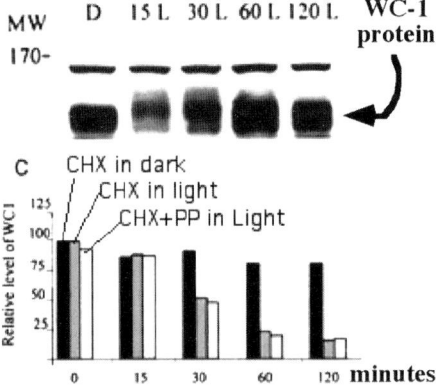

Figure 5. WC-1 levels were followed in the dark, and after exposure to light with CHX (cycloheximide) to block de novo protein synthesis, +/-λ phosphatase (PP) to follow decay of phosphorylated WC-1. From Talora et al, 1999.

7. WC-1 AND WC-2 - POSITIVE ELEMENTS IN THE CIRCADIAN FEEDBACK LOOP

Circadian rhythmicity is inextricably linked with light responses in most organisms: even though the clock will by design continue to run in the absence of light cues, it is light and dark transitions that most commonly provide the cues required to correctly phase the internal clock oscillation so that internal subjective day as experienced in the cell corresponds to external day as defined by the earth's rotation. For this reason, although it is only indirectly associated with the photobiology of Neurospora *per se*, it should be noted here that the WCC plays an essential role in circadian timekeeping in the dark, exclusive of its role in the light-induction of the known light-induced genes in *Neurospora*.

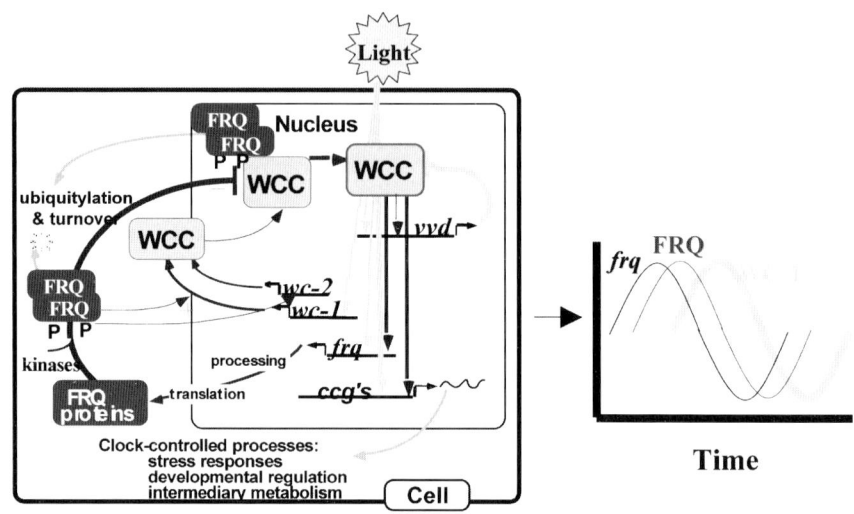

Figure 6. Known molecular components in the circadian feedback loop of Neurospora. Light lines represents effects of light, grey or dark lines actions that take place in the dark. Within the cell, the WCC drives expression of FRQ which does three things: (1) Its first and dominant action is to bind to the WCC to block its function, thus constituting a negative feedback loop that is the clock. (2) It promotes the synthesis of WC-1 posttranscriptionally and wc-2 mRNA, actions that promote robustness in the feedback loop; (3) It becomes phosphorylated which eventually leads to its turnover. When this happens, the bolus of WCC whose synthesis was promoted by FRQ is released to start the cycle again (Lee et al., 2000; Loros and Dunlap, 2001). The result of these feedback loops are daily rhythms in expression of frq mRNA, FRQ, and WC-1 whose timing within the day defines biological time; for instance, high frq mRNA doesn't just occur in mid-day, high frq defines mid-day for the organism (adapted form Dunlap, 2003).

The WCC is the positive factor that drives expression of the negative element in the negative feedback loop comprising the circadian oscillator (Crosthwaite et al., 1997; Figure 6). The simple model is this: In the transcriptional/translational feedback loops comprising the core of the cellular circadian oscillators of eukaryotes in the fungal/animal lineage, negative elements (e.g. FRQ in *Neurospora*, PER/TIM in *Drosophila*, and PER1, PER2, CRY1, and CRY2 in mammals) act to block activation by heterodimeric PAS domain-containing positive elements (e.g. WC-1/WC-2 in *Neurospora*, dCLK/CYC in *Drosophila*, and CLOCK/BMAL1 in mammals). The positive elements in turn activate expression of the negative elements, thereby giving rise to the negative feedback loop that comprises a core of the clock (Dunlap, 1998). Several predictions that follow from the anticipated role of the WCC in the clock have been confirmed: (1) defects in WC-2 should lengthen the circadian period length in the dark; (2) proteins such as WC-2 should play a role in temperature compensation (Collett et al., 2001); (3) WC-2, FRQ, and WC-1 should physically interact in solution. These all confirm central predictions of the current model for the *Neurospora* circadian clock in which FRQ acts to depress the level of it's own transcript at least in part by interfering with the action of the WCC, thereby blocking activation of the *frq* gene by the WCC (Froehlich *et al.*, 2003). These and other data are consistent with a model in which the PAS-protein heterodimers in eukaryotic clock loops are regulated by protein:protein interactions occurring in solution rather than on clock gene promoters (Bae *et al.*, 2000; Denault *et al.*, 2001). The essential role of the WCC in the operation of the circadian clock, and of heterodimers of PAS domain proteins (such as CLOCK/BMAL in mammals or CLK/CYC in *Drosophila*) in circadian feedback loops was not known until the work of Crosthwaite in 1997 (Crosthwaite *et al.*, 1997). That report provided the first precedent for the role of PAS:PAS heterodimers as activators in a circadian feedback loop, although it was followed just two weeks later by the cloning and description of the mammalian CLOCK gene (King *et al.*, 1997) in which similar conclusions in the mammalian circadian system were independently reached.

7.1 How light resets the clock

The realization that *frq* was a central component in the circadian clock mechanism and that it could be strongly induced by light quickly gave rise to a mechanism for light-induced clock resetting that has proven to be the paradigm for understanding how visible light resets not only fungal but also mammalian clocks. Briefly, we know from personal experience that our sense of time of day has less to do with the amount of ambient light on our bodies or in our eyes than it does with our internal clocks; when we fly long distances, this disconnect between ambient light and internal time sense is what we know as jet lag. Jet lag dissipates as our internal clocks are gradually reset and become synchronized with the external day night cycle. In Neurospora, the daily rhythm in *frq* expression actually defines biological time; the biological time of the organism is reset to correspond to the external day/night cycle through light induction of *frq* expression (Figure 7). Clearly, if the day starts early, light exposure late at night should advance the internal clock into the day phase whereas if day lasts longer so light hits the evening part of the cycle, it should act to delay the clock back to the end of the previous day. Figure 7 illustrates how this same light exposure can yield opposite biological responses. This mechanism was first developed based on work in Neurospora (Crosthwaite *et al.*, 1995) but has since been shown to be equally applicable to mammals where light induces the acute induction of the clock gene *Per1* in the SCN of the hypothalamus, thereby resetting the organism's clock (Shigeyoshi *et al.*, 1997).

In having an important and clearly distinct dark-only function, the WCC is probably unique among eukaryotic photoreceptors. To begin to gauge the importance of this regulation, Lewis et al (Lewis *et al.*, 2002) expressed WC-1 under a regulatable promoter and examined gene regulation using microarrays. They found many genes to be regulated directly or indirectly in the dark by WC-1, but interestingly most were not subsequently found to be light-induced. Conversely they found 22 light-induced genes that included 4 (like *frq*) that were also activated in the dark by WC-1. These data would be consistent with a role for WCC as a master regulator that acts to turn on downstream regulators that, in turn, more directly regulate target genes.

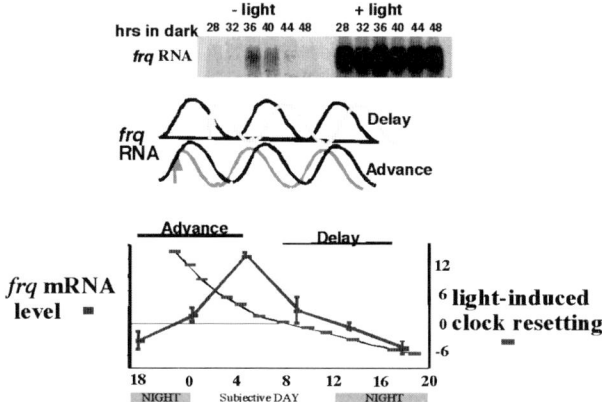

Figure 7. Light resetting of the Neurospora clock. At the top is shown a Northern: Absent light on the left, *frq* expression levels are seen to rise and fall with a circadian rhythm whereas on the right, light exposure at any time of day results in a rapid large induction of *frq*. In the middle panel is shown a schematic of how this light induction resets the clock: the dark curve represents the control rhythm of *frq* in the dark. Light exposure when *frq* is falling (evening) delays the clock whereas light exposure when *frq* is rising (late night to early morning) advances the clock. The bottom panel shows hour by hour through the day how light induction of *frq* will reset the clock, by plotting the daily rhythm in *frq* mRNA on top of the amount of clock resetting elicited by light exposure (adapted from Dunlap, 1999).

8. VIVID, A SECOND PHOTORECEPTOR THAT MODULATES LIGHT RESPONSES

Although all known light responses in Neurospora are specific for blue light, and WC-1 is required for all known light responses, it does not follow that WC-1 or the WCC act are sufficient to mediate all Neurospora photobiology. If there is to be any light-dependent modulation of a light response, it follows that there must be additional photoreceptors, and this is the case: The VIVID protein (VVD) acts to modulate responses initiated by WC-1, and as such it plays an important role in several aspects of photoresponses. Like WC-1 and WC-2, VVD is a member of the PAS protein superfamily (Heintzen *et al.*, 2001), and its action describes an autoregulatory negative feedback loop that closes outside of the core circadian oscillator (Figure 6) to affect both the clock and photoresponses. Interestingly, VVD as a photoreceptor binds an FMN chromophore (Schwerdtfeger and Linden, 2003); in response to light, it helps to modulate the organism's response to subsequent light signals, a process known as photoadaptation (Schwerdtfeger and Linden, 2001, 2003; Shrode *et al.*, 2001). As might be inferred from its place in the regulatory scheme, when *vvd* is lost all photoresponses in the organism are elevated. However, there are several important corollary effects in part deriving from this primary one. One is that photoadaptation is lost.

When Neurospora sees light and exhibits an immediate early light response, for instance the rapid induction of *al-1* gene expression, it is typically observed that the response peaks within 15-30 minutes and then decays away even when the light stays on (Figure 8). Interestingly, even though the response has decayed away, the photoresponse system remains refractory such that a second light exposure of the same fluence anytime within two hours has no additional inductive effect. Exposure to brighter light, however, can yield induction. This process by which a light exposure can influence the effects of subsequent exposures is called photoadaptation, and it requires the VVD protein as well as phosphorylation of unspecified proteins by protein kinase C. When VVD is lost, the rate of decay of light–induced gene expression is slowed, and more importantly perhaps, second exposures to light within two hours of a first exposure have additive effects (Schwerdtfeger and Linden, 2001, 2003; Shrode *et al.*, 2001). As an aside it is worth noting that not all genes are subject to photoadaptation; for instance, the clock gene *frq* is simply induced by light (Crosthwaite *et al.*, 1995) and its level does not decay over time to the same extent as other more typical genes.

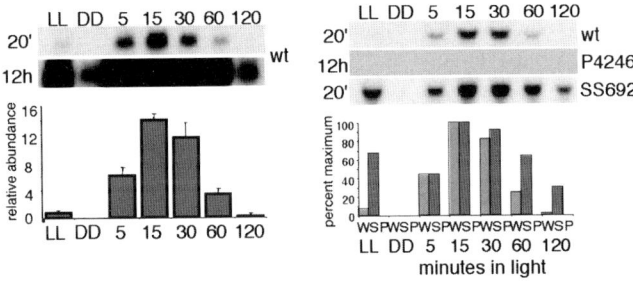

Figure 8. *vvd is rapidly light induced, and loss of vvd slows the rate of decay of light-induced transcripts. On the right, P and S stand for two alleles, P which makes no transcript and S that makes transcript but no protein; W denotes wild type. (From (Heintzen et al., 2001) with permission).*

In a second effect, because VVD is strongly induced by light and in turn acts as a repressor of light responses, VVD affects circadian entrainment, the process by which the internal clock is set by the daily light/dark cycle. *vvd* is expressed at a significant level only during the first day in constant darkness, and VVD thus probably accounts for the finding that light signals have little if any clock-resetting effects when delivered on the first day after a light-to-dark transfer. Although VVD is not required for circadian rhythmicity, its loss affects the perception of light and thereby affects clock resetting. Clock regulation of the immediate and transient repressor VVD contributes to circadian entrainment by making dark to light transitions more discrete (Heintzen *et al.*, 2001).

In a third related effect, *vvd* expression is controlled in part by the clock and VVD in turn feeds back to regulate the expression of a number of input and output genes including itself. Because of this, the effect of a brief exposure to light on gene expression or on the clock will depend on the subjective time of day when the light is seen. This is an aspect of regulation known as circadian gating, and VVD participates in it. However, loss of VVD does not result in complete loss of gating, suggesting that there are other factors involved (Heintzen *et al.*, 2001).

VVD is a small protein, basically just a short N-terminal extension preceding a LOV domain (Figure 2). Its sequence thus suggested a photoreceptor function even before Schwerdtfeger and colleagues reported chromophore binding and a photocycle (Schwerdtfeger and Linden, 2003; Schwerdtfeger *et al.*, 2003). A role of this LOV domain as a photoresponse module has also been supported by domain swap experiments in which LOV domains from various proteins were swapped into WC1 and the chimeras assayed in vivo for rescue of *frq* light induction and other related responses (Cheng *et al.*, 2003). Upon exposure of the FMN chromophore to light, VVD undergoes a classic photocycle; as with WC-1, based on the story with the LOV domain of PHOT1, the expectation is that the C4a carbon of the FMN bound by VVD undergoes a transient covalent addition to the Cys in the NCRFLQ flavin binding site, resulting in structural changes in the protein that affect its activity and ability to interact with partners (Crosson and Moffat, 2001). It is expected that VVD will interact with the transcriptional activator WCC, down-regulating its activity and thereby influencing the expression of WCC controlled genes, although this has not yet been verified biochemically.

9. COMPLEXITIES IN LIGHT REGULATORY PATHWAYS

The simplest interpretation and expectation of the data above is that the WCC would bind to LREs in the promoters of all light regulated genes and activate them in concert, in some cases being modified by the action of VVD. This model is almost certainly an oversimplified version of reality. To begin with, not surprisingly, diverse induction kinetics are seen with light induced genes, even with rapidly induced genes which can peak in expression between 15 and 30 minutes after light exposure. More telling is the observation that some genes are clearly much more sensitive to the dose of WCC than are others and, a probably related observation, some genes will tolerate mutations in the WCC that other genes cannot. For instance, Collett *et al.* (Collett *et al.*, 2002) found that point mutations in the Zn finger DNA binding domain of WC-2 abolished light induction of some target genes but allowed induction of *frq* to proceed unaffected. Similarly, mutations in WC-1 that have little effect on *frq* induction can abrogate light responses of other genes (Lee *et al.*, 2003; Cheng *et al.*, 2003). This is borne out by the finding that when expression of the WC proteins is placed under the control of inducible promoters, normal light induced expression of *frq* is seen at basal levels of induction (Lee *et al.*, 2003; Cheng *et al.*, 2002). This extreme differential dependence on WCC levels has led to at least one error in the literature that is instructive. In the MK1 allele of WC-1, there is a mutation that induces a frame shift and places a STOP codon in the N-terminal region before the LOV domain; however, due to reinitiation of translation a very low level of WC-1 is made, amounting to less than 1% of the normal amount; this is sufficient to be detected in Westerns following immunoprecipitation and it is clear that the N-terminal fragment alone is not sufficient for any light responses (Lee et al., 2003; Cheng et al., 2003). Another occasionally used allele of *wc-1* (Talora et al., 1999) is one that arose from the RIP process (Selker, 1997) which can introduce point mutations into a gene. In one notable case (Dragovic et al., 2002), a *wc-1* RIP allele, having point mutations in the N-terminal region of the gene that introduced STOP codons, was used to assess the importance of WC-1 is *frq* and operation of the clock. Not appreciating the importance of using true knockout strains, nor the differential requirements for WCC levels in light-induction, the authors assumed that this allele was a null mutant and reported that "blind" mutants of Neurospora still showed robust light responses (Dragovic et al., 2002). The error persists in the literature pending a retraction or clarification.

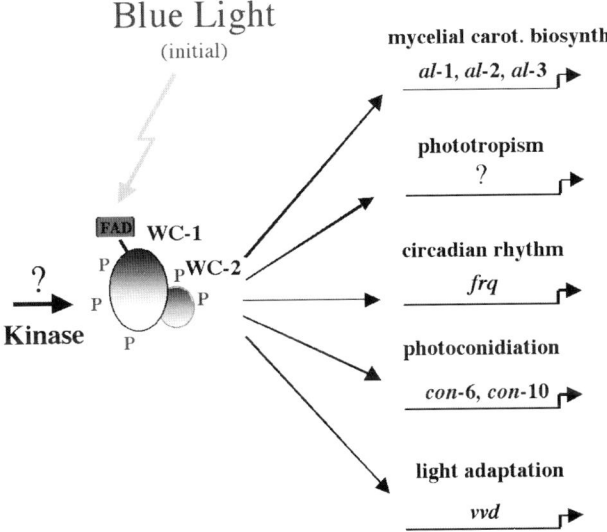

Figure 9. Summary of light responses in Neurospora (Figure courtesy of C. Schwerdtfeger; all rights reserved).

It now appears that light induced genes can be sorted into three groups based on their sensitivity to WC levels. *frq* is in a class by itself in requiring log orders less WCC than some genes. A second group includes *vivid* which itself acts as a photoreceptor as described above, and the third and most sensitive group includes the genes encoding components of the carotenoid biosynthesis pathways (*al-1, al-2,* and *al-3*) and photoconidiation (e.g. *con-6* and *con-10*) pathways. At present, the molecular basis of this sensitivity difference remains obscure. Although *frq* has two LREs in contrast to most genes with only one, it would be necessary to invoke a highly cooperative binding or activation reaction to explain the enormous differences in inducibility. Another possibility would be differences in the LRE sequences giving rise to differences in binding affinity.

A summary of the known and verified light responses in Neurospora is shown in Figure 9. It makes the point that to date there are no responses that do not require the WCC, although there is good reason to believe that this will change soon for reasons described in the next section. It is likely that one or more kinases act on the WCC to modify its activity, and that VVD in some way interacts with it to modify its function. However, in all cases, the initial light response is triggered by light that is senses by the FAD chromophore of WC-1.

10. OTHER NEUROSPORA PHOTORECEPTORS

All known light responses in *Neurospora* are specific to blue light; there are no known red or far-red responses. It has thus been a surprise when ongoing genomics efforts have turned up additional putative photoreceptors (Figure 2). The first of these was a Neurospora opsin, NOP-1, whose sequence appeared as a result of an extensive EST sequencing project aimed at identifying circadianly regulated genes (Zhu *et al.*, 2001). The gene is therefore clearly expressed, and analysis of the protein following heterologous expression showed that it did in fact bind a retinal cofactor and could undergo a photocycle. Although detailed phenotypic analysis of knockouts did identify a subtle alteration in growth, the work failed to identify a solid light-regulatory function for the protein (Bieszke *et al.*, 1999) leaving its role in photoresponses, if any, obscure. This is a theme that has been repeated as seen below. More recently, the completed genomic sequence of *Neurospora* identified several putative photoreceptors including several bacteriophytochromes and a cryptochrome (Galagan *et al.*, 2003; Borkovich *et al.*, 2004). Each of these has now been examined and a wealth of information is emerging, although as with NOP-1 there are no clear light-regulatory phenotypes associated with any of the genes. The CRY (Froehlich *et al.*, 2005a) is strongly light induced in a WC-1 mediated pathway, is circadianly regulated, and has been shown to bind FAD in vitro. It is found in both the cytoplasm and nucleus. Since CRYs are often characterized as proteins having sequences similar to DNA photolyases but lacking in photolyase function, it is of significance that complete deletion of CRY has no impact of photoreactivation repair. Disappointingly, however, there are no light or clock phenotypes that can be associated with loss of the protein.

A similar story obtains with two genes encoding proteins having extensive similarities to bacteriophytochromes (Figure 2). Based on sequence they are members of the Cph1 family of phytochromes, with two PAS and a PHY domain followed by an ATPase and a response-regulator domain, all based on the primary sequence. The genes are clearly expressed at the RNA and protein level, and are regulated although disappointingly not by blue, red, or far red light (Froehlich *et al.*, 2005b). A N-terminal fragment of 515 amino acids of PHY-2 expressed in E. coli has been shown to bind either biliverdin or PCB and to undergo spectral shifts when exposed to red or far red light, and after red exposure there is a gradual decay back to the red absorbing form (Froehlich, A., Vierstra, R., Loros, J., Dunlap, J. C., unpublished). Importantly, as with the CRY, there are no clock or light-related phenotypes that can be associated with loss of either PHY. Additionally, microarray analyses have so far failed to detect transcripts whose regulation is altered either by red or far red light, or by loss of either PHY. Based on their expression and regulation, it seems likely that the proteins have a role in the cell, but at this point there is little to suggest a clear role in the photobiology of the organism.

11. FLAVIN BINDING DOMAIN PROTEINS AS PHOTORECEPTORS IN PHOTOSYNTHETIC EUKARYOTES

The phototropins were the pioneer proteins for the class of proteins using LOV domains as light sensors (Christie et al, 1999; see Chapter 12). PHOT-1 and PHOT-2 mediate nontranscriptional effects of blue light and thereby direct diverse aspects of the photobiology of organisms. As we have seen above, the LOV domain of WC-1 is clearly mediates a transcriptional effect by strongly potentiating the transcriptional activating ability of the WCC. The structural means by which the LOV domain in VVD acts is not yet understood. However, it is clear the LOV domains, or similar domains that bind a flavin, can act as photoreceptive modules that can be hooked up to use light to regulate the activity of a large variety of proteins. Four additional examples of this are found among photosynthetic eukaryotes, in the protist Euglena and the higher plant *Arabidopsis*.

Euglena exhibits several strong blue light responses that are used to regulate its mobility in the water so that cultures can concentrate at regions in the water column where the level of light is optimal. Light is sensed by a pigment(s) whose action spectrum suggested a flavin, and fluorescence microscopy suggested the paraflagellar body as the source of the fluorescence. Based on this correlation Watanabe and colleagues (Iseki *et al.*, 2002) used gentle cell disruption techniques and gradient centrifugation to isolate a fraction enriched in paraflagellar bodies. Subsequent purification of the flavoprotein components yielded a complex containing two similar proteins of 859 and 1019 amino acids, each of which binds two FAD. The sequence of each of the two flavin binding domains in each protein established these as members of the BLUF class of flavin binding photoresponse molecules (Gomelsky and Klug, 2002). To confirm that these paraflagellar body-associated flavoproteins were functioning as photoreceptors, the investigators used RNAi directed against the two messages corresponding to the flavoproteins in the complex. RNAi treatment resulted in loss of paraflagellar body fluorescence and loss of the light step-up photophobic response but, interestingly, not the light step-down response. Sequence analysis of the flavoprotein genes identified adenyl cyclase domains adjacent to the flavin binding pockets, and subsequent biochemical analysis confirmed the presence of blue light activatable adenylate cyclase associated with the proteins in the complex (Iseki *et al.*, 2002). This complex also mediates the positive and negative phototaxis responses (Ntefidou *et al.*, 2003), but interestingly additional data strongly suggest that the paraflagellar body is not the sole intracellular location of the photoreceptor in all euglenids. The identification of this blue light-activatable adenyl cyclase provides a second example for how a flavin binding domain may be used to couple light input with the regulation of an important intracellular signalling molecule.

A third set of examples is provided by the higher plant Arabidopsis where a flavin binding LOV domain is seen controlling a third distinct aspect of intracellular regulation. In the context of circadian rhythmicity and light-regulation of flowering, Arabidopsis has enlisted the LOV domain in three distinct proteins to control light-mediated ubiquitination and protein turnover (Figure 10; see Chapter 26). *ZTL*, the founding member of the class, was isolated as a long period circadian clock mutant in which the differential between wild type and mutant period lengths increased as fluence rates decreased (Somers *et al.*, 2000). This phenotype suggested that ZTL functioned on the light input pathway to the clock. *ZTL* encodes an F-box protein containing KELCH repeats at its C-terminus and a LOV domain at its N-terminus. Proteins with these domains typically function in mediating the association between an E3 ubiquitin ligase and its substrate, and in this case the interaction is further regulated by light. ZTL binds to flavin mononucleotide (FMN), a co-factor for blue light photoreceptors, and exhibits spectral properties consistent with photoperception. These observations suggest that ZTL could be a circadian photoreceptor that transmits light signals into the clock, possibly driving the light-triggered turnover of a central clock protein such as TOC1.

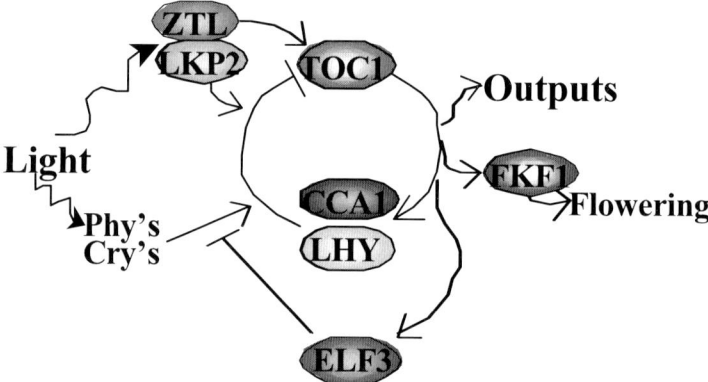

Figure 10. Placement of LOV domain-associated protein turnover photoreceptors within the circadian system in Arabidopsis. The feedback loop formed by TOC1 and LHY/CCA1 is at the core of the clock. ZTL and LKP2 are associate with light input and FKF1 with output in the regulation of CO in the context of flowering. Clock-regulated ELF3 regulates light signalling pathways. Figure courtesy of Thomas Schulz and Steven Kay; all rights reserved.

The second member of the family, *FKF1* (FLAVIN-BINDING, KELCH REPEAT, F-BOX), was identified as a deletion resulting in a late flowering phenotype (Nelson *et al.*, 2000). In contrast to *ZTL*, the *FKF1* gene is regulated by the circadian clock and loss-of-function mutants did not exhibit any circadian phenotypes, consistent for a role of FKF1 on the output side of the clock in control of flowering time and hypocotyl length. This has recently been confirmed by showing that FKF1 binds a flavin and mediates the light-dependent turnover the central flowering regulator CONSTANS (Imaizumi *et al.*, 2003). By being both light and clock regulated, FKF1 can control the day-length dependent expression of CO on which the flowering decision rests.

The last member of the family, LKP2 (LOV KELCH REPEAT PROTEIN 2) simply turned up in a database search for *Arabidopsis* PAS proteins, although it was subsequently shown that plants over-expressing *LKP2* exhibited arrhythmic circadian phenotypes for multiple outputs, and long hypocotyl and late flowering phenotypes (Schultz *et al.*, 2001). The arrhythmic phenotype was also observed in constant dark suggesting a role close to the oscillator as seen in the figure, although true loss-of-function alleles would be needed to confirm that conclusion. Similar to *ZTL*, *LKP2* transcripts are not clock-regulated. These data suggest a role for LKP2 near to ZTL on the input side of the circadian system (Figure 10). Although LKP2 can function in the dark, like its brethren ZTL and FKF1, the LOV domain of LKP2 binds FMN and displays spectral properties similar to the other two family members (Imaizumi *et al.*, 2003), strongly suggestive of a role of light in somehow influencing the activity of the protein. Taken together with the studies on ZTL and FKF1, it now appears that these three proteins constitute a gene family of blue-light photoreceptors in Arabidopsis in which the light is used to regulate ubiquitination and subsequent turnover of specific proteins important for circadian timing and/or flowering.

12. SUMMARY AND CONCLUSION

Neurospora has for a long time been useful for dissecting the role of light in biological systems. Diverse responses to light have been observed, but transcriptional responses appear to predominate, and to date responses only to blue light have been identified. These are all mediated initially by the blue light photoreceptor WC-1 that uses an FAD cofactor bound in a LOV domain in this novel multidomain protein that acts as a light-inducible transcription factor. WC-1 acts with a partner WC-2 to form a complex that initiates all blue light signalling in the organism; VVD, using FMN bound in a LOV domain, acts independently as a photoreceptor to modulate the WCC response. Unique, perhaps, among photoreceptors, the WCC also displays a prominent function in the dark by playing the central role of key transcriptional activator in the negative feedback loop at the core or the circadian system. Genomics in Neurospora has identified several other potential photoreceptors including an opsin, a cryptochrome, and two phytochromes, but to date no significant photobiological phenotypes have been associates with loss of these genes. Thus, all known photoresponses in Neurospora are mediated by LOV domain containing proteins. This domain, first described in the higher plant phototropins, has also recently been shown to provide the photoresponsive domain of a phototaxis-associated light-regulated adenyl cyclase in Euglena and a family of three Arabidopsis proteins (ZTL, FKF1, and LKP2) with roles in the regulation of flowering time and the response of the circadian clock to light.

13. REFERENCES

Arpaia, G., Cerri, F., Baima, S., Macino, G. (1999) Involvement of protein kinase C in the response of *Neurospora crassa* to blue light. *Mol Gen Genet, 262*, 314-322.

Bae, K., Lee, C., Hardin, P. E., Edery, I. (2000) dCLOCK is present in limiting amounts and likely mediates daily interactions between the dCLOCK-CYC transcription factor and the PER-TIM complex. *J Neurosci, 20*, 1746-1753.

Ballario, P. and Macino, G. (1997) White collar proteins: PASsing the light signal in *Neurospora crassa*. *Trends Microbiol, 5*, 458-462.

Ballario, P., Talora, C., Galli, D., Linden, H., Macino, G. (1998) Roles in dimerization and blue light photoresponse of the PAS and LOV domains of *Neurospora crassa* WHITE COLLAR proteins. *Molec Microbiol, 29*, 719-729.

Ballario, P., Vittorioso, P., Magrelli, A., Talora, C., Cabibbo, A., Macino, G. (1996) *White collar-1*, a central regulator of blue-light responses in *Neurospora crassa*, is a zinc-finger protein. *EMBO J, 15*, 1650-1657.

Bieszke, J. A., Braun, E. L., Bean, L. E., Kang, S., Natvig, D. O., Borkovich, K. A. (1999) The *nop-1* gene of *Neurospora crassa* encodes a seven transmembrane helix retinal-binding protein homologous to archaeal rhodopsins. *Proc Natl Acad Sci USA., 96*, 8034-8039.

Borkovich, K., Alex, L., Yarden, O., Freitag, M., Turner, G., Read, N., *et al.* (2004) Lessons from the genome sequence of Neurospora crassa: Tracing the path from genomic blueprint to multicellular organism. *Molec and Microb Rev,68*, 1-108.

Brain, R., Woodward, D., Briggs, W. (1977) Correlative studies of light sensitivity and cytochrome content in *Neurospora crassa*. *Carnegie Inst Washington Yearbk, 76*, 295-299.

Carattoli, A., Cogoni, C., Morelli, G., Macino, G. (1994) Molecular characterization of upstream regulatory sequences controlling the photoinduced expression of the *al-3* gene of *Neurospora crassa*. *Molec Microbiol, 13*, 787-795.

Carattoli, A., Kato, E., Rodriguez-Franco, M., Stuart, W. D., Macino, G. (1995) A chimeric light-regulated amino acid transport system allows the isolation of blue light regulator (*blr*) mutants of *Neurospora crassa*. *Proc Nat Acad Sci USA, 92*, 6612-6616.

Cerda-Olmedo, E. (2001) Phycomyces and the biology of light and colour. *FEMS Microbiological Reviews, 25*, 503-512.

Cheng, P., He, Q., Yang, Y., Wang, L., Liu, Y. (2003) Functional conservation of light, oxygen, or voltage domains in light sensing. *Proc Nat Acad Sci USA,100*, 5938-5943.

Cheng, P., Yang, Y., Gardner, K. H., Liu, Y. (2002) PAS Domain-Mediated WC-1/WC-2 Interaction Is Essential for Maintaining the Steady-State Level of WC-1 and the Function of Both Proteins in Circadian Clock and Light Responses of Neurospora. *Mol Cell Biol, 22*, 517-524.

Cheng, P., Yang, Y., Liu, Y. (2001) Interlocked feedback loops contribute to the robustness of the *Neurospora* circadian clock. *Proc Nat Acad Sci USA, 98*, 7408-7413.

Cheng, P., Yang, Y., Wang, L., He, Q., Liu, Y. (2003) WHITE COLLAR-1, a multifunctional *Neurospora* protein involved in the circadian feedback loops, light sensing, and transcription repression of *wc-2*. *J Biol Chem, 278*, 3801-3808.

Christie, J. M., Salomon, M., Nozue, K., Wada, M., Briggs, W. R. (1999) LOV (light, oxygen, or voltage) domains of the blue-light photoreceptor phototropin (nph1): binding sites for the chromophore flavin mononucleotide. *Proc Natl Acad Sci USA, 96*, 8779-8783.

Collett, M., Dunlap, J. C., Loros, J. J. (2001) Circadian clock-specific roles for the light response protein WHITE COLLAR-2. *Molec Cell Biol, 21*, 2619-2628.

Collett, M. A., Garceau, N., Dunlap, J. C., Loros, J. J. (2002) Light and clock expression of the *Neurospora* clock gene frequency is differentially driven by but dependent on WHITE COLLAR-2. *Genetics, 160*, 149-158.

Crosson, S. and Moffat, K. (2001) Structure of a flavin-binding plantphotoreceptor domain: insights into light mediated signal transduction. *Proc Nat Acad Sci USA, 98*, 2995-3000.

Crosthwaite, S. C., Dunlap, J. C., Loros, J. J. (1997) *Neurospora wc-1* and *wc-2*: Transcription, photoresponses, and the origins of circadian rhythmicity. *Science, 276*, 763-769.

Crosthwaite, S. C., Loros, J. J., Dunlap, J. C. (1995) Light-Induced resetting of a circadian clock is mediated by a rapid increase in *frequency* transcript. *Cell, 81*, 1003-1012.

Degli Innocenti, F. and Russo, V. E. A. (1983) Photoinduction of perithecia in *Neurospora crassa* by blue light. *Photochem Photobiol, 37*, 49-51.

Degli Innocenti, F. and Russo, V. E. A. (1984a) "Genetic analysis of blue light-induced responses in *Neurospora crassa.*" In *Blue Light Effects in Biological Systems,* Senger H. (ed) Springer Verlag, Berlin, 213-219.

Degli Innocenti, F. and Russo, V. E. A. (1984b) Isolation of new *white collar* mutants of *Neurospora crassa* and studies of their behaviour in blue light-induced formation of protoperithecia. *J Bacteriol, 159,* 757-761.

Denault, D. L., Loros, J. J., Dunlap, J. C. (2001) WC-2 mediates WC-1-FRQ interaction within the PAS protein-linked circadian feedback loop of *Neurospora crassa. EMBO Journal, 20,* 109-117.

Dragovic, Z., Tan, Y., Gorl, M., Roenneberg, T., and Merrow, M. (2002) Light reception and circadian behaviour in "blind" and "clock-less" mutants of *Neurospora. EMBO J, 21,* 3643-3651.

Dunlap, J. C. (1998) An end in the beginning. *Science, 280,* 1548-1549.

Froehlich, A., Loros, J. J., Dunlap, J. C. (2005a) A Neurospora Cryptochrome. *in preparation.*

Froehlich, A., Loros, J. J., Dunlap, J. C. (2005b) Neurospora phytochromes. *in preparation.*

Froehlich, A. C., Loros, J. J., Dunlap, J. C. (2002) WHITE COLLAR-1, a Circadian Blue Light Photoreceptor, Binding to the *frequency* Promoter. *Science, 297,* 815-819.

Froehlich, A. C., Loros, J. J., Dunlap, J. C. (2003) Rhythmic Binding of a WHITE COLLAR Containing Complex to the *frequency* Promoter is Inhibited by FREQUENCY. *Proc Nat Acad Sci USA, 100,* 5914-5919.

Galagan, J., Calvo, S., Borkovich, K., Selker, E., Read, N., FitzHugh, W., *et al.* (2003) The Genome Sequence of the Filamentous Fungus *Neurospora crassa. Nature, 422,* 859-868.

Gomelsky, M. and Klug, G. (2002) BLUF: a novel FAD-binding domain involved in sensory transduction in microorganisms. *Trends Biochem Sci., 27,* 497-500.

Hall, M. D., Bennett, S. N., Krissinger, W. A. (1993) Characterization of a newly isolated pigmentation mutant of *Neurospora crassa. Georgia . Sci, 51,* 27 (Abstr.)

Harding, R. and Melles, S. (1983) Genetic analysis of phototropism of Neurospora crassa perithecial beaks using white collar and albino mutants. *Plant Physiol, 72,* 745-749.

Harding, R. W. and Shropshire, W. J. (1980) Photocontrol of carotenoid biosynthesis. *Ann Rev Plant Physiol, 31,* 217-238.

He, Q., Cheng, P., Yang, Y., Wang, L., Gardner, K., Liu, Y. (2002) WHITE COLLAR-1, a DNA binding transcription factor and a light sensor. *Science, 297,* 840-842.

Heintzen, C., Loros, J. J., Dunlap, J. C. (2001) VIVID, Gating and the Circadian Clock: the PAS protein VVD defines a feedback loop that represses light input pathways and regulates clock resetting. *Cell, 104,* 453-464.

Hochberg, M. L. and Sargent, M. L. (1974) Rhythms of enzyme activity associated with circadian conidiation in *Neurospora crassa. J Bacteriol, 120,* 1164-1175.

Imaizumi, T., Tran, H., Swartz, T. E., Briggs, W. R., Kay, S. A. (2003) FKF1 is essential for photoperiodic-specific light signalling in Arabidopsis. *Nature, 426,* 302-306.

Iseki, M., Matsunaga, S., Murakami, A., Ohno, K., Shiga, K., Yoshida, K., *et al.* (2002) A blue-light-activated adenylyl cyclase mediates photoavoidance in Euglena gracilis. *Nature, 415,* 1047-1050.

King, D., Zhao, Y., Sangoram, A., Wilsbacher, L., Tanaka, M., Antoch, M., *et al.* (1997) Positional cloning of the mouse circadian *CLOCK* gene. *Cell, 89,* 641-653.

Klemm, E. and Ninneman, H. (1978) Correlation between absorbance changes and a physiological response induced by blue light in *Neurospora crassa. Photochem Photobiol, 28,* 227-230.

Lakin-Thomas, P. L. and Brody, S. (2000) Circadian rhythms in *Neurospora crassa. Proc Nat Acad Sci USA, 97,* 256-261.

Lauter, F.-R. (1996) Molecular Genetics of Fungal Photobiology. *J Genet, 75,* 375-386.

Lee, K., Dunlap, J. C., Loros, J. J. (2003) Roles for WHITE COLLAR-1 in Circadian and General Photoperception in *Neurospora crassa. Genetics, 163,* 103-114.

Lee, K., Loros, J. J., Dunlap, J. C. (2000) Interconnected Feedback Loops in the Neurospora Circadian System. *Science, 289,* 107-110.

Lewis, Z. A., Correa, A., Schwerdtfeger, C., Link, K. L., Xie, X., Gomer, R. H., *et al.* (2002) Overexpression of White Collar-1 (WC-1) activates circadian clockassociated genes, but is not sufficient to induce most light-regulated gene expression in Neurospora crassa. *Mol Microbiol, 45,* 917-931.

Linden, H., Ballario, P., Arpaia, G., Macino, G. (1999) Seeing the light: News in Neurospora blue light signal transduction. *Advances in Genetics, 41,* 35-54.

Linden, H., Ballario, P., Macino, G. (1997) Blue light regulation in *Neurospora crassa*. *Fungal Genet Biol, 22*, 141-150.

Linden, H. and Macino, G. (1997) White collar-2, a partner in blue-light signal transduction, controlling expression of light-regulated genes in *Neurospora crassa*. *EMBO Journal, 16*, 98-109.

Loros, J. J. and Dunlap, J. C. (2001) Genetic and molecular analysis of circadian rhythms in *Neurospora*. *Annu Rev Physiol, 63*, 757-794.

Martens, C. L. and Sargent, M. L. (1974) Conidiation rhythms of nucleic acid metabolism in *Neurospora crassa*. *J Bacteriol, 117*, 1210-1215.

Merrow, M., Franchi, L., Dragovic, Z., Gorl, M., Johnson, J., Brunner, M., et al. (2001) Circadian regulation of the light input pathway in *Neurospora crassa*. *EMBO J, 20*, 307-315.

Nelson, D. C., Lasswell, J., Rogg, I. E., Cohen, M. A., Bartel, B. (2000) FKF1, a clock-controlled gene that regulates the transition to flowering in Arabidopsis. *Cell, 101*, 331-340.

Ninneman, H. (1979) Photoreceptors for Circadian Rhythms. *Photochem Photobiol Rev, 4*, 207-265.

Ninneman, H. (1991) Photostimulation of conidiation in mutants of *Neurospora crassa*. *Photochem Photobiol, 9*, 189-199.

Ntefidou, M., Iseki, M., Watanabe, M., Lebert, M., Hader, D. P. (2003) Photoactivated adenylyl cyclase controls phototaxis in the flagellate *Euglena gracilis*. *Plant Physiol, 133*, 1517-1521.

Paietta, J. and Sargent, M. (1981) Photoreception in *Neurospora crassa*: correlation of reduced light sensitivity with flavin deficiency. *Proc Nat Acad Sci USA, 78*, 5573-5577.

Paietta, J. and Sargent, M. (1983) Modification of blue light photoresponses by riboflavin analogs in *Neurospora crassa*. *Plant Physiol, 72*, 764-766.

Paietta, J. and Sargent, M. L. (1983) Isolation and characterization of light-insensitive mutants of *Neurospora crassa*. *Genetics, 104*, 11-20.

Payen, A. (1843) Extrait d'un report adressé àM. Le MaréchalDuc de Dalmatie, ministre de la Guerre, Président do Conseil, sur une altérationextraordinaire du pain de munition. *Ann Chim Phys Troisieme Ser, 9*, 5-21.

Perkins, D. D. (1988) Photoinduced carotenoid synthesis in perithecial wall tissue of *Neurospora crassa*. *Fung Genet Newsl, 35*, 38-39.

Perkins, D. D. (1992) *Neurospora*: the organism behind the molecular revolution. *Genetics, 130*, 687-701.

Potapova, T., Levina, N., Belozerskaya, T., Kritsky, M., Chailakhian, L. (1984) Investigation of electrophysiological responses of *Neurospora crassa* to blue light. *Arch Microb, 137*, 262-265.

Ramsdale, M. and Lakin-Thomas, P. L. (2000) sn-1,2-Diacylglycerol levels in the fungus *Neurospora crassa* display circadian rhythmicity. *J Biol Chem., 275*, 27541-27550.

Rensing, L., Bos, A., Kroeger, J., Cornelius, G. (1987) Possible link between circadian rhythm and heat shock response in *Neurospora crassa*. *Chronobiol Int, 4*, 543- 549.

Roeder, P. E., Sargent, M. L., Brody, S. (1982) Circadian rhythms in *Neurospora crassa*: oscillations in fatty acids. *Biochemistry, 21*, 4909-4916.

Russo, V. (1986) Are carotenoids the blue light photoreceptor in the photoinduction of protoperithecia in *Neurospora crassa*? *Planta, 168*, 56-60.

Sargent, M. (1985) *Neurospora* in the classroom. *Neurospora Newsletter, 2*, 12-13.

Sargent, M. L. and Briggs, W. R. (1967) The effect of light on a circadian rhythm of conidiation in *Neurospora*. *Plant Physiol, 42*, 1504-1510.

Sargent, M. L., Briggs, W. R., Woodward, D. O. (1966) The circadian nature of a rhythm expressed by an invertaseless strain of *Neurospora crassa*. *Plant Physiol, 41*, 1343-1349.

Schultz, T. F., Kiyosue, T., Yanofsky, M., Wada, M., Kay, S., A. (2001) A role for LKP2 in the circadian clock of Arabidopsis. *Plant Cell, 13*, 2659-2670.

Schwerdtfeger, C. and Linden, H. (2000) Localization and light-dependent phosphorylation of White Collar-1 and 2, the two central components of blue light signalling in *Neurospora crassa*. *Eur J Biochem, 267*, 414-422.

Schwerdtfeger, C. and Linden, H. (2001) Blue light adaptation and desensitization of light signal transduction in *Neurospora crassa*. *Molecular Microbiology, 39*, 1080-1086.

Schwerdtfeger, C. and Linden, H. (2003) VIVID is a flavoprotein and serves as a fungal blue light photoreceptor for photoadaptation. *EMBO Journal, 22*, 4846-4855.

Schwerdtfeger, C., Loros, J. J., Dunlap, J. C., Linden, H. (2003) *VIVID is a flavoprotein and serves as a fungal blue light photoreceptor for photoadaptation.* Paper presented at the Fungal Genetics, Asilomar, CA.

Selker, E. U. (1997) Epigenetic phenomena in filamentous fungi. *Trends genet., 13*, 296-301.

Shigeyoshi, Y., Taguchi, K., Yamamoto, S., Takeida, S., Yan, L., Tei, H., *et al.* (1997) Light-induced resetting of a mammalian circadian clock is associated with rapid induction of the *mPer1* transcript. *Cell, 91*, 1043-1053.

Shrode, L. B., Lewis, Z. A., White, L. D., Bell-Pedersen, D., Ebbole, D. J. (2001) *vvd* Is Required for Light Adaptation Conidiation-Specific Genes of *Neurospora crassa*, but Not Circadian Conidiation. *Fungal Genet Biol, 32*, 169-181.

Siegel, R. W., Matsuyama, S., Urey, J. (1968) Induced macroconidia formation in *Neurospora crassa*. *Experiencia, 24*, 1179-1181.

Somers, D. E., Schultz, T. F., Milnamow, M., Kay, S. A. (2000) *ZEITLUPE* encodes a novel clock-associated PAS protein from Arabidopsis. *Cell, 101*, 319-329.

Talora, C., Franchi, L., Linden, H., Ballario, P., Macino, G. (1999) Role of a *white collar-1-white collar-2* complex in blue-light signal transduction. *EMBO Journal, 18*, 4961-4968.

Taylor, B. L. and Zhulin, I. B. (1999) Pas domains: Internal sensors of oxygen, redox potential, and light. *Micro and Molec Biol Rev, 63*, 479-506.

Tokugawa, Y. and Emoto, Y. (1924) Über einen, kurz nach der letzten Feuerbrunst plötzlich entwickelnden Schimmelpilz. *Japan J Bot, 2*, 175-188.

Went, F. A. F. C. (1904) Über den Einfluss des Lichtes auf Entstehung des Carotins und auf die Zersetzung der Enzyme. *Recueil des travaux botaniques neerlandais, 1*, 106-119.

Zhu, H., Nowrousian, M., Kupfer, D., Colot, H.Berrocal-Tito, G., Lai, H., *et al.* (2001) Analysis of expressed sequence tags from two starvation, time-of-day-specific libraries of *Neurospora crassa* reveals novel clock-controlled genes. *Genetics, 157*, 1057-1065.

Chapter 14

UV-B PERCEPTION AND SIGNALLING IN HIGHER PLANTS

Roman Ulm
Institute of Biology II/ Botany, University of Freiburg, Schänzlestrasse 1, 79104 Freiburg, Germany (e-mail: Roman.Ulm@biologie.uni-freiburg.de)

1. INTRODUCTION

Ultraviolet (UV) radiation is an intrinsic part of the sunlight reaching the surface of the earth. This region of the electromagnetic spectrum is by convention divided into three classes: UV-A (320-400 nm), UV-B (280-320 nm) and UV-C (<280 nm). Solar UV radiation reaching the earth is composed only of UV-A and part of UV-B, since penetration of the atmospheric ozone layer drops dramatically for wavelengths below 320 nm and declines to an undetectable level below 290 nm (Figure 1). Surface UV-B levels are determined by a number of factors, making it a highly dynamic environmental component (McKenzie *et al.*, 2003; Paul and Gwynn-Jones, 2003). In general, less than 0.5% of the sun's energy at the earth's surface consists of biologically active UV-B radiation (Blumthaler, 1993). Although only small quantities of UV-B are involved, they have significant biological effects. And indeed, all living organisms sense and respond to UV radiation as an environmental cue. It is also UV-B in the sunlight that is notorious for being a ubiquitous and potent environmental carcinogen affecting human skin cells.

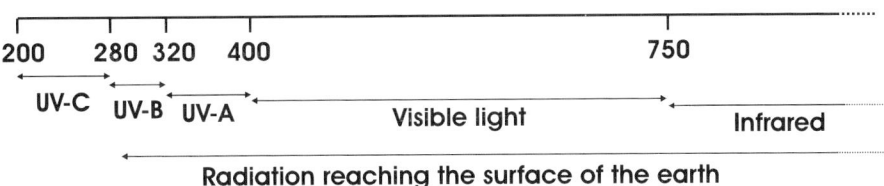

Figure 1. The solar radiation spectrum: conventional division of radiation into UV-C (200-280 nm), UV-B (280-320 nm), UV-A (320-400 nm), visible light (400-750 nm), and infrared (>750 nm). Radiation that passes the atmosphere of the earth includes ultraviolet wavelength regions from approx. 290 nm.

The effects of UV-B radiation in plants can be broadly divided into two classes, namely damage (stress) and regulation (morphogenic, non-damage). In the former case, UV-B may cause widespread cellular damage and oxidative stress, consequently activating rather general stress signal transduction pathways. In contrast, in the latter case UV-B is perceived by a specific UV-B photoreceptor yet unknown at the molecular level and transduced via distinct signalling pathways. Separation of UV-B effects into these two intrinsically different "action mechanisms" is in most cases not clear-cut as yet.

Particularly well established is the mutagenic effect on DNA and other damages conveyed to various cellular components (e.g. Jordan, 1996; Jansen et al., 1998; Britt, 2004). In general, the damaging effects are due to the high energy per photon of short wavelength UV radiation and the wide range of biologically active molecules that absorb it, including nucleic acids, aromatic amino acids and lipids. In any case, plants must withstand the harmful effects of UV radiation owing to the fact that they are reliant on sunlight for photosynthesis and cannot evade UV exposure by simply walking out of the sun. Thus, a number of studies were centred on UV-B tolerance mechanisms, as plants are directly affected by alterations in terrestrial UV-B fluence, such as those evoked by the human-made depletion of the stratospheric ozone layer. Plants defend themselves against the detrimental consequences of UV-B radiation with a variety of repair and acclimation responses. In general, UV-acclimated plants are endowed with an elevated capacity for DNA repair, mount enzymatic and non-enzymatic scavengers of reactive oxygen species (ROS), accumulate high levels of phenolic "sunscreen" pigments and adjust their growth pattern. The efficiency of these UV-B acclimation responses might be reflected in the subtle effects of elevated UV-B levels on biomass accumulation under field conditions (e.g. Searles et al., 2001). Correspondingly, the genetic interference with key acclimation processes, such as DNA repair, antioxidants or "sunscreen" accumulation results in hypersensitivity to UV radiation (Li et al., 1993; Landry et al., 1995; Conklin et al., 1996; Britt and Fiscus, 2003). On the other hand, *Arabidopsis* mutants with enhanced DNA repair or phenolics content possess higher tolerance to threatening UV-B levels (Bieza and Lois, 2001; Tanaka et al., 2002).

UV-B radiation is not only a source of stress for plants and other organisms, but is also providing critical information about the prevailing environment (Paul and Gwynn-Jones, 2003). As sessile organisms, plants in particular have evolved numerous developmental responses to deal with diverse aspects of the ever-changing surroundings. The light environment is a key factor and governs a multitude of developmental processes during their entire life cycle. To this end, plants have evolved potent perception and signalling systems to assess and interpret the whole complexity of light, including quality, quantity, direction and duration. To gather such informational signals from sunlight, plant photosensory systems employ at least three classes of photoreceptors that discriminate between different wavelengths, sensing the red/far-red (phytochromes, see Chapter 7), the blue/UV-A (cryptochromes and phototropins, see Chapters 11 and 12, respectively), and the UV-B region of the spectrum. An important aspect of UV-B is that its level reaching an organism does not necessarily correlate with wavelength in the visible range (approximately 400-700 nm, equivalent to photosynthetically active radiation) (Caldwell and Flint, 1994; Rozema et al., 1997), suggesting that its direct perception might be beneficial for proper responses. However, in contrast to considerable knowledge on visible light-sensory systems, the UV-B photoreceptor(s) remains elusive as not yet identified at the molecular level.

Even though it is difficult at present to differentiate between damage- and non-damage-mediated responses in particular cases, this chapter intends to focus on recent advances in the determination of molecular mechanisms of the regulatory (i.e. non-damage) UV-B responses, particularly through recent work conducted with the model plant *Arabidopsis thaliana*. DNA damage and repair will be discussed as a select example of UV-B-mediated damage; an important aspect with particular recent success in genetic screens for UV-B tolerance in *Arabidopsis*. It should already be noted here that the distinction between damage- and non-damage-mediated effects results in a fair amount of speculation that may only be resolved once the genuine UV-B receptor is defined. Ecological and agronomical aspects of the impact of UV-B on plants, particularly in the frame of ozone depletion research, were recently discussed, and the reader is referred to the recent literature (Tevini, 1999; Searles *et al.*, 2001; Caldwell *et al.*, 2003, and references therein). Similarly, the impact of UV-B on the photosynthetic machinery is out of scope of this chapter, and was extensively reviewed before (e.g. Vass, 1997).

2. DNA DAMAGE AND REPAIR

UV radiation can inflict a variety of damage on living organisms and it is well established as a genotoxic agent in the sunlight. DNA is considered a major cellular target for UV radiation, with peak absorption at around 260 nm determined by its component nucleotides. UV radiation induces oxidative damage (pyrimidine hydrates), DNA-protein and DNA-DNA crosslinks and most prevalently various pyrimidine dimers, in particular cyclobutane pyrimidine dimers (CPD) that constitute about 75% of UV-induced DNA lesions and pyrimidine-pyrimidinone dimers (6-4 photoproduct, 6-4PP) that make up the majority of the remainder (Britt, 2004). The deleterious effects of these photoproducts include inhibition of replication and transcription, mutation, growth arrest and cell death. In addition to the effects on survival, UV-B-mediated DNA damage seems to signal the induction of a PR protein (β-1,3-glucanase) in bean (Kucera *et al.*, 2003) and isoflavonoid synthesis in leguminous plants (Beggs *et al.*, 1985). These conclusions are drawn from the action spectra that revealed a maximal effectiveness at 260 nm and the possibility to abrogate the effect by simultaneous or subsequent irradiation with UV-A or blue light, which leads to photoreactivation as described below (Beggs *et al.*, 1985; Kucera *et al.*, 2003).

In order to prevent mutation and/or cell death, UV-induced DNA damage must be repaired before DNA replication. An alternative, in case of incomplete repair, is to tolerate unrepaired DNA lesions by a damage-tolerance pathway. This includes a process called translesion synthesis or dimer bypass, which involves polymerases that are more tolerant to aberrant template structures (Britt, 2004). Replication despite unrelieved DNA damage is responsible for the mutagenic effects of many genotoxic agents. Recent evidence with the *rev3* mutant impaired in the catalytic subunit of DNA polymerase ζ leading to DNA damage hypersensitivity provides initial genetic support for its existence in higher plants as well (Sakamoto et al., 2003). On the other hand, three repair mechanisms are employed to cope with UV-induced DNA damage: photoreactivation, excision repair or recombination. Pyrimidine dimers can be repaired by all three mechanisms, with photoreactivation as the major pathway in light, while the other UV lesions can only be repaired by excision or recombinational repair. In *Arabidopsis*, diverse genetic screens have identified components of different DNA damage repair pathways, loss of which leads to hypersensitivity to UV-B and other genotoxic stresses (Table 1, and references therein). These mutants are briefly introduced below according to the affected repair pathway.

Photoreactivation is mediated by photolyases, which harness blue/UV-A light energy to break bonds by a cyclic electron transfer mechanism, restoring the integrity of DNA. Photolyases are 55-70 kD monomeric proteins that contain two non-covalently bound prosthetic groups, flavin adenin dinucleotide (FAD) and a pterin (mostly methenyl tetrahydrofolate, MTHF). The FAD cofactor is essential both for binding to damaged DNA and catalysis. The second chromophore, a pterin, acts as a photoantenna and transfers excitation energy to the catalytic $FADH^-$ cofactor, thereby increasing the rate of repair (Sancar, 2003). Interestingly, cryptochromes and photolyases display a high degree of sequence identity and structural conservation, suggesting a similar reaction mechanism (see Chapter 11). Two photolyases are known in *Arabidopsis*, namely the CPD-specific PHR1/UVR2 and the 6-4PP-specific UVR3 (Ahmad et al., 1997; Nakajima et al., 1998; Waterworth et al., 2002). Both white light and UV-B irradiation induce PHR1/UVR2 protein abundance in *Arabidopsis* tissues, whereas the level of the UVR3 protein seems to be regulated neither by white light nor by UV-B (Waterworth et al., 2002). As expected, the photolyase mutants *uvr2* and *uvr3* in *Arabidopsis* and Norin1 in rice are highly sensitive to UV-B (Ahmad et al., 1997; Nakajima et al., 1998; Hidema et al., 2000).

Excision repair involves recognition of DNA lesions, nicking of the damaged DNA close to the damage, removal of the damaged fragment, and resynthesis with the intact strand as a blueprint to fill in the gap. Certain yeast *rad* mutants and the seven complementation groups of the heritable human disease xeroderma pigmentosum (*XPA-XPG*) are deficient in components of the nucleotide excision repair machinery. This deficiency leads to an extreme UV sensitivity and skin cancer predisposition in XP patients (e.g. Friedberg, 2001). A number of *Arabidopsis* mutants with UV-B hypersensitivity are impaired in homologues of these human and yeast proteins: UVH1 (human XPF/yeast RAD1 homologue, Fidantsef et al., 2000; Liu et al., 2000), UVH3/UVR1 (XPG/RAD2 homologue, Liu et al., 2001), UVH6 (XPD/RAD3 homologue, Liu et al., 2003), and UVR7 (ERCC1/RAD10 homologue, Hefner et al., 2003). Thus, the nucleotide excision repair pathway appears to be functionally conserved in plants and to play a crucial role in coping with DNA damage invoked by UV-B radiation.

Table 1. A non-exhaustive list of genetically defined components involved in UV-B responses in Arabidopsis

Mutant	Affected gene and homologues	Predicted or known function[a]	Ref.[b]
"Sunscreen"			
tt4	*CHS*, chalcone synthase	phenylpropanoid metabolism	1
tt5	*CHI*, chalcone isomerase	phenylpropanoid metabolism	2
fah1	ferulate hydroxylase	phenylpropanoid metabolism	3
DNA repair			
rev1	DNA polymerase ζ	Damage-tolerance, TLS	4
uvh1	XPF/RAD1 homologue	5' DNA-structure-specific endonuclease with ERCC1/RAD10, NER	5
uvh3/uvr1	XPG/RAD2 homologue	3' DNA-structure-specific endonuclease, NER	6
uvh6	XPD/RAD3 homologue	5'→3' DNA helicase, NER	7
uvr2	*PHR1*, CPD photolyase	PHR of CPD photoproducts	8
uvr3	6-4 photolyase	PHR of 6-4 photoproducts	9
uvr7	ERCC1/RAD10 homologue	5' DNA-structure-specific endonuclease with XPF/RAD1, NER	10
Oxidative stress			
vtc1	GDP-mannose pyrophosphorylase	Ascorbic acid synthesis, antioxidant	11
ssadh-1	succinic-semialdehyde dehydrogenase	γ-aminobutyrate shunt, mitochondrial	12
Signalling			
atr	Ser/Thr-protein kinase (PI3K-like)	G2-phase cell-cycle checkpoint regulation	13
uli3	put. heme and DAG binding sites	UV-B signalling intermediate	14
hy5	bZIP transcription factor	Activator of UV-B-responsive genes	15
phyB	phytochrome B	Modulator of UV-B responses	16
etr1	ethylene receptor	Perception of UV-B-responsive ethylene	17
jar1	JA-amino synthetase	Activation of JA	18
uvr8	RCC1-like	Positive regulator of phenylpropanoid synthesis	19
myb4	MYB transcription factor	Negative regulator of sinapate ester synthesis	20

[a]Abbr.: NER, nucleotide excision repair; PHR, photoreactivation; TLS, translesion synthesis
[b]References: 1 (Li et al., 1993), 2 (Li et al., 1993; Landry et al., 1995), 3 (Landry et al., 1995), 4 (Sakamoto et al., 2003), 5 (Fidantsef et al., 2000; Liu et al., 2000), 6 (Liu et al., 2001), 7 (Liu et al., 2003), 8 (Ahmad et al., 1997), 9 (Nakajima et al., 1998), 10 (Hefner et al., 2003), 11 (Conklin et al., 1996), 12 (Bouche et al., 2003), 13 (Culligan et al., 2004), 14 (Suesslin and Frohnmeyer, 2003), 15 (Ulm et al., 2004), 16 (Boccalandro et al., 2001; Jenkins et al., 2001), 17 (A-H-Mackerness, 2000), 18 (A-H-Mackerness, 2000), 19 (Kliebenstein et al., 2002), 20 (Jin et al., 2000).

Recombinational repair was found to be increased by elevated UV-B doses as measured with a recombination reporter gene in *Arabidopsis* and tobacco. A clue to the potential signal for this effect is indicated by the higher induction of recombination by UV-B in the CPD-photolyase mutant *uvr2* (Ries *et al.*, 2000a,b). This indicates that homologous recombination might be employed to purge UV-B-induced DNA lesions; however, its role in UV-B tolerance remains to be established. Nonetheless, the data suggest that elevated UV-B levels will impinge on genomic stability through the activation of recombination in plants. Similarly, elevated UV-B exposure of maize pollen can activate immobile *Mutator* transposons, thereby increasing its impact on genome integrity (Walbot, 1999). Generation of genomic variability through the activation of transposons and recombination may represent an adaptive mechanism during "genomic shock" (McClintock, 1984).

3. PHOTOMORPHOGENIC RESPONSES TO UV-B

UV-B radiation evokes diverse phenotypic responses in higher plants, amongst them hypocotyl growth inhibition, cotyledon expansion, phototropic growth, changes in stomata aperture, leaf and tendril curling and the induction of UV-protecting pigmentation (e.g. Li *et al.*, 1993; Wilson and Greenberg, 1993; Beggs and Wellmann, 1994; Ballare *et al.*, 1995b; Brosche and Strid, 2000; Mazza *et al.*, 2000; Eisinger *et al.*, 2003; Shinkle *et al.*, 2004). Several of these phenotypes were proposed to be mediated by damage-independent perception pathways, based mostly on the low threshold of the response or the action spectra that identifies an optimum wavelength response that does not correlate with the optimum for DNA damage. Thus at present, as no UV-B photoreceptor is known at the molecular level, UV-B photobiology depends on defining UV-B treatments that induce specific responses, separable from those resulting from damage. Indeed, several morphogenic responses are induced by relatively low UV-B levels, which do not hamper general plant performance.

Progress in *Arabidopsis* genetics has produced powerful tools to analyse complex physiological processes, including photomorphogenesis (see Chapter 3). However, only a relatively small part of the work on UV-B-induced photomorphogenesis has been pursued in *Arabidopsis*, and in particular, genetic approaches that were very successful in the identification of visible light photoreceptors and downstream components have not yet been widely applied. In addition to the paucity of well-defined visible phenotypes, a further complication is the confoundingly damaging effect of UV-B radiation. Regarding this latter aspect, a number of genetic screens for altered UV tolerance were carried out using genotoxic levels of UV-B, as also reflected by the identification of several mutants with defects in DNA repair components (Table 1). Nonetheless, a limited number of UV-B photomorphogenic phenotypes were identified in *Arabidopsis* that were or might be used in forward genetic screens in this model plant (Figure 2).

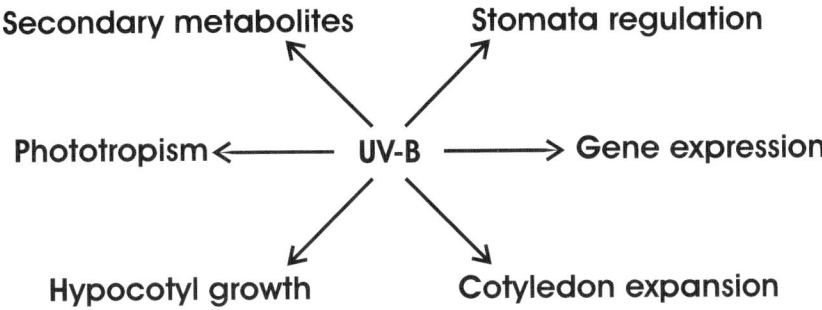

Figure 2. Photomorphogenic effects of UV-B on plants that are postulated to be photoreceptor-mediated (see references in the text).

3.1 Synthesis of "sunscreen" metabolites

Plants defend themselves against the UV-B component of incident sunlight by activating the general phenylpropanoid pathway, thereby accumulating flavonoids and sinapic esters. These "sunscreen" compounds are mostly accumulated in the epidermal cell layer and protect the underlying cells through their absorption properties in the UV-B wavelength range. In contrast, while fending off the UV-B component, these pigments allow photosynthetically active radiation to penetrate and support photosynthesis in the mesophyll and palisade tissues. It should be noted here that flavonoid accumulation is a characteristic feature of plant stress in general and, in addition to screening out harmful radiation from the sunlight, it affects numerous additional physiological functions: flavonoids may also act as antioxidants, antimicrobials, and regulators of auxin transport (e.g. Winkel-Shirley, 2002).

Initial experiments on UV-B-induced flavonoid formation in parsley cell suspension cultures and seedlings revealed an action spectrum with maximal effectiveness at about 300 nm. Moreover, the effect followed the rule of reciprocity and showed a linear fluence-response relationship (Beggs and Wellmann, 1994, and references therein). This was confirmed in several other plant systems and represents a rather extensively studied example of a UV-B photomorphogenic effect (Beggs and Wellmann, 1994; Kucera *et al.*, 2003), establishing this pathway as a UV-B-specific response mediated by the postulated UV-B photoreceptor (Beggs and Wellmann, 1994; Björn, 1999).

Likewise, phenylpropanoids are elevated in response to UV-B in *Arabidopsis* (Lois, 1994; Bharti and Khurana, 1997; Suesslin and Frohnmeyer, 2003). The induction of secondary metabolites acting as "sunscreen" pigments is physiologically significant, as failure or higher proficiency to synthesize UV-protective phenylpropanoid pigments results in UV hyper- or hyposensitivity, respectively (Li *et al.*, 1993; Landry *et al.*, 1995; Jin *et al.*, 2000; Bieza and Lois, 2001; Kliebenstein *et al.*, 2002).

The regulation of the key flavonoid biosynthesis enzyme chalcone synthase (CHS), encoded by a single gene in *Arabidopsis*, provides a well-established model system to analyse the interplay of UV-B irradiation with other parts of the light spectrum (Jenkins *et al.*, 2001). The transcriptional regulation of *CHS* relies on interactions within a network of phytochrome-, cryptochrome- and UV-B-signalling pathways. CHS gene expression is controlled by distinct UV-A/blue and UV-B photoperception systems, the former being cry1-dependent whereas the latter is independent of known photoreceptors. However, it was found that distinct UV-A and blue light pathways interact synergistically with the UV-B pathway to enhance *CHS* expression. These two synergistic interactions occur via a transient UV-A signal compared to a sustained blue light signal; interestingly in both cases independent of cry1 and cry2 (Fuglevand *et al.*, 1996; Wade *et al.*, 2001). Thus it appears that presently unknown UV-A and blue light photoreceptors mediate these synergistic interactions with the UV-B pathway. In addition, phyB is a negative regulator of the UV-B inductive pathway for *CHS* and *CHI* (Jenkins *et al.*, 2001; Wade *et al.*, 2001; Brosche and Strid, 2003); however, the generality of these interactions remains to be demonstrated.

A genetically defined component of the UV-B-induced phenylpropanoid pathway is ULI3, a mutation causing decreased UV-B-induced anthocyanin accumulation and concomitantly reduced *CHS* gene activation (Suesslin and Frohnmeyer, 2003, see Sect. 3.2 for further discussion). A more general and ubiquitously acting negative regulator of flavonoid biosynthesis is defined by the *icx1* mutant, which is also suggested to be impaired in the UV-B-inductive pathway (Wade *et al.*, 2003). However, the responsible mutation has not been identified, precluding insight into ICX1 function at the molecular level at present. Two other mutants, the UV-B hypersensitive *uvr8* and the hyposensitive *myb4* corroborate the importance of UV-B stress perception signalling to increase "sunscreen" components. It seems that UVR8 functions as a positive regulator of phenylpropanoid metabolism in response to UV-B, whereas the R2R3 MYB transcription factor MYB4 plays a negative regulatory role (Jin *et al.*, 2000; Kliebenstein *et al.*, 2002). MYB4 functions as a repressor of its target gene *C4H* encoding cinnamate 4-hydroxylase under standard growth conditions. However, UV-B-mediated down-regulation of *MYB4* expression leads to derepression of *C4H* and consequently synthesis of sinapate esters, an important UV-B protecting sunscreen (Jin *et al.*, 2000). Consistently, the *myb4* mutant has enhanced UV-B tolerance attributable to elevated production of protecting sinapate esters, whereas *MYB4*-overexpression results in UV-B sensitivity (Jin *et al.*, 2000). In contrast, a mutation in the *UVR8* gene reduces UV-B-activated *CHS* gene expression and, as a result, flavonoid accumulation (Kliebenstein *et al.*, 2002). The *uvr8* mutant is hypersensitive to UV-B stress and shows enhanced levels of pathogenesis-related PR1 and PR5 proteins. The *UVR8* gene encodes a protein with similarity to the human regulator of chromatin condensation 1 (RCC1) (Kliebenstein *et al.*, 2002), a nuclear guanine nucleotide exchange factor for the small GTPase Ran, which is implicated in diverse fundamental cellular processes, including nucleocytoplasmic transport, mitotic spindle formation, cell cycle progression and nuclear envelope assembly (Dasso, 2002). Ran GTPases and interacting proteins exist in *Arabidopsis*, indicating conservation of the Ran regulatory mechanism (Haizel *et al.*, 1997). It will be of interest to identify the mechanistic role of UVR8 in the UV-B stress response; the significance of the similarity between UVR8 and RCC1 remains to be established.

3.2 Inhibition of hypocotyl growth

Inhibition of plant stem extension by low-level UV-B radiation can be viewed as a photomorphogenic response rather than a sign of damage. For example, the maximum effectiveness in inhibition is around 300 nm for stem elongation in etiolated tomato (Ballare *et al.*, 1995a). This change in growth is a rather fast process and precedes the onset of UV-protective pigments by at least 3h (Ballare *et al.*, 1995b). In cucumber, the inhibition of hypocotyl elongation was reversible through gibberellic acid application, it took place without concomitant change in dry matter production and was due to UV-B perception in the cotyledons, all in all excluding direct damage to hypocotyls as a cause. Moreover, it was independent of stable phytochrome (Ballare *et al.*, 1991).

In *Arabidopsis*, similar to wavelength ranges in the visible light spectrum, UV-B induces hypocotyl growth inhibition (Kim *et al.*, 1998; Boccalandro *et al.*, 2001; Suesslin and Frohnmeyer, 2003). The two facets of UV-B, damage and non-damage, are apparent in the biphasic hypocotyl growth inhibition revealed in fluence response curves, where the low fluence response can be assigned as photomorphogenic, whereas the high fluence response most likely results from the stress of DNA damage (Kim *et al.*, 1998; Boccalandro *et al.*, 2001). The latter is supported by the stronger hypocotyl growth inhibition at greater UV-B fluence rates in the compound DNA repair mutants *uvr1uvr2* and *uvr1uvr3* as compared to wild type (Boccalandro *et al.*, 2001). Thus, the same phenotype seems to result from both a UV-B stress and a UV-B photomorphogenic signal. Therefore, the irradiation conditions were adjusted to a level where available repair mutants do not show an increased sensitivity (Kim *et al.*, 1998; Suesslin and Frohnmeyer, 2003). Using these conditions, the UV-B-mediated growth inhibition seems to be impaired in the constitutive photomorphogenic mutants *cop1* and *det1*, but not in *det2* (Kim *et al.*, 1998). The *DET1* and *COP1* genes encode repressors of light signalling, whereas *DET2* is required for the biosynthesis of the steroid hormone brassinolide (Fujioka *et al.*, 1997). Interestingly, DET1 binds nucleosome core particles and exists in a complex containing a homologue of the UV-damaged DNA binding protein (DDB1) that is part of histone acetyltransferase complexes in animals (Benvenuto *et al.*, 2002; Schroeder *et al.*, 2002). COP1, on the other hand, is an E3 ubiquitin ligase targeting a subset of positive regulators of photomorphogenesis for degradation (Saijo *et al.*, 2003; Seo *et al.*, 2003, see also Chapter 18). These data suggest a possible involvement of chromatin remodelling and regulated protein degradation in UV-B responses. However, it remains to be established if COP1 and DET1 are directly involved in UV-B signalling and, if they are, exactly what role they play.

This is of particular importance, because under the conditions employed, involvement of phytochrome function in establishing the UV-B phenotype was suggested (Kim *et al.*, 1998).

It is known that phytochromes do absorb UV-B wavelengths due to their protein moiety, which can cause photoconversion (Pratt and Butler, 1970). Thus, the UV-B-mediated inhibition of hypocotyl growth was analysed in photoreceptor-defective mutants and found to be independent of phyA, phyB, cry1, cry2 and phot1 photoreceptors (Suesslin and Frohnmeyer, 2003), suggesting dependence on the action of a dedicated UV-B photoreceptor. However, there is an apparent controversy regarding phyA and phyB involvement, as in similar experiments the *phyAphyB* double mutant exhibited longer hypocotyls than the wild-type seedlings under low fluences of UV-B (Kim *et al.*, 1998). The redundant action of phyA and phyB might indicate that it is due to their function as UV-B receptors or, more likely, that phytochrome might be required for co-action with a distinct UV-B receptor (Beggs and Wellmann, 1994; Mohr, 1994).

The hypocotyl phenotype of *Arabidopsis* was recently used to identify the first potentially specific UV-photomorphogenic mutants, designated *uli* mutants (*U*V-B *l*ight *i*nsensitive) (Suesslin and Frohnmeyer, 2003). It is of interest that the recovered mutants belonging to three complementation groups (*uli1* to *uli3*) have altered sensitivity specifically to UV-B, as their hypocotyl growth inhibition was similar to wild type under red, far-red, blue or UV-A light. The deduced ULI3 protein contains potential heme- and diacylglycerol-binding sites, and seems to be localized to the cytoplasm and adjacent to the plasma membrane (Suesslin and Frohnmeyer, 2003). The *ULI3* gene is preferentially expressed in the outer cell layers of above-ground tissues, and is itself transcriptionally induced by UV-B exposure. In agreement with its hyposensitive phenotype, the UV-B-mediated induction of select marker genes, *chalcone synthase* (*CHS*), *pathogenesis-related protein* (*PR-1*) and *nucleotide diphosphate kinase* (*NDPK1a*), was reduced in *uli3* mutants (Suesslin and Frohnmeyer, 2003). Thus, ULI3 defines a positive regulator of UV-photomorphogenic responses; however, its mechanism and position of action in the UV-B signalling network remain to be determined.

3.3 Cotyledon opening and expansion

Typically photomorphogenic cotyledon expansion is promoted by low-fluence rates of UV-B, and not by damaging high-fluence rates (Kim *et al.*, 1998). Similarly, in etiolated *Arabidopsis* seedlings, a daily 2.5 h exposure to low-fluence rate UV-B preceding a short red light pulse resulted in an enhanced cotyledon opening response (Boccalandro *et al.*, 2001). This positive phenotype was abrogated by higher fluence rates of UV-B, thus showed a bell-shaped fluence rate response curve. This negative effect was enhanced in DNA damage repair mutants *uvr1uvr2* and *uvr1uvr3*, indicating DNA lesions as an underlying cause (Boccalandro *et al.*, 2001). The UV-B-mediated enhancement of the cotyledon opening response to a subsequent red light pulse was dependent on phyB but independent of other known photoreceptors and DNA damage (Boccalandro *et al.*, 2001). Thus, UV-B perceived by a separate receptor system synergistically enhances the phytochrome B mediated opening response.

4. UV-B PERCEPTION

4.1 Supporting evidence and possible nature of a specific UV-B photoreceptor

UV-responses in plants could theoretically result from any number of primary UV-B events inflicted on DNA (Britt, 2004), the photosynthetic machinery (Vass, 1997), membranes (Murphy, 1983), aromatic residues in proteins (Kim *et al.*, 1992), the plant hormone auxin (Ros and Tevini, 1995), other small molecules like provitamin D (Björn and Wang, 2001) or *trans*-cinnamic acid (Braun and Tevini, 1993), phytochrome (which photoconverts in response to UV-B) (Pratt and Butler, 1970) or signal transduction via a dedicated UV-B photoreceptor as a protein-pigment complex (Figure 3). The link between such a possible primary event and a particular outcome is in most cases poorly established at present.

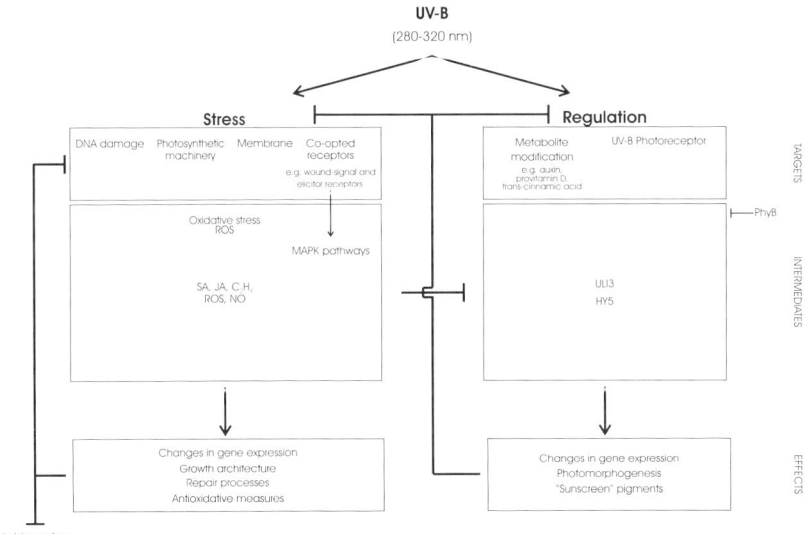

Figure 3. Speculative model of UV-B perception and signalling. UV-B effects can be divided into damaging and regulatory effects, the latter downstream of the postulated UV-B photoreceptor. Known and proposed components of UV-B perception and signal transduction are tentatively placed into these two modes of UV-action.

In regard of UV-B perception and its interaction with other light sensing mechanisms, the possibility of a specific UV-B photoreceptor is particularly intriguing. Information obtained through UV-B photoreceptor(s) may well supplement and extend that captured by the better-established photoreceptors for the visible light spectrum. A number of findings suggest that neither a known photoreceptor nor induced damage is the primary event for UV-B perception leading to photomorphogenic responses. Most work regarding the regulatory effects of UV-B radiation at the morphological level has addressed the induction of protective pigment synthesis and the inhibition of stem growth. Evidence that points to a specific UV-B perception mechanism for these developmental responses is described in the previous section. For instance, the spectral optimum for these phenotypes lies at about 300 nm in the UV-B wavelength region, and not in the germicidal UV-C (Hashimoto *et al.*, 1991; Beggs and Wellmann, 1994; Ballare *et al.*, 1995a). Another example is leaf curling that is induced by relatively low levels of UV-B but inhibited by increased UV levels (Wilson and Greenberg, 1993). Such fluence rate dependence is supportive of true photomorphogenic responses. At the molecular level, supportive evidence is derived from the UV-B-mediated transcriptional activation of *CHS* by the use of millisecond UV-B flashes in parsley cells, whereby the formation of DNA damage could be minimised (Frohnmeyer *et al.*, 1999). Moreover, the stimulation of *CHS* gene expression could be separated from pyrimidine dimer formation by virtue of the distinct wavelength sensitivities of the two processes (Frohnmeyer *et al.*, 1999). Similarly, in *Argenteum* pea plants levels of pyrimidine dimers did not correspond to transcript levels of *CHS* or *SadA* encoding a short-chain alcohol dehydrogenase (Kalbin *et al.*, 2001). It is also known that photoreactivating blue or UV-A light given together with UV-B hyperstimulates *CHS* gene expression (Wade *et al.*, 2001). A recent analysis of the genome-wide expression in the response of *Arabidopsis* to different UV-B wavelength ranges also allowed the separation of postulated UV-B photoreceptor output genes from potential stress-responsive genes, part of the former being dependent on the key photomorphogenic transcriptional activator HY5 (Ulm *et al.*, 2004, see also Sect. 6). Thus, all this evidence demonstrates a lack of correlation between DNA damage and gene expression resulting from UV-B irradiation.

Several findings also indicate that neither phytochromes nor cryptochromes function as UV-B photoreceptors. Phytochrome- and cryptochrome-deficient mutants are, for instance, proficient in activation of genes induced by low-level UV-B irradiation (Ulm *et al.*, 2004), among them *CHS* and *CHI* (Jenkins *et al.*, 2001; Brosche and Strid, 2003). Phytochrome involvement was also excluded in a number of cases by the non-reversibility of the UV-B effect by far-red light (Beggs and Wellmann, 1994; Björn, 1999, and references therein). Moreover, Ca^{2+}/calmodulin was found to play different roles in phytochrome- and UV-regulated *CHS* expression in soybean cell cultures (Frohnmeyer *et al.*, 1998). However, it is also evident that there exists an intricate interplay of the UV-B responses with blue and red light photoreceptors (Beggs and Wellmann, 1994; Mohr, 1994; Jenkins *et al.*, 2001; Brosche and Strid, 2003, see also Sect. 3.1).

Given that a bona fide UV-B photoreceptor should exist as a protein-pigment complex, possible chromophores that have been put forward include a tetrahydropterin derivative (Galland and Senger, 1988b; Björn, 1999) or a flavin (Galland and Senger, 1988a; Ensminger and Schäfer, 1992; Ballare *et al.*, 1995a).

The absorbance maxima of reduced flavins and pterins correspond to the action spectra of many UV-B responses with peaks at about 300 nm (Galland and Senger, 1988a,b). Moreover, feeding of parsley cells with riboflavin enhanced the UV-B-mediated increase of chalcone synthase and flavonoid levels (Ensminger and Schäfer, 1992). Tomato seedlings treated with compounds that perturb the normal (photo)chemistry of flavins and pterins were impaired in the UV-B-mediated inhibition of elongation (Ballare et al., 1995a). Clarification of this matter, however, has to await the identification of the UV-B receptor molecule.

It is noteworthy that in addition to the postulated specific UV-B photoreceptor perceiving low fluxes of UV-B, stress responses provoked by higher UV-B levels (or UV-C) might be mediated by activation of cell surface receptors in ligand-independent manner (Stratmann, 2003; Ulm, 2003). It is proposed that these UV-B/UV-C recognition events are similar to the situation in animals, where signal transduction cascades normally used for other cellular responses including growth factor signalling are exploited (Liu et al., 1998, and references therein). UV irradiation leads to clustering, phosphorylation and activation of several growth factor and cytokine receptor tyrosine kinases at the cell membrane. However, it is not exactly known how UV activates the cell surface receptors in mammalian cells. It has been suggested that UV irradiation might perturb the cell surface or alter receptor conformation leading to receptor multimerization, clustering and activation. Alternatively, the inhibition of a membrane-bound protein tyrosine phosphatase was proposed to prevent dephosphorylation of the receptor tyrosine kinases (Liu et al., 1998, and references therein). Whatever the precise mechanism of signal initiation, this signal transduction pathway involves MAP kinase cascades (Liu et al., 1998; Ulm, 2003, and references therein). Similarly to the mammalian system, UV exposure triggers MAP kinase pathways in tomato, tobacco and *Arabidopsis* (Stratmann et al., 2000; Miles et al., 2002; Ulm et al., 2002) and seems to co-opt cell surface receptors for the wound signalling peptide systemin and pathogen-related elicitors in tomato (Yalamanchili and Stratmann, 2002; Stratmann, 2003). This would provide a mechanistic explanation for the substantial overlap seen in wounding/herbivore and UV-B stress responses (Stratmann, 2003), and parallels the mammalian and yeast "non-DNA damage" pathways (Ulm, 2003). The hypersensitivity to UV-C in the MAP kinase phosphatase 1-deficient *mkp1* mutant indicates the essentiality of the MAP kinase pathway for genotoxic stress resistance (Ulm, 2003). It remains to be determined how these UV stress events initiated at the cell periphery relate to both DNA damage-related and UV-B photomorphogenic responses.

4.2 Possible importance of specific UV-B perception?

It is obviously valid and important to ask what advantage plants can gain by the perception of UV-B in the field. Even though the relevance of the UV-B response under natural conditions is evident, it remains a possibility that potential UV assault may be extrapolated from monitoring visible light. However, it is significant in this regard that the UV-B region of sunlight affecting plants is not tightly coupled with the remainder of the solar spectrum (Caldwell and Flint, 1994). This in turn implies that UV-B flux information may not always be anticipated from perception of other parts of the solar spectrum by phytochromes, cryptochromes or phototropins. Therefore, sensitive perception of UV-B radiation may result in a more accurate preparation before any damage occurs. Indeed, most UV-B responses point to such a prophylactic protection mechanism. Another interesting aspect is the difference in penetration of plant canopies between UV-B and photosynthetically active radiation (PAR). It is known that UV-B is scattered to a greater degree by the atmosphere than PAR. Due to this property, UV-B tends to predominate in shaded areas, resulting in a high UV-B:PAR ratio (Flint and Caldwell, 1998). Finally, it was noted that at least some characteristics of the shade-avoidance phenotype, including promotion of elongation and reduction of branching, can be induced by lowering the UV-B component (Ballare, 1999). Indeed, the UV-B-induced alterations in shoot morphology may lead to shifts in competitive balances in the field (Barnes *et al.*, 1996). However, the generality and adaptive significance of different morphological changes remains to be determined.

One recurrent effect of ambient UV-B observed in the field is its negative impact on the interaction of plants with herbivorous insects (e.g. Caldwell *et al.*, 2003). This reduction in plant tissue consumption by insects and other herbivores may be caused by UV-B-mediated changes in plant tissue chemistry or priming through the co-option of herbivore activated signalling pathways (Stratmann, 2003). Both would indicate that this UV-B effect might be an important "side product" of the plant's response to UV-B stress. In any case, these UV-B effects may cause significant changes in trophic interactions at the ecosystem level, particularly in case of further depletion of the ozone layer and the resulting elevated UV-B fluxes reaching the earth.

5. UV-B SIGNALLING

After the direct or indirect perception of UV-B by the plant, signal transduction pathways must be recruited to bring about responses such as gene activation or repression. At present, knowledge of the pathways underlying UV-B responses at the cellular as well as the organism level is only fragmentary. Signalling intermediates that have been linked to UV-B signalling in plants through mutants, transgenic plants and/or pharmacological reagents include ROS, Ca^{2+}/calmodulin, nitric oxide (NO), reversible protein phosphorylation and various plant hormones (reviewed, for example, in Brosche and Strid, 2003; Frohnmeyer and Staiger, 2003) (Figure 3).

5.1 Reactive oxygen species

In plant systems UV-B exposure leads to the generation of ROS that is partly held responsible for UV-B-mediated damage, but can also act as second messengers in signalling pathways (Hideg *et al.*, 2002). Pharmacological approaches suggest that generation of ROS is required for the induction and repression of a number of UV-B-responsive genes (A-H-Mackerness, 2000; Desikan *et al.*, 2001). In contrast, the UV-B-mediated induction of CHS gene expression does not seem to involve oxidative stress signalling (A-H-Mackerness *et al.*, 2001; Jenkins *et al.*, 2001). It is of note that ROS may be generated in an enzymatic or non-enzymatic manner. Even though a role of ROS in UV-B signalling has been supported in several biological systems, the nature and origin of these ROS have remained elusive. Recent data, however, indicate superoxide radicals as the dominant ROS in UV-irradiated spinach leaves (Hideg *et al.*, 2002).

Activation of an NADPH oxidase by UV has been demonstrated in *Arabidopsis* (Rao *et al.*, 1996; A-H-Mackerness *et al.*, 2001). There are ten NADPH oxidases (*Atrboh* genes) encoded in the *Arabidopsis* genome, two of which (AtrbohD and F) were linked by mutant analysis to stress responses, including pathogens (Torres *et al.*, 2002) and abscisic acid, a plant stress hormone (Kwak *et al.*, 2003). Similarly, reverse genetics should provide evidence regarding the involvement of particular NADPH oxidases in UV-B signalling. Interestingly, plant NADPH oxidases contain calcium-binding domains, indicating a direct regulatory role of Ca^{2+} ions, thus possibly linking Ca^{2+} release and ROS signalling. However, next to the origin of UV-B-responsive ROS production, the specific involvement of ROS in UV-B responses and in particular its potential role as a second messenger in UV-B signalling still remains to be worked out.

5.2 Plant hormones

UV-B exposure can evoke increase in the levels of salicylic acid, jasmonic acid and ethylene (Surplus *et al.*, 1998; A-H-Mackerness *et al.*, 1999). These plant hormones are implicated downstream of the oxidative stress pathway resulting from high UV-B level exposure. Mutations in components involved in the signalling or perception of these hormones (jasmonic acid-insensitive mutant *jar1* and ethylene-insensitive *etr1*), or transgenic lines engineered to break down the hormone (salicylic acid hydroxylase overexpressing *NahG* plants) indicate their involvement in gene expression of pathogenesis-related marker genes after UV exposure (A-H-Mackerness *et al.*, 1999). These results point to the involvement of JA and ethylene for *PDF1.2* and ethylene and SA for *PR* gene induction; however, none was found necessary for the repression of photosynthetic genes or *CHS*-induction. Additional evidence for the involvement of JA comes from work in tomato, where it was found that UV-B and UV-C activate proteinase inhibitor gene expression in a JA-dependent manner, similarly to wounding (Conconi *et al.*, 1996). Under conditions where UV-B radiation alone did not induce proteinase inhibitor synthesis, UV-B had a potentiating effect on proteinase inhibitor accumulation in response to wounding (Stratmann *et al.*, 2000). However, UV-B did not lead to an increase of JA under these conditions, indicating that the potentiating effect might be independent of JA (Stratmann *et al.*, 2000).

It is of note that select *Arabidopsis* mutants affected in the three hormone pathways implicated in UV-B responses demonstrate that the integrity of these pathways is a necessity for UV-B tolerance (A-H-Mackerness *et al.*, 1999; A-H-Mackerness, 2000). These pathways are similar to those initiated in response to pathogen and herbivore attack, indicating the use of shared components and a possible mechanism for cross-tolerance (A-H-Mackerness, 2000).

5.3 Calcium

Pharmacological experiments in *Arabidopsis* cell cultures indicate the involvement of calcium/calmodulin in the signalling pathway leading to *CHS* gene induction (Christie and Jenkins, 1996; Frohnmeyer *et al.*, 1998; Frohnmeyer *et al.*, 1999). Transient UV-B irradiation results in prolonged $[Ca^{2+}]_i$ in parsley cells as revealed by Fura-2 fluorescence ratio measurements (Frohnmeyer *et al.*, 1999). A pharmacological approach using the broad range Ca^{2+} channel antagonist nifedipine and parsley protoplasts generated from stable *CHS::Luc* cell lines provided evidence that elevated $[Ca^{2+}]_i$ levels are required for UV-B-activated *CHS* gene expression (Frohnmeyer *et al.*, 1999). However, the sole elevation of cytosolic calcium with an ionophore was insufficient to stimulate *CHS* expression in *Arabidopsis* cell cultures (Christie and Jenkins, 1996). It is of note that the positive regulation of *CHS* expression by Ca^{2+}/calmodulin in response to UV-B is mirrored by its negative regulatory function in the phytochrome-mediated control of *CHS* in response to red light (Bowler *et al.*, 1994; Frohnmeyer *et al.*, 1998). This difference between the induction mechanisms is also reflected by the different kinetics of *CHS* activation in response to red compared to UV-B light, as well as the requirement for *de novo* protein synthesis in the latter case (Frohnmeyer *et al.*, 1998).

5.4 Phosphorylation

The involvement of phosphorylation events in the UV-B signal transduction pathways has been demonstrated by using a variety of pharmacological substances (Christie and Jenkins, 1996; Frohnmeyer *et al.*, 1997; Frohnmeyer *et al.*, 1998). However, although the pharmacological interference is indicative, it does not pinpoint the involved protein kinases or phosphatases. One group of kinases known to be activated in response to UV-B belongs to the class of mitogen-activated protein (MAP) kinases (Stratmann, 2003; Ulm, 2003).

MAP kinase pathways are modules involved in the signal transduction of a manifold of endo- and exogenous stimuli in all eukaryotes. These three-tiered phosphorylation cascades fill a prominent role in transducing various stress signals and are employed in the UV-B response as well (Bode and Dong, 2003). Activation of plant MAP kinases was described in response to UV-C in *Arabidopsis* and tobacco (Miles *et al.*, 2002; Ulm *et al.*, 2002), and after UV-B exposure in tomato (Stratmann *et al.*, 2000). Recent work indicates shared components in the elicitor and UV-B signalling pathways culminating in the induction of MAP kinases (Yalamanchili and Stratmann, 2002). In tomato suspension-cultured cells, MAP kinases and other, molecularly unidentified signalling elements of the polypeptide wound signal systemin are suggested to be employed in the response to UV-B (Yalamanchili and Stratmann, 2002). Subsequently, it was shown that elicitors and UV-B activate the tomato MAP kinases LeMPK1 and 2, whereas LeMPK3 was activated by UV-B only, which led the authors to speculate that the specificity for the UV-B response might lie in LeMPK3 activation (Holley et al., 2003). LeMPK1 and 2 are closely related to the UV-C-activated tobacco SIPK (Miles et al., 2002) and *Arabidopsis* MPK6 (Ulm et al., 2002), whereas LeMPK3 is homologous to the tobacco WIPK and *Arabidopsis* MPK3 (Holley et al., 2003). Interestingly, *AtMPK3* is transcriptionally activated in response to UV-C (Ulm et al., 2002) and UV-B radiation (Desikan et al., 2001), at least in the latter case possibly independently of oxidative stress (Desikan et al., 2001). In contrast, UV-C activation of SIPK in tobacco was dependent on ROS production (Miles et al., 2002). Accordingly, it is known that ROS can activate MAPK pathways in plants, including MPK3 and MPK6 in *Arabidopsis* (Kovtun et al., 2000; Yuasa et al., 2001). Interestingly, a recent report linked the UV-B-inducible nucleotide diphosphate kinase NDPKIa/2 (Zimmermann et al., 1999) to oxidative stress and MPK3/6 signalling in *Arabidopsis* (Moon et al., 2003) (note: NDPKIa and NDPK2 differ in three amino acids, representing identical genes from two different *Arabidopsis* ecotypes). This NDPK complements the *gcn4* yeast mutant, binds to the *HIS4* promoter in vitro and induces *HIS4* transcription in yeast (Zimmermann et al., 1999). In addition, the expression of the *Arabidopsis HIS4* homologue *HDH* (encoding a histidinol dehydrogenase) is inducible by UV-B radiation (Zimmermann et al., 1999), altogether suggesting the existence of a UV response equivalent in plants (reviewed in Ulm, 2003). Moreover, NDPK2 was also identified as a phytochrome-interacting light signal transducer (Choi et al., 1999), pointing to an interaction between the light and UV-B/oxidative stress pathways.

Obviously, the exact set up of the UV-B activated MAP kinase pathways and their targets are presently unknown. In addition, the involvement of a much broader range of phosphorylation events in response to UV-B irradiation has to be anticipated, and thus, remains to be determined.

5.5 Nitric oxide

NO is an important second messenger with diverse physiological functions, ranging from development to defense (Wendehenne et al., 2001). Pharmacological approaches using NO scavenger or NO synthase inhibitors indicated that the up regulation of *CHS* expression by UV-B requires NO. Moreover, two compounds known to generate NO were found to activate *CHS* gene expression without UV-B (A-H-Mackerness et al., 2001). Thus, NO involvement in UV-B signalling is indicated, however, further studies to confirm the importance of this notion are needed. For instance, recent work has identified plant nitric oxide synthases, opening the way for a genetic analysis of the involvement of NO in UV-B signalling (Desikan et al., 2002; Guo et al., 2003).

6. TRANSCRIPTIONAL RESPONSE TO UV-B RADIATION

An important output of signal transduction pathways involves changes in gene expression. Therefore, the determination of the transcriptional response to a stimulus is instrumental for the characterization of signalling pathways and their interactions. Exposure of plants to UV-B radiation leads to transcriptional activation and repression of a number of genes (Jordan, 2002; Brosche and Strid, 2003, and references therein), including those that might be attributed to protection and defence, such as DNA repair (Ries *et al.*, 2000b), protective pigment synthesis (Logemann *et al.*, 2000; Jenkins *et al.*, 2001), cell cycle (Logemann *et al.*, 1995) and detoxification of reactive oxygen species (Willekens *et al.*, 1994). In most cases it is not yet possible to unequivocally classify induction of particular genes to either damage-mediated or potential UV-B photoreceptor-mediated pathways. However, an initial attempt was made by using as a criterion the level of UV-B required to first detect regulation of a specific gene (Brosche and Strid, 2003). This approach has the rationale that genes that need high levels are most likely regulated by a damage-responsive pathway, whereas genes responsive to low levels might be downstream of a more specific sensory system. Even though obviously speculative, it offers an initial means to separate different genes and pathways to provide a working basis.

With the recent development of microarray technology for the parallel analysis of thousands of transcripts in a high-throughput fashion, the description of the UV-B-transcriptome at the whole-genome level is within reach. Recent studies using array technology have already highlighted that UV-B leads to profound changes in gene expression and these changes can account to some extent for the effects observed at the physiological level (Brosche *et al.*, 2002; Casati and Walbot, 2003; Izaguirre *et al.*, 2003; Casati and Walbot, 2004; Ulm *et al.*, 2004). In principle, genome-wide expression profiling will result in a comprehensive definition of the molecular UV-B-signalling readout.

The transcriptional response to ambient UV-B in field-grown *Nicotiana longiflora* was determined with a cDNA microarray enriched in wound- and insect-responsive sequences (Izaguirre *et al.*, 2003). Approximately 20% of the 241 genes on the array were found to be differentially expressed in response to solar UV-B, representing a substantial overlap with the gene expression profile generated in response to insect herbivory (Izaguirre *et al.*, 2003). This convergence determined by the activation of common regulatory elements is particularly interesting with regard to the impact of UV-B on insect herbivory as previously documented in field studies. Namely, the experimental attenuation of the UV-B wavelength ranges of solar radiation frequently results in increased insect herbivory (reviewed in Caldwell *et al.*, 2003; Paul and Gwynn-Jones, 2003). It is thus conceivable that the negative impact of UV-B on phytophagous insects are mediated by plant defence measures related to those triggered by insect attack.

A cDNA microarray approach was also used to assess the response of maize to UV-B (Casati and Walbot, 2003, 2004). These experiments included both UV-B exclusion in the field and UV-B supplementation in the greenhouse. In addition, the response was analysed in four near isogenic maize lines that vary in flavonoid content (Casati and Walbot, 2003). Plants were grown for several weeks under filters that either transmitted the full light spectrum or selectively attenuated the UV-B component of the sunlight. 304 genes (12% of the total probe set of 2,500 genes) were identified as responsive to UV-B in at least one of the experiments or genotypes, with 268 and 36 transcripts being up- or down-regulated, respectively. These genes were classified to several functional groups, with the largest number of the up-regulated genes encoding components involved in protein synthesis. On the other hand, transcripts encoding proteins related to photosynthesis and CO_2 fixation were down-regulated by UV-B radiation. Moreover, the shielding effect of phenylpropanoid "sunscreen" pigments was confirmed at the level of gene induction (Casati and Walbot, 2003). A recent follow-up microarray approach examining 5,664 maize genes revealed that most of the UV-B-responsive genes are organ-specific and that a signal from irradiated to shielded tissues seems to exist (Casati and Walbot, 2004).

An initial small-scale array analysis of low-level UV-B-induced genes in *Arabidopsis* resulted in the identification of 70 potentially UV-B-regulated genes that showed over two-fold changes in expression level, out of 5,000 ESTs on the array (false-positive rate approx. 50%, Brosche *et al.*, 2002). A more extensive transcript profiling was performed using oligonucleotide microarrays comprising almost the full *Arabidopsis* genome (>24,000 genes) to monitor expression changes in *Arabidopsis* seedlings exposed to varying UV-B wavelength ranges (Ulm *et al.*, 2004). White-light-grown seedlings were exposed for fifteen minutes to polychromatic radiation with decreasing short-wave cut-off in the UV range and transcript levels were measured one and six hours after the start of irradiation. This analysis identified interactions between cellular responses to different UV-B ranges, indicating the presence of at least two UV-B perception and signalling pathways: one pathway is triggered by the longer wavelengths of UV-B radiation, whereas a second pathway is activated by shorter wavelengths of the UV-B spectrum, with the latter negatively interfering with the former (Ulm *et al.*, 2004). The former is suggested to represent a specific UV-B photoreceptor-mediated pathway, whereas the latter may represent indirect effects of UV-B exposure through a general stress pathway or a distinct UV-B photoreceptor. Low levels of longer wavelength UV-B radiation regulated a robust set of 107 genes one hour post-irradiation (100 activated and 7 repressed) that were postulated to comprise the UV-B photoreceptor readout by the authors (Ulm *et al.*, 2004). Interestingly, these genes include a number of transcription factors that may constitute a UV-B-regulated transcriptional network (Ulm *et al.*, 2004), similarly to the case of far-red and phyA action (Tepperman *et al.*, 2001, see also Chapter 17).

Among the UV-B-induced transcriptional regulators was the bZIP family member HY5, which is known as a key player during the transition from growth in complete darkness to growth in the light. In the dark, HY5 is destabilized and degraded by the proteasome, whereas in light HY5 is required for the expression of a number of light-responsive genes (Osterlund *et al.*, 2000, see also Chapter 18). *HY5* is transcriptionally induced after UV-B irradiation, independently of phyA/B or cry1/2, and subsequent analysis with a *hy5* null mutant identified HY5 as crucial regulator of a subset of UV-B induced genes (Ulm *et al.*, 2004). Importantly, this report links HY5, a component widely accepted to be involved in establishing photoreceptor-mediated photomorphogenic responses, to the postulated UV-B photoreceptor. Moreover, it is of interest that the regulator of HY5 protein abundance during the dark-light transition, the COP1 protein, was also put forward as a possible component of the UV-B response (Kim *et al.*, 1998). However, it remains to be established if and how they interplay in response to UV-B radiation.

Thus, the recent use of microarray technology to analyse UV-B-mediated gene expression changes on a global scale provides unprecedented access to the molecular readout of the employed pathways, not least providing a powerful tool for the discovery of gene functions.

7. CONCLUSIONS AND PERSPECTIVES

Exposure of plants to UV-B results in a multitude of rapid, intermediate- and long-term responses that are mediated by an unknown number of unidentified (photo-) sensory systems. Besides the lack of molecularly identified genuine UV-B photoreceptor(s), insight into downstream signal transduction pathways is also rather scant. Particular model pathways such as the regulation of *CHS* are explored in quite some detail; however, the generalities of these findings to UV-B-responsive gene expression changes remain to be determined. Parallel profiling of thousands of genes using microarray technology recently revealed a number of candidate genes for such an endeavour. Moreover, the generated data will allow a number of forward and reverse genetic approaches that are expected to result in the identification of components involved in UV-B perception and signalling pathways. These approaches will greatly benefit from the recent advent in *Arabidopsis* molecular genetics. Specific UV-B morphogenesis mutants will be the cornerstones of genetic dissection of responses to this fraction of the sunlight. Owing to a paucity of well-defined visible phenotypes and confounding damaging aspects, conventional genetic screens have not been very successful in isolating UV-B signal transduction mutants in plants. Screens using reporter gene fusions might be a promising approach for the future. Furthermore, reverse genetics will be invaluable in analysing the extent of involvement of components linked to UV-B by pharmacological means (ROS, NO, Ca^{2+}/calmodulin, kinases, phosphatases, etc.) or gene expression profiling. However, it is evident that as long as no UV-B photoreceptor and its loss-of-function mutant are identified, it remains difficult (if not impossible) to unambiguously separate specific UV-B responses from indirect consequences derived from concomitant general damage. A combination of approaches is eventually required to determine the informational processes that underpin responses to UV-B within the frame of a complex environment.

Acknowledgements: I would like to thank Alexander Baumann, Erzsebet Fejes and Eckard Wellmann for helpful comments on the manuscript and the European Molecular Biology Organization for support by a long-term fellowship.

8. REFERENCES

A-H-Mackerness, S. (2000) Plant responses to ultraviolet-B (UV-B: 280-320 nm) stress: what are the key regulators? *Plant Growth Reg, 32*, 27-39.

A-H-Mackerness, S., John, C. F., Jordan, B., Thomas, B. (2001) Early signaling components in ultraviolet-B responses: distinct roles for different reactive oxygen species and nitric oxide. *FEBS Lett, 489*, 237-242.

A-H-Mackerness, S., Surplus, S. L., Blake, P., John, C. F., Buchanan-Wollaston, V., Jordan, B. R., et al. (1999) Ultraviolet-B-induced stress and changes in gene expression in Arabidopsis thaliana: role of signalling pathways controlled by jasmonic acid, ethylene and reactive oxygen species. *Plant Cell Environ, 22*, 1413-1423.

Ahmad, M., Jarillo, J. A., Klimczak, L. J., Landry, L. G., Peng, T., Last, R. L., et al. (1997) An enzyme similar to animal type II photolyases mediates photoreactivation in Arabidopsis. *Plant Cell, 9*, 199-207.

Ballare, C.L. (1999) Keeping up with the neighbours: phytochrome sensing and other signalling mechanisms. *Trends Plant Sci, 4*, 97-102.

Ballare, C. L., Barnes, P. W., Flint, S. D. (1995a) Inhibition of hypocotyl elongation by ultraviolet-B radiation in de-etiolating tomato seedlings. I. The photoreceptor. *Physiol Plant, 93*, 584-592.

Ballare, C. L., Barnes, P. W., Flint, S. D., Price, S. (1995b) Inhibition of hypocotyl elongation by ultraviolet-B radiation in de-etiolating tomato seedlings. II. Time-course, comparison with flavonoid responses and adaptive significance. *Physiol Plant, 93*, 593-601.

Ballare, C. L., Barnes, P. W., Kendrick, R. E. (1991) Photomorphogenic effects of UV-B radiation on hypocotyl elongation in wild type and stable-phytochrome-deficient mutant seedlings of cucumber. *Physiol Plant, 83*, 652-658.

Barnes, P. W., Ballare, C. L., Caldwell, M. M. (1996) Photomorphogenic effects of UV-B radiation on plants: consequences for light competition. *J Plant Physiol, 148*, 15-20.

Beggs, C. J., Stolzer-Jehle, A., Wellmann, E. (1985) Isoflavonoid formation as an indicator of UV stress in bean (Phaseolus vulgaris L.) leaves. *Plant Physiol, 79*, 630-634.

Beggs, C.J. and Wellmann, E. (1994) "Photocontrol of flavonoid biosynthesis." In: *Photomorphogenesis in plants, 2nd ed.*, Kendrick R.E. and Kronenberg G.H.M. (eds.) The Netherlands: Kluwer Academic Publishers, Dordrecht, 733-751.

Benvenuto, G., Formiggini, F., Laflamme, P., Malakhov, M., Bowler, C. (2002) The photomorphogenesis regulator DET1 binds the amino-terminal tail of histone H2B in a nucleosome context. *Curr Biol, 12*, 1529-1534.

Bharti, A. K. and Khurana, J. P. (1997) Mutants of Arabidopsis as tools to understand the regulation of phenylpropanoid pathway and UVB protection mechanisms. *Photochem Photobiol, 65*, 765-776.

Bieza, K. and Lois, R. (2001) An Arabidopsis mutant tolerant to lethal ultraviolet-B levels shows constitutively elevated accumulation of flavonoids and other phenolics. *Plant Physiol, 126*, 1105-1115.

Björn, L. O. (1999) "UV-B effects: receptors and targets." In *Concepts in Photobiology: Photosynthesis and Photomorphogenesis*, Singhal G. S., Renger G., Sopory S. K., Irrgang K.-D. and Govindjee (eds.) Narosa Publishing House, New Delhi, India, 821-832.

Björn, L. O. and Wang, T. (2001) Is provitamin D a UV-B receptor in plants? *Plant Ecology, 154*, 3-8.

Blumthaler, M. (1993) "Solar UV measurements." In *UV-B radiation and ozone depletion: effects on humans, animals, plants, microorganisms, and materials*, Tevini M. (ed.) Lewis Publishers, Florida, 71-94.

Boccalandro, H. E., Mazza, C. A., Mazzella, M. A., Casal, J. J., Ballare, C. L. (2001) Ultraviolet B radiation enhances a phytochrome-B-mediated photomorphogenic response in Arabidopsis. *Plant Physiol, 126*(2), 780-788.

Bode, A. M. and Dong, Z. (2003) Mitogen-activated protein kinase activation in UV-induced signal transduction. *Sci STKE*, re2.

Bouche, N., Fait, A., Bouchez, D., Moller, S. G., Fromm, H. (2003) Mitochondrial succinic-semialdehyde dehydrogenase of the gamma-aminobutyrate shunt is required to restrict levels of reactive oxygen intermediates in plants. *Proc Natl Acad Sci USA, 100*(11), 6843-6848.

Bowler, C., Yamagata, H., Neuhaus, G., Chua, N.H. (1994) Phytochrome signal transduction pathways are regulated by reciprocal control mechanisms. *Genes Dev, 8,* 2188-2202.

Braun, J. and Tevini, M. (1993) Regulation of UV-protective pigment synthesis in the epidermal layer of rye seedlings. *Photochem Photobiol, 57,* 318-323.

Britt, A. and Fiscus, E. L. (2003) Growth responses of Arabidopsis DNA repair mutants to solar irradiation. *Physiol Plant, 118,* 183-192.

Britt, A. B. (2004) Repair of DNA damage induced by solar UV. *Photosynth Res, 81,* 105-112.

Brosche, M., Schuler, M. A., Kalbina, I., Connor, L., Strid, A. (2002) Gene regulation by low level UV-B radiation: identification by DNA array analysis. *Photochem Photobiol Sci, 1,* 656-664.

Brosche, M. and Strid, A. (2000) Ultraviolet-B radiation causes tendril coiling in Pisum sativum. *Plant Cell Physiol, 41,* 1077-1079.

Brosche, M. and Strid, A. (2003) Molecular events following perception of ultraviolet-B radiation by plants. *Physiol Plant, 117,* 1-10.

Caldwell, M. M., Ballare, C. L., Bornman, J. F., Flint, S. D., Bjorn, L. O., Teramura, A. H., *et al.* (2003) Terrestrial ecosystems, increased solar ultraviolet radiation and interactions with other climatic change factors. *Photochem Photobiol Sci, 2,* 29-38.

Caldwell, M. M. and Flint, S.D. (1994) Stratospheric ozone reduction, solar UV-B radiation and terrestrial ecosystems. *Climatic Change, 28,* 375-394.

Casati, P. and Walbot, V. (2003) Gene expression profiling in response to ultraviolet radiation in maize genotypes with varying flavonoid content. *Plant Physiol, 132,* 1739-1754.

Casati, P. and Walbot, V. (2004) Rapid transcriptome responses of maize (Zea mays) to UV-B in irradiated and shielded tissues. *Genome Biol, 5,* R16.

Choi, G., Yi, H., Lee, J., Kwon, Y. K., Soh, M. S., Shin, B., *et al.* (1999) Phytochrome signalling is mediated through nucleoside diphosphate kinase 2. *Nature, 401,* 610-613.

Christie, J. M. and Jenkins, G. I. (1996) Distinct UV-B and UV-A/blue light signal transduction pathways induce chalcone synthase gene expression in Arabidopsis cells. *Plant Cell, 8,* 1555-1567.

Conconi, A., Smerdon, M. J., Howe, G. A., Ryan, C. A. (1996) The octadecanoid signalling pathway in plants mediates a response to ultraviolet radiation. *Nature, 383,* 826-829.

Conklin, P. L., Williams, E. H., Last, R. L. (1996) Environmental stress sensitivity of an ascorbic acid-deficient Arabidopsis mutant. *Proc Natl Acad Sci USA, 93,* 9970-9974.

Culligan, K., Tissier, A., Britt, A. (2004) ATR regulates a G2-phase cell-cycle checkpoint in Arabidopsis thaliana. *Plant Cell, 16,* 1091-1104.

Dasso, M. (2002) The Ran GTPase: theme and variations. *Curr Biol, 12,* R502-508.

Desikan, R., A-H-Mackerness, S., Hancock, J. T., Neill, S. J. (2001) Regulation of the Arabidopsis transcriptome by oxidative stress. *Plant Physiol, 127,* 159-172.

Desikan, R., Griffiths, R., Hancock, J., Neill, S. (2002) A new role for an old enzyme: nitrate reductase-mediated nitric oxide generation is required for abscisic acid-induced stomatal closure in Arabidopsis thaliana. *Proc Natl Acad Sci USA, 99,* 16314-16318.

Eisinger, W. R., Bogomolni, R. A., Taiz, L. (2003) Interactions between a blue-green reversible photoreceptor and a separate UV-B receptor in stomatal guard cells. *Am J Bot, 90,* 1560-1566.

Ensminger, P.A. and Schäfer, E. (1992) Blue and ultraviolet-B light photoreceptors in parsley cells. *Photochem Photobiol, 55,* 437-447.

Fidantsef, A. L., Mitchell, D. L., Britt, A. B. (2000) The Arabidopsis UVH1 gene is a homolog of the yeast repair endonuclease RAD1. *Plant Physiol, 124,* 579-586.

Flint, S. D. and Caldwell, M.M. (1998) Solar UV-B and visible radiation in tropical forest gaps: measurements partitioning direct and diffuse radiation. *Global Change Biology, 4,* 863-870.

Friedberg, E.C. (2001) How nucleotide excision repair protects against cancer. *Nat Rev Cancer, 1,* 22-33.

Frohnmeyer, H., Bowler, C., Schäfer, E. (1997) Evidence for some common signal transduction events involved in UV B light-dependent responses in parsley protoplasts. *J Exp Bot, 48,* 739-750.

Frohnmeyer, H., Bowler, C., Zhu, J.-K., Yamagata, H., Schäfer, E., Chua, N.-H. (1998) Different roles for calcium and calmodulin in phytochrome- and UV-regulated expression of chalcone synthase. *Plant J, 13,* 763-772.

Frohnmeyer, H., Loyall, L., Blatt, M. R., Grabov, A. (1999) Millisecond UV-B irradiation evokes prolonged elevation of cytosolic-free Ca^{2+} and stimulates gene expression in transgenic parsley cell cultures. *Plant J, 20,* 109-117.

Frohnmeyer, H. and Staiger, D. (2003) Ultraviolet-B radiation-mediated responses in plants. Balancing damage and protection. *Plant Physiol, 133,* 1420-1428.

Fuglevand, G., Jackson, J. A., Jenkins, G. I. (1996) UV-B, UV-A, and blue light signal transduction pathways interact synergistically to regulate chalcone synthase gene expression in Arabidopsis. *Plant Cell, 8*, 2347-2357.

Fujioka, S., Li, J., Choi, Y. H., Seto, H., Takatsuto, S., Noguchi, T., et al. (1997) The Arabidopsis deetiolated2 mutant is blocked early in brassinosteroid biosynthesis. *Plant Cell, 9*, 1951-1962.

Galland, P. and Senger, H. (1988a) The role of flavins as photoreceptors. *J Photochem Photobiol B: Biol*, 277-294.

Galland, P. and Senger, H. (1988b) The role of pterins in the photoreception and metabolism of plants. *Photochem Photobiol, 48*, 811-820.

Guo, F. Q., Okamoto, M., Crawford, N. M. (2003) Identification of a plant nitric oxide synthase gene involved in hormonal signaling. *Science, 302*, 100-103.

Haizel, T., Merkle, T., Pay, A., Fejes, E., Nagy, F. (1997) Characterization of proteins that interact with the GTP-bound form of the regulatory GTPase Ran in Arabidopsis. *Plant J, 11*, 93-103.

Hashimoto, T., Shichijo, C., Yatsuhashi, H. (1991) Ultraviolet action spectra for the induction and inhibition of anthocyanin synthesis in broom sorghum seedlings. *J Photochem Photobiol B: Biol, 11*, 353-363.

Hefner, E., Preuss, S. B., Britt, A. B. (2003) Arabidopsis mutants sensitive to gamma radiation include the homologue of the human repair gene ERCC1. *J Exp Bot, 54*, 669-680.

Hideg, E., Barta, C., Kalai, T., Vass, I., Hideg, K., Asada, K. (2002) Detection of singlet oxygen and superoxide with fluorescent sensors in leaves under stress by photoinhibition or UV radiation. *Plant Cell Physiol, 43*, 1154-1164.

Hidema, J., Kumagai, T., Sutherland, B.M. (2000) UV radiation-sensitive norin 1 rice contains defective cyclobutane pyrimidine dimer photolyase. *Plant Cell, 12*, 1569-1578.

Holley, S. R., Yalamanchili, R. D., Moura, D. S., Ryan, C. A., Stratmann, J. W. (2003) Convergence of signaling pathways induced by systemin, oligosaccharide elicitors, and ultraviolet-B radiation at the level of mitogen-activated protein kinases in Lycopersicon peruvianum suspension-cultured cells. *Plant Physiol, 132*, 1728-1738.

Izaguirre, M. M., Scopel, A. L., Baldwin, I. T., Ballare, C. L. (2003) Convergent responses to stress. Solar ultraviolet-B radiation and Manduca sexta herbivory elicit overlapping transcriptional responses in field-grown plants of Nicotiana longiflora. *Plant Physiol, 132*, 1755-1767.

Jansen, M. A. K., Gaba, V., Greenberg, B.M. (1998) Higher plants and UV-B radiation: balancing damage, repair and acclimation. *Trends Plant Sci, 3*, 131-135.

Jenkins, G. I., Long, J. C., Wade, H. K., Shenton, M. R., Bibikova, T. N. (2001) UV and blue light signalling: pathways regulating chalcone synthase gene expression in Arabidopsis. *New Phytol, 151*, 121-131.

Jin, H., Cominelli, E., Bailey, P., Parr, A., Mehrtens, F., Jones, J., et al. (2000) Transcriptional repression by AtMYB4 controls production of UV-protecting sunscreens in Arabidopsis. *EMBO J, 19*, 6150-6161.

Jordan, B. R. (1996) "The effects of ultraviolet-B radiation on plants: a molecular perspective." In *Advances in Botanical Research*, Vol. 22, Callow J. A. (ed) Academic Press Ltd, London, 97-162.

Jordan, B. R. (2002) Molecular response of plant cells to UV-B stress. *Funct. Plant Biol., 29*, 909-916.

Kalbin, G., Hidema, J., Brosche, M., Kumagai, T., Bornman, J. F., Strid, A. (2001) UV-B-induced DNA damage and expression of defence genes under UV-B stress: tissue-specific molecular marker analysis in leaves. *Plant Cell Environ, 24*, 983-990.

Kim, B. C., Tennessen, D. J., Last, R. L. (1998) UV-B-induced photomorphogenesis in Arabidopsis thaliana. *Plant J, 15*, 667-674.

Kim, S.-T., Li, Y. F., Sancar, A. (1992) The third chromophore of DNA photolyase: Trp-277 of Escherichia coli repairs thymine dimers by direct electron transfer. *Proc Natl Acad Sci USA, 89*, 900-904.

Kliebenstein, D. J., Lim, J. E., Landry, L. G., Last, R. L. (2002) Arabidopsis UVR8 regulates ultraviolet-B signal transduction and tolerance and contains sequence similarity to human regulator of chromatin condensation 1. *Plant Physiol, 130*, 234-243.

Kovtun, Y., Chiu, W. L., Tena, G., Sheen, J. (2000) Functional analysis of oxidative stress-activated mitogen-activated protein kinase cascade in plants. *Proc Natl Acad Sci USA, 97*, 2940-2945.

Kucera, B., Leubner-Metzger, G., Wellmann, E. (2003) Distinct ultraviolet-signaling pathways in bean leaves. DNA damage is associated with β-1,3-glucanase gene induction, but not with flavonoid formation. *Plant Physiol, 133*, 1445-1452.

Kwak, J. M., Mori, I. C., Pei, Z. M., Leonhardt, N., Torres, M. A., Dangl, J. L., et al. (2003) NADPH oxidase AtrbohD and AtrbohF genes function in ROS-dependent ABA signaling in Arabidopsis. *EMBO J, 22*, 2623-2633.

Landry, L. G., Chapple, C. C., Last, R. L. (1995) Arabidopsis mutants lacking phenolic sunscreens exhibit enhanced ultraviolet-B injury and oxidative damage. *Plant Physiol, 109*, 1159-1166.

Li, J., Ou-Lee, T. M., Raba, R., Amundson, R. G., Last, R. L. (1993) Arabidopsis flavonoid mutants are hypersensitive to UV-B irradiation. *Plant Cell, 5*, 171-179.

Liu, Y., Gorospe, M., Holbrook, N. J., Anderson, C. W. (1998) "Posttranslational mechanisms leading to mammalian gene activation in response to genotoxic stress." In *DNA damage and repair, Vol. 2: DNA repair in higher eukaryotes*, Nickoloff J. A. and Hoekstra M. F. (eds) Humana Press Inc. Totowa, NJ, 263-298.

Liu, Z., Hall, J. D., Mount, D. W. (2001) Arabidopsis UVH3 gene is a homolog of the Saccharomyces cerevisiae RAD2 and human XPG DNA repair genes. *Plant J, 26*, 329-338.

Liu, Z., Hong, S. W., Escobar, M., Vierling, E., Mitchell, D. L., Mount, D. W., et al. (2003) Arabidopsis UVH6, a homolog of human XPD and yeast RAD3 DNA repair genes, functions in DNA repair and is essential for plant growth. *Plant Physiol, 132*, 1405-1414.

Liu, Z., Hossain, G. S., Islas-Osuna, M. A., Mitchell, D. L., Mount, D. W. (2000) Repair of UV damage in plants by nucleotide excision repair: Arabidopsis UVH1 DNA repair gene is a homolog of Saccharomyces cerevisiae Rad1. *Plant J, 21*, 519-528.

Logemann, E., Tavernaro, A., Schulz, W., Somssich, I. E., Hahlbrock, K. (2000) UV light selectively coinduces supply pathways from primary metabolism and flavonoid secondary product formation in parsley. *Proc Natl Acad Sci USA, 97*, 1903-1907.

Logemann, E., Wu, S. C., Schroder, J., Schmelzer, E., Somssich, I. E., Hahlbrock, K. (1995) Gene activation by UV light, fungal elicitor or fungal infection in Petroselinum crispum is correlated with repression of cell cycle-related genes. *Plant J, 8*, 865-876.

Lois, R. (1994) Accumulation of UV-absorbing flavonoids induced by UV-B radiation in Arabidopsis thaliana L. I. Mechanisms of UV-resistance in Arabidopsis. *Planta, 194*, 498-503.

Mazza, C. A., Boccalandro, H. E., Giordano, C. V., Battista, D., Scopel, A. L., Ballare, C. L. (2000) Functional significance and induction by solar radiation of ultraviolet-absorbing sunscreens in field-grown soybean crops. *Plant Physiol, 122*, 117-126.

McClintock, B. (1984) The significance of responses of the genome to challenge. *Science, 226*, 792-801.

McKenzie, R. L., Bjorn, L. O., Bais, A., Ilyasd, M. (2003) Changes in biologically active ultraviolet radiation reaching the Earth's surface. *Photochem Photobiol Sci, 2*, 5-15.

Miles, G. P., Samuel, M. A., Ellis, B. E. (2002) Suramin inhibits oxidant signalling in tobacco suspension-cultured cells. *Plant Cell Environ, 25*, 521-527.

Mohr, H. (1994). "Coaction between pigment systems." In *Photomorphogenesis in Plants - 2nd edition*, Kendrick R. E. and Kronenberg G. H. M. (eds.) Kluwer Academic Publishers, 353-373.

Moon, H., Lee, B., Choi, G., Shin, D., Prasad, D. T., Lee, O., et al. (2003) NDP kinase 2 interacts with two oxidative stress-activated MAPKs to regulate cellular redox state and enhances multiple stress tolerance in transgenic plants. *Proc Natl Acad Sci USA, 100*, 358-363.

Murphy, T. M. (1983) Membranes as targets of ultraviolet radiation. *Physiol Plant, 58*, 381-388.

Nakajima, S., Sugiyama, M., Iwai, S., Hitomi, K., Otoshi, E., Kim, S. T., et al. (1998) Cloning and characterization of a gene (UVR3) required for photorepair of 6-4 photoproducts in Arabidopsis thaliana. *Nucleic Acids Res, 26*, 638-644.

Osterlund, M. T., Hardtke, C. S., Wei, N., Deng, X. W. (2000) Targeted destabilization of HY5 during light-regulated development of Arabidopsis. *Nature, 405*, 462-466.

Paul, N. D. and Gwynn-Jones, D. (2003) Ecological roles of solar UV radiation: towards an integrated approach. *Trends Ecol Evol, 18*, 48-55.

Pratt, L. H. and Butler, W. L. (1970) Phytochrome conversion by ultraviolet light. *Photochem Photobiol, 11*, 503-509.

Rao, M. V., Paliyath, G., Ormrod, D. P. (1996) Ultraviolet-B- and ozone-induced biochemical changes in antioxidant enzymes of Arabidopsis thaliana. *Plant Physiol, 110*, 125-136.

Ries, G., Buchholz, G., Frohnmeyer, H., Hohn, B. (2000a) UV-damage-mediated induction of homologous recombination in Arabidopsis is dependent on photosynthetically active radiation. *Proc Natl Acad Sci USA, 97*, 13425-13429.

Ries, G., Heller, W., Puchta, H., Sandermann, H., Seidlitz, H.K., Hohn, B. (2000b) Elevated UV-B radiation reduces genome stability in plants. *Nature, 406*, 98-101.

Ros, J. and Tevini, M. (1995) Interaction of UV-radiation and IAA during growth of seedlings and hypocotyl segments of sunflower. *J Plant Physiol, 146*, 295-302.

Rozema, J., van de Staaij, J., Björn, L. O., Caldwell, M. (1997) UV-B as an environmental factor in plant life: stress and regulation. *Trends Ecol Evol, 12*, 22-28.

Saijo, Y., Sullivan, J. A., Wang, H., Yang, J., Shen, Y., Rubio, V., et al. (2003) The COP1-SPA1 interaction defines a critical step in phytochrome A-mediated regulation of HY5 activity. *Genes Dev, 17*, 2642-2647.

Sakamoto, A., Lan, V. T., Hase, Y., Shikazono, N., Matsunaga, T., Tanaka, A. (2003) Disruption of the AtREV3 gene causes hypersensitivity to ultraviolet B light and γ-rays in Arabidopsis: implication of the presence of a translesion synthesis mechanism in plants. *Plant Cell, 15*, 2042-2057.

Sancar, A. (2003) Structure and function of DNA photolyase and cryptochrome blue-light photoreceptors. *Chem Rev, 103*, 2203-2237.

Schroeder, D. F., Gahrtz, M., Maxwell, B. B., Cook, R. K., Kan, J. M., Alonso, J. M., et al. (2002) De-etiolated 1 and damaged DNA binding protein 1 interact to regulate Arabidopsis photomorphogenesis. *Curr Biol, 12*, 1462-1472.

Searles, P. S., Flint, S. D., Caldwell, M. M. (2001) A meta-analysis of plant field studies simulating stratospheric ozone depletion. *Oecologia, 127*, 1-10.

Seo, H. S., Yang, J. Y., Ishikawa, M., Bolle, C., Ballesteros, M. L., Chua, N. H. (2003) LAF1 ubiquitination by COP1 controls photomorphogenesis and is stimulated by SPA1. *Nature, 424*, 995-999.

Shinkle, J. R., Atkins, A. K., Humphrey, E. E., Rodgers, C. W., Wheeler, S. L., Barnes, P. W. (2004) Growth and morphological responses to different UV wavebands in cucumber (Cucumis sativus) and other dicotyledonous seedlings. *Physiol Plant, 120*, 240-248.

Stratmann, J. (2003) Ultraviolet-B radiation co-opts defense signaling pathways. *Trends Plant Sci, 8*, 526-533.

Stratmann, J. W., Stelmach, B. A., Weiler, E. W., Ryan, C. A. (2000) UVB/UVA radiation activates a 48 kDa myelin basic protein kinase and potentiates wound signaling in tomato leaves. *Photochem Photobiol, 71*, 116-123.

Suesslin, C. and Frohnmeyer, H. (2003) An Arabidopsis mutant defective in UV-B light-mediated responses. *Plant J, 33*, 591-601.

Surplus, S. L., Jordan, B. R., Murphy, A. M., Carr, J. P., Thomas, B., A-H-Mackerness, S. (1998) UV-B induced responses in Arabidopsis thaliana: role of salicylic acid and ROS in the regulation of transcripts and acidic PR proteins. *Plant Cell Environ, 21*, 685-694.

Tanaka, A., Sakamoto, A., Ishigaki, Y., Nikaido, O., Sun, G., Hase, Y., et al. (2002) An ultraviolet-B-resistant mutant with enhanced DNA repair in Arabidopsis. *Plant Physiol, 129*, 64-71.

Tepperman, J. M., Zhu, T., Chang, H. S., Wang, X., Quail, P. H. (2001) Multiple transcription-factor genes are early targets of phytochrome A signaling. *Proc Natl Acad Sci USA, 98*, 9437-9442.

Tevini, M. (1999) "UV-effects on plants." In *Concepts in Photobiology: Photosynthesis and Photomorphogenesis*, Singhal G. S., Renger G., Sopory S. K., Irrgang K.-D. and Govindjee (eds.) Narosa Publishing House, New Delhi, India, 588-613.

Torres, M. A., Dangl, J. L., Jones, J. D. (2002) Arabidopsis gp91phox homologues AtrbohD and AtrbohF are required for accumulation of reactive oxygen intermediates in the plant defense response. *Proc Natl Acad Sci USA, 99*, 517-522.

Ulm, R. (2003) Molecular genetics of genotoxic stress signalling in plants. *Topics Curr Genet, 4*, 217-240.

Ulm, R., Baumann, A., Oravecz, A., Mate, Z., Adam, E., Oakeley, E. J., et al. (2004) Genome-wide analysis of gene expression reveals function of the bZIP transcription factor HY5 in the UV-B response of Arabidopsis. *Proc Natl Acad Sci USA, 101*, 1397-1402.

Ulm, R., Ichimura, K., Mizoguchi, T., Peck, S. C., Zhu, T., Wang, X., et al. (2002) Distinct regulation of salinity and genotoxic stress responses by Arabidopsis MAP kinase phosphatase 1. *EMBO J, 21*, 6483-6493.

Vass, I. (1997) "Adverse effects of UV-B light on the structure and function of the photosynthetic apparatus". In *Handbook of Photosynthesis*, Pessarakli M. (ed) Marcel Dekker Inc, NY, 931-949.

Wade, H. K., Bibikova, T. N., Valentine, W. J., Jenkins, G. I. (2001) Interactions within a network of phytochrome, cryptochrome and UV-B phototransduction pathways regulate chalcone synthase gene expression in Arabidopsis leaf tissue. *Plant J, 25*, 675-685.

Wade, H. K., Sohal, A. K., Jenkins, G. I. (2003) Arabidopsis ICX1 is a negative regulator of several pathways regulating flavonoid biosynthesis genes. *Plant Physiol, 131*, 707-715.

Walbot, V. (1999) UV-B damage amplified by transposons in maize. *Nature, 397*, 398-399.

Waterworth, W. M., Jiang, Q., West, C. E., Nikaido, M., Bray, C. M. (2002) Characterization of Arabidopsis photolyase enzymes and analysis of their role in protection from ultraviolet-B radiation. *J Exp Bot, 53*, 1005-1015.

Wendehenne, D., Pugin, A., Klessig, D. F., Durner, J. (2001) Nitric oxide: comparative synthesis and signaling in animal and plant cells. *Trends Plant Sci, 6*, 177-183.

Willekens, H., van Camp, W., van Montagu, M., Inze, D., Langebartels, C., Sandermann, H. Jr., (1994) Ozone, sulfur dioxide, and ultraviolet-B have similar effects on mRNA accumulation of antioxidant genes in Nicotiana plumbaginifolia L. *Plant Physiol, 106*, 1007-1014.

Wilson, M. I. and Greenberg, B. M. (1993) Specificity and photomorphogenic nature of ultraviolet-B-induced cotyledon curling in Brassica napus L. *Plant Physiol, 102*, 671-677.

Winkel-Shirley, B. (2002) Biosynthesis of flavonoids and effects of stress. *Curr Opin Plant Biol, 5*, 218-223.

Yalamanchili, R. D. and Stratmann, J. W. (2002) Ultraviolet-B activates components of the systemin signaling pathway in Lycopersicon peruvianum suspension-cultured cells. *J Biol Chem, 277*, 28424-28430.

Yuasa, T., Ichimura, K., Mizoguchi, T., Shinozaki, K. (2001) Oxidative stress activates ATMPK6, an Arabidopsis homologue of MAP kinase. *Plant Cell Physiol, 42*, 1012-1016.

Zimmermann, S., Baumann, A., Jaekel, K., Marbach, I., Engelberg, D., Frohnmeyer, H. (1999) UV-responsive genes of Arabidopsis revealed by similarity to the Gcn4-mediated UV response in yeast. *J Biol Chem, 274*, 17017-17024.

Chapter 15

SIGNAL TRANSDUCTION IN BLUE LIGHT-MEDIATED RESPONSES

Vera Quecini[1] and Emmanuel Liscum
Division of Biological Science, University of Missouri, Columbia MO 65211, U.S.A., [1] *Present address: Universidade de São Paulo, Piracicaba SP 13400-290, Brazil (e-mail: LiscumM@missouri.edu)*

1. INTRODUCTION

The study of blue light (B)-mediated responses has a long history (Darwin, 1881; Sachs, 1887), yet molecular identification of the photoreceptors mediating these responses has only occurred in the last decade. Although identification of cryptochrome (cry)-, phototropin (phot)-, and other B-receptor-mediated signal transduction components have lagged even farther behind, use of mutational analyses in the model plant *Arabidopsis thaliana* has resulted in many exciting new advances in recent years. The focus of this chapter is on signalling associated with the cryptochromes and phototropins. For simplicity, subsections are divided by a response output although this in no way implies that a given response operates in isolation relative to other receptors and/or responses. Co-action and interaction between various photosignal-response systems is covered elsewhere (see Chapters 20, 21).

2. CRYPTOCHROME SIGNALING

2.1 Cryptochromes and Photomorphogenesis

The physiological and metabolic processes involved in the transition from dark-grown seedlings dependent upon seed reserves for survival to photosynthetic-competent seedlings are collectively known as photomorphogenesis (Kendrick and Kronenberg, 1994). Proper photomorphogenic development requires the cooperative action of both the phytochromes and the cryptochromes (see Chapter 20). Here we will discuss the signalling events associated with cry-dependent photomorphogenic growth responses (section 2.1.1) and those associated with alterations in gene expression involved in chloroplast biogenesis and pigment production (section 2.1.2). We defer discussion of cry influences on circadian and photoperiodic responses as they are covered in detail elsewhere (see Chapters 26, 27).

2.1.1 Cryptochrome Signalling and Photomorphogenic Growth Responses

Both cry1 and cry2 appear to contribute to the regulation of photomorphogenetic development of *Arabidopsis* under monochromatic B conditions although to different extents: cry1 being dominant under high-intensity B and cry2 functioning mainly at low B intensities (Mockler *et al.*, 1999; Mazzella *et al.*, 2001). The light-sensing capacity of cry1 and cry2 is mediated by the amino (N)-terminal PHR (photolyaserelated domain), while the ability of both receptors to transduce signals perceived in the PHR domain appears tightly coupled to the carboxyl (C)-terminal regions. The importance of these C-terminal regions, called CCT1 and CCT2, respectively for cry1 and cry2, was initially demonstrated by mutations in these regions that impair function (Ahmad *et al.*, 1995; Cashmore *et al.*, 1999; Lin and Shalitin, 2003). Recent studies have shown that transgenic plants overexpressing (OX) translational fusions of GUS with either CCT1 or CCT2 have constitutive photomorphogenic phenotypes in the dark (Wang *et al.*, 2001; Yang *et al.*, 2000, 2001). These phenotypes are compelling since they resemble the *cop/det/fus* (*constitutive photomorphogenesis/de-etiolated/fusca*) class of mutants (Serino and Deng, 2003). As will be discussed shortly, this is intriguing since several of the *COP/DET/FUS* loci have been implicated as downstream components of cry signalling pathways (Serino and Deng, 2003). Although the CCT-OX plants are hypersensitive to light, that hypersensitivity is blocked if small portions of the N-terminal region of either cry are added to the transgene construct. This has led to the conclusion that the signal transduction activity of a cry is brought about by its C-terminus and this activity is inhibited by the N-terminus in the dark but is relieved upon absorption of B (Wang *et al.*, 2001; Yang *et al.*, 2001).

Table 1. Cryptochrome-mediated responses and associated signalling components

Receptor	Response	Signaling element (biochemical identity)
cry1	photomorphogenic growth control	SUB1 (Ca^{2+}-binding EF-hand protein)
		HY5 (basic leucine zipper-type transcription factor) HFR1 (basic helix-loop-helix transcription factor) AtPP7 (ser/thr protein phosphatase)
cry1 and cry2	photomorphogenic growth control and regulation of gene expression	COP1 (E3 ubiquitin ligase)
		COP10 (E2 ubiquitin-conjugating enzyme) CNS subunits (COP9 signalosome)
cry1 and cry2	plasma membrane depolarisation	unknown (Clchannel?; other anion channel?)
cry1	regulation of gene expression	Ca^{2+}
cry3	Regulation of gene expression (?)	direct influence (?)

Beyond mutations in the receptors themselves, two general classes of loss-of-function photomorphogenic mutants have been especially useful in identifying signalling components acting downstream of the cryptochromes (Figure 1): 1) mutants with altered B responses; and 2) mutants exhibiting constitutive photomorphogenic responses in the absence of light. To date mutations in only three non-cry loci have been found by forward-genetic approaches to result in appreciable alteration of Binduced photomorphogenesis, namely *HFR1* (*LONG HYPOCOTYL IN FRA-RED*; Duek and Fankhauser, 2003), *HY5* (*LONG-HYPOCOTYL 5*; Koornneef *et al.*, 1980), and *SUB1* (*SHORT UNDER BLUE LIGHT*; Gou *et al.*, 2001) (Table 1; Figure 1). It is interesting to note that *sub1* was the only mutant identified in a B-specific screen. In contrast, *hy5* was identified in a screen for seedlings with long hypocotyls in white light (WL) (Koornneef *et al.*, 1980) and *hfr1* (also know as *rep1* and *rsf1*) mutants were identified in screens for phyA signalling intermediates with long hypocotyls in far-red light (FR) (Fairchild *et al.*, 2000; Fankhauser and Chory, 2000; Soh *et al.*, 2000; Spiegelmann *et al.*, 2000).

SUB1 has been shown to encode a Ca^{2+} binding protein containing EF hand-like calcium binding motifs in its C-terminus (Gou *et al.*, 2001). Although SUB1 contains a basic region resembling a nuclear localization signal, SUB1-GUS fusions have shown to preferentially localize to the nuclear envelope, with some fusion protein remaining in the cytoplasm (Figure 1). In addition to altered B responsiveness *sub1* mutants also exhibit an alteration in FR-dependent hypocotyl elongation, suggesting that SUB1 may define a point of crosstalk between cry1 and phyA pathways. It appears that SUB1 functions as a negative regulator of B-dependent photomorphogenesis through an inhibitory effect on the light-induced HY5 accumulation. It has been suggested, due to its localization, that the mechanism of SUB1-mediated downregulation of HY5 involves changes in the nuclear trafficking of signal transduction molecules (Guo *et al.*, 2001; Figure 1).

HY5 encodes a basic leucine zipper-type transcription factor and is constitutively localized in the nucleus (Oyama *et al.*, 1997) where it binds to the G-box (CACGTG) motif of multiple light-inducible promoters driving their expression (Chattopadhyay *et al.*, 1998; Figure 1). The abundance and activity of HY5 in the nucleus is increased with increasing fluence rate of incident light and correlates with the expression level of light-induced genes (Osterlund *et al.*, 2000). Consistent with the observed mutant phenotypes, the abundance of HY5 is regulated by not only B but also R and FR.

HFR1 is a basic helix-loop-helix (bHLH) transcription factor (Fairchild *et al.*, 2000; Spiegelman *et al.*, 2000; Soh *et al.*, 2000) in the same family as the PIF3 (PHYTOCHROME-INTERACTING FACTOR 3) and PIF4 bHLH proteins that mediate phyB signalling (Toledo-Ortiz *et al.*, 2003). While the target promoters for HFR1 have not been described, PIF3 and PIF4 have both been shown to interact with G-box elements common to a number of light-regulated genes (Martinéz-Garcia *et al.*, 2000; Toledo-Ortiz *et al.*, 2003; Figure 1). If HFR1 does interact with G-box elements this could explain both, why *hfr1* mutant phenotypes are observed across a broad wavelength range (Fankhauser and Chory, 2000; Duek and Fankhauser, 2003), and how *hfr1* and *hy5* mutants could exhibit an additive interaction (Kim *et al.*, 2002).

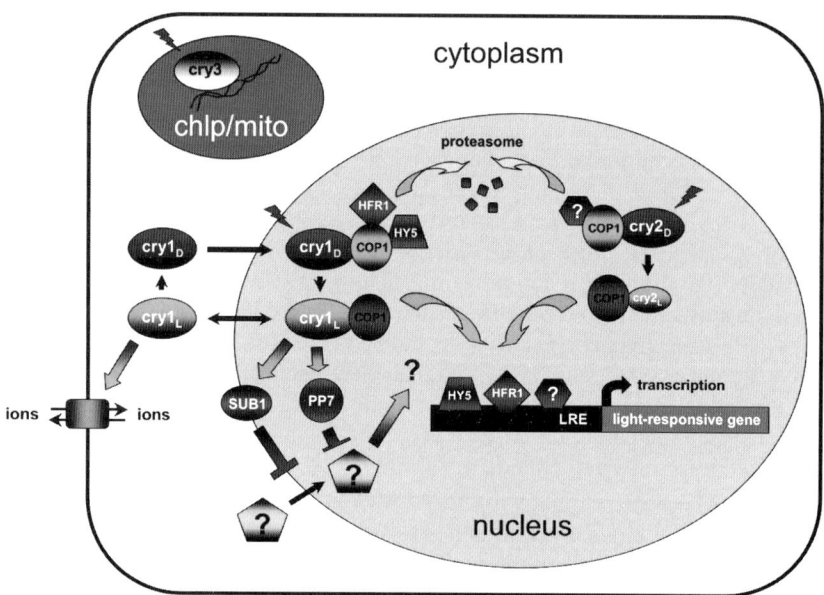

Figure 1. Depiction of the cryptochrome-associated signalling events discussed in the text. Dark and lit-states of cry2 and cry2 are shown together as they relate to intracellular localization and function. Question marks refer to inferred, but still unidentified, molecules.

In contrast to the few loci associated with cry photomorphogenic signalling by alteration in B responsiveness, a relatively large number of *COP/DET/FUS* loci have been identified whose mutants result in constitutive photomorphogenesis in the absence of light signals (Serino and Deng, 2003). Of particular interest to the current discussion of cry photomorphogenic signalling are loci that encode components of protein degradation machinery (Table 1), including the COP1 E3 ubiquitin ligase, the COP10 E2 ubiquitin-conjugating enzyme, and various subunits of the COP9 signalosome (CNS), which include COP8, COP9, COP11, FUS5, FUS11 and FUS12 (Serino and Deng, 2003). The CNS functions as a proteosome, together with COP1 and COP10, to regulate the abundance of positive transcriptional regulators of photomorphogenesis. Interestingly, HY5 has been shown to be a target for degradation by this complex in darkness (Osterlund *et al.*, 2000; Holm *et al.*, 2002; Seo *et al.*, 2003; Figure 1), providing a direct connection between the positive and negative regulators of cry responses. HFR1 has been shown to be partially required for a subset of COP1-triggered photomorphogenic responses in darkness, suggesting that HFR1 may also be a target for COP1 action (Kim *et al.*, 2002; Figure 1).

The key regulatory component of this system appears to be the cytoplasmic-nuclear partitioning of COP1; in the light COP1 accumulates in the cytoplasm rendering the CNS inactive relative to degradation of proteins involved in photomorphogenesis, while in darkness COP1 is nuclear in residence thus targeting proteins like HY5 to the CNS (Serino and Deng, 2003). It is interesting to note that both full length and CCT versions of cry1 and cry2 interact with COP1, although in a light-independent fashion (Wang *et al.*, 2001; Yang *et al.*, 2001). Based on these results it has been hypothesized that the B-activation of cry impairs the E3 ubiquitin ligase activity of COP1, thereby preventing the targeting of proteins such as HY5 to the CNS (Wang *et al.*, 2001; Yang *et al.*, 2001; Figure 1). Independent of light conditions CCT versions of cry1 and cry2 also appear capable of inhibiting COP1 activity, providing a simple explanation of the phenotypes observed in the CCT-OX plants (Lin and Shalitin, 2003).

Reversible protein phosphorylation represents a recurrent theme in signal transduction (Hunter, 1995), including pathways involved in the transduction of light signals in plants (Fankhauser and Chory, 1999; Fankhauser and Staiger, 2002). Both cry1 and cry2 are phosphorylated in response to B irradiation (see Chapter 11), and the phototropins and phytochromes are light-acted protein kinases (see Chapters 12 and 16, respectively). In the case of the cryptochromes, phosphorylation may represent a mechanism for both activation and desensitisation/degradation (Lin and Shalitin, 2003). In this context phosphorylation of cry1 by phyA could provide an explanation for the observed co-dependence of cry and phy-mediated responses (see Chapter 11). Recently, the novel serine/threonine protein phosphatase AtPP7 from *Arabidopsis* (Andreeva *et al.*, 1998) has been shown to function as a positive regulator of cry-mediated signal transduction (Møller *et al.*, 2003; Table 1). While, knock-down transgenic loss-of-function mutant AtPP7 plants had clear defects in B responsiveness, they exhibited no detectable alteration in photomorphogenesis under red or far-red monochromatic light, suggesting that AtPP7 functions exclusively in B signal transduction (Møller *et al.*, 2003). Although AtPP7 is constitutively localized in the nucleus (Andreeva and Kutuzov, 2001; Møller *et al.*, 2003) no physical interaction with cry1 has been detected. Møller and colleagues (2003) proposed that AtPP7 may function as a signalling checkpoint by dephosphorylating a nuclear localized cry1 signalling intermediate, thus sustaining or amplifying cry1 output signals (Figure 1).

2.1.2 Cryptochrome Signalling and Electrophysiological Processes

Changes in electrical potentials and currents have been associated with a number of B responses in plants, such as photomorphogenesis and phototropism, and are thought to occur through changes in ion channel and proton pump activities (Spalding and Cosgrove, 1992; Spalding, 2000; Figure 1). There is some evidence that hetero-trimeric G-proteins and/or second messengers, such as calcium, might represent intermediates in the signalling pathway from B receptor to electrical response (Spalding, 2000). Early studies in cucumber and beans showed that B-dependent inhibition of hypocotyl growth and stimulation of leaf lamina expansion is preceded by an intense membrane depolarisation of cells in the expansion zone of these tissue (Meijer, 1968; Spalding and Cosgrove, 1992). This phenomenon is extremely rapid and is preceded by an extensive activation of plasmamembrane-localized ion channels (Cho and Spalding, 1996), at least one of which is a Cl⁻ channel (Spalding and Cosgrove, 1992; Table 1). Transgenic *Arabidopsis* seedlings expressing the Ca^{2+} reporter protein aequorin have been used to demonstrate that the activation of anion channels by B occurs independent of the concentration of cytoplasmic Ca^{2+} ($[Ca^{2+}]_{cyt}$) (Lewis et al., 1997).

B has been shown to induce shrinkage of protoplasts from maize coleoptiles and has been proposed to have a causal correlation with the reduction of cell turgor pressure and inhibition of growth (Wang and Iino, 1997). It has been demonstrated that this response is dependent upon anion efflux and reduction of cell osmolarity due to the activation of anion channels, leading the authors to propose a cry-based model for anion channel-dependent growth inhibition in maize (Wang and Iino, 1997). Similar effects were observed in protoplasts from *Arabidopsis* hypocotyls (Wang and Iino, 1998), permitting genetic analyses of the receptors associated with this response. Studies employing a *cry1*-null mutant have shown that the initial fast growth-inhibition phase is independent of anion channel activity and cry1 (Parks et al., 1998). After approximately 30 min of continuous B irradiation, the onset of a genetically separable second phase of growth inhibition has been observed (Parks et al., 1998; Wang and Iino, 1998). This second phase is dependent on the activity of anion channels and mainly mediated by the cryptochromes (Parks et al., 1998; Wang and Iino, 1998; Folta and Spalding, 2001a).

By employing high-resolution analysis techniques, Folta and Spalding (2001a) have shown that both *Arabidopsis cry1* and *cry2* mutants are severely and equally impaired in B-induced plasma membrane depolarisation, which is surprising considering that the role of cry2 in hypocotyl growth inhibition during seedling establishment is less significant than that of cry1 (Lin et al., 1998). Again two distinct phases of B-mediated growth inhibition have been identified: (i) an early phototropin 1-dependent phase and (ii) a latter cry-dependent phase. Both phases have been shown to be interdependent, since the timing of the cry phase is set by the transient phot1-dependent phase, and in the absence of active cry, phot1 action delays the growth acceleration (Folta and Spalding, 2001a). The cry phase itself appears to be subdivided into an initial phase dependent on cry1/cry2 activation of anion channels and a later phase dependent solely on cry1 (Folta and Spalding, 2001a, 2001b). Interestingly, the signal transduction mechanism of cry1 and cry2 appears to change over time, since the hypocotyls of *cry2* mutants are not as long as those from *cry1* after days of B despite having indistinguishable phenotypes during the initial period of B irradiation (Folta and Spalding, 2001a). Thus, the requirement of cry2 is short-lived agreeing with its labile nature.

Folta and Spalding (2001b) have also shown that phyA is necessary for the cry1/cry2-mediated activity of anion channels and resultant membrane depolarisation in the first few seconds of growth inhibition under B. It is not clear at this point whether phyA activates the anion channels directly or indirectly via cry1/cry2. Similar to cry2, the requirement of phyA for the inhibition of hypocotyl growth appears to be transient and restrict to the initial stages of the cry-mediated phase (Folta and Spalding, 2001b). The authors have correlated the kinetics of the nuclear trafficking of phyA and cry with their changing roles in the inhibition of hypocotyl growth under B, proposing that the distinct mechanisms are brought about the changes in subcellular localization.

2.1.3 Cryptochrome Signalling and the Regulation of Gene Expression

In flowering plants, light regulation of gene expression occurs mainly at the transcriptional level (Silverthorne and Tobin, 1984; Batschauer *et al.*, 1994) through either transcriptional activation (Tobin and Kehoe, 1994) or repression (von Armin and Deng, 1996). Interestingly, opposite patterns of transcriptional regulation are often modulated by the same photoreceptor (Gilmartin *et al.*, 1990). Recent results from whole-genome array studies suggested that phot-mediated effects on gene expression are minimal, if present at all, while cry1-mediated changes are numerous (Ma *et al.*, 2001; Ohgishi *et al.*, 2002). Relative to signaling downstream of photoperception, the transcriptional regulation of photosynthesis-associated nuclear genes and genes encoding flavonoid biosynthetic enzymes by phytochromes has been proposed to involve distinct signal transduction pathways, one calcium/calmodulin-dependent and the other cGMP-dependent (Bowler *et al.*, 1994a, 1994b; Neuhaus *et al.*, 1993, 1997). B-induced transcription of *CHS* in *Arabidopsis* has been shown to involve Ca^{2+}, although not in an essential manner (Christie and Jenkins, 1996). A role for cGMP has not been reported for B-dependent changes in transcription.

In general, the presence of light-responsive elements (LRE) within the promoter sequences is essential and sufficient for photoreceptor-mediated regulation of transcription (Terzaghi and Cashmore, 1995; Millar and Kay, 1996). Transcriptional specificity or responsiveness to multiple photoreceptor systems seem to be dependent on the combinatorial interaction of distinct LREs in a given promoter context, with distinct photoreceptor-initiated signals converging on similar *cis*-acting elements (Martinez-Hernandez *et al.*, 2002; Yadav *et al.*, 2002; Figure 1). Yet, to date, no single element can be attributed to cry-regulated gene expression alone.

Cry1 and cry2 are also involved in the regulation of chloroplast-encoded genes in *Arabidopsis*, as demonstrated by a 75% reduction in plastid transcriptional activity in *cry1cry2* double mutants (Thum *et al.*, 2001). The recently characterized *Arabidopsis* cry3 is localized in chloroplasts and mitochondria and has been shown to bind DNA in a sequence independent manner (Kleine *et al.*, 2003; Figure 1). The cry3 paralog, cry DASH, from *Synechocyscitis* has been shown to function as a transcriptional repressor (Brudler *et al.*, 2003; Table 1). The evolutionary conservation of the DNA binding capacity in cry3 (Kleine *et al.*, 2003), cry DASH (Brudler *et al.*, 2003), mouse cry1 (Kobayashi *et al.*, 1998) and human cry2 (Özgür and Sancar, 2003) suggests this is an important feature of cry-mediated signalling.

3. PHOTOTROPIN SIGNALING

3.1 Phototropins and Plant Movement Responses

Phototropism, or the curvature of a plant organ towards or away from directional B, is one of the oldest known plant photoresponses (for example, see Darwin, 1880; Sachs, 1887). Yet, phototropism it is but one of several "movement" responses initiated by B in plants, that include stomatal and mesophyll chloroplast movement responses (Stone and Liscum, 2003). Over the past decade significant progress has been made in our understanding of the molecular and genetic components mediating these plant movements, not the least of which has been the identification of the phototropins as the primary photoreceptors mediating these responses. Here we will discuss the current state of knowledge relative to phototropin signalling associated with phototropism (section 3.1.1), stomatal aperture control (section 3.1.2), and chloroplast movements (section 3.1.3). In the last section (3.2) we will discuss particular electrophysiological responses associated with the phototropins, which may or may not ultimately impact one or more of the aforementioned responses.

3.1.1 Phototropins and Phototropism

In *Arabidopsis*, both phot1 and phot2 have been demonstrated to mediate the B-induced phototropism in etiolated seedlings, with phot1 responding over a broad range of fluence rates and phot2 responding to relatively high intensity B (Sakai *et al.*, 2001). Phot-mediated signal transduction is likely initiated by the light-induced formation of a flavin-C(4a) cysteinyl adduct (Crosson and Moffat, 2001; Iwata *et al.*, 2002; Crosson *et al.*, 2003) and transmitted to phot-signaling partners through the adduct state of the LOV domain (Crosson and Moffat, 2002; Crosson *et al.*, 2003; Harper *et al.*, 2003, 2004). A number of biochemical studies suggest that the C-terminal serine/threonine protein kinase domain of the phototropins is involved in transmission of phot signals (Short and Briggs, 1994; Briggs and Huala, 1999). However, it has recently been suggested that the kinase activity of the phototropins could be responsible for signal attenuation, rather than direct signal transmission (Christie *et al.*, 2002; Liscum, 2002). This signal attenuation, or desensitisation response, could result from alterations in a signalling complex (Liscum and Stowe-Evans, 2000; Liscum, 2002).

At least one component of a hypothesized phot1-complex has been identified, namely the NPH3 (NON-PHOTOTROPIC HYPOCOTYL 3) protein (Motchoulski and Liscum, 1999; Table 2; Figure 2). Mutations in the *NPH3* locus were identified in the same genetic screen that identified the *nph1/phot1* mutants (Liscum and Briggs, 1995, 1996). The *NPH3* locus encodes a novel plant-specific protein containing two recognized protein-protein interaction domains: an N-terminal BTB/POZ domain (BROAD COMPLEX, TRAMTRACK, BRIC-À-BRAC/ POXVIRUS AND ZINC FINGER) (Albagli *et al.*, 1995; Aravind and Koonin, 1999; Collins *et al.*, 2001) and a C-terminal coiled-coil domain (Lupas, 1996). Although highly hydrophilic and globular in nature, the NPH3 protein, like phot1 (Briggs and Huala, 1999), has been found to associate with the plasma membrane (Motchoulski and Liscum, 1999; Figure 2). Consistent with the mutant phenotypes, and concordance in intracellular localization, NPH3 was found to interact physically with phot1 in both a yeast two-hybrid and *in vitro* pull-down assays (Motchoulski and Liscum, 1999). This interaction occurs via the C-terminal coiled-coil-containing region of NPH3 and the N-terminal LOV domain-containing region of phot1. NPH3 has been shown to exhibit reversible phosphorylation (Motchoulski and Liscum, 1999). Although the relationship of this biochemical property to phot1 kinase activity is not entirely clear it is likely connected in some way, whether inductive or responsive, to the apparent B-induced dissociation of the phot1-NPH3 complex (Liscum, 2002).

Table 2. Phototropin-mediated responses and associated signaling components

Receptor	Response	Signaling element (biochemical identity)
phot1	phototropism	NPH3 (phot1-interacting protein; putative scaffold/adaptor protein)
		RPT2 (NPH3 paralogue)
		14-3-3 (?)
		NPH4/ARF7 (auxin-responsive transcriptional activator)
phot2	phototropism	unidentified NPH3 paralog (?)
phot1 and phot2 (?)	phototropism	indole-3-acetic acid (IAA, plant hormone)
		PIN1 (facilitator of auxin transport)
		PIN3 (facilitator of auxin transport)
		MDR1 (facilitator of auxin transport)
		PGP1 (facilitator of auxin transport)
phot1 and phot2	stomatal aperture control	H^+-ATPase
		14-3-3
		RPT2
		K^+ channel
phot1 and phot2	Chloroplast movements	cytoskeletal elements (e.g., CHUP1?)
phot1	Ca^{2+} influx from apoplast	plasma membrane Ca^{2+} channel/PACC1 (?)
phot2	Ca^{2+} flux from intracellular stores	unknown

Another apparent phototropin signalling intermediate is the NPH3 paralogue, RPT2 (ROOT PHOTOTROPISM 2) (Table 2; Figure 2). The *rpt2* mutant exhibits reduced phototropism in partially deetiolated seedlings under high intensity white light, while retaining nearly normal bending responses under low light conditions (Okada and Shimura, 1992; Sakai et al., 2000). These mutant phenotypes indicate that the RPT2 protein functions over a fluence rate range similar to that of phot2 (Sakai *et al.,* 2000, 2001). Interestingly *PHOT2* and *RPT2* show similar wavelength- and fluence ratedependencies for increase in transcript accumulation in response to light (Sakai *et al.,* 2000). Based on these observations and similarities in sequence and predicted structure of RPT2 to that of NPH3, it was proposed that RPT2 functions downstream of and physically interacts with phot2, similarly to the role proposed for NPH3 in phot1-mediated phototropism (Motchoulski and Liscum,

1999; Liscum, 2002). However, a recent study has demonstrated that RPT2, like NPH3, interacts with phot1 (Figure 2) but not phot2, suggesting that RPT2 functions in the phot1-specific pathway (Inada et al., 2004). Unlike the NPH3-phot1 interaction that is mediated by the coiled-coil-containing region of NPH3, the BTB/POZ domain of RPT2 appears to function as the phot1-interacting domain. Moreover, the BTB/POZ domain appears to mediate both RPT2 homodimerization and RPT2-NPH3 heterodimerization (Inada et al., 2004). While the mechanistic outcome of these interactions is currently unknown, these results, along with the differences in expression between *NPH3* and *RPT2*, suggest that a phot1-signalling complex can contain a variety of proteins depending upon the light condition the plant is experiencing. The variety of proteins involved in a phot1-signaling complex for phototropism could be even greater than that just discussed as a 14-3-3 protein has recently been shown to interact with phot1 in hypocotyl cells of etiolated seedlings (Kinoshita et al., 2003). Yet, to date there has been no demonstration of a direct role of the 14-3-3 protein in phototropism.

The final effector of the tropic growth response stimulated by B has long been thought to be the unequal distribution of plant hormone auxin between the proximal and the distal side of plant relative to the stimulus (Went and Thimann, 1937; Briggs et al., 1957; Friml and Palme, 2002; Friml, 2003; Table 2; Figure 2). But how is such a "lateral gradient" of auxin generated and what are molecular components and events between phototropin activation and this output? Although the picture of how a gradient of auxin is established across the diameter of the plant stem is far from complete, several aspects of auxin transport and localization have been elucidated and are providing enticing clues. For example, it is known that auxin is transported from the apical shoot region to other parts of a young seedling via both phloem-mediated routes and a directional "polar" transport system (Estelle, 1996; Friml and Palme, 2002). While auxin can passively diffuse into a cell from the apoplast, movement from the symplast to the apoplast requires an active transport system. This latter auxin efflux is facilitated by members of at least two transmembrane protein families; namely the PIN (PIN-FORMED) and MDR (MULTI-DRUG RESISTANCE)/PGP (P-GLYCOPROTEIN) families (Friml, 2003; Muday et al., 2003) (Table 2; Figure 2). A third class of proteins, the AUX (AUXIN RESISTANT 1)/LAX (LIKE-AUX1) proteins, function as auxin influx carriers to enhance the uptake from the apoplast (Parry et al., 2002). An asymmetric distribution of such facilitators and carriers within a cell provides a simple but effective means to establish polar/directional transport of auxin. It seems highly probably that phototropin signalling leads to an alteration in the activity or localization of auxin transport facilitators such that polar movement of the hormone is disrupted or modified and a lateral, rather than longitudinal, gradient of auxin forms. Noh and colleagues (2003) recently found that disruption of PIN1 polar localization in *mdr1pgp1* double mutant seedlings results in enhanced phototropic and gravitropic responses. Moreover, it has been reported that the basal localization of PIN1 within hypocotyl cells of etiolated *Arabidopsis* seedlings is disrupted by exposure to unidirectional B (Blakeslee et al., 2004). In particular, much reduced levels of PIN1 are found in the basal position of cells on the "shaded" versus "lit" side of the plant (Figure 2). While the mechanism underlying this PIN1 delocalisation is not understood, phot1 is clearly coupled to the response since *phot1*-null mutant seedlings fail to exhibit any obvious B-dependent change in PIN1 localization (Blakeslee *et al.*, 2004). These results suggest that by altering polar basipetal transport of auxin a larger lateral gradient of auxin can be established in response to tropic stimulation. It is plausible that the laterally-localized PIN3 protein (Friml *et al.*, 2002) facilitates an enhanced lateral conductance of auxin in this context (Figure 2).

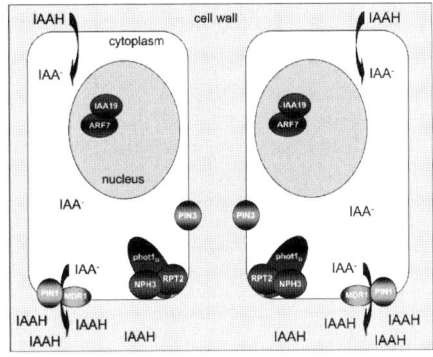

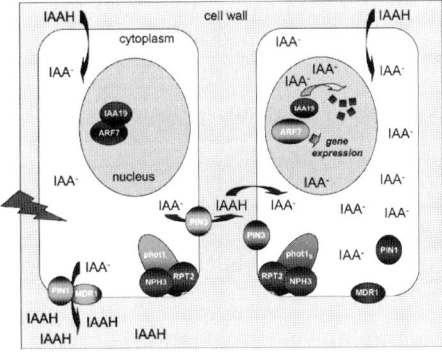

Figure 2. Depiction of the phototropin 1 signaling events associated with phototropism. In darkness (top panel) phot1 is in its resting-state, and as such, its signaling is not activated. In the absence of phot1 signaling basipetal polar auxin (IAA) transport predominates, as mediated by passive uptake of IAAH from the cell wall space and active efflux of the IAA anion via action of the PIN1-MDR1 (PGP1, not shown) complex. Also inactive in the dark is the ARF7 transcriptional machinery, as the ARF7 inhibitor, IAA19, persists. Upon unilateral B irradiation phot1 signaling is activated and a dramatic redistribution of auxin occurs. Although it is not clear at present how phot1 activates this process at least two events appear likely: 1) disruption of the the PIN1-MDR1 complex on the shaded side that thus disrupts efflux in that side, and 2) activation of laterally localized PIN3 that results in directional lateral flow of auxin. Both of these events lead to accumulation of auxin on the shaded side of the plant that in turn stimulate the degradation of IAA19 and subsequent activation of ARF7-dependent transcription.

While significant questions still remain relative to how phototropin signals influence auxin localization, it is clear, again through mutational studies, that auxin is an critical modulator of phot1-dependent phototropism (Liscum, 2002). Perhaps the strongest evidence in support of this thesis comes from studies of the *NPH4* (*NONPHOTOTROPIC HYPOCOTYL 4*) locus, identified via mutations that disrupt phototropic response to low fluence rate B (Liscum and Briggs, 1995, 1996; Table 2; Figure 2). *NPH4* was found to encode the auxin response factor ARF7 (Harper *et al.*, 2000), which had been previously shown to function as a transcriptional activator that can bind directly to auxin-responsive elements (AuxREs) in genes whose transcription is enhanced by the presence of auxin (Ulmasov *et al.*, 1999a, 1999b; Hagen and Guilfoyle, 2002; Liscum and Reed, 2002). The loss of phototropic responsiveness in the *nph4/arf7* mutant background indicates that auxin-dependent changes in transcription are an essential component in the phototropic signal-response pathway (Stowe-Evans *et al.*, 1998; Harper *et al.*, 2000; Liscum, 2002). This conclusion is further supported by the identification of mutations in the *MSG2* (*MASSUGU2*) locus encoding a second regulator of auxin-dependent gene expression (Tatematsu *et al.*, 2004; Table 2). In particular, dominant gain-of-function mutations in the *MSG2* gene, which encodes the auxin-dependent transcriptional repressor IAA19, were found to also result in aphototropic seedlings. Interestingly, MSG2/IAA19 was shown to heterodimerize with NPH4/ARF7 (Tatematsu *et al.*, 2004; Figure 2), consistent with the notion that IAA proteins function as transcriptional repressors by preventing activator ARFs from homodimerizing (Liscum and Reed, 2002). The gain-of-function alleles of *msg2/iaa19* are proposed to prevent NPH4/ARF7 function because of increased stability of the mutant iaa protein (Tatematsu *et al.*, 2004), rendering it less sensitive to auxin-stimulated degradation (Gray *et al.* 2001; Ramos *et al.* 2001; Zenser *et al.* 2001, 2003; Kepinski and Leyser, 2004). Together these studies of the *nph4/arf7* and *msg2/iaa19* mutants also suggest that the limited phot1dependent changes in gene expression observed by microarray studies of whole seedlings (Ohgishi *et al.*, 2002) likely do not accurately reflect the phot1-dependent changes that are possible.

3.1.2 Phototropins and Stomatal Aperture Control

Gas exchange in the aerial organs of terrestrial plants occurs to ensure optimal concentration of internal CO_2 for photosynthesis and prevent excessive transpiration (Salisbury and Ross, 1985). This process is regulated by the aperture of microscopic pores that are formed between a pair of highly specialized modified epidermal cells, called guard cells. Constraints on radial expansion of the guard cells allow for turgordependent expansion only along the long axis of the cells. Because each pair of guard cells is fixed in position at the two ends of their long axes, expansion results in an outward "bowing" of the cells leading to formation of the stomatal pore. Thus, changes in turgor pressure determine to what extent pores are open. For example, Bdependent increases in solute accumulation leads to uptake of H_2O and increased turgor pressure in the guard cells and thus greater stomata opening, whereas decreases in solute concentration reduces the turgor pressure, leading to stomata closure (Assmann, 1993, Dietrich *et al.*, 2001; Schroeder *et al.*, 2001). The principal solutes involved in turgor pressure changes are sucrose, K^+ and the accompanying anions, and their levels are regulated by environmental conditions, mainly B and diurnal cycles (Schroeder et al., 2001). The B effects occur through the activation of a plasma membrane H^+-ATPase whose action generates an electrical potential that in turn activates a voltage-gated K^+ channel (Schroeder et al., 2001; Figure 3).

It has recently been shown that phot1 and phot2 function redundantly to mediate perception of B signals that regulate stomatal aperture (Kinoshita et al., 2001; Figure 3). Signal transduction from the phototropins to the H^+-ATPase in the plasma membrane of the guard cells has been proposed to involve protein phosphorylation since serine/threonine photein kinase inhibitors completely impair B-induced H^+ pumping in *Vicia faba* guard cell protoplasts, as well as stomatal opening in *Commelina* (Assmann and Shimazaki, 1999; Table 2). The B-induced phosphorylation of the C-terminus of an H^+-ATPase and its association with 14-3-3 proteins has been positively correlated with proton pumping activity and stomatal opening (Kinoshita and Shimazaki, 2002). The binding of a 14-3-3 protein to the autoinhibitory C-terminal domain of the H^+-ATPase prevents its interaction with the catalytic domain leading to a high-activity state of the proton pump. The phosphoserine residues targeted to 14-3-3 protein-binding belong to the conserved consensus motifs RSXpSXP and RXY/FXpSXP in metazoans (Muslin et al., 1996; Yaffe et al., 1997), but are absent in plant H^+-ATPase isoforms. Axelsen et al., (1999) have found two potential 14-3-3 binding sites in the plant H^+-ATPases AHA2 (from *Arabidopsis*) and VHA1 and VHA2 (from *Vicia*) that are distinct from those conserved in metazoan H^+-ATPases. Using synthetic phosphopeptides designed from the C-terminus of VHA1, Kinoshita and Shimazaki (2002) have shown that 14-3-3 binding is required for functional activation of the guard cell proton pump. However, phosphorylation independent binding of a 14-3-3 protein to the H^+-ATPases has also been observed, suggesting that the role of phosphorylation is to stabilize a pre-existent less specific interaction between the 14-3-3 and several residues in the autoinhibitory region of the H^+-ATPase (Fuglsang et al., 2003; Table 2). Recently Kinoshita and colleagues (2003) have shown that phot1 and phot2 both interact with a 14-3-3 protein in guard cells of *Vicia*. The recent finding that *rpt2* mutants are defective in the stomatal response to B (Inada et al., 2004) suggests that the phototropins, a 14-3-3 protein, RPT2 and the H^+-ATPase can exist in a single complex, thus providing a potential direct connection between the B-activated kinase activity of the phototropins and phosphorylation-dependent activation of the H^+-ATPase (Figure 3).

In *Arabidopsis*, 14-3-3 proteins constitute a large and diverse protein family, with distinct localization and specificity for interacting proteins (Wu et al., 1997; Sehnke et al., 2002). Yet, the function of 14-3-3 proteins appears to be highly conserved, even in distinct organisms; namely to alter the activity state of an interacting protein (Sehnke et al., 2002). It is possible that members of the NPH3/RPT2 family act together with 14-3-3 proteins, providing specificity through combinatorial arrangement between the members of the two families of adapter/scaffold proteins. The functional characterization of the members of the 14-3-3 and NPH3/RPT2 families will provide means to investigate further their role in phot-mediated signal transduction.

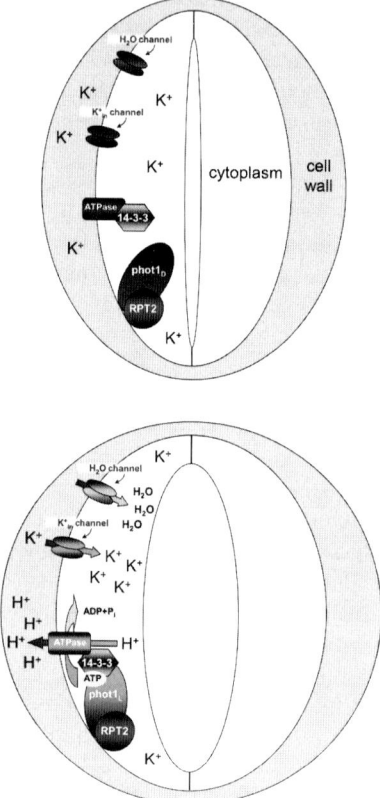

Figure 3. *Depiction of the phototropin signaling events associated with B-dependent stomatal aperture control. In darkness (or B deficient light) the phot signaling system is in its "off-state" and thus the guard cells exhibit no appreciable ion gradient across the plasma membrane (top panel). In the absence of such a gradient no water uptake occurs and a small stomatal pore is maintained. Upon exposure to B (bottom panel), the phot signaling is activated (activation is shown in only one cell for simplicity, but the mirror image processes occur in the second guard cell) whereby the H^+-ATPase is activated leading to a membrane hyperpolarization that activates inward rectifying K^+ channels. The resultant K^+ gradient in turn drives the uptake of H_2O and turgor driven cell expansion, which leads to an opening of the pore.*

3.1.3 Phototropins and Chloroplast Movement

Like all light responses chloroplast photorelocation requires the perception of a light stimulus, transmission of signal from receptor to the effectors, and the effectors themselves (Kagawa, 2003; Sato *et al.*, 2003; Wada *et al.*, 2003). B can induce both chloroplast accumulation and avoidance responses in all plant species examined to date. However, in some cryptogamous plants R is also an affective inducer of these responses (Wada *et al.*, 2003). In *Arabidopsis*, phot1 and phot2 have been shown to act redundantly in the control of low-light accumulation response, while phot2 appears solely responsible for the avoidance response (Jarillo *et al.*, 2001; Kagawa *et al.*, 2001; Sakai *et al.*, 2001). Detailed observation of the phenomenon has shown that phot1 mediates only the accumulation response even under high-fluence-rate B and phot2 mediates accumulation movements at low-fluence rates and avoidance movements at high-fluence rates (Sakai *et al.*, 2001; Sato *et al.*, 2003; Wada *et al.*, 2003). Doi and colleagues (2004) have recently shown that a *PHOT1* transgene can complement the chloroplast accumulation defect of *phot1phot2* double mutants, providing additional support for the notion that phot1 and phot2 function as redundant receptors for this response.

The nature of the signal(s) transferred from the photoreceptors to the chloroplasts remains unknown, but data from microbeam irradiation experiments provide some helpful insight into the mechanism (Kagawa and Wada, 1999). For example, irradiation of a small area of a dark-adapted cell with high intensity B stimulates chloroplast movement from anticlinal wall positions toward positions close to, but not within, the beam. Upon termination of the B irradiation the chloroplasts also move to the region where the beam had been. These results suggest that: (i) high-fluence-rate B generates both accumulation and avoidance signals; (ii) the signal for accumulation movements travels a relatively long distance, but the one for avoidance movements stays within the irradiated area; (iii) the signal for avoidance movements is dominant over the one for accumulation movements during illumination; and (iv) the duration of the signal for accumulation responses is longer than that for avoidance movements (Kagawa and Wada, 1999; Wada *et al.*, 2003). However, it remains unclear whether accumulation and avoidance responses share the same signal(s).

Chloroplast movement is ultimately accomplished through the function of the cytoskeleton, thus the signals from the photoreceptors have to be transmitted to those elements. Specific cytosolic inhibitors have been used to study the motile system responsible for chloroplast photomovement in plant cells (Kadota and Wada, 1992; Tlalka and Gabrys, 1993; Malec *et al.*, 1996). In cryptogams, such as *Physcomitrella patens*, both actin filaments and microtubules appear essential from proper chloroplast movement responses (Sato *et al.*, 2003). In contrast, in higher plants, only actin filaments appear necessary. For example, Kandasamy and Meagher (1999) have used immunofluorescence microscopy to demonstrate that chloroplasts move along actin filaments in *Arabidopsis*. Structural studies suggest that a pre-polymerised network of actin filaments is utilized (Kandasamy and Meagher, 1999). However, it remains possible that freshly polymerised actin filaments also play a role (Wada *et al.*, 2003). The availability of GFP/RFP translational fusions with actin-binding proteins (Kost *et al.*, 1998) should provide a powerful means to study the involvement of cytoskeleton in phot-mediated chloroplast relocation. Genetic support for the involvement of actin elements in phot-mediated chloroplast movements comes the identification of the *chup* mutants of *Arabidopsis* (Oikawa *et al.*, 2003). *CHUP* encodes a modular actin-binding protein that appears to play a critical role in chloroplast positioning and movement since mesophyll chloroplasts of the *chup* mutants aggregate along the bottom of the cell farthest from the adaxial surface of the leaf, independent of light condition (Oikawa *et al.*, 2003; Table 2). Although CHUP binds actin, the gross actin filament distribution is not disrupted in *chup* mutants (Oikawa *et al.*, 2003), making it unclear how this interaction may influence chloroplast movements.

3.2 Phototropin Signalling and Electrophysiological Processes

As already introduced, B can activate a variety of plasma membrane ion channels. These channel activities lead to changes in the electrical potential of the cell surface and initiation and/or alteration of signal transduction (Spalding, 2000). Both phot1 and phot2 have been demonstrated to be plasma membrane-localized (Christie *et al.*, 2002; Sakamoto and Briggs, 2002; Harada *et al.*, 2003). All three of the currently identified phot1-interacting proteins (NPH3, RPT2, and the 14-3-3 protein) are also plasma-membrane-localized (Motchoulski and Liscum, 1999; Kinoshita *et al.*, 2003; Inada *et al.*, 2004). The common intracellular localization of these proteins with B-activated ion channels suggests that changes in membrane potential and ion channel activities may represent critical phot-signalling elements. As discussed below several recent studies provide support for this hypothesis.

In guard cells, phot1 and phot2 function redundantly to increase the activity of a plasma membrane H^+-ATPase (Kinoshita *et al.*, 2001). Although this activation could potentially occur via direct influence of the phototropins, it could also occur via changes in other ion channel activities. For example, it has been suggested that B activation of plasma membrane H^+-ATPase and anion channel activities in *Arabidopsis* and *Vicia* are coupled to Ca^{2+}-channel activities (Lewis *et al.*, 1997; Shimazaki *et al.*, 1999; Table 2). For example, transgenic *phot1* mutant seedlings expressing aequorin exhibit a reduced transient increase in $[Ca^{2+}]_{cyt}$ compared to both wild-type and *cry1cry2* plants (Baum *et al.*, 1999). Babourina *et al.*, (2002) used a non-invasive ion-selective microelectrode to show that the influx of apoplastic Ca^{2+} in hypocotyls of etiolated *Arabidopsis* seedlings is completely impaired in *phot1* but not in *phot2* single mutants, indicating that in dark-grown seedlings phot1 is the sole photoreceptor mediating Ca^{2+} influx from extracellular spaces. In contrast, Harada *et al.*, (2003) have shown that both phototropins appear to influence Ca^{2+} channel activity in leaf mesophyll cells. Although the observed transient increase in $[Ca^{2+}]_{cyt}$ was equally induced by phot1 and phot2, their role is partially redundant, with phot1 being responsible for response to lower fluence rate B and phot2 for response to higher fluence rate B. Moreover, by employing specific Ca^{2+} channel blockers and Ca^{2+} chelating agents, the authors Harada and colleagues (2003) demonstrated that both phototropins are capable of inducing the Ca^{2+} influx from the apoplast through the Ca^{2+} channel in the plasma membrane. A potential phot-activated calciumpermeable influx channel may be PACC1 identified by Stoelzle and colleagues (2003) (Table 2).

4. CONCLUDING REMARKS

The isolation of the cryptochromes and phototropins in plants has allowed the discovery of orthologuous molecules in other organisms, unveiling distinct roles and providing clues about the evolution of these photoreceptors. The initial photoreaction of phototropins is the best-characterized among plant photoreceptors and downstream steps are likely to involve (possibly phosphorylation-dependent) changes in the activity of a number of plasma membrane proteins, including H-ATPases in guard cells, ion transporters and channels in leaf cells, and/or regulators of auxin transport in hypocotyl cells. Much less is known about the photochemistry and the initial steps of cryptochrome-mediated signal transduction. However, the nuclear localization of cryptochromes and the physical interaction with COP1 predict a short signal-transduction pathway. Future biochemical analyses of protein-protein interactions and use of genetics (both forward and reverse) will continue to provide new insights into phototropin and cryptochrome signalling. The use of genome-wide data collection, proteomics, and mathematical modelling will help to decipher complex regulatory interactions between phot- and cry-dependent processes and their interactions with other signalling systems. The coming years should be very informative indeed!

Acknowledgements: Unpublished works from the anchor authors' laboratory were supported by grants from the U.S. National Science Foundation (MCB-0077312 and DBI-0114992).

5. REFERENCES

Ahmad, M., Lin, C., Cashmore, A. R. (1995) Mutations throughout an *Arabidopsis* blue-light photoreceptor impair blue-light responsive anthocyanin accumulation and inhibition of hypocotyl elongation. *Plant J, 8*, 653-658.

Albagli, O., Dhordain, P., Deweindt, C., Lecocq, G., Leprince, D. (1995) The BTB/POZ domain: a new protein-protein interaction motif common to DNA- and actin-binding proteins. *Cell Growth Differ, 6*, 1193-1198.

Andreeva, A. V. and Kutuzov, M. A. (2001) Nuclear localization of the plant protein Ser/Thr phosphatase PP7. *Mol Cel Biol Res Commun, 4*, 345-352.

Andreeva, A. V., Evans, D. E., Hawes, D. R., Bennett, N., Kutuzov, M. A. (1998) PP7, a plant phosphatase representing a novel evolutionary branch of eukaryotic ser-thr protein phosphatases. *Biochem Mol Biol Int, 44*, 703-715.

Aravind, L. and Koonin, E. V. (1999) Fold prediction and evolutionary analysis of the POZ domain: structural and evolutionary relationship with the potassium channel tetramerization domain. *J Mol Biol, 285*, 1353-1361.

Assmann, S. M. (1993) Signal transduction in guard cells. *Annu Rev Cell Biol, 9*, 345-375.

Assmann, S. M. and Shimazaki, K. (1999) The multisensory guard cell. Stomatal responses to blue light and abscisic acid. *Plant Physiol, 119*, 809-815.

Axelsen, K. B., Venema, K., Jahn, T., Baunsgaard, L., Palmgren, M. G. (1999) Molecular dissection of the C-terminal regulatory domain of the plant plasma membrane H^+-ATPase AHA2: mapping of residues that when altered give rise to an activated enzyme. *Biochemistry, 38*, 7227-7234.

Babourina, O., Newman, I., Shabala, S. (2002) Blue light-induced kinetics of H and Ca fluxes in etiolated wild-type and phototropin-mutant *Arabidopsis* seedlings. *Proc Natl Acad Sci USA, 99*, 2433-2438.

Batschauer, A., Gilmartin, P. M., Nagy, F., Schäfer E. (1994) "The molecular biology of photoregulated genes." In *Photomorphogenesis in Plants - 2nd Edition*, Kendrick R. E. and Kronenberg, G. H. M. (eds) Kluwer Academic Publishers, Dordrecht, 559-599.

Baum, G., Long, J. C., Jenkins, G. I., Trewavas, A. J. (1999) Stimulation of the blue light phototropic receptor NPH1 causes a transient increase in cytosolic Ca^{2+}. *Proc Natl Acad Sci USA*, 96, 13554–13559.

Blakeslee, J. J., Bandyopadhyay, A., Peer, W. A., Makam, S. N., Murphy, A. S. (2003) Relocalization of the PIN1 auxin efflux facilitator plays a role in phototropic responses. *Plant Physiol*, 134, 28-31.

Bowler, C., Neuhaus, G., Yamagata, H., Chua, N.-H. (1994a) Cyclic GMP and calcium mediate phytochrome phototransduction. *Cell*, 77, 73-81.

Bowler, C., Yamagata, H., Neuhaus, G., Chua, N.-H. (1994b) Phytochrome signal transduction pathways are regulated by reciprocal control mechanisms. *Genes Dev*, 8, 2188-2020.

Briggs, W. R. and Huala, E. (1999) Blue-light photoreceptors in higher plants. *Ann Rev Cell Dev Biol*, 15, 33-62.

Briggs, W. R., Tocher, R. D., Wilson, J. F. (1957) Phototropic auxin redistribution in corn coleoptiles. *Science*, 126, 210-212.

Brudler, R., Hitomi, K., Daiyasu, H., Toh, H., Kucho, K.-I., Ishiura, M., et al. (2003) Identification of a new cryptochrome class: structure, function and evolution. *Mol Cell*, 11, 59-67.

Cashmore, A. R., Jarillo, J. A., Wu, Y.-J., Liu, D. (1999) Cryptochromes: blue light receptors for plants and animals. *Science*, 284, 760-765.

Chattopadhyay, S., Ang, L. H., Puente, P., Deng, X.-W., Wei N. (1998) *Arabidopsis* bZIP protein HY5 directly interacts with light-responsive promoters in mediating light control of gene expression. *Plant Cell*, 10, 673-683.

Cho, M. H. and Spalding, E. P. (1996) An anion channel on *Arabidopsis* hypocotyls activated by blue light. *Proc Natl Acad Sci USA*, 93, 8134-8138.

Christie, J. M. and Jenkins, G. I. (1996) Distinct UV-B and UV-A/blue-light mediated signal transduction pathways induce chalcone synthase gene expression in *Arabidopsis* cells. *Plant Cell*, 8, 1555-1567.

Christie, J. M., Swartz, T. E., Bogomolni, R. A., Briggs, W. R. (2002) Phototropin LOV domains exhibit distinct roles in regulating photoreceptor function. *Plant J*, 32, 205-219.

Collins, T., Stone, J. R., William, A. J. (2001) All in the family: the BTB/POZ, KRAB, and SCAN domains. *Mol Cel Biol*, 21, 3609-3615.

Crosson, S. and Moffat, K. (2001) Structure of a flavin-binding plant photoreceptor domain: Insights into light-mediated signal transduction. *Proc Natl Acad Sci USA*, 98, 2995-3000.

Crosson, S. and Moffat, K. (2002) Photoexcited structure of a plant photoreceptor domain reveals a light-driven molecular switch. *Plant Cell*, 14, 1-9.

Crosson, S., Rajagopal, S., Moffat K. (2003) The LOV domain family: photoresponsive signaling modules coupled to diverse output domains. *Biochemistry*, 42, 2-10.

Darwin, C. *The Power of Movement in Plants.* D. New York: Appleton and Company,. 1881.

Dietrich, P., Sanders, D., Hedrich, R. (2001) The role of ion channels in light-dependent stomatal opening. *J Exp Bot*, 52, 1959-1967.

Doi, M., Shigenaga, T., Emi, T., Kinoshita, T., Shimazaki K. (2004) A transgene encoding a blue-light receptor, phot1, restores blue-light responses in the *Arabidopsis photphot2* double mutant. *J Exp Bot*, 55, 517-523.

Duek, P. D. and Fankhauser, C. (2003) HFR1, a putative bHLH transcription factor, mediates both phytochrome A and cryptochrome signaling. *Plant J*, 34, 827-836.

Estelle, M. (1996) Plant tropisms: the ins and outs of auxin. *Curr Biol*, 6, 1589-1591.

Fairchild, C. D., Shumaker, M. A., Quail P. H. (2000) HFR1 encodes an atypical bHLH protein that acts in phytochrome A signal transduction. *Genes Dev*, 14, 2377-2391.

Fankhauser, C. and Chory, J. (1999) Photomorphogenesis: light receptor kinases in plants! *Curr Biol*, 9, 123-126.

Fankhauser, C. and Chory, J. (2000) RSF1, an *Arabidopsis* locus implicated in phytochrome A signalling. *Plant Physiol*, 124, 39-45.

Fankhauser, C. and Staiger, D. (2002) Photoreceptors in *Arabidopsis thaliana*: light perception, signal transduction and entrainment of the endogenous clock. *Planta*, 216, 1-16.

Folta, K. M. and Spalding, E. P. (2001a) Unexpected roles for cryptochrome 2 and phototropin revealed by high-resolution analysis of blue light-mediated hypocotyl growth inhibition. *Plant J*, 26, 471-478.

Folta, K. M. and Spalding, E. P. (2001b) Opposing roles of phytochrome A and phytochrome B in early cryptochrome-mediated growth inhibition. *Plant J, 28*, 333-340.

Friml, J. (2003) Auxin transport – shaping the plant. *Curr Opin Plant Biol, 6,* 7-12.

Friml, J. and Palme, K. (2002) Polar auxin transport: old questions and new concepts? *Plant Mol Biol, 49,* 273-284.

Friml, J., Wisniewska, J., Benková, E., Mendgen, K., Palme, K. (2002) Lateral relocation of auxin efflux regulator PIN3 mediates tropism in *Arabidopsi*s. *Nature, 415,* 806-809.

Fuglsang, A. T., Borch, J., Bych, K., Jahn, T. P., Roepstorff, P., Palmgreen, M. G. (2003) The binding site for regulatory 14-3-3 protein in the plant plasma membrane H -ATPase: involvement of a region promoting phosphorylation-independent in addition to phosphorylation-dependent C-terminal end. *J Biol Chem, 278,* 42266-42272.

Gilmartin, P. M., Sarokin, L., Memmelink, J., Chua, N.-H. (1990) Molecular light switches for plant genes. *Plant Cell, 2,* 369-378.

Gray, W. M., Kepinski, S., Rouse, D., Leyser, O., Estelle, M. (2001) Auxin regulates SCF TIR1-dependent degradation of AUX/IAA proteins. *Nature, 414,* 271-276.

Guo, H., Mockler, T., Duong, H., Lin, C. (2001) SUB1, an Arabidopsis Ca^{2+}-binding protein involved in cryptochrome and phytochrome coaction. *Science, 291,* 487-490.

Hagen, G. and Guilfoyle, T. (2002) Auxin-responsive gene expression: genes, promoters and regulatory factors. *Plant Mol Biol, 49,* 373-385.

Harada, A., Sakai, T., Okada, K. (2003) phot1 and phot2 mediate blue light-induced transient increases in the cytosolic Ca^{2+} differently in *Arabidopsis* leaves. *Proc Natl Acad Sci USA, 100,* 8583-8588.

Harper, R. M., Stowe-Evans, E. L., Luesse, D. R., Muto, H., Tatematsu, K., Watahiki, M. K., *et al.* (2000) The *NPH4* locus encodes the auxin response factor ARF7, a conditional regulator of differential growth in aerial *Arabidopsis* tissue. *Plant Cell, 12,* 757-770.

Harper, S. M., Neil, L. C., Day, I. J., Hore, P. J., Gardner, K. H. (2004) Conformational changes in a photosensory LOV domain monitored by time-resolved NMR spectroscopy. *J Am Chem Soc, 126,* 3390-3391.

Harper, S. M., Neil, L. C., Gardner, K. H. (2003) Structural basis of a phototropin light switch. *Science, 301,* 1541-1544.

Holm, M., Ma, L. G., Qu, L. J., Deng, X.-W. (2002) Two interacting bZIP proteins are direct targets of COP1-mediated control of light-dependent gene expression in *Arabidopsis*. *Genes Dev, 16,* 1249-1259.

Inada, S., Ohgishi, M., Mayama, T., Okada, K., Sakai, T. (2004) RPT2 is a signal transducer involved in phototropic response and stomatal opening by association with phototropin 1 in Arabidopsis thaliana. *Plant Cell, 16,* 887-896.

Iwata, T., Tokutomi, S., Kandori, H. (2002) Photoreaction of the cysteine S-H group in the LOV2 domain of *Adiantum* phytochrome3. *J Am Chem Soc, 124,* 11840-11841.

Jarillo, J. A., Gabrys, H., Capel, J., Alonso, J. M., Ecker, J. R., Cashmore, A. R. (2001) Phototropin-related NPL1 controls chloroplast relocation induced by blue light. *Nature, 410,* 952-954.

Kagawa, T. (2003) The phototropin family as photoreceptors for blue light-induced chloroplast relocation. *J Plant Res, 116,* 77-82.

Kagawa, T. and Wada, M. (1999) Chloroplast-avoidance response induced by high-fluence blue light in prothallial cells of the fern *Adiantum capillus-veneris* as analyzed by microbeam irradiation. *Plant Physiol, 119,* 917-924.

Kagawa, T., Sakai, T., Suetsugu, N., Oikawa, K., Ishiguro, S., Kato, T., *et al.* (2001) *Arabidopsis* NPL1: A phototropin homolog controlling the chloroplast highlight avoidance response. *Science, 291,* 2138-2141.

Kandasamy, M. K. and Meagher, R. B. (1999) Actin-organelle interaction: association with chloroplast in *Arabidopsis* leaf mesophyll cells. *Cell Motil Cytoskel, 44,* 110-118.

Kendrick, R. E. and Kronenberg, G. H. M. (eds) *Photomorphogenesis in Plants* −2nd Edition. Dordrecht: Kluwer Academic, 1994.

Kepinski, S. and Leyser, O. (2004) Auxin-induced SCF^{TIR1}-AUX/IAA interaction involves stable modification of the SCF^{TIR1} complex. *Proc Natl Acad Sci USA, 101,* 12381-12386.

Kim, Y.-M., Woo, J.-C., Song, P.-S., Soh, M.-S. (2002) HFR1, a phytochrome A-signalling component, acts in a separate pathway from HY5, downstream of COP1 in *Arabidopsis thaliana*. *Plant J, 30,* 711-719.

Kinoshita, T. and Shimazaki, K. (1999) Blue light activates the plasma membrane H$^+$-ATPase by phosphorylation of the C-terminus in stomatal guard cells. *EMBO J, 18*, 5548-5558.

Kinoshita, T., Doi, M., Suetsugu, N., Kagawa, T., Wada, M., Shimazaki, K. (2001) phot1 and phot2 mediate blue light regulation of stomatal opening. *Nature, 414*, 656–660.

Kinoshita, T., Emi, T., Tominga, M., Sakamoto K., Shigenaga A., Doi M., *et al.* (2003) Bluelight- and phosphorylation-dependent binding of a 14-3-3 protein to phototropins in stomatal guard cells of broad bean. *Plant Physiol, 133*, 1453-1463.

Kleine,T., Lockhart, P., Batschauer, A. (2003) An *Arabidopsis* protein closely related to *Synechocystis* cryptochrome is targeted to organelles. *Plant J, 35*, 93-103.

Kobayashi, Y., Kanno, S., Smit, B., van der Horst, G. T., Takao, M., Muijtjens, M., *et al.* (1998) Characterization of photolyase/blue-light receptor homologs in mouse and human cells. *Nuc Acid Res, 26*, 5086-5092.

Koornneef, M., Rolff, E., Spruit, C. J. P. (1980) Genetic control of light-inhibited hypocotyl elongation in *Arabidopsis thaliana* (L.) Heynh. *Z Pflanzenphysiol, Bd 100*, 147-160.

Kost, B., Spielhofer, P., Chua, N.-H. (1998) A GFP-mouse talin fusion protein labels plant actin filaments *in vivo* and visualizes the actin cytoskeleton in growing pollen tubes. *Plant J, 16*, 393- 401.

Lewis, B. D., Karlin-Neumann, G., Davis, R. W., Spalding, R. P. (1997) Ca^{2+}-activated anion channels and membrane depolarizations induced by blue light and cold in *Arabidopsis* seedlings. *Plant Physol, 114*, 1327-1334.

Lin, C. and Shalitin, D. (2003) Cryptochrome structure and signal transduction. *Annu Rev Plant Biol, 54*, 469-496.

Lin, C., Yang, H., Guo, H., Mockler, T., Chen, J., Cashmore A. R. (1998) Enhancement of blue-light sensitivity of *Arabidopsis* seedlings by a blue light receptor cryptochrome 2. *Proc Natl Acad Sci USA, 95*, 2682-2690.

Liscum, E and Reed, J. W (2002) Genetics of Aux/IAA and ARF action in plant growth and development. *Plant Mol Biol, 49*, 387-400.

Liscum, E. (2002) "Phototropism: mechanisms and outcome." In *The Arabidopsis Book*, Somerville C. R. and Meyerowitz E. M. (eds) American Society of Plant Biologists: Rockville (doi/10.1199/tab.0042, http://www.aspb.org/publications/arabidopsis).

Liscum, E. and Briggs, W. R. (1995) Mutations in the *NPH1* locus of *Arabidopsis* disrupt the perception of phototropic stimuli. *Plant Cell, 7*, 473- 485.

Liscum, E. and Briggs, W. R. (1996) Mutations of *Arabidopsis* in potential transduction and response components of the phototropic signaling pathway. *Plant Physiol, 112*, 291- 296.

Liscum, E. and Stowe-Evans, E. L. (2000) Phototropism: A "simple" physiological response modulated by multiple interacting photosensory-response pathways. *Photochem Photobiol, 72*, 273-282.

Lupas, A. (1996) Coiled-coils: new structures and new families. *Trends Biochem Sci, 21*, 375-382.

Ma, L., Li, J., Qu, L., Hager, J., Chen, Z., Zhao, H., *et al.* (2001) Light control of *Arabidopsis* development entails coordinated regulation of genome expression and cellular pathways. *Plant Cell, 12*, 2589-2607.

Malec, P., Rinaldi, R. A., Gabrys, H. (1996) Light-induced chloroplast movements in *Lemna trisulca* identification of the motile system. *Plant Sci, 120*, 127- 137.

Martinéz-Garcia, J. F., Huq, E., Quail, P. H. (2000) Direct targeting of light signals to a promoter element-bound transcription factor. *Science, 288*, 858-863.

Martinez-Hernandez, A., López-Uchoa, L., Argüello-Astorga, G., Herrera-Estrella, L. (2002) Functional properties and regulatory complexity of a minimal *RBCS* light-responsive unit activated by phytochrome, cryptochrome, and plastid signals. *Plant Physiol, 128*, 1223-1233.

Mazzella, M. A., Cerdán, P. D., Stateline, R. J., Casal, J. J. (2001) Hierarchical coupling of phytochromes and cryptochromes reconciles stability and light modulation of *Arabidopsis* development. *Development, 128*, 2291-2299.

Meijer, G. (1968) Rapid growth inhibition of gherkin hypocotyls in blue light. *Acta Bot Neel, 17*, 9-14.

Millar, A. J. and Kay, S. A. (1996) Integration of circadian and phototransduction pathways in the network controlling *CAB* gene transcription in *Arabidopsis*. *Proc Natl Acad Sci USA, 93*, 15491-15496.

Mockler, T. C., Guo, H., Yang, H., Duong, H., Lin, C. (1999) Antagonistic actions of *Arabidopsis* cryptochromes and phytochrome B in the regulation of floral induction. *Development, 126*, 2073-2082.

Møller, S. G., Kim, Y.-S., Kunkel, T., Chua, N.-H. (2003) PP7 is a positive regulator of blue light signaling in *Arabidopsis. Plant Cell, 15*, 1111-1119.
Motchoulski, A. and Liscum, E. (1999) *Arabidopsis* NPH3: a NPH1 photoreceptor-interacting protein essential for phototropism. *Science, 286*, 961-964.
Muday, G. K., Peer, W. A., Murphy, A. S. (2003) Vesicular cycling mechanisms that control auxin transport polarity. *Trends Plant Sci, 8*, 301-304.
Muslim, A. J., Tanner, J. M., Allen, P. M., Shaw, A. S. (1996) Interaction of 14-3-3 protein with signaling proteins is mediated by the recognition of phosphoserine. *Cell, 84*, 889-897.
Neuhaus, G., Bowler, C., Hiratsuka, K., Yamagata, H., Chua, N.-H. (1997) Phytochrome regulated repression of gene expression requires calcium and cGMP. *EMBO J, 16*, 2534-2564.
Neuhaus, G., Bowler, C., Kern, R., Chua, N.-H. (1993) Calcium/calmodulin-dependent and –independent phytochrome signal transduction pathways. *Cell, 73*, 937-952.
Noh, B., Bandyopadhyay, A., Peer, W. A., Spalding, E. P., Murphy, A. S. (2003) Enhanced gravi-and phototropism in plant *mdr* mutants mislocalizing the auxin efflux protein PIN1. *Nature, 423*, 999-1002.
Ohgishi, M., Sakai, T., Okada, K. (2002) Analysis of blue-light signaling pathways using a *cry1cry2phot1phot2* quadruple mutant. (Abstract 9-28). 11th. International Conference on *Arabidopsis* Research, Seville, Spain.
Oikawa, K., Kasahara, M., Kiyosue, T., Kagawa, T., Suetsugu, N., Takahashi, F., *et al.* (2003) CHLOROPLAST UNUSUAL POSITIONING1 is essential for proper chloroplast positioning. *Plant Cell, 15*, 2815-2825.
Okada, K. and Shimura, Y. (1992) Mutational analysis of root gravitropism and phototropism of *Arabidopsis thaliana* seedlings. *Aust J Plant Physiol, 19*, 439-448.
Osterlund, M. T., Hardtke, C. S., Wei, N., Deng, X.-W. (2000) Targeted destabilization of HY5 during light-regulated development of *Arabidopsis. Nature, 405*, 462-466.
Oyama, T., Shimura, Y., Okada, K. (1997) The *Arabidopsis HY5* gene encodes a bZIP protein that regulates stimulus-induced development of root and hypocotyl. *Genes Dev, 11*, 2983-2995.
Özgür, S. and Sancar, A. (2003) Purification and properties of human blue-light photoreceptor cryptochrome 2. *Biochemistry, 42*, 2926-2932.
Parks, B. M., Cho, M. H., Spalding, E. P. (1998) Two genetically separated phases of growth inhibition induced by blue light in *Arabidopsis* seedlings. *Plant Physiol, 118*, 608-615.
Parry, G., Marchant, A., May, S., Swarup, R., Swarup, K., James, N., *et al.* (2001) Quick on the uptake: characterization of a family of plant auxin influx carriers. *J Plant Growth Reg, 20*, 217-225.
Ramos, J. A., Zenser, N., Leyser, O., Callis, J. (2001) Rapid degradation of AUX/IAA proteins requires conserved amino acids of domain II and is proteosome-dependent. *Plant Cell, 15,* 2349-2360.
Sakai, T., Kagawa, T., Kasahara, M., Swartz, T. E., Christie, J. M., Briggs, W. R., *et al.* (2001) *Arabidopsis* nph1 and npl1: Blue light receptors that mediate both phototropism and chloroplast relocation. *Proc Natl Acad Sci USA, 98,* 6969-6974.
Sakai, T., Wada, T., Ishiguro, S., Okada, K. (2000) RPT2: a signal transducer of the phototropic response in *Arabidopsis. Plant Cell, 12*, 225-236.
Sakamoto, K. and Briggs, W. R. (2002) Cellular and subcellular localization of phototropin 1. *Plant Cell, 14*, 1723-1735.
Salisbury, F. B. and Ross, C. W., *Plant Physiology (Third Edition)*. Belmont: Wadsworth Publishing Company, 1985.
Sato, Y., Kadota, A., Wada, M. (2003) Chloroplast movement: dissection of the events downstream of photo- and mechano-perception. *J Plant Res, 1161,* 1-5.
Schroeder, J. I., Allen, G. J., Hugouvieux, V., Kwak, J. M., Waner, D. (2001) Guard cell signal transduction. *Annu Rev Plant Physiol Plant Mol Biol, 52*, 627- 658.
Sehnke, P. C., DeLille, J. M., Ferl, R. (2002) Consummating signal transduction: the roles of 14-3-3 proteins in the completion of signal-induced transitions in protein activity. *Plant Cell, 14*, S339-S354.
Seo, H. S., Yang, J.-Y., Ishikawa, M., Bolle, C., Ballesteros, M. L., Chua, N.-H. (2003) LAF1 ubiquitination by COP1 controls photomorphogenesis and is stimulated by SPA1. *Nature, 423*, 995-999.
Serino, G. and Deng, X.-W. (2003) The COP9 signalosome: regulating plant development through the control of proteolysis. *Annu Rev Plant Biol, 54*, 165-182.

Shimazaki, K., Goh, C. H., Kinoshita, T. (1999) Involvement of intracellular Ca^{2+} in blue lightdependent proton pumping in guard cell protoplasts from *Vicia faba*. *Physiol Plant*, *105*, 554-561.
Short, T. W. and Briggs, W. R. (1994) The transduction of blue light signals in higher plants. *Annu Rev Plant Physiol Plant Mol Biol*, *45*, 143-171.
Silverthorne, J. and Tobin, E. (1984) Demonstration of transcriptional regulation of specific genes by phytochrome action. *Proc Natl Acad Sci USA*, *81*, 1112-1116.
Soh, M. S., Kim, Y. M., Han, S. J., Song, P.-S. (2000) REP1, a basic helix-loop-helix protein, is required for a branch of phytochrome A pathway in *Arabidopsis*. *Plant Cell*, *12*, 2016-2074.
Spalding, E. P. (2000) Ion channels and the transduction of light signals. *Plant Cell Environ*, *23*, 665-674.
Spalding, E. P. and Cosgrove, D. J. (1992) Mechanisms of blue-light induced plasma-membrane depolarization in etiolated cucumber hypocotyls. *Planta*, *188*, 199-205.
Spiegelmann, J. J., Mindrinosa, M. N., Fankhauser, C., Richards, D., Lutes, J., Chory, J., *et al.* (2000) Cloning of the *Arabidopsis RSF1* gene using a strategy based on high density DNA arrays and denaturing high performance liquid chromatography. *Plant Cell*, *12*, 2485-2498.
Stoelzle, S., Kagawa, T., Wada, M., Hedrich, R., Dietrich, P. (2003) Blue light activates calcium-permeable channels in *Arabidopsi*s mesophyll cells via the phototropin signaling pathway. *Proc Natl Acad Sci USA*, *100*, 1456-1461.
Stone, B. B. and Liscum, E. (2003) "Photoreceptors and associated signaling III: Phototropins." In *Encyclopedia of Plant and Crop Science*, Goodman R. M. (ed) Marcel Dekker Inc, New York, *in press*.
Stowe-Evans, E. L., Luesse, D. R., Liscum, E. (2001) The enhancement of phototropin-induced phototropic curvature in Arabidopsis occurs via a photoreversible phytochrome A-dependent modulation of auxin responsiveness. *Plant Physiol*, *126*, 826-834.
Tatematsu, K., Kumagai, S., Muto, H., Sato, A., Watahiki, M. K., Harper, R. M., *et al.* (2004) *MASSUGU2* encodes Aux/IAA19, an auxin-regulated protein that functions together with the transcriptional activator NPH4/ARF7 to regulate differential growth responses of hypocotyl and formation of lateral roots in *Arabidopsis thaliana*. *Plant Cell*, *16*, 379-393.
Terzaghi, W. B. and Cashmore, A. R. (1995) Light-regulated transcription. *Ann Rev Plant Phys Plant Mol Biol*, *46*, 445-474.
Thum, K. E., Kim, M., Christopher, D. A., Mullet, J. E. (2001) Cryptochrome 1, cryptochrome 2 and phytochrome A co-activate the chloroplast *psbD* blue-light responsive promoter. *Plant Cell*, *13*, 2747-2760.
Tlalka, M. and Gabrys, H. (1993) Influence of calcium on blue-light-induced chloroplast movement in *Lemna trisulca* L. *Planta*, *189*, 491-498.
Tobin, E. and Kehoe, D. M. (1994) Phytochrome regulated gene expression. *Sem Cell Biol*, *5*, 335-346.
Toledo-Ortiz, G., Huq, E., Quail, P. H. (2003) The *Arabidopsis* basic/helix-loop-helix transcription factor family. *Plant Cell*, *15*, 1749-1770.
Ulmasov, T., Hagen, G., Guilfoyle, T. J. (1999a) Activation and repression of transcription by auxin-response factors. *Proc Natl Acad Sci USA*, *96*, 5844-5849.
Ulmasov, T., Hagen, G., Guilfoyle, T. J. (1999b) Dimerization and DNA-binding of auxin response factors. *Plant J*, *19*, 1-11.
von Arnim, A. G. and Deng, X.-W. (1994) Light inactivation of *Arabidopsis* photomorphogenic repressor COP1 involves a cell-specific regulation of its nucleocytoplasmic partitioning. *Cell*, *79*, 1035-1045.
von Sachs, J., *Lectures on the physiology of plants.* Oxford: Clarendon, 1887.
Wada, M., Kagawa, T., Sato, Y. (2003) Chloroplast movement. *Annu Rev Plant Biol, 54*, 455-468.
Wang, H., Ma, L. G., Li, J. M., Zhao, H. Y., Deng, X. W. (2001) Direct interaction of *Arabidopsis* cryptochromes with COP1 in light control development. *Science*, *294*, 154-158.
Wang, X. and Iino, M. (1997) Blue-light induced shrinkage of protoplasts from maize coleoptiles and its relationship to coleoptile growth. *Plant Physiol*, *114*, 1009-1020.
Wang, X. and Iino, M. (1998) Interactions of cryptochrome 1, phytochrome, and ion fluxes in blue-light induced shrinkage of *Arabidopsis* hypocotyl protoplasts. *Plant Physiol*, *117*, 1265-1279.
Went, F. W. and Thimann, K. V. *Phytohormones*. New York: Macmillan, 1937.
Wu, R., Rooney, M. C., Ferl, R. (1997) The *Arabidopsis* 14-3-3 multigene family. *Plant Physiol, 114*, 1421-1431.
Yadav, V., Kundu, S., Chattopadhyay, S., Negi, P., Wei, N., Deng, X.-W. (2002) Light-regulated modulation of Z-box containing promoters by photoreceptors and downstream regulatory components, COP1 and HY5, in *Arabidopsis*. *Plant J*, *31*, 741-753.

Yaffe, M. B., Rittinger, K., Volinia, S., Caron, P. R., Aitkins, A., Lefferes, H., *et al.* (1997) The structural basis for 14-3-3:phosphopeptide binding specificity. *Cell*, *96*, 961-971.

Yang, H.-Q, Tang, R. H., Cashmore, A. R. (2001) The signaling mechanism of *Arabidopsis* CRY1 involves direct interaction with COP1. *Plant Cell*, *13*, 2573-2587.

Yang, H.-Q., Wu, Y.-J., Tang, R.-H., Liu, D., Liu, Y., Cashmore, A. R. (2000) The C termini of *Arabidopsis* cryptochromes mediate a constitutive light response. *Cell*, *103*, 815-827.

Zenser, N., Ellsmore, A., Leasure, C., Callis, J. (2001) Auxin modulates the degredation of AUX/IAA proteins. *Proc Natl Acad Sci USA*, *98*, 11795-11800.

PART 4: SIGNAL TRANSDUCTION IN PHOTOMORPHOGENESIS

Chapter 16

GENERAL INTRODUCTION

Peter H. Quail
Department of Plant and Microbial Biology, University of California, Berkeley, CA 94720, and USDA-ARS Plant Gene Expression Centre, 800 Buchanan Street, Albany, CA 94710, (email: quail@nature.berkeley.edu)

In the strictest sense, the terms "signalling" or "signal transduction" are reserved to refer to those events from perception of a physical or biological signal to the first change in gene expression elicited by that signal, i.e., pre-transcriptional events (Quail, 1983; Lissemore and Quail, 1988; Quail, 1991; Quail, 1994a; Bowler and Chua, 1994; Millar *et al.*, 1994). In the plant field, however, these terms are frequently used more loosely to encompass a considerably more inclusive and less well-defined set of events between the perception of a signal and the appearance of a measurable growth or developmental change in the organism under study (Abel *et al.*, 1996; Deng and Quail, 1999; Bowler and Chua, 2000; Stepanova and Ecker, 2000; Leyser and Deng, 2000; McCarty and Chory, 2000; Chory and Wu, 2001). This is particularly true of components that have been defined genetically as having some functional role in a given process or response by virtue of a mutated locus causing a perturbation in the overt growth or developmental process being monitored. Until these components are further characterized, the potential remains that they may well function downstream of the primary changes in transcription, either as a direct or indirect consequence of those changes, in cascade fashion.

Much of the work in the photomorphogenesis field discussed here utilizes the broader definition of signalling referred to (Quail, 1994a, 2000, 2002a,b; Deng and Quail, 1999; Chory and Wu, 2000; Hudson, 2000; Smith, 2000; Moller *et al.*, 2002; Nagy and Schaefer, 2002; Gyula *et al.*, 2003; Liscum *et al.*, 2003; Parks, 2003; Chapter 15). Implicitly or otherwise, the bulk of this work is focused on defining intracellular signalling pathways that transfer the information perceived by the light-activated photoreceptor molecule to the machinery that ultimately drives visible photomorphogenic changes. By definition, then, these segments of the pathways will be cell autonomous, triggered by the photoreceptor molecules resident in the cell. However, because the visible phenotypic changes being monitored almost invariably involve changes in growth and/or development of multicellular organs, the final display of this response may require, in addition, coordinate intercellular signalling. There is clear evidence for this in the case of responses such as flowering, phototropism, and probably hook opening, where intercellular transfer of signalling molecules is necessary. This consideration has important consequences for genetic and reverse genetic approaches directed at identifying and/or testing the activity of potential signalling components. Because the phenotype is complex, observed perturbation of that process by genetic mutation could, in principle, potentially involve any component necessary for manifestation of the response, whether in the primary, intracellular segment of the pathway, or downstream machinery involved in elaboration of the cellular response, or perhaps intercellular signalling.

The chapters in this part of the book seek to assemble an analysis of current understanding of a number of major facets of signalling in photomorphogenesis. These include the definition of components involved in cytoplasmic- and nuclear-localized events in phy signalling, the role of chloroplast to nucleus signalling, the role of regulated proteolysis in photomorphogenic responses, the complexities of interactions between the multiple plant photoreceptors (phytochromes, cryptochromes and phototropins) responsible for mediating light-induced changes in growth and development, and the intersection of the light-signalling networks with those of the major plant hormones. Historically, two principal classes of research strategy have been used to identify components involved in signalling in photomorphogenesis: (a) a biochemical/molecular strategy aimed at isolation and identification of the immediate reaction-partner molecules that physically interact with the photoreceptor molecules in the signalling process, and (b) a conventional forward genetic strategy involving screening for mutants defective in photoresponsiveness. The powerful molecular technologies and reverse-genetic strategies that have evolved from the recombinant DNA era into those of the genomics era have been added to this repertoire in recent years. The application of these concepts and technologies to the development of Arabidopsis as the primary model plant system has revolutionized the field. In particular, the availability of the complete Arabidopsis genome sequence, as well as large populations of tagged, insertional mutants, high-density microarrays, the yeast two-hybrid system, and GFP-fusion proteins, have together greatly accelerated the pace at which new insights have been gained into light-signalling.

Much of the understanding we currently have of photosensory signalling pathways and transcriptional networks has come from the use of the seedling deetiolation process as a model system in *Arabidopsis* (Quail, 1994a, 2000, 2002a,b; Deng and Quail, 1999; Chory and Wu, 2000; Hudson, 2000; Smith, 2000; Moller *et al.*, 2002; Nagy and Schaefer, 2002; Gyula *et al.*, 2003; Liscum *et al.*, 2003; Parks, 2003; Chapter 15). Deetiolation is the initial prominent redirection of development that is induced upon the first exposure of seedlings to light after germination. At the visible phenotypic level this response involves a switch from the etiolated (skotomorphogenic) growth pattern of dark-grown seedlings (displayed as rapid hypocotyl elongation, closed apical hook, small, appressed cotyledons enclosing undeveloped leaf primordia, and a pale yellow colour), to the converse (photomorphogenic) pattern of normal light-grown seedlings (exhibited as inhibited hypocotyl elongation, straightened apical hook, separation and rapid expansion of cotyledons and leaves, and green colour). At the cellular level, this response involves inhibition of cell elongation in the hypocotyl, stimulation of cell expansion in the cotyledons, and induction of chloroplast development.

This response has been the focus of studies by numerous researchers because it has many conceptual and experimental advantages for investigating the primary mechanisms involved in light perception, signal transduction and transcriptional regulation. These advantages include the following: (a) The photoreceptor molecules are synthesized and accumulate exclusively in the non-photoactivated state in seedlings germinated and grown in complete darkness. Consequently, because the timing, quality and quantity of the light signal can be readily and precisely controlled, the initiation of the intracellular perception and transduction processes can likewise be triggered and manipulated, non-invasively, with exquisite precision. Because the temporal progression of events can be a powerful indicator of the sequence in which components act, monitoring the time-course of molecular and cellular responses following initial exposure to light can provide important information on the hierarchical order in which early signalling intermediates act. (b) The experimental period is short (only a few days), and large numbers of samples can be processed in parallel *in petri* plates because the seedlings are small. (c) Genetic screens of large, mutagenised populations for aberrations in visible aspects of the phenotype are relatively simple. (d) The concomitant, reciprocal, responses induced by light in hypocotyl cells (inhibition of expansion) and cotyledon cells (stimulation of expansion) provides a critical diagnostic visible marker of the normal photomorphogenic process. Perturbation of early events in the normal photosensory perception or signalling pathways are expected to affect the light-induced expansion responses of these two cell types, reciprocally. This provides a facile method of distinguishing between mutations causing a perturbation specific to light signalling events (Halliday *et al.*, 1999; Huq *et al.*, 2000), and others that more globally affect cell expansion responses *per se* (Okamoto *et al.*, 2001; Ullah *et al.*, 2001). In the latter case, shorter hypocotyls are accompanied by smaller cotyledons in the mutant than wild type, rather than larger cotyledons. This distinction is important because in the absence of further critical analysis (Ullah *et al.*, 2001; Jones *et al.*, 2003), a short hypocotyl phenotype can be erroneously interpreted to indicate direct involvement of the mutant locus in normal light signalling (Okamoto *et al.*, 2001).

It will be apparent from the following chapters that a substantial number of the components thus far defined as candidate signalling intermediates have been molecularly cloned and identified, providing clues to their functions. More significantly, it will also be apparent that these studies have led, in recent years, to major new directions in thinking about potential intracellular phy signalling pathways and mechanisms of action, as well as uncovering completely new mechanisms involved in light signalling. In particular, the widely held view a decade ago, chronicled in multiple chapters of the predecessor of this book (Kendrick and Kronenberg, 1994), was that the phy molecule is cytosolically localized both before and after photoconversion from its inactive Pr to its active Pfr form, and therefore, that one or more molecular intermediates must relay the signal from the photoreceptor into the nucleus to light responsive genes. Since then, evidence has accumulated from multiple laboratories that light induces phy molecules to rapidly translocate (within minutes) into the nucleus where they may interact with nuclear proteins, such as transcription factors, to regulate transcription (see Chapters 9 and 17). While such evidence certainly does not exclude the possibility of a cytoplasmic signalling pathway to photoresponsive genes, involving second messengers or phosphorylation cascades, for example (see Chapter 19), it does at minimum open up a potential "second front" of phy signalling.

The concept that regulated proteolysis might have a prominent role in light signalling had not yet been formulated in the predecessor to this volume (Kendrick and Kronenberg, 1994). In the intervening years, a large body of evidence has been presented establishing that members of the COP/DET/FUS group of components function in a nuclear ubiquitin/proteosome pathway to modulate the abundance of key transcription factors, such as HY5, through light-regulated proteolysis (see Chapter 18). Together with the evidence of more direct transcriptional regulation of gene expression mentioned above, these data have led to the concept that phy signalling potentially involves dual levels of control, both synthetic (transcriptional) and degradative (post-translational).

One of the themes to emerge with increasing prominence from these studies in recent years, is one of considerable spatio-temporal complexity at multiple levels of the signalling and transcriptional networks. Although this is not unanticipated for a biological system requiring a high degree of coordinate regulation in response to environmental signals, evidence has steadily accumulated of a complex web of interactions, cross-talk and feed-back regulation among light-signalling components (Chory and Wu, 2001; Nagy and Schaefer, 2002; Quail, 2002a,b; Monte et al., 2003). Analysis of mutants defective in the various photoreceptors themselves has revealed an array of interactions and cross-talk both within and between the multiple members of the phy, cry and phot families (see Chapter 20). Components involved in retrograde intracellular signalling from the chloroplast to light-regulated nuclear genes have also been uncovered through mutant analysis (see Chapter 23). This dimension can be seen at its most sophisticated, perhaps, where key early light-signalling intermediates have been recruited to function as integral components of the central oscillator of the circadian clock, thereby permitting many, if not all, light-regulated genes to be atuned to the planet's diurnal, dark-light cycles (see Chapters 17, 26 and 27). Finally, the apparent complexities are amplified yet further when the potential roles of plant hormone signalling networks in mediating light responses are considered (see Chapter 21).

The "Holy Grail" of phy research has for many decades been defining the "primary mechanism of action", that initial molecular process that constitutes the intermolecular transfer of perceived signalling information from the photoactivated photoreceptor to its immediate reaction partner(s) (Smith, 1975; Sage, 1992; Kendrick and Kronenberg, 1994; Quail, 2000; Smith, 2000). The biochemical mechanism underlying this transaction must be intrinsic to the photoreceptor molecule, and must occur regardless of the intracellular location or molecular nature of the primary signalling partner(s). Although the possibility that the eukaryotic phy molecule is a light-regulated protein kinase that signals by direct transphosphorylation of primary reaction partners has received considerable prominence over a number of years (see Chapters 5, 6 and 19), definitive evidence for this or other mechanisms of signal transfer in the plant cell has remained elusive. This central question remains a key unsettled issue in the field.

Collectively, then, the next few chapters document the dramatic advances that have been made in recent years in the area of understanding signal transduction in photomorphogenesis. They also highlight some of the challenges that remain. The complex web of interlocking strands that constitute the light signalling networks are just beginning to come into focus. The continued application of the powerful molecular, genetic, genomic, and proteomic tools that have emerged in recent times can be anticipated to accelerate this process yet further.

Acknowledgements. I thank the many colleagues who have contributed to the work from this laboratory cited here, and Ron Wells for manuscript preparation and editing. Research supported by NIH grant GM47475, DOE-BES grant DE-FG03-87ER13742, USDA CRIS 5335-21000-017-00D, and Torrey Mesa Research Institute, San Diego.

REFERENCES

Abel, S., Ballas, N., Wong, L.-M., Theologis, A. (1996) DNA elements responsive to auxin. *BioEssays*, *18*, 647-654.
Bowler, C. and Chua, N.-H. (1994) Emerging themes of plant signal transduction. *Plant Cell*, *6*, 1529-1541.
Chory, J. and Wu, D. Y. (2001) Weaving the complex web of signal transduction. *Plant Physiol*, *125*, 77-80.
Deng, X.-W. and Quail, P. H. (1999) Signalling in light-controlled development. *Seminars Cell Devel Biol*, *10*, 121-129.
Gyula, P., Schaefer, E., Nagy, F. (2003) Light perception and signalling in higher plants. *Curr Op Plant Biol*, *6*, 446-452.
Halliday, K. J., Hudson, M., Ni, M., Qin, M., Quail, P. H. (1999) *poc1*: an Arabidopsis mutant perturbed in phytochrome signaling due to a T-DNA insertion in the promoter of *PIF3*, a gene encoding a phytochrome-interacting, bHLH protein. *Proc Natl Acad Sci USA*, *96*, 5832-5837.
Hudson, M. E. (2000) The genetics of phytochrome signalling in *Arabidopsis*. *Sem Cell Dev Biol*, *11*, 475-483.
Huq, E., Kang, Y., Qin, M., Quail, P. H. (2000b) *SRL1*: A new locus specific to the phyB signalling pathway in *Arabidopsis*. *Plant J*, *23*, 1-11.
Jones, A. M., Ecker, J. R., Chen, J.-G. (2003) A re-evaluation of the role of the heterotrimeric G protein in coupling light responses in *Arabidopsis*. *Plant Physiol*, *131*, 1623-1627.
Kendrick, R. E. and Kronenberg, G. H. M. (eds.) *Photomorphogenesis in Plants, 2nd ed.* Dordrecht, Netherlands: Kluwer Academic Publishers, 1994.
Lariguet, P., Boccalandro, H. E., Alonso, J. M., Ecker, J. R., Chory, J., Casal, J. J., *et al.* (2003) A growth regulatory loop that provides homeostasis to phytochrome A signalling. *Plant Cell*, *15*, 2966-2978.
Leyser, O. and Deng, X. W. (2000) Cell signalling and gene regulation – New directions in plant signalling. *Curr Op Plant Biol*, *3*, 351-352.
Liscum, E., Hodgson, D. W., Campbell, T. J. (2003) Blue light signaling through the cryptochromes and phototropins. So that's what the blues is all about. *Plant Physiol*, *133*, 1429-1436.
Lissemore, J. L. and Quail, P. H. (1988) Rapid transcriptional regulation by phytochrome of the genes for phytochrome and chlorophyll a/b-binding protein in *Avena sativa*. *Mol Cell Biol*, *8*, 4840-4850.
Martínez-García, J. F., Huq, E., Quail, P. H. (2000) Direct targeting of light signals to a promoter element-bound transcription factor. *Science*, *288*, 859-863.
McCarty, D. R. and Chory, J. (2000) Conservation and innovation in plant signalling pathways [Review]. *Cell*, *103*, 201-209.
Millar, A. J., McGrath, R. B., Chua, N.-H. (1994) Phytochrome phototransduction pathways. *Annu Rev Genet*, *28*, 325-349.
Moller, S. G., Ingles, P. J., Whitelam, G. C. (2002) The cell biology of phytochrome signalling. *New Phytologist*, *154*, 553-590.
Monte, E., Alonso, J. M., Ecker, J. R., Zhang, Y., Li, X., Young, J., *et al.* (2003) Isolation and characterization of phyC mutants in Arabidopsis reveals complex cross talk between phytochrome signalling pathways. *Plant Cell*, *15*, 1962-1980.

Nagy, F. and Schäfer, E. (2000a) Nuclear and cytosolic events of light-induced, phytochrome-regulated signalling in higher plants. *EMBO J, 19*, 157-163.

Nagy, F. and Schäfer, E. (2002) Phytochromes control photomorphogenesis by differentially regulated, interacting signaling pathways in higher plants. *Annu Rev Plant Biol, 53*, 329-355.

Okamoto, H., Matsui, M., Deng, X. W. (2001a) Overexpression of the heterotrimeric G-protein "-subunit enhances phytochrome-mediated inhibition of hypocotyl elongation in Arabidopsis. *Plant Cell, 13*, 1639-1651.

Parks, B. M. (2003) The red side of photomorphogenesis. *Plant Physiol, 133*, 1437-1444.

Quail, P. H. (1983) "Rapid action of phytochrome in photomorphogenesis." In *Encyclopedia of Plant Physiology Vol. 16A*, Mohr H. and Shropshire W. Jr. (eds), Springer-Verlag, Berlin, 178-212.

Quail, P. H. (1991) Phytochrome: A light-activated molecular switch that regulates plant gene expression. *Annu Rev Genet, 25*, 389-409.

Quail, P. H. (1994a) "Phytochrome genes and their expression." In *Photomorphogenesis in Plants, 2nd ed.*, Kendrick R. E. and Kronenberg G. H. M. (eds) Kluwer Academic Press, Dordrecht, Netherlands, 71-104.

Quail, P. H. (1994b) Photosensory perception and signal transduction in plants. *Curr Opin Gen Dev, 4*, 652-661.

Quail, P. H. (2000) Phytochrome Interacting Factors. *Seminars Cell Devel Biol, 11*, 457-466.

Quail, P. H. (2002a) Phytochrome photosensory signalling networks. *Nature Rev Mol Cell Biol, 3*, 85-93.

Quail, P. H. (2002b) Photosensory perception and signalling in plant cells: new paradigms? *Curr Op Cell Biol, 14*, 180-188.

Sage, L. C. *Pigment of the Imagination, A History of Phytochrome Research*. San Diego, CA: Academic Press, 1992.

Smith, H. *Phytochrome and Photomorphogenesis*. New York: McGraw-Hill, 1975.

Smith, H. (2000) Phytochromes and light signal perception by plants – and emerging synthesis. *Nature, 407*, 585-591.

Stepanova, A. N. and Ecker, J. R. (2000) Ethylene signalling: from mutants to molecules. *Curr Op Plant Biol, 3*, 353-360.

Ullah, H., Chen, J. G., Young, J. C., Im, K. H., Sussman, M. R., Jones, A. M. (2001) Modulation of cell proliferation by heterotrimeric G protein in Arabidopsis. *Science, 292*, 2066-2069.

Chapter 17

PHYTOCHROME SIGNAL TRANSDUCTION NETWORK

Peter H. Quail
Department of Plant and Microbial Biology, University of California, Berkeley, CA 94720; and USDA-ARS Plant Gene Expression Center, 800 Buchanan Street, Albany, CA 94710, (email: quail@nature.berkeley.edu)

1. INTRODUCTION

Light-signal perception by the phytochrome (phy) photoreceptor molecule (Pfr formation) initiates an intracellular transduction process that culminates in the altered expression of target genes responsible for directing the adaptational changes in plant growth and development appropriate for the prevailing environment. This general overall process is summarized schematically in Figure 1 for the seedling deetiolation response, but occurs in similar fashion for a continuum of light-regulated events throughout the life cycle (Smith, 2000; Quail, 2002a).

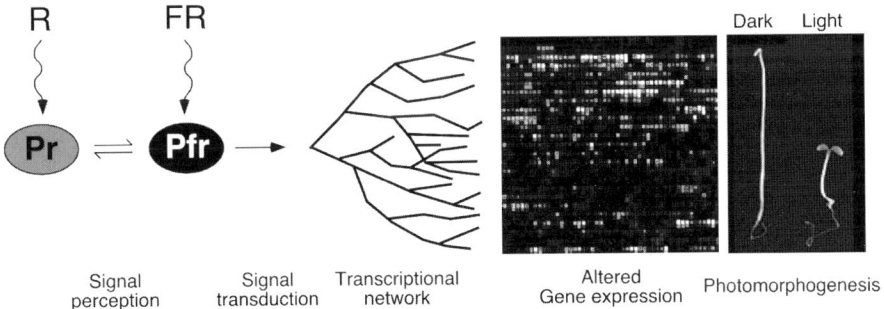

Figure 1. phy photoperception and signal transduction in seedling photomorphogenesis. phy molecules switch reversibly between their Pr and Pfr conformers upon sequential absorption of R and FR photons. Pfr formation (signal perception) triggers a signal transduction process that, via a transcriptional network, alters gene expression (displayed as a chip image from an oligonucleotide-microarray experiment) and culminates in seedling deetiolation (photomorphogenesis), manifested phenotypically as short hypocotyls and open, green, expanded cotyledons (right) compared to control seedlings held in darkness (left). (Modified from Quail, 2002a).

The central goal of a major area of current phy research is to define the molecular, cellular and biochemical mechanisms involved in this process through identifying molecular components that comprise the signalling and transcriptional networks controlled by the photoreceptor family. The principal general strategies being employed toward attaining this goal are summarized schematically in Figure 2, and include: (a) conventional, forward-genetic screens to identify mutants exhibiting aberrant visible photoresponsiveness phenotypes (morphological phenotype); (b) yeast two-hybrid screens to identify phy-interacting proteins as potential primary signalling partners; (c) molecular phylogeny analyses to identify homologs closely related to previously identified components in multigene families; (d) microarray-based expression profiling, both to define the genome-wide complement of phy-regulated genes (the molecular phenotype), and to identify the most rapidly light-responsive genes in this set as potential direct targets of phy signalling; and (e) bioinformatic analysis of the promoters of co-ordinately light-responsive genes to identify common DNA sequence elements, and, eventually, their cognate binding-proteins, potentially involved in regulating expression of those genes. Of necessity, the functional relevance to phy signalling or transcriptional regulation of any components identified by strategies (b) through (e) requires subsequent assessment by reverse-genetic methods (such as T-DNA or transposon insertion, antisense or RNAi, Tilling or Delete-a-gene technology; Henikoff and Comai, 2003), that provide targeted mutagenesis of the encoding genes, coupled with phenotypic analysis (morphological and/or molecular) for aberrant photoresponsiveness caused by the mutated component. Superimposed on these studies is the powerful cell-biological strategy of tagging phy-signalling-system proteins by fusing them to visible molecular markers, such as GUS and GFP, to permit the subcellular location, and potential colocalization of these components to be monitored (see Chapter 9).

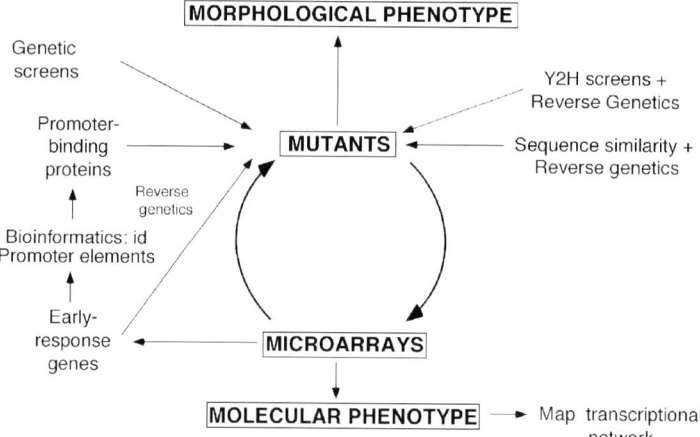

Figure 2. Schematic summary of current strategies used to dissect phy signalling and transcriptional networks. See text for details.

2. GENETICALLY-IDENTIFIED SIGNALLING-INTERMEDIATE CANDIDATES

The phenotypes of null mutants for each of the five phys in Arabidopsis have established that individual members of the family have differential, albeit partially overlapping, photosensory and/or physiological functions at various phases of the life cycle (Smith, 2000; Quail, 2002a; Monte *et al.*, 2003; Franklin *et al.*, 2003; see Chapters 7, 20 and 22). In some instances, different members of the family monitor essentially the same light signals but have predominant regulatory roles in different physiological responses. For example, whereas phyB has a predominant role in regulating seedling establishment, phyE appears to function primarily in controlling internode elongation (Devlin *et al.*, 1998; Smith, 2000). Conversely, in other instances, different members of the family monitor different light signals, but control essentially the same physiological response. For example, both phyA and phyB regulate seedling deetiolation, but, whereas phyB is activated by continuous monochromatic red light (Rc) (or red-light-rich signals), phyA is exclusively responsible for monitoring continuous monochromatic far-red light (FRc) (or far-red-light-rich signals), in controlling this response (Smith, 2000; Whitelam *et al.*, 1998; Quail *et al.*, 1995) (Figure 3).

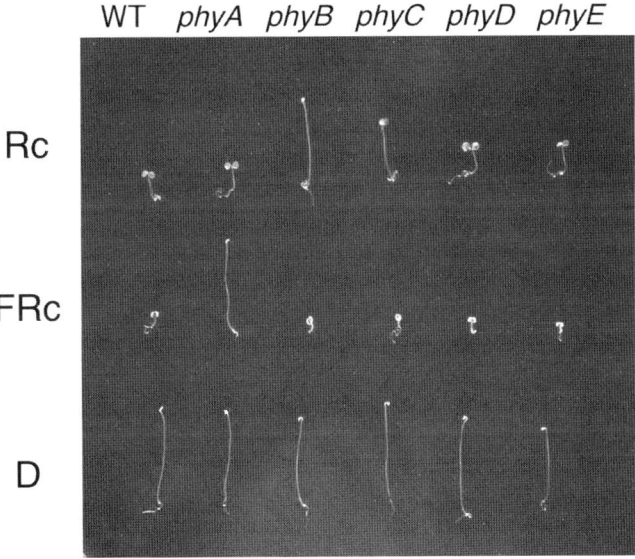

Figure 3. The phy-family mutants of Arabidopsis. Depicted are the seedling phenotypes of monogenic null mutants in each of the five phytochromes after 5-days' growth in Rc, FRc, or darkness (D). (E. Monte and P. Quail, unpublished).

Finally, in yet other instances, multiple phys appear to act additively or partly redundantly in regulating the same physiological response to the same light signal.

For example, phyB has long been considered to dominate regulation of seedling deetiolation in response to Rc signals. However, the residual, partial responsiveness of *phyB* null mutants to Rc indicates the participation of one or more other phy family members in this process (Tepperman *et al.*, 2004). The recent isolation of *phyC* mutants (Monte *et al.*, 2003; Franklin *et al.*, 2003) has shown that phyC can have a role (Figure 3), and there is evidence that other members of the family, particularly phyA, may also contribute significantly to aspects of Rc-induced deetiolation (Tepperman *et al.*, 2004). Despite these complexities, the seedling deetiolation process in Arabidopsis provides an excellent model system for dissecting the intricacies of phy signalling (Quail, 2002a). This system has been the focus of intense research efforts by a considerable number of laboratories, and has provided much of the information we currently have about the molecular and cellular aspects of the signalling process.

Conventional forward genetic screens for mutants defective in normal seedling photomorphogenesis have identified numerous non-photoreceptor loci that exhibit aberrant deetiolation (Moller *et al.*, 2002; Quail, 2002a,b; Gyula *et al.*, 2003) (Figure 4). These mutants fall into two broad classes: the *cop/det/fus* class that develop in complete darkness as if they were in the light, and those that develop normally in darkness, but display altered sensitivity to light (photodefective mutants). Most of the *cop/det/fus* class that have been molecularly cloned and characterized are considered to act downstream of the convergence of the phy and blue-light-receptor pathways, and have been shown to function either in a nuclear-localized, ubiquitin-proteosome pathway, by targeted proteolysis of the key transcription factor, HY5 (Wei and Deng, 1996; Hardtke and Deng, 2000; Serino and Deng, 2003; Seo *et al.*, 2003; Saijo *et al.*, 2003), or in the brassinosteroid pathway (Clouse, 2002; Fujioka and Yokata, 2003; Nemhauser and Chory, 2003). Because most of these mutations are recessive, the wild-type loci are considered to act negatively in darkness to suppress photomorphogenesis (see Chapter 18 for detailed discussion).

Three principal subclasses of mutants exhibiting altered responsiveness upon exposure to light have been identified: those responding aberrantly either to FRc only, to Rc only, or to both FRc and Rc (Figure 4). Those aberrant in FRc only are interpreted as representing loci involved selectively or specifically in phyA signalling, those aberrant in Rc responsiveness as involved predominantly in phyB signalling, and those aberrant in responsiveness to both wavelengths as being involved in shared phyA and phyB signalling pathways. Each of these three subclasses of mutants, in turn, contains two subclasses: those displaying reduced sensitivity to the light signal (hyposensitive), interpreted as representing loci that act positively in the pathway, and those mutants displaying enhanced sensitivity to the light signal (hypersensitive), interpreted as representing loci that act negatively in the pathway (Figure 4). These different classes of mutants suggest that early steps in the phyA and phyB pathways involve upstream intermediates dedicated to the individual photoreceptors, and that the separate pathways converge downstream in some undefined 'signal integration' process that drives later common events in photomorphogenesis and the circadian clock.

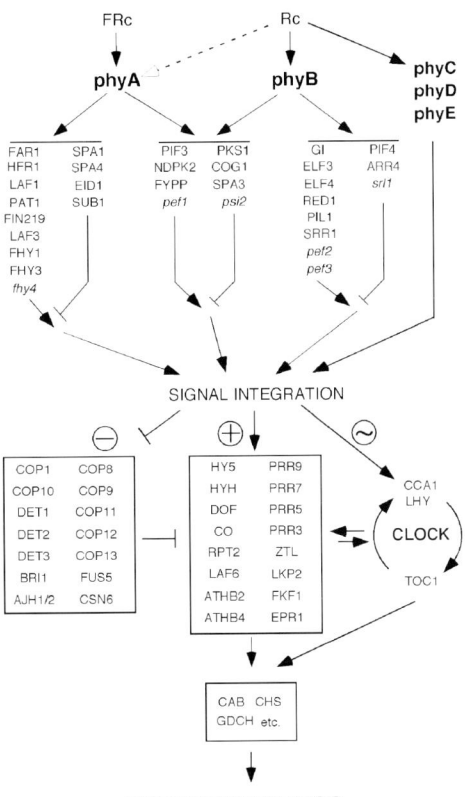

Figure 4. Simplified schematic of phy signalling pathways derived from current genetic and molecular studies of seedling photoresponsiveness. phyA is exclusively responsible for perception of FRc signals in seedling deetiolation, whereas phyB, phyC and probably phyA, are predominately responsible for Rc perception. Early signalling intermediates have been identified that are apparently specific to either phyA or phyB pathways, or are shared in a common pathway. Both positively (arrows) and negatively ('T' symbols) acting components have been identified in each case. The early pathway segments are considered to converge in a mechanistically undefined 'signal integration' process, propagated to negatively regulate the negatively-acting COP/DET/FUS group of components, and to positively regulate a core set of rapidly-responsive transcription-factor genes and circadian-clock-associated components. Other downstream genes involved in implementing different facets of photomorphogenic development are indicated. Cloned components are capitalized to indicate the encoded protein product, whereas genetically-defined, but not yet cloned, loci are lower case italicised to indicate mutant status.

A considerable number of these genetically identified loci have been cloned, while some others still await molecular definition. In addition, molecular and reverse-genetic studies have also identified components that are potentially functionally involved in phy signalling (see below). The data from these various sources are combined to provide the simplified signalling-pathway schematic in Figure 4.

How the majority of these individual components might function together in signal transduction is still largely unknown. Significantly, however, a major fraction of the components that have been cloned localize to the nucleus, suggesting possible functions in regulating gene expression (Quail, 2002a,b; Moller *et al.*, 2002; Gyula *et al.*, 2003). Consistent with this suggestion, one major subset of these components is comprised of established or predicted transcriptional regulators of various classes (FAR1, HFR1, FHY3, PIF3, PIF4, PIL1, HY5, HYH, CCA1, LHY, LAF1, COG1, ATHB2, ATHB4, EPR1) (Quail, 2002a,b; Gyula *et al.*, 2003). A second major subset (COP1, COP10, COP8, COP9, COP11, COP12, COP13, FUS5, CSN6, AJH1/2, SPA1, SPA3, SPA4, EID1, ZTL) appears to function post-translationally in regulated, proteosome-mediated proteolysis of nuclear proteins (Hardtke and Deng, 2000; Serino and Deng, 2003; Seo *et al.*, 2003; Saijo *et al.*, 2003) (see Chapter 18). These data suggest, therefore, that early phy signalling events are focused in the nucleus, and involve both synthetic (transcriptional) and degradative (post-translational) regulatory mechanisms.

3. PHYTOCHROME-INTERACTING FACTORS

Several yeast two-hybrid screens of cDNA libraries for phy-interacting proteins that may function as primary phy signalling partners have resulted in identification of three, apparently unrelated, proteins: PIF3 (phytochrome interacting factor 3) (Ni *et al.*, 1998), PKS1 (phytochrome kinase substrate 1) (Fankhauser *et al.*, 1999), and NDPK2 (nucleoside diphosphate kinase 2) (Choi *et al.*, 1999) which are capable of direct binding to phy molecules.

3.1 PIF3

PIF3 is a member of the basic helix-loop-helix (bHLH) superfamily of transcriptional regulators. Although the original yeast two-hybrid screen used only the C-terminal domain of phyB, subsequent in vitro interaction assays showed that full-length, chromophore-conjugated molecules of both phyA and phyB bind to PIF3, but only upon light-induced conversion to the biologically-active Pfr form (Ni *et al.*, 1999; Zhu *et al.*, 2000; Martinez-Garcia *et al.*, 2000) (Figure 5A). Reconversion to the Pr form resulted in rapid dissociation from PIF3, establishing that the interaction is specific to the biologically active conformer of the photoreceptor.

PIF3 localizes constitutively to the nucleus and binds in sequence specific fashion to a G-box DNA sequence, CACGTG, that is present in a variety of light-regulated promoters (Martinez-Garcia *et al.*, 2000; Quail, 2000). Strikingly, phyB can bind specifically and photoreversibly to PIF3 that is already bound to its cognate DNA binding site (Figure 5B). Together with the observed light-induced translocation of phy molecules to the nucleus (Sakamoto and Nagatani, 1996;

Kircher *et al.*, 1999; Nagy *et al.*, 2000; see Chapter 9), these data suggest that PIF3 can recruit the photoreceptor in its active form to G-box-containing promoters (Figure 5C).

Evidence that the observed phy-PIF3 interactions are relevant to phy signalling *in vivo* came initially from analysis of Arabidopsis seedlings with antisense-imposed reductions in PIF3 expression (Ni *et al.*, 1998). These seedlings exhibited strongly reduced phenotypic responsiveness to light signals perceived by phyB, and partially reduced responsiveness to signals perceived by phyA. These data were interpreted to indicate that PIF3 is functionally active in both phyA and phyB signalling pathways, consistent with its binding to both photoreceptors. RNA-blot analysis of PIF3-deficient seedlings indicated further that PIF3 is functionally necessary for phyB-induced expression of a subset of rapidly photoresponsive genes, in particular the key genes, *CCA1* and *LHY* (Martinez-Garcia *et al.*, 2000) (Figure 5D). The promoters of both of these genes contain G-box motifs and PIF3 binds to these in sequence-specific fashion, consistent with a direct role in regulating their expression. In addition, because the induction of *CCA1* and *LHY* is rapid (within 1 hour) and transient (Figure 5D), these may represent a subset of primary response genes that are directly regulated by phyB through PIF3.

Significantly, *CCA1* and *LHY* themselves encode MYB-class transcription-factor-related proteins known to function in regulating the expression of *CAB* genes (which encode a critical chloroplast component) and/or the circadian clock. Tobin and colleagues (Wang *et al.*, 1997; Wang and Tobin, 1998), first identified CCA1 as a positive regulator of light-induced *CAB* gene expression which binds to a functionally defined DNA sequence (CCA1-binding site; CBS) in the *CAB* promoter. Subsequently, CCA1 and LHY were shown to bind to a closely-similar sequence (termed an "evening element", EE) in the promoter of the *TOC1* gene, negatively regulating its expression (Harmer *et al.*, 2000). In a landmark contribution, Alabadi *et al.* (2001) showed this negative regulation of *TOC1* to be coupled in reciprocal fashion to positive regulation of *CCA1* and *LHY* gene expression by TOC1. This tight feedback loop is postulated to define the basic framework of the circadian clock in Arabidopsis (Figure 5C). It has been proposed, therefore, that PIF3 may represent the central control point through which the phy system regulates both a major branch of photomorphogenesis and the circadian oscillator, and that this regulation may be executed through a short, bifurcated transcriptional cascade using these MYB-related CCA1 and LHY transcription factors as intermediates (Quail, 2002a,b) (Figure 5C).

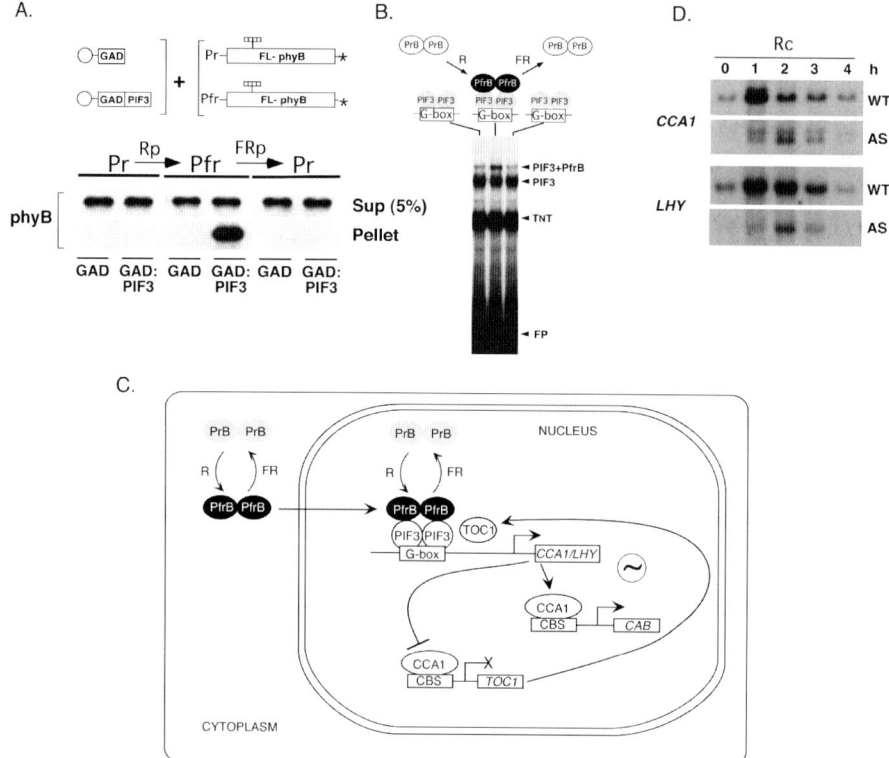

Figure 5. Postulated direct targeting of light signals to a promoter-bound bHLH factor, PIF3, that simultaneously regulates both photomorphogenic and clock genes through a short, branched transcriptional cascade. **A.** phyB binds to PIF3 specifically as the biologically active conformer Pfr. (Top). Experimental design. Recombinant Gal4-activation domain (GAD) or GAD-PIF3 fusion protein, immobilized on beads (represented by circles) were each mixed with radioactively labelled (asterisks), chromophore-ligated (represented as small striped bars), full-length (FL) phyB, which was converted either to the inactive (Pr) or active Pfr form by 5-min pulses of FR or R irradiation, respectively. (Bottom). Autoradiography of pelleted phyB obtained by centrifugation of beads after 2 hrs incubation in the dark at 4 °C, in the presence of either GAD or GAD-PIF3, after pulse irradiation with 5 min R (Rp), or with 5 min R followed immediately by 5 min of FR light (FRp). (Modified from Ni et al., 1999). **B.** phyB binds specifically in its Pfr form (PfrB) to DNA (G-box)-bound PIF3. Shown is a gel-shift assay in which PIF3 forms a complex with its target G-box DNA sequence (PIF3) and phyB forms a supershifted complex (PIF3 + PfrB) only upon R-induced conversion of the Pr form (PrB) to PfrB. Subsequent reconversion of PfrB to PrB by FR irradiation causes phyB to dissociate from the complex. TNT = non-specific complex. FP = free probe. (Modified from Martinez-Garcia et al., 2000). **C.** Model depicting the proposed direct signalling pathway from phyB to the promoters of photoresponsive genes. Light-induced conversion of PrB to PfrB triggers translocation to the nucleus where it binds to G-box-bound PIF3 and induces expression of primary target genes such as CCA1 and LHY. The encoded MYB-related transcription factors bind in turn to their cognate binding sites (here CBS for CCA1 binding site), where they either induce expression of genes such as CAB, involved in chloroplast

biogenesis, or repress expression of TOC1 ('T' symbol). TOC1 is in turn a positive regulator of CCA1 expression, creating a feedback loop that is postulated to constitute the basic framework of the circadian clock (~) in Arabidopsis cells. (Modified from Quail, 2002a). **D.** *Rapid, transient phyB-induced expression of CCA1 and LHY requires PIF3. Time-course, Northern-blot analysis of CCA1 and LHY mRNA levels in response to Rc in wild-type (WT) and PIF3-antisense (AS) seedlings expressing reduced levels of PIF3. (Modified from Martinez-Garcia et al., 2000).*

Based on the evidence that phyB binds specifically and reversibly to DNA-bound PIF3 upon light-triggered conversion to its active Pfr form (Figure 5A,B), it has been proposed that the phys may function as integral, light-switchable components of transcription-regulator complexes, directly at target promoters, following photoconversion-induced translocation into the nucleus (Martinez-Garcia *et al.*, 2000) (Figure 5C). According to this proposal, the photoreceptor could then regulate transcription either directly by functioning as a co-regulator, or indirectly by biochemically altering the transcriptional regulatory activity of PIF3 or other components of the transcriptional machinery (Figure 5C). This mechanism would target light signals directly to specific promoters permitting essentially instantaneous modulation of expression in response to fluctuations in signal content.

Despite the attractiveness of the above model of PIF3 function and mechanism of action, very recent data suggest a more complex, and possibly alternative, picture. Three laboratories have found that *pif3* mutants exhibit enhanced seedling deetiolation in Rc (hypersensitivity) (Kim *et al.*, 2003; Bauer *et al.*, 2004; E. Monte and P. Quail, unpublished) in direct contrast to the initial phenotype reported for *PIF3*-antisense seedlings, which indicated that PIF3 deficiency resulted in hyposensitivity (Ni *et al.*, 1998). It appears that the PIF3-antisense-line hyposensitivity is due to a T-DNA-induced mutation at another locus, and not to antisense suppression of *PIF3* expression (E. Monte and P. Quail, unpublished). At face value, these data suggest that PIF3 functions negatively, rather than positively, as initially reported, in the overall process of seedling deetiolation. However, *pif3* null mutants are defective in early aspects of seedling deetiolation, such as greening, upon initial exposure to Rc, suggesting a critical positive function at the initial dark-to-light transition experienced by seedlings (E. Monte and P. Quail, unpublished). Consistent with a possible transient function at the initial dark-to-light transition, the PIF3 protein displays extremely rapid light-dependent degradation over the first one hour of light exposure (Bauer *et al.*, 2004; B. Al-Sady, E. Monte, J. Tepperman and P. Quail, unpublished).

3.2 PKS1

PKS1 is a novel, constitutively cytoplasmic, protein isolated in a yeast two-hybrid screen using the C-terminal domain of phyA (Fankhauser *et al.*, 1999). *In vitro* binding assays showed that PKS1 can bind to both phyA and phyB, but with no difference in apparent affinity for the Pr and Pfr forms. Purified oat phyA preparations catalysed trans-phosphorylation of PKS1 about two-fold greater in the Pfr than in the Pr form. Because of the cytoplasmic location of PKS1, it has been suggested that the protein may function in cytoplasmic retention of phy molecules in the Pr-form (Fankhauser *et al.*, 1999). However, this is difficult to reconcile with the lack of differential binding of PKS1 to the two photochemical forms of the photoreceptor. In addition, recent evidence that a glucocorticoid receptor-phyB fusion protein is induced by dexamethosome to translocate into the nucleus in the Pr form (Huq *et al.*, 2003), seems to argue against an active cytoplasmic retention mechanism. Overexpression of PKS1 in transgenic Arabidopsis induces seedling hyposensitivity to Rc, potentially consistent with a negative functional role in phy signalling (Fankhauser *et al.*, 1999). However, transgenic seedlings expressing antisense *PKS1* sequences showed no phenotype. On the other hand, a very recent report of the isolation of a *pks1* mutant allele provides evidence that PKS1 functions specifically in a phyA-mediated very low fluence response mode, in conjunction with a related protein, PKS2, to provide homeostasis to phyA signalling (Lariguet *et al.*, 2003). The potential functional role of PKS1 phosphorylation in phy signalling is yet to be directly assessed.

3.3 NDPK2

The enzyme NDPK2 was isolated as binding to the C-terminal domain of Arabidopsis phyA in a yeast two-hybrid screen (Choi *et al.*, 1999). Subsequent *in vitro* binding assays showed that NDPK2 interacts with purified oat phyA with 1.8-fold greater apparent affinity for the Pfr than for the Pr form. Intriguingly, the extant enzymatic phosphate-exchange activity of the NDPK2 protein was enhanced about 1.7-fold by incubation with the Pfr form of oat phyA, whereas the Pr form had no effect. A loss-of-function *ndpk2* mutant displayed reduced sensitivity to both Rc and FRc as regards hook opening and cotyledon separation, but not hypocotyl elongation, consistent with a functional role in both phyA and phyB signalling. The NDPK2 protein appears to localize to both nucleus and cytoplasm, suggesting a function in either or both compartments. The molecular function in phy signalling remains unclear, although one possibility is transcriptional regulation based on yeast complementation experiments (Zimmerman *et al.*, 1999).

3.4 Other phy interactors

In addition to the non-targeted yeast two-hybrid screens described above, targeted molecular interaction studies using either quantitative yeast two-hybrid or *in vitro* binding assays with pre-selected proteins have resulted in reports of a variety of proteins that are capable of interacting with phyA and/or phyB (Moller *et al.*, 2002; Quail, 2000, 2002b; Gyula *et al.*, 2003). The complexity of the pattern of interactions determined from both targeted and non-targeted studies is summarized in Figure 6. The robustness of the evidence that these readily measurable physical interactions are functionally relevant to phy signalling *in vivo* is highly variable between studies, and is completely lacking in some cases. Therefore, it is presently unclear whether this apparent diversity of phy binding activities involving interactions of each photoreceptor with multiple, apparently unrelated proteins, indicates multiple signalling pathways radiating directly from each phy, reflects other activities of the molecules, or includes functionally irrelevant interactions.

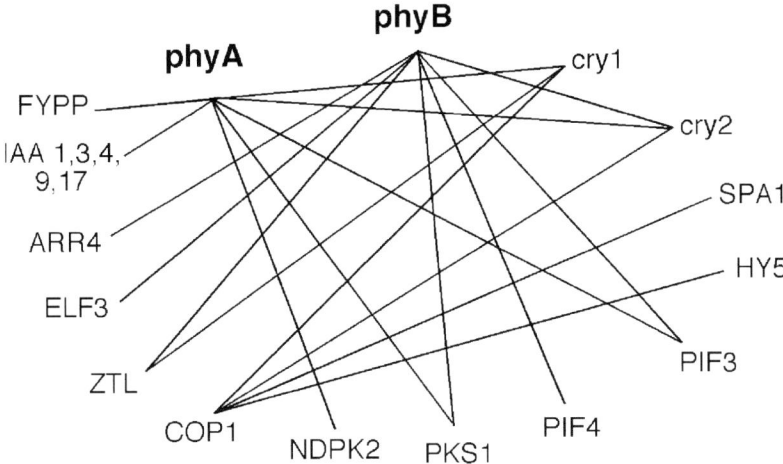

Figure 6. Molecular interaction map. Connecting lines depict physical interactions that have been reported between the phy photoreceptor molecules and various putative signalling components, as well as the blue-light photoreceptors, cry1 and cry2. (Modified from Quail, 2002b).

4. TRANSCRIPTION-FACTOR GENES ARE EARLY TARGETS OF PHY SIGNALLING

The advent of high-density microarrays for genome-wide expression profiling in Arabidopsis has provided a tool of unprecedented power for dissecting signalling and transcriptional networks. The capacity to simultaneously monitor the expression of all genes in the plant permits interrogation of the phenotype to be transferred globally to the molecular level, thereby providing a molecular phenotypic "fingerprint" of unparalleled resolution for any given growth or developmental process (Figure 2). Combined with the use of *phy*-null mutants, this technology provides the opportunity to define the complete set of genes regulated by each member of the photoreceptor family. It is then possible to define any subset of genes within this envelope controlled by specific signalling intermediates or transcriptional regulators, by systematically examining the light-induced expression profiles in mutants disrupted in these components. This strategy thus provides the potential for mapping the hierarchy, branching and interactions in the transcriptional network that elaborates photomorphogenesis, and for positioning putative upstream signalling intermediates in the circuitry linking the photoreceptor molecules to photoresponsive genes. Moreover, temporal analysis of global expression following light-signal administration permits identification of the most rapidly photoresponsive genes, thereby providing candidate primary-response genes: those responding directly to the phy signal, via pre-existing molecular components, without the need for intervention of altered expression of any of those components.

A small number of studies aimed at defining the spectrum of photoresponsive genes regulated by phyA and phyB have now used microarrays to examine the changes in expression profiles elicited by Rc- or FRc-irradiation of Arabidopsis seedlings. One set of studies employed glass-slide, spotted-cDNA arrays to measure presumptively end-point, steady-state transcript profiles after prolonged (5- to 6-day) FRc or Rc irradiation (Ma *et al.*, 2001, 2002; Wang *et al.*, 2002). The other set of studies used Affymetrix oligonucleotide microarrays to follow the time-course of changes in expression over the first 24-hours of irradiation of dark-grown seedlings (Tepperman *et al.*, 2001, 2004). Because both types of arrays were based primarily on the EST sequences available at the time of construction, each contained largely the same gene set of 6000 to 8000 genes. Although this gene set represents only about 25-30% of the total present in the Arabidopsis genome (about 26,000 genes), highly useful information that is likely representative of the genome-wide pattern has been obtained.

At a global level, the data reported indicate that 10 to 30% of the genes on the array exhibit light-induced changes (Tepperman *et al.*, 2001, 2004; Ma *et al.*, 2001; Wang *et al.*, 2002). This is illustrated for the FRc-responsive genes in wild-type seedlings following 12 hours of irradiation in Figure 7. Both sets of studies observed a high degree of overlap in the genes regulated by Rc and FRc, suggesting early convergence of the photosensory pathways regulated by these two wavelengths (Figure 8). Similarly, both sets of studies documented light-induced changes in expression of a broad array of functional categories of genes representing numerous cellular processes and pathways. This response pattern is consistent with that expected of the major redirection in development, from skotomorphogenic to photomorphogenic, that occurs during seedling deetiolation. Genes involved in photosynthesis and chloroplast biogenesis, and in an array of metabolic and biosynthetic processes predominate, reflecting the switch from heterotrophic to autotrophic growth, while genes involved in transcription, cell expansion and hormone pathways are also strongly represented.

It is noteworthy from monitoring the temporal patterns of the changes in gene expression that occur over the first 24 hours of irradiation that the light-induced transition in expression is initiated within 3 hours and essentially complete within 12 hours for the majority of photoresponsive genes (Tepperman *et al.*, 2001, 2004). This is reflected in the average time-course curves for all photoresponsive genes in wild-type seedlings depicted in Figure 9. This suggests that the steady-state expression profiles determined by end-point analysis after several days of continuous irradiation are likely to have been largely established within the first 12 hours of exposure to light. This capacity to rapidly complete the transition to photomorphogenic development might be critical to successful seedling establishment.

Comparison of the expression patterns in *phyA* and *phyB* null mutants with those of the wild type in FRc and Rc, respectively, have identified the photoresponsive genes regulated by these phy family members. These data provide robust evidence that phyA is exclusively responsible for perception of the FRc signals that induce altered gene expression during deetiolation (Tepperman *et al.*, 2001). This is depicted globally for the 12-hour time-point in Figure 7, and in Figure 9 where the mean time-course curves for both induced and repressed genes display clear responsiveness to the FRc signal in the wild type, but no detectable response in the *phyA* null mutant.

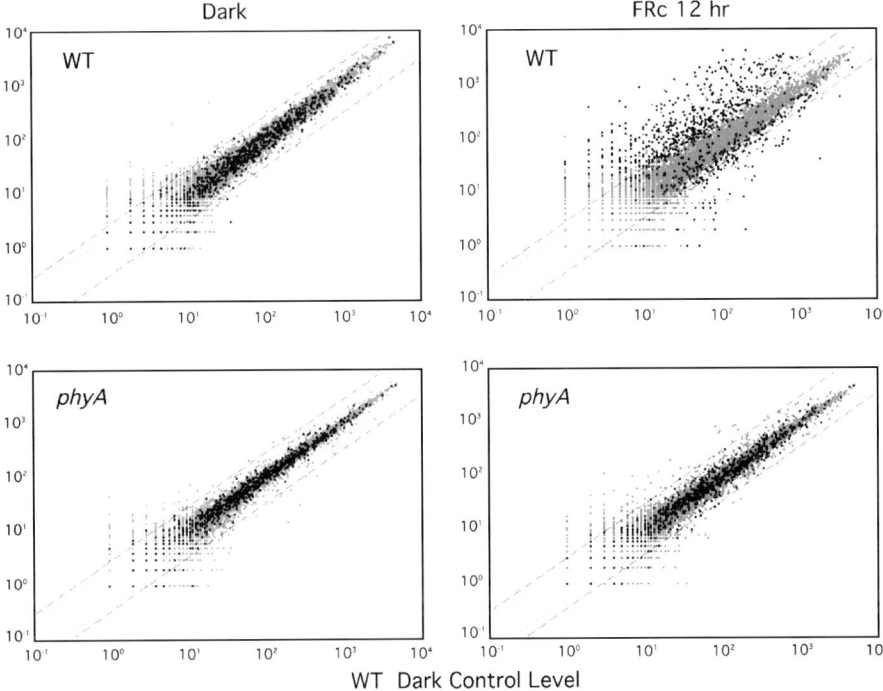

Figure 7. phyA regulates a major subset of Arabidopsis genes in response to FRc. Scatter-plots from oligonucleotide microarray analysis of transcript expression profiles in wild-type and phyA-null mutant Arabidopsis in response to a continuous far red light (FRc) signal. Expression levels of the approximately 8,200 different genes on the microarray (grey data points) are plotted for wild-type, dark-grown control seedlings at time-zero (abscissa of all plots) against those for other samples as indicated (ordinate of each plot): duplicate wild-type, dark control at time-zero (upper left), phyA-null mutant dark-control at time-zero (lower left), wild-type after 12 hours FRc (upper right), and phyA-null mutant after 12 hours of FRc (lower right). Black data points represent expression levels of the 812 genes identified as specifically phyA-regulated. Dashed lines represent two-fold deviation from unity. (Data from Tepperman et al., 2001).

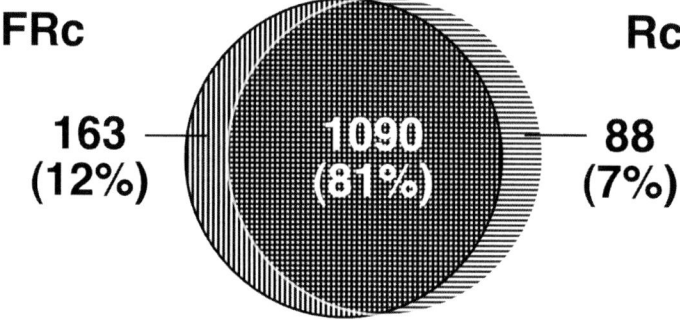

**Total Number of Genes
(1341)**

Figure 8. Genes regulated by Rc and FRc display substantial overlap. Venn diagram showing that of the total number of genes defined as FRc- and/or Rc-responsive in Arabidopsis seedlings during the first 24 hours of deetiolation (1341), 81% are regulated by both wavelengths. (Tepperman et al., 2004).

By contrast, *phyB* mutants retain a surprisingly high level of responsiveness to Rc as regards the mean expression profiles of photoresponsive genes (Tepperman *et al.*, 2004) (Figure 9). Closer consideration of these data shows that the majority (86%) of these Rc-responsive genes exhibit marginal or no dependence on phyB for Rc-regulated changes in expression, while the remainder (14%) exhibit partial to relatively clear dependence on phyB for Rc responsiveness. Similarly, closer inspection of the visible phenotype of Rc-grown *phyB* null mutants indicates that whereas hypocotyl cell elongation is essentially unresponsive to the Rc signal in the mutant (Figure 2), partial Rc-responsiveness is retained in the apical zone of the seedling, observable as hook opening and partial cotyledon expansion and greening. Given the anatomy of the Arabidopsis seedling, it is likely that the small, cytoplasmically-rich apical cells contribute the vast majority of the mRNA recovered in the whole-seedling extracts used for the microarray experiments, compared to the large, vacuole-filled, cytoplasmically-poor hypocotyl cells. These considerations lead to the following conclusions: (a) while phyB has at least a partial role in regulation of a significant number of Rc-responsive genes during deetiolation, other phy family members have a dominant role in regulating the major fraction of these genes; and (b) whereas phyB does appear to be predominantly, if not exclusively, responsible for Rc-inhibited hypocotyl elongation (and presumably the necessary underlying gene expression regulation), this photosensory function appears to be shared between phyB and one or more other phy family members in regulating Rc-induced apical zone responses (including the accompanying gene expression regulation). Collectively then, the data suggest that there are organ-specific differences in signalling activity among phy family members in regulating the gene sets responsible for Rc-induced seedling deetiolation (Tepperman *et al.*, 2004) (Figure 10).

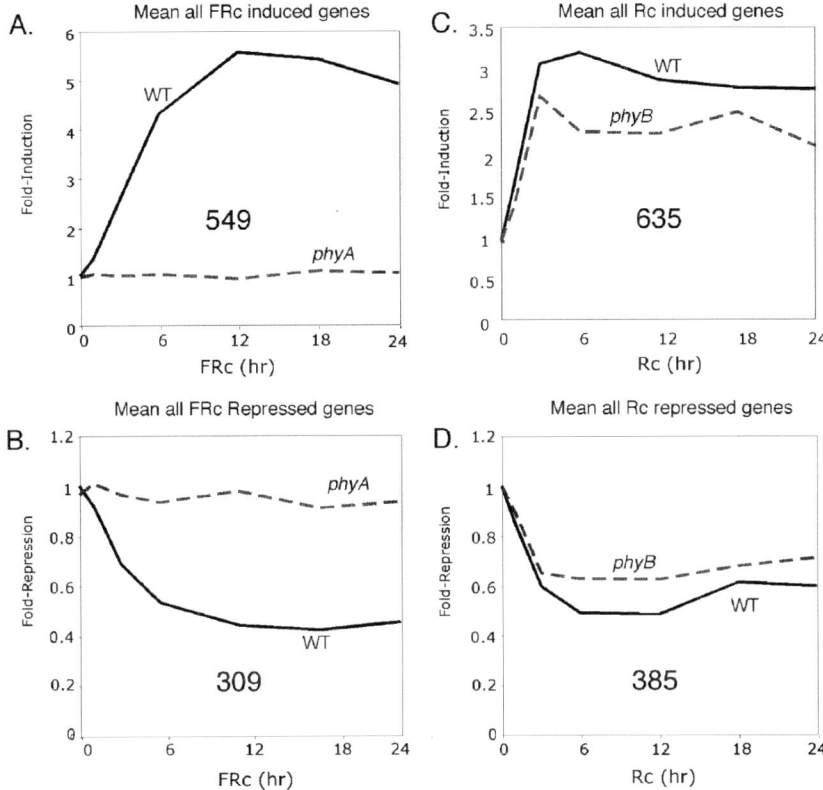

Figure 9. Mean 24-hour time-course expression profiles for all FRc- and Rc-responsive genes in Arabidopsis seedlings. *A.* and *B*. Mean time-course curves for all genes induced (*A*) or repressed (*B*) by FRc in wild-type (WT) or phyA null-mutant (phyA) seedlings. The mean time-course curves were computed by averaging the curves for all 549 induced genes (*A*) and all 309 repressed genes (*B*). *C*. and *D*. Mean time-course curves for all genes induced (*C*) or repressed (*D*) by Rc in wild-type (WT) or phyB null-mutant (phyB) seedlings. Total number of genes in each class shown in each panel. (Tepperman et al., 2004).

Analysis of the temporal patterns of light-induced expression has permitted the identification of those genes that respond most rapidly to the signal (Tepperman *et al.*, 2001, 2004). The large majority of phy-regulated genes (~90%) do not display changes in expression until 3 hours or later after the onset of the light signal ("late-response" genes). However, the remaining minority (10%) exhibit altered expression within 1 hour of the signal ("early-response" genes). Of the small group of functionally-classifiable early-response genes, a significant proportion are predicted to encode established or putative transcriptional regulators. Of these, over 70% respond to both Rc and FRc wavelengths.

The rapid responsiveness of these genes suggests that they may be integral components of a primary transcriptional network under phyA and phyB control (Figure 11). In addition, because these genes encode several different classes of factors, including zinc-finger, bZIP, MYB, bHLH, AP2-domain and homeodomain proteins, the data suggest rapid amplification and diversification of the range of gene sets that are targets of these factors. In fact, a number of these proteins already have well-documented central roles in several major known branches of the cellular activities that drive photomorphogenesis, including cell wall expansion, chloroplast biogenesis and photosynthesis, photorespiration, flavonoid biosynthesis, flowering and circadian clock regulation (Figure 11). Based on these observations, it has been suggested that these rapidly responding transcription-factor genes constitute a master-set, each with a primary role in coordinating the expression of the downstream genes that elaborate one or more major facets of light-induced development (Tepperman et al., 2001, 2004; Quail, 2002a,b).

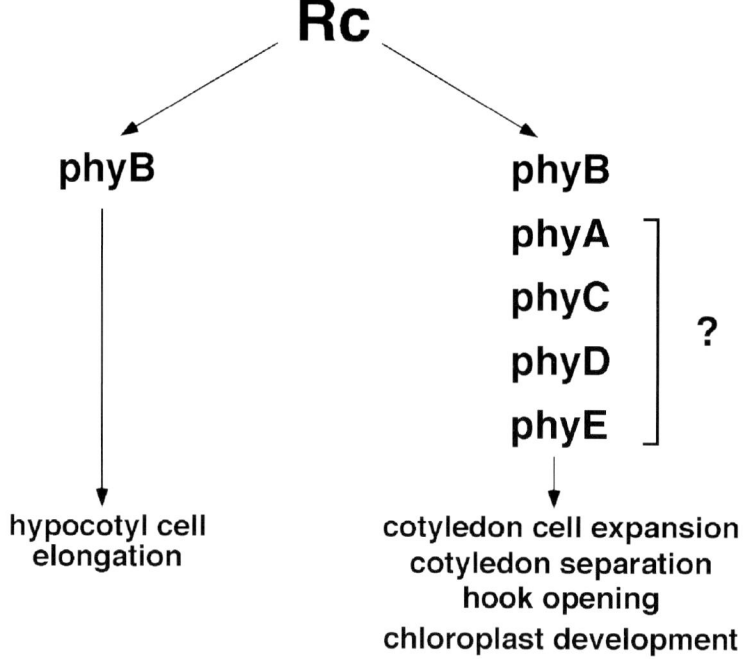

Figure 10. Schematic indicating organ-specific differences in signalling activity among phy family members in response to Rc. It is proposed that phyB dominates in regulating hypocotyl cell elongation in response to Rc, but that one or more other family members (phyA, C, D and/or E; uncertainty indicated by query), in addition to phyB, are involved in regulating various aspects of apical-zone responsiveness to Rc, including hook opening, cotyledon separation and expansion, and chloroplast development. (Tepperman et al., 2004).

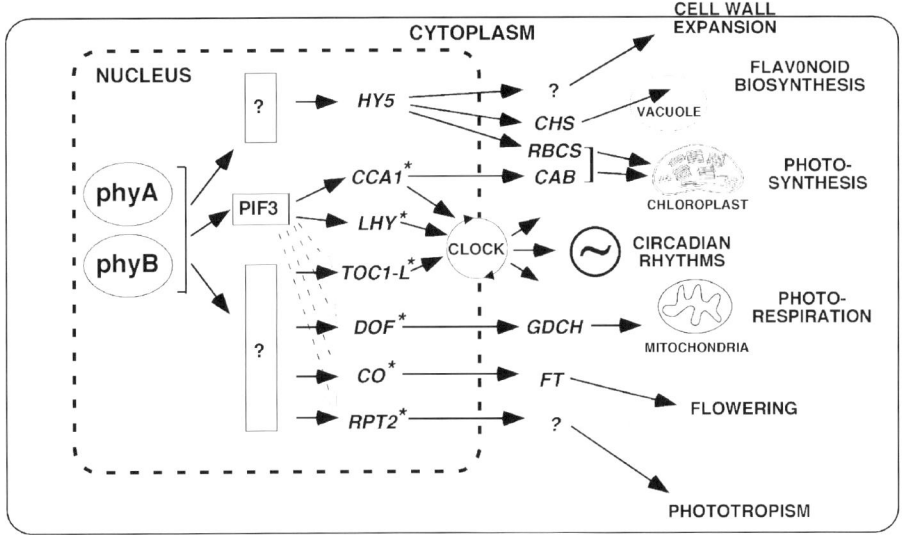

*G-box motifs in promoters

Figure 11. Simplified model of phy-regulated transcriptional network. It is proposed that a master set of rapidly responding transcription-factor genes (HY5 through RPT2 here) are primary targets of phyA and phyB signalling through constitutively present transcriptional regulators (termed signalling transcriptional regulators) that are direct recipients of incoming phyA and/or phyB signals. This master set of rapidly photoresponsive genes is proposed to encode regulators which control one or more major branches of the transcriptional network that drives the various facets of photomorphogenesis. Some of the key, established downstream genes that are known or proposed to be targets of the transcription-factor gene products listed are indicated. PIF3 is proposed to function as one such signalling transcriptional regulator, given its role in regulating CCA1 and LHY (Fig. 5). Several of the transcription-factor genes besides CCA1 and LHY also have G-box motifs in their promoters (asterisks), making them potential PIF3 targets (dashed arrows). Alternatively, other yet to be identified signalling transcriptional regulators may fulfill this role (boxed question marks). (Modified from Tepperman et al., 2001).

Importantly, *CCA1* and *LHY* are among this rapidly-responding group (Figure 11) (Tepperman *et al.*, 2001, 2004). Given the evidence that these two genes appear to be immediate targets of phy signalling through direct interaction of the photoreceptor with PIF3 bound to G-box sequences in their promoters (Figure 5) (Martinez-Garcia *et al.*, 2000), it is possible that the other transcription-factor genes in this group are also regulated in this way (Figure 11). Consistent with this possibility, many of these genes, including *TOC1-L*, *DOF*, *CO* and *RPT2*, contain G-box motifs in their promoters. In addition, the observation that phy molecules localize to multiple speckles in the nucleus after translocation from the cytoplasm (Kim *et al.*, 2000; Nagy *et al.*, 2000a,b; Mas *et al.*, 2000) is also compatible with the proposed direct regulation of multiple genes dispersed across the genome. On the other hand, other transcription-factor genes in this group, notably that encoding the well-characterized bZIP factor HY5, lack G-box motifs, suggesting that phyA and phyB signal to these genes by another pathway(s) (Figure 11).

A recent global, bioinformatic analysis of the promoters of phyA-regulated genes defined by microarray analysis has identified a series of sequence motifs potentially involved in the coordinate transcriptional regulation of these genes (Hudson and Quail, 2003). Five previously undescribed motifs were detected as enriched in the promoters of phyA-induced genes, and four novel motifs were found in phyA-repressed promoters, together with a previously described negative element, designated DE1 (Inaba *et al.*, 2000). The G-box motif, CACGTG, was prominently enriched in both induced and repressed phyA-responsive promoters compared to the remainder of the genes in the genome. However, intriguingly, two distinct consensus sequences were found in phyA-induced and phyA-repressed promoters, respectively, immediately flanking the G-box core, suggesting that different members of the same family of DNA-binding proteins (such as the bHLH and bZIP factors) might mediate phyA induction and repression of expression. Significantly, a greater abundance of G-box motifs was found in the most rapidly phyA-responsive genes, and in the promoters of phyA-regulated transcription factors, consistent with the notion that G-box-binding proteins have key functions early in the phy transcriptional network.

An initial study to begin to define the influence of several genetically-defined, phyA-pathway signalling intermediates on phyA-regulated expression profiles has been reported (Wang *et al.*, 2002). Based on the comparison of wild-type and mutant profiles after prolonged (5-day) FRc irradiation, it was concluded that FHY1, FAR1, and FHY3 likely act upstream in the phyA signalling network, whereas FIN219, SPA1 and HFR1 likely act more downstream, controlling smaller subsets of genes.

The discovery that the phy signalling intermediates, CCA1 and LHY, are integral components of the circadian clock (Alabadi *et al.*, 2001) has established that the oscillator is embedded at the apex of the phy-regulated transcriptional network, and is therefore positioned to impose oscillatory behaviour on numerous downstream genes in the light-regulated cascade. No oscillations are detectable in the central-oscillator genes *CCA1*, *LHY* and *TOC1* in dark-grown Arabidopsis seedlings (Kaczorowski and Quail, 2003; K. Franklin, G. Toledo-Ortiz and P. Quail, unpublished), apparently reflecting a steady-state equilibrium of expression established by the mutual feedback regulation exerted by these components on each other (Figure 5C). Exposure to light initiates the oscillations in this loop by rapidly inducing enhanced expression of *CCA1* and *LHY* (Kaczorowski and Quail, 2003). The characteristic biphasic profile over the first 24 hours after onset of the light-signal is consistent with the proposed clock model (Figure 5C). The data show that the initial light-induced rise in *CCA1/LHY* expression is followed by a reduction in *TOC1* expression, which is accompanied by a fall in *CCA1/LHY* expression below the initial dark control level (K. Franklin, G. Toledo-Ortiz and P. Quail, unpublished). This is consistent with the first complete loop of the oscillatory cycle (Figure 5C). Subsequently, *TOC1* expression increases while *CCA1/LHY* expression remains low, until the converse reciprocal pattern is initiated after about 12 hours, presumably reflecting the positive regulation of *CCA1/LHY* expression by the elevated levels of TOC1. This second peak of *CCA1/LHY* expression presumably represents the initiation of the second cycle that will constitute the normal circadian oscillation pattern under diurnal day-night cycles (see Chapter 26).

5. BIOCHEMICAL MECHANISM OF SIGNAL TRANSFER

The central question of the biochemical mechanism of signal transfer from the activated phy molecule to its primary signalling partner(s) has intrigued researchers for many decades. Regardless of the identity, molecular nature or subcellular location of that partner(s), molecular recognition must be followed by a biochemical event that consummates the signal transfer transaction. This mechanism remains to be defined. However, at least three non-mutually exclusive formal possibilities suggest themselves: (a) the phy molecule could function simply as a key scaffolding component in nucleating the assembly of functionally active multiprotein complexes, such as those involved in transcriptional regulation or proteolytic degradation; (b) the phy molecule could allosterically induce a conformational change in the partner(s), thereby activating its latent molecular or biochemical function; or (c) the phy molecule could catalyse the covalent modification of the partner by virtue of some enzymatic function intrinsic to the photoreceptor molecule.

The possibility that the phy molecule might function as an integral, light-switchable component of transcription-regulator complexes at target promoters was raised by the observed capacity of phyB to bind to DNA-bound PIF3 (Martinez-Garcia *et al.*, 2000; Quail, 2002a,b). However, thus far, no direct evidence that the phy molecule can directly regulate transcription at target promoters has been reported. The increase in NDPK2 enzymatic activity induced *in vitro* by the addition of oat phyA to the assay suggests that the physical binding of Pfr to this enzyme can enhance its intrinsic activity (Choi *et al.*, 1999). However, the mechanism by which this modest (1.7-fold) increase in phosphate-exchange activity might be translated into specific cellular regulatory activity remains to be clarified. Considerable attention has been focused on the attractive possibility that the phy molecule may function as a light-regulated protein kinase, and that transphosphorylation of one or more interacting partners comprises the biochemical mechanism of signal transfer (Fankhauser *et al.*, 1999). Both biochemical and evolutionary evidence have been presented in support of this proposition (see Chapter 5, 6, 19 for detailed discussions). However, a recent report that the photoactive, N-terminal domain of phyB is fully functional in the cell in the complete absence of the putative protein kinase domain, that is located in the C-terminal half of the molecule (Matsushita *et al.*, 2003), provides compelling evidence that the normal signal transfer activity of the photoreceptor does not require this postulated kinase activity, and seems unlikely therefore to involve transphosphorylation of signalling partners. The possibility that the biochemical mechanism of phy signal transfer involves a novel intermolecular transaction remains open.

Acknowledgements: I thank the many colleagues who have contributed to the work from this laboratory cited here, Jim Tepperman for figure preparation, and Ron Wells for manuscript preparation and editing. Research supported by NIH grant GM47475, DOE-BES grant DE-FG03-87ER13742, USDA CRIS 5335-21000-017-00D, and Torrey Mesa Research Institute, San Diego.

6. REFERENCES

Alabadi, D., Oyama, T., Yanovsky, M. J., Harmon, F. G., Mas, P., Kay, S. A. (2001) Reciprocal regulation between *TOC1* and *LHY/CCA1* within the Arabidopsis circadian clock. *Science, 293,* 880-883.

Bauer, D., Viczian, A., Kircher, S., Nobis, T., Nitschke, R., Kunkel, T., *et al.* (2004) Constitutive photomorphogenesis 1 and multiple photoreceptors control degradation of Phytochrome Interacting Factor 3, a transcription factor required for light signalling in *Arabidopsis*. *Plant Cell, 16,* 1433-1445.

Choi, G., Yi, H., Lee, J., Kwon, Y.-K., Soh, M. S., Shin, B., *et al.* (1999) Phytochrome signalling is mediated through nucleoside diphosphate kinase 2. *Nature, 401,* 610-613.

Clouse, S. (2002) "Brassinosteroids." In *The Arabidopsis Book*, Somerville C. R., Meyerowitz E. M. (eds) Am Soc Plant Biologists, Rockville, MD, **doi/10.1199/tab.0009,** http://www.aspb.org/publications/arabidopsis/

Fankhauser, C., Yeh, K. C., Lagarias, J. C., Zhang, H., Elich, T. D., Chory, J. (1999) PKS1, a substrate phosphorylated by phytochrome that modulates light signalling in arabidopsis. *Science, 284*(N5419), 1539-1541.

Franklin, K. A., Davis, S. J., Stoddart, W. M., Vierstra, R. D., Whitelam, G. C. (2003) Mutant analyses define multiple roles for phytochrome C in Arabidopsis photomorphogenesis. *Plant Cell, 15,* 1981-1989.

Fujioka, S. and Yokota, T. (2003) Biosynthesis and metabolism of brassinosteroids. *Annu Rev Plant Biol, 54,* 137-164.

Gyula, P., Schaefer, E., Nagy, F. (2003) Light perception and signalling in higher plants. *Curr Op Plant Biol, 6,* 446-452.

Hardtke, C. S. and Deng, X. W. (2000) The cell biology of the COP/DET/FUS proteins. Regulating proteolysis in photomorphogenesis and beyond? *Plant Physiol, 124,* 1548-1557.

Harmer, S. L., Hogenesch, J. B., Straume, M., Chang, H.-S., Han, B., Zhu, T., *et al.* (2000) Orchestrated transcription of key pathways in *Arabidopsis* by the circadian clock. *Science, 290,* 2110-2113.

Henikoff, S. and Comai, L. (2003) Single-nucleotide mutations for plant functional genomics. *Annu Rev Plant Biol, 54,* 375-401.

Hudson, M. E. (2000) The genetics of phytochrome signalling in *Arabidopsis*. *Sem Cell Dev Biol, 11,* 475-483.

Hudson, M. E. and Quail, P. H. (2003) Identification of promoter motifs involved in the network of phytochrome A-regulated gene expression by combined analysis of genomic sequence and microarray data. *Plant Physiol, 133,* 1605-1616.

Hudson, M. E., Lisch, D. R., Quail, P. H. (2003) The FHY3 and FAR1 genes encode transposase-related proteins involved in regulation of gene expression by the phytochrome A-signalling pathway. *Plant J, 34,* 453-471.

Huq, E., Al-Sady, B., Quail, P. H. (2003) Nuclear translocation of the photoreceptor phytochrome B is necessary for its biological function in seedling photomorphogenesis. *Plant J, 35,* 660-664.

Inaba, T., Nagano, Y., Reid, J. B., Sasaki, Y. (2000) DE1, a 12-base pair cis-regulatory element sufficient to confer dark-inducible and light down-regulated expression to a minimal promoter in pea. *J Biol Chem, 275,* 19723-19727.

Kaczorowski, K. A. and Quail, P. H. (2003) Arabidopsis PSEUDO-RESPONSE REGULATOR7 is a signalling intermediate in phytochrome-regulated seedling deetiolation and phasing of the circadian clock. *Plant Cell, 15,* 2654-2665.

Kim, J., Yi, H., Choi, G., Shin, B., Song, P.-S., Choi, G. (2003) Functional characterization of phytochrome interacting factor 3 in phytochrome-mediated light signal transduction. *Plant Cell, 15,* 2399-2407.

Kim, L., Kircher, S., Toth, R., Adam, E., Schaefer, E., Nagy, F. (2000) Light-induced nuclear import of phytochrome-A: GFP fusion proteins is differentially regulated in transgenic tobacco and Arabidopsis. *Plant J, 22*, 125-133.

Kircher, S., Kozma-Bognar, L., Kim, L., Adam, E., Harter, K., Schäfer, E., *et al.* (1999) Light quality-dependent nuclear import of the plant photoreceptors phytochrome A and B. *Plant Cell, 11*, 1445-1456.

Kuno, N., Moller, S. G., Shinomura, T., Xu, X. M., Chua, N.-H., Furuya, M. (2003) The novel MYB protein EARLY-PHYTOCHROME-RESPONSIVE1 is a component of a slave circadian oscillator in Arabidopsis. *Plant Cell, 15*, 2476-2488.

Lariguet, P., Boccalandro, H. E., Alonso, J. M., Ecker, J. R., Chory, J., Casal, J. J., *et al.* (2003) A growth regulatory loop that provides homeostasis to phytochrome A signalling. *Plant Cell, 15*, 2966-2978.

Ma, L., Li, J., Qu, L., Hager, J., Chen, Z., Zhao, H., *et al.* (2001) Light Control of Arabidopsis Development Entails Coordinated Regulation of Genome Expression and Cellular Pathways. *Plant Cell, 13*, 2589-2607.

Ma, L., Gao, Y., Qu, L., Chen, Z., Li, J., Zhao, H., *et al.* (2002) Genomic evidence for COP1 as a repressor of light-regulated gene expression and development in Arabidopsis? *Plant Cell, 14*, 2383-2398.

Martínez-García, J. F., Huq, E., Quail, P. H. (2000) Direct targeting of light signals to a promoter element-bound transcription factor. *Science, 288*, 859-863.

Mas, P., Devlin, P. F., Panda, S., Kay, S. A. (2000) Functional interaction of phytochrome B and cryptochrome 2. *Nature, 408*, 207-211.

Mas, P., Kim, W. Y., Somers, D. E., Kay, S. A. (2003) Targeted degradation of TOC1 by ZTL modulates circadian function in Arabidopsis thaliana. *Nature, 426*, 567-570.

Matsushita, T., Mochizuki, N., Nagatani, A. (2003) Dimers of the N-terminal domain of phytochrome B are functional in the nucleus. *Nature, 424*, 571-574.

Moller, S. G., Ingles, P. J., Whitelam, G. C. (2002) The cell biology of phytochrome signalling. *New Phytologist, 154*, 553-590.

Monte, E., Alonso, J. M., Ecker, J. R., Zhang, Y., Li, X., Young, J., *et al.* (2003) Isolation and characterization of phyC mutants in Arabidopsis reveals complex cross talk between phytochrome signalling pathways. *Plant Cell, 15*, 1962-1980.

Nagy, F., Kircher, S., Schäfer, E. (2000) Nucleo-cytoplasmic partitioning of the plant photoreceptors phytochromes. *Sem Cell Devel Biol, 11*, 505-510.

Nagy, F. and Schäfer, E. (2000b) Control of nuclear import and phytochromes. *Curr Opin Plant Biol, 3*, 450-454.

Nagy, F. and Schäfer, E. (2000a) Nuclear and cytosolic events of light-induced, phytochrome-regulated signalling in higher plants. *EMBO J, 19*, 157-163.

Nemhauser, J. L. and Chory, J. (2004) BRing it on: new insights into the mechanism of brassinosteroid action. *J Exper Biol, 55*, xxx-xxx.

Ni, M., Tepperman, J. M., Quail, P. H. (1999) Binding of phytochrome B to its nuclear signalling partner PIF3 is reversibly induced by light. *Nature, 400*, 781-784.

Ni, M., Tepperman, J. M., Quail, P. H. (1998) PIF3, a phytochrome-interacting factor necessary for normal photoinduced signal transduction, is a novel basic helix-loop-helix protein. *Cell, 95*, 657-667.

Quail, P. H. (2002b) Photosensory perception and signalling in plant cells: new paradigms? *Curr Op Cell Biol, 14*, 180-188.

Quail, P. H. (2000) Phytochrome Interacting Factors. *Seminars Cell Devel Biol, 11*, 457-466.

Quail, P. H. (2002a) Phytochrome photosensory signalling networks. *Nature Rev Mol Cell Biol, 3*, 85-93.

Saijo, Y., Sullivan, J. A., Wang, H. Y., Yang, J. P., Shen, Y. P., Rubio, V., *et al.* (2003) The COP1-SPA1 interaction defines a critical step in phytochrome A-mediated regulation of HY5 activity. *Genes Devel, 17*, 2642-2647.

Sakamoto, K. and Nagatani, A. (1996) Nuclear localization activity of phytochrome B. *Plant J, 10*, 859-868.

Schwechheimer, C. and Deng, X. W. (2001) COP9 signalosome revisited: a novel mediator of protein degradation. *Trends Cell Biol., 11*, 420-426.

Seo, H. S., Yang, J.-Y., Ishikawa, M., Bolle, C., Ballesteros, M. L., Chua, N.-H. (2003) LAF1 ubiquitination by COP1 controls photomorphogenesis and is stimulated by SPA1. *Nature, 423*, 995-999.

Serino, G. and Deng, X.-W. (2003) The COP9 signalosome: Regulating plant development through the control of proteolysis. *Annu Rev Plant Biol*, *54*, 165-182.

Smith, H. (2000) Phytochromes and light signal perception by plants - and emerging synthesis. *Nature*, *407*, 585-591.

Tepperman, J. M., Hudson, M. E., Khanna, R., Zhu, T., Chang, H.-S., Wang, X., et al. (2004) Expression profiling of *phyB* mutant demonstrates substantial contribution of other phytochromes to red-light-regulated gene expression during seedling deetiolation. *Plant J*, in press.

Tepperman, J. M., Zhu, T., Chang, H.-S., Wang, X., Quail, P. H. (2001) Multiple transcription-factor genes are early targets of phytochrome A signalling. *Proc Natl Acad Sci USA*, *98*, 9437-9442.

Wang, H., Ma, L., Habashi, J., Li, J., Zhao, H., Deng, X. W. (2002) Analysis of far-red light-regulated genome expression profiles of phytochrome A pathway mutants in Arabidopsis. *Plant J*, *32*, 723-733.

Wang, Z.-Y., Kenigsbuch, D., Sun, L., Harel, E., Ong, M. S., Tobin, E. M. (1997) A myb-related transcription factor is involved in the phytochrome regulation of an Arabidopsis *Lhcb* gene. *Plant Cell*, *9*, 491-507.

Wang, Z.-Y. and Tobin, E. M. (1998) Constitutive expression of the *CIRCADIAN CLOCK ASSOCIATED 1* (*CCA1*) gene disrupts circadian rhythms and suppresses its own expression. *Cell*, *93*, 1207-1217.

Wei, N. and Deng, X.-W. (1996) The role of the *COP/DET/FUS* genes in light control of Arabidopsis seedling development. *Plant Physiol*, *112*, 871-878.

Zhu, Y., Tepperman, J. M., Fairchild, C. D., Quail, P. (2000) Phytochrome B binds with greater apparent affinity than phytochrome A to the basic helix-loop-helix factor PIF3 in a reaction requiring the PAS domain of PIF3. *Proc Natl Acad Sci USA*, *97*, 13419-13424.

Zimmermann, S., Baumann, A., Jaekel, K., Marbach, I., Engelberg, D., Frohnmeyer, H. (1999) UV-responsive genes of arabidopsis revealed by similarity to the Gcn4-mediated UV response in yeast. *J Biol Chem*, *274*, 17017-17024.

Chapter 18

THE FUNCTION OF THE COP/DET/FUS PROTEINS IN CONTROLLING PHOTOMORPHOGENESIS: A ROLE FOR REGULATED PROTEOLYSIS

Elizabeth Strickland, Vicente Rubio, Xing Wang Deng
Department of Molecular, Cellular, and Developmental Biology, Yale University, New Haven, Connecticut, 06520-8104, USA (e-mail: xingwang.deng@yale.edu)

1. INTRODUCTION

When a seed germinates and begins to grow in the dark conditions found under the surface of the soil, it follows a developmental pattern of skotomorphogenesis that is characterized by a lengthened hypocotyl and unexpanded cotyledons (Deng, 1994; von Arnim and Deng, 1996). Once the seedling has reached the soil surface and is exposed to light, a complete developmental transition occurs to photomorphogenesis – hypocotyl growth ceases, the cotyledons open and expand, and chloroplast development occurs. In order for this transition to photomorphogenesis to occur, a seedling must perceive a light signal and change its gene expression profile to one that promotes photomorphogenesis. Experiments in the past decade have shown that a major mechanism plants use to accomplish this transition is regulated proteolysis (reviewed in Hellmann and Estelle, 2002; Sullivan *et al.*, 2003). In a simplified view, this process begins with the recognition of a light signal by a family of photoreceptors. This signal is then transduced to the nucleus where the degradation rate of key transcription factors is modulated and, thus, the gene expression profile of the plant is altered to one that allows a photomorphogenic developmental program to proceed. In fact, up to 30% of the genes in the *Arabidopsis* genome show an altered expression pattern after exposure to white light (Ma *et al.*, 2002; Tepperman *et al.*, 2001).

1.1 Genetic analysis of photomorphogenesis

A highly successful strategy for determining key players in regulating photomorphogenesis has been the identification and characterization of mutants that have alterations in this process (Wei and Deng, 1996). Several groups have carried out mutant screens in *Arabidopsis* and together have isolated a group of pleiotropic mutants referred to as the *COP/DET/FUS* genes. The first such report was the identification of *de-etiolated* mutants (*det1*, Chory *et al.*, 1989; Pepper *et al.*, 1994). A similar screen for a <u>constitutively</u> <u>photomorphogenic</u> seedling phenotype in darkness resulted in identification of the *cop* mutants (Deng *et al.*, 1991; Wei *et al.*, 1994; Kwok *et al.*, 1996). As shown in Figure 1, when grown in darkness, these mutants exhibit shortened hypocotyls and expanded cotyledons, characteristic features of photomorphogenesis. Both pleiotropic *det* and *cop* mutants were found to be allelic to the *fusca* mutants, so named for the purple colour (*fusca* meaning dark in Latin) of their seeds due to the accumulation of anthocyanins (Castle and Menke, 1994; Miséra *et al.*, 1994). All of these mutants are recessive and are caused by loss of function mutations. Thus, the *COP/DET/FUS* mutants define negative regulators of photomorphogenesis.

Nine of the *COP/DET/FUS* genes have now been cloned and extensively studied (reviewed in Wei *et al.*, 1999; Hardtke *et al.*, 2000; Schwechheimer *et al.*, 2000; Schwechheimer and Deng, 2001; Serino and Deng, 2003). These genes represent three biochemical entities: the COP1 complex (Saijo *et al.*, 2003), the COP9 signalosome (CSN, Serino and Deng, 2003) and the CDD complex, which includes both COP10 and DET1 (Yanagawa *et al.*, 2004). The three complexes are all directly involved in regulated proteolysis through the ubiquitin-proteasome system. In brief, COP1 is part of an E3 ubiquitin-protein ligase. COP10 resembles an E2 ubiquitin-conjugase variant with the ability to enhance E2 activity and form a complex with DET1 and DDB1 (Yanagawa *et al.*, 2004). Finally, the CSN is a regulator of E3 ubiquitin-protein ligases and, possibly, the 26S proteasome. This chapter will focus on recent work to describe the molecular mechanism each of these factors or complexes uses to regulate protein degradation and, thus, photomorphogenesis.

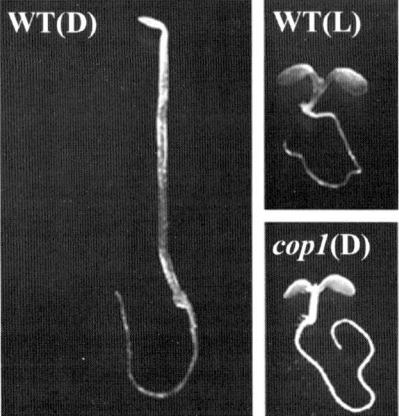

Figure 1. Dark-grown Arabidopsis *cop1* mutants show a constitutive photomorphogenic phenotype. Five days after germination, *cop1* mutants grown continuously in darkness (*cop1* (D)) show typical photomorphogenic features, such as expanded cotyledons and shortened hypocotyl, like light-grown wild type plants (WT(L)), in contrast to the skotomorphogenic phenotype of dark grown wild type plants (WT(D)). Part of the photograph adapted from previous reports (Deng et al., 1991; Deng, 1994).

2. A BRIEF SUMMARY OF THE UBIQUITIN-PROTEASOME SYSTEM

The ubiquitin-proteasome system is the major mechanism of regulated protein degradation in eukaryotic organisms (Ciechanover, 1994). It is responsible for the regulation of critical cellular processes such as cell cycle control, signal transduction, the removal of abnormal proteins and antigen presentation (Hershko and Ciechanover, 1998). The complex machinery responsible for this proteolysis is the 2 MDa 26S proteasome, which is composed of a 20S core particle containing proteolytic sites buried in its hollow, cylindrical core and a 19S regulatory particle (RP) that caps one or both ends of the core particle (Coux *et al.*, 1996; DeMartino and Slaughter, 1999). Two subcomplexes comprise the 19S complex: the "base," which contains the six ATPase-containing subunits plus two others, and the non-ATPase-containing "lid," which is composed of eight subunits (Glickman *et al.*, 1998). The Rpn10 subunit is thought to form a bridge between the base and the lid. Proteins are designated for degradation by the 26S proteasome through the covalent attachment of a polyubiquitin tag to the target protein by a cascade of enzymes: an E1 ubiquitin activating enzyme, an E2 ubiquitin conjugase and an E3 ubiquitin-protein ligase (Hershko and Ciechanover, 1998).

E3 ubiquitin-protein ligases are responsible for bringing together the substrate protein and the E2 ubiquitin conjugase to attach the polyubiquitin chain to the proteins. Several families of E3 ubiquitin-protein ligases have been identified to date (Vierstra, 2003). Among these families is the RING-finger containing E3 family. This group includes a non-cullin based superfamily (such as COP1) and a cullin based superfamily. The cullin based E3 ligase superfamily includes the SCF-like E3 ubiquitin-protein ligases. SCF E3 ubiquitin-protein ligases are comprised of four subunits: S̲KP1, c̲ullin-1/Cdc53 family member, a F̲-box protein and RBX1/ROC1/HRT1, which contains a RING finger motif (Deshaies, 1999). Together, cullin-1 and RBX1 form the core of the complex and interact with E2 ubiquitin-conjugating enzymes (Zheng *et al.*, 2002). The SKP1 subunit links the cullin and the F-box protein, which, in turn, tethers substrate proteins to the complex. Potentially, as many as 700 F-box proteins exist in *Arabidopsis*, providing a myriad of substrate recognition sites for the modularly arranged SCF-type E3 ubiquitin-protein ligases (Bachmair *et al.*, 2001; Gagne *et al.*, 2002).

3. PROPERTIES AND FUNCTIONS OF THE PLEIOTROPIC COP/DET/FUS PROTEINS

3.1 COP1

In plants, in order to undergo photomorphogenesis, a signalling pathway must exist between the molecules that recognize the light signal and those that effect the required changes in gene expression. COP1 is a central component in the signalling process mediating the overall gene expression response to light. *Arabidopsis cop1* mutant plants grown in the dark have short hypocotyls, open cotyledons and accumulate anthocyanin in the dark (Figure 1) (Deng et al., 1991). These mutant plants are constitutively photomorphogenic; wild type COP1, therefore, represses photomorphogenesis unless a light signal is present. As evidence for this, DNA microarray gene expression experiments utilizing WT and *cop1* mutant plants demonstrate that *cop1* plants grown in the dark have largely overlapping gene expression patterns with WT plants grown in the light (Ma et al., 2002).

The product of the *COP1* gene is a 76 kDa protein that has three familiar protein-protein interaction domains (Figure 2A) (Deng et al., 1992). COP1 has an N-terminal RING finger domain, a coiled-coil domain, and a C-terminal series of WD-40 repeats. The RING finger domain is similar to that found in many E3 ubiquitin-protein ligases. The coiled-coil domain may be responsible for the homo- or hetero-dimerization of COP1 (Torii et al., 1998). The WD-40-repeat domain interacts with target proteins (Torii et al., 1998; Holm et al., 2001). Significantly, the majority of COP1 isolated from dark-grown plants is part of a large (700 kDa) complex (Saijo et al., 2003). The multiple protein-protein interaction domains of COP1 presumably mediate its interactions with a dynamic assembly of protein degradation factors and substrates.

3.1.1 Nuclear localization of COP1

Interestingly, the subcellular localization of COP1 is light dependent (Figure 2B). In the dark COP1 is predominantly nuclear. Under light conditions, the total cellular amount of COP1 does not change; however, there is a great reduction in the amount of COP1 that is in the nucleus (von Arnim and Deng, 1994). This subcellular repartitioning requires the COP9 signalosome, COP10 and DET1 (von Arnim et al., 1997). Both GUS and green fluorescent protein tagged-COP1 in the nucleus are often found further localized within nuclear speckles (von Arnim et al., 1994; Ang et al., 1998). The exact nature of these nuclear speckles is unknown; however, they may represent sites of accumulation of factors involved in protein degradation and/or gene regulation. The dependence of COP1 nuclear localization on the CSN, COP10 and DET1 implies that interactions between these factors are critical for the overall control of photomorphogenesis.

In the dark, COP1 serves as a negative regulator of photomorphogenesis by degrading nuclear transcription factors required for photomorphogenesis to occur. The subcellular redistribution of COP1 is, therefore, likely a major, though not exclusive, mechanism of COP1 regulation.

3.1.2 Light regulation of COP1

To maintain its proper subcellular distribution and activity in response to light, a mechanism must be in place so that COP1 is alerted to the presence of light. To this end, two main families of photoreceptors that handle light perception have been defined (reviewed in Sullivan and Deng, 2003). These include the cryptochromes (cry1 and cry2 in *Arabidopsis*) for blue/ultraviolet-A (UV-A) light perception and the phytochromes (phyA-phyE in *Arabidopsis*) for red/far red light perception

(Ahmed and Cashmore, 1993; Nagy and Schäfer, 2002; Briggs and Christie, 2002). Each has been shown to interact with COP1 directly or indirectly.

Cryptochromes are flavin proteins that resemble mammalian photolyases (Ahmad and Cashmore, 1993). They are composed of two domains, the N-terminal of which contains a flavin adenine dinucleotide (FAD) and a pterin chromophore and the C-terminal of which is activated upon exposure to light (Ahmed and Cashmore, 1993). The C-terminal domains of both CRY1 and CRY2 have been shown to interact physically with the WD-40 domains of COP1 in both a yeast two-hybrid screen and co-immunoprecipitation experiments in both light- and dark-grown seedlings (Wang et al., 2001; Yang et al., 2001). This is interesting, in part, because, like COP1, CRY1-reporter fusion proteins localize to the nucleus in the dark and can be found in nuclear speckles (Wang et al., 2001; Yang et al., 2000). Light-activated cryptochromes may, therefore, interact with COP1 in such a way as to prevent its normal function – promoting degradation of downstream targets. Significantly, CRY1 and CRY2 mediate the accumulation of HY5 in blue light (Osterlund et al., 2000).

PhyA is the major photoreceptor responsible for photomorphogenesis in continuous far red light. Although phyA is stable and cytoplasmic in the dark, upon exposure to light it moves rapidly to the nucleus where it further localizes to nuclear bodies and is ubiquitinated and degraded (Clough and Vierstra, 1997; Kircher et al., 1999; Kircher et al., 2002; Sharrock and Clack, 2002). In transient expression assays in onion epidermal cells, it was reported that COP1 co-localizes with phyA in these nuclear bodies (Seo et al., 2004). Presumably the nuclear interaction between phyA and COP1 in *Arabidopsis* would occur before COP1 itself is depleted from the nucleus upon exposure to light. It appears that phyA plays a role in mediating nuclear depletion of COP1 under far red light (Osterlund and Deng, 1998), while COP1 mediates phyA degradation in the light (Seo et al., 2004).

3.1.3 Molecular role of COP1

In dark-grown plants, COP1's presence in the nucleus could affect gene expression in one of two ways: by directly regulating gene expression through its binding to DNA promoter elements or, alternatively, by interacting with other proteins, such as transcription factors. The absence of a noticeable DNA binding motif in COP1 and the presence of multiple protein-protein interaction domains support this later possibility.

Indeed, COP1 has been shown to interact with the transcription factor HY5 both genetically and biochemically (Ang and Deng, 1994; Ang et al., 1998). HY5 is a nuclear-localized bZIP protein that binds to light-responsive promoter elements and is a positive regulator of photomorphogenesis (Oyama et al., 1997; Chattapadahyay et al., 1998). When HY5 is present, it induces the expression of its target genes to promote photomorphogenesis. In the absence of HY5 activity, transcription of these genes does not occur. As shown in Figure 2A, COP1 interacts with the N-terminal domain of HY5 through its WD-40 domain (Holm et al., 2001). The consequence of this interaction between COP1 and HY5 is a decrease in the total amount of HY5, largely due to HY5's degradation by the ubiquitin-26S proteasome system

(Osterlund *et al.*, 2000). Thus, as a result of its nucleocytoplasmic repartitioning, COP1 controls changes in gene expression in response to light by regulating the nuclear level of HY5 and other transcription factors (Holm *et al.*, 2002).

The action of COP1 on target proteins such as HY5 may be regulated in multiple ways. Besides the regulation of COP1 nuclear abundance, another means of regulation seems to be the phosphorylation of HY5 by casein kinase II (Hardtke *et al.*, 2000). The phosphorylation of HY5 by casein kinase II inhibits the interaction between COP1 and HY5 and, consequentially, the degradation of HY5 (Hardtke *et al.*, 2000). Unphosphorylated HY5 binds target promoters more strongly than phosphorylated HY5 (Hardtke *et al.*, 2000). Together these results suggest that unphosphorylated HY5 is both more active and more rapidly degraded. The phosphorylation activity of casein kinase II may maintain a less active, but more stable, pool of HY5 for rapid activation upon exposure to light.

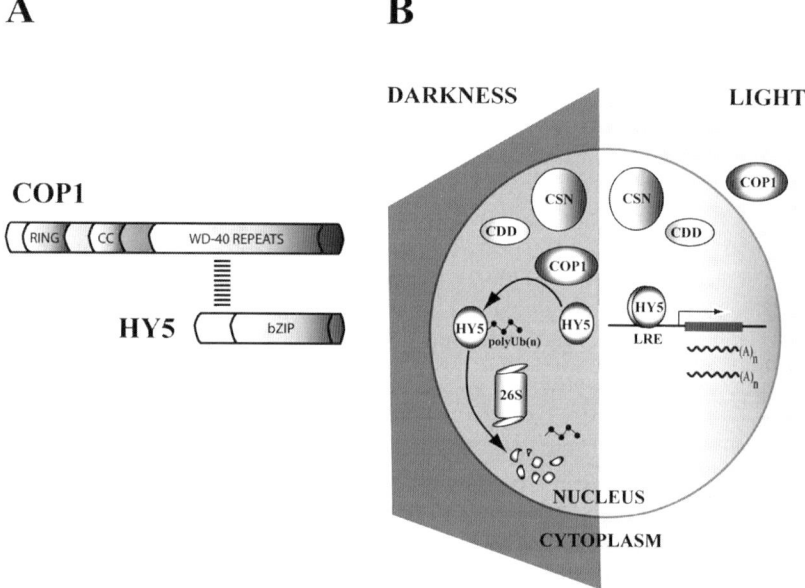

Figure 2. COP1 domain structure and nucleocytoplasmic partitioning. A. COP1 protein contains three distinct protein-protein interacting domains: an N-terminal RING finger domain (RING) conserved in E3 ubiquitin-protein ligases, a coiled-coil domain (CC) involved in COP1's dimerization and a C-terminal series of WD-40 repeats, which interact with COP1 target proteins, such as the light-responsive bZIP protein HY5 (Torii et al., 1998; Holm et al., 2001). B. In dark-grown plants, there is a CSN-dependent enrichment of COP1 in the nucleus (von Arnim et al., 1994). Together with the CDD complex (including COP10, DET1, and DDB1; Yanagawa et al., 2004) and CSN, COP1 mediates HY5 polyubiquitination and its 26S proteasome-mediated degradation (Osterlund et al., 2000). In the dark, COP1 is nuclear-enriched, leading to the degradation of HY5. HY5 binding to light responsive elements (LRE) induces gene transcription and, therefore, the transition to photomorphogenesis (Oyama et al., 1997; Chattapadahyay et al., 1998).

COP1 can be envisioned to serve as a master regulator of light-induced photomorphogenesis by targeting specific transcription factors for ubiquitination and proteasome mediated degradation. HY5 is the best characterized of these COP1 targets, but may only represent one of many such targets. The list of putative transcription factors for which there is either direct or indirect evidence of COP1 targeting them for degradation includes LAF1, a myb transcription activator that transmits phyA signals (Seo *et al.*, 2003); phytochrome interacting factor 3 (PIF3), which interacts with both phyA and phyB (Bauer *et al.*, 2004); CIP7, a positive regulator of photomorphogenesis (Yamamoto *et al.*, 1998); ATB2, a light-induced bZIP protein (Rook *et al.*, 1998); HY5's homologous partner molecule HYH (Holm *et al.*, 2002); phyA-dependent HFR1 (Kim *et al.*, 2002); flower-specific MYB21 (Shin *et al.*, 2002); and the transcription factor(s) responsible for regulating the expression of *HEMA1* and *Lhcb*, genes involved in chloroplast development (McCormac and Terry, 2002).

The pleiotropic nature of *cop1* mutants suggests that COP1-regulated proteins may include non-photomorphogenesis-related factors. Certainly, genetic evidence from *cop1* mutants demonstrates that COP1 plays a role in adult plants as *cop1* plants have defects in the timing of flowering; however, the molecular mechanisms behind these effects remain to be characterized (McNellis *et al.*, 1994). In some cases COP1 may use the familiar strategy of regulated proteolysis to oversee these cellular processes. Candidates for COP1-targeted degradation that is not directly linked to photomorphogenesis include ABI5, a bZIP transcription activator that is involved in ABA hormone signalling, and STO, a putative salt-tolerance transcription factor that may provide a link to Ca^{2+} signalling pathways (Lopez-Molina *et al.*, 2002; Holm *et al.*, 2001).

3.1.4 The E3 ubiquitin-protein ligase activity of COP1

Functionally, COP1 targets proteins for proteolysis by the ubiquitin-proteasome pathway. Because COP1 contains a RING-finger domain similar to that in many E3 ubiquitin-protein ligases, it was hypothesized to act as an E3 ubiquitin-protein ligase itself (Osterlund *et al.*, 2000). Three recent reports have demonstrated that COP1 can, in fact, serve as an E3 ubiquitin-protein ligase. In two cases, COP1 serves as an E3 ubiquitin-protein ligase toward a transcription factor. Seo, *et al* have reported that COP1 can polyubiquitinate the transcription factor LAF1 (Seo *et al.*, 2003). Similarly, Saijo, *et al.*, have reported that COP1 monoubiquitinates HY5 (Saijo *et al.*, 2003). In each report, COP1 ubiquitination of substrate protein was demonstrated in an *in vitro* assay that also included an added E1 ubiquitin activator and E2 ubiquitin-conjugating enzyme. Interestingly, in each case the E3 ubiquitin-protein ligase activity of COP1 toward its substrate is affected by the presence of SPA1, with which it physically interacts through its coiled-coil domain (Hoecker and Quail, 2001). SPA1 serves as a negative regulator of phyA signalling and is a nuclear-localized protein that contains a sequence which shares loose similarity to Ser/Thr and Tyr kinases, two putative nuclear localization sequences, a coiled-coil domain, and a C-terminal domain of WD-40-repeats (Hoecker *et al.*, 1999). In the case of LAF1, its ubiquitination by COP1 is increased in the presence of the small coiled-coil domain of SPA1 (Seo *et al.,* 2003). In contrast, the monoubiquitination of HY5 is inhibited by full length SPA1 (Saijo *et al.,* 2003). How these conflicting reports reconcile with each other remains an outstanding question. The observation that COP1 monoubiquitinates HY5 in these *in vitro* assays may suggest the requirement of additional factors lacking in these assays for HY5 polyubiquitination. In the third demonstration to date of COP1's action as an E3 ubiquitin-protein ligase, the target is not a transcription factor, but, rather, phytochrome A (Seo *et al.,* 2004).

In addition to ubiquitinating phyA and the transcription factors LAF1 and HY5, COP1 auto-ubiquitinates itself (Seo *et al.,* 2003; Saijo *et al.,* 2003). This auto-ubiquitination appears to depend upon the dimerization of COP1 through its coiled-coil domain (Torii *et al.,* 1998; Seo *et al.,* 2003). The potential physiological effect of the auto-ubiquitination of COP1 on its activity remains to be defined.

3.1.5 COP1 interactors

The activity of COP1 is likely modulated by a cohort of interacting proteins. In fact, a number of COP1 interacting proteins (CIP) have been identified through yeast two-hybrid screens. CIP1 is a cytoskeletal-associated protein that has been proposed to be involved in the nucleocytoplasmic shuttling of COP1 (Matsui *et al.,* 1995). Controlling the nuclear abundance of COP1, as the CSN and CDD complexes also seem to do, could be a highly effective way of preventing COP1 from interacting with its downstream targets.

Another COP1 interacting protein is CIP8. COP1 and CIP8 interact through their respective RING domains (Torii *et al.,* 1999). CIP8 is an E3 ubiquitin-protein ligase that is capable of monoubiquitinating HY5 (Hardtke *et al.,* 2002). The relationship between the demonstrated E3 ubiquitin-protein ligase activities of COP1 and CIP8 has yet to be determined.

COP1 interacting genes have also been identified in a genetic screen to look for modifiers of *cop1* mutants. Among the proteins discovered in this way is FIN219, which has a proposed role in phyA-mediated inactivation of COP1 and whose expression may be regulated in response to auxin (Hsieh *et al.,* 2000). Factors such as FIN219 may integrate light and hormone signalling pathways with the activity of COP1.

3.2 The COP9 signalosome

Six of the *COP/DET/FUS* genes encode proteins that form a large multisubunit complex, the COP9 signalosome (CSN). The CSN is a 550 kDa multimeric protein complex composed of eight subunits designated CSN1 through CSN8 (Table 1) (Wei *et al.,* 1994; Wei *et al.,* 1998; Deng *et al.,* 2000; Serino *et al.,* 2003). Six of the subunits were identified in the original *cop/fus* mutant screens. The remaining two subunits (CSN5 and CSN6) have two gene copies each, so they were not identified in the genetic screens (Wei *et al.,* 1994; Chamovitz *et al.,* 1996). Although the CSN was originally identified in *Arabidopsis*, homologues of the CSN subunits have been identified in human, *Drosophila melanogaster*, *Caenorhabditis elegans*, *Aspergillus nidulans* and both budding and fission yeast (Wei *et al.*, 1998; Seeger *et al.*, 1998; Mundt *et al.*, 1999; Freilich *et al.*, 1999; Wee *et al.*, 2002; Maytal-Kivity *et al.*, 2002; Busch *et al.*, 2003). The CSN, like the other COP/DET/FUS proteins, appears to play a critical role in regulating the ubiquitin-proteasome system (reviewed in Schwechheimer and Deng, 2001; Bech-Otschir *et al.*, 2002; Cope and Deshaies, 2003; Serino and Deng, 2003).

Two lines of evidence support a role for the CSN in the ubiquitin-proteasome system. First are the CSN's obvious physical interactions with and structural similarities to the components of the ubiquitin-proteasome system. Second is the clear role the biochemical activities associated with the CSN play in regulating ubiquitin-proteasome mediated protein degradation.

Table 1. CSN subunit composition and 19S RP Lid homologues

Subunit	MW (kDa) Arabidopsis	Motif	*Arabidopsis* locus	Lid homologue
CSN1	50	PCI	*COP11/FUS6*	Rpn7p
CSN2	51	PCI	*COP12/FUS12*	Rpn6p
CSN3	47	PCI	*COP13/FUS11*	Rpn3p
CSN4	45	PCI	*COP8/FUS4*	Rpn5p
CSN5	40	MPN	*AJH1* *AJH2*	Rpn11p
CSN6	35	MPN	*CSN6a* *CSN6b*	Rpn8p
CSN7	25	PCI	*COP15/FUS5*	Rpn9p
CSN8	22	PCI	*COP9/FUS7/FUS8*	Rpn12p

3.2.1 Interactions and similarities between the CSN and the ubiquitin-proteasome system

Recent studies have shown that the CSN physically interacts with other components of the ubiquitin-proteasome pathway. The CSN interacts with SCF E3 ubiquitin-protein ligases (Lyapina *et al.*, 2001; Schwechheimer *et al.*, 2001), and subunits of the CSN interact with subunits of the 19S as well as the 20S particles of the 26S proteasome (Kwok *et al.*, 1999; Fu *et al.*, 2001). This theme of extensive interactions between different complexes involved in ubiquitination and proteolysis is not limited to the CSN; the 26S proteasome has previously been shown to interact with both E2 ubiquitin-conjugases and E3 ubiquitin-protein ligases (Tongaonkar *et al.*, 2000; Xie and Varshavsky, 2000). An intriguing possibility is that the ubiquitinating enzymes, the proteasome and the CSN form a large and dynamic super-complex that holds all the needed factors for regulated proteolysis in close proximity so that the entire substrate ubiquitination and degradation process can occur at one location (Peng *et al.*, 2003). Subunits from the 26S proteasome, the CSN and SCF E3 ubiquitin-protein ligase complexes have, in fact, been shown to co-fractionate (Peng *et al.*, 2003). A centralized location for all these activities would provide an efficient means to regulate proteolysis through the multiple interactions that occur between complexes.

Each subunit of the CSN contains a sequence motif that is found in components of the 19S regulatory particle of the proteasome. Six subunits have a PCI interaction domain – a 200 amino acid domain shared among the 26S proteasome, the COP9 signalosome and the eukaryotic translation initiation factor eIF3 (Table I) (Hofmann and Bucher, 1998). The other two subunits, CSN5 and CSN6, contain a MPN domain conserved in Mpr1p and the Pad1p N-terminus in yeast (Glickman *et al.*, 1998; Hofmann and Bucher, 1998). Both of these domains are assumed to promote interactions among the complex's subunits. Interestingly, the lid subcomplex of the 19S proteasome regulatory particle contains a paralogous set of subunits to the CSN (Wei *et al.*, 1998; Henke *et al.*, 1999). Although specific interactions between equivalent subunits in each complex are similar as shown by yeast two-hybrid experiments (Fu *et al.*, 2001; Serino *et al.*, 2003), the overall topology of the complexes seem to be different in electron microscopy images (Kapelari *et al.*, 2000).

The functional consequences of the similarities between the 19S RP lid and the CSN remain unclear. One possibility is that the CSN and the lid represent evolutionarily related, but functionally diverged complexes, where the 19S lid, through its Rpn11 subunit, is responsible for deubiquitination (Verma *et al.*, 2002; Yao and Cohen, 2002), while the CSN, through its Rpn11-equivalent subunit CSN5, is responsible for RUB deconjugation (see below) (Lyapina *et al.*, 2001; Schwechheimer *et al.*, 2001; Cope *et al.*, 2002). However, the recent discovery that the CSN also possesses deubiquitination activity seems to complicate this argument (Zhou *et al.*, 2003; Groisman *et al.*, 2003). Alternatively, it has been proposed that the CSN may replace the lid subcomplex of the 26S proteasome complex to form new combinations of deubiquitinating activities or specificities (Li *et al.*, 2003).

3.2.2 Biochemical activities of the CSN

There are multiple biochemical activities associated with the CSN that appear to be directly involved in regulating the activity of the ubiquitin-proteasome system. Below we discuss each of those activities.

Derubylation

The most extensively characterized of these activities of the CSN is its role as a RUB1 isopeptidase (Lyapina *et al.*, 2001; Schwechheimer *et al.*, 2001; Zhou *et al.*, 2001; Cope *et al.*, 2002). RUB1 (also called NEDD8 in mammals) is one of the ubiquitin-like molecules identified in the past few years that are covalently attached to substrate proteins (Hochstrasser, 2000). In contrast to polyubiquitin chains, which have traditionally been thought to target substrate proteins for degradation, the role of RUB1 conjugates appears to be regulatory. The commonly known targets for RUB modification are the cullin subunits of E3 ubiquitin-protein ligases (Lyapina *et al.*, 2001; Schwechheimer *et al.*, 2001). RUB1/NEDD8 modification of cullin-1s seems to increase their SCF E3 ligase activity (Read *et al.*, 2000; Wu *et al.*, 2000). This may occur either by increasing SCF complex affinity for F-box proteins (Read *et al.*, 2000; Osaka *et al.*, 2000) and/or by helping to recruit E2 ubiquitin-conjugases to the complex (Kawakami *et al.*, 2001).

The observation that while wild type (WT) cells contain both RUB modified and unmodified cullins, CSN-impaired cells accumulate much higher levels of the rubylated cullins both in fission yeast and in *Arabidopsis*, suggested the existence of a CSN-associated RUB deconjugating activity (derubylation) in cells that modulates the total level of RUB modification (Lyapina *et al.*, 2001; Schwechheimer *et al.*, 2001). Indeed, such a biochemical activity of the CSN can be demonstrated with highly purified CSN and can be localized to a Jab1/MPN domain metalloprotease motif (JAMM domain) within CSN5 (Lyapina *et al.*, 2001; Cope *et al.*, 2002). Controlling the extent of cullin rubylation may be an important mechanism the CSN uses to dictate the rate of substrate polyubiquitination and, thus, various biological processes, including photomorphogenesis. An example of CSN derubylation of a cullin having an effect on substrate abundance occurs in the case of the cyclin-dependent kinase inhibitor p27^{kip1} in mammalian cells (Yang *et al.*, 2002). In this case, CSN derubylation activity inhibits p27^{kip1} degradation and, as a result, alters cell cycle progression.

Two issues are important to mention about CSN derubylation activity in *Arabidopsis*. First, this activity may not be limited to control of photomorphogenesis. The derubylation activity of the CSN may be a general strategy that the CSN utilizes to control multiple cellular pathways regulated by proteolysis. For example, a cycle of RUB modification and deconjugation has been suggested to be essential for SCF-mediated auxin responses (Schwechheimer *et al.*, 2001). Second, the derubylation activity of the CSN may not be responsible for all the functions attributed to the CSN. For instance, a mutated form of the CSN that appears to have normal derubylation activity still fails to rescue the adult lethality of CSN null plants (Wang *et al.*, 2002).

Deubiquitination

A second way in which the CSN may regulate the ubiquitin ligase activity of SCF E3 ubiquitin-protein ligase complexes is through its associated deubiquitinating activity. Two recent papers have demonstrated a deubiquitinating activity associated with the CSN (Zhou *et al.*, 2003; Groisman *et al.*, 2003). In the first, the CSN in yeast associates with a deubiquitinating enzyme, Ubp12p (Zhou *et al.*, 2003). The CSN seems to recruit Upb12p to the nucleus where the CSN serves as a scaffold to bring Ubp12p together with the yeast cullins Pcu1p and Pcu3p, components of the SCF ubiquitin-protein ligase complex. The result of this interaction is that the deubiquitinating enzyme Ubp12p neutralizes the overall E3 ubiquitin-protein ligase activity of the SCF complex. In the second example, from mammalian cells, the newly identified ubiquitin ligase complexes formed by the cullin 4A molecule and either the UV-damaged DNA binding protein p48 (DDB2) or the Cockayne syndrome protein (CSA) also bind the CSN (Groisman *et al.*, 2003). This larger complex contains two deubiquitinating activities: the deubiquitination of polyubiquitin chains and the cleavage of ubiquitins from substrate molecules. The former does not require the metalloprotease domain in the CSN5 subunit, while the later activity does, similar to derubylation activity. One possible role of a CSN-associated deubiquitination activity may be to reverse the autoubiquitination of cullins that normally occurs so that an accumulation of polyubiquitin chains on the cullins does not occur and, thus, to prevent the proteasome-mediated proteolysis of the cullin molecules themselves.

Protein phosphorylation

A third biochemical activity associated with the CSN is a protein kinase activity. Most SCF-type E3 ubiquitin-protein ligases preferentially ubiquitinate phosphorylated substrates, at least in organisms other than *Arabidopsis* (Deshaies, 1999). The fact that the CSN may associate with a protein kinase(s) activity provokes speculation that an additional way the CSN may regulate protein ubiquitination, and, therefore, degradation, is by affecting substrate phosphorylation. In a mammalian system, for instance, the CSN directs the phosphorylation of p53 through the interaction between CSN5 and p53 (Bech-Otschir *et al.*, 2001). The result of this interaction is the increased degradation of p53. CSN-associated phosphorylation does not, however, always result in increased substrate degradation. The CSN has been reported to phosphorylate IκBα, the NF-κB precursor p105, and the transcription factor c-Jun, (Seeger *et al.*, 1998; Naumann *et al.*, 1999). c-Jun had previously been shown to interact with Jab1, the mammalian homologue of CSN5, and, as a result, to be stabilized (Claret *et al.*, 1996). The phosphorylation activity associated with the CSN appears, then, in this case, to stabilize c-Jun against ubiquitin-proteasome degradation.

Although none of the subunits within the CSN contains a kinase activity, several protein kinases that physically associate with the CSN have now been identified. These include inositol 1,3,4-triphosphate 5/6 kinase, which binds to CSN1 (Wilson *et al.*, 2001; Sun *et al.*, 2002), protein kinase D (Uhle *et al.*, 2003), and casein kinase II (Uhle *et al.*, 2003). Interestingly, casein kinase II phosphorylation of HY5 has previously been demonstrated to affect the stability of HY5 (Hardtke *et al.*, 2000). This suggests that the CSN may use phosphorylation as a means of regulating protein degradation in plants as well as mammals.

Control of nucleocytoplasmic localization

In addition to the enzymatic activities described above, the CSN controls the nucleocytoplasmic distribution of key factors, such as COP1 (Chamovitz *et al.*, 1996). In dark-grown *csn* mutants, COP1 is not properly retained in the nucleus and cannot, therefore, ubiquitinate its nuclear target molecules (Osterlund *et al.*, 1999). Maintenance of the proper nuclear localization of COP1 may, therefore, be one way the CSN regulates the overall process of protein degradation. In at least one example, the CSN also determines the nucleocytoplasmic distribution of substrates destined for ubiquitination and subsequent degradation. For example, the CSN mediates the relocalisation of the cyclin dependent kinase inhibitor p27^{kip1} from the nucleus to the cytoplasm where it is degraded (Tomoda *et al.*, 1999). Thus, by varying its subcellular localization, the CSN controls the total cellular abundance of $p27^{kip1}$. The mechanism by which the CSN dictates the nucleocytoplasmic distribution of proteins is completely unexplored.

The CSN has, in summary, four known biochemical or cellular activities associated with it. It may use each of these either independently or in concert to regulate protein degradation. The CSN may manipulate cycles of rubylation/derubylation and ubiquitination/deubiquitination of cullins and phosphorylation of target molecules to control precisely the activity of E3 ubiquitin-protein ligase complexes. These activities, of course, may be further regulated by whether the appropriate factors may interact in the cell. The interconnectedness of each of these biochemical activities of the CSN will likely provide many future avenues of study in how the CSN functions physiologically.

3.2.3 Independent roles for CSN subunits

The discussion of the biochemical activities of the CSN so far has been limited to activities of the intact CSN complex. The structural integrity of the nuclear-localized CSN complex requires the presence of each CSN subunit (Wei et al., 1994); however, various subunits may also function independently of the larger complex. The best characterized example of this is CSN5 and its mammalian homologue Jab1 (Chamovitz and Segal, 2001). CSN5 is an essential part of the nuclear CSN complex and contains the active site for the derubylation activity of the CSN complex, yet it also exists independently in the cytoplasm (Kwok et al., 1998; Cope et al., 2002). The nucleocytoplasmic shuttling of CSN5 is particularly interesting as CSN5/Jab1 functions as an adaptor protein between $p27^{kip1}$ and the nuclear export receptor CRM1 and induces the nuclear export and subsequent degradation of $p27^{kip1}$ in mammalian cells (Tomoda et al., 1999; Tomoda et al., 2002). Whether CSN5 requires a continual association and dissociation cycle from the larger CSN complex as a condition of its physiological function or has completely non-overlapping roles in its monomer and complexed forms is an open question.

3.2.4 Non-photomorphogenic roles of the CSN

Although the COP9 signalosome was originally identified as a regulator of photomorphogenesis in *Arabidopsis*, CSN functions are not limited to photomorphogenesis. The CSN uses its influence over protein degradation to control multiple biological processes in *Arabidopsis*, including control of hormone-dependent pathways, flower development and disease resistance (reviewed in Serino and Deng, 2003), as well as many processes in other organisms. The involvement of the CSN in non-photomorphogenic processes in other organisms includes cell cycle control in mammalian cells (Tomoda et al., 1999; Bech-Otschir et al., 2001; Yang et al., 2002), various aspects of *Drosophila* development (Freilich et al., 1999; Doronkin et al., 2002; Doronkin et al., 2003; Oron et al., 2002; Suh et al., 2002) and the DNA synthesis pathway in yeast (Liu et al., 2003; Nielsen, 2003). Thus, the CSN is a highly conserved regulator of eukaryotic signal transduction and development.

3.3 The CDD complex

3.3.1 COP10

The accumulation of HY5 is not only affected in *cop1* mutants, but also in *Arabidopsis cop10* mutants (Osterlund *et al.*, 2000). Null mutants of *cop10* have a strong photomorphogenic phenotype in the dark and die after the seedling stage (Wei *et al.*, 1994). Like COP1, whose role as an E3 ubiquitin-protein ligase places it on the ubiquitin proteasome pathway, COP10 may be involved in the polyubiquitination of target proteins whose degradation is essential for photomorphogenesis to proceed. COP10 has a roughly 50% identity to the E2 ubiquitin-conjugating enzymes Ubc4/Ubc5 from *Saccharomyces cerevisiae* with an important difference: the canonical cysteine residue in the E2 ubiquitin conjugase catalytic domain is replaced by a serine in COP10 (Suzuki *et al.*, 2002). This is reminiscent of E2 ubiquitin conjugase variant (UEV) proteins, which are inactive E2-like proteins that function only in concert with a true E2 ubiquitin conjugase (Hoffmann and Pickart, 1999). COP10 may serve, therefore, as an E2 ubiquitin conjugase in cooperation with the E3 ubiquitin-protein ligase activities of COP1. In fact, in a yeast two-hybrid assay, COP10 interacts with the RING-finger domain of COP1, as well as subunits of the COP9 signalosome complex (Suzuki *et al.*, 2002). Additional factors may physically interact with COP10, as suggested by the fact that COP10 fractionates on a gel filtration column at a size much larger than the molecular weight of COP10 alone would predict (Suzuki *et al.*, 2002). COP10's presence in a complex would also explain the observation that it is nuclear-localized, even though itself does not contain a recognizable nuclear localization sequence.

3.3.2 DET1

DET1 is a nuclear-localized 62 kDa protein that negatively regulates *Arabidopsis* photomorphogenesis (Pepper *et al.*, 1994). It is conserved in other plant species (tomato) (Mustilli *et al.*, 1999) as well as in animals (hDET1 in human (Wertz *et al.*, 2004) and Abo in *Drosophila* (Berloco *et al.*, 2001)). DET1 appears to repress photomorphogenesis by regulating the mRNA levels of specific light-regulated genes (Pepper *et al.*, 1994) and is required for COP1-mediated HY5 degradation in darkness (Osterlund *et al.*, 2000). Several activities have been demonstrated for DET1. DET1 binds to the N-terminal tail of histone H2B in tomato (Benvenuto *et al.*, 2002), and DET1/Abo binds to the promoter region of histone genes in *Drosophila* (Berloco *et al.*, 2001). DET1 appears, then, somehow to be involved in chromatin regulation through its interaction with histones. Significantly, DET1 has a preference for binding non-acetylated histones (Benvenuto *et al.*,2002), and histone deacetylation has been recognized as a means of regulating overall gene expression (Strahl and Allis, 2000).

How, then, do the histone-binding activities of DET1 relate to the regulation of the ubiquitin-proteasome system in which the other COP/DET/FUS proteins are involved? The answer to this question may lie in the interaction partners of DET1. *Arabidopsis* DET1 can be isolated in a complex with the UV-damaged DNA binding protein DDB1 (Schroeder *et al.,* 2002), and human DET1 may associate with DDB1, cullin-4A, COP1, and ROC1 (Wertz *et al.,* 2004). In humans, DDB1 has been shown to form additional cullin-4A containing E3 ubiquitin-protein ligase complexes that are regulated by the COP9 signalosome (Shiyanov *et al.,* 1999; Groisman *et al.,* 2003). DET1, then, may provide a link between the ubiquitin-proteasome system of regulated protein degradation and the chromosomal regulation of gene expression.

3.3.3 COP10, DDB1, and DET1 are components of the same CDD complex

In an effort to characterize the COP10-containing complex, the COP10-containing complex was purified from cauliflower (Yanagawa *et al.,* 2004). It was shown that the COP10-containing complex includes DDB1 and DET1. This complex has now been named the CDD complex. Immunoprecipitation experiments using *Arabidopsis* seedling extracts indicated that the CDD complex physically interacts with COP1 and CSN complexes and ubiquitin-conjugating enzymes (E2s) *in vivo* (Yanagawa *et al.,* 2004). Both the recombinant COP10 and the purified CDD complex can enhance E2 ubiquitin conjugase activity *in vitro* (Yanagawa *et al.,* 2004), an activity distinct from previous characterized UEVs such as MMS2 and UEV1. Thus, the CDD complex, along with the COP1 and CSN complexes, promotes the degradation of positive regulators of photomorphogenesis, such as the transcription factor HY5, via the ubiquitin/26S proteasome system. In doing so, the CDD complex may act as a ubiquitination-promoting factor to regulate photomorphogenesis.

4. CONCLUDING REMARKS

In summary, the pleiotropic COP/DET/FUS group of proteins defines three protein complexes: the COP1 complex, the COP9 signalosome and the CDD complex. These three complexes interact to regulate the multi-faceted process of photomorphogenesis in plants. They do this by integrating light signals with other cellular processes. Together they form a part of the regulatory network controlling the activity of the ubiquitin-proteasome system. COP1 appears to be directly involved in ubiquitinating protein targets for degradation, while the CDD complex and CSN appear to have a broader regulatory role in overseeing the ubiquitination and degradation process. The ubiquitin-proteasome mediated degradation of sets of transcription factors may be the major way that the COP/DET/FUS proteins can switch the entire developmental program of *Arabidopsis* upon light stimulation. The potential role of DET1 (and, thus, the CDD complex) in the large-scale control of gene expression through chromatin regulation is very exciting. A summary diagram for the reported interactions and roles of all these factors is depicted in Figure 3. To be sure, many unanswered questions remain about how this system functions, but significant progress has been made in the past decade in defining the proteins involved in photomorphogenesis and their activities.

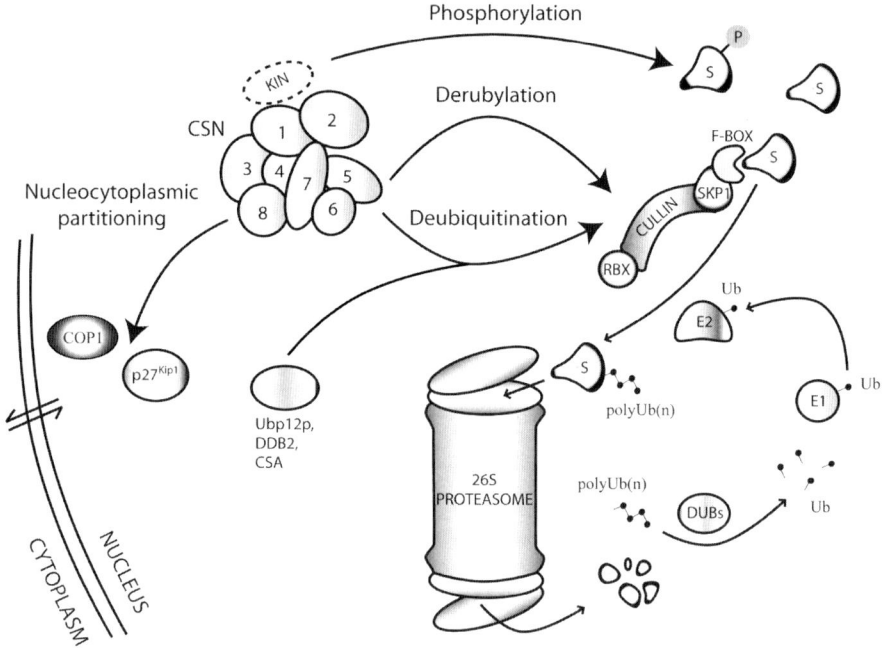

Figure 3. CSN roles in regulated-protein degradation in eukaryotes. CSN shows different activities, represented with large headed arrows, in regulated protein degradation. The former involves a cycle of substrate-specific polyubiquitination, mediated by an E1 ubiquitin-activating enzyme (E1), an E2 ubiquitin-conjugating enzyme (E2) and an E3 ubiquitin-protein ligase (E3, such as the SCF complex comprised of RBX1, a cullin, SKP1 and an F-box protein). The newly polyubiquitinated substrates are susceptible to 26S proteasome-dependent degradation. The polyubiquitin chain is then reduced to ubiquitin monomers by the deubiquitinating enzymes (DUBs) (reviewed in Vierstra, 2003). CSN-dependent derubylation of cullins, mediated by the CSN5 subunit, leads to a decrease in SCF E3 activity as has been shown in fission yeast (Yang et al., 2002). CSN plays a role in the deubiquitination of cullins together with other factors, such as yeast Ubp12p or mammalian DDB2 or CSA (Zhou et al., 2003; Groisman et al., 2003). This process seems to prevent the polyubiquitination and degradation of cullins. Phosphorylation of mammalian proteins, such as p53 or c-Jun, mediated by a kinase activity (KIN) associated with CSN, changes their susceptibility to being SCF E3 substrates (Seeger et al., 1998; Naumann et al., 1999; Bech-Otschir et al., 2001). CSN-dependent accumulation in the nucleus of Arabidopsis COP1 allows it to act as a photomorphogenic repressor (Osterlund et al., 2000). On the contrary, CSN represses mammalian $p27^{kip1}$ function by mediating its translocation to the cytoplasm where it is degraded (Tomoda et al., 1999). CSN is depicted according to its subunit contacts based on yeast two hybrid assays (Serino et al., 2003).

5. REFERENCES

Ahmed, M. and Cashmore, A. R. (1993) *HY4* gene of *A. thaliana* encodes a protein with characteristics of a blue-light photoreceptor. *Nature*, 366, 162-166.

Ang, L. Y. and Deng, X. W. (1994) Regulatory hierarchy of photomorphogenic loci: allele-specific and light-dependent interaction between the HY5 and COP1 loci. *Plant Cell*, 6, 613-628.

Ang, L. Y., Chattopadhyay, S., Wie, N., Oyama, T., Okada, K., Batschauer, A. *et al.* (1998) Molecular interaction between COP1 and HY5 defines a regulatory switch for light control of *Arabidopsis* development. *Molec Cell*, 1, 213-222.

Bachmair, A., Novatchkova, M., Potuschak, T., Eisenhaber, F. (2001) Ubiquitylation in plants: a post-genomic look at a post-translational modification. *Trend Plant Sc*, 6, 463-470.

Bauer, D., Viczián, A., Kircher, S., Nobis, T., Nitschke, R., Kunkel, T., *et al.* (2004) Constitutive photomorphogenesis 1 and multiple photoreceptors control degradation of phytochrome interacting factor 3, a transcription factor required for light signalling in *Arabidopsis*. *Plant Cell*, 16, 1433-1445.

Bech-Otschir, D., Kraft, R., Huang, X., Henklien, P., Kapelari, B., Pollman, C. *et al.* (2001) COP9 signalosome-specific phosphorylation targets p53 to degradation by the ubiquitin system. *EMBO J*, 20, 1630-1639.

Bech-Otschir, D., Seeger, M., Huang, X., Dubiel, W. (2002) The COP9 signalosome: at the interface between signal transduction and ubiquitin-dependent proteolysis. *J Cell Sc*, 115, 467-473.

Benvento, G., Formiggini, F., Laflamme, P., Malakhov, M., Bowler, C. (2002) The photomorphogenesis regulator DET1 binds the animo-terminal tail of histone H2B in a nucleosome context. *Curr Biol*, 12, 1529-1535.

Berloco, M., Fanti, L., Breiling, A., Orlando, V., Pimpinelli, S. (2001) The maternal effect gene, *abnormal oocyte* (*abo*), of *Drosophila melanogaster* encodes a specific negative regulator of histones. *Proc Natl Acad. Sci, USA*, 98, 12126-12131.

Briggs, W. R. and Christie J. M. (2002) Phototropins 1 and 2: versatile plant blue-light receptors. *Trend Plant Sc*, 7, 204-210.

Busch, S., Eckert, S. E., Krappmann, S., Braus, G. H. (2003) The COP9 signalosome is an essential regulator of development in the filamentous fungus *Aspergillus nidulans*. *Molec Microbiol*, 49, 717-730.

Castle, L. A. and Menke, D. W. (1994) A *FUSCA* gene of *Arabidopsis* encodes a novel protein essential for plant development. *Plant Cell*, 6, 25-41.

Chamovitz, D. A., Wei, N., Osterlund, M. T., von Armin, A. G., Staub, J. M., Matsui, M. *et al.* (1996) The COP9 complex, a novel multisubunit nuclear regulator involved in light control of a plant developmental switch. *Cell*, 86, 115-121.

Chamovitz, D. A. and Segal, D. (2001) JAB1/CSN5 and the COP9 signalosome: a complex situation. *EMBO Reports*, 2, 96-101.

Chattapadhyay, S., Ang, L. Y., Puente, P., Deng, X. W., Wei, N. (1998) *Arabidopsis* bZIP protein HY5 directly interacts with light-responsive promoters in mediating light control of gene expression. *Plant Cell*, 10, 673-683.

Chory, J., Peto, C. A., Feinbaum, R., Pratt, L., Ausubel, F. (1989) *Arabidopsis thaliana* mutant that develops a light-grown plant in absence of light. *Cell*, 58, 991-999.

Ciechanover, A. (1994) The ubiquitin-proteasome proteolytic pathway. *Cell*, 79, 13-21.

Claret, F. X., Hibi, M., Dhu, S., Toda, T., Karin, M. (1996) A new group of coactivators that increase the specificity of AP-1 transcription factors. *Nature*, 383, 453-457.

Clough, R. C. and Vierstra, R. D. (1997) Phytochrome degradation. *Plant Cell and Environ*, 20, 713-721.

Cope, G. A., Suh, G. S., Aravind, L., Schwarz, S. E., Zipursky, S. L., Koonin, E. V. *et al.* (2002) Role of metalloprotease motif of Jab1/Csn5 in cleavage of Nedd8 from Cul1. *Science*, 298, 608-611.

Cope, G. A. and Deshaies, R. J. (2003) COP9 Signalosome: a multifunctional regulator of SCF and other cullin-based ubiquitin ligases. *Cell*, 114, 663-671.

Coux, O., Tanaka, K., Goldberg, A. L. (1996) Structure and function of the 20S and 26S proteasomes. *Ann Rev Biochem*, 65, 801-847.

DeMartino, G. N. and Slaughter, C. A. (1999) The proteasome, a novel protease regulated by multiple mechanisms. *J Biol Chem*, 274, 22123-22126.

Deng, X. W., Caspar, T., Quail, P. H. (1991) *COP1*: a regulatory locus involved in light-controlled development and gene expression in *Arabidopsis*. *Genes Dev*, 5, 1172-1182.

Deng, X. W., Matsui, M., Wei, N., Wagner, D., Chu, A. M., Feldmann, K. A., *et al.* (1992) COP1, an *Arabidopsis* regulatory gene, encodes a protein with both a zinc-binding motif and a G_β homologous domain. *Cell, 71,* 791-801.

Deng, X. W. (1994) Fresh view of light signal transduction in plants. *Cell, 76,* 423-426.

Deng, X. W., Dubiel, W., Wei, N., Mundt, K., Colicelli, J., Kato, J. Y., *et al.* (2000) Unified nomenclature for the COP9 signalosome and its subunits: an essential regulator of development. *Trend Genet, 16,* 202-203.

Deshaies, R. J. (1999) SCF and Cullin/RING H2-based ubiquitin ligases. *Ann Rev Cell Dev Biol, 15,* 435-467.

Doronkin, S., Djagaeva, I. Beckendorf, S. K. (2002) CSN5/Jab1 mutations affect axis formation in the *Drosophila* oocyte by activating a meiotic checkpoint. *Devel, 129,* 5053-5064.

Doronkin, S., Djagaeva, I., Beckendorf, S. K. (2003) The COP9 signalosome promotes degradation of Cyclin E during early *Drosophila* embryogenesis. *Dev Cell, 4,* 699-710.

Freilich, S., Oron, E., Kapp, Y., Nevo-Caspi, Y., Orgad, S., Segal, D., *et al.* (1999) The COP9 signalosome is essential for development of *Drosophila melanogaster*. *Curr Biol, 9,* 1187-1190.

Fu, H., Reis, N., Lee, Y., Glickman, M. H., Vierstra, R. D. (2001) Subunit interaction maps for the regulatory particle of the 26S proteasome and the COP9 signalosome. *EMBO J, 20,* 7096-7101.

Gagne, J. M., Downes, B. P., Shiu, S. H., Durski, A. M., Vierstra R. D. (2002) The F-box subunit of the SCF E3 complex is encoded by a diverse superfamily of genes in *Arabidopsis*. *Proc Natl Acad Sci USA, 99,* 11519-11524.

Glickman, M. H., Rubin, D., Coux, O., Wefes, I., Pfeifer, G., Cjeka, Z., *et al.* (1998) A subcomplex of the proteasome regulatory particle required for ubiquitin-conjugate degradation and related to the COP9-signalosome and eIF3. *Cell, 94,* 615-623.

Groisman, R., Polanowska, J., Kuraoka, I., Sawada, J., Saijo, M., Drapkin, *et al.* (2003) The ubiquitin ligase activity of the DDB2 and CSA complexes is differentially regulated by the COP9 signalosome in response to DNA damage. *Cell, 113,* 357-367.

Hardtke, C. S., Gohda, K., Osterlund, M. T., Oyama, T., Okada, K., Deng, X. W. (2000) HY5 stability and activity in *Arabidopsis* is regulated by phosphorylation in its COP1 binding domain. *EMBO J, 19,* 4997-5006.

Hardtke, C. S., Okamoto, H., Stoop-Myer, C., Deng, X. W. (2002) Biochemical evidence for ubiquitin ligase activity of the *Arabidopsis* COP1 interacting protein 8 (CIP8). *Plant J, 30,* 385-394.

Hellman, H. and Estelle, M. (2002) Plant development: regulation by protein degradation. *Science, 297,* 793-797.

Henke, W., Ferrel, K., Bech-Otschir, D., Seeger, M., Schade, R., Jungblut, P., *et al.* (1999) Comparison of human COP9 signalosome and 26S proteasome 'lid.' *Molec Biol Reports, 26,* 29-34.

Hershko, A. and Ciechanover, A. (1998) The ubiquitin system. *Ann Rev Biochem, 67,* 425-479.

Hochstrasser M. (2000) Evolution and function of ubiquitin-like protein conjugation systems. *Nat Cell Biol, 2,* E153-E157.

Hoecker, U., Tepperman, J. M., Quail, P. H. (1999) SPA1, a WD-repeat protein specific to phytochrome A signal transduction. *Science, 284,* 496-499.

Hoecker U. and Quail P.H. (2001) The phytochrome A-specific signalling intermediate SPA1 interacts directly with COP1, a constitutive repressor of light signalling in *Arabidopsis*. *J Biol Chem, 276,* 38173-38178.

Hofmann, K. and Bucher, P. (1998) The PCI domain: a common theme in three multiprotein complexes. *Trend Biochem, 23,* 204-205.

Hofmann, R. M. and Pickart, C. M. (1999) Noncanonical *MMS2*-encoded ubiquitin-conjugating enzyme functions in assembly of novel polyubiquitin chains for DNA repair. *Cell, 96,* 645-653.

Holm, M., Hardtke, C. S., Gaudet, R., Deng, X. W. (2001) Identification of a structural motif that confers specific interaction with the WD40 repeat domain of *Arabidopsis* COP1. *EMBO J, 20,* 118-127.

Holm, M., Ma, L. G., Qu, L. J., Deng, X. W. (2002) Two interacting bZIP proteins are direct targets of COP1-mediated control of light-dependent gene expression in *Arabidopsis*. *Genes Dev, 16,* 1247-1259.

Hsieh, H. L., Okamoto, H., Wang, M., Ang, L. H., Matusi, M., Goodman, H., *et al.* (2000) *FIN219*, an auxin regulated gene, defines a link between phytochrome A and the downstream regulator COP1 in light control of *Arabidopsis* development. *Genes Dev, 14,* 1958-1970.

Kapelari, B., Bech-Otschir, D., Hegerl, R., Schade, R., Dumdey, R., Dubiel W. (2000) Electron microscopy and subunit-subunit interaction studies reveal a first architecture of COP9 signalosome. *J Mol Biol*, *300*, 1169-1178.

Kawakami, T., Chiba, T., Suzuki, T., Iwai, K., Yamanaka, K., Minato, N., *et al.* (2001) NEDD8 recruits E2-ubiquitin to SCF E3 ligase. *EMBO J*, *20*, 4003-4012.

Kim, Y. M., Woo, J. C., Song, P. S., Suh, M. S. (2002) HFR1, a phytochrome A-signalling component, acts in a separate pathway from HY5, downstream of COP1 in *Arabidopsis thaliana*. *Plant J*, *30*, 711-719.

Kircher, S., Kozma-Bognar, L., Kim, L., Adam, E., Harter, K., Schäfer, E. *et al.* (1999) Light quality-dependent nuclear import of the plant photoreceptors phytochrome A and B. *Plant Cell*, *11*, 1445-1456.

Kircher, S., Gil, P., Kozma-Bognar, L., Fejes, E., Speth, V., Husselstein-Muller, T., *et al.* (2002) Nucleocytoplasmic partitioning of the plant photoreceptors phytochrome A, B, C, D, and E is regulated differentially by light and exhibits a diurnal rhythm. *Plant Cell*, *14*, 1541-1555.

Kwok, S. F., Piekos, B., Miséra, S., Deng, X. W. (1996) A complement of ten essential and pleiotropic *Arabidopsis* COP/DET/FUS genes is necessary for repression of photomorphogenesis in darkness. *Plant Physiol*, *110*, 731-742.

Kwok, S. F., Solano, R., Tsuge, T., Chamovitz, D. A., Ecker, J. R., Matsui, M., *et al.* (1998) *Arabidopsis* homologs of a c-Jun coactivator are present both in monomeric form and in the COP9 complex, and their abundance is differentially affected by the pleiotropic *cop/det/fus* mutations. *Plant Cell*, *19*, 1779-1790.

Kwok, S. F., Staub, J. M., Deng, X. W. (1999) Characterization of two subunits of *Arabidopsis* 19S proteasome regulatory complex and its possible interaction with the COP9 complex. *J Molec Biol*, *285*, 85-95.

Li, L., Peng, Z., Deng, X. W. (2003) The COP9 signalosome: an alternative lid for the 26S proteasome? *Trend Cell Biol*, *13*, 507-509.

Liu, C., Powell, K. A., Mundt, K., Wu, L., Carr, A. M., Caspari, T. (2003) Cop9/signalosome subunits and Pcu4 regulate ribonucleotide reductase by both checkpoint-dependent and –independent mechanisms. *Genes Dev*, *9*, 1130-1140.

Lopez-Molina, L., Mongrand, S., Kinoshita, N., Chua, N. H. (2002) AFP is a novel regulator of ABA signalling that promotes ABI5 protein degradation. *Genes Dev*, *17*, 410-418.

Lyapina, S., Cope, G., Shevchenko, A., Serino, G., Tsuge, T., Zhou, C., *et al.* (2001) Promotion of NEDD8-CUL1 conjugate cleavage by COP9 signalosome. *Science*, *292*, 1382-1385.

Ma, L., Gao, Y., Qu, L., Li, J., Zhao, H., Deng, X. W. (2002) Genomic evidence for COP1 as a repressor of light-regulated gene expression and development in *Arabidopsis*. *Plant Cell*, *14*, 2383-2398.

Matsui, M., Stoop, C. D., von Arnim, A., Wei, N., Deng, X. W. (1995) *Arabidopsis* COP1 protein specifically interacts *in vitro* with a cytoskeleton-associated protein, CIP1. *Proc Natl Acad Sci USA*, *92*, 4239-4243.

Maytal-Kivity, V., Pick, E., Piran, R., Hofmann, K, Glickman, M. H. (2003) The COP9 signalosome-like complex in S. cerevisiae and links to other PCI complexes. *Int J Biochem Cell Biol*, *35*, 706-715.

McCormac, A. C. and Terry, M. J. (2002) Light-signalling pathways leading to the co-ordinated expression of HEMA1 and Lhcb during chloroplast development in *Arabidopsis thaliana*. *Plant J*, *32*, 549-559.

McNellis, T. W., von Armin, A. G., Araki, T., Komeda, Y., Miséra, S., Deng, X. W. (1994) Genetic and molecular analysis of an allelic series of *cop1* mutants suggests functional roles for the multiple protein domains. *Plant Cell*, *6*, 487-500.

Miséra, S., Müller, A. J., Weiland-Heidecker, U., Jürgens, G. (1994) The *FUSCA* genes of *Arabidopsis*: negative regulators of light responses. *Mol Gen Genet*, *244*, 242-252.

Mundt, K. E., Porte, J., Murray, J. M., Brikos, C., Christensen, P. U., Caspari, T., *et al.* (1999) The COP9/signalosome complex is conserved in fission yeast and has a role in S phase. *Curr Biol*, *9*, 1427-1430.

Mustilli, A. C., Fenzi, F, Ciliento, R., Alfano, F., Bowler, C. (1999) Phenotype of the tomato *high pigment-2* mutant is caused by a mutation in the tomato homolog of *DEETIOLATED1*. *Plant Cell*, *11*, 145-157.

Nagy, F. and Schäfer, E. (2002) Phytochromes control photomorphogenesis by differentially regulated, interacting signalling pathways in higher plants. *Ann Rev Plant Biol*, *53*, 329-355.

Naumann, M., Bech-Otschir, D., Huang, X., Ferrell, K., Dubiel, W. (1999) COP9 signalosome-directed c-Jun activation/stabilization is independent of JNK. *J Biol Chem, 274*, 35297-35300.
Nielsen, O. (2003) COP9 signalosome: a provider of DNA building blocks. *Curr Biol, 13*, R565-R567.
Oron, E., Mannervik, M., Rencus, S., Harari-Steinberg, O., Neumann-Silerberg, S., Segal, D., et al. (2002) COP9 signalosome subunits 4 and 5 regulate multiple pleiotropic pathways in *Drosophila melanogaster. Devel, 129*, 4399-4409.
Osaka, F., Saeki, M., Katayama, S., Aida, N., Toh-E, A., Kominami, K., et al. (2000) Covalent modifier NEDD8 is essential for SCF ubiquitin-ligase in fission yeast. *EMBO J, 19*, 3475-3484.
Osterlund, M. T. and Deng, X. W. (1998) Multiple photoreceptors mediate the light-induced reduction of GUS-COP1 from *Arabidopsis* hypocotyl nuclei. *Plant J, 16*, 201-208.
Osterlund, M. T., Ang, L. Y., Deng, X. W. (1999) The role of COP1 in repression of *Arabidopsis* photomorphogenic development. *Trend Cell Biol, 9*, 113-118.
Osterlund, M. T., Hardtke, C. S., Wei, N., Deng, X. W. (2000) Targeted destabilization of HY5 during light-regulated development of *Arabidopsis. Nature, 405*, 462-466.
Oyama, T., Shimura, Y., Okada, K. (1997) The *Arabidopsis* HY5 gene encodes a bZIP protein that regulates stimulus-induced development of root and hypocotyl. *Genes Dev, 11*, 2983-2995.
Peng, Z., Shen, Y., Feng, S., Wang, X., Chitteti, B. N., Vierstra, R. D., et al. (2003) Evidence for a physical association of the COP9 signalosome, the proteasome, and specific E3 ligases *in vivo. Curr Biol, 13*, R504-R505.
Pepper, A., Delaney, T., Washburn, T., Poole, D., Chory J. (1994) DET1, a negative regulator of light-mediated development and gene expression in *Arabidopsis* encodes a novel nuclear-localized protein. *Cell, 79*, 109-116.
Read, M. A., Brownell, J. E., Gladyshev, T. B., Hottelet, M., Parent, L. A., Coggins, M. B., et al. (2000) Nedd8 modification of CUL1 activates SCF$^{(beta(TrCP))}$-dependent ubiquitylation of I Ba. *Mol Cell Biol, 20*, 2326-2333.
Rook, F., Weisbeek, P., Smeekens, S. (1998) The light-regulated *Arabidopsis* bZIP transcription factor ATB2 encodes a protein with an unusually long leucine zipper domain. *Plant Molec Biol, 37*, 171-178.
Saijo, Y., Sullivan, J. A., Wang, H., Yang, J., Shen, Y., Rubio, V., et al. (2003) The COP1-SPA1 interaction defines a critical step in Phytochrome A-mediated regulation of HY5 activity. *Genes Dev, 17*, 2642-2647.
Schroeder, D. F., Gahrtz, M., Maxwell, B. B., Cook, R. K., Kan, J. M., Alonso, J. M., et al. (2002) Deetiolated 1 and damaged DNA binding protein 1 interact to regulate *Arabidopsis* photomorphogenesis. *Curr Biol, 12*, 1462-1472.
Schwechheimer, C. and Deng, X. W. (2000) The COP/DET/FUS proteins – regulators of eukaryotic growth and development. *Sem Cell and Dev Biol, 11*, 495-503.
Schwechheimer, C. and Deng, X. W. (2001) COP9 signalosome revisited: a novel mediator of protein degradation. *Trend Cell Biol, 11*, 420-426.
Schwechheimer, C., Serino, G., Callis, J., Crosby, W. L., Lyapina, S., Deshaies, R. J., et al. (2001) Interactions of the COP9 signalosome with the E3 ubiquitin-ligase SCF^{TIR1} in mediating auxin response. *Science, 292*, 1379-1382.
Seeger, M., Kraft, R., Ferrell, K., Bech-Otschir, D., Dumdey, R., Schade, R., et al. (1998) A novel protein complex involved in signal transduction possessing similarities to 26S proteasome subunits. *FASEB J, 12*, 469-478.
Seo, H. S., Yang, J. Y., Ishikawa, M., Bolle, C., Ballesteros, M. L., Chua, N. H. (2003) LAF1 ubiquination by COP1 controls photomorphogenesis and is stimulated by SPA1. *Nature, 423*, 995-999.
Seo, H. S., Watanabe, E., Tokutomi, S., Nagatian, A., Chua, N. H. (2004) Photoreceptor ubiquitination by COP1 E3 ligase desensitizes phytochrome A signaling. *Genes Dev, 18*, 617-622.
Serino, G. and Deng, X. W. (2003) The COP9 signalosome: regulating plant development through the control of proteolysis. *Ann Rev Plant Biol 54*, 65-82.
Serino, G., Su, H., Peng, Z., Tsuge, T., Wei, N., Gu, H., et al. (2003) Characterization of the last subunit of the *Arabidopsis* COP9 signalosome: implications for the overall structure and origin of the complex. *Plant Cell, 15*, 719-731.
Sharrock, R. A. and Clack, T. (2002) Patterns of expression and normalized levels of the five *Arabidopsis* phytochromes. *Plant Physiol, 130*, 442-456.

Shin, B., Choi, G., Yi, H., Yang, S., Cho, I., Kim, J., *et al.* (2002) AtMYB21, a gene encoding a flower-specific transcription factor, is regulated by COP1. *Plant J*, 23-32.

Shiyanov, P., Nag, A., Raychaudhuri, P. (1999) Cullin-4A associates with the UV-damaged DNA-binding protein DDB. *J Biol Chem*, 274, 35309-35312.

Strahl, B. D. and Allis, C. D. (2000) The language of covalent histone modifications. *Nature,* 403, 41-45.

Suh, G. S., Poek, B., Chouard, T., Oron, E., Segal, D., Chamovitz, D.A., *et al.* (2002) *Drosophila* JAB1/CSN5 acts in photoreceptor cells to induce glial cells. *Neuron*, 33, 35-46.

Sullivan, J. A. and Deng, X. W. (2003) From seed to seed: the role of photoreceptors in *Arabidopsis* development. *Dev Biol*, 260, 289-297.

Sullivan, J. A., Shirasu, K., Deng, X. W. (2003) The diverse roles of the ubiquitin and 26S proteasome in the life of plants. *Nature Rev Gen*, 4, 948-958.

Sun,Y., Wilson, M. P., Majerus, P. W. (2002) Inositol 1,3,4-triphosphate 5/6-kinase associates with the COP9 signalosome by binding to CSN1. *J Biol Chem*, 277, 45759-45764.

Suzuki, G., Yanagawa, Y., Kwok, S. F., Matsui, M., Deng X. W. (2002) *Arabidopsis* COP10 is an ubiquitin-conjugating enzyme variant that acts together with COP1 and COP9 signalosome in repressing photomorphogenesis. *Genes Dev*, 16, 554-559.

Tepperman, J. M., Zhu, T., Chang, H. S., Wang, X., Quail, P. H. (2001) Multiple transcription-factor genes are early targets of phytochrome A signaling. *Proc Natl Acad Sc USA*, 98, 9437-9442.

Tomoda, K., Kubota, Y., Kato J. (1999) Degradation of the cyclin-dependent-kinase inhibitor p27Kip1 is instigated by Jab1. *Nature*, 398, 160-165.

Tomoda, K., Kubota, Y., Arata, Y., Mori, S., Maeda, M., Tanaka, T., *et al.* (2002) The cytoplasmic shuttling and subsequent degradation of p27kip1 mediated by Jab1/CSN5 and the COP9 signalosome complex. *J Biol Chem*, 277, 2302-2310.

Tongaonkar, P., Chen, L., Lambertson, D., Ko, B., Madura, K. (2000) Evidence for an interaction between ubiquitin-conjugating enzymes and the 26S proteasome. *Mol Cell Biol*, 20, 4691-4698.

Torii, K. U., McNellis, T. W., Deng, X. W. (1998) Functional dissection of *Arabidopsis* COP1 reveals specific roles of its three structural modules in light control of seedling development. *EMBO J*, 17, 5577-5587.

Torii, K. U., Stoop-Myer, C. D., Okamoto, H., Coleman, J., Matsui, M., Deng, X. W. (1999) The RING finger motif of photomorphogenic repressor COP1 specifically interacts with the RING-H2 motif of a novel *Arabidopsis* protein. *J Biol Chem*, 274, 27674-27681.

Uhle, S. Medalia, O., Waldron, R., Dumdey, R., Henklein, P., Bech-Otschir, D., *et al.* (2003) Protein kinase CK2 and protein kinase D are associated with the COP9 signalosome. *EMBO J*, 22, 1302-1312.

Verma, R., Aravind, L., Oania, R., McDonald, W. H., Yates, J. R., Koonin, E. V., *et al.* (2002) Role of Rpn11 metalloprotease in deubiquitination and degradation by the 26S proteasome. *Science*, 298, 611-615.

Vierstra, R. D. (2003) The ubiquitin/26S proteasome pathway, the complex last chapter in the life of many plant proteins. *Trend Plant Sc*, 8, 135-142.

von Arnim, A. and Deng, X. W. (1994) Light inactivation of *Arabidopsis* photomorphogenic repressor COP1 involves a cell-specific regulation of its nucleocytoplasmic partitioning. *Cell*, 79, 1035-1045.

von Arnim, A. and Deng, X. W. (1996) Light control of seedling development. *Annu Rev Plant Physiol Plant Mol Biol*, 47, 215-243.

von Armin, A., Osterlund, M. T., Kwok, S. F., Deng, X. W. (1997) Genetic and developmental control of nuclear accumulation of COP1, a repressor of photomorphogenesis in *Arabidopsis*. *Plant Physiol*, 114, 779-788.

Wang, H., Ma, L. G., Li, J. M., Zhao, H. Y., Deng, X. W. (2001) Direct interaction of *Arabidopsis* cryptochromes with COP1 in light control development. *Science*, 294, 154-158.

Wang, X., Kang, D., Feng, S., Serino, G., Schwechheimer, C., Wei, N. (2002) CSN1 N-terminal-dependent activity is required for *Arabidopsis* development but not for Rub1/Nedd8 deconjugation of cullins: a structure-function study of CSN1 subunit of COP9 signalosome in *Arabidopsis*. *Mol Biol Cell*, 13, 646-655.

Wee, S., Hetfeld, B., Dubiel, W., Wolf, D. A. (2002) Conservation of the COP9/signalosome in budding yeast. *BMC Genet*, 3, 15.

Wei, N., Chamovitz, D. A., Deng, X. W. (1994) *Arabidopsis* COP9 is a component of a novel signaling complex mediating light control of development. *Cell*, 78, 117-124.

Wei, N. and Deng, X. W. (1996) The role of the *COP/DET/FUS* genes in light control of *Arabidopsis* seedling development. *Plant Physiol*, *112*, 871-878.
Wei, N., Tsuge, T., Serino, G., Dohmae, N., Takio, K., Matsui, M., et al. (1998) The COP9 complex is conserved between plants and animals and is related to the 26S proteasome regulatory complex. *Curr Biol*, *8*, 919-922.
Wei, N. and Deng, X. W. (1999) Making sense of the COP9 signalosome. *Trend Gen*, *15*, 98-103.
Wertz, I. E., O'Rourke, K. M., Zhang, Z., Dornan, D., Arnott, A., Deshaies, R. J., et al. (2004) Human deetiolated-1 regulates c-Jun by assembling a CUL4A ubiquitin ligase. *Science*, *303*, 1371-1374.
Wilson, M. P., Dun, Y., Cao, L., Majerus, P. W. (2001) Inositol 1,3,4-triphosphate 5/6-kinase is a protein kinase that phosphorylates the transcription factors c-Jun and ATF-2. *J Biol Chem*, *276*, 40998-41004.
Wu, K., Chen, A., Pan, Z. Q. (2000) Conjugation of Nedd8 to CUL1 enhances the ability of the ROC1-CUL1 complex to promote ubiquitin polymerization. *J Biol Chem*, *275*, 32317-32324.
Xie, Y., Varshavsky, A. (2000) Physical association of ubiquitin ligases and the 26S proteasome. *Proc Natl Acad S, USA*, *97*, 2497-2502.
Yao, T., Cohen, R. E. (2002) A cryptic protease couples deubiquitination and degradation by the proteasome. *Nature*, *419*, 403-407.
Yamamoto, Y. Y., Matsui, M., Ang, L. Y., Deng, X. W. (1998) Role of a COP1 interacting protein in mediating light-regulated gene expression in *Arabidopsis*. *Plant Cell*, *10*, 1083-1094.
Yang, H. Q., Wu, Y. J., Tang, R. J., Liu, D., Cashmore, A. R. (2000) The C-termini of *Arabidopsis* cryptochromes mediate a constitutive light response. *Cell*, *103*, 815-827.
Yang, H. Q., Tang, R. H., Cashmore, A. R. (2001) The signalling mechanism of *Arabidopsis* CRY1 involves direct interaction with COP1. *Plant Cell*, *13*, 2573-2587.
Yanagawa, Y., Sullivan, J. A., Komatsu, S., Gusmaroli, G., Suzuki, G., Yin, J., et al. (2004). Arabidopsis COP10 forms a complex with DDB1 and DET1 *in vivo* and enhances the activity of ubiquitin conjugating enzymes. *Genes Dev*, *18*, 2172-2181.
Yang, X., Menon, S., Lykke-Andersen, K., Tsuge, T., Xiao, D., Wang, X., et al. (2002) The COP9 signalosome inhibits $p27^{kip1}$ degradation and impedes G1-S phase progression via deneddylation of SCF Cul1. *Curr Biol*, *12*, 667-672.
Zheng, N., Schulman, B. A., Song, L., Miller, J. J., Jeffrey, P. D., Wang, P., et al. (2002) Structure of the Cul1-Rbx1-Skp1-Fbox^{Skp2} SCF ubiquitin ligase complex. *Nature*, *416*, 703-709.
Zhou, C., Seibert, V., Geyer, R., Rhee, E., Lyapina, S., Cope, G., et al. (2001) The fission yeast COP9/signalosome is invovled in cullin modification by ubiquitin-related Ned8p. *BioMed Central Biochem*, *2*, 7.
Zhou, C., Wee, S., Rhee, E., Naumann, M., Dubiel, W., Wolf, D. A. (2003) Fission yeast COP9/signalosome suppresses cullin activity through recruitment of the deubiquitylating enzyme Ubp12p. *Mol Cell*, *11*, 927-938.

Chapter 19

BIOCHEMICAL AND MOLECULAR ANALYSIS OF SIGNALLING COMPONENTS

Christian Fankhauser[1], Chris Bowler[2]
[1]*Department of Molecular Biology, University of Geneva, 30 quai E. Ansermet, 1211 Geneva 4, Switzerland (e-mail: Christian.Fankhauser@unil.ch); Current Address: Center for Integrative Genomics, University of Lausanne, 1015 Lausanne, Switzerland*
[2]*ENS/CNRS FRE 2433, Organismes Photosynthétiques et Environnement, Département de Biologie, Ecole Normale Supérieure, 46 Rue D'Ulm, 75230 Paris Cedex 05, FRANCE (e-mail: cbowler@biologie.ens.fr)*

1. INTRODUCTION

Phototransformation of phytochrome initiates a series of events that ultimately leads to physiological adaptation to new environmental conditions. The chain of events occurring in between is complex, with multiple branches and interactions that are context dependent so that they should probably be described as a signalling web rather than a signalling chain. Recent progress in the phytochrome-signalling field clearly indicates that key transduction events also occur in distinct subcellular compartments. The light regulated localisation of the photoreceptors themselves is a striking illustration of this concept (see Chapter 9). The aim of this chapter is to review the biochemical and molecular events of phytochrome signalling with an emphasis on events occurring outside of the nucleus. The large recent progress in the field prompted us to split phytochrome signalling into several chapters. Please consult chapter 17 for nuclear events and chapters 16, 18, 20 and 21 for additional considerations on phytochrome signalling. For studies performed prior to 1993 we strongly recommend reading the chapter by S. Roux in the 2nd edition of photomorphogenesis in plants (Roux, 1994).

Phototransformation occurs in the time scale of milliseconds, but the manifestation of this initial event varies largely depending on the response that is studied. This can take as little as a few seconds, as in light regulated cytoplasmic motility (Takagi *et al.*, 2003), or up to many days for photoperiodic induction of flowering (see Chapter 27). Many of these events presumably require a transcriptional cascade, although very quick responses probably do not. Such early branching in signalling events has been well characterised in other systems, e.g., for the response to pheromone in yeast (Leberer *et al.*, 1997). The presence of multiple phytochromes within any plant species (phyA-phyE in *Arabidopsis* thaliana) adds an additional level of complexity since it is well known that despite similar phototransformation reactions the different phytochromes will trigger distinct responses when exited by light (see Chapter 7). A detailed description of the different phytochrome signalling modes can be found in chapter 16. Additional levels of complexity are due to the distinct tissues and to the timing of the light cue. Although the initial phytochrome phototransformation events are believed to be the same in all tissues, the outcome of the subsequent chain of events will depend upon the organ and the timing of the event. For example, light enhances cell growth in the cotyledons but inhibits it in the hypocotyl. Moreover the effect of a saturating pulse of red light on flower induction very much depends on the timing of delivery (see Chapter 27). These examples illustrate the complexity of the signalling events occurring after photoperception.

Most of the genetic screens that have so far been performed to identify phytochrome-signalling components rely on phenotypes that take several days before they can be scored. This presumably explains the current lack of temporal resolution for phytochrome signalling events. We still know very little about the order of action of a fast growing list of phytochrome signalling-components. In this chapter we will concentrate on the components that are either constitutively cytoplasmic or that shuttle between cytoplasm and nucleus. Signalling in animals often requires second messengers such as Ca^{2+}, cyclic nucleotides, phosphorylation events or inositol phospholipids. The requirement for such molecules for phytochrome signalling has been carefully investigated, although the "jury is still out" for several of them. The involvement of phosphorylation events has been revived with the discovery of bacteriophytochromes, many of which are light regulated histidine kinases (see Chapter 6).

2. IS PHYTOCHROME A LIGHT-REGULATED PROTEIN KINASE?

Structural and evolutionary considerations clearly indicate that plant phytochromes have evolved from prokaryotic bacteriophytochromes (see Chapter 6). The majority of those bacteriophytochromes have a typical sensor kinase type configuration with the sensor domain binding the phytochrome chromophore and displaying classical phytochrome spectral properties, and a carboxy-terminal histidine kinase domain (Montgomery and Lagarias, 2002). For several bacteriophytochromes it has been demonstrated that they possess light regulated histidine kinase activity (Montgomery and Lagarias, 2002). We can therefore conclude that in prokaryotes protein kinase activity is a very early event in phytochrome signalling. A minority of bacteriophytochromes do not possess a carboxy-terminal histidine kinase domain but a PAS (Per, Arnt, Sim) domain, indicating that there are other ways to signal after the initial phototransformation event (Giraud *et al.*, 2002). This might be particularly relevant in view of plant phytochrome structure because the carboxy terminus of a typical plant phytochrome has both PAS- and histidine kinase-related domains (Montgomery and Lagarias, 2002). PAS domains can mediate protein-protein interactions and there is a large body of evidence indicating that the PAS domains of *Arabidopsis* phyA and phyB are very important for normal interaction with several signalling intermediates (Choi *et al.*, 1999; Ni *et al.*, 1999; Schaefer and Bowler, 2002; Taylor and Zhulin, 1999). Moreover in some instances these interactions are light specific (Choi et al., 1999; Ni et al., 1999). A more detailed description of factors interacting in a light specific manner with the PAS domain of phytochromes can be found in chapter 17.

Plant phytochromes are most probably not histidine kinases. Several key residues are missing in the histidine kinase-related domain (Cashmore, 1998; Elich and Chory, 1997; Reed, 1999). However oat phyA, biochemically the best characterised plant phytochrome, possesses Ser/Thr protein kinase activity in vitro (Yeh and Lagarias, 1998). Very little is known about a possible kinase activity of the other plant phytochromes. As for bacteriophytochromes, both light and the chromophore regulate the autophosphorylation activity of oat phyA (Yeh and Lagarias, 1998). Two sites of oat phyA in vitro phosphorylation have been mapped: Ser8 and Ser599 are phosphorylated by the phytochrome associated protein kinase activity (Figure 1) (McMichael and Lagarias, 1990; Song, 1999). Oat phyA purified from plants and recombinant oat phyA purified from yeast have similar in vitro kinase properties (Yeh and Lagarias, 1998). Oat phyA phosphorylation with various kinases has been used to determine structural changes between Pr and Pfr (Lapko et al., 1996; McMichael and Lagarias, 1990; Wong et al., 1986). Protein kinase A primarily phosphorylates Ser18 and Ser599 (Figure 1) (Lapko et al., 1996; Song, 1999). Interestingly oat phytochrome purified from seedlings is always phosphorylated on Ser8, and Ser599 is only phosphorylated when phyA is purified from light grown seedlings (Figure 1) (Lapko et al., 1999). These results suggest that Ser8 and Ser599 are targets of oat phyA kinase activity. Notwithstanding, in vitro studies indicate that oat phyA is not a very potent kinase in vitro. It has a high Km for ATP binding but where the ATP binding site is located within the protein is not known, indicating that we still have a lot to learn about this enzymatic activity of phyA (Wong and Lagarias, 1989).

Phosphorylation of the amino-terminus of phytochrome and conservation of this Ser-rich region in phyA from different plants prompted Stockhaus et al. (1992) to test the function of this domain by mutational analysis. A rice phyA in which the first 10 Ser (contained within the first 20 amino acids of the protein) have been mutated into Ala displays enhanced biological activity when over-expressed in tobacco (Stockhaus et al., 1992). Taken together with the in vivo and in vitro phosphorylation studies this suggests that Ser8 phosphorylation might be implicated in a desensitisation mechanism. Similar studies using oat phyA with the same Ser mutagenised into Ala, or deleting amino acids 6 to 12 (Δ6-12 mutant, which includes Ser8) have yielded similar results, again suggesting Ser8 phosphorylation as a regulatory modification (Figure 1) (Jordan et al., 1997). A careful analysis of oat phyA-mediated light responses showed that the Δ6-12 mutant is hyperactive for the VLFR when over-expressed in tobacco, shows a normal VLFR in *Arabidopsis* and shows a dominant negative suppression of the HIR in both species (Casal et al., 2002). Studies using GFP fusions indicate that the deletion of amino acids 6 to 12 (Δ6-12 mutant) alters light induced phyA nuclear speckle formation (Casal et al., 2002). Taken together these studies suggest that oat phyA auto-phosphorylates on Ser8 and that this modification fine-tunes phyA activity in vivo. The light dependent phosphorylation of Ser599 could be a regulatory step for one or several aspects of phyA light regulation (Figure 1) (Fankhauser *et al.*, 1999; Song, 1999).

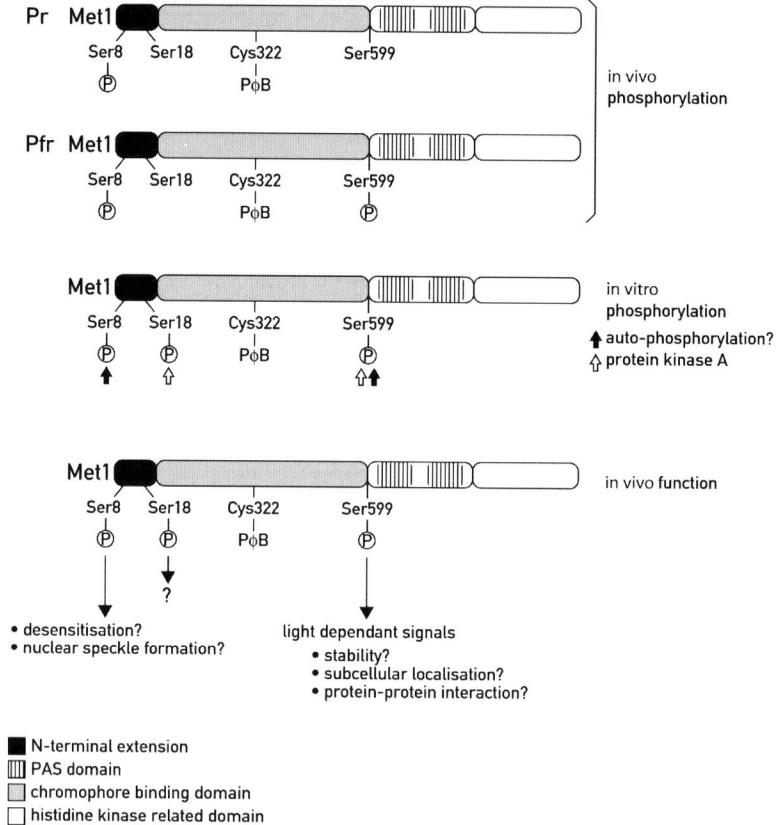

Figure 1. Oat phyA is a phosphoprotein. The phosphorylation sites of oat phyA have traditionally been labelled Ser7, Ser17 and Ser598. However this nomenclature does not take into account the initiator Met (Met1). In vivo only Ser8 is phosphorylated when phyA is purified from etiolated seedlings (Pr). When phyA is purified after a light treatment (Pfr) both Ser8 and Ser599 are phosphorylated. In vitro kinase assays have shown that Ser8, Ser18 and Ser599 are the major phosphoacceptor sites, with protein kinase A and the phyA associated kinase activity having different site preferences. A model is proposed for the possible function of these phosphorylation events.

Oat phyA can also phosphorylate several other substrates in vitro. These include the cryptochrome blue light receptors, PKS1 (phytochrome kinase substrate 1) and the IAA proteins (Ahmad *et al.*, 1998; Colon-Carmona *et al.*, 2000; Fankhauser *et al.*, 1999). The phosphorylation of some of these proteins is modulated by light in vivo, although it has not been definitively established that they are phosphorylated by phytochrome in planta. In the case of cryptochrome 2 it has even been suggested that its phosphorylation is independent of any phytochrome (Shalitin et al., 2002). The functional implications of phosphorylation of these proteins await further studies.

The biological relevance of the phyA kinase activity has still not been firmly established. A truncation of as little as the last 35 amino-acids of the carboxy terminus of oat phyA eliminates biological activity in gain of function experiments (over-expression in tobacco) (Cherry et al., 1993). It is interesting to note that this phytochrome deletion construct still multimerises and that its spectral properties are unaffected (Cherry et al., 1993). However this truncation eliminates a small part of the histidine kinase-related domain and gets very close to the putative ATP binding site present in the last 150 amino acids of phyA. These data suggest that the kinase activity of oat phyA might be important in vivo, however this construct has not been characterised for protein kinase activity and such a deletion might also have other effects. More recently the far-red insensitivity of the *Arabidopsis* ecotype Lm-2 was traced back to a missense mutation in phyA. Biochemical characterisation of this phyA variant showed that this phytochrome has reduced in vitro kinase activity but is otherwise normal (Maloof et al., 2001). This data offers strong support for the importance of the phyA kinase activity.

Structure function studies performed with phyB would actually indicate that the histidine kinase-related domain of *Arabidopsis* phyB is dispensable in vivo (Krall and Reed, 2000; Matsushita et al., 2003). This was first suggested when a *phyB* allele lacking most of the histidine kinase domain was characterised. Interestingly this *phyB* allele is only partially impaired in phyB signalling and, more surprisingly, missense alleles altering single amino-acids within the histidine kinase-related domain have a stronger phenotype than do alleles in which the histidine kinase homology domain has been entirely deleted. These phyB point mutations affect residues that are close to the putative ATP binding site (Krall and Reed, 2000). However ATP binding or protein kinase activity have not been demonstrated for phyB. These results suggest that the histidine kinase homology domain may have a regulatory function. In the wild-type protein both the PAS domain and the histidine kinase homology domain would have to be activated to achieve normal phyB activity. The missense mutants in the histidine kinase-related domain would prevent proper activation. In contrast, in the absence of this domain the need for domain activation would be eliminated (Krall and Reed, 2000). A recent publication suggests that the amino-terminal 650 amino-acids of phyB are sufficient to sustain phyB signalling as long as this protein is maintained in the nucleus (Matsushita et al., 2003). Based on this study one might conclude that phyB signalling only occurs in the nucleus and that phyB signalling does not require the carboxy terminus. Such conclusions are probably premature. It is possible that this truncated phyB acts in concert with other type II phytochromes (phyC-E) and uses their carboxy terminal domain to achieve normal signalling. The signalling activities of phyA and phyB are sufficiently different to imagine that kinase activity might be important for one but not the other phytochrome. The protein kinase domain of phytochrome has never been mapped. Based on the homology with histidine kinases it is tempting to speculate that it lies within the histidine kinase-related domain, but this has never been demonstrated. These data clearly show that the "phytochrome as a kinase" controversy is not solved yet.

3. PHOSPHORYLATION IN PHY MEDIATED SIGNALLING

Irrespective of phytochrome being a kinase or not, phosphorylation events are probably important for some aspect(s) of phytochrome signalling, although the currently available data are mainly indirect and the in vivo implication(s) of the described biochemical events are largely unclear. However given that plant phytochromes likely evolved from light regulated protein kinases (cyanobacterial phytochromes, see Chapter 6) and that phosphorylation events play central roles is most signalling cascades, it would be very surprising if phytochrome signalling in plants would be an exception.

Several biochemical methods have been utilised to study the role of protein phosphorylation: 1) In vitro kinase assays with plant extracts and various subcellular fractions 2) In vivo labelling of seedlings or protoplasts 3) In gel kinase assays with plant extracts. The first reports concerning phytochrome regulated phosphorylation were obtained from extracts. It is generally assumed that red light induction is a good indication for phytochrome regulation. In some cases the effect of red light was inhibited by a subsequent pulse of far-red light clearly indicating the involvement of phytochrome (acting in the low fluence response mode). Non reversibility by far-red light and induction by far-red light itself is however not incompatible with the involvement of phytochrome, acting in the very low fluence response mode and/or the far-red high irradiance response. Datta *et al.*, (1985) were the first to show red/far-red reversible phosphorylation of proteins from etiolated pea nuclear extracts. Interestingly phosphorylation was Ca^{2+} dependent and the use of inhibitors suggested the involvement of calmodulin (Datta *et al.*, 1985). Another study reported red- and far-red-light induced phosphorylation of proteins in oat nuclear extracts. G-protein modulators affected the phosphorylation of these proteins suggesting a cross-talk between phosphorylation and G-protein-coupled signalling events (Romero *et al.*, 1991). Studies using oat coleoptiles identified a number of proteins that were phosphorylated extremely rapidly after red light stimulation (Otto and Schäfer, 1988). Using a parsley cell suspension culture Harter and co-workers showed that a number of rapid red light stimulated (and red/far-red reversible) phosphorylation events occur in the cytoplasm (Harter *et al.*, 1994). Interestingly using different light sources they showed that light induces several signalling events involving protein kinases and phosphatases. These studies indicate that following phytochrome-mediated light perception, rapid phosphorylation events occur in several species and in several cell compartments (nucleus and cytoplasm). Unfortunately, the identity of the kinases and their substrates remains unknown. Similar experiments performed with extracts or intact tissues of wild type compared to phytochrome mutants would be a nice complement. A first report using *Arabidopsis* extracts and an in-gel kinase assay yielded some interesting results (Malec *et al.*, 2002). This study identifies a 50 kD protein kinase. Both far-red and blue light activated this protein kinase but red light did not. The mechanism of light activation and the identity of this protein kinase are unknown. The kinase activity was not light induced in phyA, phyB and cry1 mutant backgrounds, possibly suggesting the requirement for phytochrome/cryptochrome co-action for light regulation of this protein kinase (Malec *et al.*, 2002).

The plant G-box binding factors (G-boxes are regulatory sequences commonly encountered in light regulated promoters) belonging to the bZIP class of transcription factors are known substrates of light regulated protein kinase activity (Menkens *et al.*, 1995). Several bZIP transcription factors from this family might be involved in light regulated transcription (Menkens *et al.*, 1995). In some instances the DNA binding activity of a G-box binding factor was found to be regulated by phosphorylation-dephosphorylation events. Casein kinase II (CK2) is a likely protein kinase candidate (Klimczak *et al.*, 1992). A cytoplasmic protein kinase regulates the activity of CPRF2 (common plant regulatory factor 2), another member of this transcription factor family. CPRF2 phosphorylation is light regulated but does not alter its DNA binding capacity (Wellmer *et al.*, 1999). However this transcription factor is cytoplasmic in the dark and phytochrome-regulated phosphorylation triggers nuclear import of CPRF2 (Kircher *et al.*, 1999). Phosphorylation of this class of transcription factors might therefore affect their function at different regulatory levels. The functional relevance of most of these findings has unfortunately not been tested yet. However plants with reduced levels of CK2 activity display mis-regulation of light-activated genes, but it should be emphasised that it is not known if this is due to altered phosphorylation of a G-box binding factor (Lee *et al.*, 1999). Interestingly plants with elevated CK2 levels show alterations in circadian regulated gene expression. It has been proposed that this phenotype is due to altered phosphorylation of the circadian clock component CCA1, a Myb class transcription factor (Sugano *et al.*, 1999). CCA1 was initially identified as a protein that binds to the promoter element of a light regulated gene. Given the close connections between light signalling and regulation of the circadian clock this is an interesting observation.

NDPK (nucleoside diphosphate kinase) has been identified as another substrate of red light stimulated protein kinase activity (Hamada *et al.*, 1996). This phosphorylation was observed in etiolated pea stems, was rapid and red/far-red reversible. The light regulated phosphorylation was observed both in membrane and soluble fractions but the nature of the kinase(s) is unknown (Ogura *et al.*, 1999). Interestingly NDPK2 has been shown to interact directly with phytochrome A (Choi *et al.*, 1999). phyA in the Pfr form enhances γ-phosphate exchange activity of NDPK2, but it is currently not known if regulation of the enzymatic activity of NDPK2 is due to phosphorylation by phytochrome (Choi *et al.*, 1999). An *ndpk2* loss of function mutant in *Arabidopsis* shows reduced cotyledon opening in response to red and far-red light indicating a role for this protein in phytochrome signalling (Choi *et al.*, 1999). Unfortunately the function of the red light-stimulated phosphorylation of NDPK is currently unknown. A protein phosphatase 2A, another protein interacting with phyA, provides further evidence for a role of phosphorylation/dephosphorylation events in phytochrome signalling. This cytoplasmic protein interacts with both phyA and phyB and is capable of dephosphorylating oat phyA in vitro (Kim *et al.*, 2002). Genetic evidence indicates that this protein may play a role in the phytochrome-mediated control of flowering time (Kim *et al.*, 2002).

4. G-PROTEINS

A number of influential papers published during the 90s strongly suggest a role for Ca^{2+}, G-proteins and calmodulin in phytochrome signalling. The involvement of these second messengers was suspected by analogy to other signalling cascades and in particular to signalling downstream of the visual pigment rhodopsin. G-proteins are extremely well studied and a number of specific inhibitors and activators have been characterised, enabling pharmacological studies to investigate the role of these intermediates in phytochrome mediated signalling. G proteins are typically active in their GTP binding form and can be inactivated upon hydrolysis of GTP into GDP. They typically possess low GTPase activity and GAP (GTPase Activating Proteins) enzymes enhance this activity thereby returning the G-protein to its inactive state. Such G-proteins exist either as monomers (small G-proteins) or as heterotrimeric G-proteins where the α subunit is the G-protein and the two other subunits are known as β and γ. Heterotrimeric G-proteins typically transduce signals from G-protein coupled receptors, that are 7 transmembrane proteins. Receptor activation promotes the GDP to GTP exchange of the α subunit. In the GTP bound form the α subunit dissociates from βγ and is typically active, however in some cases the βγ heterodimer also possesses independent signalling activity (Leberer *et al.*, 1997). When the α subunit hydrolyses GTP into GDP it re-associates with βγ and is inactivated. When GTP-γ-S, a non hydrolysable GTP analogue, is used the system is locked in the activated state. Constitutive activation can also be achieved with cholera toxin. In contrast GDP-β-S or pertussis toxin can trap the system in the off state. Moreover structure function studies allowed the identification of specific Gα mutants that are either constitutively active or inactive.

Rhodopsin is the best studied example of this class of receptors (Meng and Bourne, 2001). G-protein coupled receptors represent the biggest gene family in animals and there are large families coding for the Gα, Gβ and Gγ subunits (Meng and Bourne, 2001). In contrast the fully sequenced *Arabidopsis* genome uncovered single genes coding for a Gα and a Gβ and two genes coding for Gγ subunits (Jones, 2002). Moreover there is currently no characterised G-protein coupled 7 transmembrane receptor (Jones, 2002). The *Arabidopsis* GRC1 protein is a predicted 7 transmembrane receptor that might be G protein coupled. The ligand of GRC1 is currently unknown. Over-expression of GRC1 alters the cell cycle, abolishes seed dormancy and shortens time to flowering (Colucci *et al.*, 2002). A role for G proteins has been established for *Arabidopsis* growth and development, but there is still very little known about the mode of activation of plant heterotrimeric G-proteins (Jones, 2002). Interestingly eliminating GPA1, the Gα subunit of *Arabidopsis* also leads to alterations of the cell division cycle (Jones, 2002). This genomic view highlights some differences between plant and animal G-protein mediated signalling.

One of the earliest reports suggesting a role for G-proteins in phytochrome signalling was obtained with etiolated wheat protoplasts. In this system light induces cellular swelling, and the use of a number of agonists and antagonists indicated a possible role for G-proteins (Bossen *et al.*, 1990). Subsequent work showed that activating G-proteins with cholera toxin could mimic the effect of red light regulated gene expression (Romero *et al.*, 1991; Romero and Lam, 1993). Interestingly Clark *et al.* (1993) reported phytochrome regulation of monomeric GTP binding proteins associated with the nuclear envelope of peas. A direct role for these nuclear envelope bound G-proteins in phytochrome modulated gene expression has however not been established yet (Clark *et al.*, 1993). All these papers suggested a role for G-proteins in diverse aspects of phytochrome signalling.

A series of elegant papers gave strong support to this hypothesis (Neuhaus *et al.*, 1993; Bowler *et al.*, 1994a; Bowler *et al.*, 1994b; Wu *et al.*, 1996; Kunkel *et al.*, 1996; Neuhaus *et al.*, 1997). All those publications used a "biochemical complementation" approach. The tomato aurea mutant is a phytochrome chromophore biosynthetic mutant (Terry and Kendrick, 1996). This mutant is largely impaired in phytochrome signalling and has hypocotyl cells that are large enough to allow micro-injection. Several phenotypes can be used to monitor phytochrome responses such as anthocyanin accumulation and chloroplast development that are almost entirely absent when the mutant is shifted into the light for 48 hours. In addition, light-inducible reporter genes can be injected to easily follow gene expression responses.

When purified oat phyA is micro-injected into aurea hypocotyl cells "biochemical complementation" can be observed in a cell autonomous fashion (Neuhaus *et al.*, 1993). Gain-of-function studies using GTP-γ-S or cholera toxin suggested a role for G-proteins, and loss of function experiments co-injecting phyA with GDP-β-S or pertussis toxin yielded similar conclusions. Using the same strategy phytochrome signalling was divided into three branches downstream of the Gα step, with the different pathways controlling each other (Bowler *et al.*, 1994a; Bowler *et al.*, 1994b). One branch requires cGMP only and leads to *CHS* expression. A second branch requiring Ca^{2+} and calmodulin leads to *CAB* expression and the third branch requires both cGMP and Ca^{2+} and allows regulation of genes such as *FNR* and *AS1* (Figure 2). In agreement with this branched pathway Wu *et al.* (1996) reported that Ca^{2+} and cGMP target distinct light responsive promoter elements (Wu *et al.*, 1996).

Because aurea mutants are defective for all phytochromes, it was of interest to test whether purified phyB could also complement this mutant. Recombinant yeast-produced phyB could activate both the *CAB* and *FNR* genes but not the *CHS* gene (Figure 2) (Kunkel *et al.*, 1996). Taken together these experiments strongly support the G-protein signalling hypothesis, although strong genetic support for this proposal is still lacking.

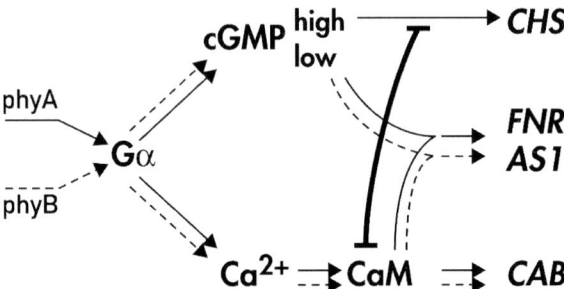

Figure 2. A model for G protein function in phytochrome signalling. This model is mainly based on the tomato micro-injection studies.

In a first attempt to gain genetic support for the G protein model Okamoto et al. (2001) used a gain of function approach in *Arabidopsis*. They moderately over-expressed either a wild-type or a constitutively activated form of the single *Arabidopsis* Gα subunit. Although the seedlings developed normally in the dark, light-dependent hypocotyl growth inhibition was greater in the seedlings with elevated Gα protein levels (Okamoto et al., 2001). The same result was observed when seedlings were grown in continuous blue, red and far-red light, suggesting that Gα acts downstream of cry1, phyB and phyA (the major photoreceptors regulating inhibition of hypocotyl growth in response to blue, red and far-red light, respectively). In agreement with this idea the Gα effect requires phyA in far-red light and phyB in red light. However cry1 is not required for the Gα effect in blue light suggesting a possible role for phytochrome or redundancy between the two *Arabidopsis* cryptochromes (Okamoto et al., 2001). Moreover in far-red light the Gα effect requires the signalling component FHY1 but not FHY3. These studies offer genetic support for the pharmacological experiments, however somewhat unexpectedly the seedlings over-expressing Gα have smaller cotyledons, reduced chlorophyll accumulation and reduced *CAB* expression in the light. All these phenotypes are opposite to what one might have expected based on the increased inhibition of hypocotyl growth and the pharmacological experiments (Okamoto et al., 2001).

Furthermore a recent study by Jones et al. (2003) indicates that heterotrimeric G proteins play no direct role in the red and far-red mediated inhibition of *Arabidopsis* hypocotyl growth (Jones et al., 2003). Null mutants of the α and the β subunits and a Gα Gβ double mutant were used to reach this conclusion. Moreover in these experiments ectopic expression of the wild-type and constitutively active Gα subunit did not lead to altered light sensitivity (Jones et al., 2003). The different results concerning Gα gain of function might relate to the fact that Okamoto et al (2001) used lines that over-express 2-3 fold whereas the lines from Jones et al. (2003) were in the 6-10 fold range. The loss of function experiments are more problematic for the G-protein theory. Based on the currently available and very complete sequence information it is very unlikely that there are other classical Gα or Gβ subunits encoded in the *Arabidopsis* genome (Jones *et al.*, 2003). It is difficult to rule out the possibility that some other G protein-like molecule mediates phytochrome effects. For example, a recent study has revealed a novel type of nitric oxide synthase (NOS) in plants that has similar biochemical activity as the animal NOS enzymes, but whose sequence is entirely unrelated (Guo *et al.* 2003). It should also be noted that the genetic experiments tested phytochrome mediated inhibition of hypocotyl growth whereas the pharmacological experiments investigated gene expression, anthocyanin accumulation and chloroplast development. We clearly know that phytochrome signalling is a highly branched pathway, and so G-proteins might therefore only affect a subset of the phytochrome responses.

5. RAPID ION FLUXES

The role of rapid ion fluxes as second messengers of the phytochrome response has been a topic of intense research, although this aspect of signalling has received little attention during the last decade. We strongly recommend the chapter by Roux (1994) in the previous edition of this book who carefully describes earlier work. As discussed in the previous section Ca^{2+} is required for a subset of phytochrome mediated responses in "biochemical complementation" assays of the tomato aurea mutant. It is therefore not surprising that reports describing different phytochrome responses in different organisms have not always found an important function for Ca^{2+} (Roux, 1994). The available phytochrome mutants in several plant species now allow a re-evaluation of the previously reported phytochrome mediated ion fluxes. Moreover with the molecular identification of loci affecting phytochrome signalling (particularly in *Arabidopsis*) genetic support can back up previous work.

In *Arabidopsis* hypocotyls a rapid and transient membrane depolarisation occurs in response to blue light. This early event occurring within seconds of the light treatment has been linked to light mediated growth control (Parks *et al.*, 2001). This rapid and transient blue light induced membrane depolarisation is due to the activation of an anion channel (Parks *et al.*, 2001). Inhibition of this initial event either by mutating cry1 or by pharmacological inhibition of anion channels (using 5-nitro-2 (3-phenylpropylamino)-benzoic acid, NPPB) affects hypocotyl growth similarly but very importantly not additively. These studies indicate that upon blue light perception the cryptochromes activate an anion channel (Parks *et al.*, 1998). This channel activation then affects cryptochrome mediated hypocotyl growth control that starts about 30-60 minutes after light perception (Figure 3). The blue light growth response is complex and has been dissected genetically. It requires the action of the phototropins during the initial 30 minutes. The phototropin mediated growth phase does not require membrane depolarisation (Folta and Spalding, 2001). This initial phase is followed by a prolonged cryptochrome controlled hypocotyl growth (Figures 3 and 4).

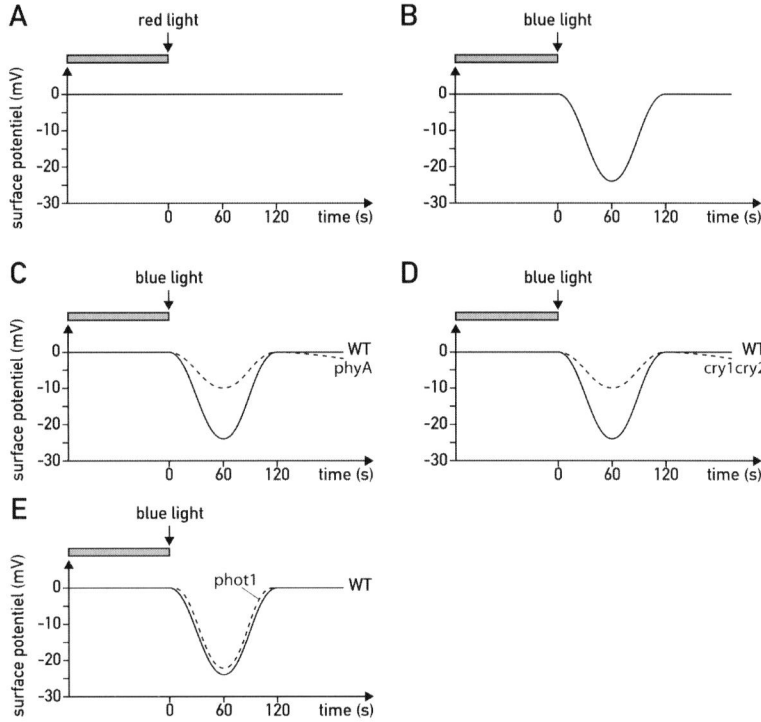

Figure 3. Light induced membrane depolarisation. This figure is based on the work presented by E. Spalding and co-workers (Folta and Spalding, 2001; Parks et al., 1998; Parks and Spalding, 1999). It represents the membrane depolarisation of Arabidopsis hypocotyl cells upon a light treatment in the wild type and various mutant backgrounds The grey bar represents darkness.

As for a number of other cryptochrome responses this growth control is modulated by the action of the phytochromes. This has been demonstrated in *Arabidopsis* where *phyA* mutants are strongly impaired in blue light mediated membrane depolarisation (Folta and Spalding, 2001). The *phyA* phenotype for this response is as strong as the phenotype of the *cry1cry2* double mutant (Figure 3). Incidentally the membrane depolarisation defect of *phyA* mutants is probably the fastest phy phenotype described to date in *Arabidopsis* (Folta and Spalding, 2001). *phyA* mutants also have a hypocotyl growth phenotype similar to the one of *cry* mutants. The phyA modulation of the cryptochrome growth phase lasts for a few hours only (Parks *et al.*, 2001). It is unlikely that the effect is exclusively indirect via the cryptochromes because the *phyAcry1* double mutant has a stronger hypocotyl growth and a stronger membrane depolarisation phenotype than either single mutant and than the *cry1cry2* double mutant (Folta and Spalding, 2001). phyB modulates the cryptochrome growth response very differently. *phyB* mutants show rather normal membrane depolarisation and the *phyB* mutant suppresses the early growth phenotypes of *cry1* and *phyA* mutants. However this suppression of the growth phenotype is not related to the transient membrane depolarisation because the *phyAphyB* double mutant behaves similarly to the *phyA* single mutant (Folta and Spalding, 2001). Taken together these results indicate that upon light perception a balance of growth inhibitory and growth promoting factors controls hypocotyl growth (Parks *et al.*, 2001). The light effects that are visible within seconds (membrane depolarisation) and that are controlled by both phyA and the cryptochromes indicate that upon light perception both classes of photoreceptors mediate very rapid signalling responses at the plasma membrane.

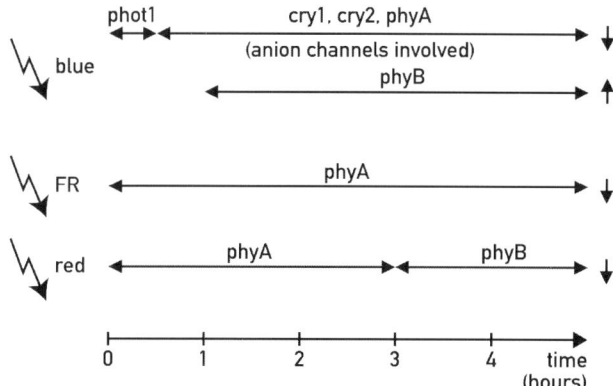

Figure 4. Light modulated hypocotyl growth in Arabidopsis requires the sequential action of several photoreceptors. This figure is based on the work from E. Spalding and co-workers and adapted from a model presented in (Parks et al., 2001). The arrows on the right pointing up represent enhancement of growth and pointing down represent inhibition of growth.

Inhibition of *Arabidopsis* hypocotyl growth by red and far-red light is slower than in response to blue light. The response occurs within about 15 minutes and is not preceded by a transient membrane depolarisation (Parks and Spalding, 1999). The use of phytochrome mutants shows that the far-red growth response is strictly phyA-dependent, the response to red light depends mainly on phyA during the first 3 hours of de-etiolation and then is predominantly controlled by phyB (Figures 3 and 4). This rather slow phytochrome effect has been taken as an indication that nuclear events are involved in this growth regulation (Parks and Spalding, 1999). In contrast, in the moss Physcomitrella red light induces a very rapid membrane depolymerisation response (visible within 10 to 15 seconds) (Ermolayeva *et al.*, 1996). This response is red/far-red reversible and requires Ca^{2+} in the external medium. The use of inhibitors lead to the conclusion that K^+ and Cl^- channels are important for the response but that photosynthesis is not required (Ermolayeva *et al.*, 1996). These studies indicate that although rapid red light induced membrane depolarisation responses have not been identified in *Arabidopsis* they do exist in other photosynthetic organisms. The identification of such responses in *Arabidopsis* would allow a detailed genetic description of rapid phytochrome responses.

The growth effects of light can also be observed is isolated protoplasts from a number of species. This has for instance been shown with protoplasts from etiolated wheat leaves that swell within 3 minutes in response to red light (Bossen *et al.*, 1990). The direct effects of phytochromes and cryptochromes could recently be assessed with protoplasts from *Arabidopsis* hypocotyls and pea stems isolated from various mutant backgrounds. Such protoplasts swell in response to red light. This swelling response is clearly phytochrome dependent. In pea, phyA and phyB play the prevalent role because the response is almost totally abolished in a *phyAphyB* double mutant (Long and Iino, 2001). The response is red/far-red reversible and requires 30 minutes of irradiation with red light before growth is initiated (Long and Iino, 2001). This response is much slower than in wheat leaf protoplasts (Bossen *et al.*, 1990). However the two experiments can hardly be compared since in one case the protoplasts come from etiolated leaves and in the other case from stems of red light-grown pea seedlings. Swelling of pea protoplasts requires uptake of K^+ and Cl^- and Ca^{2+} is required in the medium (Long and Iino, 2001). A similar study performed with *Arabidopsis* protoplasts shows that in this species phyA plays the prevalent role for red light induced swelling. A *phyAphyB* double mutant no longer responds to red light indicating that phyA and phyB are the major phytochromes mediating this response (Wang and Iino, 1998). The swelling is red/far-red reversible although it is generally accepted that in *Arabidopsis* phyA mediates very low fluence responses and high irradiance responses but not low fluence response. This is therefore an exception, but it is not the only one (Stowe-Evans *et al.*, 2001). In contrast to pea, the dependency of Ca^{2+} for protoplast swelling is not very obvious in *Arabidopsis* (Wang and Iino, 1998).

Protoplasts also respond to blue light that triggers rapid and transient shrinking. In *Arabidopsis* this is strictly cry1-dependent, but the phytochromes also modulate this response since a *phyAphyB* double mutant lacks the response similarly to the *cry1* mutant (Wang and Iino, 1998). A careful set of experiments demonstrated that Pfr has to be present about 30 minutes before the blue light stimulus to induce the formation of an essential short lived signalling intermediate. In the absence of this signalling intermediate the cryptochrome mediated blue light response cannot take place. The induction of this intermediate is red/far-red reversible. Interestingly NPPB inhibits the blue light response similarly to what has been found for the cryptochrome mediated growth control of *Arabidopsis* hypocotyls (Wang and Iino, 1998). These two responses are therefore very similar 1) cry1 is essential 2) phytochromes modulate the response 3) NPPB inhibits the response. By changing the bath solution ion concentration and using various inhibitors the following model for this transient shrinking response has been proposed. Light induced protoplast shrinking requires secretion of K^+ and Cl^-, the swelling then occurs by uptake of these same ions from the medium. In contrast to the red light swelling response Ca^{2+} does not seem to be required (Wang and Iino, 1998). This is another illustration of branching in phytochrome signalling because on the one hand Ca^{2+} is required for red light induced swelling and on the other hand phytochrome modulation of the blue light induced growth response does not depend on Ca^{2+}. Interestingly phyA and phyB control both the red light swelling and the modulation of the cry-mediated shrinking in a totally red/far-red reversible way (Long and Iino, 2001; Wang and Iino, 1998).

6. CYTOPLASMIC MOVEMENTS

Light has some very rapid effects on cell growth that require re-organisation of the cytoskeleton. In addition light induces rapid movement of organelles, in particular chloroplasts, that are mediated by microfilaments and microtubules (Wada et al., 2003). Several classical experiments show that phytochrome is important for these events. This has for instance been documented for Mougeotia chloroplast movements (Haupt and Häder, 1994). In the fern *Adiantum* both red and blue light control chloroplast location. A recent publication shows that an unusual phytochrome controls the red light response (Kawai et al., 2003). Similarly, in the moss *Physcomitrella* red and blue light control such photo-movements (Sato et al., 2001). Microtubule and microfilament inhibitors were used to show that these movements depend on both actin and tubulin. Interestingly blue light modulated the chloroplast movement, which was microtubule and microfilament dependent. In contrast the red light induced phytochrome response involves only the microtubules (Sato et al., 2001). In higher plants chloroplast movements are mediated by blue light. The phototropins are required for this response in *Arabidopsis* (Wada et al., 2003).

Cortical microtubules reorient in response to light in the growing zone of maize coleoptiles (Fischer and Schopfer, 1997). This response is closely correlated with changes in elongation and is phytochrome controlled. After a red light pulse the microtubules begin relocalisation within 10 minutes and the response is completed after 1 hour (Fischer and Schopfer, 1997). Such a fast effect of phytochrome on the cytoskeleton, which ultimately leads to light modulated growth control was anticipated, however the molecular details are still unknown. An even much faster response has been documented in the aquatic angiosperm *Vallisneria gigantea* (Takagi et al., 2003). Cytoplasmic mobility was monitored by infrared microscopy in epidermal leaf cells. Red light was effective in increasing the cytoplasmic mobility and these movements were regulated in a red/far-red reversible manner. The sites of increased motility did not correlate with the presence of organelles. When the light was applied on single cells with microbeam irradiation the movements were only induced in the irradiated cell. Moreover the increased motility only occurred where the cytoplasm was directly irradiated (subcellular autonomous response) (Takagi et al., 2003). The changes in movement already occurred 2.5 sec after the irradiation. Photobiological characterisation of the response is consistent with the involvement of a type II phytochrome (light stable phytochrome). Pharmacological studies indicate that those movements require actin filaments but not microtubules. Removal of Ca^{2+} leads to increased motility and the plasma membrane ATPase inhibitor vanadate or an inhibitor of photosynthesis did not inhibit the light response. These studies clearly show the presence of extremely rapid phytochrome responses that occur in the cytoplasm (Takagi et al., 2003). The following model could explain these findings. The Pr to Pfr transformation evokes changes in cytosolic Ca^{2+} levels, this leads to a modulation of activity of a Ca^{2+} dependent cytoskeletal factor, such as the motor protein myosin. In addition to these cell autonomous phytochrome responses "non nuclear" events must also be at the basis of phytochrome mediated intercellular signalling. Micro-beam irradiation of tobacco cotyledons carrying a CAB:Luciferase transgene were used to look for phytochrome mediated intercellular signalling. Expression of the reporter construct in cells that had not been irradiated could be measured within minutes (Schutz and Furuya, 2001). Interestingly the response was red/far-red reversible only at the site of irradiation. This indicates that a hormonal or an electrophysiological signal (changes in membrane potential or ion fluxes) could very rapidly spread from the site of irradiation and influence gene expression in cells that have never seen light.

7. FORWARD AND REVERSE GENETICS

During the last decade classic and reverse genetics have become increasingly more important. The striking difference between an etiolated and a light grown seedling inspired Maarten Koornneef's screen for long hypocotyl mutants (*hy*) in 1980 (Koornneef *et al.*, 1980). Hypocotyl length is the easiest phenotype to screen for, but numerous other morphological or molecular phenotypes can be used as a basis for genetic screens (Devlin *et al.*, 1998; Li *et al.*, 1994). The genetic approach has been very successful and the recent progress made in cloning the altered genes in such mutants has greatly enhanced our molecular understanding of phytochrome signalling. Those mutants can be classified into mutants that display an altered light sensitivity but develop normally in the dark, and mutants that display light independent de-etiolation (*det*) or constitutive photomorphogenesis (*cop*) (Chory *et al.*, 1989; Deng *et al.*, 1991). Many mutants of this second class are allelic to *fus* mutants that were initially identified based on enhanced anthocyanin accumulation (Miséra *et al.*, 1994). This class of mutants is covered in chapter 18. Among altered light sensitivity mutants both decreased light sensitivity (*hy* mutants for instance) and increased light sensitivity mutants have been identified (Quail, 2002). It is however important to note that mutants in this class have light specific phenotypes, in other words their morphology in the dark is wild-type. We will not cover all these mutants in this chapter but will concentrate on those gene products that have enhanced our understanding of events occurring in the cytoplasm and at the plasma membrane (Table 1). Nuclear events will be covered in chapter 17. This separation is however somewhat artificial since several signalling components can be found both in the nucleus and the cytoplasm. As a note of caution we would like to emphasise that the vast majority of subcellular localisation studies have been performed with over-expressed GFP fusions, so the results may not completely reflect to the properties of the endogenous protein.

Genetic screens for mutants specifically hyposensitive to far-red light identified a large number of loci (Wang and Deng, 2003). Since phyA is the major photoreceptor mediating de-etiolation in far-red light those loci identify gene products that are required for efficient phyA signalling. A number of these genes have been cloned. The majority of them encode nuclear proteins but PAT1 and FIN219 are exclusively cytoplasmic, LAF6 is chloroplastic and FHY1 can be found both in the nucleus and the cytoplasm (Bolle *et al.*, 2000; Hsieh *et al.*, 2000; Moller *et al.*, 2001;Desnos *et al.*, 2001. PAT1 is a member of the plant specific GRAS family. These proteins have been postulated to regulate gene expression and are involved in a variety of developmental responses including root patterning (SCR), shoot meristem differentiation (HAM) and gibberellin signalling (RGA and GAI) (Stuurman *et al.*, 2002). SCR, RGA and GAI all have nuclear localisation signals but PAT1 does not (Bolle *et al.*, 2000). The effect of PAT1 is specific to phyA signalling since *pat1* mutants do not show altered gibberellin sensitivity or defective root development. The interpretation of the *pat1* phenotype is somewhat complicated by the fact that it is not a loss of function allele. A truncated form of PAT1 is expressed in the *pat1* mutant leading to a dominant negative effect. The phenotype of a true null allele is currently unknown. The presence of several highly related PAT1 homologues suggests possible genetic redundancy. As for most genetically identified phytochrome signalling components the exact role of PAT1 is currently unknown (Bolle *et al.*, 2000).

Table 1. Short summary of phytochrome-signalling components identified by genetic and/or reverse genetic approaches that are not always localized in the nucleus. The references to all those signaling components can be found in section 19.7.

Name	homologies/ function	localization	signalling
NDPK2	nucleoside di-phosphate kinase	Nuc and Cyto	phyA/phyB
PKS1	no known function	PM and Cyto	phyA
PKS2	homologous to PKS1	PM and Cyto	phyA
PAT1	GRAS family member	Cyto	phyA
FIN219	GH3 domain protein	Cyto	phyA
FHY1	no known function, NLS, NES	Nuc and Cyto	phyA
LAF6	ABC transporter	Chl	phyA
CR88	HSP90	Chl	phyB
SRR1	no known function, NLS	Nuc and Cyto	phyB
ARR4	B-type response regulator	Nuc and Cyto	phyB
SUB1	EF hand Ca^{2+} binding	Cyto	phyA, cry

The *fin219* mutant affects a subset of the phyA responses. Hypocotyl elongation is not inhibited to the same extent as in the wild-type, cotyledon opening and expansion do not seem to be much affected, and germination and chlorophyll accumulation after a far-red light treatment are unaffected in this mutant (Hsieh *et al.*, 2000). Global gene expression profiling is also consistent with the idea that FIN219 acts late in the phyA pathway (Wang *et al.*, 2002). The mutation is semi-dominant and caused by an epigenetic modification that diminishes expression of *FIN219* in the light. Over-expression of FIN219 leads to enhanced far-red light sensitivity. FIN219 codes for an auxin induced cytoplasmic GH3-like protein (Hsieh *et al.*, 2000). GH3 proteins were initially identified from soybean as early auxin induced genes, but their function is mostly unknown. FIN219 is allelic to JAR1, a gene involved in jasmonic acid (JA) signalling (Staswick *et al.*, 2002). Biochemical characterisation of recombinant JAR1 indicates that this protein specifically adenylates jasmonic acid (Staswick *et al.*, 2002). Based on the *jar1* phenotype this suggests that jasmonic acid adenylation is important for a subset of JA responses. Among the nine *jar1* alleles not a single one affects far-red induced inhibition of hypocotyl elongation. Moreover contrary to the nine *jar1* alleles the *fin219* mutant is not affected in jasmonic acid signalling (Staswick *et al.*, 2002). These findings cast some doubt on the genetic basis for the far-red light insensitivity in the *fin219* allele. On the other hand this work may represent a very interesting link between light and auxin signalling. The interaction between light and this hormone have long been proposed and various reports support this notion, however the work from Hsieh and co-workers represents a potential molecular link (Hsieh *et al.*, 2000).

Based on the characterisation of the *fhy1* mutant one can conclude that FHY1 acts early on in phyA signalling (Desnos *et al.*, 2001). The *fhy1* mutant is allelic to *pat3* (Zeidler *et al.*, 2001). *fhy1* mutants are strongly defective for all phyA mediated responses, moreover large scale expression profiling is consistent with an early role for FHY1 in phyA signalling (Wang *et al.*, 2002). FHY1 is a limiting factor for far-red light perception since over-expression of FHY1 leads to a phyA-dependent enhancement of light sensitivity. *FHY1* expression is rapidly down-regulated by light (white, far-red, and red). Moreover the *FHY3* gene product controls *FHY1* expression (Desnos *et al.*, 2001). FHY3 is a nuclear protein that also acts early in phyA signalling (Wang and Deng, 2002). *FHY1* codes for a small protein without any overall similarity to proteins of known function. However both NLS and NES sequences can be recognised, and a close FHY1 homologue is present in the *Arabidopsis* genome (Desnos *et al.*, 2001). Subcellular localisation of FHY1 was assessed with a FHY1 promoter driven FHY1-GFP fusion protein. This construct could complement the *fhy1* mutant indicating that the fusion protein is functional. This GFP fusion was present in both the nucleus and to a lesser extent in the cytoplasm of dark-grown hypocotyl cells. Light treatments diminished the overall amount of the FHY1-GFP fusion protein, and interestingly this decrease was slower in far-red than in other light conditions, suggesting light dependent degradation. FHY1 is therefore an early component of phyA signalling that could act both in the nucleus and the cytoplasm but we currently have no idea about its biochemical function (Desnos *et al.*, 2001).

The *LAF6* locus is required for a subset of phyA responses and is particularly important under high fluence rates (Moller *et al.*, 2001). *LAF6* codes for a member of the ATP-binding cassette transporters (ABC1). It is highly homologous to a *Synechocystis* ABC protein and to a chloroplast encoded ABC protein from a unicellular alga. LAF6-GFP fusion proteins localise to the periphery of plastids in onion epidermal cells and *Arabidopsis* seedlings. Interestingly protoporphyrinogen IX, one of the chlorophyll biosynthesis precursors, over accumulates in the *laf6* mutant. Together with chlorophyll biosynthesis inhibitor experiments this study suggests that LAF6 is required for the transport and/or proper distribution of protoporphyrinogen IX. This chlorophyll precursor may be required for proper communication between plastids and nucleus (Moller *et al.*, 2001). Such intercompartmental signalling is known to be important for light signalling (Strand *et al.*, 2003) and is extensively reviewed in chapter 21. Mutation in a second chloroplast localised protein also potentially affects phytochrome signalling. The *cr88* mutant is selectively impaired in red light inhibition of hypocotyl elongation, suggesting a role in phyB signalling (Cao *et al.*, 2003). Moreover the mutant shows reduced accumulation of light induced transcripts and delayed plastid development. The phenotype results from a missense mutation in a conserved motif of a chloroplast-targeted HSP90 protein. The available data is not sufficient to place this protein directly into the phyB pathway (Cao *et al.*, 2003). It is quite surprising that the two mutants affected in chloroplast localised proteins differentially affect phytochrome signalling although it is now clear that there are multiple signals involved in plastid to nucleus communication (Strand *et al.*, 2003). The different phenotypes might therefore result from differential effects on those pathways.

Reverse genetic studies have identified NDPK2 and PKS1 as two cytoplasmic players in phytochrome signalling (Choi *et al.*, 1999; Fankhauser *et al.*, 1999). Both proteins were identified on the basis of direct interaction with the carboxy-terminal domain of phyA (see above). phyA in a Pfr specific manner modulates NDPK2 activity in vitro. GFP fusion studies indicate that NDPK2 can be found both in the nucleus and the cytoplasm (Choi *et al.*, 1999). A loss of function allele shows slightly reduced sensitivity to both red and far-red light indicating that NDPK2 is important for normal phyA and phyB signalling (Choi *et al.*, 1999). In addition to in vitro interaction with phyA and phyB, PKS1 is also a substrate of the oat phyA protein kinase activity. PKS1 phosphorylation is also light regulated in vivo. Based on over-expression studies it was postulated that PKS1 mainly plays a role in phyB signalling (Fankhauser *et al.*, 1999). More recently loss of function alleles of both PKS1 and PKS2, the closest PKS1 homologue in *Arabidopsis*, have been described. Interestingly a proper balance between PKS1 and PKS2 is required for normal phyA signalling in the VLFR mode (Lariguet *et al.*, 2003). PKS1 and PKS2 appear to be dispensable for phyA signalling in the HIR mode and for phyB signalling. The presence of two additional PKS1 homologues in *Arabidopsis* could be an indication of additional gene redundancy. PKS1-GFP fusion proteins localise to the cytoplasm and are enriched at the plasma membrane. Both PKS1 and PKS2 show complex light regulation patterns. *PKS1* mRNA is very rapidly light induced, and a single 3 minute far-red light pulse is sufficient for a massive induction indicating that a VLFR is sufficient to induce *PKS1* expression. Interestingly *PKS1* expression is light induced in the hypocotyl and root elongation zones. These are the cells where elongation is modulated by light (Lariguet *et al.*, 2003). In addition to transcriptional regulation PKS1 protein levels are controlled in a light specific manner. PKS1 is more stable in far-red than other light conditions, indicating that phyA is also important for this post-transcriptionnal mechanism (Lariguet *et al.*, 2003). PKS1 homologues are also present in rice, but we currently don't have any idea about the biochemical function of this protein. The bHLH class transcription factor PIF3 was also identified as a phytochrome interacting factor in a yeast two-hybrid screen (Ni *et al.*, 1998). The role of PIF3 and related bHLH class proteins is extensively described in chapter 17.

A number of screens have also identified mutants specifically affected in phyB signalling (Quail, 2002). Hypocotyl growth inhibition defects specific for red light is generally used as a selection criteria. Again most cloned genes code for nuclear proteins but there are exceptions. The *srr1* mutant de-etiolates normally in far-red and blue light but is hyposensitive to red light. In addition it has a reduced end of day far-red response, longer petioles and flowers early particularly in short days, similar to *phyB* mutants (Staiger *et al.*, 2003). The phenotype of a *phyBsrr1* double mutant indicates that SRR1 has a function in addition to phyB signalling. SRR1 plays an important role for normal circadian clock function. *srr1* mutants are affected in multiple clock outputs in constant light and constant darkness suggesting that SRR1 is required for normal central oscillator function rather than light input (Staiger *et al.*, 2003). However the SRR1 gene is not under circadian control but it is light induced. A GFP fusion protein localises both to the nucleus and the cytoplasm indicating a possible function in both cell compartments. The primary amino-acid sequence gives no clues to a possible biochemical function of SRR1 but interestingly there are homologues of SRR1 is most eukaryotes. The function of these genes is currently unknown (Staiger *et al.*, 2003).

The response regulator gene ARR4 is a protein that localises to both the cytoplasm and the nucleus and plays a role in phyB signalling (Sweere *et al.*, 2001). This protein specifically interacts with the amino-terminus of phyB but not of phyA. ARR4 modulates the dark-reversion rate of phyB (Pfr to Pr transformation in the absence of light). This regulatory mechanism is of great interest since phyB dark reverts much faster than phyA. Modulation of dark reversion will affect the pool of active phyB which will have several physiological implications. Consistent with its function in phyB signalling ARR4 is induced by red light but not by far-red light. The induction is red/far-red reversible and abolished in a phyB mutant background, consistent with phyB regulation. In addition ARR4 gene expression is induced by cytokinins suggesting a possible hormonal regulation of phytochrome activity (Sweere *et al.*, 2001).

In contrast to the genes discussed so far SUB1 is important for both phytochrome and cryptochrome signalling (Guo *et al.*, 2001). Loss of function *sub1* mutants show enhanced sensitivity to both far-red and blue light but respond normally to red light. *sub1* mutants also show stronger light induction of two genes in response to far-red and blue light. Double mutant analysis with phyA shows that the enhanced far-red sensitivity of the sub1 mutant strictly depends on phyA. This is a very clear genetic indication that SUB1 directly modulates phyA signalling (Guo *et al.*, 2001). SUB1-

GUS fusions can be found throughout the cytoplasm but are enriched at the nuclear periphery. SUB1 codes for a protein with EF hand Ca^{2+} binding motifs and recombinant SUB1 does bind Ca^{2+}. This result is very significant because it represents direct genetic evidence for the involvement of Ca^{2+} in phytochrome signalling (Guo et al., 2001). This study also indicates that Ca^{2+} is important for normal cryptochrome signalling. The convergence between genetic and biochemical/pharmacological studies is a long awaited result. It should however be emphasised that this is the only example so far.

The progress made using genetic approaches has been quite spectacular if as a criteria one uses the number of cloned genes which when mutated alter photomorphogenesis. It must however be said that we currently stand in front of a rather large number of pieces in a puzzle and that we have very little understanding of the order of action of these components. This might in part be due to the fact that the genetic screens that have been performed mainly score phenotypes that take several days before they can be analysed. The vast number of implicated loci suggests that a rather large number of events can be modified leading to a hypocotyl growth phenotype. We now need to apply methods that can inform us on the order of action of these gene products such as expression profiling.

8. INTERACTIONS WITH INTERNAL CUES (GROWTH REGULATORS, CIRCADIAN CLOCK)

Light affects plant growth and development in a tissue specific and timing dependent manner. Light regulated growth control is a classic example. Early in development far-red light will strongly inhibit hypocotyl growth, however once a seedling is properly de-etiolated far-red light will have the opposite effect. In this particular case this change in growth response results from an early phyA dominated phase to a later phyB dominated phase. However in many other cases the effects will not simply depend on the dominating phytochrome species but on the cellular context. We still know little about the interactions between these endogenous cues and phytochrome signalling. In the previous section we have highlighted a few examples where light and hormonal signalling interact with each other (ARR4 and cytokinin, FIN219 and auxin)(Hsieh et al., 2000; Sweere et al., 2001). A number of additional examples could be described but in most cases the evidence for the interaction is still very indirect. Some of the most direct evidence has been uncovered for the brassinosteroid (BR)/light interaction. In *Arabidopsis* BR is important for normal etiolated growth (Li et al., 1996). The requirement for BR is somewhat species dependent since in pea this hormone is less important for etiolated development (Symons et al., 2002). The level of BR is controlled both by biosynthesis and by catabolism. The *Arabidopsis BAS1* gene is particularly interesting in this context. *BAS1* codes for a cytochrome P450 enzyme that presumably hydroxylates BR biosynthetic intermediates thereby reducing the levels of active BR (Neff et al., 1999). Plants over-expressing BAS1 show enhanced light responses and plants with reduced BAS1 levels have reduced light responses. Double mutant analysis is consistent with a role for BAS1 downstream of phyA and cry1 (Neff *et al.*, 1999). In a second recent publication it has been proposed that BR levels have an influence on the mode of phyA signalling. Plants with reduced BR levels show enhanced phyA-mediated VLFR responses but reduced HIR responses (Luccioni *et al.*, 2002). The different modes of phyA signalling have to be properly co-ordinated for proper development suggesting that local BR levels might participate in fine tuning phyA responses. A molecular link between light and BR has also been uncovered in pea. The expression of the small GTPase Pra2 correlates very well with the elongation zone of etiolated pea epicotyls (Nagano *et al.*, 1995). Light down-regulates the expression of Pra2 in a phytochrome and cryptochrome regulated manner (Inaba *et al.*, 1999). Pra2 interacts in vitro with the DDWF1 protein in a GTP dependent manner (Kang *et al.*, 2001). DDWF1 presumably acts in the BR biosynthetic pathway and has the same tissue specific and light-regulated expression pattern as Pra2. Moreover both proteins co-localise in the endoplasmic reticulum. Experiments performed with plants that either under-express or over-express Pra2 suggest that this GTPase modulates BR content and thereby light regulated hypocotyl growth (Kang *et al.*, 2001). These data are of interest concerning light-hormone interaction and concerning the role of small G proteins in photomorphogenesis. For a more extensive discussion about this topic we recommend recent reviews (Halliday and Fankhauser, 2003; Moller *et al.*, 2002).

The circadian clock is a second very important endogenous cue controlling light signalling in general and phytochrome mediated events in particular. In fact light signalling controls the phase of the clock since light is the most effective external cue to reset the clock. In turn the circadian clock controls a number of light signalling events. Clock control of the phytochrome response has been uncovered very early on. One example that comes to mind is the control of flowering time in the short day plant *Chenopodium* (Mohr, 1982). Red light night breaks will either promote or inhibit flowering in a circadian pattern. In other words it is not only the quantity and quality of given light that matters, it is also a question of timing. Similar experiments have recently shown a circadian pattern for red light-induced *CAB* expression (Millar and Kay, 1996). A large number of studies have lead to a so-called gating hypothesis. This hypothesis proposes that the circadian clock determines if light signals can be delivered further down the signalling cascade (see Chapter 26). The circadian clock basically determines the window of opportunity for light signals to induce responses. Such gating is presumably very important, for example in response to far-red light a plant has to distinguish between shade induced far-red light and far-red rich light occurring every dawn and dusk. The very close connection between regulation of the circadian clock and phytochrome signalling has also become apparent since a number of phyB signalling mutants are also required for circadian clock control (Huq *et al.*, 2000; Liu *et al.*, 2001; Staiger *et al.*, 2003). The conceptual distinction between light input, central oscillator and clock controlled output pathways is presumably rather misleading because these three parts of the clock are presumably very closely interconnected. For a more detailed review of the circadian clock please consult Chapter 26.

9. CONCLUSIONS

During the last ten years enormous progress has been made concerning our molecular understanding of phytochrome signalling events. The majority of our knowledge has been obtained with the use of genetics. Various screening methods have been used to identify mutants affected in their light perception. The available physical maps, BAC and YAC contigs and recent sequencing of the *Arabidopsis* genome considerably speeds up map based cloning of such mutants. The large collections of insertional mutants are also speeding up reverse genetic approaches. This is an important asset because in vitro results (i.e. interactions) can now be assessed for in vivo relevance relatively quickly. It should however be said that despite a large number of cloned phytochrome signalling components we still know very little about the order of action and the precise role of most of them. Since most genetic screens have relied on hypocotyl growth, a phenotype that takes several days before it can be scored, we have very little temporal resolution. New genetic screens that could identify phenotypes that only take minutes or hours would probably be informative but more technically challenging. Our understanding of the initial light induced event(s) is actually still very poor. A better biochemical/structural characterisation of the phytochromes is certainly needed to go any further. A second aspect of phytochrome signalling that is still very unclear is the relationship between the genetic model and models based on biochemical/pharmacological approaches. Genetic screens have uncovered very little evidence to support data obtained using these other techniques. The reasons for this apparent discrepancy might be multiple but one important difference might be the timing of the studied responses. Genetic screens often uncovered signalling components that are needed for long-term phytochrome effects (e.g., hypocotyl length). In contrast many pharmacological studies have targeted rapid and transient responses such ion fluxes, membrane depolarisation etc. This is certainly only part of the answer. During the last 10 years phytochrome signalling has shifted from a cytoplasmic view to a nuclear view. The rapid light-regulated nuclear translocation of these photoreceptors had a great impact on our models for phytochrome signalling. It is now absolutely clear for phyB that nuclear translocation is an essential step for a number of phytochrome responses (Huq *et al.*, 2003; Matsushita *et al.*, 2003). This is not contradictory with phytochrome signalling also occurring in the cytoplasm. Cytoplasmic events could modulate the timing of phytochrome nuclear translocation. In addition very rapid phytochrome responses probably rely on cytoplasmic signalling only (Takagi *et al.*, 2003).

10. REFERENCES

Ahmad, M., Jarillo, J. A., Smirnova, O., Cashmore, A. R. (1998) The CRY1 blue light photoreceptor of *Arabidopsis* interacts with phytochrome A *in vitro*. *Mol Cell, 1*, 939-948.

Bolle, C., Koncz, C., Chua, N. H. (2000) PAT1, a new member of the GRAS family, is involved in phytochrome A signal transduction. *Genes Dev, 14*, 1269-1278.

Bossen, M. E., Kendrick, R. E., Vredenberg, W. J. (1990) The involvement of a G-protein in phytochrome-regulated, Ca^{2+}-dependent swelling of etiolated wheat protoplasts. *Physiol Plant, 80,* 55-62.

Bowler, C., Neuhaus, G., Yamagata, H., Chua, N.-H. (1994a) Cyclic GMP and calcium mediate phytochrome phototransduction. *Cell, 77,* 73-81.

Bowler, C., Tamagata, H., Neuhaus, G., Chua, N.-H. (1994b) Phytochrome signal transduction pathways are regulated by reciprocal control mechanisms. *Genes and Development, 8,* 2188-2202.

Cao, D., Froehlich, J. E., Zhang, H., Cheng, C. L. (2003) The chlorate-resistant and photomorphogenesis-defective mutant cr88 encodes a chloroplast-targeted HSP90. *Plant J, 33,* 107-118.

Casal, J. J., Davis, S. J., Kirchenbauer, D., Viczian, A., Yanovsky, M. J., Clough, R. C., et al. (2002) The serine-rich N-terminal domain of oat phytochrome a helps regulate light responses and subnuclear localization of the photoreceptor. *Plant Physiol, 129,* 1127-1137.

Cashmore, A. R. (1998) Higher-plant phytochrome: "I used to date histidine, but now I prefer serine". *Proc Natl Acad Sci USA, 95,* 13358-13360.

Cherry, J. R., Hondred, D., Walker, J. M., Keller, J. M., Hershey, H. P., Vierstra, R. D. (1993) Carboxy-terminal deletion analysis of oat phytochrome A reveals the presence of separate domains required for structure and biological activity. *Plant Cell, 5,* 565-575.

Choi, G., Yi, H., Lee, J., Kwon, Y. K., Soh, M. S., Shin, B., et al. (1999) Phytochrome signalling is mediated through nucleotide diphosphate kinase 2. *Nature, 401,* 610-613.

Chory, J., Peto, C., Feinbaum, R., Pratt, L., Ausubel, F. (1989) *Arabidopsis* thaliana mutant that develops as a light-grown plant in the absence of light. *Cell, 58,* 991-999.

Clark, G. B., Memon, A. R., Tong, C. G., Thompson, G. A., Jr., Roux, S. J. (1993) Phytochrome regulates GTP-binding protein activity in the envelope of pea nuclei. *Plant J, 4,* 399-402.

Colon-Carmona, A., Chen, D. L., Yeh, K. C., Abel, S. (2000) Aux/IAA proteins are phosphorylated by phytochrome in vitro. *Plant Physiol, 124,* 1728-1738.

Colucci, G., Apone, F., Alyeshmerni, N., Chalmers, D., Chrispeels, M. J. (2002) GCR1, the putative *Arabidopsis* G protein-coupled receptor gene is cell cycle-regulated, and its overexpression abolishes seed dormancy and shortens time to flowering. *Proc Natl Acad Sci USA, 99,* 4736-4741.

Datta, N., Chen, Y.-R., Roux, S. J. (1985) Phytochrome and calcium stimulation of protein phosphorylation in isolated pea nuclei. *Biochem Biophys Res Comm, 128,* 1403-1408.

Deng, X.-W., Caspar, T., Quail, P. (1991) cop1: a regulatory locus involved in light-controlled development and gene expression in *Arabidopsis*. *Genes and Development, 5,* 1172-1182.

Desnos, T., Puente, P., Whitelam, G. C., Harberd, N. P. (2001) FHY1: a phytochrome A-specific signal transducer. *Genes Dev, 15,* 2980-2990.

Devlin, P. F., Patel, S. R., Whitelam, G. C. (1998) Phytochrome E Influences Internode Elongation and Flowering Time in *Arabidopsis*. *Plant Cell, 10,* 1479-1488.

Elich, T. D. and Chory, J. (1997) Phytochrome: if it looks and smells like a histidine kinase, is it a histidine kinase? *Cell, 91,* 713-716.

Ermolayeva, E., Hohmeyer, H., Johannes, E., Sanders, D. (1996) Calcium-dependent membrane depolyrisation activated by phytochrome in the moss *Physcomitrella patens*. *Planta, 199,* 352-358.

Fankhauser, C., Yeh, K. C., Lagarias, J. C., Zhang, H., Elich, T. D., Chory, J. (1999) PKS1, a substrate phosphorylated by phytochrome that modulates light signalling in *Arabidopsis*. *Science, 284,* 1539-1541.

Fischer, K. and Schopfer, P. (1997) Separation of Photolabile-Phytochrome and Photostable-Phytochrome Actions on Growth and Microtubule Orientation in Maize Coleoptiles (A Physiological Approach). *Plant Physiol, 115,* 511-518.

Folta, K. M. and Spalding, E. P. (2001) Opposing roles of phytochrome A and phytochrome B in early cryptochrome-mediated growth inhibition. *Plant J, 28,* 333-340.

Folta, K. M. and Spalding, E. P. (2001) Unexpected roles for cryptochrome 2 and phototropin revealed by high- resolution analysis of blue light-mediated hypocotyl growth inhibition. *Plant J, 26,* 471-478.

Giraud, E., Fardoux, J., Fourrier, N., Hannibal, L., Genty, B., Bouyer, P., et al. (2002) Bacteriophytochrome controls photosystem synthesis in anoxygenic bacteria. *Nature, 417,* 202-205.

Guo, H., Mockler, T., Duong, H., Lin, C. (2001) SUB1, an *Arabidopsis* Ca2+-binding protein involved in cryptochrome and phytochrome coaction. *Science, 291,* 487-490.

Guo, F. Q., Okamoto, M., Crawford, N. M. (2003) Identification of a plant nitric oxide synthase gene involved in hormonal signaling. *Science, 302,* 100-103.

Halliday, K. J. and Fankhauser, C. (2003) Phytochrome-hormonal signalling networks. *New Phytologist*, *157*, 449-463.

Hamada, T., Tanaka, N., Noguchi, T., Kimura, N., Hasunuma, K. (1996) Phytochrome regulates phosphorylation of a protein with characteristics of a nucleoside diphosphate kinase in the crude membrane fraction from stem sections of etiolated pea seedlings. *J Photochem Photobiol, B 33*, 143-151.

Harter, K., Frohnmeyer, H., Kircher, S., Kunkel, T., Mühlbauer, S., Schäfer, E. (1994) Light induces rapid changes of the phosphorylation pattern in the cytosol of evacuolated parsley protoplasts. *Proc Natl Acad Sci USA*, *91*, 5038-5042.

Haupt, W. and Häder, D. P. (1994) "Photomovement." In *Photomorphogenesis in plants, 2nd edition*, Kendrick R. E. and Kronenberg G. H. M. (eds.) Kluwer Academic Press, Dordrecht, 707-732.

Hsieh, H. L., Okamoto, H., Wang, M., Ang, L. H., Matsui, M., Goodman, H., *et al.* (2000) FIN219, an auxin-regulated gene, defines a link between phytochrome A and the downstream regulator COP1 in light control of *Arabidopsis* development. *Genes Dev*, *14*, 1958-1970.

Huq, E., Al-Sady, B., Quail, P. H. (2003) Nuclear translocation of the photoreceptor phytochrome B is necessary for its biological function in seedling photomorphogenesis. *Plant J*, *35*, 660-664.

Huq, E., Tepperman, J. M., Quail, P. H. (2000) GIGANTEA is a nuclear protein involved in phytochrome signaling in *Arabidopsis*. *Proc Natl Acad Sci USA*, *97*, 9789-9794.

Inaba, T., Nagano, Y., Sakibara, T., Sasaki, Y. (1999) Identification of a cis-regulatory element involved in phytochrome down- regulated expression of the pea small GTPase gene pra2. *Plant Physiol*, *120*, 491-500.

Jones, A. M. (2002) G-protein-coupled signalling in *Arabidopsis*. *Curr Opin Plant Biol*, *5*, 402-407.

Jones, A. M., Ecker, J. R., Chen, J. G. (2003) A re-evaluation of the role of the heterotrimeric G protein in coupling light responses in *Arabidopsis*. *Plant Physiol*, *131*, 1623-1627.

Jordan, E. T., Marita, J. M., Clough, R. C., Vierstra, R. D. (1997) Characterization of regions within the N-terminal 6-kilodalton domain of phytochrome A that modulate its biological activity. *Plant Physiol*, *115*, 693-704.

Kang, J. G., Yun, J., Kim, D. H., Chung, K. S., Fujioka, S., Kim, J. I., *et al.* (2001) Light and brassinosteroid signals are integrated via a dark-induced small G protein in etiolated seedling growth. *Cell*, *105*, 625-636.

Kawai, H., Kanegae, T., Christensen, S., Kiyosue, T., Sato, Y., Imaizumi, T., *et al.* (2003). Responses of ferns to red light are mediated by an unconventional photoreceptor. *Nature*, *421*, 287-290.

Kim, D. H., Kang, J. G., Yang, S. S., Chung, K. S., Song, P. S., Park, C. M. (2002) A phytochrome-associated protein phosphatase 2A modulates light signals in flowering time control in *Arabidopsis*. *Plant Cell*, *14*, 3043-3056.

Kircher, S., Wellmer, F., Nick, P., Rügner, A., Schäfer, E., Harter, K. (1999) Nuclear import of the parsley bZIP transcription factor CPRF2 is regulated by phytochrome photoreceptors. *J Cell Biol*, *144*, 201-211.

Klimczak, L. J., Schindler, U., Cashmore, A. R. (1992) DNA binding activity of the *Arabidopsis* G-box binding factor GBF1 is stimulated by phosphorylation by casein kinase II from broccoli. *Plant Cell*, *4*, 87-98.

Koornneef, M., Rolff, E., Spruit, C. J. P. (1980) Genetic control of light inhibited hypocotyl elongation in *Arabidopsis* thaliana (L.) *Heynh Z Pflanzenphysiol*, *100*, 147-160.

Krall, L. and Reed, J. W. 2000. The histidine kinase-related domain participates in phytochrome B function but is dispensable. *Proc Natl Acad Sci USA*, *97*, 8169-8174.

Kunkel, T., Neuhaus, G., Batschauer, A., Chua, N.-H., Schäfer, E. (1996) Functional analysis of yeast-derived light responses in *Arabidopsis* A and B phycocyanobilin adducts. *Plant J*, *10*, 625-636.

Lapko, V. N., Jiang, X.-Y., Smith, D. L., Song, P.-S. (1999) Mass spectroscopic characterization of oat phytochrome A: isoforms and post-translational modifications. *Protein Sci*, *8*, 1032-1044.

Lapko, V. N., Wells, T. A., Song, P. S. (1996) Protein kinase A-catalyzed phosphorylation and its effect on conformation in phytochrome A. *Biochemistry*, *35*, 6585-6594.

Lariguet, P., Boccalandro, H. E., Alonso, J. M., Ecker, J. R., Chory, J., Casal, J. J., *et al.* (2003) The balance between PKS1 and PKS2 provides homeostasis for phytochrome A signalling in *Arabidopsis*. *Plant Cell*, *15*, 2966-2978.

Leberer, E., Thomas, D. Y., Whiteway, M. (1997) Pheromone signalling and polarized morphogenesis in yeast. *Curr Opin Genet Dev*, *7*, 59-66.

Lee, Y., Lloyd, A. M., Roux, S. J. (1999) Antisense expression of the CK2 alpha-subunit gene in *Arabidopsis*. Effects on light-regulated gene expression and plant growth. *Plant Physiol, 119*, 989-1000.

Li, H.-M., Altschmied, L., Chory, J. (1994) *Arabidopsis* mutants define downstream branches in the phototransduction pathway. *Genes Dev, 8*, 339-349.

Li, J., Nagpal, P., Vitart, V., McMorris, T. C., Chory, J. (1996) A role for brassinosteroids in light-dependent development of *Arabidopsis*. *Science, 272*, 398-401.

Liu, X. L., Covington, M. F., Fankhauser, C., Chory, J., Wagner, D. R. (2001) ELF3 encodes a circadian clock-regulated nuclear protein that functions in an *Arabidopsis* PHYB signal transduction pathway. *Plant Cell, 13*, 1293-1304.

Long, C. and Iino, M. (2001) Light-dependent osmoregulation in pea stem protoplasts. photoreceptors, tissue specificity, ion relationships, and physiological implications. *Plant Physiol, 125*, 1854-1869.

Luccioni, L. G., Oliverio, K. A., Yanovsky, M. J., Boccalandro, H. E., Casal, J. J. (2002). Brassinosteroid mutants uncover fine tuning of phytochrome signalling. *Plant Physiol, 128*, 173-181.

Malec, P., Yahalom, A., Chamovitz, D. A. (2002) Identification of a light-regulated protein kinase activity from seedlings of *Arabidopsis* thaliana. *Photochem Photobiol, 75*, 178-183.

Maloof, J. N., Borevitz, J. O., Dabi, T., Lutes, J., Nehring, R. B., Redfern, J. L., *et al.* (2001) Natural variation in light sensitivity of *Arabidopsis*. *Nat Genet, 29*, 441-446.

Matsushita, T., Mochizuki, N., Nagatani, A. 2003. Dimers of the N-terminal domain of phytochrome B are functional in the nucleus. *Nature, 424,* 571-574.

McMichael, R. W. and Lagarias, J. C. 1990. Phosphopeptide mapping of avena phytochrome phosphorylated by protein kinases *in vitro*. *Biochem, 29*, 3872-3878.

Meng, E. C. and Bourne, H. R. (2001) Receptor activation: what does the rhodopsin structure tell us? *Trends Pharmacol Sci, 22*, 587-593.

Menkens, A. E., Schindler, U., Cashmore, A. R. (1995) The G-box: a ubiquitous regulatory DNA element in plants bound by the GBF family of bZIP proteins. *TIBS*, 506-510.

Millar, A. J., Kay, S. A. (1996) Integration of circadian and phototransduction pathways in the network controlling CAB gene transcription in *Arabidopsis*. *Proc Natl Acad Sci USA, 93*, 15491-15496.

Miséra, S., Müller, A. J., Weiland-Heidecker, U., Jürgens, G. (1994) The *FUSCA* genes of *Arabidopsis*: negative regulators of light responses. *Mol Gen Genet, 244*, 242-252.

Mohr, H., *Lectures on photomorphogenesis.* Springer Verlag, 1982.

Moller, S. G., Ingles, P. J., Whitelam, G. C. (2002) The cell biology of phytochrome signalling. *New Phytologist, 154*, 553-590.

Moller, S. G., Kunkel, T., Chua, N. H. (2001). A plastidic ABC protein involved in intercompartmental communication of light signalling. *Genes Dev, 15*, 90-103.

Montgomery, B. L. and Lagarias, J. C. (2002) Phytochrome ancestry: sensors of bilins and light. *Trends Plant Sci, 7*, 357-366.

Nagano, Y., Okada, Y., Narita, H., Asaka, Y., Sasaki, Y. (1995) Location of light-repressible, small GTP-binding protein of the YPT/rab family in the growing zone of etiolated pea stems. *Proc Natl Acad Sci USA, 92*, 6314-6318.

Neff, M. M., Nguyen, S. M., Malancharuvil, E. J., Fujioka, S., Noguchi, T., Seto, H., Tsubuki, M., Honda, T., Takatsuto, S., Yoshida, S., and Chory, J. 1999. BAS1: A gene regulating brassinosteroid levels and light responsiveness in *Arabidopsis*. *Proc Natl Acad Sci USA, 96*, 15316-15323.

Neuhaus, G., Bowler, C., Hiratsuka, K., Yamagata, H., Chua, N. H. (1997) Phytochrome-regulated repression of gene expression requires calcium and cGMP. *Embo J, 16,* 2554-2564.

Neuhaus, G., Bowler, C., Kern, R., Chua, N. H. (1993) Calcium/calmoduling-dependent and -independent phytochrome signal transduction pathways. *Cell, 73*, 937-952.

Ni, M., Tepperman, J. M., Quail, P. H. (1999) Binding of phytochrome B to its nuclear signalling partner PIF3 is reversibly induced by light. *Nature, 400*, 781-784.

Ni, M., Tepperman, J. M., Quail, P. H. (1998) PIF3, a phytochrome-interacting factor necessary for normal photoinduced signal transduction, is a novel basic helix-loop-helix protein. *Cell, 95*, 657-667.

Ogura, T., Tanaka, N., Yabe, N., Komatsu, S., Hasunuma, K. (1999) Characterization of protein complexes containing nucleoside diphosphate kinase with characteristics of light signal transduction through phytochrome in etiolated pea seedlings. *Photochem Photobiol, 63*, 397-403.

Okamoto, H., Matsui, M., Deng, X. W. (2001) Overexpression of the heterotrimeric G-protein alpha-subunit enhances phytochrome-mediated inhibition of hypocotyl elongation in *Arabidopsis*. *Plant Cell, 13*, 1639-1652.

Otto, V. and Schäfer, E. (1988) Rapid phytochrome-controlled protein phosphorylation and dephosphorylation in *Avena sativa*. *Plant Cell Physiol*, 29, 1115-1121.
Parks, B. M., Cho, M. H., Spalding, E. P. (1998) Two genetically separable phases of growth inhibition induced by blue light in *Arabidopsis* seedlings. *Plant Physiol*, 118, 609-615.
Parks, B. M., Folta, K. M., Spalding, E. P. (2001) Photocontrol of stem growth. *Curr Opin Plant Biol*, 4, 436-440.
Parks, B. M. and Spalding, E. P. (1999) Sequential and coordinated action of phytochromes A and B during *Arabidopsis* stem growth revealed by kinetic analysis. *Proc Natl Acad Sci USA*, 96, 14142-14146.
Quail, P. H. (2002) Phytochrome photosensory signalling networks. *Nat Rev Mol Cell Biol*, 3, 85-93.
Reed, J. W. (1999) Phytochromes are Pr-ipatetic kinases. *Curr Opin Plant Biol*, 2, 393-397.
Romero, L. C., Biswal, B., Song, P.-S. (1991) Protein phosphorylation in isolated nuclei from etiolated Avena seedlings. Effects of red/far-red light and cholera toxin. *FEBS Lett*, 282, 347-350.
Romero, L. C. and Lam, E. (1993) Guanine nucleotide binding protein involvement in early steps of phytochrome-regulated gene expression. *Proc Natl Acad Sci USA*, 90, 1465-1469.
Roux, S. J. (1994) "Signal transduction in phytochrome responses." In *Photomorphogenesis in plants 2nd edition*, Kendrick R. E. and Kronenberg G. H. M. (eds) Kluwer Academic Press, Dordrecht, 187-209.
Sato, Y., Wada, M., Kadota, A. (2001) Choice of tracks, microtubules and/or actin filaments for chloroplast photo-movement is differentially controlled by phytochrome and a blue light receptor. *J Cell Sci*, 114, 269-279.
Schäfer, E. and Bowler, C. (2002) Phytochrome-mediated photoperception and signal transduction in higher plants. *EMBO Reports*, 3, 1042-1048.
Schutz, I. and Furuya, M. (2001) Evidence for type II phytochrome-induced rapid signalling leading to cab::luciferase gene expression in tobacco cotyledons. *Planta*, 212, 759-764.
Shalitin, D., Yang, H., Mockler, T. C., Maymon, M., Guo, H., Whitelam, G. C., *et al.* (2002) Regulation of *Arabidopsis* cryptochrome 2 by blue-light-dependent phosphorylation. *Nature*, 417, 763-767.
Song, P. S. (1999) Inter-Domain Signal Transmission within the Phytochromes. *J of Biochem and Mol Biol*, 32, 215-225.
Staiger, D., Allenbach, L., Salathia, N., Fiechter, V., Davis, S. J., Millar, *et al.* (2003) The *Arabidopsis SRR1* gene mediates phyB signalling and is required for normal circadian clock function. *Genes Dev*, 17, 256-268.
Staswick, P. E., Tiryaki, I., Rowe, M. L. (2002) Jasmonate response locus JAR1 and several related *Arabidopsis* genes encode enzymes of the firefly luciferase superfamily that show activity on jasmonic, salicylic, and indole-3-acetic acids in an assay for adenylation. *Plant Cell*, 14, 1405-1415.
Stockhaus, J., Nagatani, A., Halfter, U., Kay, S., Furuya, M., Chua, N. H. (1992) Serine-to-alanine substitutions at the amino-terminal region of phytochrome A result in an increase in biological activity. *Genes Dev*, 6, 2364-2372.
Stowe-Evans, E. L., Luesse, D. R., Liscum, E. (2001) The enhancement of phototropin-induced phototropic curvature in *Arabidopsis* occurs via a photoreversible phytochrome A-dependent modulation of auxin responsiveness. *Plant Physiol*, 126, 826-834.
Strand, A., Asami, T., Alonso, J., Ecker, J. R., Chory, J. (2003) Chloroplast to nucleus communication triggered by accumulation of Mg-protoporphyrinIX. *Nature*, 421, 79-83.
Stuurman, J., Jaggi, F., Kuhlemeier, C. (2002) Shoot meristem maintenance is controlled by a GRAS-gene mediated signal from differentiating cells. *Genes Dev*, 16, 2213-2218.
Sugano, S., Andronis, C., Ong, M. S., Green, R. M., Tobin, E. M. (1999) The protein kinase CK2 is involved in regulation of circadian rhythms in *Arabidopsis*. *Proc Natl Acad Sci USA*, 96, 12362-12366.
Sweere, U., Eichenberg, K., Lohrmann, J., Mira-Rodado, V., Bäurle, I., Kudla, J., *et al.* (2001) Interaction of the Response Regulator ARR4 with Phytochrome B in Modulating Red Light Signalling. *Science*, 294, 1108-1111.
Symons, G. M., Schultz, L., Kerckhoffs, L. H., Davies, N. W., Gregory, D., Reid, J. B. (2002) Uncoupling brassinosteroid levels and de-etiolation in pea. *Physiol Plant*, 115, 311-319.
Takagi, S., Kong, S. G., Mineyuki, Y., Furuya, M. (2003) Regulation of actin-dependent cytoplasmic motility by type II phytochrome occurs within seconds in *Vallisneria gigantea* epidermal cells. *Plant Cell*, 15, 347-364.
Taylor, B. L. and Zhulin, I. B. (1999) PAS domains: internal sensors of oxygen, redox potential, and light. *Microbiol Mol Biol Rev*, 63, 479-506.

Terry, M. J. and Kendrick, R. E. (1996) The aurea and yellow-green-2 mutants of tomato are deficient in phytochrome chromophore synthesis. *J Biol Chem*, *271*, 21681-21686.

Wada, M., Kagawa, T., Sato, Y. (2003) Chloroplast movement. *Annu Rev Plant Biol*, *54*, 455-468.

Wang, H. and Deng, X. W. (2002) *Arabidopsis* FHY3 defines a key phytochrome A signalling component directly interacting with its homologous partner FAR1. *Embo J*, *21*, 1339-1349.

Wang, H. and Deng, X. W. (2003) Dissecting the phytochrome A-dependent signalling network in higher plants. *Trends Plant Sci*, *8*, 172-178.

Wang, H., Ma, L., Habashi, J., Zhao, H., Deng, X. W. (2002) Analysis of far-red light-regulated genome expression profiles of phytochrome A pathway mutants in *Arabidopsis*. *Plant J*, *32*, 723-733.

Wang, X. and Iino, M. (1998) Interaction of cryptochrome 1, phytochrome, and ion fluxes in blue-light-induced shrinking of *Arabidopsis* hypocotyl protoplasts. *Plant Physiol*, *117*, 1265-1279.

Wellmer, F., Kircher, S., Rügner, A., Frohnmeyer, H., Schäfer, E., Harter, K. (1999) Phosphorylation of the parsley bZIP transcription factor CPRF2 is regulated by light. *J Biol Chem*, *274*, 29476-29482.

Wong, Y.-S. and Lagarias, J. C. (1989) Affinity labelling of Avena phytochrome with ATP analogs. Proc. Natl. Acad. Sci. 86: 3469-3473.

Wong, Y. S., Cheng, H. C., Walsh, D. A., and Lagarias, J. C. 1986. Phosphorylation of *Avena* phytochrome *in vitro* as a probe of light-induced conformational changes. *J Biol Chem*, *261*, 12089-12097.

Wu, Y., Hiratsuka, K., Neuhaus, G., Chua, N. H. (1996) Calcium and cGMP target distinct phytochrome-responsive elements. *Plant J*, *10*, 1149-1154.

Yeh, K. C. and Lagarias, J. C. (1998) Eukaryotic phytochromes: light-regulated serine/threonine protein kinases with histidine kinase ancestry. *Proc Natl Acad Sci USA*, *95*, 13976-13981.

Zeidler, M., Bolle, C., Chua, N. H. (2001) The phytochrome A specific signalling component PAT3 is a positive regulator of *Arabidopsis* photomorphogenesis. *Plant Cell Physiol*, *42*, 1193-1200.

Chapter 20

THE PHOTORECEPTOR INTERACTION NETWORK

Jorge José Casal
IFEVA, Faculty of Agronomy, University of Buenos Aires and CONICET, 1417-Buenos Aires, Argentina (e-mail: casal@ifeva.edu.ar)

1. INTRODUCTION

1.1 Light signals and photoreceptors

Plants are exposed to stressful conditions including the aggressive transition between belowground and aboveground environments, the presence of competitors, and seasons with extreme temperatures or shortage of water. Specific combinations of irradiance, spectral composition, duration of the daily photoperiod and angle of incidence of light are tightly linked to these threats and very often anticipate their impending occurrence (Figure 1). When an organ emerges from the soil, it experiences a transition between full darkness and daily light cycles. The ratio between red light and far-red light (R:FR) is inversely related to the presence, size and position of neighbour plants that can compete for resources, because green leaves reflect and transmit much more far-red than red light (see Chapter 22). The photoperiod correlates with time of the year. The angle of incidence depends on the position of neighbours and soil aggregates or stones. The perception and transduction of these signals by the network formed by phytochromes, cryptochromes, phototropins and their downstream signalling elements leads to changes in plant growth and development that favour adjustment to the associated stressful conditions.

In addition to the light signals, plants are also exposed to noise from the light environment. This noise is caused by fluctuations that do not carry ecologically relevant information. Some of these fluctuations are in the range of magnitude of the light signals. This is the case, for instance, of the reductions in R:FR that take place at the beginning and the end of the photoperiod due to atmospheric reasons and not to the presence neighbours. Furthermore, some fluctuations can be signals at a given stage and noise at another. The photoperiod is indicative of season but a seedling that emerges from the soil has to respond to light no matter whether the day is short or long.

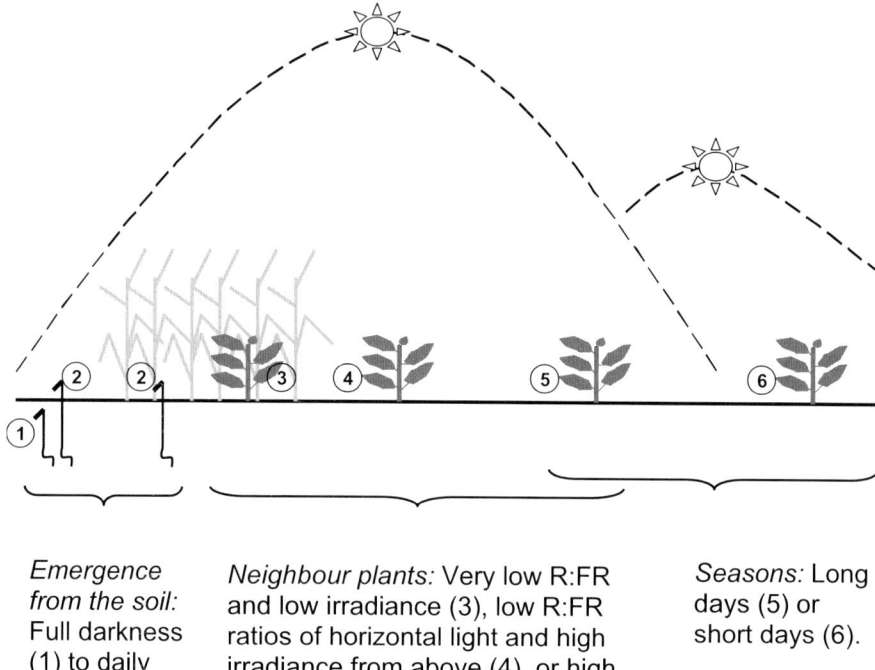

Figure 1. Plants are exposed to light signals that provide information about organ position above or below ground, presence, size and density of neighbour plants and season.

The evolutionary significance of the occurrence of different photoreceptors is partially accounted for by presence of the divergent signals described above. Phytochrome A (phyA), for instance, is the only photoreceptor able to perceive the difference between belowground darkness and the low R:FR experienced beneath a deep canopy (Yanovsky et al., 1995). However, as a result of its photochemical properties and signalling cascade, phyA is not a good receptor of changes in R:FR or light direction. Conversely, phytochrome B (phyB) is a good sensor of changes in R:FR but it is not involved in the response to very low R:FR compared to darkness (Quail et al., 1995). In this chapter, we will describe an additional consequence of the occurrence of multiple photoreceptors: the formation of a network that provides a refined response to the signals against the background noise.

1.2 Shared and specific control of light responses by different photoreceptors

Phytochromes (see Chapters 16, 17), cryptochromes (see Chapter 11) and phototropins (see Chapter 12) are key players in plant growth and development by controlling the timing and/or intensity of seed germination, seedling de-etiolation, shade-avoidance reactions, flowering, tuber/bulb formation and senescence. These effects involve alterations in organogenesis, growth rate, organ orientation and cellular function. Photoreceptors also control "behavioural" responses such as the movement of the occlusive cells of the stomata and the cellular orientation of the chloroplasts.

Each of the photoreceptors tends to have a bearing, either direct or indirect via modification of the response mediated by other photoreceptors, in the control of most of the aforementioned developmental responses. However, the relative impact of the various photoreceptors depends on the species, the developmental context and the specific process. This hierarchy of photoreceptors is one of the key determinants of the most effective light signal(s) controlling a given photomorphogenic event. For instance, small changes in photoperiod perceived by cryptochrome 2 (cry2) are very important in the control of flowering and not in the control of shade-avoidance reactions.

Compared to other responses, seedling de-etiolation provides a simpler model for the study of signalling interaction downstream the photoreceptors because it is initiated by a single signal – i.e. the transition between full darkness and light, which simultaneously activates all the photoreceptors. Flowering of a long-day plant, on the contrary, can be initiated or accelerated by long photoperiods and by low R:FR. Long, compared to short photoperiods keep photoreceptors like cry2 and phyA in their active stage but low, compared to high R:FR shift phyB, phytochrome D (phyD) and phytochrome E (phyE) to their inactive form. This interaction among signals adds a layer of complexity to the interactions among photoreceptors controlling flowering.

2. PHOTORECEPTOR INTERACTION DURING DE-ETIOLATION

2.1 Multiple photoreceptors control de-etiolation

The signal that initiates de-etiolation is the transition between full darkness and daily light cycles. The physiological responses that occur during de-etiolation include the inhibition of stem growth and the display of the foliage and the organization of the photosynthetic apparatus. In Arabidopsis and other dicots, morphological changes involve inhibition of hypocotyl growth and expansion and unfolding of the cotyledons. In this species, hypocotyl growth is inhibited mainly by phyA (Whitelam *et al.*, 1993), phyB (Reed *et al.*, 1993) and cry1 (Ahmad and Cashmore, 1993). cry2 (Lin *et al.*, 1998), phyD (Aukerman *et al.*, 1997), phyC (Franklin *et al.*, 2003a; Monte *et al.*, 2003) and phot1 (Folta and Spalding, 2001b) have more modest effects. We will first describe the main patterns of interaction among these photoreceptors and then the consequences of such interactions.

2.2 Redundancy

2.2.1 The potential action of a photoreceptor can be hidden by the action of others

phyA has a unique role in the inhibition of hypocotyl growth under the shade of dense canopies (Yanovsky et al., 1995) because it appears to be the only photoreceptor having different activity under prolonged far-red light compared to darkness (Quail et al., 1995). Under white light, the other photoreceptors become active and the *phyA* mutant shows little difference with the wild type, suggesting that phyA plays a negligible role (Whitelam et al., 1993; Figure 2). However, if the *phyA phyB* double mutant is compared with the *phyB* mutant or the *phyA phyB cry1* triple mutant is compared to the *phyB cry1* double the consequences of the presence or absence of phyA are readily evident (Figure 2). This indicates that phyA is able to control hypocotyl growth under white light but it bears a lower hierarchy than phyB or cry1 in this action.

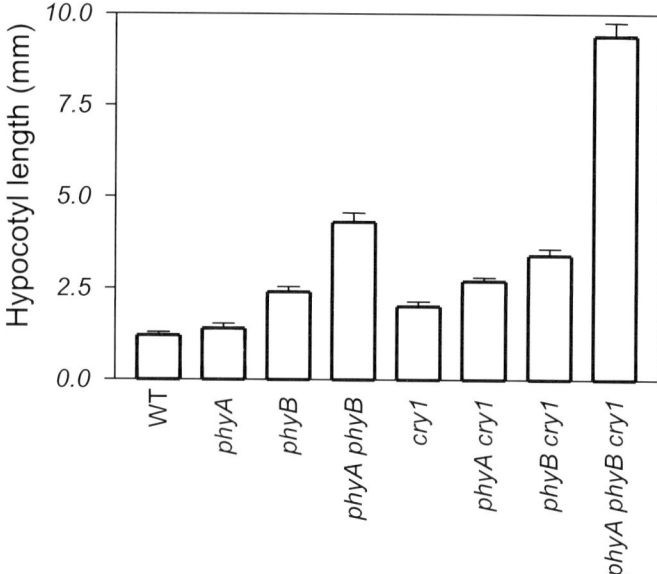

Figure 2. Redundancy. Length of the hypocotyl in Arabidopsis seedlings grown under white-light photoperiods. Note that the phyA *mutant shows little difference with the wild type (WT) but the* phyA phyB *and* phyA phyB cry1 *mutants are significantly taller than the* phyB *and* phyB cry1 *mutants, respectively. This indicates that phyA can control hypocotyl growth under these conditions but this effect is hidden by the action of other photoreceptors. Error bars are standard errors*

There are several examples where elimination of one or more photoreceptors reveals the role played by remaining members of the family. Single *phyA*, *phyD* or *phyE* mutants have no cotyledon-expansion phenotype under continuous red light but when combined with the *phyB* mutation, which in itself reduces cotyledon expansion, double, triple and quadruple mutants uncover the role of phyA, phyD and phyE in addition to phyB in the control of this response (Franklin *et al.*, 2003b). The *phyC* mutant of Arabidopsis shows reduced inhibition of hypocotyl growth under red or blue light and this effect is particularly obvious in the *phyA* mutant background (Franklin *et al.*, 2003a; Monte *et al.*, 2003). In tomato, the *PHYB* gene sub-family is composed of two members –*PHYB1* and *PHYB2*. The *phyB2* mutant has no obvious de-etiolation phenotype under red light but the *phyB1 phyB2* double mutant shows taller hypocotyls and reduced cotyledon growth, anthocyanin and chlorophyll synthesis compared to the *phyB1* single mutant (Weller *et al.*, 2000).

The *cry2* mutant of Arabidopsis shows little difference with the wild type at high fluence rates of white light. However, the double *cry1 cry2* mutant is taller than the *cry1* mutant indicating that cry2 is able to control hypocotyl growth under the latter light conditions (Mazzella *et al.*, 2001; Mazzella and Casal, 2001).

2.2.2 Definition of redundancy

The aforementioned observations share the occurrence of a potential role of a given photoreceptor, hidden by the action of others and uncovered by mutation of their genes. Genes whose products are capable of compensating for the loss of activity of other gene are said to display redundant (or partially redundant) genetic functions (Pickett and Meeks-Wagner, 1995). Plant photoreceptors show only "generic redundancy" (Nowak *et al.*, 1997) because they have overlapping functions under certain conditions (e.g. inhibition of hypocotyl growth under white light) but not others (e.g. inhibition of hypocotyl growth under far-red light).

The definition of redundancy applies only to the level of analysis. For instance, if two photoreceptors are redundant in the control of hypocotyl growth under white light we cannot conclude that they exert this control via the same molecular or cellular pathways. We can only predict that their signalling fluxes converge at or before the level of analysis.

2.2.3 The mechanisms of redundancy

Redundancy can have two origins. One is simply the relationship between signal and response (Figure 3). If the action of one photoreceptor (Y) is enough to mediate the maximum level of response (or a large proportion of this maximum response), a loss of function mutation at the other photoreceptor (Z) is predicted to have little effect. Once the major photoreceptor gene is mutated the role of the other(s) becomes evident. Another possibility (that does not exclude the former one) is based on the occurrence of negative regulation exerted by at least some element of one pathway on a target element of the other (Pickett and Meeks-Wagner, 1995) (Figure 4). In the example, the photoreceptor Z contributes to the response but it simultaneously mediates negative regulation of signalling downstream Y. The combined action of both photoreceptors is therefore less than that predicted by the additive action of each one in the absence of the other.

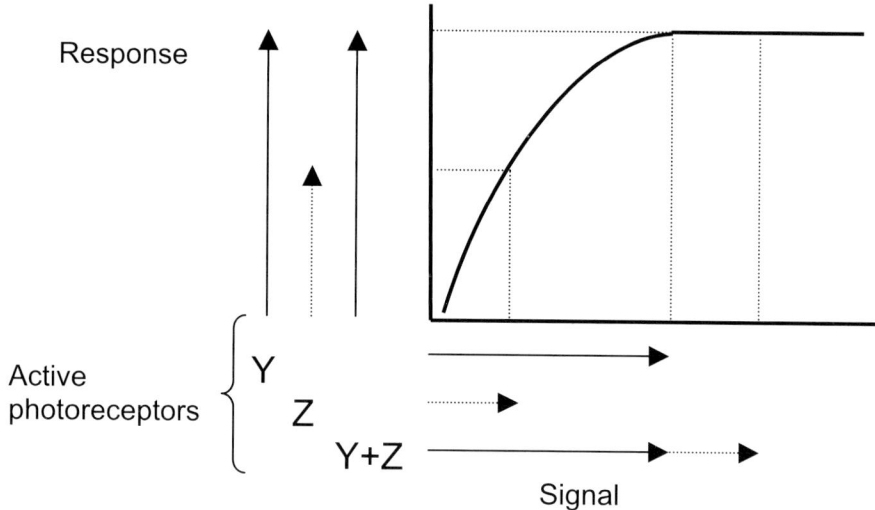

Figure 3. When two photoreceptors (Y and Z) share the control of a given process, they become redundant if their combined action yields a signal higher than that required to saturate the response.

2.2.4 Redundant photoreceptors are not equally important

The cases of redundancy often reveal a hierarchy of the different photoreceptors (Mazzella *et al.*, 2001). For instance, for hypocotyl growth inhibition in Arabidopsis, the *phyB* or *cry1* single mutations have respectively larger effects than the *phyA* or *cry2* single mutations. This situation could reflect differences in photoreceptor abundance (see Chapter 8). However, even for a given developmental stage and environmental condition the hierarchy depends on the response under consideration. cry2 has weak effects on hypocotyl growth and stronger effects on long-term *Lhcb1*2* expression (Mazzella *et al.*, 2001).

2.3 Synergism between phytochromes and cryptochromes

2.3.1 Blue light-mediated responsivity amplification towards phytochrome

The observation that light protocols combining red and blue light have larger effects than those of each waveband in isolation can be traced back to early studies of plant photomorphogenesis. Hans Mohr and co-workers rationalised this idea and proposed a model where blue light perceived by a specific blue-light photoreceptor (phytochtomes absorb blue light but their activity is obviously not specific for short wavebands) enhance the capacity of the system to respond to active phytochrome (Mohr, 1986, 1994). The basic photobiological protocol consisted of a period of several hours under blue light terminated with either a red light or a far-red light pulse. A pulse of red light is predicted to activate phytochrome but in many species this light treatment has at most weak effects on seedling morphology when compared to a pulse of far-red light or to darkness. The difference between these light/dark conditions is expressed if the seedlings are previously exposed to prolonged blue light. In other words, the maximum effect requires both blue light and terminal red light. This phenomenon is called "responsivity amplification" towards phytochrome.

In some cases, responsivity amplification to phytochrome can be achieved by light pre-treatments perceived by phytochromes themselves (Mohr, 1986, 1994) or by UV-B radiation (Boccalandro *et al.*, 2001).

One of the most provocative issues of the model proposed by Mohr *et al.*, is that all the effect of the blue-light photoreceptor involved convergence to phytochrome signalling, as phytochrome was the "effector proper". This aspect of the model has not received strong support from available results with Arabidopsis but the ultimate experimental test that involves the analysis of cryptochrome-mediated effects in a quintuple *phyA phyB phyC phyD phyE* null mutant is still pending.

2.3.2 cry1 amplifies responsivity towards phyB

Despite this open question, much has been learnt in recent years with respect to the role played by different photoreceptors in the synergism between blue and red light. By using a protocol adapted from that designed by Mohr *et al.*,, in combination with the *cry1*, *phyA* and *phyB* mutants the synergism can be assigned mainly to phyB and cry1. The unfolding of the cotyledons depends on the combination of a daily pulse of red light to activate phytochrome and exposure to blue light (Figure 5).

In seedlings exposed to the optimum combination, i.e. blue light followed by a pulse of red light, the *phyB* mutation is equivalent to the absence of the red light pulse and the *cry1* mutation is almost equivalent to the absence of blue light (Casal and Boccalandro, 1995; Figure 5). The *phyA* mutation does not affect the synergism between blue light and red light. phyB, however, is not the only phytochrome showing synergistic interaction with cry1. In the absence of phyB, phyD but not phyE is able to interact synergistically with cry1 in the control of hypocotyl growth in Arabidopsis (Hennig *et al.*, 1999a). In addition, synergism between cry1 and phyB2 has been demonstrated for anthocyanin synthesis in tomato (Weller *et al.*, 2001).

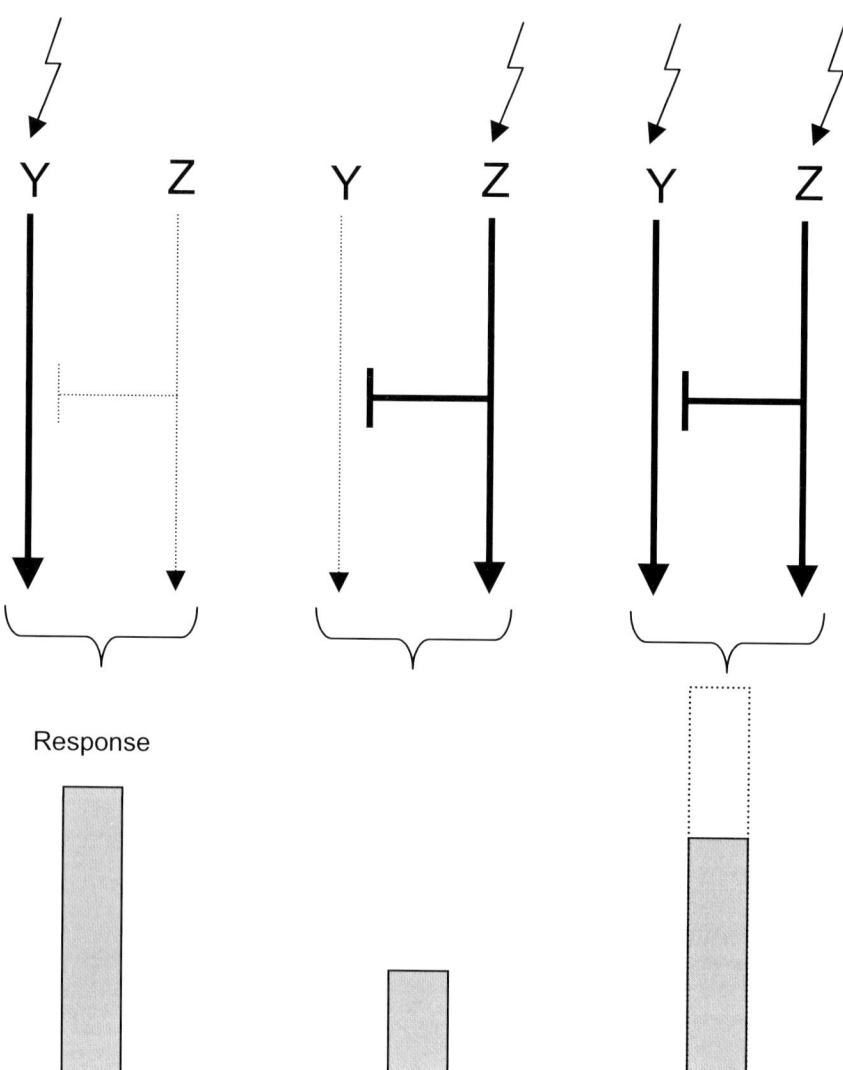

Figure 4. Negative interactions between two pathways controlling a given response create redundancy, i.e. both pathways together provide a response smaller than the sum of the responses of each pathway in isolation. Lines terminated in "T" or arrowheads indicate negative or positive regulation, respectively. Solid and dotted lines indicate active or inactive pathways, respectively.

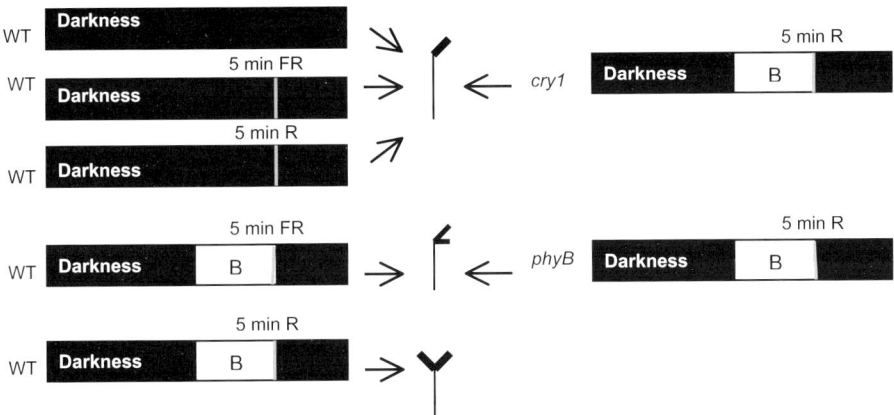

Figure 5. Synergism between cry1 and phyB. Wild type (WT) seedlings of Arabidopsis exposed daily to the indicated light protocols show maximum cotyledon unfolding only if a pulse of red light is combined with exposure to blue light (B). The effects of blue light terminated with far-red light (FR), or red light (R) without a blue light pre-treatment are small. The WT response is larger than predicted by the sum of the residual effects in the phyB and cry1 mutants.

2.3.3 The synergism between cry1 and phyB is conditional

The use of different light protocols in combination with genetic tools demonstrates the conditional nature of the synergism between cry1 and phyB in Arabidopsis (Casal and Mazzella, 1998). Etiolated seedlings were grown under:

1. a background of continuous red light to activate phyB with continuous blue light to activate cry1,

2. a background of continuous red light with a short (3 h) daily supplement of blue light to activate cry1 only during these few hours

3. or a background of far-red light to maintain very low levels of active phyB with continuous blue light.

The interaction is evaluated genetically by using plants with wild type levels of both photoreceptors (*PHYB CRY1*), null levels of phyB (*phyB CRY1*), null levels of cry1 (*PHYB cry1*) or null levels of both (*phyB cry1* double mutant), and comparing *PHYB* versus *phyB* either in the *CRY1* or the *cry1* backgrounds, as well as *CRY1* versus *cry1* either in the *PHYB* or *phyB* backgrounds.

Under the most favourable conditions for both photoreceptors (continuous red light and blue light) phyB and cry1 operate additively, i.e. the action of phyB (*PHYB* versus *phyB*) is independent of the *CRY1* or *cry1* background and vice versa (Figure 6). When blue light is added to the red light background only for 3 h phyB and cry1 operate synergistically, i.e. the action of phyB (*PHYB* versus *phyB*) is stronger in the *CRY1* than *cry1* background and vice versa (Figure 6). A similar picture is observed when blue light is added to a background of far-red light. Since short blue light is sub-optimal for cry1 and far-red light is sub-optimal for phyB, the common theme is that the synergism between phyB and cry1 is observed only when the activity of one of them is not optimal (Casal and Mazzella, 1998).

2.3.4 Other manifestations of synergism between phytochromes and cryptochromes

The enhanced action of a waveband by combination with another provided photobiological evidence for the occurrence of interactions among the photoreceptors. These interactions were later resolved in terms of specific photoreceptors with the aid of mutants. In addition, these genetic tools revealed interactions that had not been predicted by photobiological experiments.

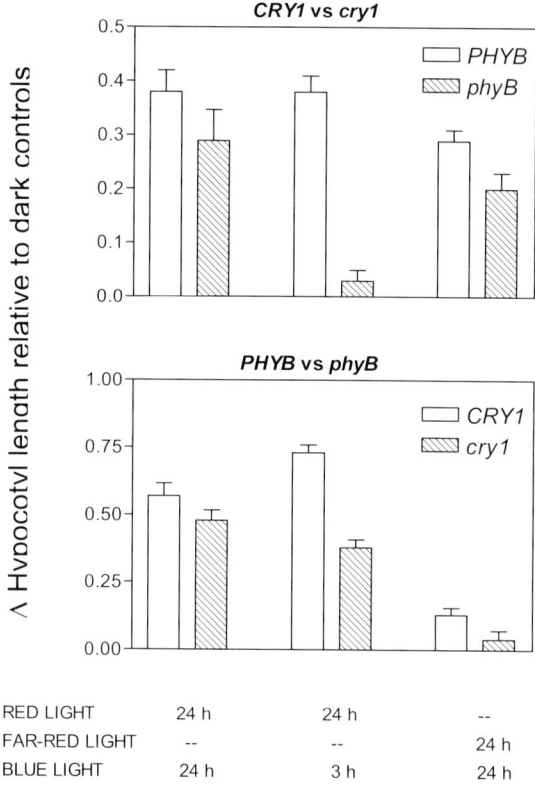

Figure 6. The synergism between phyB and cry1 is conditional. The inhibition of hypocotyl growth by cry1 (CRY1 vs cry1) and phyB (PHYB vs phyB) was calculated for the PHYB and phyB or the CRY1 and cry1 backgrounds, respectively. The synergism is observed only if light conditions are sub-optimal for cry1 action (only 3 h blue light per day) or for phyB (a background of far-red instead of red light). All the genotypes are in the phyA mutant background to avoid inhibition of hypocotyl growth by far-red light. Redrawn after Casal and Mazzella (1998).

The seedling phenotype of the *cry2* mutant can be detected at very low fluence rates of blue light (Lin *et al.*, 1998). Under higher irradiances of blue or white light *cry2* shows wild-type inhibition of hypocotyl growth, but compared to the *cry1* single mutant the *cry1 cry2* mutant is taller. This effect of *cry2* in the *cry1* background is no longer detectable in the absence of either phyA or phyB. In other words, cry2 is redundant with cry1 but it requires phyA and phyB indicating a synergism between cry2 and phytochromes (Mazzella *et al.*, 2001).

The *phyA* mutant shows deficient hypocotyl-growth inhibition under blue light (Whitelam *et al.*, 1993). Ahmad *et al.*, (2002) observed no hypocotyl growth inhibition in the *cry1 cry2* double mutant under monochromatic light sources within the range of 390 to 530 nm. Taken together, these results suggest that the response to blue light is primarily mediated by cryptochromes and that phyA has only a modulatory role activity of cryptochrome with no other photoreceptor having a significant primary role at the fluence range tested (Ahmad *et al.*, 2002). However, we do observe effects of blue light in the *cry1 cry2* double mutant that would be consistent with a more direct role of phyA as they are no longer present in the *phyA cry1 cry2* triple mutant (Mazzella and Casal, unpublished). The use of monochromatic versus broad-band blue light could perhaps account for this discrepancy. The rapid inhibition of hypocotyl growth that begins after approximately 30 min of blue light and the previous anion channel activation necessary for this response require cry1, cry2 and phyA suggesting that these photoreceptors could synergistically activate early events out of the nucleus (Folta *et al.*, 2001a).

Cotyledon unfolding induced by hourly pulses of far-red light compared to darkness is stronger in the Cape Verde Islands (Cvi) than in the Landsberg *erecta* (L*er*) accession of Arabidopsis (Botto *et al.*, 2003). The genetic loci involved in this polymorphism have been mapped by using recombinant inbred lines derived from a cross between both accessions. One of these loci is *CRY2* because cotyledon unfolding under pulses or far-red light is enhanced in transgenic lines by ectopic expression of the $CRY2^{Cvi}$ allele and not of the $CRY2^{Ler}$ allele (Botto *et al.*, 2003). The $CRY2^{Cvi}$ allele bears methionine instead of valine at position 367, which is a highly conserved residue in plant cryptochromes. This mutation has been characterised because it also renders early flowering plants irrespective of daylength (El-Assal *et al.,*, 2001). The observation that $CRY2^{Cvi}$ enhances cotyledon unfolding under hourly pulses of far-red light is surprising because it implies cry2 activity in the absence of blue light. The *phyA* mutation is epistatic to the $CRY2^{Cvi}$ allele indicating that cry2 is somehow able to modulate some phyA-mediated responses (Botto *et al.*, 2003). $CRY2^{Cvi}$ is a gain-of-function allele (El-Assal *et al.*, 2001), the single *cry1* or *cry2* loss-of-function mutants show apparently normal cotyledon unfolding under pulses of far-red light but the *cry1 cry2* double mutant is slightly impaired in this response, indicating a positive relationship between cryptochrome action and phyA action under these light conditions (Botto *et al.*, 2003).

2.4 Synergistic or antagonistic interaction between phyA and phyB

The interactions between phyA and phyB can be synergistic or antagonistic, depending on light conditions. As described above for blue light pre-treatments, exposure to several hours of far-red light is also able to amplify the response to a pulse of red light in etiolated seedlings (Figure 7).

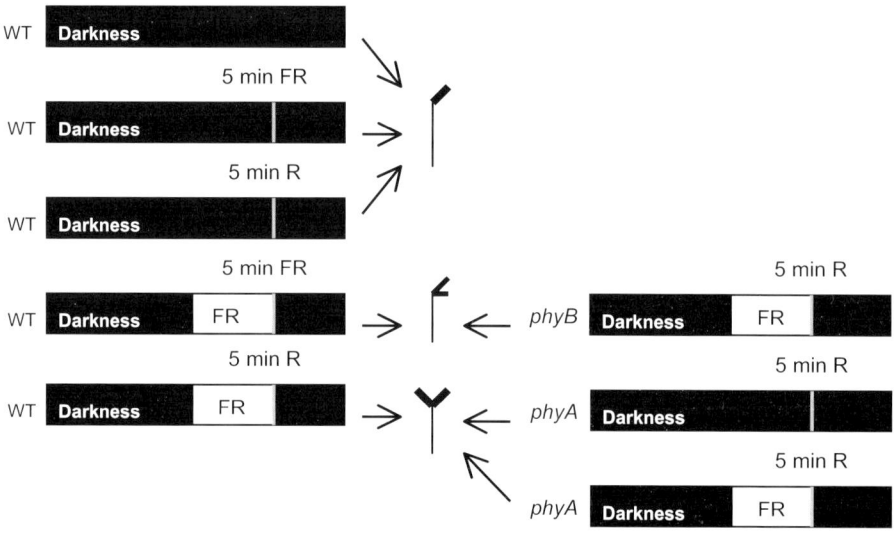

Figure 7. Synergistic or antagonistic interaction between phyA and phyB. Wild type (WT) seedlings of Arabidopsis exposed daily to the indicated light protocols show maximum cotyledon unfolding only if a pulse of red light (R) is combined with a far-red light (FR) pre-treatment (3 h). The phyB *mutant fails to respond to the R pulse. The* phyA *mutant shows a constitutive response to R, not enhanced by the FR pre-treatment.*

The *phyB* mutant is unable to respond to the red light pulse even after a far-red light pre-treatment (Figure 7). Since the difference between full darkness and far-red light is perceived by phyA, these observations can be described as a synergistic interaction between phyA and phyB (Casal and Boccalandro, 1995; Hennig *et al.,* 1999b). In accordance with this view, prolonged far-red light fails to amplify the response to red light in the *phyA* mutant. However, *phyA* does not require a far-red light pre-treatment to exhibit morphological responses to a pulse of red light (Figure 7). Since this effect noted in the *phyA* null mutant is absent in the *phyA phyB* double mutant, the observations can be interpreted as a negative regulation of the phyB-mediated response by phyA in wild type seedlings (Casal and Boccalandro, 1995; Mazzella *et al.,* 1997; Cerdán *et al.,* 1999; Hennig *et al.,* 1999b).

There are two different pathways of phyA signalling that can be dissected based on several criteria (Casal *et al.*, 2003). The frequency of excitation of phyA with far-red light defines two discrete phases of response. The very-low fluence response (VLFR) pathway, which saturates with infrequent excitation, and the high-irradiance response (HIR) pathway, which becomes evident under sustained excitation (Casal *et al.*, 2000; see Chapter 2). The VLFR also saturates with low fluence rates and can be initiated by red light or far-red light whereas the HIR requires higher fluence rates and is specific for far-red light. The HIR requires residues of the PAS2 domain of phyA (Yanovsky *et al.*, 2002), and cis-acting regions of target promoters that are dispensable for the VLFR (Cerdán *et al.*, 2000). A serine-rich N-terminal domain of phyA reduces VLFR but enhances HIR (Casal *et al.*, 2002).

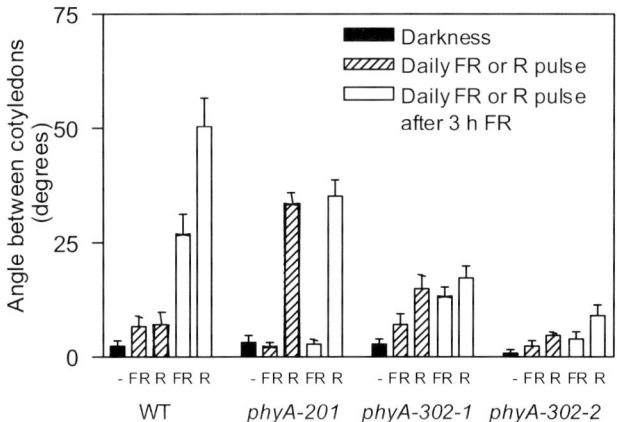

Figure 8. Synergistic or antagonistic interaction between phyA and phyB. Seedlings of Arabidopsis were daily exposed to either a red (R) or a far-red (FR) pulse in factorial combination with or without a 3 h FR pre-treatment and dark controls were included (protocol as in Figure 7). The phyA-302 *mutants do not show the constitutive response to R observed in the* phyA-201 *(null) mutant because they retain the negative effect of the VLFR on phyB-mediated signalling. In the* phyA-302 *mutants the response to R is not amplified by the FR pre-treatment because these mutants lack a HIR. After Yanovsky et al., 2002.*

The occurrence of synergistic and antagonistic interactions between phyA and phyB can be accounted for on the basis of these two pathways of phyA signalling. In seedlings exposed to prolonged far-red light, the HIR pathway of phyA becomes active and enhances the subsequent response to active phyB established by the red-light pulse. In the absence of prolonged far-red light, the red light pulse is enough to activate the VLFR but not the HIR pathway of phyA. Then, the VLFR pathway would negatively regulate phyB signalling. The behaviour of the *phyA-302* mutant, which bears a glutamic to lysine substitution at position 777 (PAS 2 domain) and retains almost normal VLFR but completely lacks HIR, is consistent with this model (Yanovsky *et al.*, 2002). The *phyA-302* mutants fail to show strong phyB-mediated cotyledon unfolding (expressed by the difference between the red and far-red light pulse) observed either in the wild type exposed to a far-red light pre-treatment or in the *phyA-201* null mutant even in the absence of pre-treatment (Figure 8). Compared to the *phyA* null mutant, *phyA-302* mutants fail because their residual VLFR pathway is enough to negatively regulate the phyB-mediated response. Compared to the wild type exposed to far-red light pretreatments *phyA-302* mutants fail because they are unable to produce the HIR necessary to amplify the subsequent effect of phyB (Yanovsky *et al.*, 2002).

Both synergistic and antagonistic regulation of phyB mediated responses by phyA do not involve changes in PHYB protein abundance indicating that phyA action would target downstream components of phyB signalling (Cerdán *et al.*, 1999). FHY1 (Desnos *et al.*, 2001) operates both in the VLFR and HIR pathways and the *fhy1* mutation affects both the synergistic and antagonisic regulation by phyA (Cerdán *et al.*, 1999). FHY3 (Wang and Deng, 2002) affects only the HIR and the synergism between phyA and phyB (Yanovsky *et al.*, 2000). These observations indicate that the negative regulation of phyB signalling is exerted by signalling element(s) downstream phyA and not by phyA itself (Figure 9). Some loci as *EVE1/DIM1* and *DET2* (Luccioni *et al.*, 2002) and *SPA1* (Hoecker *et al.*, 1999; Baumgardt *et al.*, 2002) reduce VLFR and enhance the phyB-mediated response. Other loci like *VLF1-5*, *VLF7* and *CP3* affect the VLFR but not the phyB-mediated response (Yanovsky *et al.*, 1997; Botto *et al.*, 2003). These two sets of genes are predicted to regulate phyA signalling respectively upstream or downstream the elements of this signalling pathway that in turn regulate phyB signalling (Figure 9).

While phyA in the VLFR mode negatively regulates phyB signalling, phyB appears to interfere with phyA mediated responses (Short *et al.*, 1999; Hennig *et al.*, 2001). However, these two effects occur at different levels. phyB does not affect phyA abundance, spectral activity or light-induced degradation rate but is likely to interact unproductively with phyA partners. This idea is supported by the interference caused even by mutant phyB impaired in its signalling ability (Short *et al.*, 1999; Hennig *et al.*, 2001).

2.5 Synergism between phyB and phyC

Seedlings of the *phyC* mutant show reduced inhibition of hypocotyl growth and promotion of cotyledon expansion by red light. These effects are observed in the presence but not in the absence of phyB (Franklin *et al.*, 2003a; Monte *et al.*, 2003). Since the effect of both photoreceptors acting together is larger than their additive effects in the absence of the other photoreceptor, phyB and phyC operate synergistically.

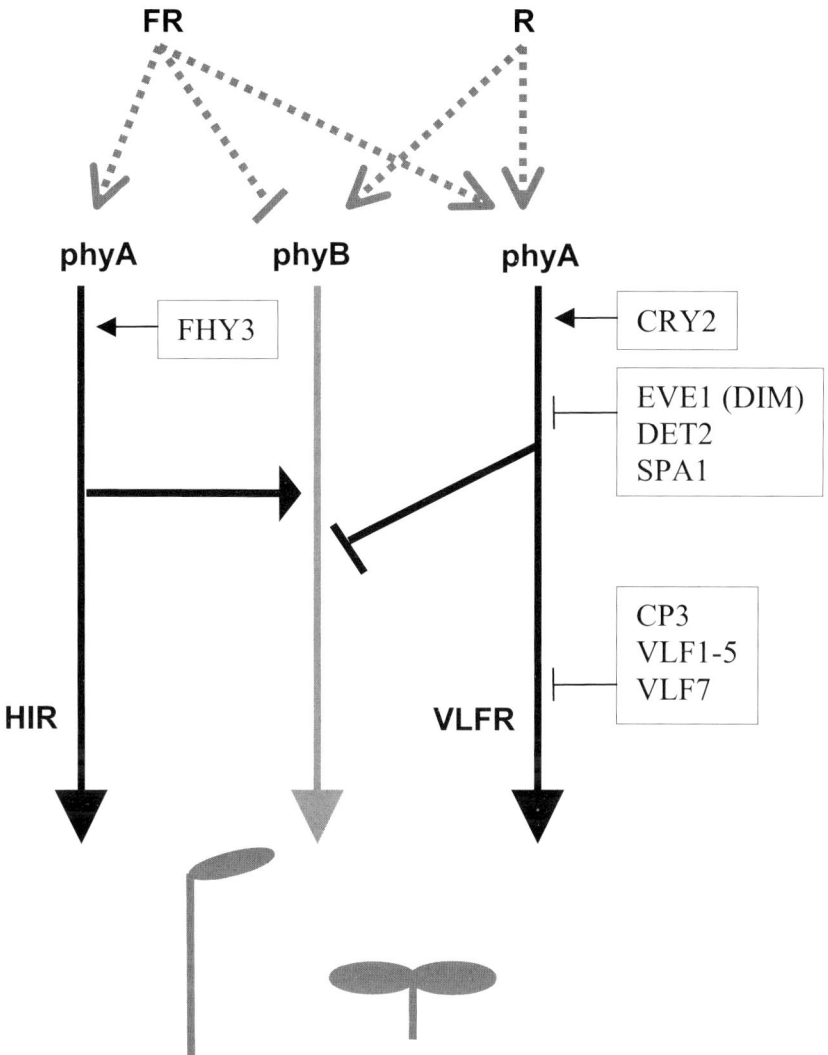

Figure 9. Schematic representation of the interactions between phyA and phyB during de-etiolation of Arabidopsis seedlings exposed to red (R) or far-red light (FR). The predicted place of action of several loci affecting VLFR is indicated.

2.6 Interactive signalling under sunlight reduces noise/signal ratio

When grown under sunlight, the single *phyA* mutant of Arabidopsis is not significantly different from the wild type, the *phyB* mutant shows reduced inhibition throughout the whole range of irradiances and the phenotype of the *cry1* mutant increases with sunlight irradiance (Figure 10). The phenotype of the double and triple mutants is not that predicted from the additive effects of single mutations and this is indicative of interactions (Mazzella and Casal, 2001).

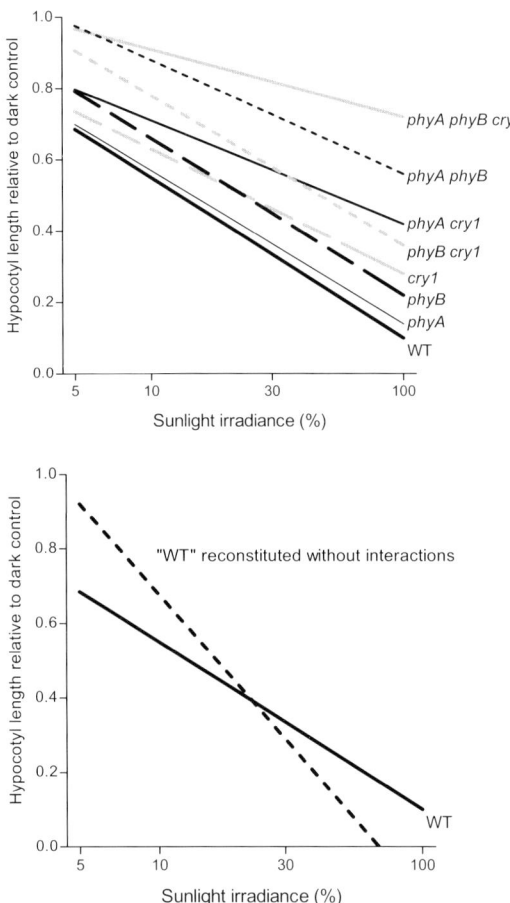

Figure 10. Photoreceptor interactions reduce the impact of irradiance on hypocotyl growth inhibition during de-etiolation of Arabidopsis seedlings under sunlight. The top panel shows hypocotyl length in different photoreceptor mutants. In the lower panel the wild type (WT) is compared with the theoretical WT obtained by adding the effects of phyA, phyB and cry1 without interactions. Redrawn after Mazzella and Casal, 2001.

The *phyA phyB* double mutant is taller than the *phyB* mutant (Figure 10). Since the *phyA* mutation increases the phenotype in the *phyB* but not in the *PHYB* background, there is redundancy between phyA and phyB. This redundancy can be accounted for by the aforementioned antagonism between phyA in the VLFR mode and phyB. During most part of the photoperiod, the irradiances represented in Figure 10 are well above the minimum required for VLFR but this should not preclude the operation of phyA in the VLFR mode.

The *phyA cry1* double mutant is taller than the wild type even at the lowest irradiances; i.e. a condition where neither the *cry1* nor the *phyA* single mutants had a hypocotyl length phenotype (Figure 10). This indicates that phyA and cry1 are redundant, particularly at low irradiances. However, compared to the *phyA phyB cry1* triple mutant the *phyA phyB* double mutant is shorter at high but not at low irradiances. Taken together, these observations indicate that cry1 can operate at low irradiances if phyB is active but it does not require phyB at higher irradiances; a scenario that is fully consistent with the conditional synergism between these two photoreceptors described for experiments under controlled conditions (see 2.3.3). One explanation for the finding that under sunlight the synergism between phyB and cry1 requires the absence of phyA could be based on the fact that both phyA in the HIR mode and cry1 can operate synergistically with phyB. The redundancy between phyA and cry1 could occur specifically at the level of their interaction with phyB (Mazzella and Casal, 2001).

The difference between the wild type and the *phyA phyB cry1* triple mutant provides an estimation of the actions and interactions among phyA, phyB and cry1. In order to evaluate the impact of the interactions a theoretical "wild type without interactions" among these three photoreceptors can be calculated. The inhibition of hypocotyl growth caused by phyA in the absence of phyB and cry1 is the difference in length between the *phyA phyB cry1* triple mutant and the *phyB cry1* double mutant. A similar procedure can be followed to calculate the effects of phyB in the absence of phyA and cry1 and the effects of cry1 in the absence of phyA and phyB. If all these effects are subtracted from the hypocotyl length of the *phyA phyB cry1* triple mutant we obtain a "wild type" reconstituted without interactions among phyA, phyB and cry1. The results are shown in the lower panel of Figure 10 and indicate that the interactions among these three photoreceptors reduce hypocotyl length at low irradiances and increase hypocotyl length at high irradiances. In other words, although the extreme variation of the environment between 5 and 100% of sunlight affects de-etiolation, this influence is significantly less than it could be in the absence of interactions. Interactive signalling provides de-etiolation homeostasis.

A comparable type of analysis has been conducted for seedlings de-etiolating under different photoperiods of sunlight (Mazzella and Casal, 2001). Again, the conclusion is that in the presence of interactions the effect of photoperiod is much weaker than in the absence of interactions. Hypocotyl growth inhibition under red light plus far-red light mixtures with intermediate R:FR is larger in the wild type than the inhibitory effects of phyA alone plus phyB alone (differences in length between the *phyA phyB* and *phyB* mutants, and between the *phyA phyB* and *phyA* mutants, respectively, calculated after Smith *et al.*, 1997). This suggests, that the synergistic interaction between phyA and phyB could enhance hypocotyl growth inhibition at intermediate R:FR that are sub-optimal for either phyA or phyB. The signal that initiates de-etiolation is the change from full darkness to the occurrence of daily light cycles. The interactions appear to reduce the impact of variations in irradiance, photoperiod or R:FR; i.e. fluctuations that are not central to the signal that initiates de-etiolation. In other words, the interactions among phyA, phyB and cry1 provide de-etiolation homeostasis.

3. PHOTORECEPTOR INTERACTION DURING ADULT PLANT BODY SHAPE FORMATION

3.1 Redundant control of normal progression of vegetative development by phytochromes and cryptochromes

Plants cannot survive prolonged dark periods that exhaust carbohydrate reserves. The analysis of multiple photoreceptor mutants has revealed a second role of light; i.e. even in the presence of light for photosynthesis growth does not proceed normally when the major photoreceptors are not active. Organogenesis is very slow in the *phyA phyB cry1 cry2* mutant of Arabidopsis (Mazzella *et al.*, 2001). The only single mutation that reduces the rate of leaf production is *phyB* and in the *phyB* background specific combinations of *phyA*, *cry1* and *cry2* mutations further reduce this rate. Even more impressive is the case of tomato, where the *phyA phyB1 phyB2 cry1* mutant is effectively lethal (Weller *et al.*, 2001). Cotyledon expansion is slow but it achieves a relatively normal size in this quadruple mutant of tomato. However, subsequent development is arrested and most seedlings develop into white plantets with rudimentary leaves and very short internodes (Weller *et al.*, 2001).

The relative significance of a given photoreceptor in the control of a response depends not only on the light conditions but also on other environmental cues like temperature (Mazzella *et al.*, 2000; Halliday *et al.*, 2003a; b). For instance, phyB or cry1 are required to arrest the growth of the internodes and maintain a typical rosette in Arabidopsis. The double *phyB cry1* mutant shows elongated internodes at 20 °C during the vegetative phase. This phenotype disappears at 6 °C (Mazzella *et al.*, 2000).

3.2 The response to R:FR

Low R:FR caused by the presence of neighbours that reflect or transmit far-red light (Ballaré *et al.*, 1987) initiate a series of responses that include accelerated stem growth in dicots (Morgan and Smith, 1976) and reduced tillering in grasses (Deregibus *et al.*, 1983). In Arabidopsis, the petioles show enhanced elongation and the leaves adopt a more erect position. These effects are mediated by phyB and secondarily by phyD and phyE, which are shifted to their inactive forms by low R:FR. The *phyB phyD phyE* triple mutant shows a constitutive phenotype characteristic of low R:FR (Franklin *et al.*, 2003b).

phyA does not appear to play a major direct role in the perception of the low R:FR signal that initiates the changes described above. Actually, the *phyA* mutant tends to respond better than the wild type to intense reductions in R:FR (Johnson *et al.*, 1994). This can be accounted for by the fact that far-red light enhances rather than reduces phyA activity via the HIR pathway and this contrasts with the situation for phyB, phyD and phyE. However, the *phyA* mutant fails to perceive the early warning signal provided by far-red light reflected by neighbour plants before mutual shading is established in growing canopies (Casal, 1996). A more detailed analysis of this phenomenon can be achieved by delivering different R:FR as a brief pulse only at the end of the white light photoperiod in order to manipulate the status of phyB without causing a HIR of phyA (the latter requires prolonged exposures to far-red light). The response to "end-of-day" far-red light is normal in the *phyA* mutant. However, if mixtures of red and far-red light are used to provide intermediate R:FR values, the stem growth promotion caused by lowering the R:FR is more sensitive in the wild type than in the *phyA* mutant (Casal 1996). In other words, the *phyA* mutant requires stronger reductions in R:FR to induce a response. This situation resembles the antagonism between phyA and phyB observed during de-etiolation (Figure 9). Following with the argument, the presence of phyA would reduce the inhibition of stem growth caused by phyB. In the wild type, a small reduction in R:FR, and consequently in active phyB, would be enough to promote stem growth because phyA is already antagonising the inhibition caused by phyB. In the *phyA* mutant the inhibition caused by phyB is stronger and a small reduction in R:FR would not be enough to alleviate the inhibition of stem growth caused by phyB. Transgenic pants with high levels of phyA are more sensitive to small reductions in R:FR than the wild type (Casal, 1996). This fine regulation of the effect of mild reductions in phyB Pfr by phyA might help to maintain a high sensitivity to R:FR during daytime, when this signal indicates the presence of neighbours (Figure 1), compared to the end of the day, when the activity of phyA is predicted to decay due to the reduced irradiance and the R:FR is lower due to atmospheric reasons.

4. PHOTORECEPTOR INTERACTION IN PHOTOTROPISM

4.1 Phototropins perceive the unilateral stimulus

The orientation of the shoot of young seedlings responds to the direction of blue light. The response to a short pulse of unilateral blue light, termed first positive curvature, is bell-shaped with no response below 0.01 $\mu mol.m^{-2}$ (450 nm), a maximum around 0.5 $\mu mol.m^{-2}$, and no curvature at 10 $\mu mol.m^{-2}$ in Arabidopsis (Janoudi, 1997). The response to prolonged exposures to blue light is denominated second positive curvature. The gradient of blue light is perceived by phototropins (see Chapter 12). The first positive curvature and the second positive curvature response to low fluence rates of blue light are mediated by phototropin 1, whereas the second positive curvature response to higher fluences is redundantly mediated by both phototropin 1 and 2 (Liscum and Briggs, 1995; Huala *et al.*, 1997; Sakai *et al.*, 2001).

4.2 Phytochromes enhance the responses mediated by phototropins

Light exposure causes adaptation of the phototropic response in etiolated seedlings, which includes desensitisation and curvature enhancement. Unilateral red light is unable to induce phototropic curvature in etiolated Arabidopsis seedlings but a pulse of red light (which can be given from above) increases the maximum first positive curvature to subsequent unilateral blue light without altering the range of effective fluences. The enhancement induced by red light is mediated by phyA and phyB (Parks *et al.*, 1996; Janoudi *et al.*, 1992; 1997a, b). The *phyA phyB* double mutant shows reduced first positive curvature and reduced sensitivity (increased time threshold) for the second positive curvature even in the absence of red light. This indicates that blue light perceived by phytochromes is able to modulate the phototropic response mediated by phototropins. The enhancement caused by red light perceived by phyA could be mediated by a modulation of auxin responsiveness (Stowe-Evans *et al.*, 2001; see Chapter 15).

4.3 The role of cryptochromes

cry1 and cry2 affect phototropic responses (Ahmad *et al.*, 1998). However, the role of cry1 and cry2 is as positive modulators of the action of phototropins both in the first-positive phototropism (Lascève *et al.*, 1999) and second-positive phototropism when induced by low fluence rates (<1 $\mu mol.m^{-2} s^{-1}$, Whippo *et al.*, 2003), rather than photoreceptors of the blue light gradient inducing phototropism. A detailed kinetic analysis indicates that at fluence rates of blue light above 1 $\mu mol.m^{-2} s^{-1}$ the lag time required for the phototropic response is increased and the magnitude of the curvature is attenuated by the action of phototropins themselves and cryptochromes (Whippo *et al.*, 2003). Since phototropic bending depends on a balance between growth promotion and growth inhibition on the flanks of the hypocotyl, the negative effect of phototropins and cryptochromes at high fluence rates is likely the consequence of hypocotyl growth inhibition by these photoreceptors (Whippo *et al.*, 2003).

5. PHOTORECEPTOR INTERACTION IN CLOCK ENTRAINMENT

A number of photoreceptors, including cry1, cry2, phyA, phyB, phyD and phyE, are involved in the light input to the circadian clock (Devlin and Kay, 2000; Yanovsky *et al.*, 2001, see Chapter 26). These photoreceptors show some degree of redundancy, particularly under blue light. Interestingly, the *cry1* mutant shows a longer period of expression of a chlorophyll a/b-binding gene than the wild type under red light and the *cry2* mutation (that per se is not effective under these conditions) further extends the period in the *cry1* mutant background (Devlin and

Kay, 2000). In Arabidopsis, the phase shift of the leaf movement rhythm induced by several hours of far-red light anticipating the end of the night is mediated by phyA but does not show the aforementioned cry1-dependency (Yanovsky *et al.,* 2001). The phyA-mediated effect of red light on gene expression could occur via the VLFR pathway, which can be affected by cryptochromes (see 2.3.4), whereas the phyA-mediated effect of far-red light on leaf angle occurs via the HIR pathway, which is not obviously affected by cryptochromes.

6. PHOTORECEPTOR INTERACTION CONTROLLING FLOWERING

6.1 Different light signals control the transition between vegetative and reproductive growth

There are two light signals that exert a major influence on the transition of the apex between vegetative and reproductive development. One is day length, which provides an indication of the season. In short-day plants, flowering requires or it is accelerated by short days, whereas long day plants respond in the opposite direction and day neutral plants are insensitive to photoperiod (see Chapter 27). The other is the R:FR, inversely related to canopy density, which tends to accelerate flowering. In the long-day plant Arabidopsis, flowering can be accelerated by long days or by low R:FR compared to short days with a high R:FR. There is interaction between these two light signal as lowering the R:FR has a larger effect if the day is short than if it is long. This situation is reflected on the interactions among photoreceptors.

6.2 Roles of cry2, cry1 and phyA in the photoperiodic response

In Arabidopsis, the difference between short and long days of white light with a high R:FR is perceived primarily by cry2 (Guo *et al.,* 1998). In the Landsberg *erecta* background, cry1 also participates in this response as the *cry1* mutant flowers later than the wild type and the *cry1 cry2* double mutant flowers later than the *cry2* single mutant, particularly when time is measured on the biological scale provided by the number of leaves (Bagnall *et al.,* 1996; Mazzella *et al.,* 2001). In the Columbia background, the contribution of cry1 is not that obvious (Bagnall *et al.,* 1996) and, if the plants are grown under continuous blue light, the *cry1 cry2* mutant can flower simultaneously (Mockler *et al.,* 1999) or later (Mockler *et al.,* 2003) than the *cry2* single mutant.

In pea, phyA is required to perceive day extensions with a high R:FR (Weller *et al.,* 1997). In Arabidopsis plants grown under long days with high R:FR, the *phyA* mutant flowers simultaneously with the wild type but the *phyA cry2* or *phyA cry1 cry2* mutants flower slightly earlier than *cry2* or *cry1 cry2*, respectively (Mazzella *et al.,* 2001). This effect of the *phyA* mutation is already noted in plants grown under short days in the *cry2* (Mazzella *et al.,* 2001), *phyE* or *phyD* (Halliday *et al.,* 2003b) backgrounds. This suggests some sort of negative interaction between phyA and cry2 that affects flowering without been necessarily involved in the perception of the daylength signal. A positive regulation of flowering by phyA becomes obvious under low R:FR or under continuous far-red light (Johnson *et al.*, 1994; Mockler *et al.*, 2003). This could simply reflect the fact that far-red light is required for the HIR of phyA but other interpretations cannot be ruled out.

6.3 Roles of phyB, phyD and phyE in the response to low R:FR

Flowering is promoted by low R:FR of the white light because this condition shifts the photoequilibrium from the Pfr (active) to the Pr (inactive) form of phyB, phyD and phyE. This conclusion is based on the observation that the *phyB* (Goto *et al.*, 1991; Reed *et al.*, 1993), *phyD* (Devlin *et al.*, 1999) and *phyE* (Devlin *et al.*, 1998; Halliday, 2003b) mutations accelerate flowering and the triple *phyB phyD phyE* mutant flowers early irrespectively of high or low R:FR (Franklin *et al.*, 2003b; see Chapter 22). The hierarchy of these three photoreceptors is temperature dependent because at cooler conditions the *phyB* mutant is late, rather than early flowering and phyD and phyE become more important in the repression of flowering (Halliday *et al.*, 2003a, b).

6.4 Integration of the responses to photoperiod and R:FR

The larger effect of long compared to short days at high compared to low R:FR is genetically reflected on the epistatic effect of the *phyB* mutation on *cry2* (Mockler *et al.*, 1999; Mazzella and Casal, 2001). Furthermore, the *cry2* mutation flowers later than the wild type under continuous red plus blue light but not under red or blue light alone, and the *phyB* mutation accelerates flowering under red plus blue but not under blue light alone (Guo *et al.*, 1998). These observations are consistent with a model where a major role of cry2 on flowering would be to alleviate the repression imposed by phyB. These photoreceptors converge to control the stability of CONSTANS, a positive regulator of flowering (Valverde *et al.*, 2003; see Chapter 27).

7. POINTS OF CONVERGENCE IN THE PHOTORECEPTOR SIGNALLING NETWORK

7.1 The occurrence of interactions is an emergent property of the signalling network

The interactions described in previous sections of this chapter are predicted to have their origin in the convergent action of different photoreceptors. Convergence occurs if a signalling element is shared by two photoreceptors or if the status of a signalling element of one photoreceptor is affected by the action of another photoreceptor, which does not involve the affected element among its own downstream players (Figure 11). The possibilities of variation around this theme are immense because the effects can be positive or negative, and the connectivity between the pathways can occur at multiple points and in both directions (from Z to X and from X to Z). In addition, a single photoreceptor can initiate divergent signalling cascades as evidenced by genetic and biochemical studies, further increasing the points of potential interaction with other photoreceptors.

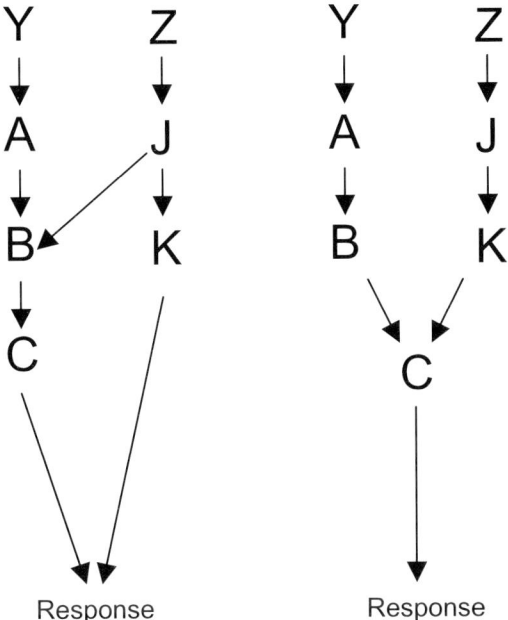

Figure 11. Signalling by a given photoreceptor (Z) converges with another (Y) if at least one of the elements of its signalling pathway (including Z itself) affects the activity of at least one of the elements of the signalling pathway of the other (including Y as a possible target) (left); or at least one element is common to both pathways (right).

The relative significance of each of these connections could be affected by the cellular and developmental context, or the time frame of the analysis. Thus, although some specific patterns of genetic interaction appear to be repeated in different developmental contexts, they cannot be considered as the only possible ones. The conditional nature of the interactions appears to be the rule.

Uncovering signalling convergence is only the first step in accounting for the genetic interactions. It is necessary to demonstrate that the specific connection under study is necessary for a given genetic interaction observed for physiological responses. A status that has not been reached for most of the points of convergence uncovered so far.

7.2 Direct convergence: Physical interaction between photoreceptor pigments

The occurrence of physical interaction between phytochromes and cryptochromes has been documented by different experimental approaches:

1. The PHYA and CRY1 apoproteins show in vitro interaction in the yeast two-hybrid system (Ahmad *et al.*, 1998).

2. Immunoprecipitation with anti-phyB of extracts from transgenic plants overexpressing cry2, followed by detection with anti-cry2 antibody yielded a band that evidenced co-immunoprecipitation (Más *et al.*, 2000).

3. BY-2 protoplasts co-transfected with a fusion protein of CRY2 and red fluorescent protein and a fusion of PHYB and green fluorescent protein revealed that these photoreceptors co-localise in nuclear speckles (see Chapter 9) upon illumination (Más *et al.*, 2000).

4. Fluorescent resonance energy transfer (FRET) microscopy applied to protoplasts co-transfected with CRY2-red fluorescent protein and PHYB-green fluorescent protein indicates direct physical interaction between phyB and cry2 in nuclear speckles (Más *et al.*, 2000).

7.3 Convergence in the control of transcription: HFR1

The control of transcription is a target where the action of different photoreceptors is likely to merge. The action of different photoreceptors ultimately leads to common changes in transcriptome (Ma *et al.*, 2001). The patterns of convergence, however, are not clearly established. Plants mutant for the basic Helix-Loop-Helix transcription factor HFR1 show deficient responses to both far-red light and blue light (see Chapters 17, 19). The *hfr1* phenotype under far-red light is due to impaired phyA signalling. The blue light phenotype is not fully accounted for by the action of phyA because the *phyA hfr1* double mutant is taller than the *phyA* single mutant. HFR1 participates both in phyA and cry1 signalling (Duek *et al.*, 2003). The *hfr1* mutant shows wild-type morphology under red light. Thus, the wavelength-dependency of the phenotype correlates with the pattern of *HFR1* expression because mRNA levels are high under far-red light or blue light and low under red light (Duek *et al.*, 2003). Although branches of phyA and cry1 signalling converge at or upstream the transcriptional control of *HFR1* expression, this convergence does not necessarily play a role in the interactions between phytochrome and cryptochrome during the control of hypocotyl growth of etiolated Arabidopsis, actually the main interaction appears to be between cry1 and phyB rather than phyA (Duek *et al.*, 2003). This situation is useful to illustrate that although convergence is necessary to account for interaction, it does not per se lead to interaction.

7.4 Post-transcriptional convergence accounts for the interaction between phyB and phyC

The *phyC* mutant is impaired in de-etiolation responses to red light and shows early flowering under short days (Franklin *et al.*, 2003a; Monte *et al.*, 2003). The *phyB* mutation is epistatic to *phyC* and causes a significant reduction in phyC levels (Hirschfeld *et al.*, 1998). Thus, the synergism between phyB and phyC can be accounted for by the positive effect of phyB on the levels of phyC.

7.5 Convergence in the control of protein stability: COP1

COP1 is a nuclear repressor of photomorphogenesis that targets to degradation positive light signalling elements such as HY5 (Osterlund *et al.*, 2000; see Chapter 18). At least one of the mechanisms by which COP1 is inactivated is via light-induced translocation to the cytosol. The latter effect is mediated by different photoreceptors such as phyA, phyB and cry1 indicating that some signalling branches of these photoreceptors converge at or upstream COP1 (Osterlund *et al.*, 1998). Partitioning of COP1 to the cytosol by a given photoreceptor could facilitate the action of a second photoreceptor if they are sequentially activated. This could contribute to the synergism observed under light protocols that alternate wavebands that initiate the action of specific photoreceptors (e.g. "responsivity amplification" experiments). This possibility, however, remains to be evaluated.

7.6 Photoreceptor sub-cellular partitioning

In etiolated tobacco seedlings, a pulse of red light induces nuclear import of phyB if the seedlings are pre-treated with prolonged red light or blue light (a far-red light pre-treatment is not effective; see Chapter 9). If the blue-light pre-treatment is replaced by irradiation with monochromatic light at 695 nm, which is predicted to establish a proportion of Pfr similar to that provided by blue light, the subsequent pulse of red light is not effective. This suggests that a specific blue-light photoreceptor could enhance phyB migration to the nucleus (Gil *et al.*, 2000). Repetition of these experiments in different mutant backgrounds in Arabidopsis will help to elucidate whether this phenomenon is involved in the observed interaction between phyB and cry1 during de-etiolation.

7.7 SUB1

SUB1 is a Ca^{2+}-binding protein that localises throughout the cytosol and is apparently enriched in the region surrounding the nucleus (Guo *et al.*, 2001). The *sub1* mutant is hipersensitive to low fluences of blue, a light signal that is perceived by phyA, cry1 and cry2 (Guo *et al.*, 2001). Since *sub1* is also hypersensitive to far-red light and the *phyA* mutation is epistatic to *sub1* both under blue and far-red light, SUB1 appears to be a negative regulator of phyA signalling. SUB1 negatively regulates HY5 abundance and the *hy5* mutation is epistatic to *sub1* (Guo et al., 2001) suggesting that HY5, a positive player in phyA-mediated signalling, could be the point of convergence between SUB1 and phyA actions. The role played by cryptochromes in this scenario is not simple as low fluence rates of blue light are also perceived by cry1 and cry2 but the *cry1* and *cry2* mutations are not epistatic to *sub1*. The *cry2 sub1* is as hypersensitive to blue and far-red light as the *sub1* mutant. The *cry1 sub1* mutant is also similar to the *sub1* mutant at low fluence rates and only at higher fluence rates the double mutant shows increased hypocotyl length and reaches the level of the *cry1* mutant. Based on these results, Guo et al., (2001) have proposed a model where the main role of cry1 and cry2 at low fluence rates is to down regulate SUB1 and in this way enhance phyA signalling.

8. OVERVIEW

The aim of this final section is to present the main general conclusions that emerge from the analysis of the specific cases.

8.1 Redundancy

Photomorphogenic responses are generally controlled by multiple photoreceptors that can often be activated simultaneously. Under optimal light conditions the effect of these photoreceptors tends to be less than predicted by an additive model based on the contribution of each photoreceptor in the absence of the others. In some cases this redundancy could be accounted for by the convergence of different photoreceptors to the control of a given process whose response to light saturates with less than the maximum signalling capacity of the system. In others, redundancy appears to have its origin in negative regulation between two photoreceptor pathways.

8.2 Hierarchical action

Although photomorphogenic responses are generally controlled by multiple photoreceptors, one or two tend to have the largest impact. This hierarchy is strongly context dependent (specific process, developmental status, temperature, etc.). The hierarchy defines the light signals more effective to control a given response.

8.3 Synergism

Particularly under weak light signals, the effects of some photoreceptors tend to be higher than predicted by an additive model based on the contribution of each photoreceptor in the absence of the others. Although multiple photoreceptors can control a given response simultaneously, their participation in the synergism is often very selective.

8.4 Sensitivity and homeostasis

Redundancy reduces the effect of optimal light conditions for photoreceptor activity. Synergism between photoreceptors is more important under sub-optimal light conditions and enhances the response when signals are weak. Thus, the interplay between redundancy and synergism provides a buffer that appears to favour the balance between sensitivity and response to noise from the light environment.

8.5 Connectivity

It is becoming clear that plant photoreceptors create a signalling network with numerous points of convergence or divergence of specific pathways. Some of the points of convergence are already known but they do not necessarily account for the interactions observed in physiological experiments. Uncovering the molecular and cellular basis of the interactions among photoreceptors will help to elucidate the functional structure of the signalling network.

9. REFERENCES

Ahmad, M. and Cashmore, A. R. (1993) *HY4* gene of *Arabidopsis thaliana* encodes a protein with characteristics of a blue-light photoreceptor. *Nature, 366*, 162-166.

Ahmad, M., Grancher, N., Heil, M., Black, R. C., Giovani, B., Galland, P., Lardemer, D. (2002) Action spectrum for cryptochrome-dependent hypocotyl growth inhibition in *Arabidopsis*. *Plant Physiology, 129*, 774-785.

Ahmad, M., Jarillo, J. A., Smirnova, O., Cashmore, A. R. (1998) Cryptochrome blue-light photoreceptors of *Arabidopsis* implicated in phototropism. *Nature, 392*, 720-723.

Ahmad, M., Jarrillo, J. A., Smirnova, O., Cashmore, A. A. (1998) The CRY1 blue light photoreceptor of *Arabidopsis* interacts with phytochrome A *in vitro*. *Molecular Cell, 1*, 939-948.

Aukerman, M. J., Hirschfeld, M., Wester, L., Weaver, M., Clack, T., Amasino, R. M., *et al.* (1997) A deletion in the *PHYD* gene of the Arabidopsis Wassilewskija ecotype defines a role for phytochrome D in red/far-red light sensing. *Plant Cell, 9*, 1317-1326.

Bagnall, D. J., King, R. W., Hangarter, R. P. (1996) Blue-light promotion of flowering is absent in *hy4* mutants of *Arabidopsis*. *Planta, 200*, 278-280.

Ballaré, C. L., Sánchez, R. A., Scopel, A. L., Casal, J. J., Ghersa, C. M. (1987) Early detection of neighbour plants by phytochrome perception of spectral changes in reflected sunlight. *Plant, Cell and Environment, 10*, 551-557.

Baumgardt, R.-L., Oliverio, K. A., Casal, J. J., Hoecker, U. (2002) SPA1, a component of phytochrome A signal transduction, regulates the light signalling current. *Planta, 215*, 745-753.

Boccalandro, H. E., Mazza, C. A., Mazzella, M. A., Casal, J. J., Ballaré, C. L. (2001) UV-B enhances phytochrome B mediated cotyledon opening in Arabidopsis. *Plant Physiology, 126*, 780-788.

Botto, J. F., Alonso Blanco, C., Garzarón, I., Sánchez, R. A., Casal, J. J. (2003) The Cvi allele of cryptochrome 2 enhances cotyledon unfolding in the absence of blue light in Arabidopsis. *Plant Physiology, 133*, 1547-1556.

Casal, J. J. (1996) Phytochrome A enhances the promotion of hypocotyl growth caused by reductions of phytochrome B Pfr levels in light-grown *Arabidopsis thaliana*. *Plant Physiology, 112*, 965-973.

Casal, J. J. and Boccalandro, H. (1995) Co-action between phytochrome B and HY4 in *Arabidopsis thaliana*. *Planta, 197*, 213-218.

Casal, J. J., Davis, S. J., Kirchenbauer, D., Viczian, A., Yanovsky, M. J., Clough, R. C., *et al.* (2002) The serine-rich N-terminal domain of oat phytochrome A helps regulate light responses and subnuclear localization of the photoreceptor. *Plant Physiology, 129*, 1127-1137.

Casal, J. J., Luccioni, L. G., Oliverio, K. A., Boccalandro, H. E. (2003) Light, phytochrome signalling and photomorphogenesis in Arabidopsis. *Photochemical and Photobiological Sciences*, *2*, 625-636.

Casal, J. J. and Mazzella, M. A. (1998) Conditional synergism between cryptochrome 1 and phytochrome B is shown by the analysis of *phyA*, *phyB* and *hy4* simple, double and triple mutants in *Arabidopsis*. *Plant Physiology*, *118*, 19-25.

Casal, J. J., Yanovsky, M. J., Luppi, J. P. (2000) Two photobiological pathways of phytochrome A activity, only one of which shows dominant negative suppression by phytochrome B. *Photochem and Photobiol*, *71*, 481-486.

Cerdán, P. D., Staneloni, R. J., Ortega, J., Bunge, M. M., Rodriguez-Batiller, J., Sánchez, R. A., et al. (2000) Sustained but not transient phytochrome A signalling targets a region of a *Lhcb1*2* promoter that is not necessary for phytochrome B action. *Plant Cell*, *12*, 1203-1211.

Cerdán, P. D., Yanovsky, M. J., Reymundo, F. C., Nagatani, A., Staneloni, R. J., Whitelam, G. C., et al. (1999) Regulation of phytochrome B signalling by phytochrome A and FHY1 in *Arabidopsis thaliana*. *Plant Journal*, *18*, 499-507.

Deregibus, V. A., Sanchez, R. A., Casal, J. J. (1983) Effects of light quality on tiller production in *Lolium spp.* . *Plant Physiology*, *72*, 900-902.

Desnos, T., Puente, P., Whitelam, G., Harberd, N. (2001) FHY1: a phytochrome A-specific signal transducer. *Genes and Development*, *15*, 2980-2990.

Devlin, P., Patel, S. R., Whitelam, G. C. (1998) Phytochrome E Influences Internode Elongation and Flowering Time in Arabidopsis. *Plant Cell*, *10*, 1479-1488.

Devlin, P. F. and Kay, S. A. (2000) Cryptochromes are required for phytochrome signaling to the circadian clock but not for rhythmicity. *Plant Cell*, *12*, 2499-2509.

Devlin, P. F., Robson, P. R. H., Patel, S. R., Goosey, L., Sharrock, R. A., Whitelam, G. C. (1999) Phytochrome D acts in the shade-avoidance syndrome in Arabidopsis by controlling elongation growth and flowering time. *Plant Physiology*, *119*, 909-915.

Duek, P. D. and Fankhauser, C. (2003) HFR1, a putative bHLH transcription factor, mediates both phytochrome A and cryptochrome signalling. *Plant Journal*, *34*, 827-836.

El-Assal, S.-D., Alonso Blanco, C., Peeters, A. J. M., Raz, V., Koornneef, M. (2001) A QTL for flowering time in *Arabidopsis* reveals a novel allele of *CRY2*. *Nature Genetics*, *29*, 435-439.

Folta, K. M. and Spalding, E. P. (2001a) Opposing roles of phytochrome A and phytochrome B in early cryptochrome-mediated growth inhibition. *Plant Journal*, *28*, 333-340.

Folta, K. M. and Spalding, E. P. (2001b) Unexpected roles for cryptochrome 2 and phototropin revealed by high-resolution analysis of blue light-mediated hypocotyl growth inhibition. *Plant Journal*, *26*, 471-478.

Franklin, K. A., Davis, S. J., Stoddart, W. M., Vierstra, R. D., Whitelam, G. C. (2003a) Mutant analyses define roles for phytochrome C in *Arabidopsis* photomorphogenesis. *Plant Cell*, *15*, 1981-1989.

Franklin, K. A., Praekelt, U., Stoddart, W. M., Bilingham, O. E., Halliday, K. J., Whitelam, G. C. (2003b) Phytochromes B, D, and E act redundantly to control multiple physiological responses in Arabidopsis. *Plant Physiology*, *113*, 1340-1346.

Gil, P., Kircher, S., Adam, E., Bury, E., Kozma-Bognar, L., Schäfer, E. et al. (2000) Photocontrol of subcellular partitioning of phytochrome-B: GFP fusion protein in tobacco seedlings. *The Plant Journal*, *22*, 135-145.

Goto, N., Kumagai, T., Koornneef, M. (1991) Flowering responses to light-breaks in photomorphogenic mutants of Arabidopsis thaliana, a long-day plant. *Physiologia Plantarum*, *83*, 209-215.

Guo, H., Mockler, T., Duong, H., Lin, C. (2001) SUB1, an *Arabidopsis* Ca^{2+}-binding protein involved in cryptochrome and phytochrome coaction. *Science*, *19*, 487-490.

Guo, H., Yang, H., Mockler, T. C., Lin, C. (1998) Regulation of flowering time by *Arabidopsis* photoreceptors. *Science*, *279*, 1360-1363.

Halliday, K. J., Salter, M. G., Thingnaes, E., Whitelam, G. C. (2003a) Phytochrome control of flowering is temperature sensitive and correlates with expression of the floral integrator *FT*. *Plant Journal*, *33*, 875-885.

Halliday, K. J., Whitelam, G. C. (2003b) Changes in photoperiod or temperature alter the functional relatioships between phytochromes and reveal roles for phyD and phyE. *Plant Physiology*, *131*, 1913-1920.

Hennig, L., Funk, M., Whitelam, G. C., Schäfer, E. (1999a) Functional interaction of cryptochrome 1 and phytochrome D. *Plant Journal*, *20*, 289-294.

Hennig, L, Poppe, C., Sweere, U, Martin A, Schäfer E. (2001) Negative interference of endogenous phytochrome B with phytochrome A function in Arabidopsis. *Plant Physiology, 125,* 1036-1044.

Hennig, L., Poppe, C., Unger, S., Schäfer, E. (1999b) Control of hypocotyl elongation in *Arabidopsis thaliana* by photoreceptor interaction. *Planta, 208,* 257-263.

Hirschfeld, M., Tepperman, J. M., Clack, T., Quail, P. H., Sharrock, A. R. (1998) Coordination of phytochrome levels in *phyB* mutants of Arabidopsis as revealed by apoprotein-specific monoclonal antibodies. *Genetics, 149,* 523-535.

Hoecker, U., Tepperman, J. M., Quail, P. H. (1999) SPA1, a WD-repeat protein specific to phytochrome A signal transduction. *Science, 284,* 496-499.

Huala, E., Oeller, P. W., Liscum, E., Han, I-S., Larsen, E., Briggs, W. R. (1997) Arabidopsis NPH1: A protein kinase with a putative redox-sensing domain. *Science, 278,* 2120-2123.

Janoudi, A.-K., Gordon, W. R., Wagner, D., Quail, P. H., Poff, K. L. (1997) Multiple phytochromes are involved in red-light-induced enhancement of first-positive phototropism in *Arabidopsis thaliana*. *Plant Physiology, 113,* 975-979.

Janoudi, A.-K., Konjevic, R., Whitelam, G., Gordon, W., Poff, K. L. (1997) Both phytochrome A and phytochrome B are required for the normal expression of phototropism in *Arabidopsis thaliana* seedlings. *Physiologia Plantarum, 101,* 278-282.

Janoudi, A.-K. and Poff, K. L. (1992) Action spectrum for enhancement of phototropism by *Arabidopsis thaliana* seedlings. *Photochemistry and Photobiology, 56,* 655-659.

Johnson, E., Bradley, M., Harberd, P., Whitelam, G. C. (1994) Photoresponses of light-grown phyA mutants of *Arabidopsis*. Phytochrome A is required for the perception of daylength extensions. *Plant Physiology, 105,* 141-149.

Lascève, G., Leymarie, J., Olney, M. A., Liscum, E., Christie, J. M., Vavasseur, Briggs, W. R. (1999) Arabidopsis contains at least four independent blue-light-activated signal transduction pathways. *Plant Physiology, 120,* 606-614.

Lin, C., Yang, H., Guo, H., Mockler, T., Chen, J., Cashmore, A. (1998) Enhancement of blue-light sensitivity of Arabidopsis seedlings by blue light receptor cryptochrome 2. *Proceedings of the National Academic of Sciences, 95,* 2686-2690.

Liscum, E. and Briggs, W. R. (1995) Mutations in the NPH1 locus of arabidopsis disrupt the perception of phototropic stimuli. *Plant Cell, 7,* 473-485.

Luccioni, L. G., Oliverio, K. A., Yanovsky, M. J., Boccalandro, H., Casal, J. J. (2002) Brassinosteroid mutants uncover fine tunning of phytochrome signaling. *Plant Physiology, 178,* 173-181.

Ma, L., Li, J., Qu, L., Hager, J., Chen, Z, Zhao, H., *et al.* (2001) Light control of *Arabidopsis* development entails coordinated regulation of genome expression and cellular pathways. *Plant Cell, 13,* 2589-2607.

Más, P., Devlin, P. F, Panda, S., Kay, S. A. (2000) Functional interaction of phytochrome B and cryptochrome 2. *Nature, 408,* 207-211.

Mazzella, M. A., Alconada Magliano, T. M., Casal, J. J. (1997) Dual effect of phytochrome A on hypocotyl growth under continuous red light. *Plant, Cell and Environment, 20,* 261-267.

Mazzella, M. A., Bertero, D., Casal, J. J. (2000) Temperature-dependent internode elongation in vegetative plants of *Arabidopsis thaliana* lacking phytochrome B and cryptochrome 1. *Planta, 210,* 497-501.

Mazzella, M. A. and Casal, J. J. (2001) Interactive signalling by phytochromes and cryptochromes generates de-etiolation homeostasis in *Arabidopsis thaliana*. *Plant, Cell and Environment, 24,* 155-162.

Mazzella, M. A., Cerdán, P. D., Staneloni, R., Casal, J. J. (2001) Hierarchical coupling of phytochromes and cryptochromes reconciles stability and light modulation of *Arabidopsis* development. *Development, 128,* 2291-2299.

Mockler, T., Guo, H., Yang, H., Duong, H., Lin, C. (1999) Antagonistic actions of Arabidopsis cryptochromes and phytochrome B in the regulation of floral induction. *Development, 126,* 2073-2082.

Mockler, T., Yu, X-H., Parikh, D., Cheng, Y.-C., Dolan, S., Lin, C. (2003) Regulation of photoperiodic flowering by *Arabidopsis* photoreceptors. *Proceedings of the National Academy of Sciences USA, 100,* 2140-2145.

Mohr, H. (1986) "Coaction between pigment systems." In *Photomorphogenesis in Plants*, Kendrick, R. E. and Kronenberg, G. H. M. (eds) Marthinus Nijhoff Publishers, Dordrecht, The Netherlands, 547-564.

Mohr, H. (1994) "Coaction between pigment systems." In *Photomorphogenesis in Plants*, Kendrick, R. E. and Kronenberg, G. H. M. (eds) Kluwer Academic Publishers, Dordrecht, The Netherlands, 353-373.

Monte, E., Alonso, J. M., Ecker, J. R., Zhang, Y., Li, X., Young, J., et al. (2003) Isolation and characterization of *phyC* mutants in Arabidopsis reveals complex cross talk between phytochrome signalling pathways. *Plant Cell, 15*, 1962-1989.

Morgan, D. C. and Smith, H. (1976) Linear relationship between phytochrome photoequilibrium and growht in plants under natural radiation. *Nature, 262*, 210-212.

Nowak, M. A., Boerlijst, M. C., Cooke, J., Smith, J. M. (1997) Evolution of genetic redundancy. *Nature, 388*, 167-171.

Osterlund, M. K. and Deng, X.-W. (1998) Multiple photoreceptors mediate the light induced reduction of GUS-COP1 from Arabidopsis hypocotyl nuclei. *Plant Journal, 16*, 201-208.

Osterlund, M. T., Hardtke, N. W., Deng, X. W. (2000) Targeted destabilization of HY5 during light-regulated development of *Arabidopsis*. *Nature, 405*, 462-466.

Parks, B. M., Quail, P. H., Hangarter, R. P. (1996) Phytochrome A regulates red-light induction of phototropic enhancement in Arabidopsis. *Plant Physiology, 110*, 155-162.

Pickett, F. B. and Meeks-Wagner, D. R. (1995) Seeing double: appreciating genetic redundancy. *Plant Cell, 7*, 1347-1356.

Quail, P. H., Boylan, M. T., Parks, B. M., Short, T. W., Xu, Y., Wagner, D. (1995) Phytochromes: Photosensory perception and signal transduction. *Science, 268*, 675-680.

Reed, J. W., Nagpal, P., Poole, D. S., Furuya, M., Chory, J. (1993) Mutations in the gene for the red/far-red light receptor phytochrome B alter cell elongation and physiological responses throughout Arabidopsis development. *Plant Cell, 5*, 147-157.

Sakai, T., Kagawadagger, T., Kasahara, M., Swartz, T. E., Christie, J. M., Briggs, W. R., et al. (2001) *Arabidopsis* nph1 and npl1: Blue light receptors that mediate both phototropism and chloroplast relocation. *Proceedings of the National Academy of Sciences USA, 98*, 6969-6974.

Short, T. W. (1999) Overexpression of *Arabidopsis* phytochrome B inhibits phytochrome A function in the presence of sucrose. *Plant Physiology, 119*, 1497-1505.

Smith, H., Xu, Y., Quail, P. H. (1997) Antagonistic but complementary actions of phytochromes A and B allow optimum seedling de-etiolation. *Plant Physiology, 114*, 637-641.

Stowe-Evans, E. L., Luesse, D. R., Liscum, E. (2001) The enhancement of phototropin-induced phototropic curvature in Arabidopsis occurs via photoreversible phytochrome A-dependent modulation of auxin responsiveness. *Plant Physiology, 126*, 826-834.

Valverde, F., Mouradov, A., Soppe, W., Ravenscroft, D., Samach, A., Coupland, G. (2003) Photoreceptor regulation of CONSTANS protein and the mechanism of photoperiodic flowering. *Science, 303*, 1003-1006.

Wang, H. and Deng, X. W. (2002) Arabidopsis FHY3 defines a key phytochrome A signalling component directly interacting with its homologous partner FAR1. *The EMBO Journal, 21*, 1339-1349.

Weller, J. D., Murfet, I. C., Reid, J. B. (1997) Pea mutants with reduced sensitivity to far-red light define an important role for phytochrome A in day-length detection. *Plant Physiology, 114*, 1225-1236.

Weller, J. L., Perrotta, G., Schreuder, M. E. L., van Tuinen, A., Koornneef, M., Giuliano, G., et al. (2001) Genetic dissection of blue-light sensing in tomato using mutants deficient in cryptochrome 1 and phytochromes A, B1 and B2. *Plant Journal, 25*, 427-440.

Weller, J. L., Schreuder, M. E., Smith, H., Koornneef, M., Kendrick, R. E. (2000) Physiological interactions of phytochromes A, B1 and B2 in the control of development in tomato. *Plant Journal, 24*, 345-356.

Whippo, C. W. and Hangarter, R. P. (2003) Second positive phototropism results from coordinated co-action of the phototropins and cryptochromes. *Plant Physiology, 132,* 1499-1507.

Whitelam, G. C., Johnson, E., Peng, J., Carol, P., Anderson, M. L., Cowl, J. S., et al. (1993) Phytochrome A null mutants of *Arabidopsis* display a wild-type phenotype in white light. *Plant Cell, 5*, 757-768.

Yanovsky, M. J., Casal, J. J., Luppi, J. P. (1997) The *VLF* loci, polymorphic between ecotypes Landsberg erecta and Columbia dissect two branches of phytochrome A signalling pathways that correspond to the very-low fluence and high-irradiance responses of phytochrome. *Plant Journal, 12*, 659-667.

Yanovsky, M. J., Casal, J. J., Whitelam, G. C. (1995) Phytochrome A, phytochrome B and HY4 are involved in hypocotyl growth responses to natural radiation in *Arabidopsis*: weak de-etiolation of the *phyA* mutant under dense canopies. *Plant, Cell & Environment, 18*, 788-794.

Yanovsky, M. J., Luppi, J. P., Kirchenbauer, D., Ogorodnikova, O. B., Sineshchekov, V. A., Adam, E., et al. (2002) Missense mutation in the PAS2 domain of phytochrome A impairs subnuclear localization and a subset of responses. *Plant Cell*, *14*, 1591-1603.

Yanovsky, M. J., Mazzela, M. A., Whitelam, G. C., Casal, J. J. (2001) Re-setting of the circadian clock by phytochromes and cryptochromes in *Arabidopsis*. *Journal of Biological Rhythms*, *16*, 523-530.

Yanovsky, M. J., Whitelam, G. C., Casal, J. J. (2000) *fhy3-1* retains inductive responses of phytochrome A. *Plant Physiology*, *123*, 235-242.

Chapter 21

INTERACTION OF LIGHT AND HORMONE SIGNALLING TO MEDIATE PHOTOMORPHOGENESIS

Michael M. Neff, Ian H. Street, Edward M. Turk and Jason M. Ward
Department of Biology, Washington University, One Brookings Drive, St. Louis, MO 63130, USA (e-mail: mneff@biology2.wustl.edu)

1. INTRODUCTION

When taking a reductionist view of experimental results we often think of signal transduction as a linear chain of events leading from a specific stimulation to a change in growth and development. This reductionist view is certainly helpful as one dissects the fine points of mechanistic interactions between individual signalling components. Sometimes, however, this method of reasoning becomes muddled by data that cannot be supported by the models at hand. This scenario has been repeated many times during studies of photomorphogenic and hormone responses in plants. Though molecular genetic studies in Arabidopsis have allowed us to uncover many of the signalling components involved in both photomorphogenic and hormone signal transduction, fitting them into a model that describes plant development has proven difficult. This is no surprise given the plethora of both photomorphogenic photoreceptors and phytohormones.

It has become clear that plant photomorphogenesis is better thought of as a web of interactions between multiple receptor-mediated signal transduction events leading to a fine-tuning of plant development in response to a constantly changing light environment. Given that plants can only change their environment by altering their growth, it is no surprise that such a web of photomorphogenic interactions evolved. One of the main mechanisms by which a plant or a portion of a plant grows towards a more optimal light environment is via stem elongation, often coupled with cell expansion. This expansion/elongation response is first seen in the hypocotyls and epicotyls of developing seedlings after germination and involves an integration of the phytochromes, cryptochromes and phototropins. Upon exposure to light at the soil/air interface, hypocotyl cells stop elongating whereas the apical meristem and cotyledons expand, become photosynthetic and go on to develop as juvenile plants competing for optimal light conditions.

Many plant hormones regulate the same cell division and expansion growth responses modulated by photomorphogenesis. Gibberellins, auxins and brassinosteroids are generally considered to be growth-stimulating hormones. Ethylene and cytokinins are primarily considered to be growth-inhibiting hormones. These phytohormones can also act in concert or antagonistically with one another. In addition, there are cases where a growth-promoting hormone can act as an inhibitor and vice versa. This chapter focuses on photomorphogenic responses modulated by these phytohormones. As a reductionist approach, we have grouped the discussions by hormone. We also address the positive and negative interactions that occur with other phytohormones within each sub-chapter.

2. GIBBERELLINS

Initially characterized as a metabolite of the rice-infecting fungus *Gibberella fujikuroi*, gibberellins (GAs) have been studied for over 70 years. This "foolish seedling" disease on rice causes elongated, pale plants and also results in a low seed set, thus lowering food production. GAs affect many aspects of plant growth and development from seed germination to flowering. Over 100 GAs have been identified to date (for reviews see Hedden and Phillips, 2000; Richards *et al.*, 2001). Though many plant GAs have been identified, most are biosynthetic intermediates or catabolites of active GAs (for review see Hedden, 1997). The alteration of GA pathways in crop plants is one of the keys to the "Green revolution" (for review see Salamini, 2003).

GA application causes plants to become etiolated, pale and flower early, similar to loss-of-function alleles of *phyB* (Reed *et al.*, 1996). The inhibition of GA biosynthesis causes a severe dark-green-dwarf phenotype as do semi-dominant mutations in GA signalling components such as *gai* (Dill *et al.*, 2001). Other GA signalling mutations, such as *spy* alleles, confer the opposite phenotype, resembling plants with constitutive GA response (for review see Olszewski *et al.*, 2002). Two main factors in how a plant responds to GA are based on the level of the hormone and tissue sensitivity (for review see Kamiya and Garcia-Martinez, 1999). Depending on the plant species and habit of growth (e.g. long-day vs. short-day), GA has many affects on plant development. In general, however, GAs are growth-promoting hormones essential for normal plant development. This section focuses specifically on the role GAs play in photomorphogenic responses.

2.1 Gibberellin biosynthetic genes and seed germination

The interaction between GAs and light has been suggested by studies of seed germination in various plant species (for review see Peng, 2002). Lettuce, Arabidopsis, pea and many grasses require a red-light pulse to germinate (for review see Garcia-Martinez and Gil, 2001). Exogenous GA compensates for this red-light requirement, suggesting that red light may involve up-regulation of GA biosynthesis (for review see Yamaguchi and Kamiya, 2001). The red-light pulse is also far-red reversible, implicating phytochromes in this process (Figure 1). Evidence suggests some GA biosynthetic enzymes are under light control as well as feed-back regulation by GA levels themselves (for reviews see Hedden, 1997; Hedden and Phillips, 2000; Ogawa *et al.*, 2003).

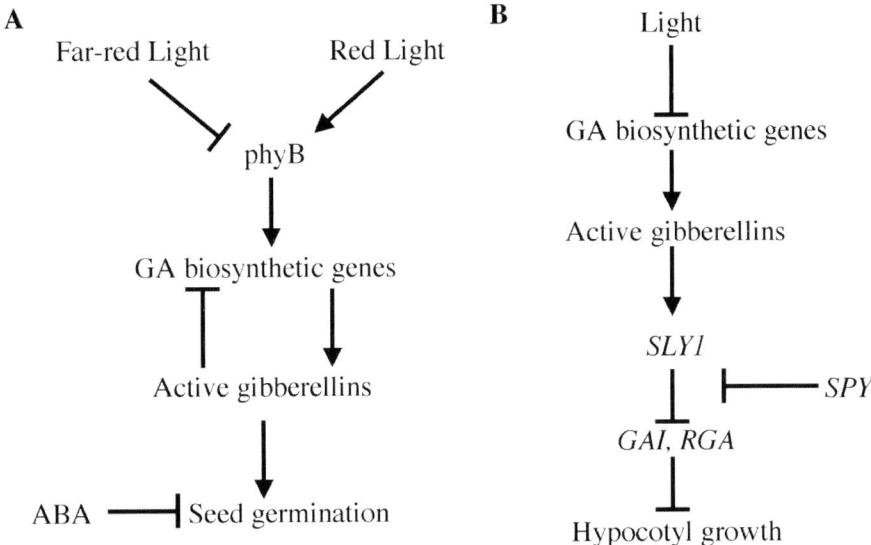

Figure 1. Generalized models of seed germination (A) and hypocotyl elongation (B) illustrating the mode of gibberellin action in Arabidopsis.

In seeds, phyB is the main photoreceptor that regulates active GA levels to induce germination. However, phyA and at least one other photoreceptor also have a role (for review see Garcia-Martinez and Gil, 2001). Evidence for this hypothesis comes from classic as well as modern experiments. Indeed, the existence of phytochrome was postulated based on the red/far-red light reversible control of lettuce seed germination (Toyomasu *et al.*, 1998). Modern molecular biology firmly establishes the importance of GAs in seed germination and that several GA biosynthetic genes are light regulated. GAs promote cell elongation and cell division and are therefore important in plant growth post-germination as well (for reviews see Garcia-Martinez and Gil, 2001; Richards *et al.*, 2001).

Five Arabidopsis GA biosynthetic genes (*ga1-ga5*) were initially found by EMS mutagenesis (Koornneef and van der Veen, 1980). *GA4* encodes a GA 3β-hydroxylase, which catalyses the final step in active GA biosynthesis, converting GA_9 and GA_{20} to GA_4 and GA_1, respectively (Cowling *et al.*, 1998). In seeds, *GA4* transcript accumulation appears to be regulated both by GA levels and light (Cowling *et al.*, 1998; Garcia-Martinez and Gil, 2001). There is another redundant, though differentially regulated Arabidopsis gene, *GA4H*, which shows strong light regulation at the transcriptional level (Yamaguchi *et al.*, 1998). *GA4* is down regulated by GA_4, a feedback mechanism that influences levels of other GA biosynthesis genes. phyB seems to mediate up-regulation of *GA4H* though not *GA4*, which may be under the regulation of another Arabidopsis phytochrome (Yamaguchi *et al.*, 1998). In Arabidopsis, these two genes are differentially regulated, with red light inducing expression at different times during seed germination.

Although it is clear that there is light-induced expression of several GA biosynthetic genes, the biological significance of this is unresolved. Presumably, in seed germination, the induction of GA biosynthesis by light is significant, though the full mechanism for this process has yet to be worked out, as the process is molecularly complex, not simply involving photoreceptors and GA biosynthesis (for review see Ogawa *et al.*, 2003). Some of the GA signalling components involved in the germination process are also antagonized by the plant hormone, abscisic acid (ABA), thus adding another layer of complexity to the regulation of seed germination (Gualberti *et al.*, 2002; Zentella *et al.*, 2002).

2.2 Gibberellins and de-etiolation

The interactions between GAs and light signalling change after germination. In plants where phyB is involved in GA mediated germination, the role of phyB seems to change post-germination by reducing the sensitivity of the emerging hypocotyl to GA. Though the light regulation of GA biosynthesis is clear, how this translates into GA-mediated growth responses requires plants to sense GA levels and respond accordingly.

After germination, there is evidence that light modulation of GA levels is involved in the de-etiolation response of hypocotyls, particularly during the first 24 hours of seedling development (O'Neill *et al.*, 2000; Reid *et al.*, 2002). Active GA levels in the first days post-germination are low while levels of GA catabolites are high, suggesting activity of GA-inactivating enzymes such as GA2ox (Hedden and Phillips, 2000; Garcia-Martinez and Gil, 2001). This reduction of active GA levels may be part of the de-etiolation response (Figure 1). After the pool of active GA is reduced, evidence suggests that sensitivity to GA is also lowered in latter stages of seedling development (Reed *et al.*, 1996). Thus the level and sensitivity to active GAs are both important factors in plant growth.

The post-germination role of GA-biosynthetic genes has been explored in *Pisum sativum* (Pea). The pea homologue of *GA4*, *PsGA3ox1*, is down-regulated within 2 hours of 7-day-old pea seedlings being transferred to blue, red or far-red light (Reid *et al.*, 2002). In addition, the level of active GA (GA_1) dramatically decreases upon exposure to light (Reid *et al.*, 2002). Since phytochromes promote de-etiolation and GAs promote etiolation, the influence of phytochromes would be predicted to counteract GA action. In the pea system, evidence suggests that phyA and a blue light photoreceptor play a role in down-regulating GA_1 biosynthesis enzymes as well as up-regulating the GA inactivating enzymes, *PsGA2ox1* and *PsGA2ox2*, thus promoting photomorphogenesis (Reid *et al.*, 2002).

A role for cry1 is also supported in a microarray experiment showing that, in a *cry1* background, several GA biosynthesis genes are up-regulated compared to the wild type (Folta et al., 2003). Although GA$_1$ levels drop within the first 4 hours after transfer to light, active GA levels recover to etiolated seedling levels after 24-72 hours of exposure to light, presumably representing the transition to adult growth and development. phyB also inhibits hypocotyl growth in the light, though in pea, it apparently does not regulate GA biosynthesis (Reid et al., 2002; Symons and Reid, 2003). There is evidence to suggest that phyB may alter GA sensitivity of the hypocotyl in Arabidopsis and the same might hold true for other plant species (Reed et al., 1996).

Gibberellins also play a role in the repression of photomorphogenesis when plants are grown in the dark (Alabadí et al., 2004). GA inhibitors and ga1-3 mutant seedlings grown in the dark have light grown seedling phenotypes such as shorter hypocotyls and expanded, un-hooked cotyledons (Alabadí et al., 2004). Gibberellins thus not only affect light grown development, but also the mechanism by which photomorphogenesis is repressed in the absence of light.

2.3 The SPY and PHOR1 genes

The Arabidopsis mutant *spindly* (*spy*) was identified based on the ability to germinate in the presence of the GA biosynthesis inhibitor paclobutrazol. Mutations at this locus are recessive and confer adult phenotypes similar to *phyB-null* alleles and wildtype plants repeatedly treated with GA (Jacobsen and Olszewski, 1993; Wilson and Somerville, 1995; for review see Halliday and Fankhauser, 2003). The *SPY* gene encodes an O-linked N-acetylglucosamine transferase, involved in protein modifications. SPY is a negative regulator of GA signalling, as loss-of-function alleles resemble plants treated with GA (Jacobsen et al., 1996) (Figure 1). The proteins modified by SPY are unknown, though SPY appears to have O- linked N-acetylglucosamine transferase activity (for review see Filardo and Swain, 2003). The specific role *SPY* plays in GA signalling is also currently unknown. Genetic interactions with *spy-4* and *ga1-2* and *gai* (a semidominant mutation) suggest that *SPY* is epistatic to both, acting as a signalling component downstream of GA biosynthesis. However, a precise pathway position for SPY does not exist (Jacobsen et al., 1996).

The *spy* phenotype is similar to loss-of-function *phyB* mutants. Recently, a role for *SPY* in photomorphogenesis has been uncovered. *SPY* negatively regulates GA signalling in far-red light, inhibiting hypocotyl growth (Tseng et al., 2004) (Figure1). In red light, an interacting protein, *GIGANTEA,* negatively regulates *SPY*, thus *SPY* is differentially regulated in red and far-red light (Tseng et al., 2004). This series of experiments also implicates *SPY* as a negative regulator of the GA flowering time pathway as well as having a role in the long day flowering pathway.

It is possible that SPY is involved in the post-translational modification of GA signalling component proteins, though no such modification has yet been reported. A hint that SPY might modify GA signalling components is that both are localized to the nucleus (Swain et al., 2002). Another O-linked N-acetylglucosamine transferase has been identified in the Arabidopsis genome, and the loss of both genes is lethal at the embryonic stage (Hartweck et al., 2002). Though originally identified as a GA signalling component, SPY may have broader functions in plant development other than modifying GA action (Swain *et al.*, 2001; for review see Filardo and Swain, 2003).

phyB and GA are also important in later stages of the plant life cycle. An example of this is seen in potato tuberization, which normally occurs under short day conditions. Anti-sense *phyB* plants lose the photoperiodic control over tuberization (for review see Zhao and Chory, 2001). GA has been shown to be important as a negative regulator of tuberization. A potential link between phyB and GA signalling in the tuberization process is *PHOR1*, an *ARMADILLO/β-CATENIN*-like gene that shows photoperiodic regulation as well as responses to GAs (Amador *et al.*, 2001). The tuberization response to light is perceived in the leaves, and travels via an unknown long-distance signal to the stolon, which then develops into another shoot under growth in long days, or into a tuber in short day conditions. PHOR1 is differentially regulated in short-day photoperiods, showing a post-dusk peak, which is absent under non-inductive conditions. Thus PHOR1 is likely a positive signalling component (Amador *et al.*, 2001). Anti-sense lines of *PHOR1* are less sensitive to GA and nuclear localization of PHOR1 is GA dependent, suggesting this protein is a GA signalling component (Amador *et al.*, 2001; Zhao and Chory, 2001).

The role of PHOR1 in the interaction between light perception and GA signalling is complex. *phyB* loss-of-function mutants have the opposite phenotype of *PHOR1* antisense lines (which resemble GA deficient mutants, darker green and shorter stature) (Amador *et al.*, 2001). Homologues of PHOR1 are present in other plant species and determining the role that this family of proteins plays in other plants will dramatically increase our understanding of how light influences GA signalling.

PHOR1 also represents a photoperiodic gene, implicating the circadian clock in the modulation of GA signalling (see Chapter 27). GA and light also affect the timing of flowering. The interaction of these two pathways is complex, as it appears that GA and light signalling pathways act independently, at least with regard to Arabidopsis flowering (Blazquez and Weigel, 1999; Blazquez *et al.*, 2002). It is apparent, however, that in some plant species, photoperiod dramatically affects GA levels (Lee and Zeevaart, 2002). The affects of GA can be dependent on photoperiod and thus flowering in many plant species, and several reviews address this idea more fully than discussed here (for review see Garcia-Martinez and Gil, 2001; Simpson and Dean, 2002).

2.4 A possible role for protein degradation

GAs have been implicated in many plant processes, though their mode of action have only recently been uncovered. GA action is mediated in part by the de-stabilization of proteins in the presence of active GA (for review see Sun, 2000). Some of these transcription factors are also involved in the GA biosynthesis feedback mechanism discussed above (Dill *et al.*, 2001). All of these GA responsive transcription factors have a conserved amino acid sequence known as the DELLA domain. The DELLA domain is apparently involved in the GA-dependent degradation of such proteins, thus letting the growth response of the plant move forward (Dill *et al.*, 2001). Thus GA, in part, acts through a de-repression mechanism (Figure 1). Protein degradation does not occur in the *gai* loss-of-function mutant, a plant that is phenotypically similar to GA-deficient dwarfs. However, other DELLA domain proteins are able to suppress the *ga1-3* dwarf phenotype (Dill and Sun, 2001).

The degradation of DELLA domain proteins in some cases is carried out by the ubiquitination pathway (Fu *et al.*, 2002; Itoh *et al.*, 2003; McGinnis *et al.*, 2003; Sasaki *et al.*, 2003). Several F-box containing proteins have been cloned from rice and Arabidopsis. The *GID-2* mutant of rice and *SLEEPY (SLY1)* in Arabidopsis are both F-box containing genes known to be E3 ubiquitin ligase subunits involved in the degradation of specific proteins (see Chapter 18; McGinnis *et al.*, 2003; Sasaki *et al.*, 2003). These F-box proteins lend even more credibility to the hypothesis that GA action is mediated through the degradation of DELLA domain proteins (Figure 1). It has recently been demonstrated that *SLY1* is involved in the degradation of *GAI* and *RGA* and acts as an E3 ubiquitin ligase (Dill *et al.*, 2004; Fu *et al.*, 2004). The de-repression/degradation pathway of GA signalling has not been connected to other signalling pathways, and this area of research will be important in the determination of how plants integrate environmental signals into hormone responses.

Although the presumed GA receptor has not been identified, there are hints that a receptor-mediated signal transduction pathway regulates GA action. Heterotrimeric G-proteins play a role in GA responses, and are perhaps part of the early steps in GA perception and signal transduction pathways (Hooley, 1998; Assmann, 2002). The *d1* mutants in rice and *gpa* in Arabidopsis have reduced sensitivity to GA, and reduced stature as adults (Ashikari, 1999; Ueguchi-Tanaka *et al.*, 2000; Ullah *et al.*, 2002). Although heterotrimeric G-proteins have been observed to have alterations in GA signalling/responses, they have not been linked to specific downstream targets. In perhaps the simplest scheme, the G-protein signal negatively regulates DELLA domain containing proteins, thus allowing plant growth to occur. Some evidence has been proposed linking photomorphogenic responses with G-protein signalling, though little evidence supports this hypothesis (for review see Assmann, 2002).

2.5 Interactions with other hormone signalling pathways

Like many other plant hormones, GAs are likely to not operate alone. For example, there is evidence to suggest that GA mediated DELLA protein regulation is modulated by auxin and ethylene (Achard *et al.*, 2003; Saibo *et al.*, 2003). Auxin may also promote root growth in part through GAs (Fu and Harberd, 2003). Together these papers suggest that GAs work in concert with other hormones during plant development. This cross-talk may also play a role during modulation of photomorphogenesis.

3. AUXIN

The growth-promoting phytohormone auxin has been studied for over 100 years, yet much of the progress in understanding its role in plant development has occurred within the past ten years. Auxin is responsible for a number of responses throughout a plant's life cycle, including cotyledon/leaf expansion, hypocotyl/stem elongation, gravitropism, phototropism, apical dominance, fruit development, leaf abscission, formation of lateral roots, and inhibition of root elongation. Many of these auxin responses are stimulated by environmental signals including light. As photomorphogenic and auxin-related genes are identified, it is becoming evident that these signal transduction pathways are interconnected.

The connection between light and auxin signalling has been studied since the 19th century when Darwin observed a transmitted signal that caused the coleoptiles of canary grass to grow towards the light (Darwin, 1881). This "transmitted signal" that Darwin observed was later termed auxin. It has been shown that polar auxin transport is responsible for this phototropic response. A number of models have been proposed to explain how auxin is transported in young seedlings, and how this auxin gradient throughout the coleoptile (for monocotyledonous species) or stem/hypocotyl (for dicotyledonous species) causes phototropic responses (for review see Firn, 1994; Liscum and Stowe-Evans, 2000; Friml and Palme, 2002).

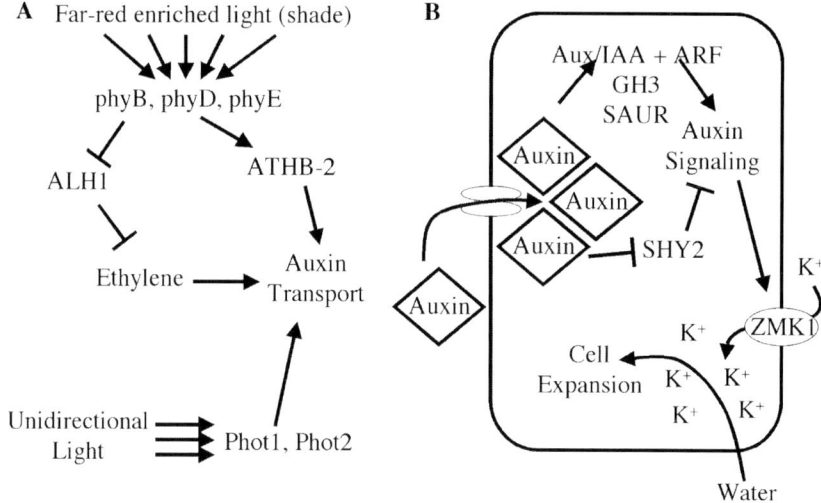

Figure 2. A model in which changes in light cause cell expansion through auxin signalling. A change in light direction or quality is sensed by photoreceptors which starts a signal transduction pathway leading to the transport of auxin. Auxin accumulation leads to activation of auxin responsive genes which starts a signalling cascade causing an accumulation of potassium followed by cell expansion.

A new model is emerging which describes the interactions between light and auxin signalling pathways. It is becoming clear that auxin transport plays a key role in plant responses to changes in light direction (phototropism) and quality (shade avoidance) (for review see Liscum and Stowe-Evans 2000; Morelli and Ruberti, 2000; see Chapters 12, 15, 22). In this model auxin is transported to the cells that will expand. Following this auxin transport and accumulation, auxin responsive genes are induced. These auxin responsive genes are responsible for signalling pathways that ultimately lead to cell expansion (Figure 2). However, this does not exclude the auxin responsive genes from being directly affected by light (Tian *et al.*, 2002), possibly as a means of fine tuning these growth responses during plant development.

3.1 Auxin transport

Auxin is synthesized in the apical region of seedlings and transported downward through the hypocotyl creating a concentration gradient (Sanchez-Bravo *et al.*, 1992). This gradient is important for hypocotyl elongation during growth and light alters the levels of auxin and auxin-binding proteins while also affecting transport (Walton and Ray, 1981; Iino, 1982; Jones *et al.*, 1991; Behringer and Davies, 1992). Auxin is either transported throughout the plant via the phloem or through influx and efflux carriers. Movement through phloem is rapid and non-directional, whereas, polar auxin transport via efflux carriers is slower and directional (for review see Friml and Palme, 2002). Many of the auxin carriers and auxin response components have been identified through the use of auxin efflux inhibitors, such as 1-naphthylphthalamic acid (NPA). Using Arabidopsis seedlings, Jensen *et al.* have shown that NPA is unable to inhibit hypocotyl elongation in the dark (Jensen *et al.*, 1998). However NPA does inhibit hypocotyl elongation in light grown seedlings, and this inhibition requires the presence of phyA, phyB, and cry1, suggesting that polar auxin transport is regulated in a light-dependent manner.

The 560 kD, calossin-like protein, BIG/DOC1/TIR3, which has been identified in genetic mutant screens for both photomorphogenic and auxin transport components, may represent a link between polar auxin transport and photomorphogenesis (Li *et al.*, 1994; Ruegger *et al.*, 1997; Gil *et al.*, 2001). *BIG/DOC1/TIR3* transcript accumulation is light regulated, and the *doc1* mutation causes light induced genes to be expressed in the dark, suggesting an involvement in photomorphogenesis. The *tir3* mutant has many phenotypes suggesting it is involved in auxin efflux transport, including lack of lateral roots, reduced apical dominance, and decreased elongation in a number of organs. Also in the presence of NPA, *tir3* mutants have altered localization of the auxin efflux carrier PIN1, suggesting that BIG/DOC1/TIR3 may be a NPA-binding protein.

Mutations in *BIG/DOC1/TIR3* have also been identified in genetic screens for altered shade-avoidance response (*asa1*) and cytokinin sensitivity (*umb1*) (Kanyuka *et al.*, 2003). The *asa1* and *umb1* mutants have altered phenotypes in response to cytokinins, ethylene, NPA, and GA, and subtle phenotypes in response to auxin, ABA, and brassinosteroids. In addition the *Asa1* mutation can suppress the long hypocotyl phenotype of the *phyAphyB* double mutant. Taken together it appears that BIG/DOC1/TIR3 is not only involved in light signalling and polar auxin transport, but in a number of hormone signalling pathways. It is possible that BIG/DOC1/TIR3 represents a link between photomorphogenesis and many of these hormone-signalling pathways.

3.2 Auxin and phototropism

Directional growth of a plant organ, called tropism, can occur in response to distinct stimuli such as gravity and light. Tropism in response to gravity and light require directional auxin transport, leading to unequal auxin accumulation and ultimately differential cell expansion in the responding tissue. Phototropism occurs when a young seedling bends at the coleoptile/stem toward or away from a light source. Blue or UV-A wavelengths are most effective in eliciting the phototropic response. In Arabidopsis two blue light photoreceptors, called phototropins have been identified which mediate these responses (see Chapter 12). Upon perception of light by the phototropins, auxin is transported away from the light source. As auxin accumulates, the auxin response pathway initiates cell elongation, causing the coleoptile/stem to bend toward the light (for review see Liscum and Stowe-Evans, 2000; see Chapter 15).

The cell expansion driving the bending of the stem during phototropic responses requires an osmotic motor. In *Zea mays*, a voltage-dependent K^+ channel, called ZMK1, may be the osmotic motor responsible for cell expansion during phototropic curvature (Philippar *et al.*, 1999; Fuchs *et al.*, 2003). The auxin responsive *ZMK1* gene mediates K^+ uptake into the coleoptile, which precedes the blue light induced phototropic response. Transcript levels of *ZMK1* increase in response to auxin and the expression pattern changes upon blue light- mediated auxin redistribution in the coleoptile. This suggests that as auxin accumulates in response to blue light, *ZMK1* transcript levels increase, which leads to uptake of K^+ followed by cell expansion (Fuchs *et al.*, 2003).

Other phototropic components linking light and auxin signalling have been identified in Arabidopsis. As is evident during phototropism, light perception and signalling are critical for polar auxin transport throughout the stem of a plant. The *NPH4* gene provides a molecular connection between light and auxin signalling during phototropism. The *nph4* mutant lacks hypocotyl and root phototropic responses (Liscum and Briggs, 1996). In addition NPH4 plays roles in other growth responses, such as gravitropism, apical hook maintenance, and leaf expansion (Stowe-Evans *et al.*, 1998). The *NPH4* gene encodes a transcriptional activator, ARF7, which is a member of the auxin responsive factor family of proteins involved in auxin signalling (Harper *et al.*, 2000). In addition, the *nph4* mutation confers altered expression levels of auxin responsive genes. Together, these results suggest that auxin signalling through NPH4 is essential for phototropic bending (see Chapter 15).

3.3 Auxin and shade avoidance

Plants use changes in light quality and intensity to determine their location relative to neighbouring plants. When plants grow near each other they compete for light. As space within the canopy fills, red light is used for photosynthesis whereas far-red light reflects off of neighbouring plants and passes through the canopy, causing a reduction in the red:far-red ratio. This alteration in the ratio of red:far-red light is perceived via phytochromes leading to rapid changes in plant growth in an avoidance response. Plants can react within minutes to a reduction in the red:far-red ratio by changing growth patterns, such as stem elongation, increased apical dominance, reduced leaf development and branching, and accelerated flowering (for review see Ballare, 1999; see Chapter 22).

Arabidopsis *phyB* mutants exhibit constitutive shade avoidance phenotypes suggesting that this response is mediated by phyB. However characterization of other photoreceptor mutants suggests that phyD and phyE also contribute to the shade avoidance response. Downstream of the photoreceptors, a homeodomain-leucine-zipper protein encoded by *ATHB-2* is involved in growth responses to shade. *ATHB-2* is induced by changes in the red:far-red ratio, and over-expression inhibits cotyledon/leaf expansion and enhances hypocotyl elongation; phenotypes exhibited during the shade avoidance response (Carabelli *et al.*, 1996; Steindler *et al.*, 1999). *ATHB-2* overexpression also results in a decrease in vascular tissue and fewer lateral roots. This root phenotype can be rescued by treating with exogenous auxin, suggesting that ATHB-2 is involved in both the shade avoidance response as well as in auxin transport or signalling.

In conjunction with auxin, it appears that the hormone ethylene is also involved in the shade avoidance response. Ethylene is rapidly produced by Arabidopsis rosette leaves in response to decreased light intensities (Vandenbussche *et al.*, 2003b). This increase in ethylene affects auxin inducible genes in the hypocotyl, as an ethylene precursor increases the expression of an auxin-responsive reporter gene (Vandenbussche *et al.*, 2003b). The *ALH1* gene may represent a link between the auxin and ethylene signalling pathways during the shade avoidance response. Like *phyB* mutants, the *alh1* mutation confers constitutive shade avoidance phenotypes, which can be reverted with auxin transport inhibitors. In addition the *alh1* mutant overproduces ethylene, and the roots are hyper-responsive to gravity, which suggests altered auxin response (Vandenbussche *et al.*, 2003a). Taken together it appears that auxin and ethylene signalling pathways may cross-talk during the shade avoidance response. This cross-talk may be linked by ALH1 as well as polar auxin transport.

Just as polar auxin transport is important for differential phototropic and gravitropic growth, it also appears to mediate the shade avoidance response. A model has been proposed describing how a reduction in the red:far-red ratio alters auxin transport, and this change in auxin transport causes altered growth, such as stem elongation and reduced leaf expansion (for review see Morelli and Ruberti, 2002). This model states that during normal light conditions auxin is transported through a central route, possibly through the vascular tissue, into the roots. Transport through this central route is rapid meaning less auxin accumulation in the hypocotyl/stem. Thus the hypocotyl/stem is proposed to be short when grown in normal light conditions, based on correlation of auxin levels and hypocotyl elongation (Romano *et al.*, 1995; Gray *et al.*, 1998). When there is a reduction in the red:far-red ratio of light, the model proposes that auxin is no longer transported through this central route, allowing auxin to accumulate in the outer cell layers of the stem (for review see Morelli and Ruberti, 2002). It is this auxin accumulation in the stem that is proposed to trigger the elongation exhibited during shade avoidance.

3.4 Auxin responsive genes involved in photomorphogenesis

During the phototropic and shade avoidance responses, light affects polar auxin transport causing changes in plant growth. In addition to light affecting polar auxin transport, it also affects a number of auxin responsive genes. The *Aux/IAA*, *GH-3* and *SAUR* gene families are rapidly induced by auxin. The most extensively studied group, the *Aux/IAA* genes are very short-lived, nuclear proteins (for review see Reed, 2001). Through interactions between conserved domains, the *Aux/IAA* gene products form homo- and hetero-dimers with other *Aux/IAA* proteins and with auxin response factors (ARF) (for review see Liscum and Reed, 2002). It is through dimerization that these proteins regulate transcription of auxin signalling genes. The Aux/IAA proteins act as repressors of transcription, while the ARF proteins can be either activators or repressors of transcription. Many of these auxin-responsive transcriptional regulators are involved in light signalling. NPH4, a member of the ARF family of transcription factors, is required for the phototropic response (see above). Other members of the *Aux/IAA* family seem to be involved in light signalling as well (see Chapter 15).

The *shy2/IAA3* (short hypocotyl2) mutant was identified as a semi-dominant, gain-of-function suppressor of the *phyB* mutant long-hypocotyl phenotype (Reed *et al.*, 1998). The *shy2/IAA3* mutant displays a short hypocotyl, flowers early, and develops leaves in both the light and the dark. In addition to these photomorphogenic phenotypes, *shy2/IAA3* also has reduced auxin response and is caused by a mutation in a member of the auxin-induced *Aux/IAA* family of proteins. The *shy2/IAA3* mutant has reduced lateral root formation and decreased response to gravity (Tian and Reed, 1999). SHY2, a negative regulator of auxin signalling, modulates the transcript levels of many auxin response genes, and its expression levels are regulated by light as well as auxin (Tian *et al.*, 2002).

Another Aux/IAA protein, IAA7, was identified as an auxin-resistant, gain-of-function mutant, *axr2-1/IAA7*. The *axr2-1/IAA7* mutant has many phenotypes characteristic with altered auxin response including auxin-resistant root growth and agravitropism (Wilson *et al.*, 1990). In addition to these phenotypes, the loss-of-function mutant has a long hypocotyl in white light, but elongates normally in the dark, suggesting that it is involved in light signalling (Nagpal *et al.*, 2000).

In addition to the nuclear *Aux/IAA* proteins, there are cytoplasmic, auxin responsive genes that also appear to be involved in light signalling. The far-red-insensitive mutant, *FIN219*, identified as a suppressor of COP1, has a long-hypocotyl phenotype in far-red light. *FIN219* encodes a cytoplasm-localized member of the GH3 group of auxin-induced genes, and its expression is rapidly induced by auxin (Hsieh *et al.*, 2000). Recently *FIN219* has been shown to be allelic to *JAR1*, which is involved in protection against a number of stresses (Staswick *et al.*, 2002). Other *JAR1* alleles respond normally to far-red light, suggesting that this auxin-induced gene may not be involved in light signalling, but instead may represent a link between jasmonic acid and auxin signalling.

However other members of this GH3 group of auxin responsive genes, *DFL1* and *AtGH3a* may link the light and auxin signalling pathways. The activation-tagging mutant, *dfl1-D*, has a short hypocotyl in all light conditions, but elongates normally in the dark (Nakazawa *et al.*, 2001). The *DFL1* transcript is induced in response to auxin, and the *dfl1-D* mutant has phenotypes consistent with reduced auxin response although these mutant phenotypes have not been confirmed with a loss-of-function allele. *AtGH3a* transcript levels are induced by auxin as well as far-red light treatment (Tanaka *et al.*, 2002). Expression of the *AtGH3a* gene is regulated by phyB, and is reduced in the auxin resistant mutant, *axr2*, suggesting that one way phyB may regulate gene expression is by altering auxin levels.

Many auxin responsive genes confer photomorphogenic phenotypes when mutated. There are also photomorphogenic mutants that have phenotypes consistent with altered auxin response. In fact disruption of the *Physcomitrella patens* cryptochrome genes alters expression levels of auxin responsive genes, thus altering responses to this hormone (Imaizumi *et al.*, 2002). In addition to the photoreceptors, a number of transcription factors that are important for photomorphogenic responses may also be involved in auxin signalling. The homeodomain-leucine zipper *ATHB-2* alters auxin response during shade avoidance. In addition to ATHB-2, the *hy5* mutant also has phenotypes suggesting it is involved in auxin response as well as photomorphogenesis. The *hy5* mutant, which was originally identified based on its altered hypocotyl length in response to white light, plays a key role in photomorphogenic signal transduction (Koornneef, *et al.*, 1980). In addition to conferring altered responsiveness to light, the *hy5* mutation causes altered responses to gravity and touch, suggesting that HY5, a bZIP transcription factor, may also modulate auxin signalling (Oyama *et al.*, 1997).

3.5 Auxin and protein degradation

A common theme in both photomorphogenesis and auxin signalling is the regulation of transcription factors by ubiquitin-mediated protein degradation. Ubiquitin-mediated degradation is a multiple step process whereby the ubiquitin is activated through its interaction with an ubiquitin-activating enzyme, E1. Following activation the ubiquitin is passed to the E2 or ubiquitin-conjugating enzyme where it is ligated, with the help of the E3-ubiquitin protein ligase, to the protein being marked for degradation (for review see Kepinski and Leyser, 2002). During photomorphogenesis transcription factors such as HY5, HYH and PIF3 are marked for degradation via interactions with the E3 ubiquitin protein ligase, COP1 (see Chapter 18). During auxin signalling the short-lived Aux/IAA proteins are degraded through ubiquitin-mediated protein degradation. Many of the Aux/IAA proteins have been identified as dominant, gain-of-function mutants (Tian and Reed, 1999;

Nagpal et al., 2000). Upon cloning it seems that many of the mutations are single amino acid changes in the conserved domain II, which is important for protein destabilization. The Aux/IAA proteins are degraded through interactions between this domain II and the E3 ubiquitin protein ligase SCF^{TIR1} (Gray et al., 2001).

It is poorly understood how SCF^{TIR1}-mediated Aux/IAA protein degradation is regulated. A common point of regulation during signal transduction phosphorylation, and Aux/IAA proteins are phosphorylated by phyA *in vitro* is (Colon-Carmona et al., 2000). This is of interest as dominant, stabilizing mutations in several Aux/IAA proteins cause photomorphogenic development in the dark (Tian and Reed, 1999; Nagpal et al., 2000). Upon ubiquitination the protein is degraded by the COP9 signalosome (CSN), and the CSN represents another link between the light and auxin signalling pathways. Components of the CSN have been identified based on their photomorphogenic phenotypes in the dark (see Chapter 18). Upon further examination, partial loss-of-function CSN mutants also have reduce auxin response, suggesting CSN may play a role in Aux/IAA degradation (Schwechheimer et al., 2001). In addition CSN has been shown to interact with SCF^{TIR1}, suggesting that ubiquitin-mediated protein degradation is essential in both light and auxin signalling pathways, with these two pathways sharing components.

3.6 Interaction of auxin with other hormone signalling pathways

Plants often use hormones, such as auxin to respond to changes in light. As we identify components of both light and auxin signalling pathways, it is clear that the two are interconnected. In fact other hormones appear to be involved as well, and these hormone-signalling pathways are acting together to respond to changes in light. For example, ethylene, which is produced by the plant in response to changes in light intensity, affect auxin signalling in the stem (Vandenbussche et al., 2003a; Vandenbussche et al., 2003b). In addition to affecting each other, hormone pathways share common components. The *BIG* gene seems to be involved in light signalling as well as a number of hormone signalling pathways (Gil et al., 2001; Kanyuka et al., 2003). In all, many of the hormone response pathways seem to be interconnected with each other and with light signalling to generate photomorphogenic responses.

4. BRASSINOSTEROIDS

In 1968 it was discovered that three extracts from an evergreen Japanese plant called Isonuki (*Distylium racemosum*) contained plant growth promoting activity (Marumo, 1968). The activity was observed using the rice (*Oryza sativa*) lamina inclination assay, which involves swelling of the adaxial cells in the joint between the leaf blade (lamina) and sheath of etiolated rice seedlings and subsequent leaf bending. The angle between the lamina and the sheath increases in a dose-dependent response that provides a sensitive bioassay for growth-promoting activity. In 1970 a bioassay utilizing elongation of the excised pinto bean (*Phaseolus vulgaris*) second internode was used to demonstrate that an oil fraction from extracts of rape (*Brassica napus* L.) pollen also contained a growth-promoting substance termed "brassin" (Mitchell, 1970). In 1979 the growth-promoting compound was finally identified after the Herculean effort to obtain 40kg of bee collected rape pollen, from which 4mg of the growth-promoting substance was purified (Grove *et al.*, 1979). X-ray analysis of the crystallized compound revealed the structure to be a steroid with similarity to ecdysone, an insect molting steroid hormone, as well as to testosterone and oestradiol, mammalian sex steroid hormones (Voigt *et al.*, 2001). The structure is unique among natural steroids in that it contains a B-ring lactone. This compound was named BRASSINOLIDE because it was identified from *BRASSIca Napus* and contains a lactone, which is commonly given the suffix of OLIDE.

To date, over 40 naturally occurring brassinolide analogues have been identified and are collectively termed brassinosteroids. They are defined as steroids that carry an oxygen moiety at C-3 and additional ones at one or more of the C-2, C-6, C-22 and C-23 carbon atoms (Bishop and Yokota, 2001). Although many of the analogues have growth-promoting activity, brassinolide is the most bioactive brassinosteroid.

The brassinolide biosynthetic pathway has been largely determined by feeding deuterium labelled precursors to cultured cells of *Catharanthus roseus* followed by GC-MS analysis of the resulting metabolites (Fujioka *et al.*, 2000). Similar experiments have also been performed on *Catharanthus roseus* seedlings, *Oryza sativa* (rice) seedlings and cultured cells, and *Nicotiana tabacum* (tobacco) seedlings and cultured cells (Choi *et al.*, 1996). The pathway has since been largely confirmed in Arabidopsis by identification of the corresponding genes (Noguchi *et al.*, 2000). The entire genetic pathway and corresponding molecular structures can be viewed on the web at www.brassinosteroid.com.

Brassinosteroids are essential to plant development and have been discovered in every multi-cellular plant species tested, but have not been identified in other organisms, which makes this class of steroids an exclusive phytohormone. A hormone is a small signal molecule that triggers cells to develop into the appropriate tissues and organs at the appropriate time. Animal hormones are synthesized in one tissue and transmitted to the entire organism to affect those tissues with the appropriate receptors, while brassinosteroids appear to be synthesized and perceived ubiquitously (Friedrichsen *et al.*, 2000). In contrast to mammalian steroid hormone receptors, which are transported from the cytoplasm to the nucleus upon hormone binding (Beato, 1995), the putative brassinolide receptor is bound to the plasma membrane (Friedrichsen *et al.*, 2000). Yet plant and animal steroid hormone signalling can each lead to alterations in transcription, as well as cellular responses that are independent of transcription (Bishop and Koncz, 2002). This is likely due to shared as well as unique signalling components among plants and animals (for review see Thummel and Chory, 2002). Other excellent reviews on brassinosteroid biology include (Nemhauser and Chory, 2004; Bishop and Koncz, 2002; Clouse, 2002; Friedrichsen and Chory, 2001; Mussig and Altmann, 2001).

4.1 Brassinosteroid-deficient mutants

Brassinosteroids were first suspected to be involved in photomorphogenesis when their application to bean (Phaseolus vulgaris L. 'Pinto') plants led to the observation that they act to ameliorate the inhibitory effect of white light on internode elongation (Krizek and Mandava, 1983). In 1996 the identification and analysis of the Arabidopsis *DET2* and *CPD* genes provided the first data to support a connection between brassinosteroids and photomorphogenesis (Li *et al.*, 1996; Szekeres *et al.*, 1996). Loss of DET2 or CPD in Arabidopsis resulted in dark-grown seedlings that are not green, but have some light grown phenotypes: short and thick hypocotyls, accumulated anthocyanins, open and partially-expanded cotyledons, developed primary leaf buds, and de-repression of light-responsive genes.

Analysis of additional brassinosteroid-deficient and -insensitive mutants and plants treated with the biosynthesis inhibitor, Brz2000, has reproduced these brassinosteroid-deficient phenotypes in dark-grown seedlings (Fankhauser and Chory, 1997; Szekeres and Koncz, 1998; Asami and Yoshida, 1999; Clouse and Feldmann, 1999; Mussig and Altmann, 1999; Nagata *et al.*, 2000). These observations have led to the proposal that light alters either the concentration of brassinosteroids or the responsivity of cells to these steroids. This hypothesis assumes that brassinosteroids act as downstream mediators of light signal transduction and their removal results in disruption of a downstream negative regulator. However, it should be noted that brassinosteroids could act as negative regulators of the light-signal pathway in addition to or rather than functioning as downstream mediators.

Identification of a brassinosteroid-deficient mutant in a recent far-red light screen strengthens the hypothesis that brassinosteroids are involved in photomorphogenesis. A mutant with enhanced very-low-fluence-responses (a far-red specific response mediated by phyA), termed *eve1*, is defective in DWF1/DIM, a gene involved in brassinosteroid biosynthesis (Luccioni *et al.*, 2002). Luccioni and colleagues demonstrated that both the *eve1/dwf1/dim* mutant, and the *det2* mutant, have enhanced very-low-fluence-responses and reduced high-irradiance-responses when chlorophyll and anthocyanin accumulation are measured in either continuous or hourly pulses of far-red light.

Brassinosteroid-deficient mutants grown in the light have pleiotropic phenotypes, the most obvious of which include severe dwarfism, reduced fertility, prolonged life, and dark green leaves. Although most of these phenotypes are likely due to loss of the cell elongation properties of brassinosteroids, the dark green leaf phenotype is consistent with the hypothesis that brassinosteroids act to negatively regulate photomorphogenesis. While the above reports provided data to suggest a connection between brassinosteroids and photomorphogenesis, the genetic connection is weak. It is important to identify the genetic components that link these two pathways in order to gain an understanding of the mechanism and ultimately the significance of such a connection.

4.2 Brassinosteroids and gene expression

Analysis of gene expression in response to applied brassinolide demonstrates that *PIF3* mRNA is down-regulated within 15 minutes and reduced by two-thirds within 2 to 3 hours (Goda *et al.*, 2002). This potent repression occurs at physiological concentrations of brassinolide (0.1 to 100 nM). PIF3 is a basic helix-loop-helix transcription factor that is localized to the nucleus, interacts with biologically active phytochromes (Ni *et al.*, 1999; Fairchild *et al.*, 2000; Zhu *et al.*, 2000) and binds to light-regulated promoters through the G-box sequence motif (CACGTG) (Martinez-Garcia *et al.*, 2000). Loss-of-function *pif3* mutants also have subtle though significant hypocotyl phenotypes further implicating this gene in photomorphogenesis (Kim *et al.*, 2003; Bauer *et al.*, 2004).

Transcript accumulation of *Lhcb1.3* (Jansson, 1999) and *rbcS-1A* (Krebbers *et al.*, 1998), which both contain the G-box sequence motif in their promoter and are regulated by PIF3, was also shown to be down-regulated by brassinolide after a lag period of approximately 1 hour. *Lhcb1* encodes a light-harvesting chlorophyll *a/b*-binding protein and *rbcS-1A* encodes the small subunit of ribulose-1,5-bisphosphate carboxylase. Transcript levels of both are reduced by two-thirds 24 hours after treatment with brassinolide, at which time the chlorophyll content of these plants is about one-half that of mock-treated controls, without a significant change in the chlorophyll *a/b* ratio. These results suggest that brassinosteroids modulate photomorphogenesis by affecting *PIF3* mRNA expression and supports the notion that brassinosteroids act as negative regulators of photomorphogenesis. It is noteworthy that regulation of *PIF3* expression provides a mechanism by which brassinosteroids regulate the light-signalling pathway in addition to or rather than functioning as downstream mediators of light signal transduction.

A 1-hour treatment of 100µM CHX (an inhibitor of cytosolic protein synthesis) increased the expression of *PIF3* and prevented its repression by subsequent brassinolide treatment. This result suggests that *PIF3* expression is regulated by a short-lived repressor acting at the transcriptional or post-transcriptional level, and that brassinolide is upstream of this repressor. Analysis of *PIF3* expression in response to applied brassinolide and CHX provides further evidence that brassinosteroids modulate photomorphogenesis, and refines the hypothesis to include regulation of the light-signalling pathway.

4.3 Further genetic connections between brassinosteroids and light

BAS1/CYP72B1 is a brassinosteroid inactivating enzyme and a modulator of photomorphogenesis (Neff *et al.*, 1999; Turk *et al.*, 2003). This Arabidopsis gene encodes a cytochrome P450 that inactivates brassinosteroids by carbon-26 hydroxylation. The link between CYP72B1 and photomorphogenesis has been established by analysis of *cyp72b1-1* (a null allele) seedling growth in response to applied brassinolide and various intensities of white light or dark. *cyp72b1-1* is less responsive than the wild type to increasing intensities of brassinolide in white light, though there is no difference in the dark. This light-dependent phenotype suggests that *CYP72B1* is involved in photomorphogenesis.

Further support for this hypothesis is based on the observation that *cyp72b1-1* is less responsive than the wild type to sub-saturating intensities of white light even in the absence of applied brassinolide. This null mutant is also less responsive to sub-saturating intensities of red, blue, and far-red light; suggesting that *CYP72B1*-mediated inactivation of brassinosteroids is involved in multiple photomorphogenic pathways. That the *cyp72b1-1* mutant has normal hypocotyl growth in the dark suggests that hypocotyl elongation is not simply impaired by this mutation. Seedling growth response assays to applied brassinolide in red, blue, and far-red light demonstrate that loss of CYP72B1 causes an increase in brassinolide sensitivity in far-red light and over-expression of CYP72B1 has the opposite affect. On the other hand, blue and red light do not affect brassinolide sensitivity of the mutants. These data suggest that far-red light modulates brassinosteroid levels in part by the action of CYP72B1.

DDWF1 is a cytochrome P450 involved in brassinosteroid biosynthesis in pea (*Pisum sativum*). This protein interacts with Pra2, a dark-inducible, phytochrome-repressed small G protein, in yeast-two hybrid analysis (Kang *et al.*, 2001). These proteins were also shown to co-localize on the endoplasmic reticulum. This data suggests a mechanism in which the photomorphogenic and brassinosteroid pathways communicate through a light dependent decrease in brassinosteroid biosynthesis and a concomitant decrease in brassinosteroid signalling, resulting in de-repression of photomorphogenesis. A seemingly contradictory report measured the levels of brassinosteroids in pea and found no difference or a slightly increased levels of brassinosteroids in response to light (Symons and Reid, 2003). In addition, three brassinosteroid biosynthetic mutants in pea are not de-etiolated in the dark (Symons *et al.*, 2002). It was later reported that Pra2 is a Rab GTPase predominantly localized on Golgi and endosomes and may function in vesicle transport (Inaba *et al.*, 2002). Identification of the analogous proteins in Arabidopsis followed by genetic and biochemical studies will be needed to resolve this potential mechanism.

4.4 Brassinosteroids and light signalling: three speculative models

Protein degradation mediated by the COP9 signalasome/26S proteosome is a key player in the photomorphogenic pathway through the repression of the HY5 transcription factor in a light dependent manner (see Chapter 18; Hardtke and Deng, 2000) (Figure 3). The 26S proteosome is also a key player in the brassinosteroid pathway through the repression of the BES1 and BZR1 signalling components in a brassinolide dependent manner. It is therefore tempting to speculate that one molecular mechanism of interaction between the photomorphogenic and brassinosteroid pathways is at the point of protein degradation by the 26S proteosome.

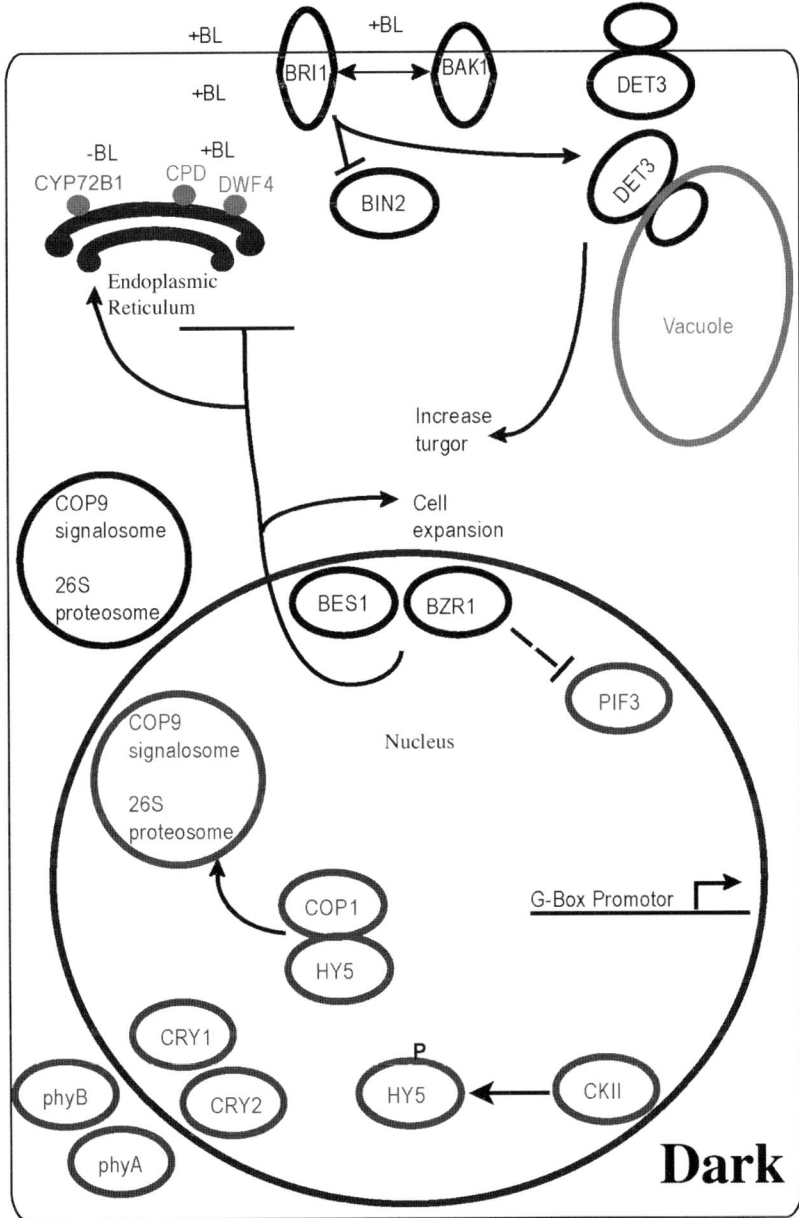

Figure 3a. *Model of a hypocotyl cell in the dark including proteins of the brassinosteroid and photomorphogenic pathways. +BL represents active brassinosteroids. –BL represents inactive brassinosteroids.*

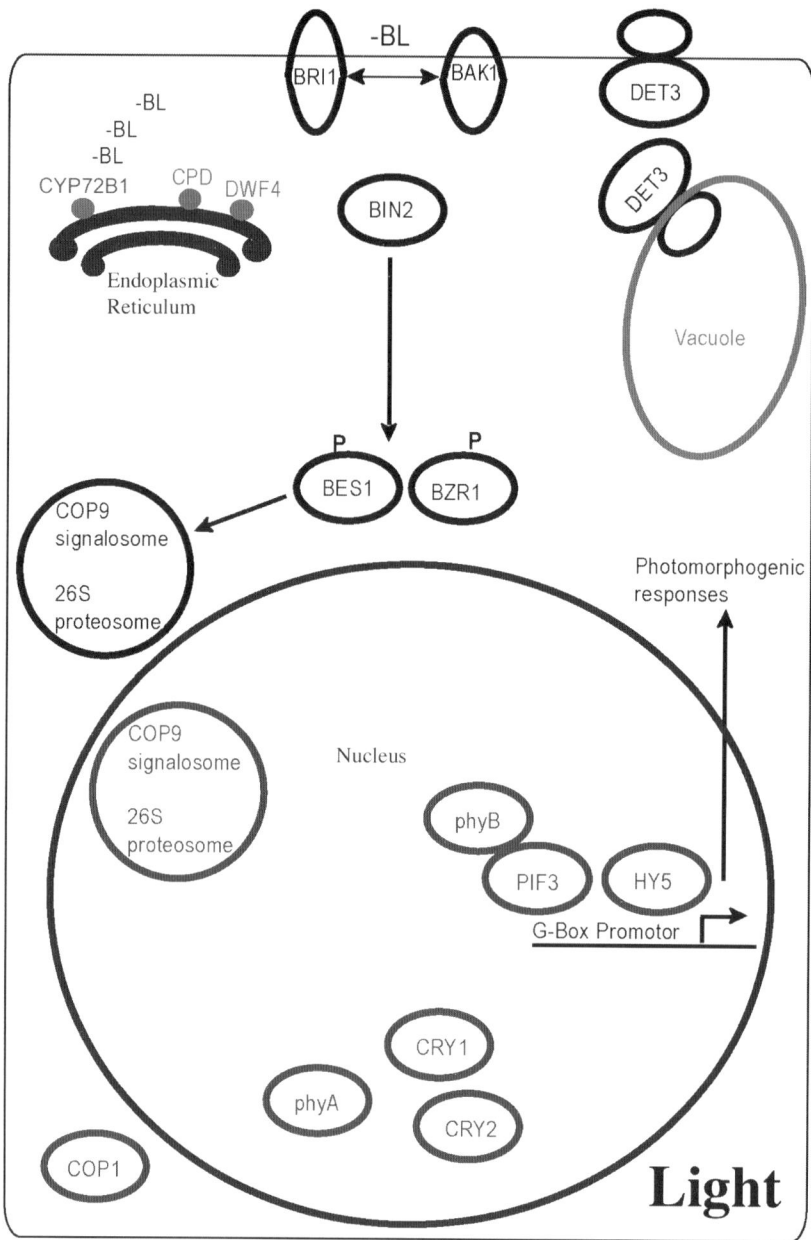

Figure 3b. Model of a hypocotyl cell in the light including proteins of the brassinosteroid and photomorphogenic pathways.

This hypothesis is strengthened by transcriptional repression of the PIF3 photomorphogenic transcription factor upon brassinolide application, coupled with loss of repression by application of a protein synthesis inhibitor. These results suggest that a short-lived protein, which is downstream of brassinolide and possibly degraded by the 26S proteosome, represses PIF3 transcription. In fact COP1 (a mediator of the 26S proteosome) is required for PIF3 protein accumulation in the dark (Bauer et al., 2004). This speculative model will require the identification of photomorphogenic repressors, which are regulated by brassinolide, in order to be resolved.

Hormone inactivation mediated by BAS1/CYP72B1 is another possible molecular mechanism linking the photomorphogenic and brassinosteroid pathways (Figure 3). CYP72B1 mRNA expression is reduced by light as well as by the loss of brassinosteroid biosynthesis or signalling. Transcript accumulation of this gene is elevated in darkness or by brassinolide application or increased brassinolide biosynthesis (such as in a DWARF4 overexpression background) (Choe et al., 2001). This suggests that CYP72B1 is part of the brassinosteroid-receptor mediated feedback loop that modulates brassinosteroid pools, but also suggests that this enzyme is a modulator of photomorphogenesis. This hypothesis is strengthened by a CYP72B1-GUS translational-fusion protein, driven by the genes native promoter, which shows that while the mRNA level is elevated in the dark and far-red light, the protein level is only elevated in far-red light (Turk et al., 2003). This suggests that far-red light is repressing the putative protein instability/degradation in the dark, which would hypothetically increase the overall brassinosteroid inactivation, thus decreasing brassinosteroid signalling, and its resulting photomorphogenic repression. Although loss of CYP72B1 results in some blue and red light mutant phenotypes, the CYP72B1-GUS fusion does not accumulate in those wavelengths of light. This suggests that brassinosteroid levels are differentially modulated by CYP72B1 depending on the wavelength of light. A careful examination of CYP72B1 mRNA and protein levels in the various photoreceptor null backgrounds and under the various wavelengths of light will be necessary to further examine this mechanism.

Differential sensitivity to brassinosteroids depending on the light conditions is a third possible mechanism linking the photomorphogenic and brassinosteroid pathways (Turk et al., 2003) (Figure 3). Mutants defective in brassinosteroid biosynthesis (e.g. det2 and cpd) or perception (e.g. bri1) are short in the dark as well as in the light. This demonstrates that brassinosteroids are required for the hypocotyl to elongate. In fact, addition of brassinolide to the growth media has a stimulatory affect on hypocotyl elongation of light grown seedlings. A seemingly contradictory phenomenon is the ability of exogenous brassinolide to inhibit hypocotyl elongation in the dark at concentrations that have a stimulatory effect in the light. This suggests that the brassinolide response is nearly saturated in the dark, and as a consequence, increases in concentration of this growth-promoting hormone act to inhibit or "poison" the response. This response dichotomy may be due to differential brassinolide sensitivity between the two environments. An analogous mechanism is the dichotomy of response to auxin between the shoot and the root in response to gravity (Ottenschlager et al., 2003). Auxin accumulates in the lower portion of the root to inhibit cell elongation and affect downward bending, while auxin accumulation in the shoot stimulates cell elongation to cause upward bending. Based on this analogy, a working model of brassinosteroid action in seedlings should include a mechanism in which dark-grown seedlings are more sensitive to brassinosteroids than the light grown seedlings. Upon illumination the seedling could become less sensitive to brassinosteroids, allowing a rapid shutting off of the growth promoting effect in the hypocotyl. The level of brassinosteroids could also be decreased to complement the change in sensitivity. No change or even an increase in the level of brassinosteroids would also be compatible with a change-in-sensitivity model.

5. ETHYLENE

The gaseous hormone ethylene has been studied for more than one hundred years. Though not chemically identified at the time, early observations of plant responses to products of combustion implicated a gaseous compound in regulating certain aspects of development including early shedding of urban trees, accelerated flowering in pineapples and the ripening of oranges (for review see Arshad and Fankenberger 2002). Though most commonly known for its role in fruit ripening, ethylene also affects seed germination, seedling development, root development, leaf abscission, floral senescence, disease resistance, responses to wounding, and responses to abiotic stresses caused by ozone and UV-B light (for reviews see Johnson and Ecker 1998; Wang et al., 2002). Dark-grown Arabidopsis seedlings treated either with ethylene or the precursor ACC (1-aminocyclopropane-1-carboxylic acid) respond dramatically with an increased curvature in the apical hook, a decrease in hypocotyl and root elongation and an increase in hypocotyl width. This effect, referred to as the "triple response", has been the basis of screens for mutants that are altered in their responses to ethylene. The identification and characterization of many ethylene insensitive and constitutive "triple response" mutants in Arabidopsis has led to a well-supported model of ethylene perception and signal transduction (for reviews see Johnson and Ecker 1998; Wang et al., 2002).

One component of the triple response seen in dark-grown Arabidopsis seedlings treated with ethylene is the inhibition of hypocotyl elongation. In contrast, ethylene enhances hypocotyl elongation of light-grown seedlings (Smalle et al., 1997). This response is most dramatic when seedlings are grown on nutrient-poor media though still visible with standard growth media. Arabidopsis mutants that are insensitive to ethylene are also less responsive to ethylene treatment in the light with the degree of insensitivity to ethylene treatment in the light correlating in most cases with the degree of insensitivity to ethylene treatment in the dark. These results suggest that ethylene signal transduction in light-grown seedlings is genetically similar to the dark-grown signalling pathways though the "output" is different. Treatment with silver nitrate ($AgNO_3$), an inhibitor of ethylene responses, abolished the hypocotyl response to ethylene in the light. Treatment with $AgNO_3$ also abolished auxin-induced hypocotyl elongation for seedlings grown on nutrient-poor media:

suggesting interplay between ethylene and auxin signalling pathways (Smalle *et al.*, 1997).

5.1 Genetic connections between ethylene and photomorphogenesis

The above response was the basis for a mutant screen on nutrient-poor media, resulting in the identification of the *alh1* mutant that has a slightly longer hypocotyl than the wild type in white light (Vandenbussche *et al.*, 2003a). In white light, hypocotyl responses to the ethylene precursor, ACC, are saturated at lower concentrations in the *alh1* mutant than the wild type. Light-grown *alh1* seedlings also have phenotypes reminiscent of constitutive auxin and ethylene responses suggesting a degree of constitutive ethylene signalling in this mutant. Dark-grown *alh1* seedlings are significantly less sensitive to ACC than the wild type, having a constitutive hook-response that is likely to be caused by the four-fold overproduction of ethylene in these growth conditions. However, *alh1* hypocotyls are not short in the dark, demonstrating only a partial "triple response" in this mutant. ACC inhibits dark-grown wildtype hypocotyl elongation yet has much less of an affect on the *alh1* mutant.

In the light, *alh1* hypocotyls are also much less sensitive to auxin than the wild type, suggesting cross-talk between these two hormone pathways during hypocotyl responses to white light. Light-grown roots of the *alh1* mutant are more responsive to gravitropic stimuli than the wild type, demonstrating an involvement of *alh1* in this auxin response (Vandenbussche *et al.*, 2003a). Though the *alh1* mutant uncovers the complexities of interactions between light, ethylene and auxin (Figure 2), this interpretation should be made with caution since the mutation has not been cloned and the semi-dominant nature of the mutation does not indicate whether this allele is hyper-morphic or hypo-morphic. Interactions between ethylene and auxin signalling have been uncovered by additional mutants in Arabidopsis. The *hls1* mutant was identified based on its hookless seedling phenotype when grown in the dark either in the presence or absence of ethylene (Guzman and Ecker, 1990). The *HOOKLESS* gene has been cloned and mRNA accumulation is increased with ethylene treatments while reduced in an ethylene insensitive mutant (Lehman *et al.*, 1996). Altering auxin transport or levels phenocopies the *hls1* hypocotyl phenotype. Expression patterns of auxin-response genes are also altered in the *hls1* mutant. Together these results suggest that the *HLS1* gene integrates auxin and ethylene response pathways (Lehman *et al.*, 1996). The auxin-resistant mutant, *axr4-1*, is less sensitive to ACC than the wild type at intermediate concentrations though *axr4-1* responds the same as the wild type at high and lower doses, further demonstrating possible interactions between auxin and ethylene signalling in Arabidopsis (Hobbie and Estelle, 1995).

The asa1 (attenuated shade avoidance 1) mutant was identified as a suppressor of the constitutive shade avoidance phenotypes seen in a phyAphyB double mutant. The asa1 mutation is caused by a single-base insertion leading to the introduction of 8 novel amino acids and a premature stop codon in the calossin-like protein, BIG (Kanyuka *et al.*, 2003). BIG is required for polar auxin transport and affects the expression of light-regulated genes (Gil *et al.*, 2001). Mutations in BIG have been identified in auxin (Ruegger *et al.*, 1997) and cytokinin (Kanyuka *et al.*, 2003) related mutant screens as well as mutant screens targeting genes involved in photomorphogenesis (Li *et al.*, 1994; Kanyuka *et al.*, 2003). Multiple mutations in BIG confer aberrant photomorphogenic phenotypes in addition to altered responses to the hormones ethylene, cytokinin, GA and auxin (Kanyuka *et al.*, 2003). Indeed, studies of the BIG gene suggest a complex, non-linear interaction between photomorphogenesis and multiple hormone pathways.

5.2 Ethylene mutants and shade-avoidance

The Arabidopsis *etr1-1* mutation is dominant and confers a blocking of many ethylene-signalling responses (Bleecker *et al.*, 1988; Chang and Meyerowitz, 1995; Ecker, 1995). Tobacco plants transformed with the *etr1-1* mutant gene from Arabidopsis are insensitive to ethylene in a dark-grown seedling triple response assay demonstrating that this mutant gene can be used to block ethylene signalling in heterologous species (Knoester *et al.*, 1998). These ethylene-insensitive, *Tetr*, tobacco plants also fail to reduce their growth in response to crowding, suggesting a role for ethylene in neighbour-perception and shade avoidance in tobacco plants (Knoester *et al.*, 1998). However, these studies leave open the question of whether this affect of ethylene is due to an overall reduced light level or changes in the red/far-red ratio as perceived by the crowded plants. Further studies with the ethylene-insensitive *Tetr* tobacco have shown that these plants are deficient in shade-avoidance responses to low ratios of red/far-red light when grown either in monoculture or in a mixed population with wildtype plants. In both cases, these ethylene-insensitive plants have a clear competitive disadvantage against the wild type (Pierik *et al.*, 2003).

The red/far-red absorbing phytochromes control shade-avoidance in plants. Under natural growth conditions, this is probably the most important process regulated by this family of photoreceptors (Smith, 1995). Genetic and physiological studies of the short-day, photoperiodic grass *Sorghum bicolor* demonstrate interplay between phytochrome B and circadian-regulated ethylene production (Finlayson *et al.*, 1998). When *Sorghum* plants are grown in heavy shade, circadian-ethylene production is enhanced. Under these growth conditions, wildtype plants resemble the constitutive shade-avoidance phenotypes conferred by mutations in *phyB*. One of the mechanisms controlling this process appears to be via phyB-mediated transcriptional regulation of the ethylene biosynthesis enzyme ACC oxidase. This conclusion is based on observations that *Sorghum phyB* mutations confer enhanced circadian ethylene production and enhanced ACC oxidase transcript accumulation (Finlayson *et al.*, 1999).

This connection between phytochromes, ethylene and shade avoidance is also seen in Arabidopsis as ACC oxidase transcript accumulation is regulated by phyA (Tepperman *et al.*, 2001). Furthermore, in Arabidopsis ethylene production is light regulated and a *phyB*-null mutant has elevated levels of ethylene production relative to the wild type. Interestingly, when the *alh1* mutant is grown in dim light, this mutation confers leaf phenotypes that are similar to those seen in *phyB* null mutants (Vandenbussche *et al.*, 2003b). Taken together with the *Sorghum* results described above, it appears that multiple phytochrome pathways intersect with ethylene-mediated signal transduction (Figure 2).

5.3 Ethylene and fruit ripening

Ethylene is well known as a hormone involved in fruit ripening. Tomato fruit ripening is also thought to be modulated by phytochrome. *PHYA* transcript accumulation is dramatically up regulated (approximately 10-fold) during the ripening process in these fruits. In addition, red/far-red light treatments of ripening tomato fruit implicate phytochromes in accumulation of the carotenoid, lycopene (Alba *et al.*, 2000). However, these studies demonstrate that altering red/far-red light treatments do not alter ethylene production in ripening tomato. These results suggest an uncoupling of ethylene production and phytochrome accumulation during the ripening of tomato fruit (Alba *et al.*, 2000) though other possibilities, such as phytochrome-mediated alteration of ethylene sensitivity, have not been ruled out.

The interactions between ethylene signalling and photomorphogenic pathways are clearly complex. This complexity is exacerbated by interactions with other hormone signalling pathways, such as auxins. Mutations in the *BIG* gene suggest interactions between photomorphogenesis and the hormones ethylene, auxin, GA and cytokinin (Ruegger *et al.*, 1997; Gil *et al.*, 2001; Kanyuka *et al.*, 2003). One possible mechanism linking at least ethylene, auxin and GA responses during seedling development involves the DELLA class of nuclear growth repressor proteins (Achard *et al.*, 2003).

6. CYTOKININS

Cytokinins are a class of plant hormones involved in plant cell division, originally identified as factors stimulating this response in tissue culture (Skoog and Miller, 1957). Cytokinins, together with auxins, regulate or modulate a wide variety of developmental responses including seed germination, seedling development, apical dominance and leaf greening and senescence (for review see, Hutchison and Kieber, 2002; Sheen, 2002). There are multiple examples of interactions between photomorphogenic development and cytokinin signal transduction. For example, certain light-dependent phenotypes can be mimicked by high concentrations of exogenous cytokinin applied to dark-grown Arabidopsis seedlings. These responses include changes in gene expression, some aspects of chloroplast development, hypocotyl growth inhibition, induction of cotyledon and leaf development and phenotypically mimicking some aspects of the *deetiolation* mutant *det1* (Chory *et al.*, 1994). The Arabidopsis *amp1* (altered meristem program) mutant, which is allelic with the constitutive photomorphogenesis mutant, *cop2*, confers some aspects of light grown morphology when grown in the dark and has been shown to overproduce cytokinins (Chin-Atkins *et al.*, 1996; Wei and Deng, 1996). However, other *det/cop* mutants, such as the brassinosteroid mutant *det2* (Li *et al.*, 1996; Li and Chory, 1997), contain normal levels of cytokinin.

Cytokinin-mediated inhibition of hypocotyl elongation does not appear to interact directly with photomorphogenic signalling pathways, instead working indirectly through the modulation of ethylene levels (Cary *et al.*, 1995; Su and Howell, 1995). This interaction between cytokinin and ethylene can also be extended to modulating hypocotyl agravitropic responses to red light. Red light causes a loss of negative gravitropic growth in hypocotyls of Arabidopsis seedlings. However, the addition of cytokinin restores the negative gravitropism response of seedlings grown in red light (Golan *et al.*, 1996).

Studies with the *lip1* mutant in pea further implicate a connection between cytokinins and photomorphogenesis. The *lip1* mutation confers phenotypes of light-grown plants when grown in the dark. The addition of cytokinin inhibits this response, phenocopying dark-grown wildtype plants grown in the presence of cytokinin (Seyedi *et al.*, 2001). However, etiolated growth of both the wild type and *lip1* is significantly inhibited by cytokinin applications. Dark-grown *lip1* mutants have reduced levels of phyA. Application of cytokinin increased the levels of phyA in etiolated *lip1* mutants while decreasing levels of the chlorophyll biosynthetic enzyme protochlorophyllide oxidoreductase in both the mutant and the wild type. These results demonstrate that interactions between cytokinin signalling and photomorphogenic pathways, though variable in nature, are present in multiple plant species.

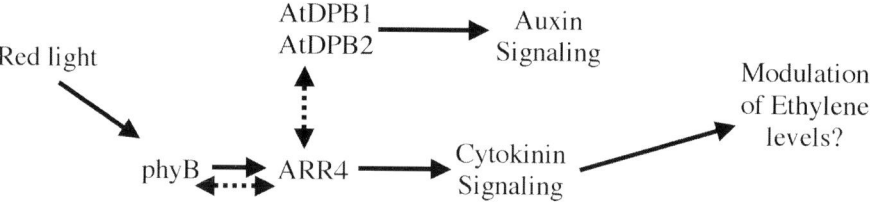

Figure 4. A model incorporating multiple hormone signalling pathways with potential interactions between the cytokinin response regulator ARR4 and phyB as well as the auxin response genes AtDPB1/ AtDPB2. Double arrowed, dashed lines represent possible physical interactions.

The Arabidopsis ARR4 response regulator is likely to represent a direct interaction between cytokinin signalling and phytochrome-mediated photomorphogenesis (Sweere *et al.*, 2001). ARR4, an early response component in cytokinin signalling, is induced by red light via phyB signalling. ARR4 also interacts with the amino terminus of phyB, stabilizing the active Pfr-form. In addition, over-expression of ARR4 confers increased sensitivity to red light (Sweere et al., 2001). ARR4 also interacts with the auxin response genes AtDPB1 and AtDPB2 suggesting a mechanism of interaction between auxin and cytokinin signalling pathways (Yamada et al., 1998). These results support the notion of direct interactions between some aspects of photomorphogenesis and cytokinin signalling and may begin to address the complex interactions between multiple hormone signalling pathways and light-mediated developmental responses in plants (Figure 4). However, this is clearly the tip of the iceberg, as the ARR4 studies do not shed light on cytokinin-mediated induction of photomorphogenic processes in dark-grown plants.

7. SUMMARY

Though this chapter has addressed the role of gibberellin, auxin, brassinosteroid, ethylene and cytokinin hormone signalling pathways as they interact with photomorphogenesis, it is quite likely that other hormones will also be involved with these critical plant responses. Clearly it will be quite difficult to build a simple model describing the complex interaction between multiple photoreceptor-mediated light signalling pathways and each of these phytohormones. Given the diversity of responses to both light and hormones across plant species, it is unlikely that a single, unifying model will ever be supported. However, detailed analysis of each interaction between hormone and photoreceptor as well as the complications caused by interactions amongst the different hormones (and between the multiple photoreceptor signalling pathways) will give us a better handle on how plants exquisitely refine their development in the face of an ever-changing light environment.

Acknowledgement. The authors would like to thank the National Science Foundation (grant no. 0114726), the Department of Energy (grant no. DE-FG02-02ER15340), the Monsanto Corporation (grant no. 46011J) and Washington University for funding our research on photomorphogenesis and hormone signal transduction.

8. FURTHER READING

Bishop, G. J. and Koncz, C. (2002) Brassinosteroids and plant steroid hormone signalling. *Plant Cell, 14*, Suppl: S97-110.
Clouse, S. D. (2002) Brassinosteroids. Plant counterparts to animal steroid hormones? *Vitam Horm, 65*, 195-223.
Friedrichsen, D. and Chory, J. (2001) Steroid signalling in plants: from the cell surface to the nucleus. *Bioessays, 23*, 1028-1036.
Friml, J. and Palme, K. (2002) Polar auxin transport - old questions and new concepts? *Plant Mol Biol, 49*, 273-284.
Halliday, K. and Fankhauser, C. (2003) Phytochrome-hormonal signalling networks. *New Phytologist, 157*, 449-463.
Hutchison, C. E, Kieber, J.J. (2002) Cytokinin signalling in Arabidopsis. *Plant Cell, 14*, S47-S59.
Johnson, P. R., Ecker, J. R. (1998) The ethylene gas signal transduction pathway: a molecular perspective. *Ann. Rev. Genet., 32*, 227-254.

Liscum, E., Stowe-Evans, E. L. (2000) Phototropism: a "simple" physiological response modulated by multiple interacting photosensory-response pathways. *Photochem Photobiol, 72*, 273-282.

Mussig, C. and Altmann, T. (1999) Physiology and molecular mode of action of brassinosteroids. *Plant Physiol Biochem, 37*, 363-372.

Nemhauser, J. L. and Chory, J. (2004) BRing it on: new insights into the mechanism of brassinosteroid action. *J Exp Bot, 55*, 265-270.

Olszewski, N. E., Sun, T., Gubler, F. (2002) Gibberellin Signalling: Biosynthesis, Catabolism and Response Pathways. *Plant Cell Supplement*, S61-S80.

Richards, D. E., King, K. E., Ait-ali, T., Harberd, N. P. (2001) How Gibberellin Regulates Plant Growth and Development: A Molecular Genetic Analysis of Gibberellin Signalling. *Annu Rev Plant Physiol Plant Mol Biol, 52*, 67-88.

Sheen, J. (2002) Phosphorelay and transcription control in cytokinin signal transduction. *Science, 296*, 1650-1652.

Thomas, S. G. and Sun, T. (2004). Update on Gibberellin Signalling. A Tale of the Tall and the Short *Plant Phys, 135*, 668-676.

Thummel, C. S. and Chory, J. (2002) Steroid signalling in plants and insects--common themes, different pathways. *Genes Dev, 16*, 3113-3129.

Tseng, T., Salome, P. A., McClung, C. R., Olszewski, N. E. (2004) SPINDLY and GIGANTEA Interact and Act in *Arabidopsis thaliana* Pathways Involved in Light Responses, Flowering, and Rhythms in Cotyledon Movements. *Plant Cell, 16*, 1550-1583.

Wang, K.L., Li, H., Ecker, J. R. (2002) Ethylene biosynthesis and signalling networks. *Plant Cell, 14*, S131-S151.

9. REFERENCES

Achard, P., Vriezen, W. H., van der Straeten, D., Harberd, N. P. (2003) Ethylene Regulates Arabidopsis Development via the Modulation of DELLA Protein Growth Repressor Function. *Plant Cell, 15*, 2816-2825.

Alabadí, D., Gil, J., Blázquez, M. A., García-Martínez, J. L. (2004) Gibberellins Repress Photomorphogenesis in Darkness. *Plant Physiol, 134*, 1050-1057.

Alba, R., Cordonnier-Pratt, M. M., Pratt, L. H. (2000) Fruit-localized phytochromes regulate lycopene accumulation independently of ethylene production in tomato. *Plant Physiol, 123*, 363-370.

Amador, V., Monte, E., Garcia-Martinez, J. L., Prat, S. (2001) Gibberellins signal nuclear import of PHOR1, a photoperiod-responsive protein with homology to Drosophila armadillo. *Cell, 106*, 343-354.

Arshad, M. and Fankenberger, W. T., *Ethylene, Agricultural Sources and Applications*, New York: Kluwer Academic, 2002.

Asami, T. and Yoshida, S. (1999) Brassinosteroid biosynthesis inhibitors. *Trends Plant Sci, 4*, 348-353.

Ashikari, M., Wu, J., Yano, M., Sasaki, T., Yoshimura, A. (1999) Rice gibberellin-insensitive dwarf mutant gene Dwarf 1 encodes the -subunit of GTP-binding protein. *Proc Natl Acad Sci USA, 96*, 10284-10289.

Assmann, S. M. (2002) Heterotrimeric and unconventional GTP binding proteins in plant cell signalling. *Plant Cell, 14* , Suppl, S355-373.

Ballare, C. L. (1999) Keeping up with the neighbours, phytochrome sensing and other signalling mechanisms. *Trends Plant Sci, 4*, 97-102.

Bauer, D., Viczian, A., Kircher, S., Nobis, T., Nitschke, R., Kunkel, T. *et al.* (2004) Constitutive photomorphogenesis 1 and multiple photoreceptors control degradation of phytochrome interacting factor 3, a transcription factor required for light signalling in Arabidopsis. *Plant Cell, 16*, 1433-1445.

Beato, M., Herrlich, P., Schutz, G. (1995) Steroid hormone receptors-Many actors in search of a plot. *Cell, 83*, 851-857.

Behringer, F. J. and Davies, P. J. (1992) Indole-3-acetic acid levels after phytochrome-mediated changes in the stem elongation rate of dark- and light-grown *Pisum* seedlings. *Planta, 188*, 85-92.

Bishop, G. J. and Koncz, C. (2002) Brassinosteroids and plant steroid hormone signalling. *Plant Cell, 14*, Suppl, S97-110.

Bishop, G. J. and Yokota, T. (2001) Plants steroid hormones, brassinosteroids, current highlights of molecular aspects on their synthesis/metabolism, transport, perception and response. *Plant Cell Physiol, 42*, 114-120.

Blazquez, M. A., Trenor, M., Weigel, D. (2002) Independent control of gibberellin biosynthesis and flowering time by the circadian clock in *Arabidopsis*. *Plant Physiol, 130*, 1770-1775.

Blazquez, M. A. and Weigel, D. (1999) Independent regulation of flowering by phytochrome B and gibberellins in Arabidopsis. *Plant Physiol, 120*, 1025-1032.

Bleecker, A. B., Estelle, M. A., Somerville, C., Kende, H. (1988) Insensitivity to ethylene conferred by a dominant mutation in *Arabidopsis thaliana*. *Science, 241*, 1086-1089.

Carabelli, M., Morelli, G., Whitelam, G., Ruberti, I. (1996) Twilight-zone and canopy shade induction of the Athb-2 homeobox gene in green plants. *Proc Natl Acad Sci USA, 93*, 3530-3535.

Cary, A., Liu, W., Howell, S. (1995) Cytokinin action is coupled to ethylene in its effects on the inhibition of root and hypocotyl elongation in Arabidopsis thaliana seedlings. *Plant Physiol, 107*, 1075-1082.

Chang, C. and Meyerowitz, E. M. (1995) The ethylene hormone response in Arabidopsis, A eukaryotic two-component signalling system. *Proc Natl Acad Sci USA, 92*, 4129-4133.

Chin-Atkins, A. N., Craig, S., Hocart, C. H., Dennis, D. S., Chaudhury, A. M. (1996) Increased endogenous cytokinin in the Arabidopsis *amp1* mutant corresponds with de-etiolation responses. *Planta, 198*, 549-556.

Choe, S., Fujioka, S., Noguchi, T., Takatsuto, S., Yoshida, S., Feldmann, K. A. (2001) Overexpression of DWARF4 in the brassinosteroid biosynthetic pathway results in increased vegetative growth and seed yield in Arabidopsis. *Plant J, 26*, 573-582.

Choi, Y. H., Fujioka, S., Harada, A., Yokota, T., Takatsuto, S., Sakurai, A. (1996) A brassinolide biosynthetic pathway via 6-deoxocastasterone. *Phytochemistry, 43*, 593-596.

Chory, J., Reinecke, D., Sim, S., Washburn, T., Brenner, M. (1994) A role for cytokinins in de-etiolation in Arabidopsis. *det* mutants may have an altered response to cytokinins. *Plant Physiol, 104*, 339-347.

Clouse, S. D. (2002) Brassinosteroids. Plant counterparts to animal steroid hormones? *Vitam Horm, 65*, 195-223.

Clouse, S. D. and Feldmann, K. A. (1999) "Molecular genetics of brassinosteroid action." In *Brassinosteroids, Steroidal Plant Hormones*. Sakurai A., Yokota, T. and Clouse, S. D. (eds) Springer-Verlag, Tokyo, 163-190.

Colon-Carmona, A., Chen, D. L., Yeh, K. C., Abel, S. (2000) Aux/IAA proteins are phosphorylated by phytochrome in vitro. *Plant Physiol, 124*, 1728-1738.

Cowling, R. J., Kamiya, Y., Seto, H., Harberd, N. P. (1998) Gibberellin dose-response regulation of GA4 gene transcript levels in Arabidopsis. *Plant Physiol, 117*, 1195-1203.

Darwin, C. (1881) Das Bewegungsvermögen der Pflanzen, von Charles Darwin mit Unterstützung von Francis Darwin. Aus dem Englischen übers. von J. Victor Carus. Schweizerbart, Stuttgart

Dill, A., Jung, H. S., Sun, T. P. (2001) The DELLA motif is essential for gibberellin-induced degradation of RGA. *Proc Natl Acad Sci USA, 98*, 14162-14167.

Dill, A., Sun, T. (2001) Synergistic derepression of gibberellin signalling by removing RGA and GAI function in Arabidopsis thaliana. *Genetics, 159*, 777-785.

Dill, A., Thomas, S. G., Hu, J., Steber, C. M., Sun, T. (2004) The Arabidopsis F-box Protein SLEEPY1 Targets Gibberellin Signalling Repressors for Gibberellin-Induced Degradation. *Plant Cell, 16*, 1392-1405.

Ecker, J. R. (1995) The ethylene signal transduction pathway in plants. *Science, 268*, 667-675.

Fairchild, C. D., Schumaker, M. A., Quail, P. H. (2000) HFR1 encodes an atypical bHLH protein that acts in phytochrome A signal transduction. *Genes Dev, 14*, 2377-2391.

Fankhauser, C. and Chory, J. (1997) Light control of plant development. *Annu Rev Cell Dev Biol, 13*, 203-229.

Filardo, F. F. and Swain, S. M. (2003) SPYing on GA signalling and Plant Development. *J Plant Growth Regul, 22*, 163-175.

Finlayson, S. A., Jung, I.-J., Mullet, J. E., Morgan, P. W. (1999) The mechanism of rhythmic ethylene production in Sorghum. The role of phytochrome B and simulated shading. *Plant Physiol, 119*, 1083-1089.

Finlayson, S. A., Lee, I.-J., Morgan, P. W. (1998) Phytochrome B and the regulation of circadian ethylene production in sorghum. *Plant Physiol, 116*, 17-25.

Firn, R. D. (1994) "Phototropism." In *Photomorphogenesis in Plants, 2nd ed.* Kluwer Academic, Dordrecht, 659-680.

Folta, K. M., Pontin, M. A., Karlin-Neumann, G., Bottini, R., Spalding, E. P. (2003) Genomic and physiological studies of early cryptochrome 1 action demonstrate roles for auxin and gibberellin in the control of hypocotyl growth by blue light. *Plant J, 36,* 203-214.

Friedrichsen, D. and Chory, J. (2001) Steroid signalling in plants, from the cell surface to the nucleus. *Bioessays, 23,* 1028-1036.

Friedrichsen, D. M., Joazeiro, C. A., Li, J., Hunter, T., Chory, J. (2000) Brassinosteroid-insensitive-1 is a ubiquitously expressed leucine-rich repeat receptor serine/threonine kinase. *Plant Physiol, 123,* 1247-1256.

Friml, J. and Palme, K. (2002) Polar auxin transport - old questions and new concepts? *Plant Mol Biol, 49,* 273-284.

Fu, X., Harberd, N. P. (2003) Auxin promotes Arabidopsis root growth by modulating gibberellin response. *Nature, 421,* 740-743.

Fu, X., Richards, D. E., Ait-Ali, T., Hynes, L. W., Ougham, H., Peng, J., Harberd, N. P. (2002) Gibberellin-mediated proteasome-dependent degradation of the barley DELLA protein SLN1 repressor. *Plant Cell, 14,* 3191-3200.

Fu, X., Richards, D. E., Fleck, B., Xie, D., Burton, N., Harberd, N. P. (2004) The Arabidopsis Mutant sleepy1^{gar2-1} Protein Promotes Plant Growth by Increasing the Affinity of the SCF^{SLY1} E3 Ubiquitin Ligase for DELLA Protein Substrates. *Plant Cell, 16,* 1406-1418.

Fuchs, I., Philippar, K., Ljung, K., Sandberg, G., Hedrich, R. (2003) Blue light regulates an auxin-induced K+-channel gene in the maize coleoptile. *Proc Natl Acad Sci USA, 100,* 11795-11800.

Fujioka, S., Noguchi, T., Watanabe, T., Takatsuto, S., Yoshida, S. (2000) Biosynthesis of brassinosteroids in cultured cells of Catharanthus roseus. *Phytochemistry, 53,* 549-553.

Garcia-Martinez, J. L., Gil, J. (2001) Light Regulation of Gibberellin Biosynthesis and Mode of Action. *J Plant Growth Regul, 20,* 354-368.

Gil, P., Dewey, E., Friml, J., Zhao, Y., Snowden, K. C., Putterill, J., *et al.* (2001) BIG, a calossin-like protein required for polar auxin transport in *Arabidopsis. Genes Dev, 15,* 1985-1997.

Goda, H., Shimada, Y., Asami, T., Fujioka, S., Yoshida, S. (2002) Microarray analysis of brassinosteroid-regulated genes in Arabidopsis. *Plant Physiol, 130,* 1319-1334.

Golan, A., Tepper, M., Soudry, E., Horwitz, B. A., Gepstein, S. (1996) Cytokinin, acting through ethylene, restores gravitropism to Arabidopsis seedlings grown under red light. Plant Physiol, 112, 901-904.

Gray, W. M., Kepinski, S., Rouse, D., Leyser, O., Estelle, M. (2001) Auxin regulates SCF(TIR1)-dependent degradation of AUX/IAA proteins. *Nature, 414,* 271-276.

Gray, W. M., Ostin, A., Sandberg, G., Romano, C. P., Estelle, M. (1998) High temperature promotes auxin-mediated hypocotyl elongation in Arabidopsis. *Proc Natl Acad Sci USA, 95,* 7197-7202.

Grove, M. D., Spencer, G. F., Rohwedder, W. K., Mandava, N., Worley, J. F., Warthen, J. D. Jr., Steffens, G. L., Flippen-Anderson, J. L., Cook, J. C. Jr. (1979) Brassinolide, a plant growth-promoting steroid isolated from Brassica napus rape pollen. *Nature, 281,* 216-217.

Gualberti, G., Papi, M., Bellucci, L., Ricci, I., Bouchez, D., Camilleri, C., *et al.* (2002) Mutations in the Dof zinc finger genes DAG2 and DAG1 influence with opposite effects the germination of Arabidopsis seeds. *Plant Cell, 14,* 1253-1263.

Guzmán, P. and Ecker, J. R. (1990) Exploiting the triple response of Arabidopsis to identify ethylene-related mutants. *Plant Cell, 2,* 513-523.

Halliday, K. and Fankhauser, C. (2003) Phytochrome-hormonal signalling networks. *New Phytologist, 157,* 449-463.

Hardtke, C. S. and Deng, X. W. (2000) The cell biology of the COP/DET/FUS proteins. Regulating proteolysis in photomorphogenesis and beyond? *Plant Physiol, 124,* 1548-1557.

Harper, R. M., Stowe-Evans, E. L., Luesse, D. R., Muto, H., Tatematsu, K., Watahiki, M. K., *et al.* (2000) The NPH4 locus encodes the auxin response factor ARF7, a conditional regulator of differential growth in aerial Arabidopsis tissue. *Plant Cell, 12,* 757-770.

Hartweck, L. M., Scott, C. L., Olszewski, N. E. (2002) Two O-linked N-acetylglucosamine transferase genes of Arabidopsis thaliana L. Heynh. have overlapping functions necessary for gamete and seed development. *Genetics, 161,* 1279-1291.

Hedden, P. (1997) GIBBERELLIN BIOSYNTHESIS, Enzymes, Genes and Their Regulation. *Annu Rev Plant Physiol Plant Mol Biol, 48,* 431-460.

Hedden, P. and Phillips, A. L. (2000) Gibberellin metabolism, new insights revealed by the genes. *Trends Plant Sci, 5*, 523-530.

Hobbie, L. and Estelle, M. (1995) The *axr4* auxin-resistant mutants of Arabidopsis thaliana define a gene important for root gravitropism and lateral root initiation. *Plant J, 7*, 211-220.

Hooley, R. (1998) Plant hormone perception and action, a role for G-protein signal transduction? *Philos Trans R Soc Lond B Biol Sci, 353*, 1425-1430.

Hsieh, H. L., Okamoto, H., Wang, M., Ang, L. H., Matsui, M., Goodman, H., *et al.* (2000) FIN219, an auxin-regulated gene, defines a link between phytochrome A and the downstream regulator COP1 in light control of Arabidopsis development. *Genes Dev, 14*, 1958-1970.

Hutchison, C. E., Kieber, J. J. (2002) Cytokinin signalling in Arabidopsis. *Plant Cell, 14*, S47-S59.

Iino, M. (1982) Inhibitory action of red light on the growth of the maize mesocotyl, evaluation of the auxin hypothesis. *Planta*, 156, 388-395.

Imaizumi, T., Kadota, A., Hasebe, M., Wada, M. (2002) Cryptochrome light signals control development to suppress auxin sensitivity in the moss *Physcomitrella patens*. *Plant Cell, 14*, 373-386.

Inaba, T., Nagano, Y., Nagasaki, T., Sasaki, Y. (2002) Distinct localization of two closely related Ypt3/Rab11 proteins on the trafficking pathway in higher plants. *J Biol Chem, 277*, 9183-9188.

Itoh, H., Matsuoka, M., Steber, C. M. (2003) A role for the ubiquitin-26S-proteasome pathway in gibberellin signalling. *Trends Plant Sci, 8*, 492-497.

Jacobsen, S. E., Binkowski, K. A., Olszewski, N. E. (1996) SPINDLY, a tetratricopeptide repeat protein involved in gibberellin signal transduction in Arabidopsis. *Proc Natl Acad Sci USA, 93*, 9292-9296.

Jacobsen, S. E. and Olszewski, N. E. (1993) Mutations at the SPINDLY locus of Arabidopsis alter gibberellin signal transduction. *Plant Cell, 5*, 887-896.

Jansson, S. (1999) A guide to the Lhc genes and their relatives in *Arabidopsis*. *Trends Plant Sci, 4*, 236-240.

Jensen, P. J., Hangarter, R. P., Estelle, M. (1998) Auxin Transport Is Required for Hypocotyl Elongation in Light-Grown but Not Dark-Grown *Arabidopsis*. *Plant Physiol, 116*, 455-462.

Johnson, P. R. and Ecker, J. R. (1998) The ethylene gas signal transduction pathway, a molecular perspective. *Ann Rev Genet, 32*, 227-254.

Jones, A. M., Cochran, D. S., Lamerson, P. M., Evans, M. L., Cohen, J. D. (1991) Red light-regulated growth. I. Changes in the abundance of indoleacetic acid and a 22-kilodalton auxin-binding protein in the maize mesocotyl. *Plant Physiol, 97*, 352-358.

Kamiya, Y. and Garcia-Martinez, J. L. (1999) Regulation of gibberellin biosynthesis by light. *Curr Opin Plant Biol, 2*, 398-403.

Kang, J. G., Yun, J., Kim, D. H., Chung, K. S., Fujioka, S., Kim, J. I., *et al.* (2001) Light and brassinosteroid signals are integrated via a dark-induced small G protein in etiolated seedling growth. *Cell, 105*, 625-636.

Kanyuka, K., Praekelt, U., Franklin, K. A., Billingham, O. E., Hooley, R., Whitelam, G. C., *et al.* (2003) Mutations in the huge Arabidopsis gene BIG affect a range of hormone and light responses. *Plant J, 35*, 57-70

Kepinski, S. and Leyser, O. (2002) Ubiquitination and auxin signalling, a degrading story. *Plant Cell, 14*, Suppl, S81-95.

Kim, J., Yi, H., Choi, G., Shin, B., Song, P.-S., Choi, G. (2003) Functional characterization of phytochrome interacting factor 3 in phytochrome-mediated light signal transduction. *Plant Cell, 15*, 2399-2407.

Knoester, M., van Loon, L. C., van den Heuvel, J., Hennig, J., Bol, J. F., Linthorst, H. J. M. (1998) Ethylene-insensitive tobacco lacks nonhost resistance against soil-born fungi. *Proc Natl Acad Sci USA, 95*, 1033-1973.

Koornneef, M. and van der Veen, J. H. (1980) Induction and analysis of gibberellin sensitive mutants in Arabidopsis thaliana (L.). *Heynh Theor Appl Genet, 58*, 257-263.

Koornneef, M., Rolff, E., Spruit, C. J. P. (1980) Genetic Control of Light-inhibited Hypocotyl Elongation in Arabidopsis thaliana (L.). *Heynh Z Pflanzenphysiol, 100S*, 147-160

Krebbers, E., Seurinck, J., Herdies, L., Cashmore, A., Timko, M. (1998) Four genes in two diverged subfamilies encode the ribulose-1,5-bisphosphate carboxylase small subunit polypeptides of Arabidopsis thaliana. *Plant Mol Biol, 11*, 745-759.

Krizek, D. T. and Mandava, N. B. (1983) Influence of spectral quality on the growth response of intact bean plants to brassinosteroid, a growth-promoting steroidal lactone. I. Stem elongation and morphogenesis Phaseolus vulgaris. *Physiologia plantarum, 57*, 317-323.

Lee, D. J. and Zeevaart, J. A. (2002) Differential regulation of RNA levels of gibberellin dioxygenases by photoperiod in spinach. *Plant Physiol, 130,* 2085-2094.

Lehman, A., Black, R., Ecker, J. R. (1996) *HOOKLESS1*, an ethylene response gene, is required for differential cell elongation in the Arabidopsis hypocotyl. *Cell, 85,* 183-194.

Li, H. M., Altschmied, L., Chory, J. (1994) Arabidopsis mutants define downstream branches in the phototransduction pathway. *Genes Dev, 8,* 339-349.

Li, J. and Chory, J. (1997) A putative leucine-rich repeat receptor kinase involved in brassinosteroid signal transduction. *Cell, 90,* 929-938.

Li, J., Nagpal, P., Vitart, V., McMorris, T. C., Chory, J. (1996) A role for brassinosteroids in light-dependent development of Arabidopsis. *Science, 272,* 398-401.

Liscum, E. and Briggs, W. R. (1996) Mutations of Arabidopsis in potential transduction and response components of the phototropic signalling pathway. *Plant Physiol, 112,* 291-296.

Liscum, E. and Reed, J. W. (2002) Genetics of Aux/IAA and ARF action in plant growth and development. *Plant Mol Biol, 49,* 387-400.

Liscum, E., Stowe-Evans, E. L. (2000) Phototropism, a "simple" physiological response modulated by multiple interacting photosensory-response pathways. *Photochem Photobiol, 72,* 273-282.

Luccioni, L. G., Oliverio, K. A., Yanovsky, M. J., Boccalandro, H. E., Casal, J. J. (2002) Brassinosteroid mutants uncover fine tuning of phytochrome signalling. *Plant Physiol, 128,* 173-181.

Martinez-Garcia, J. F., Huq, E., Quail, P. H. (2000) Direct targeting of light signals to a promoter element-bound transcription factor. *Science, 288,* 859-863.

Marumo, S., Hattori, H., Nanoyama, Y., Munakata, K. (1968) The presence of novel plant growth regulators in leaves of *Distylium racemosum*. *Agricultural Biological Chemistry, 32,* 528-529.

McGinnis, K. M., Thomas, S. G., Soule, J. D., Strader, L. C., Zale, J. M., Sun, T. P., *et al.* (2003) The Arabidopsis SLEEPY1 gene encodes a putative F-box subunit of an SCF E3 ubiquitin ligase. *Plant Cell, 15,* 1120-1130.

Mitchell, J. W., Mandava, N., Worley, J. F., Plimmer, J. R., Smith, M. V. (1970) Brassins, A new family of plant hormones from rape pollen. *Nature, 225,* 1065-1066.

Morelli, G. and Ruberti, I. (2002) Light and shade in the photocontrol of Arabidopsis growth. *Trends Plant Sci, 7,* 399-404.

Mussig, C. and Altmann, T. (1999) Physiology and molecular mode of action of brassinosteroids. *Plant Physiol Biochem, 37,* 363-372.

Mussig, C. and Altmann, T. (2001) Brassinosteroid signalling in plants. *Trends Endocrinol Metab, 12,* 398-402.

Nagata, N., Min, Y. K., Nakano, T., Asami, T., Yoshida, S. (2000) Treatment of dark-grown Arabidopsis thaliana with a brassinosteroid-biosynthesis inhibitor, brassinazole, induces some characteristics of light-grown plants. *Planta, 211,* 781-790.

Nagpal, P., Walker, L. M., Young, J. C., Sonawala, A., Timpte, C., Estelle, M., *et al.* (2000) AXR2 encodes a member of the Aux/IAA protein family. *Plant Physiol, 123,* 563-574.

Nakazawa, M., Yabe, N., Ichikawa, T., Yamamoto, Y. Y., Yoshizumi, T., Hasunuma, K., *et al.* (2001) DFL1, an auxin-responsive GH3 gene homologue, negatively regulates shoot cell elongation and lateral root formation, and positively regulates the light response of hypocotyl length. *Plant J, 25,* 213-221.

Neff, M. M., Nguyen S, M., Malancharuvil, E. J., Fujioka, S., Noguchi, T., Seto, H., *et al.* (1999) BAS1, A gene regulating brassinosteroid levels and light responsiveness in Arabidopsis. *Proc Natl Acad Sci USA, 96,* 15316-15323.

Nemhauser, J. L. and Chory, J. (2004) BRing it on, new insights into the mechanism of brassinosteroid action. *J Exp Bot, 55,* 265-270.

Ni, M., Tepperman, J. M., Quail, P. H. (1999) Binding of phytochrome B to its nuclear signalling partner PIF3 is reversibly induced by light. *Nature, 400,* 781-784.

Noguchi, T., Fujioka, S., Choe, S., Takatsuto, S., Tax, F. E., Yoshida, S., *et al.* (2000) Biosynthetic pathways of brassinolide in Arabidopsis. *Plant Physiol, 124,* 201-209.

O'Neill, D. P., Ross, J. J., Reid, J. B. (2000) Changes in gibberellin A(1) levels and response during de-etiolation of pea seedlings. *Plant Physiol, 124,* 805-812.

Ogawa, M., Hanada, A., Yamauchi, Y., Kuwahara, A., Kamiya, Y., Yamaguchi, S. (2003) Gibberellin biosynthesis and response during Arabidopsis seed germination. *Plant Cell, 15,* 1591-1604.

Olszewski, N., Sun, T. P., Gubler, F. (2002) Gibberellin signalling, biosynthesis, catabolism, and response pathways. *Plant Cell, 14,* Suppl, S61-80.

Ottenschlager, I., Wolff, P., Wolverton, C., Bhalerao, R. P., Sandberg, G., Ishikawa, H., *et al.* (2003) Gravity-regulated differential auxin transport from columella to lateral root cap cells. *Proc Natl Acad Sci USA, 100*, 2987-2991.

Oyama, T., Shimura, Y., Okada, K. (1997) The Arabidopsis HY5 gene encodes a bZIP protein that regulates stimulus-induced development of root and hypocotyl. *Genes Dev, 11*, 2983-2995.

Peng Ja, H. N. P. (2002) The role of GA-mediated signalling in the control of seed germination. *Current Opinion in Plant Biology, 5*, 376-381.

Philippar, K., Fuchs, I., Luthen, H., Hoth, S., Bauer, C. S., Haga, K., *et al.* (1999) Auxin-induced K+ channel expression represents an essential step in coleoptile growth and gravitropism. *Proc Natl Acad Sci USA, 96*, 12186-12191.

Pierik, R., Visser, E. J. W., De Kroon, H., Voesenek, L. A. C. J. (2003) Ethylene is required in tobacco to successfully compete with proximate neighbors. *Plant Cell Envir, 26*, 1229-1234.

Reed, J. W. (2001) Roles and activities of Aux/IAA proteins in Arabidopsis. *Trends Plant Sci, 6*, 420-425.

Reed, J. W., Elumalai, R. P., Chory, J. (1998) Suppressors of an Arabidopsis thaliana phyB mutation identify genes that control light signalling and hypocotyl elongation. *Genetics, 148*, 1295-1310.

Reed, J. W., Foster, K. R., Morgan, P. W., Chory, J. (1996) Phytochrome B affects responsiveness to gibberellins in Arabidopsis. *Plant Physiol, 112*, 337-342.

Reid, J. B., Botwright, N. A., Smith, J. J., O'Neill, D. P., Kerckhoffs, L. H. (2002) Control of gibberellin levels and gene expression during de-etiolation in pea. *Plant Physiol, 128*, 734-741.

Richards, D. E., King, K. E., Ait-ali, T., Harberd, N. P. (2001) How Gibberellin Regulates Plant Growth and Development, A Molecular Genetic Analysis of Gibberellin Signalling. *Annu Rev Plant Physiol Plant Mol Biol, 52*, 67-88.

Romano, C. P., Robson, P. R., Smith, H., Estelle, M., Klee, H. (1995) Transgene-mediated auxin overproduction in Arabidopsis, hypocotyl elongation phenotype and interactions with the hy6-1 hypocotyl elongation and axr1 auxin-resistant mutants. *Plant Mol Biol, 27*, 1071-1083.

Ruegger, M., Dewey, E., Hobbie, L., Brown, D., Bernasconi, P., Turner, J., *et al.* (1997) Reduced naphthylphthalamic acid binding in the tir3 mutant of Arabidopsis is associated with a reduction in polar auxin transport and diverse morphological defects. *Plant Cell, 9*, 745-757.

Saibo, N. J., Vriezen, W. H., Beemster, G. T., van der Straeten, D. (2003) Growth and stomata development of Arabidopsis hypocotyls are controlled by gibberellins and modulated by ethylene and auxins. *Plant J, 33*, 989-1000.

Salamini, F. (2003) Plant Biology. Hormones and the green revolution. *Science, 302*, 71-72.

Sanchez-Bravo, J., Ortuno, A. M., Botia, J. M., Acosta, M., Sabater, F. (1992) The Decrease in Auxin Polar Transport Down the Lupin Hypocotyl Could Produce the Indole-3-Acetic Acid Distribution Responsible for the Elongation Growth Pattern. *Plant Physiol, 99*, 108-114.

Sasaki, A., Itoh, H., Gomi, K., Ueguchi-Tanaka, M., Ishiyama, K., Kobayashi, M., *et al.* (2003) Accumulation of phosphorylated repressor for gibberellin signalling in an F-box mutant. *Science, 299*, 1896-1898.

Schwechheimer, C., Serino, G., Callis, J., Crosby, W. L., Lyapina, S., Deshaies, R. J., *et al.* (2001) Interactions of the COP9 signalosome with the E3 ubiquitin ligase SCFTIRI in mediating auxin response. *Science, 292*, 1379-1382.

Seyedi, M., Selstam, E., Timko, M. P., Sundqvist, C. (2001) The cytokinin 2-isopentenyladenine causes partial reversion to skotomorphogenesis and induces formation of prolamellar bodies and protochlorophyllide 657 in the *lip1* mutant of pea. *Physiol Plant, 112*, 261-272.

Sheen, J. (2002) Phosphorelay and transcription control in cytokinin signal transduction. *Science, 296*, 1650-1652.

Simpson, G. G. and Dean, C. (2002) Arabidopsis, the Rosetta stone of flowering time? *Science, 296*, 285-289.

Skoog, F. and Miller, C. O. (1957) Chemical regulation of growth and organ formation in plant tissues cultured *in vitro*. *Symp Soc Exp Biol, 11*, 118-131.

Smalle, J., Haegman, M., Kurepa, J., van Montague, M., Straeten, D. V. (1997) Ethylene can stimulate Arabidopsis hypocotyl elongation in the light. *Proc Nat Acad Sci USA, 94*, 2756-2761.

Smith, H. (1995) Physiological and ecological function within the phytochrome family. *Annu Rev Plant Physiol Plant Mol Biol, 46*, 289-315.

Staswick, P. E., Tiryaki, I., Rowe, M. L. (2002) Jasmonate response locus JAR1 and several related Arabidopsis genes encode enzymes of the firefly luciferase superfamily that show activity on jasmonic, salicylic, and indole-3-acetic acids in an assay for adenylation. *Plant Cell, 14*, 1405-1415.

Steindler, C., Matteucci, A., Sessa, G., Weimar, T., Ohgishi, M., Aoyama, T., *et al.* (1999) Shade avoidance responses are mediated by the ATHB-2 HD-zip protein, a negative regulator of gene expression. *Development, 126*, 4235-4245.

Stowe-Evans, E. L., Harper, R. M., Motchoulski, A. V., Liscum, E. (1998) NPH4, a conditional modulator of auxin-dependent differential growth responses in Arabidopsis. *Plant Physiol, 118*, 1265-1275.

Su, W. and Howell, S. (1995) The effects of cytokinin and light on hypocotyl elongation in Arabidopsis seedlings are independent and additive. *Plant Physiol, 108*, 1423-1430.

Sun, T. (2000) Gibberellin signal transduction. *Curr Opin Plant Biol, 3*, 374-380.

Swain, S. M., Tseng, T. S., Olszewski, N. E. (2001) Altered expression of SPINDLY affects gibberellin response and plant development. *Plant Physiol, 126*, 1174-1185.

Swain, S. M., Tseng, T. S., Thornton, T. M., Gopalraj, M., Olszewski, N. E. (2002) SPINDLY is a nuclear-localized repressor of gibberellin signal transduction expressed throughout the plant. *Plant Physiol, 129*, 605-615.

Sweere, U., Eichenberg, K., Lohrmann, J., Mira-Rodado, V., Bäurle, I., Kudla, J., *et al.* (2001) Interaction of the response regulator ARR4 with the photoreceptor phytochrome B in modulating red light signalling. *Science, 294*, 1108-1111.

Symons, G. M. and Reid, J. B. (2003) Hormone levels and response during de-etiolation in pea. *Planta, 216*, 422-431.

Symons, G. M., Schultz, L., Kerckhoffs, L. H. J., Davies, N. W., Gregory, D., Reid, J. B. (2002) Uncoupling brassinosteroid levels and de-etiolation in pea. *Physiologia Plantarum, 115*, 311-319.

Szekeres, M. and Koncz, C. (1998) Biochemical and genetic analysis of brassinosteroid metabolism and function in Arabidopsis. *Plant Physiol Biochem, 36*, 145-155.

Szekeres, M., Nemeth, K., Koncz-Kalman, Z., Mathur, J., Kauschmann, A., Altmann, T., *et al.* (1996) Brassinosteroids rescue the deficiency of CYP90, a cytochrome P450, controlling cell elongation and de-etiolation in Arabidopsis. *Cell, 85*, 171-182.

Tanaka, S., Mochizuki, N., Nagatani, A. (2002) Expression of the AtGH3a gene, an Arabidopsis homologue of the soybean GH3 gene, is regulated by phytochrome B. *Plant Cell Physiol, 43*, 281-289.

Tepperman, J. M., Zhu, T., Chang, H. S., Wang, X., Quail, P. H. (2001) Multiple transcription-factor genes are early targets of phytochrome A signalling. *Proc Natl Acad Sci USA, 98*, 9437-9442.

Thummel, C. S. and Chory, J. (2002) Steroid signalling in plants and insects - common themes, different pathways. *Genes Dev, 16*, 3113-3129.

Tian, Q. and Reed, J. W. (1999) Control of auxin-regulated root development by the Arabidopsis thaliana SHY2/IAA3 gene. *Development, 126*, 711-721.

Tian, Q., Uhlir, N. J., Reed, J. W. (2002) Arabidopsis SHY2/IAA3 inhibits auxin-regulated gene expression. *Plant Cell, 14*, 301-319.

Toyomasu, T., Kawaide, H., Mitsuhashi, W., Inoue, Y., Kamiya, Y. (1998) Phytochrome regulates gibberellin biosynthesis during germination of photoblastic lettuce seeds. *Plant Physiol, 118*, 1517-1523.

Tseng, T., Salomé, P. A., McClung, C. R., Olszewski, N. E. (2004) SPINDLY and GIGANTEA Interact and Act in *Arabidopsis thaliana* Pathways Involved in Light Responses, Flowering, and Rhythms in Cotyledon Movements. *Plant Cell, 16*, 1550-1563.

Turk, E. M., Fujioka, S., Seto, H., Shimada, Y., Takatsuto, S., Yoshida, S., *et al.* (2003) CYP72B1 inactivates brassinosteroid hormones, an intersection between photomorphogenesis and plant steroid signal transduction. *Plant Physiol, 133*, 1643-1653.

Ueguchi-Tanaka, M., Fujisawa, Y., Kobayashi, M., Ashikari, M., Iwasaki, Y., Kitano, H., *et al.* (2000) Rice dwarf mutant d1, which is defective in the alpha subunit of the heterotrimeric G protein, affects gibberellin signal transduction. *Proc Natl Acad Sci USA, 97*, 11638-11643.

Ullah, H., Chen, J. G., Wang, S., Jones, A. M. (2002) Role of a heterotrimeric G protein in regulation of Arabidopsis seed germination. *Plant Physiol, 129*, 897-907.

Vandenbussche, F., Smalle, J., Le, J., Saibo, N. J., De Paepe, A., Chaerle, L., *et al.* (2003a) The Arabidopsis mutant alh1 illustrates a cross talk between ethylene and auxin. *Plant Physiol, 131*, 1228-1238.

Vandenbussche, F., Vriezen, W. H., Smalle, J., Laarhoven, L. J., Harren, F. J., van der Straeten, D. (2003b) Ethylene and auxin control the Arabidopsis response to decreased light intensity. *Plant Physiol*, *133*, 517-527.

Voigt, B., Whiting, P., Dinan, L. (2001) The ecdysteroid agonist/antagonist and brassinosteroid-like activities of synthetic brassinosteroid/ecdysteroid hybrid molecules. *Cell Mol Life Sci*, *58*, 1133-1140.

Walton, J. and Ray, P. (1981) Evidence for Receptor Function of Auxin Binding Sites in Maize. *Plant Physiol*, *68*, 1334-1338.

Wang, K. L., Li, H., Ecker, J. R. (2002) Ethylene biosynthesis and signalling networks. *Plant Cell*, *14*, S131-S151.

Wei, N. and Deng, X. W. (1996) The role of the COP/DET/FUS genes in light control of Arabidopsis seedling development *Plant Physiol*, *112*, 871-878.

Wilson, A. K., Pickett, F.B., Turner, J. C., Estelle, M. (1990) A dominant mutation in Arabidopsis confers resistance to auxin, ethylene and abscisic acid. *Mol Gen Genet*, *222*, 377-383.

Wilson, R. N. and Somerville, C. R. (1995) Phenotypic Suppression of the Gibberellin-Insensitive Mutant (gai) of Arabidopsis. *Plant Physiol*, *108*, 495-502.

Yamada, H., Hanaki, N., Imamura, A., Ueguchi, C., Mizuno, T. (1998) An Arabidopsis protein that interacts with the cytokinin-inducible response regulator, ARR4, implicated in the His-Asp phosphorelay signal transduction. *FEBS Lett*, *436*, 76-80.

Yamaguchi, S. and Kamiya, Y. (2001) Gibberellins and Light-Stimulated Seed Germination. *J Plant Growth Regul*, *20*, 369-376.

Yamaguchi, S., Smith, M. W., Brown, R. G., Kamiya, Y., Sun, T. (1998) Phytochrome regulation and differential expression of gibberellin 3beta-hydroxylase genes in germinating Arabidopsis seeds. *Plant Cell*, *10*, 2115-2126.

Zentella, R., Yamauchi, D., Ho, T. H. (2002) Molecular dissection of the gibberellin/abscisic acid signalling pathways by transiently expressed RNA interference in barley aleurone cells. *Plant Cell*, *14*, 2289-2301.

Zhao, Y. and Chory, J. (2001) A link between the light and gibberellin signalling cascades. *Dev Cell*, *1*, 315-316.

Zhu, Y., Tepperman, J. M., Fairchild, C. D., Quail, P. H. (2000) Phytochrome B binds with greater apparent affinity than phytochrome A to the basic helix-loop-helix factor PIF3 in a reaction requiring the PAS domain of PIF3. *Proc Natl Acad Sci USA*, *97*, 13419-13424.

PART 5: SELECTED TOPICS

Chapter 22

THE ROLES OF PHYTOCHROMES IN ADULT PLANTS

Keara A. Franklin and Garry C. Whitelam
Department of Biology, University of Leicester, Leicester, LE1 7RH, United Kingdom
(e-mail: gcw1@leicester.ac.uk)

1. INTRODUCTION

All living organisms perceive environmental signals and use the acquired information to modify their behaviour or development. As obligatorily sessile organisms, plants need to be especially plastic in their development in order to optimise their growth in response to predictable and unpredictable environmental changes. Being photoautotrophic organisms that depend upon photosynthesis for their survival, plants are especially sensitive to variations in the light environment. It is therefore not surprising that plants monitor a wide range of light signals, including the amount, quality and direction of ambient light and use the information to modulate their growth and development. By interaction with the endogenous circadian oscillator light signals also provide plants with a means to monitor the length of the day (photoperiod). This allows plants to anticipate the light/dark cycle, as well as providing a means to monitor the changing seasons. Light signals are perceived via specialised information-transducing photoreceptors. In the higher plants, three such families have been identified and characterised. These are the phytochromes (see Chapter 7) that absorb light predominantly in the red (R) and far-red (FR, i.e. wavelengths beyond 700nm) regions of the spectrum and the cryptochromes (see Chapter 11; Cashmore *et al.*, 1999) and phototropins (see Chapter 12; Briggs and Huala, 1999) that absorb light maximally in the UV-A/blue regions of the spectrum. Plants are also likely to possess one or more specialised UV-B photoreceptors (see Chapter 14), although the identity of any such photoreceptors remains elusive. The phytochromes are unique in being reversibly photochromic photoreceptors. Higher plants contain multiple, discrete but related phytochromes, the apoproteins of which are encoded by a small family of divergent genes (see Quail, 1994). All of the higher plant phytochromes are thought to share the same basic structure, being of a dimer of identical ~120 kDa polypeptides. Each of the monomers carries a single, covalently-linked linear tetrapyrrole chromophore (phytochromobilin), attached via a thioether bond to a conserved cysteine residue in the N-terminal globular domain of the protein (Furuya and Song, 1994). The more elongated, non-chromophorylated, C-terminal domain of the protein is involved in dimerization (Edgerton and Jones, 1992) and may perform regulatory functions. The phytochromes can exist in either of two relatively stable isoforms: a R light-absorbing form, Pr, with an absorption maximum at about 660 nm, or a FR light-absorbing form, Pfr, with an absorption maximum at about 730 nm. It is the Pfr form of phytochrome that is generally considered to be biologically active, whilst Pr is considered to be inactive. The absorption spectra of Pr and Pfr show considerable overlap throughout the visible light spectrum, thus *in vivo* phytochromes will exist in an equilibrium mixture of the two forms under almost all irradiation conditions.

The size of the phytochrome family varies among different plant species. All higher plants studied, as well as several lower plants, possess a family of discrete phytochromes. In the angiosperms, there appear to be three major phytochrome types, phytochromes A, B, and C, the apoproteins of which are encoded by the *PHYA, PHYB,* and *PHYC* genes (Mathews and Sharrock, 1997). Phylogenetic analyses indicate that these genes are well separated from one another in the earliest flowering plants, indicating that the gene duplications from which they arose occurred near the origin of flowering plants (Mathews *et al.*, 1995; Mathews and Sharrock, 1997). In dicotyledonous plants, additional *PHY* genes are found, presumably the products of more recent gene duplication events. In particular, the dicots are characterized by the possession of *PHYB*-like pairs of genes that are considered to have arisen independently in different taxa (Mathews *et al.*, 1995). In the model plant species *Arabidopsis thaliana,* five apophytochrome encoding genes (*PHYA -PHYE*) have been characterised (Sharrock and Quail, 1989; Clack *et al.*, 1994). The protein products of the Arabidopsis *PHYB* and *PHYD* genes share ~80% amino acid sequence identity and *PHYB* and *PHYD* are considered to be the result of a gene duplication in a recent progenitor of the Cruciferae (Mathews and Sharrock, 1997). The PHYB and PHYD proteins are slightly more related to PHYE (~55% identity) than they are to either the PHYA or PHYC proteins (~47% identity). Thus, the *PHYB*, *PHYD*, and *PHYE* genes are considered to form a more recently subgroup of the Arabidopsis *PHY* gene family (Goosey *et al.*, 1997; see Chapter 7 for further details).

The *PHYA* gene has been shown to encode the apoprotein of phytochrome A, the well-characterised phytochrome that predominates in etiolated seedlings and that is subject to relatively rapid proteolytic degradation upon photoconversion to Pfr (Quail, 1994). The other *PHY (B-E)* genes encode the apoproteins of lower abundance phytochromes that appear to less susceptible to proteolysis in the Pfr form. Counterparts of *PHYA, PHYB* and other *PHY* genes have been isolated from several other plant species (Quail, 1994).

Phytochromes adopt distinct and overlapping regulatory roles throughout photomorphogenesis, acting both independently and in co-operation with other photoreceptors. Physiological responses regulated by phytochromes include seed germination and seedling establishment through to the regulation of mature plant architecture and the onset of reproduction. Here we focus on the role of phytochromes in the perception of light quality changes and the resultant modulation of growth and development of adult plants in natural light environments.

2. THE NATURAL LIGHT ENVIRONMENT

The solar radiation that reaches the earth's atmosphere has a spectral photon distribution similar to that of a black-body radiator at a surface temperature of 5,800°K. Prior to reaching the earth's surface this radiation is significantly attenuated within the atmosphere. In particular, longer wavelength regions of the visible spectrum and the FR region are strongly absorbed by oxygen and water vapour and short-wavelength radiation is selectively attenuated by the ozone layer (Smith 1975). A typical spectral energy distribution of incident scattered daylight is shown in Figure 1. As long as the sun is more than about 10° above the horizon, then the spectral energy distribution of daylight remains fairly constant, being only modestly influenced by cloud cover or haze. In relation to the phytochromes, a useful parameter the describe the natural light environment is the ratio of photon irradiance in the R, to that in the FR (R:FR ratio). Of course, this parameter is directly related to the properties of phytochrome and is often precisely defined as follows:

$$\text{R:FR ratio} = \frac{\text{Photon irradiance between 660 and 670 nm}}{\text{Photon irradiance between 725 and 735 nm}}$$

The R:FR ratio of scattered daylight is typically around 1.15 and this value varies little with prevailing weather conditions or the time of the year (Smith 1982).

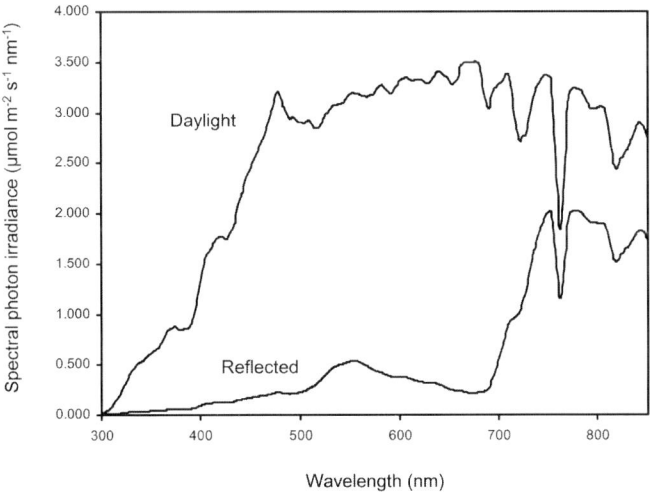

Figure 1. The spectral photon distribution of daylight and the radiation reflected from a stand of wheat seedlings.

The daylight spectrum shown in Figure 1 was obtained on a clear, sunny day in Leicester and the R:FR ratio was 1.14. The progression of the sun across the sky leads to daily fluctuations the spectral quality of daylight. At solar elevations of less that 10°, as seen at dawn and dusk, the increasing path length through the earth's atmosphere leads to enhanced absorption and scattering, as well as refraction of the solar beam by the atmosphere, resulting in preferential enhancement of longer wavelengths. This, together with an increased contribution of scattered skylight, leads to a twilight spectrum that is relatively enriched in the blue and the FR regions, but relatively poor in the orange-red regions. Thus, the onset of dusk can be associated with a significant drop in R:FR ratio from about 1.15 to about 0.7-0.8.

The spectral energy distribution of daylight is very dramatically altered by vegetation. The photosynthetic pigments, chlorophylls and carotenoids, absorb light over most of the visible spectrum, although some of the green light is reflected or transmitted. This is, of course, why leaves appear green to our eyes. Radiation in the FR region is very poorly absorbed by vegetation and consequently, the light that is transmitted through or reflected from, vegetation displays a significantly reduced R:FR ratio compared with daylight (Figure 1). Reported R:FR ratios underneath vegetational canopies are typically in the range 0.09-0.7 (Smith, 1982). The spectrum of light reflected from a stand of wheat seedlings is shown in Figure 1. Here the R:FR ratio has been reduced to 0.169. The reduction in R:FR ratio of the light transmitted through, or reflected from vegetation, can be detected by nearby plants as a change in the relative proportions of Pr and Pfr and so provides a unique and unambiguous signal that potential competitors are nearby (see below). Furthermore, the extent of the reduction in R:FR ratio is quantitatively related to the density and proximity of the neighbouring vegetation (Smith and Whitelam 1997). The only other circumstance where the natural radiation environment shows significant alteration in R:FR ratio is underwater. Water displays strong absorption bands in the FR at about 730 nm and the in the infra-red region. There, as daylight passes through water there is selective attenuation of FR such that with increasing depth there is an increase in R:FR ratio. Of course, many natural waters contain organic material and this can lead to significant attenuation of the blue and R regions.

3. R:FR RATIO AND SHADE AVOIDANCE

In all situations where plants grow in close proximity to one another there will be constraints on the availability of radiant energy to drive photosynthesis and as a consequence, there will be competition for this key resource. Success in this competition will be a crucial element in the survival of the individual plant. In simple terms, plants have evolved two principal strategies to enhance their survival prospects under such conditions; they may tolerate shading by other plants or they may avoid shading. Many angiosperms, including crop plants, have evolved a remarkable capacity to avoid vegetational shade. One of the key elements of shade avoidance is the ability of plants to acquire information about the proximity of potentially competing vegetation and then to initiate appropriate modulations of growth and development in order to minimise the likelihood of shading. It is well established that the reduction in R:FR ratio of the radiation within plant communities is a key environmental cue enabling plants to detect neighbouring vegetation. The reversibly photochromic phytochromes, with absorption maxima in the R and the FR, are superbly adapted to perceive alterations in R:FR ratio.

It is probable that the earliest report establishing the link between R:FR ratio, the phytochromes and the avoidance of vegetational shade was that of Cumming in 1963. This study showed that germination of seeds of the weed species *Chenopodium rubrum* was modulated by R:FR ratio and it was speculated that this could provide a mechanism to limit germination in areas of intense shade. Our current concepts on the role of R:FR ratio in initiating the altered patterns of growth and development of plants that are regarded as the 'shade avoidance syndrome' (Smith and Whitelam, 1997) are derived from the pioneering work of Harry Smith and colleagues initiated during the 1970s. The starting point for the development of these ideas was a series of detailed quantitative measurements of natural radiation spectra, above and below plant canopies, and their description in terms of R:FR ratio (e.g. Holmes and Smith, 1975). These natural variations in R:FR ratio were then related to the photoequilibrium (i.e. Pfr/Pr+Pfr) status of phytochrome, with the relationship being a rectangular hyperbola (Holmes and Smith, 1977). The nature of this relationship ensures that reductions in R:FR ratio below a value of about 1.0 (i.e. typical of daylight) lead to the greatest reductions in Pfr/Pr+Pfr. Thus, the phytochrome system is exquisitely sensitive to the variations in R:FR ratio that characterise the terrestrial radiation environment.

In order to establish the importance of the R:FR ratio signal in the perception of vegetational shade light, a whole series of physiological experiments were performed in which plants were grown in artificial light environments where R:FR ratio was varied. This was achieved using the output of white fluorescent tubes to provide a constant background of photosynthetically active radiation to which varying amounts of supplementary FR light was added (Morgan and Smith, 1976, 1978, 1981). These experiments led to the characterisation of a whole range of growth and developmental responses to reductions in R:FR ratio that showed a close correlation with the growth responses of plants subjected to vegetational shading in the natural environment. Furthermore, the analysis of a range of plant species that are naturally adapted to grow in the presence of vegetational shade has revealed that they show much weaker responses to low R:FR ratio compared with plant species that normally grow in more open conditions (Morgan and Smith, 1979). The principal responses of typical shade avoiding species to a reduction in R:FR ratio is a reallocation of resources preferentially into stem and petiole elongation at the expense of leaf and storage organ development. This is illustrated in Figure 2, showing the shade avoidance response in mustard (*Sinapis alba*) seedlings following prolonged exposure to a low R:FR ratio artificial light environment. In addition to the obvious increase in internode extension these seedlings display additional responses to low R:FR ratio, including reduced leaf expansion and a reduced accumulation of chlorophyll. A common response of many dicotyledonous seedlings to low R:FR ratio is the reduction in leaf thickness (McLaren and Smith, 1978). A further facet of the increased stem elongation at the expense of leaf development is an increase in apical dominance leading to reduced branching in dicots and reduced tillering in grasses (Casal *et al.*, 1986).

*Figure 2. The responses of mustard (*Sinapis alba*) seedlings to low R:FR ratio. The two seedlings were grown under white light from fluorescent tubes providing equal photosynthetically active radiation (400-700 nm). The seedling on the right received supplementary FR such that the R:FR ratio was reduced.*

The acceleration of elongation growth in response to a reduction in R:FR ratio is extremely rapid. Using linear voltage displacement transducers, Smith and colleagues were able to make real-time measurements of the elongation growth of individual seedlings (e.g. Morgan *et al.*, 1980) and record the effects reductions in R:FR ratio. By using fibre optic light guides to provide supplementary FR irradiation, it was possible to reduce the R:FR ratio incident upon a single internode of a mustard seedling and record changes in extension rate. The acceleration of growth in response to added FR was observed after a lag phase of only around 10 minutes, with growth rate increasing by up to five-fold within 30 minutes of the addition of supplementary FR (Morgan *et al.*, 1980; Child and Smith, 1987). The deceleration of growth upon switch off of the supplementary FR light was equally rapid, displaying a lag of only about 6 minutes, with growth rate returning to the pre-stimulation rate within about 15 minutes (Child and Smith, 1987). The use of transducers and fibre optic light guides enabled the participation of the phytochromes in the growth responses to low R:FR ratio to be confirmed. Thus, it was established that whilst the addition of supplementary FR to a mustard internode led to increase extension growth, it was shown that growth rate could be restored to the pre-stimulation level by the addition of supplementary R in addition to the FR (Morgan *et al.*, 1980). This demonstration of the classical R/FR reversibility of the response established the operation of the phytochrome system.

The observation that a reduction in the R:FR ratio incident upon an internode could lead to a rapid change in growth rate is consistent with the concept of early neighbour detection (Ballaré *et al.*, 1987, 1990). This concept arose from the observation that the R:FR ratio received by a sensor with a geometry approximating to that of a stem and positioned within or close to a plant canopy was significantly reduced as a result of the selective reflection of FR (Ballaré et al., 1987). Furthermore, this effect was detectable in developing stands of plants before any actual shading, assessed as a reduction in incident photosynthetic light, was detectable. The notion that this reflected light signal might provide an early warning of presence of potential competitors, as so initiate shade avoidance responses before actual shading, was tested experimentally by rendering individual internodes 'blind' to the reflected FR. This was achieved by growing seedlings of *Datura ferox* L. and *Sinapis alba* L. in stands of different densities and covering individual internodes with annular cuvettes containing $CuSO_4$ solutions of varying concentrations (Ballaré *et al.*, 1990). Since $CuSO_4$ solutions absorb strongly in the FR region of the spectrum, but lead to little attenuation of visible wavebands, the R:FR perceived by individual internodes fitted with the cuvettes could be accurately manipulated. This approach revealed that in even-aged canopies, whereas stem elongation in control seedlings fitted with cuvettes containing water increased with increasing planting density, this response was greatly attenuated in those seedlings fitted with $CuSO_4$-containing cuvettes. Furthermore, the increase in stem elongation in response to laterally propagated low R:FR ratio reflected light occurred well before the leaves of the canopy were subjected to a drop in light availability (Ballaré *et al.*, 1990). These findings have led to the suggestion that the ability to respond to the perceived threat of shading and thereby initiate escape responses before canopy closure is an essential competitive strategy in rapidly growing populations (Ballaré *et al.*, 1990).

Another very obvious component of the shade avoidance syndrome in many plant species is the marked acceleration of flowering (e.g. Halliday *et al.*, 1994) induced by exposure to low R:FR ratio conditions (see Figure 2). Unlike the increase in stem elongation, the early flowering response is only initiated by prolonged exposure to low R:FR ratio. Temporary shading is a frequent phenomenon in the natural environment, and impending competition for sunlight by neighbouring plants may be quickly overcome by a rapid increase in elongation growth allowing the plant to resume a normal growth habit. A too-rapid switch in development to flowering would not prove advantageous to the plant under these conditions. Low R:FR ratio- induced flowering would only be beneficial to the plant in a situation where the plant is unable to overtop competing vegetation. In such cases, it could be argued that the precocious transition from vegetative to reproductive development may be the best strategy for success since it may increase the likelihood of survival to reproduction under the otherwise unfavourable conditions of vegetational shade (Botto and Smith, 2002; Donohue *et al.*, 2001; Dudley and Schmitt, 1995).

It has been argued that the ability to modulate plant architecture in response to the perceived threat of shading represents one of the most radical adaptive strategies available to higher plants. The adaptive value the elongation response to low R:FR ratio has been assessed in a number of ecological investigations. These studies suggest that increased elongation growth in response to low R:FR can confer a high relative fitness in dense stands of plants (Schmitt 1997). Transgenic tobacco plants that constitutively over-express an oat *PHYA* cDNA were used in these studies. The overexpression of the *PHYA* cDNA has previously been shown to result in the suppression of normal shade avoidance elongation growth responses through persistence of a normally transient phyA-mediated inhibition of elongation in FR-rich light environments (McCormac *et al.*, 1991,1992). When planted in dense stands, the transgenic plants were unable of to elongate in response to the low R:FR ratio signals and this resulted in a decrease in fitness compared with wild-type tobacco plants, as measured by dry biomass accumulation (Schmitt *et al.*,

1995, Robson *et al.,* 1996). Whilst the ability to increase stem elongation may provide a fitness benefit in dense stands, it has been suggested that the reallocation of resources towards elongation growth, in the absence of competition, may reduce overall fitness and increase the risk of mechanical damage to stems (Casal and Smith 1989). This notion is supported by the observation that when the constitutively elongated *ein* mutant of *Brassica rapa* was grown in a low density mixed population, decreases in both dry biomass and number of reproductive structures were observed (Schmitt *et al.,* 1995). The disadvantage of increased stem elongation in the absence of competition is further supported by observations showing that elongated *lh* mutants of cucumber display increased mechanical damage when grown individually the field (Schmitt 1997; Casal *et al.*, 1994).

4. ROLES OF DIFFERENT PHYTOCHROMES IN SHADE AVOIDANCE

The perception of the R:FR ratio signal is intimately associated with the reversible photochromicity of phytochrome and represents a major role for the phytochromes in the light responses of the adult plant. The roles of individual members of the phytochrome family in regulating these responses to R:FR ratio have been largely inferred from studies of mutant plants that are deficient in one or more of the phytochromes. The long hypocotyl (*lh*) mutant of cucumber was the first phytochrome-deficient mutant to be characterised in relation to shade avoidance responses to reduced R:FR ratio. A combination of spectrophotometric and immunochemical analyses of the phytochrome status of both etiolated and light grown *lh* plants provided evidence that whereas the mutant possessed wild-type levels of a light-labile phytochrome pool predominating in etiolated tissues (i.e. phyA), it was deficient in a small phytochrome pool that was not subject to light induced proteolysis. More specifically it was shown that a polypeptide species reactive with a monoclonal antibody raised against a recombinant fragment of tobacco PHYB was absent in extracts of *lh* seedlings (Adamse *et al.*, 1988; López-Juez *et al.*, 1992).

*Figure 3. Phenotype of the long hypocotyl (*lh*) mutant of cucumber and its near isogenic wild type.*

It had previously been established that *lh* mutant seedlings displayed a number of aberrant responses to light (e.g. Adamse *et al.*, 1987). Most strikingly it was observed that light-grown *lh* seedlings resemble wild type seedlings displaying the shade avoidance syndrome (López-Juez *et al.*, 1990; Ballaré *et al.*, 1991). Thus, light-grown *lh* mutant seedlings display increased axis elongation, reduced leaf expansion and reduced chlorophyll accumulation (Figure 3). Significantly, it was shown that constitutively elongated *lh* seedlings show no further elongation responses to a reduction in R:FR ratio, achieved by the provision of supplementary FR during the photoperiod (e.g. Adamse *et al.*, 1988; López-Juez *et al.*, 1990; Ballaré *et al.*, 1991). The *lh* mutant was also reported to lack responses to brief irradiations with FR light given at the end of the photoperiod. These end-of-day (EOD) FR treatments are known to mimic responses to daytime reductions in R:FR ratio, presumably due to the reduction in the amount of Pfr present at the light-to-dark transition. These observations led to the conclusion that phyB was responsible for mediating shade avoidance responses in cucumber. The subsequent analysis of phyB-deficient mutants in other species, most notably the phyB-null mutants of *Arabidopsis*, confirmed the striking similarity between the phenotypes of such mutants and the phenotypes of wild type plants displaying the shade avoidance syndrome (e.g. Nagatani *et al.*, 1991; Somers *et al.*, 1991; Devlin *et al.*, 1992; Reed *et al.*, 1993). This too lent support to the notion that phyB mediates responses to vegetational shade.

Despite initial suggestions that phyB-deficient mutants showed no responses to supplementary FR during the photoperiod or to EOD FR, it is now apparent that many such responses are detectable in this class of mutants. For instance, the hypocotyls of light-grown cucumber *lh* seedlings, although already elongated, do show a small but significant additional elongation response to supplementary FR (Whitelam and Smith, 1991; Smith *et al.*, 1992). Of course this residual response could indicate that the *lh* mutation is leaky, and that *lh* seedlings do produce some functional phyB, or they could indicate that phytochromes other than the phyB-like species absent in *lh*, are also able to mediate responses to R:FR ratio. This question has been addressed by analysis of null alleles of the Arabidopsis *phyB* mutant. These mutants, although constitutively elongated and early flowering also retain small but significant residual shade avoidance responses to supplementary FR given during the photoperiod and to EOD FR treatments (e.g. Whitelam and Smith, 1991; Goto *et al.*, 1991; Robson *et al.*, 1993; Halliday *et al.*, 1994; Devlin *et al.*, 1996). Most obviously, both daytime reduction in R:FR ratio and EOD FR treatments induce an early flowering response in phyB-null mutants (Whitelam and Smith, 1991; Goto *et al.*, 1991; Halliday *et al.*, 1994; Devlin *et al.*, 1996; Franklin *et al.*, 2003a). These observations provide a very clear indication that phyB is not the sole mediator of the shade avoidance syndrome in Arabidopsis.

Analysis of the *tri* mutant of tomato (Kendrick *et al.*, 1997) provides compelling evidence that phyB is not the sole, or even the predominant, mediator of the shade avoidance syndrome in all plants. This mutant has been shown to be deficient in a homologue of phyB (van Tuinen *et al.*, 1995; Kerckhoffs *et al.*, 1996). However, unlike many other phyB-deficient mutants, light-grown *tri* seedlings do not obviously resemble the shade avoidance syndrome of wild type plants. Furthermore, *tri* seedlings show more-or-less normal responses to both supplementary FR during the photoperiod and EOD FR (e.g. Kerckhoffs *et al.*, 1992). The observation that phyB is not necessary for the shade avoidance syndrome in tomato is consistent with the notion that phyB does not play a significant role in these responses. This finding raises the possibility that there is redundancy among the phytochromes of tomato with respect to the shade avoidance syndrome.

The analysis of phytochrome-deficient mutants in Arabidopsis has enabled a thorough analysis of the roles of different members of the phytochrome family in the perception of R:FR ratio. As for other plant species, the phenotypic similarity of light-grown Arabidopsis *phyB* mutants to wild-type plants that have been exposed to low R:FR ratio (Figure 4) indicates a predominant role for this photoreceptor in the regulation of shade avoidance responses (Whitelam and Devlin, 1997; Robson *et al.*, 1993). Some shade avoidance responses of phyB mutants to low R:FR or EOD FR treatments, for instance petiole elongation, are severely attenuated (Nagatani *et al.*, 1991). However, other responses to low R:FR ratio, such as reduction in leaf area and the acceleration of flowering, are clearly retained in mutants null for phyB (Robson *et al.*, 1993; Halliday *et al.*, 1994). This clearly indicates the action of phytochromes other than phyB in the control of shade avoidance responses.

Figure 4. The constitutively elongated and early flowering phenotype of the phyB *mutant of Arabidopsis.*

Through the analysis of *phyA* mutants, and *phyAphyB* double mutants, it is apparent that phyA is not necessary for display of the shade avoidance syndrome in Arabidopsis (Yanovsky *et al.*, 1995; Devlin *et al.*, 1996; Whitelam and Devlin, 1997). In fact, at least during seedling establishment, the action of phyA in plants exposed to low R:FR ratio antagonises that of phyB in the control of elongation growth (Yanovsky *et al.*, 1995; Smith *et al.*, 1997). Consequently, *phyA* mutants display such exaggerated elongation responses to low R:FR ratio that many of them die. This suggests that a possible role for phyA in de-etiolating seedlings is to limit some of the shade avoidance responses.

In response to a low R:FR ratio or to EOD FR treatments *phyAphyB* double mutants display an acceleration of flowering and a promotion of the elongation of internodes between rosette leaves (Devlin *et al.*, 1996; Figure 5). The responses of the *phyAphyB* double mutant to EOD FR treatments are reversible by subsequent R treatment, indicating that one or more of phytochromes C, D and E control flowering time and internode elongation (Devlin *et al.*, 1996).

The discovery of a naturally occurring mutation within the *PHYD* gene of the Wassilewskija (Ws) ecotype of Arabidopsis provided an opportunity to study the role of phyD in seedlings response to low R:FR ratio and EOD FR treatment. Adult plants of the monogenic *phyD* mutant showed a more or less wild-type phenotype and wild-type responses low R:FR ratio and EOD FR treatments (Aukerman *et al.*, 1997; Devlin *et al.*, 1999). However, compared with the *phyB* monogenic mutant, the *phyBphyD* double mutant displayed elongated petioles and was early flowering. These phenotypes are reminiscent of the shade avoidance syndrome and suggest that phyD plays a role, in conjunction with phyB, in mediating shade avoidance responses. The high degree of sequence conservation between phytochromes B and D (Mathews and Sharrock, 1997) and the similar patterns of expression of the *PHYB* and *PHYD* promoters (Goosey *et al.*, 1997) are consistent with a similarity in function of these phytochromes. However, the absence of a detectable mutant phenotype in monogenic *phyD* seedlings suggests that there is redundancy of function. Despite their elongated and early flowering phenotype, mutants that are doubly null for phytochromes B and D, such as the *phyAphyB phyD* mutant (Figure 5) still retain an early flowering response to EOD FR treatments, as well as displaying increased elongation of internodes. These observations clearly indicate the participation of other phytochromes in the shade avoidance responses.

Figure 5. The responses of mutants deficient in multiple phytochromes to end-of-day FR light treatments.

The retained shade avoidance response of Arabidopsis *phyAphyB* double mutants was exploited in a screen to identify new photoreceptor mutants. This screen led to the identification of a *phyE* mutant (Devlin et al., 1998). The screen from which the *phyE* mutant was isolated derived from the observation that, in response to EOD FR treatments *phyAphyB* double mutants show a pronounced acceleration of flowering and a promotion elongation of the internodes between rosette leaves (Devlin et al., 1996). Compared with the parental line, the *phyAphyBphyE* triple mutant is constitutively early flowering and constitutively produces internodes between rosette leaves (Devlin et al., 1998). Thus, the phenotype of *phyAphyBphyE* strongly resembles that of *phyAphyB* grown under EOD FR conditions (Figure 5) and so phyE is implicated in the control of the responses. However, the absence of a detectable mutant phenotype in monogenic *phyE* seedlings suggests that there is redundancy of function. The severe attenuation of the flowering response to EOD FR in the *phyAphyBphyE* triple mutant and the retention of a significant flowering response in the *phyAphyBphyD* triple mutant suggest that when phyB is absent, phyE plays a more dominant role than phyD in this response. In a *phyAphyB* double mutant background, phyD deficiency leads to increased petiole elongation, whereas in the same background phyE deficiency leads to increased internode elongation (see Figure 5). This observation suggests that in the photoregulation of elongation growth, phyE plays a role that is distinct from that of phyD. This distinction could reflect the differential spatial patterns of activity of the *PHYD* and *PHYE* promoters (Goosey et al., 1997).

Any possible involvement of phyC in the perception of the low R:FR ratio signal has been excluded following the observation that *phyBphyDphyE* triple mutant plants were blind to the reduced R:FR ratio signal and to EOD FR treatments (Franklin et al., 2003a). Furthermore, the isolation and characterisation of *phyC* null mutants has confirmed that whilst phyC may play a role in early seedling development, phyC-deficiency does not lead to aberrant shade avoidance responses, nor does it modify the constitutive shade avoidance phenotype of the *phyB* mutant when combined with phyB-deficiency (Franklin et al., 2003b).

From the foregoing, it is clear that in Arabidopsis at least, the perception of changes in R:FR ratio and the initiation of shade avoidance responses involves the actions of phytochromes B, D and E (Table 1). These three phytochromes represent the most recently evolved members of the phytochrome family, forming a distinct subgroup (Mathews and Sharrock, 1997). It has therefore been speculated that competition for light, manifested as a reduction in R:FR ratio, may have provided the selective pressure for their evolution (Devlin et al., 1998).

4.1 Roles for phytochrome A in adult plants

The roles of phytochromes B, D and E in regulating internode and petiole elongation and flowering in Arabidopsis in response to reduced R:FR ratio are well established (Devlin et al., 1998, 1999; Franklin et al., 2003a). It is also clear that phyA plays a role in the modulation of plant responses to low R:FR ratio signals. Such a role seems to be inconsistent with the fact that phyA is subject to rapid proteolytic degradation upon photoconversion to Pfr, and so is present at much lower levels in light-grown plants compared with etiolated seedlings, and the observations that when grown under artificial white light, or under natural daylight conditions, *phyA* mutants appear to display a more-or-less wild type phenotype. Taken together this might suggest that the role of phyA is restricted to the early stages of seedling de-etiolation. In etiolated seedlings a major role for phyA, and the basis upon which *phyA* mutants were identified, is mediating FR High Irradiance Responses (HIR), a response mode mediated solely by phyA, usually observed under laboratory conditions, and showing an action maximum in the FR region of the spectrum (see Smith and Whitelam, 1990). The FR-HIR response mode is thought to be only strongly displayed by etiolated seedlings and to be diminished as phyA levels are reduced following exposure to light (e.g. Beggs *et al.*, 1980). Nevertheless, phyA plays an important role in the regulation of hypocotyl elongation in light-grown seedlings in response to alterations in R:FR ratio. For light-grown Arabidopsis seedlings grown in the laboratory under low R:FR ratio conditions, it has been shown (Johnson *et al.*, 1994) that hypocotyls of *phyA* display an enhanced elongation response compared with wild-type seedlings. This led to the suggestion that under prolonged irradiation with supplementary FR, phyA action in wild-type seedlings led to an inhibition of hypocotyl elongation that was antagonising the mainly phyB-mediated shade avoidance response to low R:FR ratio which leads to increased hypocotyl elongation. The notion that phyA action may antagonise the action of phyB in the regulation of elongation growth under low R:FR ratios is supported by analysis of the effect of such conditions on hypocotyl elongation in *phyB* mutants and in *phyAphyB* double mutants. Under low R:FR ratio conditions the hypocotyls of *phyAphyB* double mutants are significantly more elongated than those of the monogenic *phyB* mutant. This indicates that the action of phyA in the monogenic *phyB* mutant reduces the long hypocotyl phenotype that is caused by the phyB deficiency (Johnson *et al.*, 1994). The growth inhibitory action of phyA under low R:FR ratios has been successfully exploited in transgenic approaches where constitutive over-expression of *PHYA* can lead to the 'elimination' of shade avoidance reactions (see Robson *et al.*, 1996).

The action of phyA in FR-rich light environments to antagonise the phyB-mediated shade avoidance response is not restricted to laboratory conditions. Yanovsky *et al.*, (1995) have elegantly shown that phyA action in the FR-HIR is of fundamental importance in seedling establishment under natural vegetational shade conditions. In dense vegetational shade conditions, in contrast to wild-type seedlings, Arabidopsis *phyA* mutants display impaired de-etiolation, characterised by extreme hypocotyl elongation. As a result, significant number of *phyA* mutant seedlings do not become established and they die. This dramatic demonstration indicates that the phyA-mediated FR HIR is essential for seedling survival under some circumstances.

Additional roles for phyA in modulating the development of the mature light-grown plant have also been demonstrated. For example, *phyA* mutants of the long-day plants Arabidopsis and pea flower later under long day conditions that their respective wild types, indicating a role for phyA in daylength perception (Johnson *et al.*, 1994; Lin, 2000; Weller *et al.*, 1997). Phytochrome A also plays a role in the maintenance of the rosette habit of light-grown Arabidopsis plants (Table 1).

Table 1. Summary of the physiological roles of the different phytochromes

Phytochrome	Mutant combinations studied	Functional Roles Revealed	Reference
phyA	*phyA*	Germination	Shinomura *et al.*, 1994, 1996
	phyA *phyAphyB*	Seedling de-etiolation	Nagatani *et al.*, 1993, Parks and Quail, 1993, Whitelam *et al.*, 1993, Reed *et al.*, 1994
	phyAphyB *phyAphyBphyD* *phyAphyBphyE* *phyAphyBphyDphyE*	Mature plant architecture	Devlin *et al.*, 1996, 1998, 1999, Franklin *et al.*, 2003a
	phyA	Promotion of flowering	Johnson *et al.*, 1994, Neff and Chory, 1998
phyB	*phyB*	Germination	Shinomura *et al.*, 1994, 1996
	phyB	Seedling de-etiolation	Koornneef *et al.*, 1980, Somers *et al.*, 1991
	phyB	Shade avoidance	Somers *et al.*, 1991
	phyB	Inhibition of flowering	Goto *et al.*, 1991
phyC	*phyC* *phyCphyD* *phyAphyCphyD* *phyBphyCphyD*	Seedling de-etiolation, leaf development	Franklin *et al.*, 2003b, Monte *et al.*, 2003
phyD	*phyD* *phyBphyD*	Seedling development	Aukerman *et al.*, 1997
	phyD *phyBphyD* *phyAphyBphyDphyE*	Shade avoidance	Devlin *et al.*, 1999, Franklin *et al.*, 2003a
phyE	*phyAphyBphyE*	Germination	Hennig *et al.*, 2002
	phyBphyE *phyAphyBphyE*	Shade avoidance	Devlin *et al.*, 1998
	phyAphyBphyE	Maintenance of rosette habit	Devlin *et al.*, 1998

Whereas wild-type plants structure their leaves in a compact rosette, internode elongation and loss of rosette habit were observed in *phyAphyB* double and *phyAphyBphyD* triple mutant plants following EOD FR treatments (Devlin *et al.*, 1996, 1999; Figure 5). The *phyAphyBphyE* triple mutant has constitutively elongated internodes, leading to the proposal that maintenance of the rosette phenotype is regulated by phyA, B and E acting redundantly (Devlin *et al.*, 1998). This proposal is supported by the more recent observation that the *phyBphyDphyE* triple mutant does not display elongated internodes whereas the *phyAphyBphyDphyE* quadruple mutant does (Franklin *et al.*, 2003a). This same study also reported that the *phyAphyBphyDphyE* quadruple mutant displays a pronounced increase in leaf length/width ratio compared with the *phyBphyDphyE* triple mutant, suggesting a significant role for phyA in inhibiting leaf elongation in light-grown plants (Franklin *et al.*, 2003a)

5. MOLECULAR MECHANISMS CONTROLLING SHADE AVOIDANCE RESPONSES

5.1 The acceleration of flowering

Although the photoreceptors involved in shade avoidance have been well characterised, little is known of the molecular signalling events that couple detection of R:FR ratio with changed patterns of growth and development. Some progress is being made in defining some of the components that are involved in triggering the acceleration of flowering by low R:FR ratio in Arabidopsis, a response regulated by the action of phyB, phyD and phyE. In addition to light quality, many other environmental cues control flowering time in Arabidopsis, especially those that indicate the passage of winter and the onset of spring and summer. These signals are transduced by multiple regulatory pathways that converge to regulate the developmental fate of the shoot apical meristem via altered expression of meristem identity genes such as *LFY* (see Simpson and Dean, 2002). The expression of meristem identity genes is regulated by a number of floral pathway integrators, for instance FT and SOC1, that are themselves controlled by the action of transcriptional regulators, such as FLC and CO (Simpson and Dean, 2002). For example, the photoperiodic pathway, leading to the promotion of flowering under longs days, is thought to involve a photoreceptor-derived signal coinciding with high levels of CO in order to activate *FT* expression (Yanovsky and Kay, 2002). The levels of CO are regulated by the circadian oscillator, and the requirement for the photoreceptor-derived signal to coincide with appropriate levels of CO provides a mechanism for discriminating between long days and short days (see Chapter 27 for further details).

Phytochrome B-deficiency, and by implication low R:FR ratio, was reported to regulate expression of the floral meristem identity gene *LFY*, independently of both CO and FT, suggesting that the light quality pathway is not only separate from the photoperiodic pathway, but is also independent of one of the key floral pathway integrators (Blázquez and Weigel, 1999). More recently, two independent studies have provided evidence that although the R:FR ratio-dependent light quality pathway regulating flowering is independent of CO, it does involve FT (Halliday *et al.*, 2003; Cerdán and Chory, 2003). Halliday *et al.*, (2003) showed that the well-established early flowering phenotype of the *phyB* mutant correlates with increased expression of *FT* and demonstrated that this effect was temperature conditional. Thus, the early flowering of *phyB* mutants compared with wild-type plants, observed at, say, 22°C was completely abolished by growing the plants at 16°C. The early flowering of the *phyAphyBphyD* triple mutant was also abolished by growing the plants at 16°C. In both cases, growth of the mutant plants at 16°C also abolished the elevated expression of *FT*. This, together with the finding that wild type plants grown at 16°C still display a marked acceleration of flowering by low R:FR ratio, suggests that phyE may be the predominate regulator of the flowering response to R:FR ratio at this temperature. This was confirmed by the finding that in the *phyAphyBphyD* triple mutant background, the additional loss of phyE triggered both an acceleration of flowering and a rise in *FT* expression in plants grown at 16°C (Halliday *et al.*, 2003).

A role for FT in the regulation of flowering by phyB has also been suggested by Cerdán and Chory (2003), who also observed a correlation between the early flowering of the *phyB* mutant and elevated expression of *FT*. This study also identified an additional possible component of the pathway linking phyB action to altered *FT* expression. From a screen for mutants displaying defective flowering time, a recessive mutation, *pft1*, was isolated. Following growth under both long-day and short-day conditions the *pft1* mutant was late flowering and completely suppressed the early-flowering phenotype of *phyB*, suggesting that PFT1 is essential for phyB regulation of flowering time. The increased petiole length of *phyB* mutants was largely unaffected in *pft1phyB* double mutants. This result, together with the finding that the effect of *pft1* on the photoperiodic induction of flowering was relatively minor in both wild-type and phytochrome mutant backgrounds, suggests that the main role of PFT1 is to regulate flowering time downstream of phyB in a photoperiod-independent pathway. The *pft1* mutation also leads to significant impairment of the flowering response to EOD FR light treatments (Cerdán and Chory, 2003), suggesting that PFT1 also acts downstream of phyD and phyE, since these two phytochromes act with phyB to mediate the EOD FR effect. The phenotype of the *pft1* mutant is reminiscent of that of some natural Arabidopsis accessions that are essentially unresponsive to low R:FR with respect to flowering time, but display normal elongation growth responses to low R:FR ratio (Botto and Smith, 2002). The *PFT1* gene was isolated and found to encode a nuclear-localised protein of 836 amino acids, with a domain structure similar to that of some transcriptional activators (Cerdán and Chory, 2003).

The low R:FR ratio signal, and indeed phyB-deficiency, can fully correct the late flowering of several induced mutations in components of the autonomous pathway that confer a vernalisation requirement (Bagnall, 1993; Halliday *et al.*, 1994). The normal function of these autonomous pathway components is to limit the expression of *FLC* (Sheldon *et al.*, 1999) and the induced mutations have elevated *FLC* expression. It is therefore possible that phyB/D/E may also control *FLC* levels, although such regulation has not been reported.

5.2 Early events in R:FR ratio signalling

A number of proteins, including PIF3, a basic Helix-Loop-Helix (bHLH) transcription factor (Ni *et al.*, 1998, 1999), that interact physically with phyB, have been isolated. The interaction between PIF3 and phyB is reported to provide a mechanism by which phyB can more-or-less directly regulate the expression of target genes. Thus, complexes of PIF3 and phyB have also been demonstrated to bind to regulatory elements in the promoter regions of known light responsive genes (Martínez-García *et al.*, 2000). A role for PIF3, or any of the phyB-interacting factors, in shade avoidance has yet to be demonstrated.

The expression of at least one transcriptional regulator is controlled by phyB and phyE, in a manner that is similar to way these phytochromes regulate shade avoidance responses. The gene *ATHB-2* (also known as *HAT4)*, and the related gene *ATHB-4*, was the first to be shown to be reversibly regulated by changes in the R:FR ratio in light-grown plants (Carabelli *et al.*, 1993, 1996). *ATHB-2* transcript levels are low in tissues that have been maintained in high R:FR ratio white light, but are rapidly and substantially increased following transfer to low R:FR ratios or EOD FR treatments. The analysis of mutants null for multiple phytochromes revealed that the response to reduced R:FR ratio was controlled by phytochromes B and E acting in a functionally redundant manner (Franklin *et al.*, 2003a). The products of the *ATHB-2* and *ATHB-4* genes are members of a large class of proteins that carry a homeodomain linked to a leucine zipper motif (Ruberti *et al.*, 1991). A combination of DNA-binding studies and transient expression assays has shown that these proteins act as transcriptional regulators. ATHB-2 acts as a negative regulator of gene expression (Steindler *et al.*, 1999).

The possible involvement of ATHB-2 in shade avoidance responses has been inferred from the phenotypes of transgenic plants with increased or reduced *ATHB-2* expression (Schena and Davies, 1992; Steindler *et al.*, 1999). Transgenic seedlings overproducing ATHB-2 had longer hypocotyls and petioles and smaller leaves than wild-type plants. Moreover, seedlings with reduced levels of ATHB-2 had shorter hypocotyls and larger leaves (Steindler *et al.*, 1999). The phenotypes of transgenic plants over-expressing *ATHB-2* are reminiscent of those displayed by wild-type plants grown under low R:FR ratio conditions. This, together with the R:FR ratio-dependent regulation of *ATHB-2* gene expression by phytochromes suggests a role for ATHB-2 in the regulation of shade avoidance (Carabelli *et al.*, 1996; Steindler *et al.*, 1999). Anatomical studies of the hypocotyls of transgenic plants with altered expression of *ATHB-2* revealed changes in both the orientation of cell expansion and the production of the secondary vascular tissue. Similar changes were observed in wild-type seedlings grown under low R:FR ratio conditions (Steindler *et al.*, 1999). It is proposed that these cellular events reflect changes in auxin transport mediated by R:FR ratio-dependent changes in the expression of *ATHB-2* (Morelli and Ruberti, 2002).

A role for auxin transport in aspects of the shade avoidance responses has also been inferred from phenotypic analysis of the recessive mutant designated *asa1*, for *a*ttenuated *s*hade *a*voidance (Kanyuka *et al.*, 2003). This mutant was isolated following mutagenesis of the Arabidopsis *phyAphyB* mutant and screening for individuals that fail to display the constitutive shade avoidance phenotype of the parent. The *asa1* mutation overcomes the elongated, early flowering phenotype caused by phyB-deficiency and reduces the sensitivity of the plant to low R:FR signals. The *ASA1* gene was found to encode *BIG* (= *DOC1, TIR3, UMB, GA6*), a massive calossin-like protein involved in polar auxin transport and other aspects hormone signalling (Kanyuka *et al.,* 2003). Given the very pleiotropic nature of *asa1* mutant phenotype, it is unlikely that ASA1 plays a very specific role in the transduction of the low R:FR ratio signal. Nevertheless, the mutant phenotype may indicate a requirement for normal polar auxin transport for the normal display of shade avoidance responses.

Recently, transcriptomics approaches have been used in attempts to identify possible molecular components of the early R:FR ratio signalling cascade. Analysis of gene expression, using Affymetrix oligoarrays, representing about 8,000 genes, revealed that several genes show substantially altered expression (> 3fold change in expression) in response to a 1 h exposure to low R:FR ratio conditions (Salter *et al.,* 2003). As expected *ATHB-2* was among those genes displaying the most marked changes in expression in response to brief low R:FR ratio. However, the greatest change in transcript level in response to low R:FR ratio (> 30-fold increase in transcript abundance) was observed for a gene identified as *PIL1*, for *PIF3-like 1*. This gene encodes a basic helix-loop-helix protein that has 37% identity with the transcriptional regulator *PIF3* (Ni *et al* 1998). Analysis of a *pil1* knockout mutant indicated that PIL1 is required for the display of the normal elongation growth responses associated with the rapid shade avoidance responses to low R:FR ratio (Salter *et al.,* 2003).

PIL1 was initially isolated as a protein that interacts with the circadian clock protein TOC1 (Makino *et al.,* 2002). The increased expression of *PIL1* in response to a low R:FR ratio treatment is gated by the circadian clock, with the greatest increase in *PIL1* transcript occurring in response to treatments given at subjective dawn (Salter *et al.,* 2003). The regulation of elongation growth by low R:FR ratio treatment is also gated by the circadian clock. Thus, a transient low R:FR ratio treatment given at subjective dawn results in an inhibition of hypocotyl extension compared with seedlings maintained in high R:FR ratio conditions. This response probably reflects the increased levels and/or action of phyA because the growth inhibition was absent in seedlings of the *phyA* mutant (Salter *et al.,* 2003). Phytochrome A protein levels show both diurnal and circadian regulation, with the highest levels being detected at the end of the subjective night (Sharrock and Clack, 2002). A transient reduction in low R:FR ratio given at subjective dusk leads to a marked (25-35%) increase in hypocotyl length, measured following return to high R:FR ratio for 24 hours. These data are consistent with observations showing the elongation of Arabidopsis hypocotyls to be under circadian control with a daily arrest of growth at dawn and a period of rapid elongation at dusk (Dowson-Day and Millar, 1999). Seedlings of the *phyB* mutant displayed an attenuated elongation response to reductions in R:FR ratio, which was also gated by the circadian clock (Salter *et al.,* 2003).

The increased expression of *PIL1* in response to a low R:FR ratio treatment at dusk coincides with maximal levels of *TOC1* gene expression (Strayer *et al.,* 2000) and results in a significant promotion of hypocotyl elongation. The inability of *toc1-2* mutant seedlings to elongate following a dusk reduction in R:FR ratio confirms the functional dependence of TOC 1 protein for this response (Salter *et al.*, 2003).

6. REFERENCES

Adamse, P., Jaspers, P. A. P. M., Bakker, J. A., Kendrick, R. E, Koornneef, M. (1988) Photophysiology and phytochrome content of long-hypocotyl and wild-type cucumber seedlings. *Plant Physiology, 87*, 264-268.

Adamse, P., Jaspers, P. A. P. M., Kendrick, R. E., Koornneef, M. (1987.) Photomorphogenetic responses of long hypocotyl mutant of *Cucumis sativus. Journal of Plant Physiology, 127*, 481-491.

Aukerman, M. J., Hirschfeld, M., Wester, L., Weaver, R., Clack, T., Amasino, R. M., *et al.* (1997) A deletion in the *PHYD* gene of the Arabidopsis Wassilewskija ecotype defines a role for phytochrome D in red/far-red light sensing. *Plant Cell, 9*, 1317-132.

Bagnall, D. J. (1993) Light quality and vernalization interact in controlling late flowering *Arabidopsis* ecotypes and mutants. *Annals of Botany, 71*, 75-83.

Ballaré, C. L., Casal, J. J., Kendrick, R. E. (1991) Responses of light-grown wild-type and long-hypocotyl mutant cucumber seedlings to natural and stimulated shade light. *Photochemistry and Photobiology, 54*, 819-826.

Ballaré, C. L., Sánchez, R. A., Scopel, A. L., Casal, J. J., Ghersa, C. M. (1987) Early detection of neighbour plants by phytochrome perception of spectral changes in reflected sunlight. *Plant, Cell and Environment, 10*, 551-557.

Ballaré, C. L., Scopel, A. L., Sánchez, R. A. (1990) Far-red radiation reflected from adjacent leaves: an early signal of competition in plant canopies. *Science, 247*, 329-332.

Beggs, C. J., Holmes, M. G., Jabben, M., Schäfer, E. (1980) Action spectra for the inhibition of hypocotyl growth by continuous irradiation in light, dark-grown *Sinapis alba* L. seedlings. *Plant Physiology, 66*, 615-618.

Blázquez, M. A. and Weigel, D. (1999) Independent regulation of flowering by phytochrome B and gibberellins in Arabidopsis. *Plant Physiology, 120*, 1025-32.

Botto, J. F. and Smith, H. (2002) Differential genetic variation in adaptive strategies to a common environmental signal in Arabidopsis accessions: phytochrome-mediated shade avoidance. *Plant Cell and Environment, 25*, 53-63.

Briggs, W. R. and Huala, E. (1999) Blue-light photoreceptors in higher plants. *Annual Review of Cell and Developmental Biology, 15*, 33-62.

Carabelli, M., Morelli, G., Whitelam, G. C., Ruberti, I. (1996) Twilight-zone and canopy shade induction of the *ATHB-2* homeobox gene in green plants. *Proceedings of the National Academy of Sciences USA, 93*, 3530-3535.

Carabelli, M., Sessa, G., Ruberti, I., Morelli, G. (1993) The *Arabidopsis ATHB-2* and -*4* genes are strongly induced by far-red-rich light. *Plant Journal, 4*, 469-479.

Casal, J. J., Ballaré, C. L., Tourn, M., Sánchez, R. A. (1994) Anatomy, growth and survival of a long-hypocotyl mutant of *Cucumis sativus* deficient in phytochrome B. *Annals of Botany, 73*, 569-575.

Casal, J. J., Sánchez, R. A., Deregibus, V. V. (1986) The effect of plant density on tillering: The involvement of R/FR ratio and the proportion of radiation intercepted per plant. *Environmental and Experimental Botany, 26*, 365-371.

Casal, J. J., Smith, H. (1989) The function, action and adaptive significance of phytochrome in light-grown plants. *Plant Cell and Environment, 12*, 855-862.

Cashmore, A. R., Jarillo, J. A., Wu, Y. J., Liu, D. (1999) Cryptochromes: blue light receptors for plants and animals. *Science, 284*, 760-765.

Cerdán, P. D. and Chory, J. (2003) Regulation of flowering time by light quality. *Nature, 423*, 881-885.

Child, R. and Smith, H. (1987) Phytochrome action in light-grown mustard: Kinetics, fluence-rate compensation and ecological significance. *Planta, 172*, 219-229.

Clack, T., Mathews, S., Sharrock, R. A. (1994) The phytochrome apoprotein family in *Arabidopsis* is encoded by five genes: the sequences and expression of *PHYD* and *PHYE*. *Plant Molecular Biology, 25*, 413-427.

Cumming, B. G. (1963) The dependence of germination on photoperiod, light quality, and temperature in *Chenopdium* spp. *Canadian Journal of Botany,* 41, 1211-1233.

Devlin, P. F., Halliday, K. J., Harberd, N. P., Whitelam, G. C. (1996) The rosette habit of *Arabidopsis thaliana* is dependent upon phytochrome action: novel phytochromes control internode elongation and flowering time. *Plant Journal, 10,* 1127-1134.

Devlin, P. F., Patel, S. R., Whitelam, G. C. (1998) Phytochrome E influences internode elongation and flowering time in *Arabidopsis*. *Plant Cell, 10,* 1479-1487.

Devlin, P. F., Robson, P. R. H., Patel, S. R., Goosey, L., Sharrock, R. A., Whitelam, G. C. (1999) Phytochrome D acts in the shade-avoidance syndrome in *Arabidopsis* by controlling elongation and flowering time. *Plant Physiology, 119,* 909-915.

Devlin, P. F., Rood, S. B., Somers, D. E., Quail, P. H., Whitelam, G. C. (1992) Photophysiology of the *elongated internode* (*ein*) mutant of *Brassica rapa*: *ein* mutant lacks a detectable phytochrome B-like protein. *Plant Physiology, 100,* 1442-1447.

Donohue, K., Pyle, E. H., Messiqua, D., Heschel, M. S., Schmitt, J. (2001) Adaptive divergence in plasticity in natural populations of *Impatiens capensis* and its consequences for performance in novel habitats. *Evolution, 55,* 692-702.

Dowson-Day, M. J. and Millar, A. J. (1999) Circadian dysfunction causes aberrant hypocotyl elongation patterns in Arabidopsis. *Plant Journal, 17,* 63-71.

Dudley, S. A. and Schmitt, J. (1995) Genetic differentiation in morphological responses to simulated foliage shade between populations of *Impatiens capensis* from open and woodland sites. *Functional Ecology,* 9, 655-666.

Edgerton, M. D. and Jones, A. M. (1992) Localization of protein-protein interactions between the subunits of phytochrome. *Plant Cell, 4,* 161-171.

Franklin, K. A., Praekelt, U., Stoddart, W. M., Billingham, O. E., Halliday, K. J., Whitelam, G. C. (2003a) Phytochromes B, D and E act redundantly to control multiple physiological responses in Arabidopsis. *Plant Physiology, 131,* 1340-1346.

Franklin, K. A., Davis, S. J., Stoddart, W. M., Vierstra, R. D., Whitelam, G. C. (2003b) Mutant analyses define multiple roles for phytochrome C in *Arabidopsis thaliana* photomorphogenesis. *Plant Cell, 15,* 1981-1989.

Furuya, M. and Song, P-S. (1994) "Assembly and properties of holophytochrome." In: *Photomorphogenesis in Plants, 2nd ed.,* Kendrick, R. E and Kronenberg, G. H. M. (eds), Kluwer Academic Publishers, Dordrecht, Netherlands, 105-140.

Furuya, M. and Schäfer, E. (1996) Photoperception and signalling of induction reactions by different phytochromes. *Trends in Plant Science, 1,* 301-307.

Gilbert, I. R., Jarvis, P. G., Smith, H. (2001) Proximity signal and shade avoidance differences between early and late successional trees. *Nature, 411,* 792-795.

Goosey, L., Palecanda, L., Sharrock, R. A. (1997) Differential patterns of expression of the Arabidopsis *PHYB, PHYD,* and *PHYE* phytochrome genes. *Plant Physiology, 115,* 959-969.

Goto, N., Kumagai, T., Koornneef, M. (1991) Flowering responses to light-breaks in photomorphogenic mutants of *Arabidopsis thaliana,* a long-day plant. *Physiologia Plantarum, 83,* 209-215.

Halliday, K. J., Koornneef, M., Whitelam, G. C. (1994) Phytochrome B and at least one other phytochrome mediate the accelerated flowering response of *Arabidopsis thaliana* L. to low red/far-red ratio. *Plant Physiology, 104,* 1311-1315.

Halliday, K. J., Salter, M. G., Thingnaes, E., Whitelam, G. C. (2003) Phytochrome control of flowering is temperature sensitive and correlates with expression of the floral integrator FT. *Plant Journal, 33,* 875-885.

Hennig, L., Stoddart, W. M., Dieterle, M., Whitelam, G. C., Schäfer, E. (2002) Phytochrome E controls light-induced germination of Arabidopsis. *Plant Physiology, 128,* 194-200.

Holmes, M. G. and Smith, H. (1975) The function of phytochrome in plants growing in the natural environment. *Nature, 254,* 512-514.

Holmes, M.G. and Smith, H. (1977) The function of phytochrome in the natural environment. II. The influence of vegetation canopies on the spectral energy distribution of natural daylight. *Photochemistry and Photobiology, 25,* 539-545.

Johnson, E., Bradley, J. M., Harberd, N. P., Whitelam, G. C. (1994) Photoresponses of light-grown *phyA* mutants of *Arabidopsis*: phytochrome A is required for the perception of daylength extensions. *Plant Physiology, 105,* 141-149.

Kanyuka, K., Praekelt, U., Billingham, O., Franklin, K. A., Hooley, R., Whitelam, G. C., Halliday, K. J. (2003) Mutations in the huge Arabidopsis gene *BIG* affect a range of hormone and light responses. *Plant Journal, 35*, 57-70

Kendrick, R. E., Kerckhoffs, L. H. J., van Tuinen, A., Koornneef, M. (1997) Photomorphogenic mutants of tomato. *Plant, Cell and Environment, 20*, 746-751.

Kerckhoffs, L. H. J., Kendrick, R. E., Whitelam, G. C., Smith, H. (1992) Extension growth and anthocyanin responses of photomorphogenic tomato mutants to changes in the phytochrome photoequilibrium during the daily photoperiod. *Photochemistry and Photobiology, 56*, 611-616.

Kerckhoffs, L. H. J., van Tuinen, A., Hauser, B. A., Cordonnier-Pratt, M. M., Nagatani, A., Koornneef, M., et al.. (1996) Molecular analysis of *tri* mutant alleles in tomato indicates the *TRI* locus is the gene encoding the apoprotein of phytochrome B1. *Planta, 199*, 152-157.

Konstantin, K., Praekelt, U., Franklin, K. A., Billingham, O. E., Hooley, R., Whitelam, G. C., Halliday, K. J. (2003) Mutations in the huge *Arabidopsis* gene *BIG* affect a range of hormone and light responses. *Plant Journal, 35*, 57-70.

Koornneef, M., Rolff, E., Spruitt, C. J. P. (1980) Genetic control of light-inhibited hypocotyl elongation in *Arabidopsis thaliana* L. Heyhh. *Zeitschrift für Pflanzenphysiologie, 100*, 147-160.

Lin, C. (2000) Photoreceptors and regulation of flowering time. *Plant Physiology, 123*, 39–50.

López-Juez, E., Buurmeijer, W. F., Heeringa, G. H., Kendrick, R. E., Wesselius, J. C. (1990) Response of light-grown wild-type and long hypocotyl mutant cucumber plants to end-of-day far-red light. *Photochemistry and Photobiology, 52*, 143-149.

López-Juez, E., Nagatani, A., Tomizawa, K-I., Deak, M., Kern, R., Kendrick, R. E., et al.. (1992) The cucumber long hypocotyl mutant lacks a light-stable PHYB-like phytochrome. *Plant Cell, 4*, 241-251.

Makino, S., Matsushika, A., Kojima, M., Yamashino, T., Mizuno, T. (2002) The APRR1/TOC1 quintet implicated in circadian rhythms of *Arabidopsis thaliana*: I. Characterization with APRR1-overexpressing plants. *Plant and Cell Physiology, 43*, 58-69.

Martínez-García, J. F., Huq, E., Quail, P. H. (2000) Direct targeting of light signals to a promoter element-bound transcription factor. *Science, 288*, 859-863.

Mathews, S., Lavin, M., Sharrock, R. A. (1995) Evolution of the phytochrome gene family and its utility for phylogenetic analyses of angiosperms. *Annals of the Missouri Botanical Garden 82*, 296321.

Mathews, S., Sharrock, R.A. (1997) Phytochrome gene diversity. *Plant Cell and Environment, 20*, 666-671.

McLaren J. S. and Smith, H. (1978) The function of phytochrome in the natural environment. VI. Phytochrome control of the growth and development of *Rumex obtusifolius* under simulated canopy light environments. *Plant, Cell and Environment, 1*, 61-67.

McCormac, A. C., Cherry, J. R., Hershey, H. P., Vierstra, R. D., Smith, H. (1991) Photoresponses of transgenic tobacco plants expressing an oat phytochrome gene. *Planta, 185*, 162-170.

McCormac, A. C., Whitelam, G. C., Smith, H. (1992) Light grown plants of transgenic tobacco expressing an introduced oat phytochrome A gene under the control of a constitutive viral promoter exhibit persistent growth inhibition by far-red light. *Planta, 188*, 173-181.

Monte, E., Alonso, J. M., Ecker, J. R., Zhang, Y., Li, X., Young, J., et al. (2003) Isolation and characterization of *phyC* mutants in *Arabidopsis* reveals complex crosstalk between phytochrome signalling pathways. *Plant Cell, 15*, 1962-1980.

Morelli, G. and Ruberti, I. (2002) Light and shade in the photocontrol of Arabidopsis growth. *Trends in Plant Science, 7*, 399-404.

Morgan, D. C. and Smith, H. (1976) Linear relationship between phytochrome photoequilibrium and growth in plants under simulated natural radiation. *Nature, 262*, 210-212.

Morgan, D. C. and Smith, H. (1978) The function of phytochrome in the natural environment. VII. The relationship between phytochrome photo-equilibrium and development in light-grown *Chenopodium album* L. *Planta, 132*, 187-193.

Morgan, D. C. and Smith, H. (1979) A systematic relationship between phytochrome-controlled development and species habitat for plants grown in simulated natural radiation. *Planta, 145*, 253259.

Morgan, D. C., O'Brien, T., Smith, H. (1980) Rapid photomodulation of stem extension in light-grown *Sinapis alba* L. Studies on kinetics, site of perception and photoreceptor. *Planta, 150*, 95-101.

Nagatani, A., Chory, J., Furuya, M. (1991) Phytochrome B is not detectable in the *hy3* mutant of *Arabidopsis*, which is deficient in responding to end-of-day far-red light treatments. *Plant and Cell Physiology, 32*, 1119-1122.

Nagatani, A., Reed, J. W., Chory, J. (1993) Isolation and initial characterisation of *Arabidopsis* mutants that are deficient in functional phytochrome A. *Plant Physiology, 102*, 269-277.

Neff, M. M. and Chory, J. (1998) Genetic interaction between phytochrome A, phytochrome B and cryptochrome 1 during *Arabidopsis* development. *Plant Physiology, 118,* 27-36.

Ni, M., Tepperman, J. M., Quail, P. H. (1999) Binding of phytochrome B to its nuclear signalling partner PIF3 is reversibly induced by light. *Nature, 400*, 781-784.

Ni, M., Tepperman, J. M., Quail, P. H. (1998) PIF3, a phytochrome-interacting factor necessary for normal photoinduced signal transduction, is a novel basic helix-loop-helix protein. *Cell, 95*, 657667.

Parks, B. M. and Quail, P. H. (1993) *hy8*, a new class of Arabidopsis long hypocotyl mutants deficient in functional phytochrome A. *Plant Cell, 3,* 39-48.

Quail, P. H. (1994) "Phytochrome genes and their expression." In: *Photomorphogenesis in Plants, 2nd ed.*, Kendrick R.E and Kronenberg G.H.M. (eds) Kluwer Academic Publishers, Dordrecht, Netherlands, 71-104.

Quail, P. H. (2002) Photosensory perception and signalling in plant cells: new paradigms? *Current Opinion in Plant Biology, 14*, 180-188.

Reed, J. W., Nagatani, A., Elich, T., Fagan, M., Chory, J. (1994) Phytochrome A and phytochrome B have overlapping but distinct functions in *Arabidopsis* development. *Plant Physiology, 104*, 1139-1149.

Reed, J. W., Nagpal, P., Poole, D. S., Furuya, M., Chory, J. (1993) Mutations in the gene for red/far-red light receptor phytochrome B alter cell elongation and physiological responses throughout *Arabidopsis* development. *Plant Cell, 5,* 147-157.

Robson, P. R. H., McCormac, A. C., Irvine, A. S., Smith, H. (1996) Genetic engineering of harvest index in tobacco through overexpression of a phytochrome gene. *Nature Biotechnology, 14*, 995-998.

Robson, P. R. H., Whitelam, G. C., Smith, H. (1993) Selected components of the shade-avoidance syndrome are displayed in a normal manner in mutants of *Arabidopsis thaliana* and *Brassica rapa* deficient in phytochrome B. *Plant Physiology, 102*, 1179-1184.

Ruberti, I., Sessa, G., Lucchetti, S., Morelli, G. (1991) A novel class of plant proteins containing a homeodomain with a closely linked leucine zipper motif. *EMBO Journal, 10*, 1787-1791.

Salter, M. G., Franklin, K. A., Whitelam, G. C. (2003) Gating of the rapid shade avoidance response by the circadian clock in plants. *Nature, 426*, 680-683.

Schena, M. and Davis, R. W. (1992) HD-Zip proteins: members of an *Arabidopsis* homeodomain protein superfamily. *Proceedings of the National Academy of Sciences USA, 89*, 3894-3898.

Schmitt, J., McCormac, A. C., Smith, H. (1995) A test of the adaptive plasticity hypothesis using transgenic and mutant plants disabled in phytochrome-mediated elongation responses to neighbours. *American Naturalist, 146*, 937-953.

Schmitt, J. (1997) Is photomorphogenic shade avoidance adaptive? Perspectives from population biology. *Plant Cell Environ.* 20, 826-830.

Sharrock, R. A. and Clack, T. (2002) Patterns of expression and normalized levels of the five Arabidopsis phytochromes. *Plant Physiology, 130*, 442- 456.

Sharrock, R. A. and Quail, P.H. (1989) Novel phytochrome sequences in *Arabidopsis thaliana*: Structure, evolution, and differential expression of a plant regulatory photoreceptor family. *Genes and Development, 3*, 1745-1757.

Sheldon, C. C., Burn, J. E., Perez, P. P., Metzger, J., Edwards, J. A., Peacock, W. J., *et al.* (1999) The *FLF* MADS box gene: a repressor of flowering in *Arabidopsis* regulated by vernalization and methylation. *Plant Cell, 11*, 445-458.

Shinomura, T., Nagatani, A., Chory, J., Furuya, M. (1994) The induction of seed germination in *Arabidopsis thaliana* is regulated principally by phytochrome B and secondarily by phytochrome A. *Plant Physiology, 104*, 363-371.

Shinomura, T., Nagatani, A., Manzawa, H., Kubota, M., Watanabe, M., Furuya, M. (1996) Action spectra for phytochrome A and B-specific photoinduction of seed germination in *Arabidopsis thaliana*. *Proceedings of the National Academy of Science USA, 93*, 8129-8133.

Simpson, G. G. and Dean, C. (2002) *Arabidopsis*, the Rosetta stone of flowering time? *Science, 296*, 285-289.

Smith, H. *Phytochrome and Photomorphogenesis.* UK: McGraw-Hill, 1975.

Smith, H. (1982) Light quality, photoperception and plant strategy. *Annual Review of Plant Physiology, 33*, 481-518.

Smith, H., Turnbull, M., Kendrick, R. E. (1992) Light-grown plants of the cucumber long hypocotyl mutant exhibit both long-term and rapid growth responses to irradiation with supplementary far-red light. *Photochemistry and Photobiology, 56*, 607-610.

Smith, H. and Whitelam, G. C. (1990) Phytochrome, a family of photoreceptors with multiple physiological roles. *Plant Cell and Environment, 13*, 695-707.

Smith, H. and Whitelam, G. C. (1997) The shade avoidance syndrome: multiple responses mediated by multiple phytochromes. *Plant Cell and Environment, 20*, 840-844.

Smith, H., Xu, Y., Quail, P. H. (1997) Antagonistic but complementary actions of phytochromes A and B allow optimum seedling de-etiolation. *Plant Physiology, 114*, 995-998.

Somers, D. E., Sharrock, R. A., Tepperman, J. M., Quail, P. H. (1991) The *hy3* long hypocotyl mutant of *Arabidopsis* is deficient in phytochrome B. *Plant Cell, 3*, 1263-1274.

Steindler, C., Matteucci, A., Sessa, G., Weimar, T., Ohgishi, M., Aoyama, T., et al.. (1999) Shade avoidance responses are mediated by the ATHB-2 HD-zip protein, a negative regulator of gene expression. *Development, 126*, 4235-4245.

Strayer, C., Oyama, T., Schultz, T. F., Raman, R., Somers, D. E., Mas, P., et al.. (2000) Cloning of the Arabidopsis clock cone *TOC1*, an autoregulatory response regulator homolog. *Science, 289*, 768-771.

van Tuinen, A., Kerckhoffs, L. H. J., Nagatani, A., Kendrick, R. E., Koornneef, M. (1995) A temporarily red light-insensitive mutant of tomato lacks a light-stable, B-like phytochrome. *Plant Physiology, 108*, 939-947.

Weller, J. L., Murfet, I. C., Reid, J. B. (1997) Pea mutants with reduced sensitivity to far-red light define an important role for phytochrome A in day-length detection. *Plant Physiology, 114*, 1225-1236.

Whitelam, G. C. and Devlin, P. F. (1997) Roles of different phytochromes in *Arabidopsis* photomorphogenesis. *Plant Cell and Environment, 20*, 752-758.

Whitelam, G. C., Johnson, E., Peng, J., Carol, P., Anderson, M. C., Cowl, J. S., et al. (1993) Phytochrome A null mutants of *Arabidopsis* display a wild-type phenotype in white light. *Plant Cell, 5*, 757-768.

Whitelam, G. C. and Smith, H. (1991) Retention of phytochrome-mediated shade avoidance responses in phytochrome-deficient mutants of *Arabidopsis,* cucumber and tomato. *Journal of Plant Physiology, 39*, 119-125.

Yanovsky, M. J., Casal, J. J., Whitelam, G. C. (1995) Phytochrome A, phytochrome B and HY4 are involved in hypocotyl growth responses to natural radiation in *Arabidopsis*: weak de-etiolation of the *phyA* mutant under dense canopies. *Plant, Cell and Environment, 18*, 788-794.

Yanovsky, M. J. and Kay, S. A. (2002) Molecular basis of seasonal time measurement in *Arabidopsis. Nature, 419*, 308-312.

Chapter 23

A ROLE FOR CHLOROPHYLL PRECURSORS IN PLASTID-TO-NUCLEUS SIGNALING

Robert M. Larkin[1]* and Joanne Chory[2]

[1] MSU-DOE Plant Research Laboratory and Department of Biochemistry and Molecular Biology, Michigan State University, East Lansing, MI 48824, USA;
[2] Howard Hughes Medical Institute and Plant Biology Laboratory, The Salk Institute for Biological Studies, La Jolla, CA 92037, USA
*Corresponding author: phone 517/432-4619; e-mail larkinr@msu.edu

1. INTRODUCTION

The photoautotrophic lifestyle of plants is absolutely dependent on chloroplasts, which are the products of an endosymbiotic relationship between an ancient eukaryote and a predecessor of modern cyanobacteria. As the endosymbiotic relationship between eukaryotic and cyanobacterial cells gave rise to modern photosynthetic organisms, the overwhelming majority of the ancient cyanobacterial genome was lost or transferred to the nucleus of the eukaryotic host (Herrmann et al., 2003). Currently chloroplast genomes of higher plants are known to encode 60 to 80 proteins, and more than 3500 nuclear genes are predicted to encode chloroplast proteins in *Arabidopsis thaliana* (Martin and Hermann, 1998; Arabidopsis Genome Initiative, 2001). As expected, the nucleus plays a dominant role in chloroplast development, but the expression of nuclear genes that encode chloroplast proteins is also dependent on the functional and developmental state of the plastid. Plastids send signals to the nucleus that are essential for proper expression of nuclear genes that encode proteins with functions related to photosynthesis, coordinating expression of chloroplast and nuclear genomes, and proper leaf morphogenesis (Rodermel, 2001; Surpin et al., 2002; Rodermel and Park, 2003; Gray et al., 2003). Because plastids perform essential metabolic and biosynthetic functions, it is not surprising that the functional state of the plastid has a dramatic influence on gene expression and developmental decisions, but little is known about the molecular mechanisms that plastids use to communicate with other cellular compartments.

Mayfield and Taylor (1984) used maize seedlings in which chloroplast development was arrested at an early developmental stage to provide the first evidence that the developmental state of the plastid has a powerful effect on the expression of nuclear genes that encode particular chloroplast proteins. Subsequently, a number of studies have indicated that proper expression of nuclear genes that encode proteins with functions related to photosynthesis is dependent on normal chloroplast development in diverse monocotyledonous and dicotyledonous plants (Oelmüller, 1989; Gray et al., 2003; Strand et al., 2003). More recent experiments indicate that there are a number of distinct plastid-to-nucleus signaling pathways. Each of these signaling pathways is likely essential for proper metabolism under certain developmental and environmental conditions (Rodermel et al., 2001; Mochizuki et al., 2001; Mullineaux and Karpinski, 2002; Surpin et al., 2002; Rodermel and Park, 2003; Gray et al. 2003; Pfannschmidt, 2003).

The physiological significance of certain plastid signaling pathways seems clear. For example, certain pathways fine tune the expression of nuclear genes that encode components of the photosynthetic machinery to particular light environments (Pfannschmidt, 2003); other pathways induce the expression of nuclear-encoded antioxidant defence proteins as reactive oxygen species accumulate within the chloroplast (Mullineaux and Karpinski, 2002). It is likely that plastid signaling pathways linking chloroplast development to other cellular processes are important as seedlings emerge from underneath the soil and/or ground cover and begin the transition from heterotrophic to photoautotrophic growth. Examples of plastid-to-nucleus signaling pathways that might be important during photomorphogenesis include pathways that link plastid differentiation to leaf morphogenesis (Rodermel, 2001) and pathways that couple the expression of nuclear genes to chloroplast development (reviewed in Oelmüller, 1989; Rodermel, 2001; Surpin et al., 2002; Rodermel and Park 2003; Gray et al., 2003). Plastids appear to use different pathways for influencing leaf development and coordinating the expression of nuclear genes encoding proteins that are active in photosynthesis with chloroplast development (Rodermel, 2001). There is now substantial evidence that the cell uses more than one signal to monitor chloroplast development, and one of these signals appears to be accumulation of particular chlorophyll precursors. In this chapter, we review the data supporting a role for certain chlorophyll precursors in intracellular communication, we discuss current models for chlorophyll precursor signaling, and we discuss future directions of this field.

2. CHLOROPHYLL BIOSYNTHETIC MUTANT, INHIBITOR, AND FEEDING STUDIES

Tetrapyrroles are the intermediates and end products of the chlorophyll, heme, and phytochromobilin biosynthetic pathway (Figure1). Tetrapyrroles are best known for their importance in metabolism, but there is precedence for heme also functioning as a ligand for factors that regulate gene expression in yeast, animal, and bacterial cells (Chen and London, 1995; Ogawa et al., 2001; Zhang and Hach, 1999; O'Brian and Thony-Meyer, 2002). Mg-protoporphyrin IX monomethyl ester (Mg-ProtoMe), a chlorophyll precursor that bears a striking resemblance to heme, was first suggested to act as a regulator of nuclear gene expression in *Chlamydomonas reinhardtii* (Johaningmeier and Howell, 1984). Using mutations and inhibitors that were previously shown to affect accumulation of the chlorophyll precursor Mg-ProtoMe, these researchers reported an inverse correlation between Mg-ProtoMe accumulation and the levels of the light-harvesting chlorophyll a*b*-binding prot ein of photosystem II (Lhcb) mRNA and ribulose 1, 5-bisphosphate carboxylase/oxygenase small subunit RBCS mRNA (Johanningmeier and Howell, 1984; Johanningmeier, 1988: Jasper *et al.*, 1991). Protoporphyrin IX (Proto) and protochlorophyllide (Pchlide) appeared to be less effective regulators of nuclear transcription than Mg-ProtoMe (Johanningmeier and Howell, 1984).

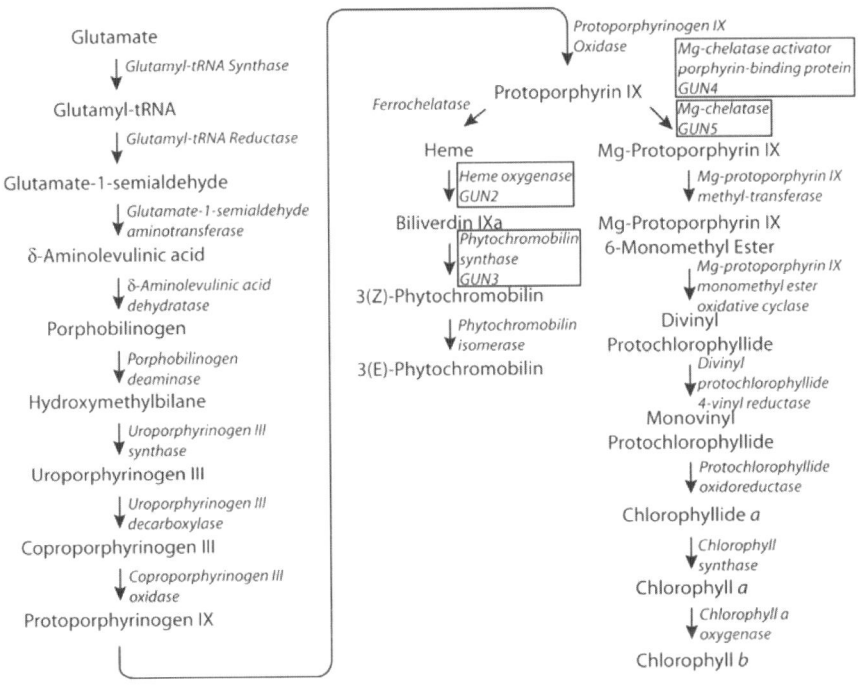

Figure 1. The plastid-localized tetrapyrrole biosynthetic pathway. The reactions in which GUN2, GUN3, GUN4, and GUN5 gene products participate are boxed. Reduction of the 4-vinyl group can occur before or after the reaction catalysed by protochlorophyllide oxidoreductase. Synthesis of common precursors (i.e., glutamate through protoporphyrin IX) and chlorophyll biosynthetic reactions are adapted from Beale (1999). Synthesis of 3(E)-phytochromobilin from protoporphyrin IX is adapted from Terry et al. (1993).

Considering the precedence for heme-regulated gene expression, it is reasonable to propose that certain chlorophyll precursors, which bear a striking resemblance to heme, might regulate the expression of particular nuclear genes in algal and plant cells. Unfortunately, serious technical difficulties have hampered efforts to examine the influence of Mg-ProtoMe and other related tetrapyrroles on the regulation of nuclear gene expression in plants and algae. For example, Mg-ProtoMe is extremely hydrophobic and is not taken up well by plants. Also, Mg-ProtoMe and other chlorophyll precursors are excellent photosensitising agents. Excited triplet states of these porphyrins generate reactive oxygen species via collisions with molecular oxygen, and misregulation of the chlorophyll biosynthetic pathway intermediates can cause lethal photooxidative damage. The photosensitising properties of Mg-ProtoMe and other chlorophyll precursors are especially troubling because some of the most interesting effects of plastid-to-nucleus signaling pathways appear to be directed at light-induced nuclear genes that encode proteins with functions related to photosynthesis. Many of these plastid-regulated genes exhibit minimal expression in the dark. Nonetheless, a number of researchers overcame these formidable technical barriers and subsequently reported additional experimental support for a role for certain chlorophyll precursors in the regulation of nuclear gene expression.

Kropat *et al.* (1997) found that chlorophyll precursor feeding does not appear to have toxic effects on *Chlamydomonas* cells in the dark. In contrast, feeding chlorophyll precursors to cells in the light causes lethal photooxidative damage. Kropat *et al.* (1997) further reported that feeding Mg-protoporphyrin IX (Mg-Proto) and Mg-ProtoMe to *Chlamydomonas* cells induced nuclear *HSP70* genes in the dark (Kropat *et al.*, 1997; 2000). Mg-Proto and Mg-ProtoMe mediated-induction of the *HSP70A* promoter appeared to require a light-responsive promoter element. Low concentrations of Mg-Proto induced *HSP70* genes and induction kinetics were rapid (Kropat *et al.*, 1997). Interestingly, *Chlamydomonas* cells take up Proto and convert it to Mg-Proto in the chloroplast. Because Proto feeding does not induce nuclear *HSP70* gene expression in the dark, Kropat *et al.* (2000) proposed that Mg-Proto and Mg-ProtoMe are exported from the plastid in the light and that export of Mg-Proto and Mg-ProtoMe is required for activation of *HSP70* gene transcription.

Plants do not take up Mg-Proto and Mg-ProtoMe, but it is possible to perturb endogenous levels of these Mg-porphyrins by feeding plants δ-aminolevulinic acid (ALA), which is a small hydrophilic tetrapyrrole precursor. ALA feeding can flood the plastid tetrapyrrole biosynthetic pathway with intermediates and cause accumulation of chlorophyll precursors (Granick, 1959). ALA feeding was reported to repress Lhcb mRNA levels in cress and in *Arabidopsis* seedlings (Kittsteiner *et al.*, 1991; Vinti *et al.*, 2000). In these experiments, Lhcb mRNA levels were monitored after a brief, nonphotosensitising pulse of red light, during continuous illumination with nonphotosensitising far-red light or in complete darkness. Because the accumulation of plastid tetrapyrroles correlated with reduced levels of Lhcb mRNA under all of these nonphotosensitising conditions, these experiments suggest that accumulation of Mg-porphyrins and/or other plastid-derived tetrapyrroles may regulate nuclear transcription in plants.

Certain inhibitor treatments appear to more specifically promote Mg-Proto and Mg-ProtoMe accumulation compared to ALA feeding. For example, the Rüdiger laboratory found that thajaplicin and amitrole treatments that cause Mg-ProtoMe and Mg-Proto accumulation, respectively, specifically inhibit *Lhcb* and/or *RBCS* expression (Oster *et al.*, 1996; La Rocca *et al.*, 2001). In these experiments, *Lhcb* and *RBCS* mRNA levels were reduced after Mg-Proto or Mg-ProtoMe levels were induced in a nonphotosensitising white light fluence, nonphotosensitising far-red light, or complete darkness. These results further support the hypothesis that Mg-Proto/Mg-ProtoMe accumulation might affect nuclear gene expression. Because the ALA-feeding and inhibitor studies were carried out under nonphotosensitising conditions, this work suggests that tetrapyrrole accumulation and not tetrapyrrole-induced photooxidative stress affected nuclear gene expression in these experiments. These results also imply that Mg-ProtoMg-P rotoMe signaling may differ between plants and *Chlamydomonas* because buildup of Mg-ProtoMg-ProtoMe in *Chlamydomonas* chloroplasts does not affect nuclear gene expression in the dark (Kropat *et al.*, 2000), but Mg-porphyrin accumulation in plant plastids has dramatic affects on nuclear gene expression in dark (Vinti *et al.*, 2000; La Rocca *et al.*, 2001).

The plastid signaling studies carried out under nonphotosensitising conditions are important to consider when interpreting other experiments that utilized the photosensitising properties of tetrapyrroles to evaluate the influence of chloroplast development on nuclear transcription. For example, many plastid-to-nucleus signaling studies have employed norflurazon, a photobleaching herbicide. Norflurazon inhibits the synthesis of carotenoids, which quench excited triplet states of chlorophyll. Therefore, when norflurazon-treated seedlings are grown in bright light, substantial amounts of reactive oxygen species are produced, and the resulting photooxidative damage blocks chloroplast development at an early stage (Oelmüer, 1989). Cells containing these developmentally arrested plastids also exhibit severe repression of photosynthesis-related genes that reside in the nucleus. Recently, Strand *et al.* (2003) reported that Mg-Proto accumulates in norflurazon-treated *Arabidopsis* seedlings. Mutations and inhibitor-treatments that block Mg-Proto accumulation in norflurazon-treated *Arabidopsis* seedlings were found to derepress *Lhcb* expression (Strand *et al.*, 2003). As outlined below, a number of mutations that led to decreased accumulation of Mg-Proto resulted in maintenance of *Lhcb* gene expression, even in the presence of norflurazon. Moreover, Strand *et al.* showed that feeding Mg-Proto, but not porphobilinogen, protoporphyrin IX, or heme to *Arabidopsis* protoplasts results in a decrease in *Lhcb* gene expression. Thus, accumulation of Mg-Proto is both necessary and sufficient to activate this signaling pathway.

The results described above provide a compelling evidence that Mg-Proto and Mg-ProtoMe accumulation regulates nuclear transcription in plants and algae. But it is also important to consider whether these particular Mg-porphyrin-induced effects on gene expression might occur during normal growth and development or whether buildup of these Mg-porphyrins is specific to certain unnatural precursor feeding and inhibitor treatments. Actually, Mg-Proto and Mg-ProtoMe levels change dramatically during the diurnal cycle. Mg-Proto and Mg-ProtoMe have been reported to transiently increase 50- to 100-fold at dawn (Pöperl *et al.*, 1998; Papenbrock *et al.*, 1999). When analysed, the ALA feeding and the inhibitor treatments described above caused only a 5- to 15-fold increase in Mg-Proto levels (La Rocca *et al.*, 2001; Strand *et al.*, 2003). Therefore, the experimental conditions commonly used to perturb Mg-Proto and Mg-ProtoMe levels do not induce these porphyrins to accumulate above levels routinely managed by plants during normal growth.

Mg-Proto transiently increases 5-fold and Mg-ProtoMe accumulates to a lesser degree after *Chlamydomonas* cells are transferred from the dark to the light. Therefore, a transient burst of Mg-Proto and Mg-ProtoMe at dawn appears to be conserved in both plants and algae. Moreover, this transient increase in *Chlamydomonas* Mg-Proto and Mg-ProtoMe levels appears to be essential for induction of a nuclear *HSP70* gene after the dark-to-light shift (Kropat *et al.*, 2000). By analogy, the Mg-ProtoMg-ProtoMe buildup at dawn might also affect nuclear transcription in plants. The Arabidopsis *cs* mutant, which contains a mutation in a gene encoding a subunit of the Mg-Proto-producing enzyme Mg-chelatase, exhibits reduced induction of *HSP70* expression after a dark-to-light shift. Surprisingly, however, the Arabidopsis *cch1* mutant, which contains a mutation in a different Mg-chelatase subunit gene, does not exhibit reduced induction kinetics of *HSP70* expression after the dark-to-light shift (Brusslan and Peterson, 2002). These results suggest that Mg-Proto might participate in *HSP70* expression in plants. However, more work will be required to confirm a role for Mg-Proto in the induction of *HSP70* gene expression in plants and to determine why *HSP70* expression is impaired in *cs* mutants but not in cch1 mutants. Exactly how the transient peak of Mg-ProtoMg-ProtoMe affects the plant transcri ptome and whether additional bursts of Mg-ProtoMg-ProtoMe regulate gene expr ession during particular phases of plant development (e.g., photomorphogenesis) remains to be determined.

3. PLASTID-TO-NUCLEUS SIGNALING MUTANTS INHIBIT MG-PORPHYRIN ACCUMULATION

Susek and Chory (1992) developed a reporter gene-based screen in *Arabidopsis thaliana* to isolate mutants in which *Lhcb* expression is uncoupled from chloroplast development. Because the normal coordinated expression of the nuclear and chloroplast genomes is uncoupled in these mutants, they are referred to as *gun* for *genomes uncoupled*. Mutations defining 5 loci were isolated from this screen, *gun1*-*gun5* (Susek *et al.*, 1993; Mochizuki *et al.*, 2001). The recessive nature of these mutations suggests that they affect repressive signaling pathways. Double mutant studies indicate that these five *gun* mutants define two partially redundant signaling pathways: one pathway is defined by *gun1*, the other by *gun2*, *gun3*, *gun4* and *gun5* (Mochizuki *et al.*, 2001).

GUN2, *GUN3*, *GUN4* and *GUN5* have been cloned, and all four of these genes are essential for proper tetrapyrrole biosynthesis in the plastid (Mochizuki *et al.*, 2001; Larkin *et al.*, 2003). The identity of the *GUN1* gene is not known. Besides the intracellular signaling defect, the only phenotype reported for *gun1* mutants is enhanced light sensitivity during deetiolation (Mochizuki *et al.*, 1996). *gun1* mutants do not exhibit defects in tetrapyrrole metabolism (Vinti *et al.*, 2000; Mochizuki *et al.*, 2001), which is consistent with *GUN1* and *GUN2*-*GUN5* encoding components of two distinct signaling pathways. Microarray studies and experiments utilizing plastid translation inhibitors provide additional support to the conclusion that these five *gun* mutants define two different pathways (Gray *et al.*, 2003; Strand *et al.*, 2003).

gun2 and *gun3* are alleles of previously described genes, *HY1* and *HY2* (Mochizuki *et al.*, 2001). *HY1*/*GUN2* encodes heme oxygenase (Davis *et al.*, 1999; Muramoto *et al.*, 1999), and *HY2*/*GUN3* encodes phytochromobilin synthase (Kohchi *et al.*, 2001), both of which participate in phytochromobilin synthesis. Heme oxygenase and phytochromobilin synthase gene mutants in *Arabidopsis*, tomato, and pea produce lower levels of chlorophyll and chlorophyll precursors than do wild type. The reduced chlorophyll accumulation in these mutants is currently believed to result largely from feedback inhibition aimed at glutamyl-tRNA reductase, the second enzyme in the plastid tetrapyrrole biosynthetic pathway. Heme inhibits glutamyl-tRNA reductase, and heme appears to accumulate in heme oxygenase and phytochromobilin synthase gene mutants (Cornah et al., 2003; Franklin et al., 2003). Because of well-established feedback inhibition from the heme/phytochromobilin branch of the plastid tetrapyrrole pathway, gun2 and gun3 were predicted to contain lower levels of Mg-Proto and Mg-ProtoMe compared to wild type (Mochizuki et al., 2001), and these predictions have been confirmed for gun2. gun2 mutants accumulate lower levels of Mg-Proto than wild type when grown in standard conditions or after a Mg-Proto-inducing norflurazon treatment (Mochizuki et al., 2001; Strand et al., 2003). GUN5 encodes the ProtoMg-Proto-binding subunit of Mg-chelatase (Mochizuki et al., 2001), and GUN4 is a previously uncharacterised gene that encodes a ProtoMg-Proto-binding protein that activates Mg-chelatase (Larkin et al., 2003). Therefore, gun4 and gun5 mutants contain lower levels of chlorophyll because of reduced Mg-chelatase activity (Mochizuki et al., 2001; Larkin et al., 2003). Like gun2 mutants, gun5 mutants accumulate lower levels of Mg-Proto than do wild type in the presence or absence of norflurazon (Strand et al., 2003). gun4 alleles are pale green or do not accumulate chlorophyll (Vinti et al., 2000; Mochizuki et al., 2001; Larkin et al., 2003). The paleness of these phenotypes is at least partially due to the substantial reduction or loss of the dramatic stimulatory effect of GUN4 on Mg-chelatase (Larkin et al., 2003). Therefore, it is very likely that norflurazon-treated gun4 seedlings also contain lower Mg-Proto and Mg-ProtoMe levels than norflurazon-treated wild-type seedlings. Thus, gun2-gun5 mutants appear to uncouple Lhcb expression from chloroplast development in photobleached seedlings by inhibiting the accumulation of Mg-Proto. Mutations affecting four additional genes that encode plastid tetrapyrrole biosynthetic enzymes or enzyme subunits were recently reported to cause gun phenotypes (Strand et al., 2003).

The GUN2-GUN5 pathway probably influences nuclear gene expression only during periods of Mg-Proto and Mg-ProtoMe accumulation, and it might be possible to block chloroplast development without causing Mg-ProtoMg-ProtoMe buildup. For example, after blocking plastid development with a plastid translation inhibitor, the gun phenotype was observed in the gun1 but not in the gun2, gun4, or gun5 mutants (Gray et al., 2003). This result may be explained by the inability of plastid translation inhibitors to induce Mg-ProtoMg-ProtoMe accumulation. Thus, this observation provides additional support for the GUN1 gene participating in a signaling pathway that does not monitor the accumulation of specific Mg-porphyrins. Instead, the GUN1 pathway probably senses some other molecular indicator of plastid development that is triggered by both photobleaching herbicides and plastid translation inhibitors. In fact, gun1 mutants have previously been shown to uncouple Lhcb transcription from chloroplast development after chloroplast development was blocked with three mechanistically distinct approaches, including inhibition of plastid translation (Susek et al., 1993).

The *long after far-red 6* (*laf6*) mutant was identified because of defects in phytochrome A signal transduction, but *laf6* also contains elevated levels of protoporphyrin IX (Proto). Interestingly, LAF6 encodes a soluble ATP-binding cassette protein that localizes in the chloroplast periphery. LAF6 was proposed to help retain Proto within the chloroplast, and accumulation of Proto in the cytoplasm was proposed to affect phytochrome A signal transduction in *laf6* mutants (Møller *et al.*, 2001). Although *laf6* mutants exhibit defects in plastid tetrapyrrole metabolism, *laf6* is not a *gun* mutant (Åstrand and J.C., unpublished).

4. MECHANISM OF MG-PROTO/MG-PROTOME SIGNALING

The mechanism by which Mg-ProtoMg-ProtoMe regulates nuclear gene expression is not understood. A popular model suggests that Mg-Proto andór Mg-ProtoMe are exported from the plastid where they participate in a signaling pathway localized in the cytoplasm that affects gene expression in the nucleus (Kropat *et al.*, 1997; Kropat *et al.*, 2000; La Rocca *et al.*, 2001; Mochizuki *et al.*, 2001; Møller *et al.*, 2001; Larkin *et al.*, 2003; Strand *et al.*, 2003; Figure 2). This model is attractive because the plastid appears to be the major site of cellular tetrapyrrole synthesis and other tetrapyrroles are routinely exported from plastids. For example, heme is exported from plastids (Thomas and Weinstein, 1990), and plastids are thought to be the major source of cellular heme synthesis (Cornah *et al.*, 2002). The limited heme biosynthesis in plant mitochondria has been proposed to rely entirely on plastid-derived protoporphyrinogen IX (Beale, 1999; Brusslan and Peterson, 2002). Phytochromobilin and chlorophyll degradation products are also exported from plastids (Terry *et al.*, 1993; Takamiya *et al.*, 2000). This model is also attractive because, as mentioned above, there is precedence for heme functioning as a ligand for regulators of gene expression. An alternative model suggests that the cell monitors flux through the chlorophyll biosynthetic pathway and somehow, possibly via a signaling function of GUN5, uses this information to regulate nuclear gene expression (Mochizuki *et al.*, 2001; Brusslan and Peterson, 2002). This sort of scenario might resemble the situation in *Bradyrhizobium japonicum* in which the iron response regulator Irr represses heme biosynthesis in response to catalytic activity of ferrochelatase (Qi and O'Brian, 2002).

The only proteins known to bind Mg-Proto in vivo are GUN4 and GUN5. Dissociation constants (K_d) for binding to Mg-deuteroporphyrin IX, a more water-soluble derivative of Mg-Proto, have been determined for *Synechocystis* relatives of GUN4 and GUN5. A K_d value of 0.26 ±0.029 µM was estimated for a *Synechocystis* GUN4 relative (Larkin *et al.*, 2003), and a K_d value of 2.43 ±0.46 µM was estimated for *Synechocystis* GUN5 (Karger *et al.*, 2001). Thus, GUN4 binds Mg-deuteroporphyrin IX and probably Mg-Proto with a higher affinity than GUN5 in *Synechocystis*. These two components of the chlorophyll biosynthetic pathway have been localized to the chloroplast stroma and the thylakoid and envelope membranes (Gibson, *et al.*, 1996; Nakayama *et al.*, 1998; Larkin *et al.*, 2003). Interestingly, GUN4 appears to be monomeric in the stroma but associated with large complexes in thylakoid and envelope membranes. The level of GUN4 is extremely low in the thylakoids compared to the levels in the stroma and envelope (Larkin *et al.*, 2003), but there is precedence for localization of chlorophyll biosynthetic enzymes in both thylakoid and envelope membranes (Block *et al.*, 2002).

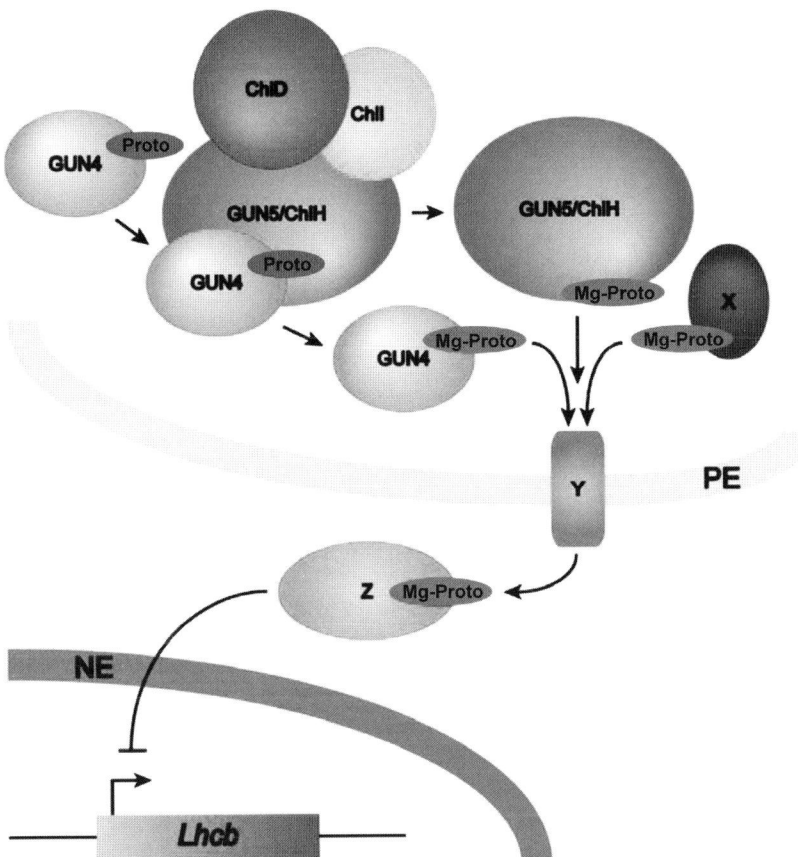

Figure 2. Model for Mg-Proto/ Mg-ProtoMe signalling. GUN5, which is also known as ChlH, two other Mg-chelatase subunits named ChlI ChlD insert Mg^{2+} into the Proto ring with assistance from GUN4. GUN4 binds the substrate and product of the Mg-chelatase reaction, Proto and Mg-Proto. Under conditions of Mg-Proto accumulation, Mg-Proto may be guided to a Tetrapyrrole transporter (Y) by GUN4, GUN5 or by another Mg-Proto-binding protein (X). After export from the plastid, a cytoplasmic factor (Z) binds Mg-Proto. Factor Z affects a signalling pathway that represses Lhcb transcription when chloroplast development is blocked; Mg-Proto regulates the activity of factor Z. Mg-ProtoMe signalling may function in a similar manner. This Proto and Mg-Proto trafficking probably takes place in the plastid envelope, but for the sake of clarity, porphyrin trafficking is not shown to be associated with the envelope. The details of the model are explained in the text. NE, nuclear envelope; PE plastid envelope.

A small GUN4 complex was purified from *Arabidopsis* thylakoids using ion exchange and immunoaffinity chromatography; the thylakoid complex was found to contain GUN4 and GUN5 (Larkin, *et al.*, 2003). The envelope GUN4 complex may also contain GUN5, but this has not been demonstrated. The chloroplast envelope is the expected location for proteins that participate in Mg-Proto/Mg-ProtoMe export; however, a role for GUN4 and GUN5 in Mg-Proto/Mg-ProtoMe trafficking remains to be established even though it is difficult to imagine that extremely lipophilic and photosensitising molecules like Mg-Proto/Mg-ProtoMe are ever permitted to diffuse freely within the cell. Because of the potentially toxic nature of these chlorophyll precursors, particular proteins should be required to target Mg-Proto and Mg-ProtoMe to particular enzymes in the chlorophyll biosynthetic pathway and prevent collisions between these Mg-porphyrins and molecular oxygen. Mg-ProtoMg-ProtoMe binding proteins might also be very important for inhibiting unproductive degradation of these intermediates by catabolic enzymes in the plastid (Whyte and Castelfranco, 1993). Therefore, it is possible that, in addition to catalytic functions, GUN5 and/or GUN4 perform Mg-Proto/Mg-ProtoMe scavenging functions during Mg-Proto/Mg-ProtoMe buildup that are somewhat analogous to the scavenging/photoprotective functions performed by protochlorophyllide oxidoreductase during Pchlide accumulation (Reinbothe *et al.*, 1996). If GUN5 and/or GUN4 turn out to participate in Mg-Proto/Mg-ProtoMe trafficking for the chlorophyll biosynthetic pathway, it seems reasonable to hypothesize that GUN5 and/or GUN4 might also participate in Mg-Proto/Mg-ProtoMe trafficking for a plastid-to-nucleus signaling pathway.

5. PLASTID AND LIGHT SIGNALING PATHWAYS APPEAR TO INTERACT

As mentioned earlier, plastid-to-nucleus signaling pathways that couple nuclear gene expression to chloroplast development severely repress the expression of nuclear genes that encode proteins with functions related to photosynthesis when chloroplast development is blocked. These repressive signaling pathways mediate their effects at the transcriptional level (Oelmüller, 1989; Gray *et al.*, 2003). Several laboratories have identified plastid-responsive promoter elements by carrying out promoter deletion studies in norflurazon-treated transgenic plants. Interestingly, in all plastid-responsive promoters studied to date, plastid and light signaling pathways utilize common promoter elements (Bolle *et al.*, 1994; Lübberstedt *et al.*, 1994; Bolle *et al.*, 1996; Kusnetsov *et al.*, 1996; Puente *et al.*, 1996; McCormac *et al.*, 2001; Martinez-Hernandez *et al.*, 2002; Strand *et al.*, 2003). Light and plastid signals also appear to utilize common promoter elements in *Chlamydomonas* (Kropat *et al.*, 1997; Hahn and Kück, 1999). However, the observation that repressive plastid signals can inhibit the limited dark expression of light-inducible nuclear genes (Sullivan and Gray, 1999; La Rocca *et al.*, 2001) and the observation that light-grown *gun* mutants partially express chloroplast-dependent genes in the absence of chloroplast development (Susek *et al.*, 1993; Mochizuki *et al.*, 2001; Strand *et al.*, 2003) suggests that the plastid and light signaling pathways are distinct pathways that converge upstream of transcription. Plastid-derived signals have been suggested to gate the light signaling pathway (McCormack and Terry, 2001). Because the plastid signal is required for the overexpression of nuclear photosynthesis genes in dark-grown *cop1* and *lip1* mutants, it was suggested that the light and plastid signaling pathways might interact downstream of the COP1 and the LIP1 proteins (Sullivan and Gray, 1999). COP1 is an E3 ubiquitin ligase that targets nuclear proteins for turnover in the dark (Seo *et al.*, 2003), and LIP1 is a pea relative of COP1 (Sullivan and Gray, 2000).

6. CONCLUSIONS AND PERSPECTIVES

A number of ALA feeding, inhibitor, and genetic studies suggest that Mg-Proto and Mg-ProtoMe accumulation within the plastid represses a number of nuclear genes with functions related to photosynthesis. How Mg-ProtoMg-ProtoMe buildup might regulate nuclear gene expression is an open question. The simplest and most attractive model is that Mg-ProtoMg-ProtoMe exits the plastid and interacts with cytosolic signaling pathways. Whether Mg-porphyrins exit the plastid using a plastid envelope ABC transporter, which would be similar to the heme transporters in bacteria (Köster, 2001) and an imal mitochondria (Shirihai *et al.*,2000), or some other export mechanism remains to be established. Because of the lipophilic and photosensitising nature of Mg-Proto and Mg-ProtoMe, these molecules probably utilize carrier proteins (e.g., GUN4 and GUN5) to move within the plastid, and additional porphyrin-binding proteins will likely be required if extraplastidic transport occurs. The nature of interactions between these Mg-porphyrins and cytosolic signaling pathways is also an open question. Heme-binding transcription factors have been described in bacteria, animals and yeast (Zhang and Hach, 1999; Ogawa *et al.*, 2001; Qi and O'Brian, 2002), but whether a similar factor might participate in the plastid-to-nucleus signaling defined by the *GUN2-GUN5* pathway remains to be determined. Proto appears to influence phytochrome A signaling, but the interaction between Proto and the phytochrome A signaling pathway does not appear to involve the *GUN* pathways. Thus, plastid tetrapyrroles may affect multiple signaling pathways.

Additional plastid signaling pathways that may be important during photomorphogenesis include the pathway defined by *GUN1* and the pathway(s) that link chloroplast development to leaf morphogenesis (Rodermel, 2001). The molecular nature of these pathways is not known, although the leaf morphogenesis pathways and the *GUN* pathways appear to be distinct (Rodermel, 2001). In contrast, the *GUN1* and the *GUN2-GUN5* pathways are partially redundant (Mochizuki *et al.*, 2001), although this redundancy is not understood in molecular terms. Identification of additional genes that participate in *GUN* pathways will be critical for understanding both the molecular nature of these pathways and the interactions between them.

Probably the most important issue to resolve is how plastid-to-nucleus signaling pathways participate in photomorphogenesis and whether these signaling pathways influence gene expression and development at other stages of the plant life cycle and during environmental stress. Recently, Richly *et al.* (2003) reported that the *GUN1* and *GUN5* pathways appear to be embedded in a larger network that regulates the nuclear chloroplast transcriptome under a large number of different conditions. Thus, the *GUN* pathways may make important contributions to transcription regulation decisions throughout the plant life cycle.

7. FURTHER READING

Gray, J. C., Sullivan, J. A., Wang, J. H., Jerome, C. A., MacLean, D. (2003) Coordination of plastid and nuclear gene expression. *Philosophical Transactions: Biological Sciences. The Royal Society* 358, 135-144.
Rodermel, S. (2001) Pathways of plastid-to-nucleus signalling. *Trends Plant Science,* 6, 471-478.
Rodermel, S. and Park, S. (2003) Pathways of intracellular communication: Tetrapyrroles and plastid-to-nucleus signalling. *Bioessays,* 25, 631-636.

8. REFERENCES

Arabidopsis Genome Initiative (2000) Analysis of the genome sequence of the flowering plant *Arabidopsis thaliana. Nature,* 408, 796-815.
Beale, S. I. (1999) Enzymes of chlorophyll synthesis. *Photosynthesis Research,* 60, 43-73.
Block, M. A., Tewari, A. K., Albrieux, C., Marechal, E., Joyard, J. (2002) The plant S-adenosyl-L-methionine:Mg-protoporphyrin IX methyltransferase is located in both envelope and thylakoid chloroplast membranes. *European Journal of Biochemistry,* 269, 240-248.
Bolle, C., Kusnetsov, V. V., Herrmann, R. G., Oelmüller, R. (1996) The spinach *AtpC* and *AtpD* genes contain elements for light-regulated, plastid-dependent and organ-specific expression in the vicinity of the transcription start sites. *The Plant Journal,* 9, 21-30.
Bolle, C., Sopory, S., Lubberstedt, T., Klosgen, R. B., Herrmann, R. G., Oelmüller, R. (1994) The role of plastids in the expression of nuclear genes for thylakoid proteins studied with chimeric β-glucuronidase gene fusions. *Plant Physiology,* 105, 1355-1364.
Brusslan, J. A. and Peterson, M. P. (2002) Tetrapyrrole regulation of nuclear gene expression. *Photosynthesis Research,* 71, 185-194.
Chen, J. J. and London, I. M. (1995) Regulation of protein synthesis by heme-regulated eIF-2 alpha kinase. *Trends in Biochemical Sciences,* 20, 105-108.
Cornah, J. E., Roper, J. M., Pal Singh, D., Smith, A. G. (2002) Measurement of ferrochelatase activity using a novel assay suggests that plastids are the major site of haem biosynthesis in both photosynthetic and non-photosynthetic cells of pea (*Pisum sativum* L.) *Biochemical Journal,* 362, 423-432.
Cornah, J. E., Terry, M. J., Smith, A. G. (2003) Green or red: what stops the traffic in the tetrapyrrole pathway? *Trends in Plant Science,* 8, 224-230.
Davis, S. J., Kurepa, J., Vierstra, R. D. (1999) The *Arabidopsis thaliana HY1* locus, required for phytochrome-chromophore biosynthesis, encodes a protein related to heme oxygenases. *Proceedings of the National Academy of Sciences USA,* 96, 6541-6546.
Franklin, K. A., Linley, P. J., Montgomery, B. L., Lagarias, J. C., Thomas, B., Jackson, S. D., *et al.* (2003) Misregulation of tetrapyrrole biosynthesis in transgenic tobacco seedlings expressing mammalian biliverdin reductase. *The Plant Journal,* 35, 717-728.
Gibson, L. C., Marrison, J. L., Leech, R. M., Jensen, P. E., Bassham, D. C., Gibson, M., *et al.* (1996) A putative Mg chelatase subunit from *Arabidopsis thaliana* cv C24. Sequence and transcript analysis of the gene, import of the protein into chloroplasts, and in situ localization of the transcript and protein. *Plant Physiology,* 111, 61-71.
Granick, S. (1959) Magnesium porphyrins formed by barley seedlings treated with δ-aminolevulinic acid. *Plant Physiology,* 34, XVIII.

Gray, J. C., Sullivan, J. A., Wang, J. H., Jerome, C. A., MacLean, D. (2003) Coordination of plastid and nuclear gene expression. *Philosophical Transactions: Biological Sciences. The Royal Society*, 358, 135-144.

Hahn, D. and Kück, U. (1999) Identification of DNA sequences controlling light- and chloroplast-dependent expression of the *lhcb1* gene from *Chlamydomonas reinhardtii*. *Current Genetics*, 34, 459-466.

Herrmann, R. G., Maier, R. M., Schmitz-Linneweber, C. (2003) Eukaryotic genome evolution: rearrangement and coevolution of compartmentalized genetic information. *Philosophical Transactions: Biological Sciences. The Royal Society*, 358, 87-97.

Jasper, F., Quednau, B., Kortenjann, M., Johanningmeier U. (1991) Control of *cab* gene expression in synchronized *Chlamydomonas reinhardtii* cells. *Journal of Photochemistry and Photobiology B: Biology*, 11, 139-150.

Johanningmeier, U. and Howell, S. H. (1984) Regulation of light-harvesting chlorophyll-binding protein mRNA accumulation in *Chlamydomonas reinhardtii*. Possible involvement of chlorophyll synthesis precursors. *Journal of Biological Chemistry*, 259, 13541-13549.

Johanningmeier, U. (1988) Possible control of transcript levels by chlorophyll precursors in *Chlamydomonas*. *European Journal of Biochemistry*, 177, 417-424.

Karger, G. A., Reid, J. D., Hunter, C. N. (2001) Characterization of the binding of deuteroporphyrin IX to the magnesium chelatase H subunit and spectroscopic properties of the complex. *Biochemistry*, 40, 9291-9299.

Kittsteiner, U., Brunner, H., Rüdiger, W. (1991) The greening process in cress seedlings. II. Complexing agents and 5-aminolevulinate inhibit accumulation of *cab* messenger RNA coding for the light-harvesting chlorophyll *ab* protein. *Physiologia Plantarum*, 81, 190-196.

Kohchi, T., Mukougawa, K., Frankenberg, N., Masuda, T., Yokota, A., Lagarias, J. C. (2001) The Arabidopsis *HY2* gene encodes phytochromobilin synthase, a ferredoxin-dependent biliverdin reductase. *Plant Cell*, 13, 425-436.

Köster, W. (2001) ABC transporter-mediated uptake of iron, siderophores, heme and vitain B_{12}. *Research in Microbiology*, 152, 291-301.

Kropat, J., Oster, U., Rüdiger, W., Beck, C. F. (1997) Chlorophyll precursors are signals of chloroplast origin involved in light induction of nuclear heat-shock genes. *Proceedings of the National Academy of Sciences USA*, 94, 14168-14172.

Kropat J., Oster U., Rüdiger W., Beck C. F. (2000) Chloroplast signalling in the light induction of nuclear *HSP70* genes requires the accumulation of chlorophyll precursors and their accessibility to cytoplasm/nucleus. *The Plant Journal*, 24, 523-531.

Kusnetsov, V., Bolle, C., Lubberstedt, T., Sopory, S., Herrmann, R.G., Oelmüller, R. (1996) Evidence that the plastid signal and light operate via the same cis-acting elements in the promoters of nuclear genes for plastid proteins. *Molecular & General Genetics*, 252, 631-639.

Larkin, R. M., Alonso, J. M., Ecker, J. R., Chory, J. (2003) *GUN4*, a regulator of chlorophyll synthesis and intracellular signaling. *Science*, 299, 902-906.

La Rocca, N., Rascio, N., Oster, U., Rüdiger, W. (2001) Amitrole treatment of etiolated barley seedlings leads to deregulation of tetrapyrrole synthesis and to reduced expression of *Lhc* and *RbcS* genes. *Planta*, 213, 101-108.

Lübberstedt, T., Oelmüller, R., Wanner, G., Herrmann, R. G. (1994) Interacting cis-elements in the plastocyanin promoter from spinach ensure regulated high-level expression. *Molecular & General Genetics*, 242, 602-613.

Martin, W. and Herrmann, R. G. (1998) Gene transfer from organelles to the nucleus: how much, what happens, and why? *Plant Physiology*, 118, 9-17.

Martinez-Hernandez, A., Lopez-Ochoa, L., Arguello-Astorga, G., Herrera-Estrella, L. (2002) Functional properties and regulatory complexity of a minimal RBCS light-responsive unit activated by phytochrome, cryptochrome, and plastid signals. *Plant Physiology*, 128, 1223-1233.

Mayfield, S. P. and Taylor, W. C. (1984) Carotenoid-deficient maize seedlings fail to accumulate light-harvesting chlorophyll *a/b* binding protein (LHCP) mRNA. *European Journal of Biochemistry*, 144, 79-84.

McCormac, A. C., Fischer, A., Kumar, A. M., Soll, D., Terry, M. J. (2001) Regulation of *HEMA1* expression by phytochrome and a plastid signal during de-etiolation in *Arabidopsis thaliana*. *The Plant Journal*, 25, 549-561.

Mochizuki, N., Brusslan, J. A., Larkin, R., Nagatani, A., Chory, J. (2001) Arabidopsis *genomes uncoupled 5* (*GUN5*) mutant reveals the involvement of Mg-chelatase H subunit in plastid-to-nucleus signal transduction. *Proceedings of the National Academy of Sciences USA*, 98, 2053-2058.

Mochizuki, N., Susek, R., Chory, J. (1996) An intracellular signal transduction pathway between the chloroplast and nucleus is involved in de-etiolation. *Plant Physiology*, 112, 1465-1469.

Møller, S.G., Kunkel, T., Chua, N.-H. (2001) A plastidic ABC protein involved in intercompartmental communication of light signaling. *Genes & Development*, 15, 90-103.

Mullineaux, P. and Karpinski, S. (2002) Signal transduction in response to excess light: getting out of the chloroplast. *Current Opinion in Plant Biology*, 5, 43-48.

Muramoto, T., Kohchi, T., Yokota, A., Hwang, I., Goodman, H. M. (1999) The Arabidopsis photomorphogenic mutant *hy1* is deficient in phytochrome chromophore biosynthesis as a result of a mutation in a plastid heme oxygenase. *Plant Cell*, 11, 335-348.

Nakayama, M., Masuda, T., Bando, T., Yamagata, H., Ohta, H., Takamiya, K. (1998) Cloning and expression of the soybean *chlH* gene encoding a subunit of Mg-chelatase and localization of the Mg^{2+} concentration-dependent ChlH protein within the chloroplast. *Plant Cell Physiology*, 39, 275-284.

O'Brian, M. R. and Thony-Meyer, L. (2002) Biochemistry, regulation and genomics of heme biosynthesis in prokaryotes. *Advances in Microbial Physiology*, 46, 257-318.

Oelmüller, R. (1989) Photooxidative destruction of chloroplasts and its effect on nuclear gene expression and extraplastidic enzyme levels. *Journal of Photochemistry and Photobiology B: Biology*, 49, 229-239.

Oster, U., Brunner, H., Rüdiger, W. (1996) The greening process in cress seedlings. V. Possible interference of chlorophyll precursors, accumulated after thujaplicin treatment, with light-regulated expression of *Lhc* genes. *Journal of Photochemistry and Photobiology B: Biology*, 36, 255-261.

Ogawa, K., Sun, J., Taketani, S., Nakajima, O., Nishitani, C., Sassa, S., et al. (2001) Heme mediates derepression of Maf recognition element through direct binding to transcription repressor Bach1. *EMBO J.*, 20, 2835-2843.

Papenbrock, J., Mock, H.-P., Kruse, E., Grimm, B. (1999) Expression studies in tetrapyrrole biosynthesis: inverse maxima of magnesium chelatase and ferrochelatase activity during cyclic photoperiods. *Planta*, 208, 264-273.

Pfannschmidt, T. (2003) Chloroplast redox signals: how photosynthesis controls its own genes. *Trends in Plant Science*, 8, 33-41.

Pöpperl, G., Oster, U., Rüdiger, W. (1998) Light-dependent increase in chlorophyll precursors during the day-night cycle in tobacco and barley seedlings. *Journal of Plant Physiology*, 153, 40-45.

Puente, P., Wei, N., Deng, X.-W. (1996) Combinatorial interplay of promoter elements constitutes the minimal determinants for light and developmental control of gene expression in Arabidopsis. *EMBO Journal*, 15, 3732-3743.

Qi, Z. and O'Brian, M. R. (2002) Interaction between the bacterial iron response regulator and ferrochelatase mediates genetic control of heme biosynthesis. *Molecular Cell*, 9, 155-162.

Reinbothe, S., Reinbothe, C., Apel, K., Lebedev, N. (1996) Evolution of chlorophyll biosynthesis-the challenge to survive photooxidation. *Cell*, 86, 703-705.

Richly, E., Dietzmann, A., Biehl, A., Kurth, J., Laloi, C., Apel, K., et al. (2003) Covariations in the nuclear chloroplast transcriptome reveal a regulatory master-switch. *EMBO Reports*, 4, 491-498.

Rodermel, S. (2001) Pathways of plastid-to-nucleus signaling. *Trends in Plant Science*, 6, 471-478.

Rodermel, S. and Park, S. (2003) Pathways of intracellular communication: Tetrapyrroles and plastid-to-nucleus signaling. *Bioessays*, 25, 631-636.

Seo, H. S., Yang, J. Y., Ishikawa, M., Bolle, C., Ballesteros, M. L., Chua, N.-H. (2003) LAF1 ubiquitination by COP1 controls photomorphogenesis and is stimulated by SPA1. *Nature*, 424, 995-999.

Shirihai, O. S., Gregory, T., Yu, C., Orkin, S. H., Weiss, M. J. (2000) ABC-me: a novel mitochondrial transporter induced by GATA-1 during erythroid differentiation. *EMBO Journal*, 19, 2492-2502.

Strand, Å., Asami, T., Alonso, J., Ecker, J. R., Chory, J. (2003) Chloroplast to nucleus communication triggered by accumulation of Mg-protoporphyrinIX. *Nature*, 421, 79-83.

Sullivan, J. A. and Gray, J. C. (1999) Plastid translation is required for the expression of nuclear photosynthesis genes in the dark and in roots of the pea *lip1* mutant. *Plant Cell*, 11, 901-910.

Sullivan, J. A. and Gray, J. C. (2000) The pea *light-independent photomorphogenesis1* mutant results from partial duplication of *COP1* generating an internal promoter and producing two distinct transcripts. *Plant Cell*, 12, 1927-1938.

Surpin, M., Larkin, R. M., Chory, J. (2002) Signal transduction between the chloroplast and the nucleus. *Plant Cell,* S327-S338.
Susek, R. E., Ausubel, F. M., Chory, J. (1993) Signal transduction mutants of Arabidopsis uncouple nuclear *CAB* and *RBCS* gene expression from chloroplast development. *Cell, 74,* 787-799.
Susek, R. and Chory, J. (1992) A tale of two genomes: role of a chloroplast signal in coordinating nuclear and plastid genome expression. *Australian Journal of Plant Physiology, 19,* 387-399.
Takamiya, K. I., Tsuchiya, T., Ohta, H. (2000) Degradation pathway(s) of chlorophyll: what has gene cloning revealed? *Trends in Plant Science, 5,* 426-431.
Terry, M. J., Wahleithner, J. A., Lagarias, J. C. (1993) Biosynthesis of the plant photoreceptor phytochrome. *Archives of Biochemistry and Biophysics, 306,* 1-15.
Thomas, J. and Weinstein, J. D. (1990) Measurement of heme efflux and heme content in isolated developing cotyledons. *Plant Physiology, 94,* 1414-1423.
Vinti, G., Hills, A., Campbell, S., Bowyer, J. R., Mochizuki, N., Chory, J., *et al.* (2000) Interactions between *hy1* and *gun* mutants of *Arabidopsis*, and their implications for plastid/nuclear signaling. *The Plant Journal, 24,* 883-894.
Whyte, B. J. and Castelfranco, P. A. (1993) Breakdown of thylakoid pigments by soluble proteins of developing chloroplasts. *Biochemical Journal, 290,* 361-367.
Zhang, L. and Hach, A. (1999) Molecular mechanism of heme signaling in yeast: the transcriptional activator Hap1 serves as the key mediator. *Cellular and Molecular Life Sciences, 56,* 415-426.

Chapter 24

PHOTOMORPHOGENESIS OF FERNS

Takeshi Kanegae[1] and Masamitsu Wada[1,2]

[1]*Department of Biological Sciences, Graduate School of Science, Tokyo Metropolitan University, Minami-Osawa 1-1, Hachioji, Tokyo 192-0397, Japan;* [2]*Division of Photobiology, National Institute for Basic Biology, Myodaiji, Okazaki 444-8585, Japan*
(e-mail: wada-masamitsu@c.metro-u.ac.jp)

1. INTRODUCTION

During the 10 years since the publication of the 2nd edition of this book, the molecular photobiology of ferns progressed very rapidly. In particular, 10 photoreceptors of *Adiantum capillus-veneris*, namely 5 cryptochromes, 3 phytochromes, and 2 phototropins have been cloned and sequenced (Figure 1). The functions of two photoreceptors, phytochrome3 and phototropin2, have been elucidated as described below in detail. Ten years ago, isolation of proteins from fern gametophytes was almost impossible because sufficient tissue for extraction could not be obtained. In 1988 we decided to extract phytochrome protein from young leaves of *Adiantum capillus-veneris*. Young leaves were collected and the tissue accumulated for 2 years in a freezer. Using the frozen tissue we extracted phytochrome and measured the difference spectrum of the phytochrome solution (Oyama *et al.*, 1990). We also detected a phytochrome signal in dark-grown living tissue of *Adiantum* (Oyama *et al.*, 1990). But although the extracted phytochrome was likely a mixture of several different species of phytochrome molecules, their spectra were likely the same. However, with the rapid progression of the techniques of molecular biology, now we can clone and sequence fern phytochrome genes rather easily referring to the homologous sequences of phytochromes from other plants.

Analysis of mutant genes also becomes possible in fern mutants using fluorescence differential display (FDD) (Uchida *et al.*, 1998), etc. Thus, we decided to screen mutants from *Adiantum* and obtained several kinds of photomorphogenesis-related mutants such as *rap2* (red light aphototropic 2) and *bhc-07* and *bhc-08* (blue light high fluence-dependent chloroplast movement). Fern gametophytes are good material to screen mutants because of their haploid character and also because each individual is very small, with from 1 to several tens of cells when screened. When spores are mutagenised by ethyl methanesulfonate (EMS) or a heavy ion beam, the germinated gametophytes are in the haploid generation; thus the mutant gametophytes should show a phenotype even if the phenomenon is controlled by a recessive gene. On the other hand, if the mutated genes are critical for survival, the mutations are lethal and cannot be rescued unless we use the diploid generation for mutant screening.

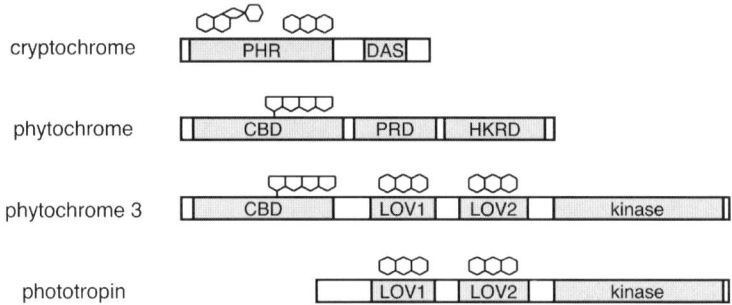

Figure 1. Photoreceptors in Adiantum

The molecular biology of ferns has progressed well in the last decade as described above, however improvement of the molecular biological techniques applicable to cell biological or physiological analyses of fern photomorphogenesis is not yet sufficient. Stable transformation of fern has not yet been established, although transient expression by particle bombardment is possible. We do not yet know the reason why foreign gene integration is not readily obtainable in fern gametophyte cells. Although stable transformation is elusive at this time, gene silencing by RNA interference (RNAi) is easily obtained in comparison with other higher plant systems. It is reported that double stranded RNAs can enter into fern spores during imbibition and gene silencing can be observed (Klink and Wokniak, 2001; Stout *et al.*, 2003). It is particularly worth noting that, in *Adiantum*, double stranded DNAs introduced into gametophyte cells trigger sequence-dependent gene silencing (Kawai-Toyooka *et al.*, 2004). Because of the similarity to RNAi, we call this phenomenon DNA interference (DNAi). DNAi will become quite a useful technique when established, because particle bombardment of PCR products of target genes is enough to induce gene silencing. Gene targeting by homologous recombination is also a powerful tool for analyzing gene functions. Recently, homologous recombination has become a routine method of targeted disruption/replacement of the genomic sequences in the moss *Physcomitrella patens* (Schaefer, 2002). Based on the similarity of their physiological and morphological features between moss and fern gametophytes, we look forward to successful targeted gene disruption being applied to ferns.

In this chapter, findings on fern photomorphogenesis and chloroplast photorelocation movement at the molecular level in the last ten years are summarized. Detailed discussion of the various physiological and cell biological phenomena in fern gametophytes can be found in the relevant chapter in the 2nd edition or in other reviews (Wada and Sugai, 1995; Wada and Kadota, 1989). Information on chloroplast photorelocation movements is to be found in several recent reviews (Kagawa and Wada, 2002; Kagawa, 2003; Wada *et al.*, 2003; Sato *et al.*, 2003).

2. PHOTORECEPTORS IN *ADIANTUM*

2.1 Cryptochromes

From screening of an *Adiantum* genomic library and consequent isolation of cDNAs corresponding to all the genes of interest, we have identified five cryptochrome genes in *Adiantum*. We named these genes as *Adiantum* cryptochrome 1 (*AcCRY1*), 2 (*AcCRY2*), 3 (*AcCRY3*), 4 (*AcCRY4*) and 5 (*AcCRY5*) (Figure 2) (Kanegae and Wada, 1998; Imaizumi *et al.*, 2000). AcCRY1, 2, 3, 4 and 5 encode polypeptides of 637, 679, 718, 699, and 487 amino acids respectively, and their predicted molecular masses are 72-, 76-, 80-, 79-, and 56-kDa, respectively. Their deduced amino acid sequences indicate that all AcCRYs have a photolyase-related domain (PHR) in their amino terminal part. Except for AcCRY5, *Adiantum* cryptochromes also contain a carboxy-terminal extension as do other plant cryptochromes. These C-terminal extensions include the DAS domain (for DQXVP-acidic-STAES: see Lin, 2002). Although AcCRY5 lacks a C-terminal extension, recombinant AcCRY5 did not complement a photolyase-deficient *E. coli* strain, suggesting that AcCRY5 might still be a cryptochrome rather than a photolyase (Imaizumi *et al.*, 2000). Other AcCRYs also lacked the photolyase activity in the *phr⁻ E. coli*, as was the case for Arabidopsis CRYs (Hoffman *et al.*, 1996; Lin *et al.*, 1995; Malhotra *et al.*, 1995).

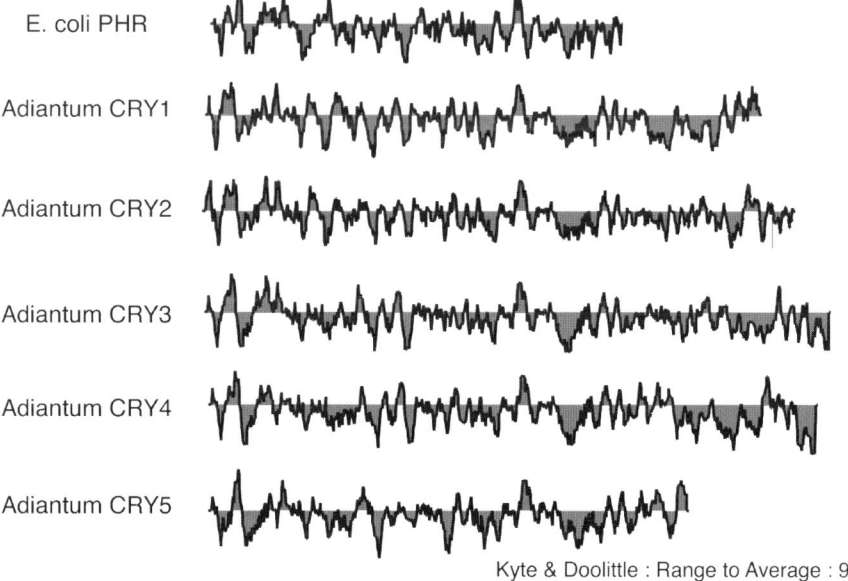

Figure 2. Hydrophilicity / hydrophobicity plot of Adiantum CRYs

AcCRY1, *2* and *3* mRNAs were shown to be constitutively expressed throughout the different developmental stages and their expression levels are likely not affected by the different light conditions. On the other hand, *AcCRY4* gene expression appears to be down regulated by light and the *AcCRY5* transcript appears to accumulate mainly in sporophyte tissues. During spore germination, the *AcCRY5* mRNA level rapidly increased up to 300- to 400-fold within the first 12 hr after the onset of red and blue light irradiation, whereas amount of *AcCRY4* mRNA was reduced by ~50-fold during the 24 hr after red light treatment. Further analysis with far-red light irradiation demonstrated that the light regulation of *AcCRY4* and *AcCRY5* gene expression is partly under the control of phytochrome. Examination of the intracellular distribution of GUS-AcCRY fusion proteins indicated that GUS-AcCRY1, GUS-AcCRY2 and GUS-AcCRY5 are localized primarily in the cytoplasm under the various light conditions tested. GUS-AcCRY3 was also distributed in the cytoplasm under blue light, but tended to accumulate in the nucleus in the dark and in red light. GUS-AcCRY4 was localized predominantly in nucleus under all light conditions investigated. Our previous physiological observations indicated that the blue light receptor involved in the inhibition of spore germination (Furuya *et al.*, 1997) and induction of cell division (Wada and Furuya, 1978) is located in or close to the nucleus. Together with their gene expression profiles and the GUS-AcCRYs' nucleocytoplasmic distribution, AcCRY3 and/or AcCRY4 are likely candidates for the photoreceptor mediating blue light-dependent inhibition of spore germination (Furuya *et al.*, 1997) and induction of cell division in protomemal cells (Wada and Furuya, 1978).

2.2 Phototropins

Phototropin (formerly NPH1 for Non-Phototropic Hypocotyl) was first cloned and sequenced in 1997 (Huala *et al.*, 1997), and identified as a blue light receptor which mediates a weak light-induced phototropic response in Arabidopsis seedlings (Christie *et al.*, 1998). Soon after, phot2 (formerly NPL1, NPH1 like 1) was found as a homolog although the function was not known (Jarillo *et al.,* 1998). NPH1 and NPL1 were renamed phototropin1 (phot1) and phototropin2 (phot2), respectively (Briggs *et al.*, 2001), as they were photoreceptor of phototropism, although phototropins were later shown to be photoreceptors for chloroplast photoorientation movement (Jarillo *et al.*, 2001; Kagawa *et al.*, 2001; Sakai *et al.*, 2001), stomata opening (Kinoshita *et al.*, 2001), and leaf expansion (Sakai *et al.*, 2001; Sakamoto and Briggs, 2002) etc. Phototropin has a protein kinase domain in the C-terminus and two chromophore binding domains (LOV1 and 2) in its N terminus half (see Chapter 12 and reviews by Briggs and Christie, 2002).

Adiantum PHOT1 (formerly *NPH1*) was first cloned by RT-PCR with cDNA prepared from gametophytes as templates with degenerate primers between LOV1 and protein kinase subdomain X. The resulting PCR product was re-amplified with a set of nested degenerate primers between LOV2 and kinase subdomain IX. Finally, both 3'- and 5'-RACE (Rapid Amplification of cDNA Ends) cloning were carried out. Sequence analysis of these clones showed that the complete *Adiantum PHOT1* cDNA clone is 3,492bp in length and encodes a protein of 1,092 amino acids with a predicted molecular weight of 122-kDa. Southern blot analysis showed that *AcPHOT1* is likely to be a single-copy gene. Thin layer chromatography of the chromophore released from the AcPHOT1 LOV2 fusion protein showed that the chromophore attached to the LOV2 domain is flavin mononucleotide (FMN) (Nozue *et al.*, 2000).

Adiantum PHOT2 was cloned by RT-PCR using sporophyte RNA extracted from young wild-type leaves with degenerate primers which were designed for LOV2 and kinase consensus regions with reference to Arabidopsis and other plant phototropins including *AcPHOT1*. The full-length cDNA sequence was determined by 5'-, 3'-RACE. The cDNA clone is 3,412 bp in length and encodes a protein of 1,019 amino acids and the predicted molecular weight is ca. 114-kDa. The genomic DNA sequence of 11.4 kbp consists of 23 exons and 22 introns and intron insertion positions are conserved among formerly identified phototropin genes (Kagawa *et al.*, 2004). *AcPHOT1* gene has a similar exon and intron structure (Kagawa and Wada, unpublished data). *Adiantum* phototropins, Acphot1 and Acphot2, were named according to the homology to Arabidopsis phot1 and phot2, respectively, based on a phylogenic tree. The homology between Atphot1 and Acphot1 at the amino-acid level is 52%, and that between Atphot2 and Acphot2 is 59%. The *AcPHOT1* and *AcPHOT2* genes are expressed both in sporophytes and gametophytes.

2.3 Phytochromes

To date quite a few fern phytochrome gene sequences have been registered in the databases. These data reveals that fern phytochromes are encoded by a small gene family as in seed plants (Figure 3). In 1993, we first reported the isolation of a phytochrome cDNA (*FP1*) from *Adiantum* (Okamoto *et al.*, 1993). Using this phytochrome sequence as a probe, we screened the *Adiantum* genomic library and obtained four different genome segments. We named these fragments *Adiantum* phytochrome 1 (*AcPHY1*), 2 (*AcPHY2*), 3 (*AcPHY3*) and 4 (*AcPHY4*) (Wada *et al.*, 1997). The nucleotide sequence of phytochrome 1 is an exact match to the previously isolated cDNA sequence (*FP1*); thus we renamed *FP1 AcPHY1*.

Our earlier spectrophotometric studies of *Adiantum* phytochrome using intact leaves demonstrated a photoreversible absorbance change in the far-red region (Oyama *et al.*, 1990). The difference spectra also indicated that the amount of *Adiantum* phytochrome decreased when the plants were transferred to continuous white light and increased when the plants were transferred to the darkness, but both the rate of decay and the rate of accumulation were slower than those of angiosperms. These measurements were performed using whole tissues grown under darkness or light, so that the multiple phytochrome species described above were likely be included; hence to obtain meaningful data elucidating the physiological function of each phytochrome species it is necessary to examine the molecular characteristics of each *Adiantum* phytochrome at the beginning.

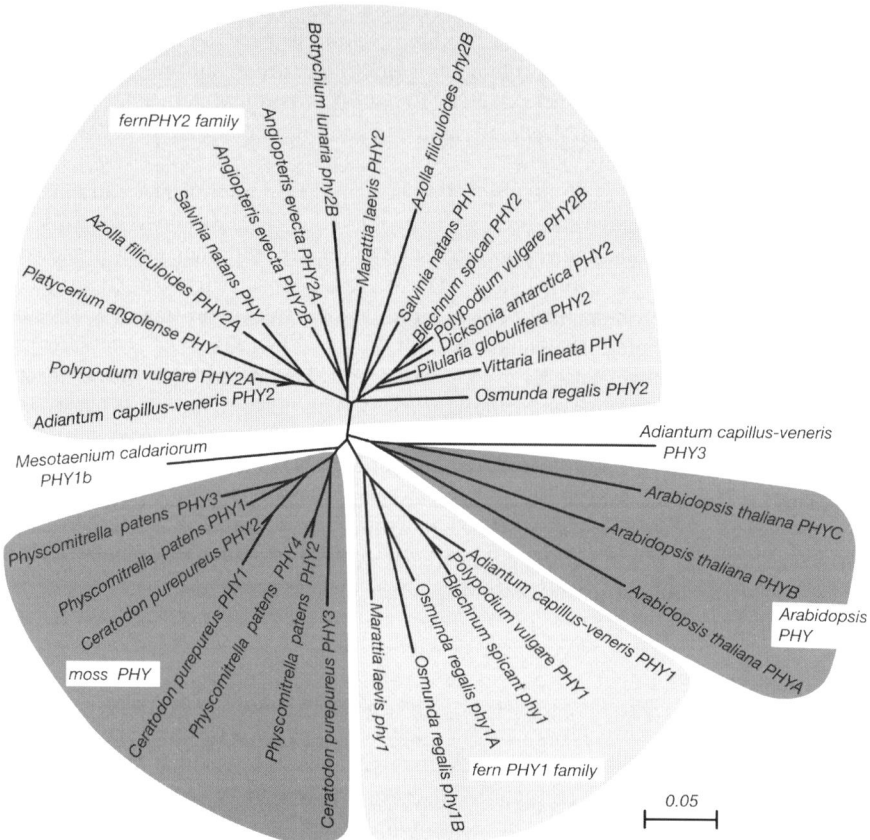

Figure 3. Phylogenetic tree showing the fern phytochrome gene lineages

2.3.1 Phytochrome 1

AcPHY1 encodes a polypeptide of 1,118 amino acids (predicted molecular weight is 124-kDa) and the deduced amino acid sequence exhibits the domain structure similar to that of the conventional phytochromes.

The structural domain order for AcPHY1 is as follows; from amino-terminus, a photosensory domain (PSD) including the chromophore-binding domain (CBD), a PAS (Per-Arnt-Sim)-related domain (PRD) and a histidine kinase-related domain (HKRD) (Yeh and Lagarias, 1998). As in seed plant phytochromes, the histidine residue conserved in the bacterial histidine kinases is not found in AcPHY1. Intron insertion positions in the *AcPHY1* gene are also similar to those of the seed plant phytochromes. *In situ* hybridisation experiments demonstrate that *AcPHY1* mRNA is predominantly localized to the abaxial rather than adaxial cells of the petiole of leaf crosiers (Okamoto *et al.*, 1997b). It is possible that this asymmetric distribution of *AcPHY1* mRNA may play a role in the differential cell growth between abaxial and adaxial cells, leading to leaf uncoiling. Interestingly, *AcPHY1* mRNA is localized primarily in the nucleus in some of the cells of the light-grown leaf crosiers, while hybridisation signals were observed uniformly in both nucleus and cytoplasm in dark-grown tissues. There was little difference in the amount of *AcPHY1* transcript between light-grown and dark-grown tissues, suggesting that *AcPHY1* gene expression is not regulated by light. By contrast, western blot analysis using anti-AcPHY1 polyclonal antibody indicates that AcPHY1 polypeptides are more abundant in dark-grown than light-grown plants. This result is similar to that for angiosperm PHYA, a light-labile phytochrome. However, constitutive expression and nuclear retention of *AcPHY1* mRNA in light-grown leaves suggests an alternative explanation; i.e. synthesis of AcPHY1 polypeptides is regulated by the export of *AcPHY1* mRNA from the nucleus.

We also analysed the photoresponses of transgenic Arabidopsis expressing AcPHY1 polypeptides (Okamoto *et al.*, 1997a). Spectroscopic analysis of the transgenic lines examined indicated both that *Adiantum* phytochrome can incorporate the chromophore from the host plants and that substantial Acphy1 holoprotein was accumulated in the transgenic plants. We then investigated inhibition of hypocotyl elongation of the transgenic lines under red and far-red light. Slightly longer hypocotyls than those of wild-type plants were observed in all of the transgenic lines under continuous far-red light, suggesting a certain degree of interference with endogenous phyA action. However, no significant difference was detected between the hypocotyl lengths of transgenic lines and those of wild-type plants under continuous red light. As a result, transgenic Arabidopsis expressing AcPHY1 did not show any remarkable difference from wild-type plant with respect to the light repression of hypocotyl elongation. Our observations suggest that AcPHY1 might be too divergent from angiosperm phytochromes to act efficiently in distantly related plant species as Arabidopsis.

2.3.2 Phytochrome 2

Using the nucleotide sequence of the *AcPHY2* genome segment, we isolated full-length *AcPHY2* cDNA by 5'- and 3'-RACE methods (Nozue *et al.*, 1998a). *AcPHY2* encodes a protein of 1,140 amino acids and has a deduced molecular mass of 127-kDa. Like AcPHY1, the AcPHY2 polypeptide is comprised of three domains; PSD (including the CBD), PRD and HKRD, and the composition and order of the domains are same as those of conventional phytochromes. As with AcPHY1, the conserved histidine phosphorylation site in the sensor histidine-kinase domain of bacterial two-component signalling systems is not found in HKRD of AcPHY2. During sequencing the *AcPHY2* genome segment, we encountered a rather long intron after exon 2. This intron is longer than 2.5Kbp, but the finding of a cDNA corresponding to the *AcPHY2* gene suggests that this intron is processed normally. This second intron contains a part of a *Ty3/gypsy*-type retrotransposon element (*ARET-2*). Interestingly, this type of retrotransposon was also found downstream of another phytochrome-like genome sequence (*AcPHY4*), as we will discuss later.

As described above, structural similarity with conventional phytochromes suggests that AcPHY2 as well as AcPHY1 might show a serine/ threonine kinase activity like that of *Avena* PHYA and *Mesotaenium* PHY1b (Yeh and Lagarias, 1998). It is also of a great interest whether the signalling mechanisms and its components of these *Adiantum* photoreceptors are related to those of conventional phytochromes. To answer these questions, more experimental evidence is needed. However, some homologous sequences for phytochrome signalling components identified so far in higher plants, such as COP1 and a PIF3-like basic Helix-Loop-Helix protein, have been found in our *Adiantum* EST data (Yamauchi, Sutoh, Kanegae, Horiguchi, Matsuoka, Fukuda, and Wada, unpublished data).

2.3.3 Phytochrome 3

AcPHY3 was cloned as one of the phytochromes obtained from the genomic library. It appears to be intronless, containing uninterrupted long open reading frame of 4,395bp. *AcPHY3* encodes a polypeptide of 1,465 amino acids and its predicted molecular mass is 161-kDa. AcPHY3 is a chimeric protein with the chromophore-binding domain of phytochrome at its N-terminal end and a full-length phototropin at its C-terminal end. AcPHY3 expressed in yeast binds phycocyanobilin, a relative of the native phytochrome chromophore phytochromobilin, and the bound form is red/far-red photoreversible, demonstrating that Acphy3 acts as a phytochrome (Nozue *et al.*, 1998b). The LOV domains from AcPHY3 were found to bind FMN (Christie *et al.*, 1999), suggesting that Acphy3 may also function as a blue light receptor as well as a phytochrome.

Given the *AcPHY3* gene structure without any introns, although the fern phototropin genes have 22 introns, it is plausible to think that *PHY3* might have been generated as a result of retrotransposition. The most likely scenario is that a phototropin mRNA was reverse transcribed and inserted into a phytochrome locus. We have not so far found an oligo(A) tract at its 3' end, an observation that strengthens the case for reverse transcription, however. If this speculation is correct, the *PHY3* gene should be found only in the ferns that evolved after the occurrence of this retrotransposition event.

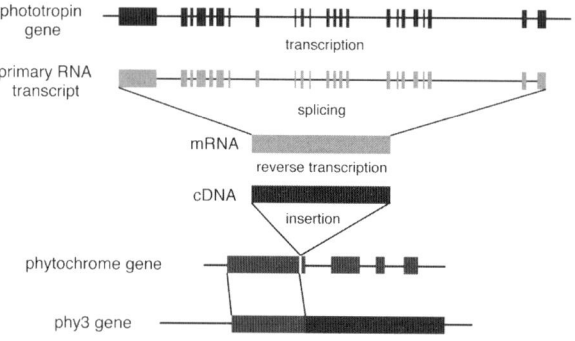

Figure 4. How was PHY3 made?

AcPHY3-homologous sequences were also found in ferns such as *Dryopteris filix-mas*, *Onoclea sensibilis*, and *Hypolepis punctata*, that are within the 'polypodiaceous' ferns sensu lato, i.e. advanced ferns (Kawai *et al.*, 2003). However, *AcPHY3*-homologous sequences could not be detected in the more primitive ferns such as *Osmunda japonica* (Osmundaceae) and *Lygodium japonicum* (Schizaeaceae). Based on the fossil record, the Osmundaceae, the oldest extant leptosporangiate fern family, arose during the Permian, the Schizaeaceae during the Triassic. Other more advanced fern families including the 'polypodiaceous' ferns appeared thereafter from the Jurassic to the Cretaceous. It is reasonable to think that the original *PHY3* appeared before the Jurassic in a polypodiaceous fern ancestor.

2.3.4 Phytochrome 4

In addition to these three *Adiantum* phytochrome genes, we have isolated one more genome segment that contains a protein coding sequence (CDS) very similar to CBD of well-known phytochromes. This CDS encodes a polypeptide of 424 amino acids, and corresponds to a part of the first exon of the CDS in *AcPHY1* and *AcPHY2*. Although the amino terminus region corresponding to 253 and 277 amino acids of AcPHY1 and AcPHY2, respectively, are missing, this CDS includes a phytochrome chromophore-binding region and shows very high similarity to AcPHY1 and AcPHY2 through that region (80.0% and 91.5% identity, respectively). Thus we tentatively refer to this CDS as *Adiantum PHY4*. Surprisingly, this *AcPHY4* CDS is followed by a full-length retrotransposon (*ARET-1*) that has long terminal repeats (LTRs) (Nozue *et al.*, 1997). *ARET-1* is 8,284 bp long and has all of the domains and elements characteristic of a retrotransposon. *ARET-1* belongs to a *Ty3/gypsy* type LTR-retrotransposon and this type of retrotransposon was also found in the second intron of *AcPHY2* (*ARET-2*) described above. Multiple stop codons found in all three open reading frames of *ARET-1* suggest that this retrotransposon element might not be active. It is not known whether these *ARET*s have some role in *Adiantum* phytochrome function. However, there is a great interest in influence of such retrotransposons on genome organization and, if any, on gene expression in *Adiantum*.

3. MUTANT ANALYSES

3.1 Methods of mutant selection

A number of mutants of photomorphogenesis and chloroplast distribution have been selected from *Adiantum capillus-veneris*. *Adiantum* dry spores were treated with EMS overnight or irradiated by a heavy ion beam of nitrogen or neon. The mutagenised spores were sown aseptically on the surface of a solidified agar medium and cultured under various light conditions, such as red light from one direction to select mutants of phototropic response under red light, or strong blue light from top to select mutants deficient in the avoidance response of chloroplasts.

Spores were sterilized by treatment with commercial bleach and sown as fairly dense lines on the surface of agar medium in a glass Petri dish of 3cm in diameter and covered by a small piece of cover glass (3 x 18 mm). Polarized red light was irradiated from above through the cover glass with the electrical vector parallel to the lines of spores. Since wild-type protonemata grow perpendicular to the plane of vibration of the electrical vector of the polarized red light, about half of the protonemata grow in parallel towards the right-hand side of the spore lines and the other half grow towards the left-hand side. However, mutants deficient in the red light-dependent tropic response grow in a random direction, so these mutants are easily detected. Red light aphototropic mutants were also detected by another method. Mutagenised spores were sown on an agar medium at low density and irradiated from one side to let them grow on an agar medium towards the red light source. Mutant protonemata could be detected because of their growth in random directions (Kadota and Wada, 1999).

The protonemata cultured under polarized red light on an agar surface were then irradiated with strong white light to induce an avoidance response of chloroplasts after pre-irradiation of weak white light from top to establish a pre-condition of chloroplast accumulation on the top side of the protonemata (Figure 5). Chloroplasts of wild-type cells move to both side of the cell, but in mutant cells chloroplasts cannot move away, and instead stay at the top side of the cells. Those mutants deficient in the avoidance response are easily found because most of the cells (i.e. wild-type in avoidance response) looks transparent but the mutant cells are fairly green depending upon the chloroplasts position at the cell surface.

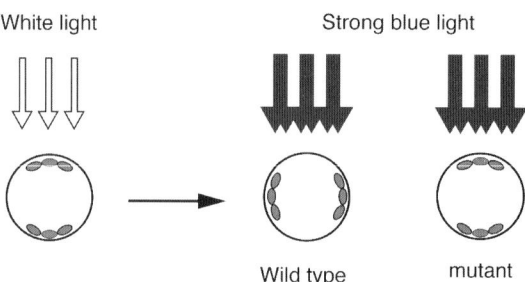

Cross section of a protonemal cell

Figure 5. Screening for mutants deficient in the chloroplast avoidance movement

Similarly, mutants showing deficient movement of chloroplasts to the dark position could also be screened in the cultures. The cells of gametophytes of just a few cells were irradiated with weak white light from the top, and then were put in the dark for two days to let the chloroplasts move to the dark position, that is to the anticlinal walls of the gametophytes. Most of the cells look transparent because of the dark positioning, but the mutant cells are green because the chloroplasts cannot move to the anticlinal walls from the periclinal walls.

Mutant screening does not require a lot of time. If protonemal cells are sown densely, tens of these cells can be seen in a single microscope field, and hundreds of cells can be observed within a few minutes. Many kinds of mutants can easily be screened if we are well informed about the physiological behaviour of these cells. Morphological mutants such as SK (Suka-suka, meaning vacant or transparent in Japanese) are also not difficult to obtain by similar mutant screening (Figure 6).

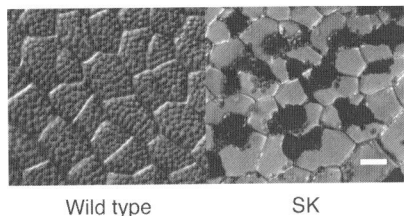

Figure 6. Chloroplast distribution of SK mutant. Bar=30μm

3.2 Red light aphototropic mutants

All *Adiantum* red-light aphototropic (*rap*) mutants tested were deficient both in red light-induced phototropism and the red light-induced chloroplast accumulation response in protonemal cells, whereas the same responses were normal under blue light. One of the *rap* mutants, *rap2* is also deficient in both of these physiological responses in leaves, indicating that the same photoreceptor is involved in these red-light responses both in gametophytes and sporophytes. However, the signal-transduction pathways downstream must be different in the phototropic response of tip-growing protonemal cells and the phototropically differential growth in multicellular petioles. *AcPHY3* genomic sequences of ten *rap* mutants generated by EMS all show defects in the *AcPHY3* gene as shown in Figure 7.

As genetic analysis of mutated genes or stable transformation for complementation testing is not possible in *Adiantum* at the moment, a transient-expression assay in the *rap2* mutant was performed by co-transfer with cDNA encoding *AcPHY3* and GFP driven by the cauliflower mosaic virus (CaMV) 35S promoter with a particle bombardment system (Kawai *et al.*, 2003). The chloroplast accumulation response was rescued in the transformed *rap2* cell when it was irradiated with a red light microbeam, whereas chloroplasts failed to gather in cells transformed only with GFP or in untransformed cells used as controls. Thus, AcPHY3 functions as the photoreceptor for red light-induced chloroplast movement. This transient assay system using mutant gametophytes is very convenient and trustworthy, because the neighbouring cells in the same gametophytes are always good controls for the assay. However, for testing the tropistic growth of protonemal cells it did not work well, because the damage from particle bombardment itself was severe and the bombarded protonemal cells died.

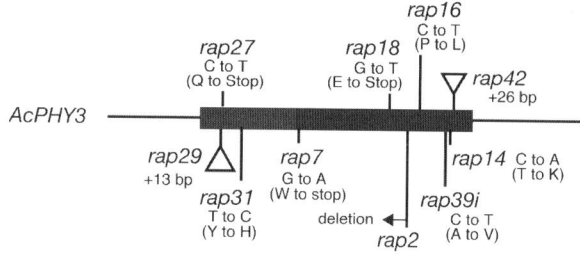

Figure 7. Positions of AcPHY3 mutations in rap mutants

3.3 Mutants deficient in the chloroplast avoidance response

Adiantum gametophytes deficient in the chloroplast avoidance response under strong white light were selected. Eleven putative *bhc* (blue high light-dependent chloroplast movement) mutants thus obtained were tested more critically with strong microbeam irradiation and two (*bhc-07* and *bhc-08*) among them were selected as genuine mutants. Chloroplasts gathered into the area irradiated even with a strong blue microbeam (10 Wm^{-2}) (under a background of red light for observation). As we thought that *AcPHOT2* might be the mutated gene, the *AcPHOT2* gene from these two mutants was sequenced and a lesion in this gene was found in both mutants.

The *AcPHOT2* gene in *bhc-07* has deletions of 2 nucleotides (TA) in the 4th exon and of one nucleotide (T) in the 4th intron whereas *bhc-08* has a deletion of 26 nucleotides (TA...AT) and an addition of 2 extra nucleotides (AG) in the first exon, resulting in a loss of the start codon (Figure 8). When prothallia of *bhc-07* and *bhc-08* mutants were co-transfected with cDNA of *AcPHOT2* and GFP both driven with the CaMV 35S promoters, the cells with GFP fluorescence showed an avoidance response under strong blue microbeam (15 Wm^{-2}) whereas the neighbouring cells without GFP fluorescence showed only chloroplast accumulation and no avoidance response under the same irradiation treatment. The cells expressing only GFP by introduction of GFP cDNA alone did not recover any avoidance movement under strong blue light. Thus, *AcPHOT2* is shown to be the photoreceptor for the chloroplast avoidance response in *Adiantum* as in the case of Arabidopsis phot2 (Kagawa *et al.*, 2004).

Figure 8. Positions of AcPHOT2 mutation in bhc mutants

3.4 Dark position-deficient mutants

When fern gametophytes are kept in darkness for one to two days, all of the chloroplasts move to anticlinal walls from periclinal walls. This phenomena was well known in the 19th century in many plants, although not many papers on the movement in darkness were published comparing to those of the accumulation and avoidance responses. Senn (1908) did some experiments on this movement in the dark and concluded that the phenomena is caused by chemotactic agents such as salts like sulphate or organic compounds migrating from cell to cell through the anticlinal walls, where chloroplasts move by chemotaxis. However, the real reason why chloroplasts move to the anticlinal position in the dark is not known. To learn the genes involved in this response we have obtained 6 mutants deficient in dark positioning from *Adiantum* gametophytes. To date, two candidate genes have been obtained from these mutants by FDD (Motoyama, Kuno, Uchida, and Wada, unpublished data).

4. FUNCTION OF PHYTOCHROME3

4.1 Phytochrome3-dependent chloroplast movement

Phytochrome-dependent chloroplast movement in *Adiantum* was originally described by Yatsuhashi *et al.* (1985), who used red light-grown single-celled protonemata kept under continuous irradiation with red light for 3 hours. The response was cancelled by simultaneous irradiation with far-red light indicating the involvement of phytochrome. At that time, red/far-red photoreversibility by short pulse irradiation was not detected. However, many years later, dark-adapted two-dimensional gametophytes were found to be very sensitive to a red light pulse for the induction of chloroplast movement (Figure 9) (Kagawa and Wada, 1994), and red/far-red photoreversibility could easily be observed. The reason for the different sensitivities of the different stages is not known but even two-celled protonemata show higher sensitivity than do single-celled protonemata (Yatsuhashi *et al.*, 1987). Different expression levels of Acphy3 is one of the possibilities to explain these different sensitivities to red light, but a detailed study on Acphy3 expression at different developmental stages has not yet been carried out.

Chloroplast movement is induced by polarized red light (Yatsuhashi *et al.*, 1985) and shows typical dichroic effect, i.e. polarized red light vibrating parallel to cell axis is more effective in inducing chloroplast movement than red light vibrating perpendicular to cell axis is, indicating that the Acphy3 molecules must be arranged in some way being anchored to the plasma membrane or to the cytoskeleton in the cortical layer. However Acphy3 has no membrane-spanning domain, at least based on its amino acid sequence, and the Acphy3 protein appears to be water-soluble.

In the *Adiantum* prothallus, chloroplast movement can be induced by red light as well as by blue light when irradiated with the light above the threshold intensity. However, when both blue and red lights under threshold intensity for induction of chloroplast movement were sequentially given, chloroplast movement was induced (Kagawa and Wada, 1996), indicating that blue and red light share their signalling pathways and function additively to induce the response. The order of the red- and blue-light irradiation is not important. Acphy3 functions as a red light photoreceptor as shown (Kawai *et al.*, 2003) but there is no evidence to date as to whether it functions as a blue light receptor as well.

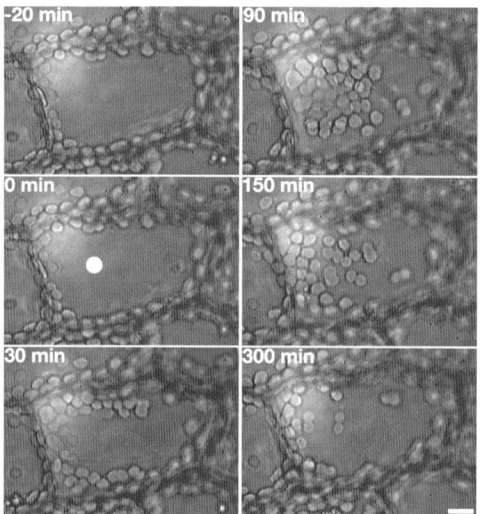

Figure 9. Chloroplast movement induced by a microbeam irradiation. Dark-adapted prothallial cell was irradiated with one minute with a red microbeam (30 Wm^{-2}). Bar=20µm

There are two known possibilities for blue light photoreceptor(s) activated in the above experiment and sharing a signalling factor with Acphy3. The first is the phototropin family (Acphot1 and/or Acphot2) and the other is Acphy3 itself. The former are plausible candidates but if Acphy3 also functions as the blue light receptor in this case, it is very interesting, because both red light and blue light can be absorbed by one photoreceptor and the signal processed through the same signal-transduction pathway.

4.2 Phytochrome3-dependent phototropism

The physiology and cell biology of Acphy3-dependent phototropism in protonemal cells has been described in detail in the 2nd issue of this book (Wada and Sugai, 1995), although Acphy3 was not known to be a photoreceptor mediating this response at that time. We briefly describe the results known so far (see more detail in the previous reviews). The photoreceptive site for phototropic response was localized to the tip area of protonemata by partial irradiation with polarized red light (Wada *et al.*, 1981). As in the case of chloroplast movement, a dichroic effect on phototropism under polarized red light was clearly shown (Kadota et al., 1982). The dichroic effect could be observed when a cell was irradiated on the flank of the subapical part of the cell (Wada et al., 1981). Even if a part of cell flank was irradiated, however, many different organelles were irradiated simultaneously, meaning that to resolve the real photoreceptive site in the cell was not possible. Hence, Wada et al. (1983) centrifuged the protonemal cells and made very vacuolated cells with almost no organelles at the cell tip. When the subapical part of the vacuolated protonemata was irradiated for a short period of time with a red microbeam, and then the organelles centrifuged back to the apical part in darkness, the cells showed phototropic response towards the irradiated side, showing that the photoreceptor must exist near or on the plasma membrane (Wada et al., 1983). We now know the photoreceptor for this phenomenon, but we still do not know how Acphy3 molecules are arranged at the plasma membrane. What is clear is that Acphy3 are not membrane-spanning molecules, and some other protein(s) or post-translational modification of Acphy3 itself must serve to connect Acphy3 to the membrane.

Phytochrome-mediated phototropism in young fern leaves was found for the first time in *Adiantum cuneatum* (Wada and Sei, 1994), although a phototropic response under white light has been previously reported (Prankerd, 1922). The tropic response was restricted to a region around 6-8 mm from the tip of crozier. Leaves can bend towards red light irrespective of the direction of the crozier. The greatest bending rate is about 10 degrees/hr, which occurs between 3-5 hr after the induction of the tropic response. A red light-induced phototropic response was also observed in young leaves of *Adiantum capillus-veneris* (Wada and Sei, 1994). Acphy3 was found to be the photoreceptor for this response (Kawai et al., 2003). It is very interesting that the mechanism for phototropism of tip-growing protonemata and that of the differential growth of phototropically responding leaves must be completely different, but both tissues nevertheless use the same Acphy3 as the photoreceptor.

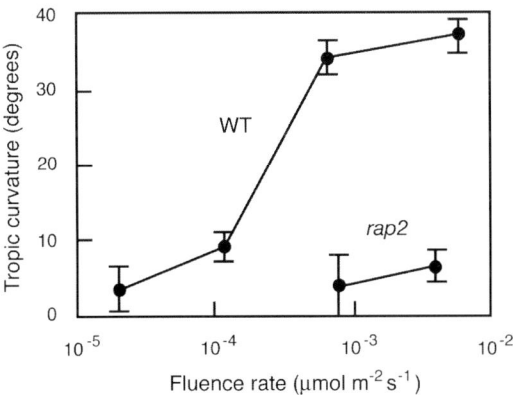

Figure 10. Phototropic response of young leaves of wild-type and rap sporophytes under white-light conditions (Kawai et al., 2003).

The invention of phy3 in advanced ferns must have given some adaptive advantages to ferns (Kawai *et al.*, 2003). We compared sensitivity of the phototropic responses of young leaves of *rap2* sporophytes with that of wild-type leaves under low white-light conditions (7.8 x 10^{-4} to 4.0 x 10^{-3} µmol m^{-2} s^{-1}) (Figure 10). The wild-type leaves showed very high photosensitivity in low light in contrast to the leaves of *rap2* mutant. Given that similar results were obtained in other Acphy3-deficient mutants, it is likely that the photosensory function of Acphy3 substantially enhances the phototropic response in *Adiantum* to white light. Based on the fact that advanced fern families have *PHY3* and increased greatly in species number in the Cretaceous, it is probable that an ancestor of these ferns obtained a functional phy3 accidentally and spread its habitat into the area under a canopy where light conditions are not optimal for other plants to live. Hence, because the ancestor became very sensitive to light and made it possible to live under the canopy.

5. FUNCTION OF PHOTOTROPIN2

5.1 Phototropin2-dependent chloroplast movement

The transient expression of *AcPHOT2* by a gene delivery system using *bhc-07* and *bhc-08* mutants showed that Acphot2 is the photoreceptor for the chloroplast avoidance response in *Adiantum* gametophytes (Kagawa *et al.*, 2004). The same strategy must now be applied to learn whether Acphot1 mediates an accumulation response or not. However, to date we do not have *Acphot1Acphot2* double mutant lines or even an *Acphot1* single mutant. Hence, we cannot test for *AcPHOT1* function at the moment. The transient expression assay of *AcPHOT2* using *bhc-07* or *bhc-08* mutants provides a big advantage for studying the function of Acphot2 domains (Figure 11). It takes only a short time to get results in comparison with similar experiments with stable transgenic plants of Arabidopsis. Moreover, chloroplast movement is easily and clearly observed in fern gametophytes because of their simple organization.

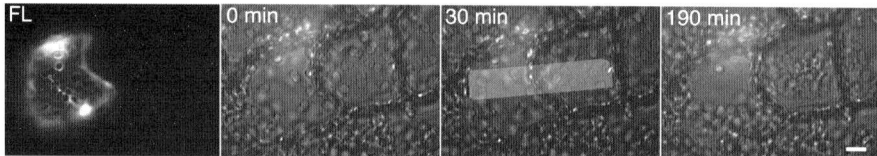

Figure 11. Phenotype recovery of a bhc mutant by introduction of AcPHOT2. A band across the cells indicated by the lighter rectangle was irradiated with a microbeam. The fluorescent image indicates the cell which expressed GFP as a marker for gene transfer. Bar=20µm

Using the transient assay, the role of the LOV domains in the N-terminus was tested (Figure 12) (Kagawa *et al.*, 2004). When a modified construct of *AcPHOT2* with deletion of 204 amino acid residues from N-terminus (dN204) or deletion including the whole LOV1 domain from the N-terminus (dN468) was bombarded into and expressed in the mutant cells, the modified *AcPHOT2* was still functional for the chloroplast avoidance movement in both cases. However, *AcPHOT2* without both LOV domains lost the function. Even if the LOV2 was replaced by LOV1, the construct did not function. Thus, LOV2 functions as the photoreceptive domain but not LOV1. There are 3 serine residues in the N-terminus. One is at the N-terminus end, the other two are in between LOV1 and LOV2. Autophosphorylation of these three serine residues are thought to be important for the phototropin function. However even when the part of the genes encoding these three serine residues was cut off, the cDNA construct still functioned in the avoidance response (Kagawa *et al.*, 2004), suggesting that the autophosphorylation must not be necessary at least for photoactivation leading to the avoidance response.

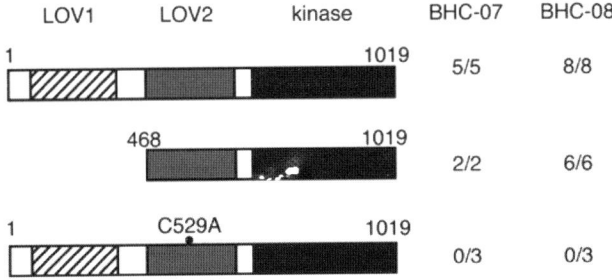

Figure 12. Phenotype complementation of bhc mutants by modified AcPHOT2 transfer (number of cells complemented) / (number of cells tested)

Similar experiments using deletion constructs at the C-terminus of Acphot2 were also performed (Kagawa *et al.*, 2004). Deletion of less than 20 residues (dC9 and dC20) from C-terminus did not affect the Acphot2 function but a deletion of more than 40 residues (dC40, dC49 and dC383) eliminated the Acphot2 function, suggesting that some residues between 20th and 40th amino acids may have a key role in Acphot2 function. When 529th cysteine in the LOV2 domain that becomes cys-adduct with FMN after blue light irradiation as the primary photoresponse, or 709th lysine in the kinase domain which is one of the conserved residues of ser/ thr kinase, was changed into alanine, the mutated Acphot2s both lost their function.

Not just the important domains and/or amino acids of the Acphot2 gene, but also key portions of any other genes could be analysed easily and rapidly by this transient assay using *Adiantum* gametophytes, provided that we can get mutants which show a clear phenotype.

5.2 Physiological estimation of the lifetime of phot signals

The nature of a signal transferred from phototropin to chloroplasts is not known. However, the speed of transfer and/or lifetime of the signal can be estimated by analysing physiological responses. When a part of a dark-adapted cell is irradiated with a small microbeam of weak blue or red light, chloroplasts at the cell periphery start to move some time after switching on the light. We can calculate roughly the speed of the signal transfer from the distance between the microbeam site and the chloroplast and the time lag between the instant that the light is switched and the start of chloroplast movement. Our earlier experiments showed that the speed of signal transfer is different for red light and blue light. Under red light the distance and the lag time are roughly proportional, and the speed of signal transfer is about 0.5µm /min (this is rough estimate because the rate varies among chloroplasts). Under blue light, on the contrary, almost all chloroplast move within 10 min irrespective of the distance between the microbeam and the chloroplasts although there is some lag time, again irrespective of the distance between microbeam and chloroplasts (Kagawa and Wada, 1996). These results indicate that transfer of the signal is very strong under blue light although the cells need to prepare for start of chloroplast movement.

When a part of a dark-adapted cell was irradiated with large microbeam (e. s. 20µm in diameter) of strong blue light, chloroplasts started to move towards the light-irradiated area but they stopped at the edge of the beam and could not enter into the beam because of the avoidance response. When the blue light was switched off, the chloroplasts at the edge of the beam entered into the formerly irradiated area. This chloroplast behaviour means that under strong blue light, both the signals for accumulation and avoidance responses were raised simultaneously and the former signal could be transferred over a long distance but the latter signal could not (Kagawa and Wada, 1999).

When a part of a cell with chloroplasts at the cell surface was irradiated with a microbeam of strong blue light for a short period, chloroplasts moved away from the irradiated area. The chloroplasts continued moving for some time even in the dark after switching off the microbeam, but then stopped and started moving back towards the original location. If the amount of signal that is generated depends upon the fluence of blue light, and if the degradation of the signal is constant, the lifetime of the signal can be estimated from the chloroplast behaviour, namely a period between the time of switching-off of blue light and the time that the chloroplasts stop or begin to move backwards. We estimated the half-life of the Acphot2 signal to be either 67 s or 167 s by the time the chloroplasts stop or begin moving backward, respectively (Kagawa et al., 2004).

6. GERMINATION-RELATED GENES

Spore germination has been studied in ferns for many years, because 1) results can be obtained rapidly (usually within a week), and 2) measuring percent spore germination is easy, although criteria for germination differ between different investigators. Usually germination is detected when a part of rhizoid or protonemal cell emerges from the spore coat. However, the germination process starts within spores long before rupture of the spore coat by cell expansion. Furuya et al. (1997) observed that a light effect on the first cell division occurred in spores could be detected and found that red light promoted the cell division but that blue light inhibited the division antagonistically. These light effects were mediated by phytochrome and a blue light receptor, both of which regulate the entry into the S phase during the first cell cycle in fern spores (Uchida and Furuya, 1997). Given that blue light advances the timing of cell division and red light inhibits the division in protonemal cells (Wada and Furuya, 1978), the light effects on the first cell division within the spores are just the opposite from those on cell division in the protonemal cells. It is interesting that the photoreceptive site of blue light in both spores and protonemal cells is the same, namely in or close to the nucleus based on the results obtained by microbeam irradiation, but the function is exactly the opposite. To explain this discrepancy we need to know which blue light photoreceptor(s) mediates these phenomena, whether cryptochrome, phototropin or some other photoreceptor? If the functional photoreceptors controlling cell division in spores and in protonemata are the same and their localization is also the same, it is quite puzzling why and how the function of one photoreceptor switches during development.

As red and blue light regulate the transition from G1 phase to S phase, genes expressed over time in *Adiantum* spores was investigated. As cyclin-A has been shown to be involved in the progression of S phase in animal cells, *Adiantum* cyclin, *CycAc1*, was cloned and the expression pattern studied (Uchida et al., 1996). CycAc1 was not found in dormant spores or during G1 phase, but its mRNA became detectable after the termination of S phase; hence it is not the key gene controlling the entrance into S phase. These workers then applied a newly developed FDD to find genes involved in the transition into S phase and found a gene regulated by phytochrome among 8000 displayed bands (Uchida et al., 1998). The cDNA has a homology with cell wall-associated extensions and was named *AcExt1*. The mRNA of *AcExt1* was detectable in spores 4 hr after red light irradiation. The induction of expression of *AcExt1* mRNA by red light is photoreversible by far-red light, but irreversible by blue light. There must be many genes expressed as is *AcExt1* during fern spore germination, so that it is very difficult to identify any real key gene controlling the transition from G1 to S phase.

7. CONCLUDING REMARKS

Fern gametophytes are good materials for analysing physiological phenomena at the cellular and/or subcellular levels because of their simple organization, not surrounded by any other tissue, and because of light-dependency of so many physiological phenomena. In addition, its haploid character is quite advantageous for molecular analyses of gene function. The diploid phase or even haploid duplication of very similar genes that redundantly mediate one phenomena make it difficult and troublesome to analyse the function of these genes. As mentioned before, cloning of various genes based on homology to those sequenced in other organisms is very easy. Moreover EST libraries make it quite promising even in *Adiantum* gametophytes and partial and full-length genes are easily obtained.

The future of fern photomorphogenesis studies using molecular biological techniques is promising even in comparison with Arabidopsis. As described above, we can analyse important domains and/or key amino acids in genes for their function by transient expression of modified gene constructs in the mutants of these genes. We need to establish various resources for ferns in near future, such as full-length cDNA libraries, DNAi instead of RNAi, and stable transformation through *Agrobacterium*. We first have to overcome some problems, but if we are successful, the function of various genes will be clarified rapidly, and accordingly the mechanisms of fern photomorphogenesis will be elucidated.

8. REFERENCES

Briggs, W. R., Beck, C., Cashmore, A. R., Christie, J. M., Hughes, J., Jarillo, J. A., *et al.* (2001) The phototropin family of photoreceptors. *Plant Cell*, *13*, 993-997.

Briggs, W. R. and Christie, J. M. (2002) Phototropins 1 and 2: versatile plant blue-light receptors. *Trends in Plant Science*, *7*, 204-210.

Christie, J. M., Reymond, P., Powell, G. K., Bernasconi, P., Raibekas, A. A., Liscum, E., *et al.* (1998) *Arabidopsis* NPH1: A flavoprotein with the properties of a photoreceptor for phototropism. *Science*, *282*, 1698-1701.

Christie, J. M., Salomon, M., Nozue, K., Wada, M., Briggs, W.R. (1999) LOV (light, oxygen, or voltage) domains of the blue-light photoreceptor phototropin (nph1): Binding sites for the chromophore flavin mononucleotide. *Proceedings of the National Academy of Sciences, USA*, *96*, 8779-8783.

Furuya, M., Kanno, M., Okamoto, H., Fukuda, S., Wada, M. (1997) Control of mitosis by phytochrome and a blue-light receptor in *Adiantum* spores. *Plant Physiology*, *113*, 677-683.

Hoffman, P. D., Batschauer, A., Hays, J. B. (1996) *PHH1*, A novel gene from *Arabidopsis thaliana* that encodes a protein similar to plant blue-light photoreceptors and microbial photolyases. *Molecular and General Genetics*, *253*, 259-265.

Huala, E., Oeller, P. W., Liscum, E., Han, I. S., Larsen, E., Briggs, W. R. (1997) *Arabidopsis* NPH1: a protein kinase with a putative redox-sensing domain. *Science*, *278*, 2120-2123.

Imaizumi, T., Kanegae, T., Wada, M. (2000) Cryptochrome nucleocytoplasmic distribution and gene expression are regulated by light quality in the fern *Adiantum capillus-veneris*. *Plant Cell*, *12*, 81-96.

Jarillo, J. A., Ahmad, M., Cashmore, A. R. (1998) NPL1 (Accession No AF053941): A second member of the NPH1 serine/threonine kinase family of Arabidopsis (PGR 98-100) *Plant Physiology*, *117*, 719.

Jarillo, J. A., Gabrys, H., Capel, J., Alonso, J. M., Ecker, J. R., Cashmore, A. R. (2001) Phototropin-related NPL1 controls chloroplast relocation induced by blue light. *Nature*, *410*, 952-954.

Kadota, A. and Wada, M. (1999) Red Light-aphototropic (rap) mutants lack red light-induced chloroplast relocation movement in the fern *Adiantum capillus-veneris*. *Plant and Cell Physiology*, *40*, 238-247.

Kadota, A., Wada, M., Furuya, M. (1982) Phytochrome-mediated phototropism and different dichroic orientation of Pr and Pfr in protonemata of the fern *Adiantum capillus-veneris* L. *Photochemistry and Photobiology*, *35*, 533-536.

Kagawa, T. (2003) The phototropin family as photoreceptors for blue light-induced chloroplast relocation. *Journal of Plant Research*, *116*, 77-82.

Kagawa, T., Kasahara, M., Abe, T., Yoshida, S., Wada, M. (2004) Functional analysis of phototropin2 using fern mutants deficient in blue light-induced chloroplast avoidance movement. *Plant and Cell Physiology*, *45*, 416-426.

Kagawa, T., Sakai, T., Suetsugu, N., Oikawa, K., Ishiguro, S., Kato, T., *et al.* (2001) *Arabidopsis* NPL1: A phototropin homolog controlling the chloroplast high-light avoidance response. *Science*, *291*, 2138-2141.

Kagawa, T. and Wada, M. (1994) Brief irradiation with red or blue light induces orientation movement of chloroplast in dark-adapted prothallial cells of the fern *Adiantum*. *Journal of Plant Research*, *107*, 389-398.

Kagawa, T. and Wada, M. (1996) Phytochrome- and blue-light-absorbing pigment-mediated directional movement of chloroplasts in dark-adapted prothallial cells of fern *Adiantum* as analysed by microbeam irradiation. *Planta*, *198*, 488-493.

Kagawa, T. and Wada, M. (1999) Chloroplast avoidance response induced by blue light of high fluence rate in prothallial cells of the fern *Adiantum* as analyzed by microbeam irradiation. *Plant Physiology*, *119*, 917-923.

Kagawa, T. and Wada, M. (2002) Blue light induced chloroplast relocation. *Plant and Cell Physiology*, *43*, 367-371.

Kanegae, T. and Wada, M. (1998) Isolation and characterization of homologues of plant blue-light photoreceptor (cryptochrome) genes from the fern *Adiantum capillus-veneris*. *Molecular and General Genetics*, *259*, 345-353.

Kawai, H., Kanegae, T., Christensen, S., Kiyosue, T., Sato, Y., Imaizumi, T., et al. (2003) Responses of ferns to red light are mediated by an unconventional photoreceptor. *Nature*, *421*, 287-290.

Kawai-Toyooka, H., Kuramoto, C., Orui, K., Motoyama, K., Kikuchi, K., Kanegae, T., et al. (2004) DNA interference: a simple and efficient gene-silencing system for high-throughput functional analysis in the fern *Adiantum*. *Plant and Cell Physiology*, *45*, 1648-1657.

Klink, V. P. and Wolniak, S. M. (2001) Centrin is necessary for the formation of the motile apparatus in spermatids of *Marsilea*. *Molecular Biology of the Cell*, *12*, 761-776.

Kinoshita, T., Doi, M., Suetsugu, N., Kagawa, T., Wada, M., Shimazaki, K. (2001) phot1 and phot2 mediate blue light regulation of stomatal opening. *Nature*, *414*, 656-660.

Lin, C. (2002) Blue light receptors and signal transduction. *Plant Cell*, *14* (supplement.), S207-S225.

Lin, C., Robertson, D. E., Ahmad, M., Raibekas, A. A., Jorns, M. S., Dutton, P. L., et al. (1995) Association of flavin adenine dinucleotide with the *Arabidopsis* blue light receptor CRY1. *Science*, *269*, 968-970.

Malhotra, K., Kim, S. -T., Batschauer, A., Dawut, L., Sancar, A. (1995) Putative blue-light photoreceptors from *Arabidopsis thaliana* and *Sinapis alba* with a high degree of sequence homology to DNA photolyase contain the two photolyase cofactors but lack DNA repair activity. *Biochemistry*, *34*, 6892-6899.

Nozue, K., Christie, J. M., Kiyosue, T., Briggs, W. R., Wada, M. (2000) Isolation and characterization of a fern phototropin (Accession No. AB037188), a putative blue-light photoreceptor for phototropism. (PGR00-039) *Plant Physiology*, *122*, 1457.

Nozue, K., Fukuda, S., Kanegae, T., Wada, M. (1998a) Isolation of second phytochrome cDNA from *Adiantum capillus-veneris* (Accession No. AB016232) (PGR-174) *Plant Physiology*, *118*, 712.

Nozue, K., Kanegae, T., Imaizumi, T., Fukuda, S., Okamoto, H., Yeh, K-. C., et al. (1998b) A phytochrome from the fern *Adiantum* with features of the putative photoreceptor NPH1. *Proceedings of the National Academy of Sciences USA*, *95*, 15826-15830.

Nozue, K., Kanegae, T., Wada, M. (1997) A full length *Ty3/gypsy*-type retrotransposon in the fern *Adiantum*. *Journal of Plant Research*, *110*, 495-499.

Okamoto, H., Hirano, Y., Abe, H., Tomizawa, K., Furuya, M., Wada, M. (1993) The deduced amino acid sequence of phytochrome from *Adiantum* includes consensus motifs present in phytochrome B from seed plants. *Plant and Cell Physiology*, *34*, 1329-1334.

Okamoto, H., Sakamoto, K., Tomizawa, K., Nagatani, A., Wada, M. (1997a) Photoresponses of transgenic Arabidopsis overexpressing the fern *Adiantum capillus-veneris* PHY1. *Plant Physiology*, *115*, 79-85.

Okamoto, H., Silverthorne, J., Wada, M. (1997b) Spatial patterns of phytochrome expression in young leaves of the fern *Adiantum capillus-veneris*. *Plant and Cell Physiology*, *38*, 1397-1402.

Oyama, H., Yamamoto, K. T., Wada, M. (1990) Phytochrome in the fern *Adiantum capillus-veneris* L.: Spectrophotometric detection in vivo and partial purification. *Plant and Cell Physiology*, *31*, 1229-1238.

Prankerd, T. L. (1922) On the irritability of the fronds of *Asplenium bulbiferum*, with special reference to graviperception. *Proceedings of the Royal Society*, *B93*, 143-152.

Sakai, T., Kagawa, T., Kasahara, M., Swartz, T. E., Christie, J. M., Briggs, W. R., et al. (2001) *Arabidopsis* Nph1 and npl1: Blue-light receptors that mediate both phototropism and chloroplast relocation. *Proceedings of the National Academy of Sciences USA*, *98*, 6969-6974.

Sakamoto, K. and Briggs, W. R. (2002) Cellular and subcellular localization of phototropin 1. *Plant Cell*, *14*, 1723-1735.

Sato, Y., Kadota, A., Wada, M. (2003) Chloroplast movement: Dissection of events downstream of photo- and mechano-perception. *Journal of Plant Research*, *116*, 1-5.

Schaefer, D. G. (2002) A new moss genetics: Targeted mutagenesis in *Physcomitrella patens*. *Annual Review of Plant Biology*, *53*, 477-501.

Senn, G. *Die Gestalts und Lageveränderungen der Pflanzenchromatophoren*, Leipzig, Germany: Engelmann, 1908.
Stout, S. C., Clark, G. B., Archer-Evance, S., Roux, S. J. (2003) Rapid and efficient suppression of gene expression in a single-cell model system, *Ceratopteris richardii*. *Plant Physiology*, *131*, 1165-1168.
Uchida, K. and Furuya, M. (1997) Control of the entry into S phase by phytochrome and blue light receptor in the first cell cycle of fern spores. *Plant and Cell Physiology*, *38*, 1075-1079.
Uchida, K., Muramatsu, T., Jamet, E., Furuya, M. (1998) Control of expression of a gene encoding an extensin by phytochrome and a blue light receptor in spores of *Adiantum capillus-veneris* L. *Plant Journal*, *15*, 813-819.
Uchida, K., Muramatsu, T., Tachibana, K., Kishimoto, T., Furuya, M. (1996) Isolation and characterization of the cDNA for an A-like cyclin in *Adiantum capillus-veneris* L. *Plant and Cell Physiology*, *37*, 825-832.
Wada, M. and Furuya, M. (1978) Effects of narrow-beam irradiations with blue and far-red light on the timing of cell division in *Adiantum* gametophytes. *Planta*, *138*, 85-90.
Wada, M. and Kadota, A (1989) Photomorphogenesis in lower green plants. *Annual Review of Plant Physiology and Plant Molecular Biology*, *40*, 169-191.
Wada, M., Kadota, A., Furuya, M. (1981) Intracellular photoreceptive site for polarotropism in protonema of the fern *Adiantum capillus-veneris* L. *Plant and Cell Physiology*, *22*, 1481-1488.
Wada, M., Kadota, A., Furuya, M. (1983) Intracellular localization and dichroic orientation of phytochrome in plasma membrane and/or ectoplasm of a centrifuged protonema of fern *Adiantum capillus-veneris*. *Plant and Cell Physiology*, *24*, 1441-1447.
Wada, M., Kagawa, T., Sato, Y. (2003) Chloroplast movement. *Annual Review of Plant Biology*, *54*, 455-468.
Wada, M., Kanegae, T., Nozue, K., Fukuda, S. (1997) Cryptogam phytochromes. *Plant, Cell and Environment*, *20*, 685-690.
Wada, M. and Sei, H. (1994) Phytochrome-mediated phototropism in *Adiantum cuneatum* young leaves. *Journal of Plant Research*, *107*, 181-186.
Wada, M. and Sugai, M. (1995) Photobiology of fern. In R. E. Kendrick and G. H. M. Kronenberg (Eds.), *Photomorphogenesis in plants* (2nd ed., pp. 783-802) Dordrecht, Boston, London: Kluwer Academic Publishers.
Yatsuhashi, H., Hashimoto, T., Wada, M. (1987) Dichroic orientation of photoreceptors for chloroplast movement in *Adiantum* protonemata. Non-helical orientation. *Plant Science*, *51*, 165-170.
Yatsuhashi, H., Kadota, A., Wada, M. (1985) Blue- and red-light action in photoorientation of chloroplasts in *Adiantum* protonemata. *Planta*, *165*, 43-50.
Yeh, K.-C. and Lagarias, J. C. (1998) Eukaryotic phytochromes: light-regulated serine/threonine protein kinases with histidine kinase ancestry. *Proceedings of the National Academy of Sciences, USA*, *95*, 13976-13981.

Chapter 25

PHOTOMORPHOGENESIS OF MOSSES

Tilman Lamparter
Freie Universtität Berlin,Pflanzenphysiologie, Königin Luise Straße 12-16, 14195 Berlin, Germany (e-mail:lamparte@zedat.fu-berlin.de)

1. INTRODUCTION

The habitus and life cycle of mosses differ from seed plants in many aspects. Mosses are small, they have no roots, are free of vascular bundles and form only few different cell types. Their life cycle (Figure 1) is dominated by the haploid gametophyte. The gametophyte plant or gametophore consists of a stem, leaves with either one or two cell layers, and rhizoid cells. Gametes are formed in multicellular organs, so-called gametangia. Egg-producing gametangia are termed archegonia, whereas sperms are produced in so-called antheridia. The eggs are fertilized within the archegonium on the mother plant, and the diploid sporophyte remains permanently to and partially dependent on the gametophyte mother plant. Meiosis and spore formation takes place in the sporangium, an organ which grows on top of the sporophyte. After dispersion of the spores by wind, a single cell of cylindrical shape emerges from the spore, which divides serially and produces a filament, the so-called protonema. Although the protonemal stage is often regarded as an obscure and unimportant step in moss development, mosses can be kept for infinite time as protonemata. This contrasts with fern protonemal cells (see Chapter 12), which after a few cell divisions will start to form prothallia. The growing zone of a protonema is restricted to the apical region of the tip cell. This type of growth reminds of pollen tubes and root hairs of seed plants and of filamentous fungi. In mosses, basal protonemal cells can divide laterally and produce new tip cells, which either form a new filament or differentiate into a three-faced apical cell to form the gametophore. In this way, a more or less dense net of moss protonemata is formed, from which the three-dimensional gametophyte plants emerge.

Besides the different habitus and life cycle, there are many features that mosses have in common with seed plants. Mosses stand at the base of the land-plant evolution. Like all land plants, they have chlorophyll a and b, and lack other pigments known from algae and bacteria. In a large scale EST programme performed with the moss *Physcomitrella patens*, it was found that about 63 % of moss genes have homologs in *Arabidopsis* (Nishiyama et al., 2003). All photoreceptor types that are known from seed plants, namely phytochromes, cryptochromes and phototropins, have moss orthologs, and there are many parallels between both groups with respect to light regulation.

Because of their simple growth, mosses are ideal model organisms for questions related to developmental biology, cell biology and photobiology. Many studies are focussed on the protonematal stage, especially the protonemal tip cell. Since in one experiment, several hundred tip cells that are all grown under identical conditions can be treated simultaneously, it may be easy to obtain many parallels for quantitative studies. Therefore, moss protonemata are ideal systems to monitor growth and photomorphogenesis at the cellular level. The identification of mutants of protonemata or gametophores is rather simple and does not require subcrossings, because in these stages the moss is haploid. In addition, it is rather easy to isolate protoplasts from mosses, and to regenerate these into new protonemal filaments (Grimsley *et al.*, 1977). Protoplasts are directly accessible for PEG-mediated DNA transfer (Schaefer *et al.*, 1991), which has become a routine technique for moss transformation. Other transformation techniques such as particle gun bombardment (Sawahel *et al.*, 1992) or microinjection (Brücker *et al.*, 2000) have also been established.

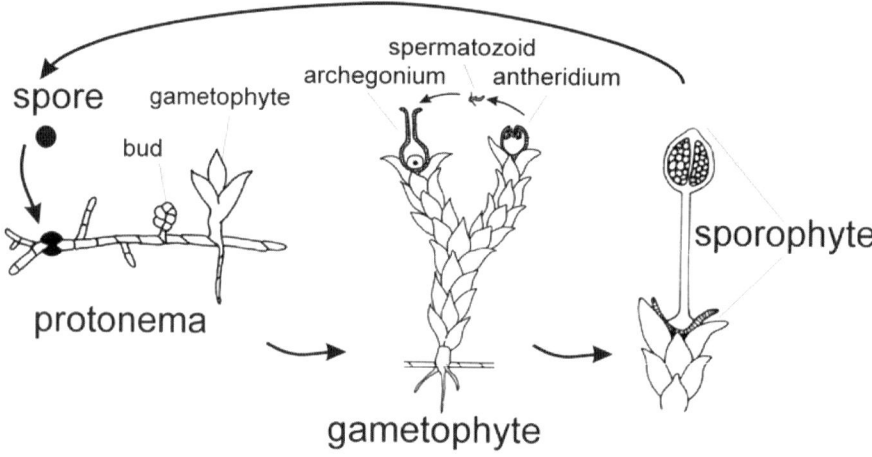

Figure 1. Life cycle of mosses. The germinating spore produces a protonemal tip cell, which divides serially to produce a protonemal filament. Side branches are formed by a lateral cell division of basal cells. Gametophyte plants, which are formed by a three-faced apical cell, produce antheridia and archegonia. For fertilization, the spermatozoid cell swims to the egg cell. The diploid sporophyte grows on the mother plant. Meiosis occurs during spore formation. Not drawn to scale.

The significance of mosses for plant science has further increased since Schaefer and Zryd found that in *Physcomitrella patens* the rate of homologous recombination is very high (Schaefer and Zryd, 1997). This feature allows efficient gene-knockout and might allow gene modification (e.g. site directed mutagenesis) in the future. After initial studies, in which homologous recombination was demonstrated for different genes at the DNA level, the function of particular genes was addressed by gene targeting experiments (Girke *et al.*, 1998; Hofmann *et al.*, 1999; Girod *et al.*, 1999; Hofmann *et al.*, 1999; Brun *et al.*, 2001; Imaizumi *et al.*, 2002; Koprivova *et al.*, 2002; Koprivova *et al.*, 2002; Zank *et al.*, 2002; Li *et al.*, 2002; Meiri *et al.*, 2002; Schipper *et al.*, 2002; Sakakibara *et al.*, 2003; Olsson *et al.*, 2003). In one of the first studies, an FtsZ gene was knocked out, which yielded filaments with impaired chloroplast division (Strepp *et al.*, 1998). In bacteria, the tubulin-like FtsZ protein plays an important role in cell division.

Largely initiated by these methodological achievements, three different groups have started *Physcomitrella* EST projects. The first project runs since 1999 and was aimed at sequencing 100,000 ESTs as a basis to find genes with novel functions (Rensing *et al.*, 2002). This project is largely financed by the agrochemical company BASF. Not all sequences of this programme are available in public databases, but are accessible upon request. In the second EST project, financed by the British BBSRC and US agencies, about 30,000 Physcomitrella and 1,500 *Ceratodon* ESTs were deposited in public databases. Finally, a Japanese group sequenced the ends of 80,000 ESTs from libraries enriched for full length cDNA (Nishiyama *et al.*, 2003). Altogether, *Physcomitrella* ESTs cover more than 15,000 different genes.

2. EFFECTS OF LIGHT ON MOSS DEVELOPMENT

As in seed plants, light affects the development of mosses throughout the life cycle. Most studies focussed on phototropism, chloroplast orientation and the initiation of side branches of protonemal filaments. Less is known about light effects in gametophores, and no light response is known for the sporophyte. In most cases, *Funaria hygrometrica*, *Physcomitrella patens* or *Ceratodon purpureus* were used as research objects. An overview is given in Table 1.

2.1 Spore germination

The effect of light on spore germination has been analysed for different species. In all cases, germination was dependent on light. In *Funaria,* germination is most sensitive to light in the red region of the spectrum and the red light effect can be reversed by far-red light (Bauer and Mohr, 1959), thus phytochrome is involved in this regulation. Red-far-red reversibility of germination has since been demonstrated for spores of *Ceratodon, Dicranum scoparium* (Valanne, 1966) and *Physcomitrella* (Schild, 1981; Cove *et al.*, 1978).

Table 1. Some light effects of mosses

Light effect	Species	photoreceptor	
Induction of spore germination	*Funaria hygrometrica, Ceratodon purpureus, Dicranium scoparium, Physcomitrella patens*	phytochrome	Bauer and Mohr 1959; Valanne, 1966; Schild, 1981; Cove *et al.*, 1978
Formation of protonemal side branches	*Physcomitrella patens, Funaria hygrometrica*	phytochrome and blue light receptor	Kagawa *et al.*, 1997; Imaizumi *et al.*, 2002
Phototropism and polarotropism of protonemal tip cell	*Physcomitrium turbinatum, Funaria hygrometrica, Ceratodon purpureus, Physcomitrella patens, Pottia intermedia*	phytochrome	Bünning and Etzold, 1958; Nebel, 1968; Jenkins and Cove, 1983b; Hartmann *et al.*, 1983; Demkiv *et al.*, 1998; Esch *et al.*, 1999
Chloroplast orientation	*Funaria hygrometrica, Ceratodon purpureus, Physcomitrella patens*	blue light receptor and / or phytochrome, depending on species	Voerkel, 1933; Kagawa *et al.*, 1997; Kadota *et al.*, 2000
Inhibition of gravitropism of protonemal tip cell	*Ceratodon purpureus, Pottia intermedia*	phytochrome and blue light receptor	Cove *et al.*, 1978; Jenkins *et al.*, 1986; Young and Sack, 1992; Lamparter *et al.*, 1996; Demkiv *et al.*, 1997a; Lamparter *et al.*, 1998b; Kern and Sack, 1999
Induction of chlorophyll synthesis in protonemal tip cell	*Ceratodon purpureus*	phytochrome and blue light receptor	Lamparter *et al.*, 1997
Increase of protoplast regeneration, direction of outgrowth	*Ceratodon purpureus*	phytochrome and blue light receptor	Cove *et al.*, 1996
Modulation of auxin sensitivity	*Physcomitrella patens*	cryptochrome	Imaizumi *et al.*, 2002

2.2 Cell differentiation

In protonemata of many moss species, two different cell types, caulonemata and chloronemata, can be distinguished. Chloronemata are densely packed with large chloroplasts and the cross walls between adjacent cells are perpendicular to the filament axis. Caulonemal cells have fewer chloroplasts that contain less chlorophyll than those in chloronemal cells. The cross walls of caulonemata are oblique. Differentiation of chloronemal to caulonemal cells requires auxin and is a light dependent process. If the medium is supplemented with carbohydrates, moss filaments can grow in darkness. Under such conditions, only caulonemal filaments are produced, whereas in the light, both cell types may be present, with the chloronemata as dominating cell type (Cove and Lamparter, 1998; Imaizumi *et al.*, 2002). Dark growing filaments form only very few side branches. The initiation of side branches is strongly increased by light, as shown for *Funaria* (Demkiv *et al.*, 1997b), *Physcomitrella* (Cove and Ashton, 1988) and *Ceratodon* (Kagawa *et al.*, 1997). As far as analysed, blue light receptors and phytochrome are involved in this induction (Kagawa *et al.*, 1997; Imaizumi *et al.*, 2002). The formation of buds, which later grow out to the three-dimensional gametophore, is dependent on the presence of cytokinin (Bopp, 1963; Hahn and Bopp, 1968; Wang *et al.*, 1980). Because of its clear-cut "all or nothing" effect, bud formation is often used as model for cytokinin studies (Saunders and Hepler, 1982; Saunders and Hepler, 1983; Featherstone *et al.*, 1990; Hahm and Saunders, 1991; Schumaker and Gizinski, 1993; Schumaker and Gizinski, 1995; von Schwartzenberg *et al.*, 1998; Gonneau *et al.*, 2001; Schulz *et al.*, 2001). Bud formation is also light dependent (Klebs, 1893; Hartmann and Jenkins, 1984; Imaizumi *et al.*, 2002).

2.3 Phototropism and polarotropism

The growth direction of the tip cell is determined by light and gravity. In darkness, tip cells grow negatively gravitropically, i.e. upwards (Jenkins *et al.*, 1986; Schwuchow *et al.*, 1995; Khorkavtsiv and Kardash, 2002; Schwuchow *et al.*, 2002). The growth direction in the light is dependent on species, growth - and light conditions. In general, the tip cell grows towards the light (positive phototropism, see Figure 2 a), but in particular cases, the tip cell can also grow away from the light (negative phototropism). Whereas in shoots of seed plant seedlings, phototropism is mediated by the blue light photoreceptor phototropin, this effect is regulated by phytochrome in many moss species (Nebel, 1968; Jenkins and Cove, 1983b; Hartmann *et al.*, 1983; Demkiv *et al.*, 1998; Bünning and Etzold, 1958). A similar response termed polarotropism can be induced by polarized light. Typically, filaments align their growth direction vertical to the E-vector of polarized light (see Figure 2 b). A tropic response can also be induced by partial irradiation with a microbeam. In general, the tip cell bends towards the irradiated side (Esch *et al.*,

1999), but high light intensities induce an avoidance response (Lamparter et al., 2004)(see Figure 3).

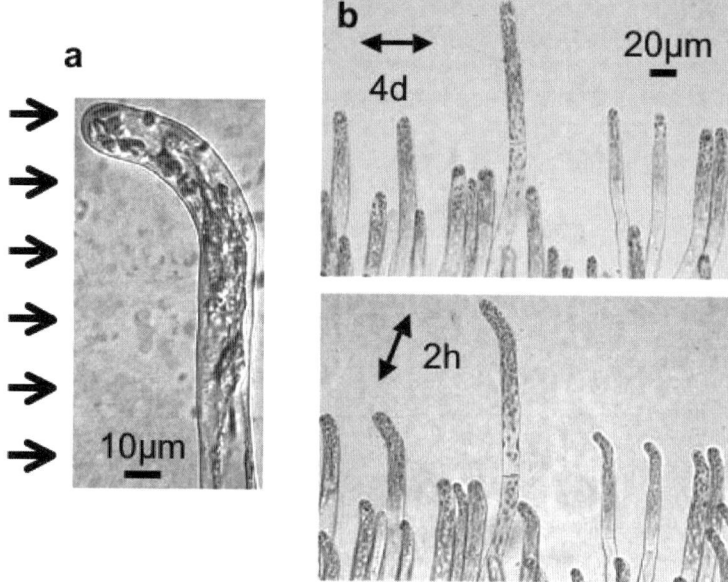

Figure 2. (a) Phototropism of a protonemal tip cell of Ceratodon purpureus (strain wt4). The cell was irradiated with monochromatic red light of 4 µmol m^{-2} s^{-1} from the left side for 2 h. Before this light treatment, the filaments were kept in unilateral low fluence white light (1 µmol m^{-2} s^{-1}) on a horizontally oriented agar dish for 4 days. This treatment leads to a parallel alignment of the filaments. In darkness, filaments may be aligned on vertically oriented agar dishes by their negative gravitropic growth. (b) Polarotropism of Ceratodon filaments. The filaments were grown for 4 days on a horizontally oriented agar dish with polarized red light (1 µmol m^{-2} s^{-1}) from above (top panel). Ceratodon filaments grow perpendicular to the E-vector (arrow). Filaments that arise from the inoculation zone grow in two opposite directions, only the apical region of one growth direction is shown. Note that the filaments are topped by a cover slip to avoid bending towards the light source (i.e. away from the growth medium). The lower panel shows the change of growth direction 2 h after turning the E-vector of polarized light.

2.4 Lights effects on gravitropism

Light affects the gravitropic response of moss protonemata (Cove et al., 1978; Jenkins et al., 1986; Cove and Knight, 1987; Young and Sack, 1992; Lamparter et al., 1996; Demkiv et al., 1997a; Lamparter et al., 1998b; Kern and Sack, 1999). Such a response is known from multicellular seed plant seedlings (Feldman and Briggs, 1987; Liscum and Hangarter, 1993). In *Ceratodon*, blue and red light act in a different manner: blue light inverts the gravitropic response, and the filaments grow downwards (Lamparter et al., 1998b), whereas red light - mediated by phytochrome - inhibits gravitropism. The red light effect is difficult to analyse with wild type filaments, because any red light irradiation induces also phototropism, but the effect on gravitropism could be unravelled by studies on class 2 mutants (see

3.1.2) which are specifically defective in the phototropic response (Lamparter et al., 1996).

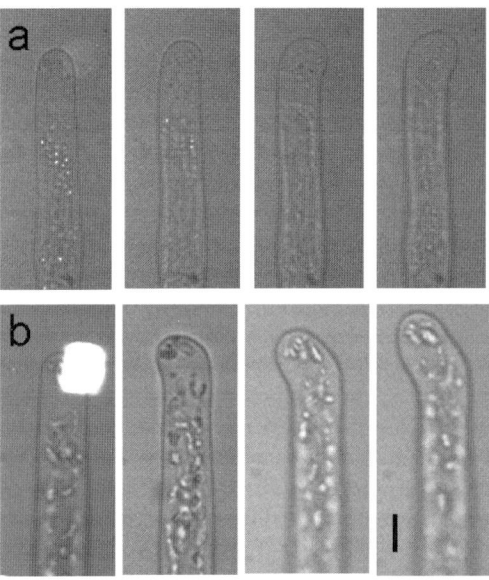

Figure 3. Microbeam irradiation of Ceratodon *tip cell. The right side of the apical dome was irradiated for 15 min. a) light intensity 3 μmol m^{-2} s^{-1}, such a treatment induces a positive response b) light intensity 100 μmol m^{-2} s^{-1}, this treatment induces a negative response. Images were taken at t=0, 15 30 and 60 min (after Lamparter et al., 2004).*

2.5 Chloroplast movement

Light-induced chloroplast movement is found in almost any plant species, also in mosses. Voerkel (1933) had already shown that chloroplast movement in leaves of *Funaria* is induced by blue light, whereas red light was inactive. The same pattern was found for *Ceratodon* (Kagawa et al., 1997). The situation for these two species is comparable with seed plants, where chloroplast movement is also only induced by blue light, mediated by phototropins, and not by red light (but see Dong et al., 1995). In *Physcomitrella* however, chloroplast movement is triggered both by blue and red light. Depending on the light intensity, either an accumulation or an avoidance response is observed. Phytochrome was shown to be involved in the red light effect (Kadota et al., 2000). Photoreceptor specificity in *Physcomitrella* is comparable with the green algae *Mougeotia scalaris* (Kraml, 1994) and *Mesotaenium caldariorum* (Herrmann and Kraml, 1997), and the fern *Adiantum capillus-veneris* (see Chapter 23).

2.6 Chlorophyll synthesis

In angiosperms, several steps in the synthesis of chlorophyll are light regulated. One of the last steps, the conversion from protochlorophyllide to chlorophyll, is in itself light dependent. It is because of this that angiosperms do not produce chlorophyll in the dark (Porra, 1997). Mosses however, are able to synthesize chlorophyll in the dark. Nevertheless, the chlorophyll content of dark-adapted *Ceratodon* tip cells is lower than that of light-grown cells (Figure 4). Chlorophyll synthesis in *Ceratodon* is under the control of phytochrome and a separate blue light photoreceptor (Lamparter *et al.*, 1997). In addition, the shape and intracellular distribution of chloroplasts is also light dependent: plastids of dark-adapted tip cells are smaller than those of cells grown in red light and light-grown cells are usually more densely packed with chloroplasts.

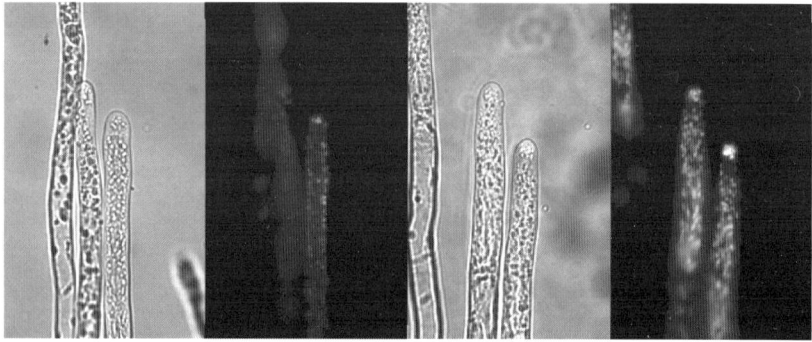

Figure 4. Transmission and chlorophyll fluorescence images of protonemal Ceratodon *tip cells. The left two panels show dark-grown cells, the right two panels show cells from a culture that has been irradiated for 24 h with red light (1 µmol $m^{-2}s^{-1}$). Dark grown cells contain chlorophyll. The chlorophyll level is light regulated via phytochrome and a blue light receptor (see also Lamparter et al., 1997).*

2.7 Protoplast regeneration

The effect of light on *Ceratodon* protoplast regeneration has been studied in detail in order to understand the formation of cell polarity (Cove *et al.*, 1996). Both the regeneration rate and the position of outgrowth are light dependent. Most protoplast grow out towards the light, but around 30% of the entire population grow out to the opposite direction. Of interest is the observation that the direction of outgrowth appears to be controlled almost equally by both red and blue light. This contrasts with the phototropic response of filaments, where blue light induces only a subtle change of growth direction.

3. DIFFERENT PHOTORECEPTORS IN MOSSES

Several phytochrome genes have been sequenced from *Ceratodon* and *Physcomitrella*, and two cryptochrome genes were found in *Physcomitrella*. The EST database contains a fragment of a putative phototropin gene, and the sequencing of four *Physcomitrella* phototropin genes has recently been reported. It therefore seems that mosses contain all types of photoreceptors known from seed plants. Since to date none of the moss genomes is fully sequenced, there might be more photoreceptor genes or genes that encode for unexpected types of photoreceptors. In the following, the results for phytochrome and the blue light photoreceptors shall be treated separately.

3.1 Phytochromes

3.1.1 Phytochrome genes and proteins

The first moss phytochrome gene, *CerpuPhy1*[1] from *Ceratodon*, was also the first example for phytochrome domain swapping (Thümmler *et al.*, 1992). The deduced protein is homologous to known plant phytochromes in the N-terminal half, but divergent in the remaining C-terminal portion (see Figure 5). The protein has a predicted molecular size of 145 kDa, significantly larger than common plant phytochromes, which range from 123 to 129 kDa. The C-terminal portion of CerpuPhy1 has similarities to protein kinases, and serine/ threonine kinase activity could be achieved *in vitro* for the truncated kinase domain in a fibroblast expression system (Thümmler *et al.*, 1995). Although common plant phytochromes show serine/ threonine kinase activity (Yeh and Lagarias, 1998), their sequence is not related to other known kinases of this type. In this context, the discovery of CerpuPhy1 highlighted the discussion from a different point of view. However, the biological function of CerpuPhy1 is enigmatic. It appears to be expressed at very low levels (Pasentsis *et al.*, 1998). A polyclonal antibody, Apc1, which is derived against the N-terminal region of CerpuPhy1, failed to detect the 145 kDa protein in extracts from protonemata but did recognize a smaller polypeptide (Lamparter *et al.*, 1995). The protein which was immunoprecipitated with this antibody cross-reacted with the universal monoclonal antibody Z-3B1, which has its epitope in the C-terminal region of conventional phytochromes. These data indicated the presence of another phytochrome gene which encodes for a conventional phytochrome in *Ceratodon*. Using the 5' part of *CerpuPhy1* as a probe, a second phytochrome was identified by Southern blots and finally cloned and sequenced (Hughes *et al.*, 1996). Indeed, this gene encodes for a conventional phytochrome, which is homologous over its entire length to known seed plant orthologs. The first *Physcomitrella* phytochrome gene isolated also encodes a normal plant phytochrome (Kolukisaoglu

---

[1] The nomenclature used here may differ from the cited articles. The abbreviations stand for the species name and contains a number for the phytochrome gene. The numbering indicates the chronological order of sequencing, but has no functional meaning. Italic abbreviations stand for the gene, the proteins are given in normal letters.

et al., 1993). PCR and Southern blotting have revealed further phytochrome genes from both species, in total there are now three fully sequenced phytochrome genes in *Ceratodon* and four in *Physcomitrella* (Mittmann *et al.*, 2004). Of three phytochrome-homologous ESTs present in the public databases, two are identical with *PhypaPhy2* and *4*, but the third probably represents a fifth phytochrome gene.

A phylogenetic tree based on N-terminal protein sequences of moss phytochromes and selected phytochromes of other plants is given in Figure 6. According to this analysis, two groups of moss phytochromes can be distinguished, one group consisting of CerpuPhy1, CerpuPhy2, PhypaPhy1 and PhypaPhy3, the other of CerpuPhy3, PhypaPhy2 and PhypaPhy4. Obviously, *CerpuPhy1* and *CerpuPhy2* arose from a late gene duplication and subsequent domain re-arrangement to form *CerpuPhy1*. It is thus no surprise that phytochrome genes that are similar to *CerpuPhy1* have not been found in other species. The *Physcomitrella* pairs *PhypaPhy1/3* and *PhypaPhy2/4* have probably also arisen from a duplication event after the divergence of the two groups of moss phytochrome genes. The simplest explanation for this arrangement is a duplication of the entire *Physcomitrella* genome. This assumption is consistent with the finding that the genome of *Ceratodon* is smaller (ca 250 Mbp, Lamparter *et al.*, 1998a) than that of *Physcomitrella* (ca. 500 Mbp, Rensing *et al.*, 2002). The high similarity between both *Physcomitrella* cryptochrome genes (see below) and other gene pairs (Markmann-Mulisch *et al.*, 2002) supports this genome duplication theory.

One of the moss phytochrome genes, *CerpuPhy2*, has been expressed in a recombinant *Saccharomyces cerevisiae* system. As for other phytochromes, it autocatalytically attaches the chromophores PCB, PEB and PΦB and behaves as a homodimer (Zeidler *et al.*, 1998). The PΦB adduct gives a difference spectrum identical to that of partially purified phytochrome from *Ceratodon* filaments (Lamparter *et al.*, 1995). This implies that the natural chromophore of *Ceratodon* is PΦB, as in seed plants. Spectral properties were reminiscent of B-type angiosperm phytochromes (Zeidler *et al.*, 1998; Sineshchekov *et al.*, 2000).

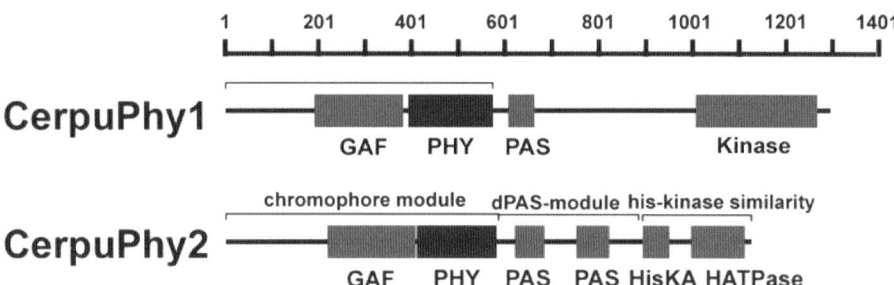

Figure 5. Domain arrangement of Ceratodon *phytochromes. CerpuPhy2 and other moss phytochromes have a conventional domain arrangement. The C-terminal region (aminoacid 600 to 1300) of CerpuPhy1 with the histidine kinase domain is unusual and has so far not been found in other phytochromes.*

The polyclonal antibody Apc1 (see above) was used for various studies (Lamparter *et al.*, 1995). Since CerpuPhy1 is not detectable on immunoblots, the antibody probably recognizes CerpuPhy2, because of the high sequence similarity between both phytochromes in the N-terminus. The immobilized antibody also precipitated photoreversible phytochrome from the crude extract of *Ceratodon* filaments. However, around 50% of the spectral activity remained in the supernatant (Lamparter *et al.*, 1995). This could be related to CerpuPhy3.

Spectrally detectable phytochrome in *Ceratodon* is light regulated, in dark adapted filaments, the concentration is higher than in filaments grown in the light (Lamparter *et al.*, 1995). In a phytochrome-chromophore deficient class 1 mutant (see 3.1.2), the content of apophytochrome was equally high in both dark- and light-grown filaments (Esch and Lamparter, 1998). Thus, the phytochrome chromophore is required for the regulation of phytochrome content. Most likely, CerpuPhy2 is slowly degraded in its far-red absorbing Pfr form. This is reminiscent of angiosperm PhyA, although PhyA is degraded at a much faster rate (Clough and Vierstra, 1997). Evidence for translational light control of phytochrome synthesis was also presented (Pasentsis *et al.*, 1998). Since in this study, light induced depletion of phytochrome was inhibited by DCMU, photosynthesis could also play a role in the regulation of phytochrome abundance.

Soluble and membrane fractions of *Ceratodon* extracts were tested for their cross-reactivity with Apc1. The majority of the phytochrome was found in the soluble fraction, but about 10% were detected in the pelletable membrane fraction (Podlowski, 1994; Esch and Lamparter, 1998). In immunocytological assays, phytochrome was detected in the cytosol of protonemal tip cells (T. Lamparter, unpublished).

3.1.2 Mutants

An obvious phytochrome effect of mosses is the phototropic response of the tip cell. Mutants with defects in phototropism, designated *ptr* mutants, have been isolated from *Physcomitrella* (Cove *et al.*, 1978) and *Ceratodon* (Lamparter *et al.*, 1996). In the *Physcomitrella ptr* mutants, both phototropism and polarotropism are equally affected (Jenkins and Cove, 1983a), whereas other phytochrome responses are regulated normally (Cove and Lamparter, 1998). Based on genetic crosses, it was found that at least 3 different complementation groups contribute to the phototropic response. In *Ceratodon* two sets of mutants that have different phenotypes in continuous red light were isolated (Figure 7). Class 1 mutants grow upwards (negatively gravitropic) and produce only very low levels of chlorophyll, whereas class 2 mutants grow in random directions, and produce normal, wild-type levels of chlorophyll. Protoplast fusion experiments demonstrate that representatives of either class complement each other (Lamparter *et al.*, 1998a). It is not known, however, whether each mutant class contains several complementation groups.

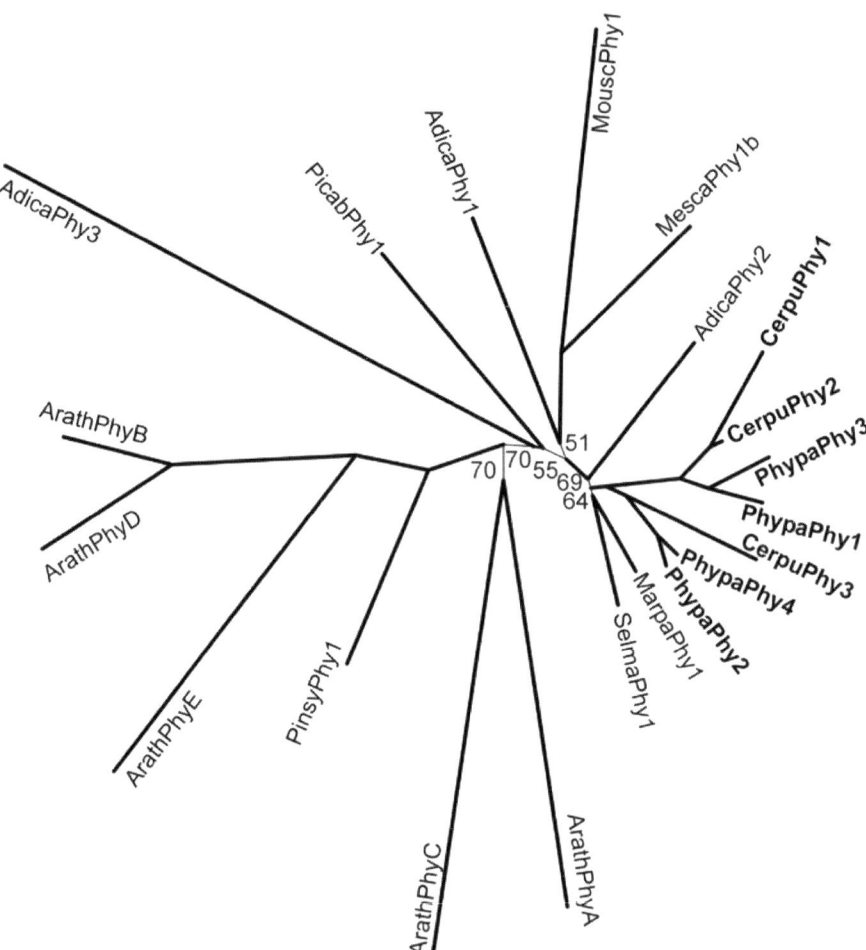

Figure 6. Phylogenetic tree of phytochromes from mosses (bold letters) and other plants. Protein sequences of the N-terminal chromophore module (ca. 600 aminoacids) were aligned with ClustalX (Thompson et al., 1997) using default parameters. The tree was constructed with the PHYLIP program package (Felsenstein, 2000) using the PROTDIST and FITCH algorithms. For bootstrapping, 100 datasets of sequences were generated, bootstrap values were usually above 90%. Branches with bootstrap values < 75% are drawn as thin lines, in those cases the % values are indicated. The following abbreviations were used: Adica: *Adiantum capillus veneris,* Arath: *Arabidopsis thaliana,* Cerpu: *Ceratodon purpureus,* Mesca: *Mesotaenium caldariorum,* Marpo: *Marchantia polymorpha,* Mousc: *Mougeotia scalaris,* Phypa: *Physcomitrella patens,* Picab: *Picea abies,* Pinys: *Pinus sylvestris,* Selma: *Selaginella martensii* (from Lamparter and Brücker, 2004).

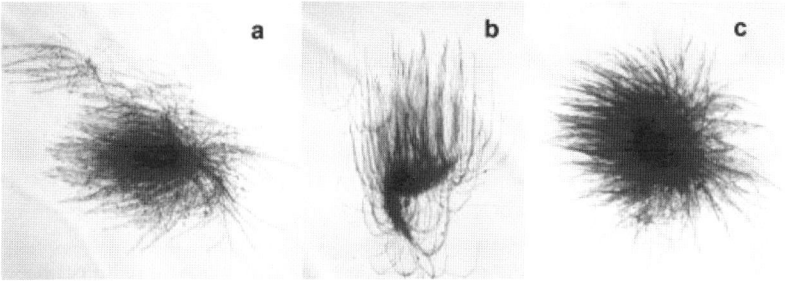

Figure 7 Phenotype of Ceratodon *wild type and mutants under continuous red light. After inoculation, the filaments were grown on vertically oriented agar dishes for 5 days under unilateral red light (660 nm, 4 µmol m^{-2} s^{-1}) from the left. (a) wild type wt3 (b) class 1 mutant* ptr1 *(c) class 2 mutant* ptr103 *(after Lamparter and Brücker, 2004).*

Ceratodon Class 1 mutants
In class 1 mutants all phytochrome responses are lost. These mutants are defective in the biosynthesis of the phytochrome chromophore, because the phototropic response can be rescued by bilins (Lamparter *et al.*, 1996). Rescue studies have been performed with more than 30 independently isolated mutant strains. All lines were rescued by biliverdin and phycocyanobilin (Esch and Lamparter, 1998). Photoreversibility assays performed with 19 arbitrarily chosen mutants showed that they contained low but still measurable quantities of photoactive phytochrome. The highest level was around 15% of the wild type. Despite this residual spectral activity, all mutants were totally aphototropic and there was no indication for a physiological action of phytochrome with respect to other responses (Esch and Lamparter, 1998). This finding implies that phytochrome action in mosses has a rather high threshold of holoprotein / apoprotein ratio. Biliverdin is a precursor in the biosynthesis of phytochromobilin, the natural phytochrome chromophore of land plants, and phycocyanobilin can functionally replace phytochromobilin. The results of the feeding experiments suggest that the class 1 mutants are defective at the point of biliverdin formation. The relevant enzyme, heme oxygenase (HO), has initially been characterized in animals where it plays an important role in heme degradation and signalling processes. The first plant HO was found in *Arabidopsis* by mutant analyses (Muramoto *et al.*, 1999).

Before the *Ceratodon* HO-gene was sequenced, HOs from rat and *Arabidopsis* were tested as to whether or not they could rescue the aphototropic phenotype of class 1 mutants. For this purpose, the respective genes were cloned into an expression vector and placed directly into mutant tip cells by means of microinjection (Brücker *et al.*, 2000). This method had been previously established for the microinjection of low and high molecular dye components, phycocyanobilin, and GFP expression constructs into mutant and wild type cells. Injection of the HO-plasmids into cells of the class 1 mutant *ptr116* resulted in a clear rescue of the mutant phenotype. The filaments formed during the days subsequent to the insertion grew towards unilateral red light and produced wild-type levels of chlorophyll. Microinjection is a powerful tool for the analysis of functional aspects of gene and metabolite activity at the cellular level, in the present case it allowed for a rapid assessment of the effect of heme oxygenase expression on phytochrome mutant phenotypes (Brücker et al., 2000).

One HO gene, *CpHO1*, was later sequenced from *Ceratodon* cDNA and genomic DNA. The gene product has an N-terminal plastid target and shows considerable sequence similarity with plant and cyanobacterial HOs. It was found that in *ptr116*, codon 31 is mutated to a stop codon. Another class 1 mutant had lost the entire gene (Brücker et al., 2004). These data suggest that the sequence is the site of the class 1 mutations, although other HO genes must exist in *Ceratodon* because all mutants contain residual photoactive phytochrome. Constructs with and without plastid target were then used for microinjection rescue experiments (Brücker et al., 2004). GFP fusions showed that the full length protein is indeed targeted to the plastid, whereas the truncated protein without the target remained in the cytosole and nucleus. Both the cytosolic and the plastidic version could rescue phototropism and chlorophyll synthesis in class 1 mutants (Brücker et al., 2004).

The strong phenotype of class 1 mutants could also be used to establish homologous recombination in *Ceratodon*. After transformation of *ptr116* with a wild type gene, several phototropic lines with high chlorophyll content were isolated. Molecular analyses showed that the rescue resulted from gene replacement events (Brücker et al., 2004). *Ceratodon* has several advantages over *Physcomitrella* for studies on photomorphogenesis. *Ceratodon* filaments grow for infinite time in darkness (important for dark controls), the phototropic response is more simple than that of *Physcomitrella*, and *Ceratodon* has fewer phytochrome genes. Therefore, the technical advance that arose from the mutant studies might help for future molecular studies with *Ceratodon*.

Ceratodon class 2 mutants
The class 2 mutants still show a weak phototropic response. Quite interestingly, the polarotropic response of the mutants is rather strong, but still weaker than that of the wild type (Esch et al., 1999). A microbeam treatment did not induce a tropic response in these mutants (Lamparter and Wada, unpublished). Other phytochrome controlled processes seem normally regulated. Some class 2 mutants have larger cells and are characterized by faster growth speed and a higher rate of cell division (Esch et al., 1999). Since it is difficult to isolate mutant genes in mosses, the nature of the lesion that confers the class 2 phenotype is as yet unclear.

Physcomitrella phytochrome knockout mutants
All four fully sequenced phytochromes in *Physcomitrella* were knocked out by insertional mutagenesis (Mittmann et al., 2004) and the mutants tested for their phototropic response. The bending direction of dark-adapted *Physcomitrella* filaments is dependent on the fluence rate. Red light around 0.15 $\mu mol\ m^{-2}s^{-1}$ induces a slightly negative curvature, with 1.5 $\mu mol\ m^{-2}s^{-1}$ red light the response is positive, and with 15 $\mu mol\ m^{-2}s^{-1}$, again a negative response is obtained. In both the

PhypaPhy1 and *PhypaPhy2* knockouts, the positive response was slightly weakened. In the *PhypaPhy3* knockout, high fluence rates induced a positive instead of a negative response. The strongest effect was found for the *PhypaPhy4* knockout. With 1.5 µmol m^{-2}s^{-1}, the filaments showed a bending response, but the direction was random and the average angle was zero. The bending can thus also be explained by the red-light induced loss of the gravitropic response (see above). These results imply that all four phytochromes participate in the phototropic response of *Physcomitrella*, and that *PhypaPhy4* is particularly important for positive phototropism (Mittmann et al., 2004).

3.1.3 Light direction and polarization

Most plants respond in various ways to the light direction by responses like phototropism, phototaxis, chloroplast orientation or solar leaf tracking. In seed plants, chloroplast orientation is controlled by phototropin (Kagawa, 2003), but in many cryptogam species, this response is also mediated by phytochrome. Similarly, phototropism is mediated by phototropin in seed plants (Kagawa and Wada, 2002), but may be mediated by phytochrome in cryptogam species. Light direction sensing is qualitatively different from other light responses, because vectorial information must be preserved through the signal transduction cascade. If single cells respond to the light direction, signal transduction does most likely not pass through the nucleus, because in the nucleus, gradual information will be lost. Therefore, these vectorial responses can not be based on differential gene expression. In contrary, many phytochrome effects in *Arabidopsis* include translocation of phytochrome into the nucleus and the regulation of gene expression. In the present sub-chapter I want to address the implications that arise from vectorial responses of single cryptogam cells, particularly in the context of phytochrome-controlled phototropism of moss protonemata. An overview about general aspects of light direction sensing is given in chapter 7.2 of the 2nd edition of "Photomorphogenesis in Plants" (Kraml, 1994).

Principles that arise from physiological studies of vectorial effects of various organisms may be summarized as follows: (i) Light attenuation, light refraction and / or the intracellular orientation of photoreceptors are important elements of gradient formation and light direction sensing. (ii) Action dichroism is often found in the case of cellular effects. Classical examples for action dichroism of phytochrome responses are chloroplast rotation of *Mougeotia* (Haupt, 1969) and the polarotropism of fern protonemata (Etzold, 1965). In mosses, action dichroism has been demonstrated for chloroplast photo-orientation (Kadota et al., 2000), polarotropism (Hartmann et al., 1983; Esch et al., 1999) and side branch formation (Kagawa et al., 1997).

Action dichroism is generally explained by oriented photoreceptors, which are localized close to the plasma membrane. In the case of phytochrome, the transition dipole moment of the Pr and Pfr form are thought to be parallel and vertical to the cell surface, respectively (see Figure 8). Studies on purified seed plant phytochromes showed that the transition dipole moment of the chromophore undergoes large vectorial changes upon photoconversion (Sundqvist and Björn, 1983; Ekelund et al., 1985; Tokutomi et al., 1992). However, the plasma membrane association of phytochrome has as yet not been unambiguously demonstrated. In general, phytochrome is extracted as soluble protein, and is detected in the cytosol or in the nucleus. In extracts of seed plants, several percent of phytochrome are always found in association with pelletable membrane fractions (Rubinstein et al., 1969; Marme, 1974; Napier and Smith, 1987; Lamparter et al., 1992). Although this fact is known since decades, it is as yet unclear if phytochrome is bound to membranes *in vivo*, to which compartment phytochrome is bound, and if this association is important for a particular function of phytochrome. Nevertheless, there are recent reports about *in vivo* stains of phytochrome at the cell periphery of *Arabidopsis* (T. Kunkel and E. Schäfer, personal communication).

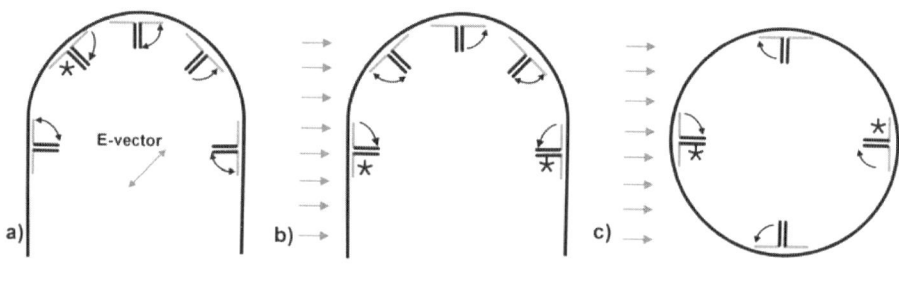

Figure 8. Cartoon to show the proposed orientation of dichroically arranged phytochrome molecules in the tip cell. a) and b): longitudinal section; c) cross section. Phytochrome molecules are drawn as dimers. Both the transition dipole moment of Pr and Pfr are presented, which are parallel and vertical to the cell surface, respectively. Arrows indicate the preferred direction of photoconversion under red polarized (a) or unilateral (b, c) light. The asterisks indicate high Pfr concentrations.

The intracellular distribution of phytochrome in cryptogam species has been addressed by several groups, with similar outcome: the large portion of spectrally active phytochrome in green algae and mosses is extracted as soluble protein (Lindemann et al., 1989; Kidd and Lagarias, 1990; Lamparter et al., 1995), and in immunological stains, phytochrome is found in the cytosole (Hanstein et al., 1992; Kidd and Lagarias, 1990, in *Ceratodon*: T. Lamparter, unpublished). As mentioned above, around 10% of extracted phytochrome 2 appear to be associated with pelletable membranes in *Ceratodon*. The intracellular distribution of phytochrome in cryptogams seems comparable with seed plants, although nuclear translocation has as yet not been shown for cryptogams. The necessity for an association of phytochrome to the plasma membrane has been questioned by several scientists, and other explanations for light direction sensing were proposed. However, the transition dipole moment of the active photoreceptor must have a defined orientation with respect to the surface of the cell, and the plasma membrane is the only compartment in the plant cell which has such an orientation.

With the present knowledge of phytochrome action and polarotropism in mosses, several scenarios are possible: (i) One particular, plasma-membrane associated phytochrome is responsible for vectorial effects. In the fern *Adiantum*, mutant analyses and molecular studies have identified PHY3 as photoreceptor for red-light controlled phototropism. This exceptional phytochrome has a C-terminal phototropin-like domain structure. In seed plants, phototropin is attached to the plasma membrane, and the same might be proposed for PHY3 (see Chapter 12 for phototropin and 24 for *Adiantum*). It seems however that PHY3 homologs are restricted to the sub-group of higher ferns and do not exist in mosses (Kawai *et al.*, 2003). PhypaPhy4 of *Physcomitrella,* which probably plays the most important role in the phototropic response of that species (see above), belongs to the group of conventional phytochromes. (ii) A sub-fraction of one or several phytochromes could be associated to the plasma membrane. This would be the case if a limited number of binding sites exists in the plasmamembrane.

For *Arabidopsis*, it has been reported that phytochrome molecules can be labelled with phycoerythrobilin (PEB), a phycobiliprotein chromophore of cyanobacteria and red algae. Phycoerythrobilin assembles with apo-phytochromes *in vitro* and *in vivo* to adducts with a high fluorescence quantum yield (Murphy and Lagarias, 1997). Phytochrome of *Ceratodon* can also be labelled by PEB. Protonemal tip cells of the class 1 mutant *ptr116* gave a strong signal, whereas the signal of wild type cells was rather weak, because the natural chromophore competes with PEB. Presumably, PEB is incorporated by any type of phytochrome in the cell. Therefore, this stain gives an impression about the intracellular distribution of all phytochrome molecules. Confocal laser scanning microscopy showed that the bulk of phytochrome is located in the cytosole (see Figure 9), consistent with the immunostain mentioned above. However, with fluorescence correlation spectroscopy, a subfraction of phytochrome with lower mobility was detected at the cell periphery (Böse *et al.*, 2004) (Figure 10). Fluorescence correlation spectroscopy allows to measure both concentration and mobility of fluorescing particles in femtoliter volume elements *in vitro* and in living cells. Although it is as yet unclear, which type of phytochrome is located at the cell periphery, this subfraction is probably important for the polarotropic response.

Which parameter determines the growth direction of a tip cell under polarized light or unilateral light? Many data imply that the position of outgrowth is defined by the highest Pfr level within the cell. If tip cells are irradiated with a microbeam on one half of the apical region, in general, the cell grows towards the irradiated side, the side of the higher Pfr concentration. If the cells are irradiated with polarized light, the final growth direction is vertical to the electric vector (Figure 2). In this case, the highest Pfr level also coincides with the growth position, as outlined schematically in Figure 8, because photoconversion of Pr into Pfr is high if the transition dipole moment of Pr is parallel to the E-vector of polarized light.

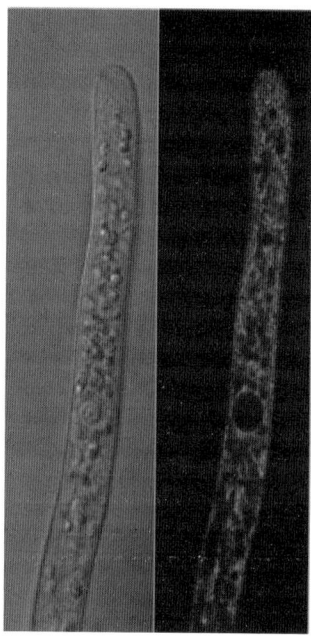

Figure 9. Intracellular phytochrome distribution in a tip cell of the class 1 mutant ptr116 monitored by PEB labelling. Left: transmission image, right: fluorescence of the PEB-phytochrome adduct. Before the fluorescence assay, cells were kept for 24 h on agar-medium containing 6 µM PEB at 20°C (after Lamparter and Brücker, 2004).

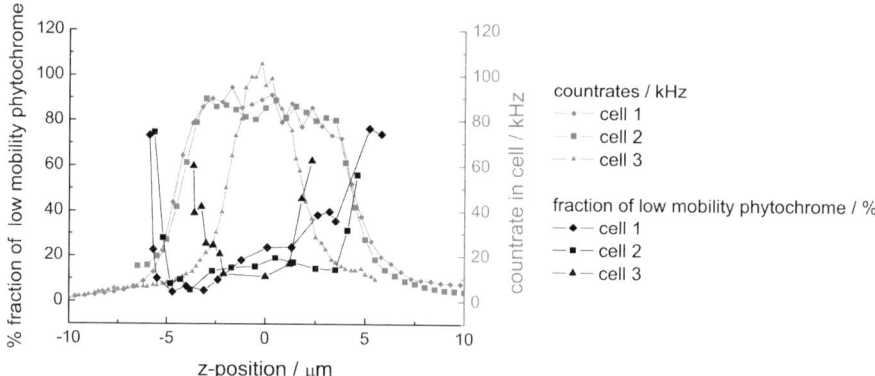

Figure 10. Fluorescence correlation spectrometry (FCS) of PEB-phytochrome in Ceratodon ptr116 tip cells. The FCS measurements were performed in the region of the apical dome at the indicated points along the Z-direction (along the cross section). The intensity profile (countrates) reflect the shape of the cells. Individual FCS traces (not shown) were fitted with a slow and a fast component. The relative fraction of the slow component is indicated by the black symbols. In all three examples, the relative contribution of this component was rather high at the cell periphery (Böse et al. 2004).

Under non-polarized, unilateral light, the intracellular Pfr pattern is more complex. In this case, the highest rate of photoconversion is given if the transition dipole moment is perpendicular to the light direction. However, there are two Pfr maxima within the cross section of the cell, namely at the light-directed and at the light-avoiding side (see Figure 8). Light attenuation or light refraction may lead to different Pfr "amplitudes" on both sides. In dark adapted *Ceratodon* cells, light attenuation plays only a minor role, because the intracellular concentration of screening pigments is low. For a clear positive phototropic response, light has to be refracted at the surface of *Ceratodon* cells. When filaments are grown within agar, only 50 to 70% bend towards the light source, the remaining filaments turn away from the light, as if each cell finds it difficult to distinguish between light-directed and light-avoiding side. In these experiments, the difference in optical density between the surrounding medium and the cell is rather weak, and the effect of light refraction will therefore be also weak. If filaments are grown on cellophane, almost the entire population bends towards the light. Due to the big optical difference between the surrounding air and the cell, the effect of light refraction is rather strong.

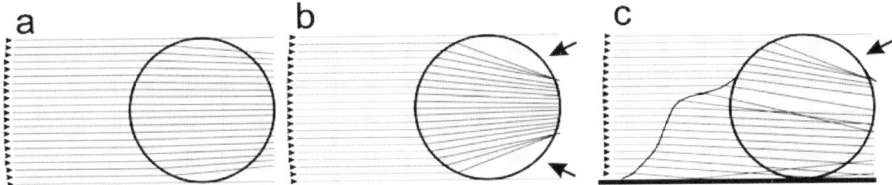

Figure 11. Proposed light path during the cross section of the tip cell. a) Cell grown within agar. The optical density of the cell is only slightly higher than that of the surrounding aqueous medium. b) Cell grown in air. Due to the difference in optical density between air and cell, light is refracted at the surface of the cell. The arrows point to the shaded areas on the light-avoiding side. c) Cell grown on top of cellophane-covered agar medium. In this case, a water film extends to both sides of the filament. This leads to a different optical situation as in b), but there is still one shaded area on the light-avoiding side (after Lamparter et al. 2004).

A combination of light refraction and dichroically oriented phytochrome molecules will lead to an intracellular Pfr gradient, from which the cell can "calculate" the light direction. Pfr levels on the flanks of the cell are always very low, because the light direction is parallel to the transition dipole moment of Pr. High Pfr levels are found on the light-directed and light-avoiding side, but in the shaded area of the light-avoiding side, the Pfr level is probably also low. Therefore, the Pfr level is higher on the light directed side (Esch *et al.*, 1999). However, strong light will saturate photoconversion into Pfr also within the shaded area. Under these conditions, light direction sensing requires a Pfr-independent input. Indeed, a Pfr-independent reaction was revealed by microbeam studies (Lamparter *et al.*, 2004). When such experiments are performed with light fluences that are expected to saturate the Pr to Pfr photoconversion, the tip cell bends towards the irradiated side. When much higher light fluences are used, the tip cell bends to the other side

(Figure 3). The inhibitory effect which was revealed by these experiments might be important to overcome the saturation effect of Pr to Pfr photoconversion.

3.2 Cryptochromes and phototropin

In a comprehensive work, the function of cryptochrome was analysed in *Physcomitrella* (Imaizumi *et al.*, 2002). Two similar cryptochrome genes were found by rapid amplification of cDNA ends (RACE). Both genes encode for almost identical proteins, termed CRY1a and CRY1b. They differ only in aminoacid 80, which is threonine in CRY1a and methionine in CRY1b. Phylogenetic analyses showed that the moss cryptochromes are closely related to cryptochromes of the fern *Adiantum*, but also quite similar to other plant orthologs (Figure 12). In a transient expression assay with *Physcomitrella* protoplasts, GUS fusion proteins were found to be targeted to the nucleus, irrespective of the light conditions. The physiological role of both cryptochromes was analysed with single- and double-knockout mutants that had been generated by insertional mutagenesis. In most tested responses, the double mutant *cry1acry1b* had a stronger phenotype than either of the single mutants *cry1a* or *cry1b*. The formation of side branches was reduced in the mutants under blue light, but not under red or white light. The mutant effect on bud formation and development of gametophore plants under different light conditions was different: under white light, the mutants produced more gametophores than the wild type. In contrary, under blue light, mutants produced fewer gametophores than the wild type. This reduction is probably an indirect effect: since the mutants produce less side branches in the blue, and side branch formation is the first step of bud formation, the overall capacity for gametophore production is reduced. The experiments with white light imply that cryptochrome inhibits the differentiation into gametophores.

Further studies showed that the cryptochrome signal alters the sensitivity of the moss to auxin. Physiological differences between wild type and mutants were enhanced upon addition of the auxin analogue 1-naphthalene acedic acid (NAA). Feeding of NAA induces the differentiation of chloronemal to caulonemal filaments. The mutants responded stronger to the addition of NAA in this respect. It therefore seems that the cryptochrome signal inhibits the sensitivity of the filaments to auxin. A transient transformation assay, in which the GUS reporter gene was expressed under the control of an auxin-inducible promoter, confirmed this suggestion: in cells of the *cry1acry1b* double mutant, the induction was saturated at lower NAA concentrations than in wild type cells.

These studies clearly showed the involvement of cryptochromes in some blue light responses of *Physcomitrella*. Since the double mutant still responds to blue light, other blue light receptors must be present in *Physcomitrella*. As outlined above, there are other blue light effects in mosses, such as photomovement of chloroplasts. Although the blue light photoreceptor for this response has not been directly addressed in mosses, results from ferns and *Arabidopsis* strongly suggest that phototropin controls this response in mosses. Four phototropin genes have been sequenced in *Physcomitrella* (Sato *et al.*, 2003). Thus, a similar redundancy as in the case of phytochromes and cryptochromes is anticipated for the *Physcomitrella* phototropins[2].

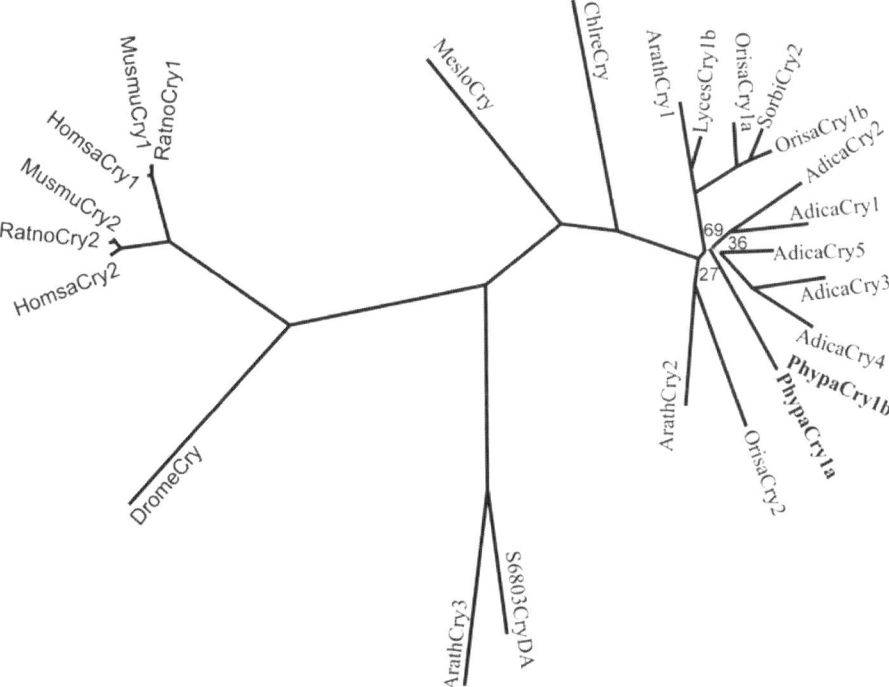

Figure 12. Phylogenetic tree of selected cryptochrome protein sequences. Alignment and phylogenetic analyses were performed as in Figure 6. Abbreviations for species names: Adica: Adiantum capillus veneris, *Arath:* Arabidopsis thaliana, *Chlre:* Chlamydomonas reinhardtii, *Drome:* Drosophila melanogaster, *Homsa:* Homo sapiens, *Lyces:* Lycopersicum esculentum, *Meslo:* Mesorhizobium loti, *Musmu:* Mus musculus, *Orisa:* Oriza sativa, *Phypa:* Physcomitrella patens, *Ratno:* Rattus norvegicus, *S6803: Synechocystis PCC 6803 (Cry DASH).*

4. SIGNAL TRANSDUCTION

As in seed plants, there is a close interrelation between light and hormone responses in mosses. Branching, bud formation, gravitropism and chloroplast division are responses which are influenced by light and hormones; one example for the effect of the cryptochrome signal on auxin sensitivity has been outlined above. In vectorial responses, the role of Ca^{2+} and of the cytoskeleton system have been particularly analysed. These data shall be discussed in the following paragraphs.

---

[2] In a recently published article by Kasahara *et al.* (2004) the authors showed that *Physcomitrella* phototropins control light-induced chloroplast movement. Phototropins appear also as downstream signal transduction components of phytochrome – mediated chloroplast movement in *Physcomitrella*. Because the article appeared during preparation of the present review, its content was not included.

4.1 Ca^{2+}

Phytochrome was found to trigger a transient plasmamembrane depolarisation of dark adapted *Physcomitrella* protonemal cells. The response was dependent on the presence of Ca^{2+} in the extracellular space (Ermolayeva *et al.*, 1996). Voltage clamp and ion flux measurements suggested that this depolarisation is partly caused by phytochrome-induced opening of voltage-gated Ca^{2+} channels (Ermolayeva *et al.*, 1997). These findings were discussed in the context of side branch formation, although other cellular effects such as chloroplast relocation or phototropism might also be regulated via Ca^{2+}. Another light effect on Ca^{2+} was found with protonemal cells in which the Ca^{2+} reporter aequorin was expressed. In this case, a transient increase in intracellular Ca^{2+} was induced by a pulse of blue light, but not by red light (Russell *et al.*, 1998). This effect was independent on extracellular Ca^{2+}.

Tip cell growth of *Ceratodon* is dependent on extracellular Ca^{2+} and seizes when Ca^{2+} chelators are added to the medium. An artificial increase of intracellular Ca^{2+} by the Ca^{2+} ionophore A23187 reduces the growth rate and decreases the ratio of cell length / cell width (Herth *et al.*, 1990). In tip cells, a clear tip to base Ca^{2+} gradient is detected by the hydrophobic, membrane bound Ca^{2+} fluorophore chlorotetracycline (CTC). Phototropic stimulation causes the gradient to shift towards the illuminated side before the change of the growth direction can be observed. This finding implies that Ca^{2+} plays a signal transducing role in the phototropic stimulus. Quite interestingly, ionophores for monovalent cations such as monensin and nigericin, result in a loss of endogenous Ca^{2+} and an inhibition of tip growth. After removal of the ionophores, tip growth is recovered. If during the inhibition the cell is stimulated by unilateral red light, the light direction is memorized and the tip grows towards the previously illuminated side. Therefore, Ca^{2+} is not required for the formation of an intracellular Pfr gradient. Since the red light effect remains reversible by far-red throughout the treatment, the Pfr gradient must persist during the entire inhibition, which can be prolonged to 2 h (Hartmann and Weber, 1988; Herth *et al.*, 1990; Meske *et al.*, 1996).

4.2 Cytoskeleton

Inhibitor studies showed that the vectorial light responses phototropism and chloroplast orientation are dependent on at least one intact cytoskeleton system. The tubulin-based microtubule (MT) and the actin-based microfilament (MF) systems have distinct roles in each process. Tip growth in *Ceratodon* persists under drugs that inhibit the MT system, but the tip to base Ca^{2+} gradient is gradually translocated out of the apex and the cells lose their response to gravity (Herth *et al.*, 1990; Meske *et al.*, 1996). Phototropism is not inhibited by anti-MT drugs, in contrary: the curvature induced by side illumination is stronger than under natural conditions because there is no gravitropic counter-reaction. It seems that the MT system defines the position of the apex and of the intracellular Ca^{2+} gradient in dark-growing tip cells and is required for the perception of gravity. Under MF blocking agents, tip growth ceases and phototropism is therefore no longer possible. However, the light direction is memorized if the cells are illuminated during MF inhibition: after drug removal, tip growth is recovered and the growth direction is defined by the light direction of the previous treatment. This finding clearly shows that active phytochrome molecules can not be associated with the MF cytoskeleton. During drug treatment, the Ca^{2+} gradient is retained; illumination results in a shift towards the irradiated side. Thus, actin is not required for the shift of the Ca^{2+} gradient.

Actin bundles converge towards the apex, but are absent from the central area of the tip thus forming a collar-like structure. Phototropic stimulation results in a re-orientation of actin strands towards the irradiated side before the new growth direction is manifested. Thus, actin re-orientation is one step in the signal transduction of the phototropic response of *Ceratodon*. The position of the actin collar structure is probably defined by the position of the highest Ca^{2+} concentration in the cell (Meske *et al.*, 1996; Meske and Hartmann, 1995; Herth *et al.*, 1990).

The role of MT and MF was also intensely studied in the context of chloroplast photorelocation. Both phytochrome and a blue light photoreceptor, probably phototropin, induce chloroplast movements in protonemal *Physcomitrella* cells (Kadota *et al.*, 2000). Partial irradiation with low light intensities results in an attraction response, i.e. chloroplasts move towards the light beam. High light intensities cause an avoidance response, during which the chloroplasts move away from the light. The blue-light- and red-light-induced movements are qualitatively different in several respects: with blue light, chloroplasts move more or less directly towards the light beam (or away from it, if the light is strong), whereas with red light, chloroplasts make more back and force movements. Red-light-induced movements are inhibited by MT inhibitors, but blue-light induced movements are still possible when either MT or MF is disrupted. Only when both cytoskeleton elements are simultaneously inhibited, blue-light induced movements are completely blocked (Sato *et al.*, 2001). These experiments showed for the first time that chloroplast movement can be mediated by both the MF and the MT system. In other species examined so far, chloroplast movement was considered to be exclusively MF based.

A memory effect was found also here. The cells were irradiated with blue or red light during MT and MF inhibition. After wash out of the inhibitors, chloroplast movement could be partially restored. The chloroplasts moved towards the previously irradiated site. This experiment suggests that neither cytoskeleton is required for the orientation of blue-light photoreceptor or phytochrome within the cell and that a structure which is independent on the cytoskeleton is able to store the spatial light information for some time.

5. SUMMARY

Mosses and seed plant share many light effects, such as phototropism, chloroplast relocation, modulation of chlorophyll synthesis or modulation of gravitropic sensitivity. Also, both groups have the same repertoire of photoreceptors, namely phytochromes, cryptochromes and phototropins. Photoreceptor specificity may however differ. For example, phototropism is controlled by phytochrome in mosses, but by phototropin in seed plants. Besides tip cell phototropism, also other cellular, vectorial effects are known from moss protonemata that are controlled by phytochrome. These effects show us that active phytochrome molecules are localised close to the plasmalemma in an oriented fashion. This contrasts with the nuclear action of phytochrome in *Arabidopsis*. Although the bulk of moss phytochrome is extracted as soluble protein and detected in the cytosole, biophysical techniques showed that a subfraction of phytochrome is immobilized at the cell periphery. Since targeted gene knockout is a routine technique for *Physcomitrella*, and site directed mutagenesis has been performed for a *Ceratodon* nuclear gene, molecular tools are available that will help to untangle photoreceptor action at the molecular level.

6. REFERENCES

Bauer, L. and Mohr, H. (1959) Der Nachweis des reversiblen Hellrot-Dunkelrot Reaktionssystems bei Laubmoosen. *Planta, 54*, 68-73.

Bopp, M. (1963) Development of the protonema and bud formation in mosses. *J. Linn. Soc. Lond. Bot., 58*, 305-309.

Böse, G., Schwille, P., Lamparter, T. (2004) The mobility of phytochrome within protonemal tip cells of the moss *Ceratodon purpureus*, monitored by fluorescence correlation spectroscopy. *Biophys J, 87*, 2013-2021.

Brücker, G., Mittmann, F., Hartmann, E., Lamparter, T. (2004) Repair of a mutated heme oxygenase locus by targeted gene replacement in the moss *Ceratodon purpureus*. *Planta, in press*. (DOI 10.1007/s00425-004-1411-6)

Brücker, G., Zeidler, M., Kohchi, T., Hartmann, E., Lamparter, T. (2000) Microinjection of heme oxygenase genes rescues phytochrome-chromophore-deficient mutants of the moss *Ceratodon purpureus*. *Planta, 210*, 529-535.

Brun, F., Gonneau, M., Doutriaux, M. P., Laloue, M., Nogue, F. (2001) Cloning of the PpMSH-2 cDNA of *Physcomitrella patens*, a moss in which gene targeting by homologous recombination occurs at high frequency. *Biochimie, 83*, 1003-1008.

Bünning, E. and Etzold, H. (1958) Über die Wirkung von polarisiertem Licht auf keimende Sporen von Pilzen, Moosen und Farnen. *Ber Dt Bot Ges, 71*, 304-306.

Clough, R. C. and Vierstra, R. D. (1997) Phytochrome degradation. *Plant Cell Environ, 20*, 713-721.

Cove, D., Schild, A., Ashton, N. W., Hartmann, E. (1978) Genetic and physiological studies of the effect of light on the development of the moss, *Physcomitrella patens*. *Photochem Photobiol, 27*, 249-254.

Cove, D. J. and Ashton, N. W. (1988) Growth regulation and development in *Physcomitrella patens*: an insight into growth regulation and development in bryophytes. *Botanical Journal of the Linnean Society, 98*, 247-254.

Cove, D. J. and Knight, C. D. (1987) Gravitropism and phototropism in the moss *Physcomitrella patens*. In: *Developmental Mutants of Higher Plants*, Thomas H. and Grierson, D. (eds) Cambridge University Press, Cambridge, UK, 181-196.

Cove, D. J. and Lamparter, T. (1998) "Light and moss development." In: *Microbial responses to light and time*, Caddick M.X., Baumberg, S. Hodgson, D.A. and Phillips-Jones, M.K. (eds) Cambridge University Press, Cambridge, UK, 125-141.

Cove, D. J., Quatrano, R. S., Hartmann, E. (1996) The alignment of the axis of asymmetry in regenerating protoplasts of the moss, *Ceratodon purpureus*, is determined independently of axis polarity. *Development, 122*, 371-379.

Demkiv, O. T., Khorkavtsiv, Y. D., Kardash, A. P., Chaban, K. I. (1997a) Interactions between light and gravitation in moss protonema tropisms. *Russian J Plant Physiol, 44*, 177-182.

Demkiv, O. T., Kordyum, E. L., Khorkavtsiv, Y., Kardash, O. R., Chaban, C. (1997b) Behaviour of amyloplasts in photo- and gravitropism of the moss protonema. *J Gravit Physiol, 4*, 75-76.

Demkiv, O. T., Kordyum, E. L., Khorkavtsiv, Y., Kardash, O. R., Chaban, C. (1998) Gravi- and photostimuli in moss protonema growth movements. *Adv Space Res, 21*, 1191-1195.

Dong, X. J., Takagi, S., Nagai, R. (1995) Regulation of the orientation movement of chloroplasts in epidermal cells of *Vallisneria*: Cooperation of phytochrome with photosynthetic pigment under low-fluence-rate light. *Planta, 197*, 257-263.

Ekelund, N. G. A., Sundqvist, C., Quail, P. H., Vierstra, R. D. (1985) Chromophore rotation in 124-kdalton *Avena sativa* phytochrome as measured by light-induced-changes in linear dichroism. *Photochem Photobiol, 41*, 221-223.

Ermolayeva, E., Hohmeyer, H., Johannes, E., Sanders, D. (1996) Calcium-dependent membrane depolarisation activated by phytochrome in the moss *Physcomitrella patens*. *Planta, 199*, 352-358.

Ermolayeva, E., Sanders, D., Johannes, E. (1997) Ionic mechanism and role of phytochrome-mediated membrane depolarisation in caulonemal side branch initial formation in the moss *Physcomitrella patens*. *Planta, 201*, 109-118.

Esch, H., Hartmann, E., Cove, D., Wada, M., Lamparter, T. (1999) Phytochrome-controlled phototropism of protonemata of the moss *Ceratodon purpureus*: physiology of wild type and class 2 *ptr* mutants. *Planta, 209*, 290-298.

Esch, H. and Lamparter, T. (1998) Light regulation of phytochrome content in wild-type and aphototropic mutants of the moss *Ceratodon purpureus*. Photochem Photobiol, 67, 450-455.

Etzold, H. (1965) Der Polarotropismus und Phototropismus der Chloronemen von *Dryopteris filix-mas* (l.) Shott. *Planta, 64*, 254-280.

Featherstone, D. R., Cove, D. J., Ashton, N. W. (1990) Genetic analysis by somatic hybridization of cytokinin overproducing developmental mutants of the moss, *Physcomitrella patens*. *Mol Gen Genet, 222*, 217-224.

Feldman, L. J. and Briggs, W. R. (1987) Light-regulated gravitropism in seedling roots of maize. *Plant Physiol, 83*, 241-243.

Felsenstein, J. (2004) PHYLIP (Phylogeny Inference Package) University of Washington, Seattle, Department of Genetics, distributed by the author.

Girke, T., Schmidt, H., Zahringer, U., Reski, R., Heinz, E. (1998) Identification of a novel delta 6-acyl-group desaturase by targeted gene disruption in *Physcomitrella patens*. *Plant J, 15*, 39-48.

Girod, P. A., Fu, H., Zryd, J. P., Vierstra, R. D. (1999) Multiubiquitin chain binding subunit MCB1 (RPN10) of the 26S proteasome is essential for developmental progression in *Physcomitrella patens*. *Plant Cell, 11*, 1457-1472.

Gonneau, M., Pagant, S., Brun, F., Laloue, M. (2001) Photoaffinity labelling with the cytokinin agonist azido-CPPU of a 34 kDa peptide of the intracellular pathogenesis-related protein family in the moss *Physcomitrella patens*. *Plant Mol Biol, 46*, 539-548.

Grimsley, N. H., Ashton, N. W., Cove, D. J. (1977) Production of somatic hybrids by protoplast fusion in moss, *Physcomitrella patens*. *Mol Gen Genet, 154*, 97-100.

Hahn, S. H. and Saunders, M. J. (1991) Cytokinin increases intracellular Ca^{2+} in Funaria: detection with Indo-1. *Cell Calcium, 12*, 675-681.

Hahn, H. and Bopp, M. (1968) A cytokinin test with high specificity. *Planta, 83*, 115-118.

Hanstein, C., Grolig, F., Wagner, G. (1992) Immunolocalization of cytosolic phytochrome in the green alga *Mougeotia*. *Botanica Acta, 105*, 55-62.

Hartmann, E. and Jenkins, G. (1984) "Photomorphogenesis of mosses and liverworts." In: *The experimental biology of bryophytes*, Dyer A.F. and Ducket J.G. (eds) Academic Press, London, 203-228.

Hartmann, E., Klingenberg, B., Bauer, L. (1983) Phytochrome mediated phototropism in protonemata of the moss *Ceratodon purpureus* BRID. *Photochem Photobiol, 38*, 599-603.

Hartmann, E. and Weber, M. (1988) Storage of the phytochrome-mediated phototropic stimulus of moss protonematal cells. *Planta, 175*, 39-49.

Haupt, W. (1969) Über den Dichroismus von Phytochrom 660 und Phytochrom 730 bei Mougeotia. *Z Pflanzenphysiol, 62*, 287-298.

Herrmann, H. and Kraml, M. (1997) Time-dependent formation of Pfr-mediated signals for the interaction with blue light in Mesotaenium chloroplast orientation. *J Photochem Photobiol B, 37*, 60-65.

Herth, W., Reiss, H.-D., Hartmann, E. (1990) "Role of calcium ions in tip growth of pollen tubes and moss protonema cells." In *Tip growth in plant and fungal cells*, Heath I. B. (ed) Academic Press, London, 91-118.

Hofmann, A. H., Codon, A. C., Ivascu, C., Russo, V. E., Knight, C., Cove, D., *et al.* (1999) A specific member of the Cab multigene family can be efficiently targeted and disrupted in the moss Physcomitrella patens. *Mol Gen Genet, 261*, 92-99.

Hughes, J., Lamparter, T., Mittmann, F. (1996) Cerpu;PHY0;2, a "normal" phytochrome in *Ceratodon*. *Plant Physiol, 112*, 446.

Imaizumi, T., Kadota, A., Hasebe, M., Wada, M. (2002) Cryptochrome light signals control development to suppress auxin sensitivity in the moss *Physcomitrella patens*. *Plant Cell, 14*, 373-386.

Jenkins, G. I., Courtice, G. R., Cove, D. J. (1986) Gravitropic responses of wild-type and mutant strains of the moss *Physcomitrella patens*. *Plant Cell Environ, 9*, 637-644.

Jenkins, G. I. and Cove, D. J. (1983a) Phototropism and polarotropism of primary chloronemata of the moss *Physcomitrella patens*: responses of mutant strains. *Planta, 159*, 432-438.

Jenkins, G. I. and Cove, D. J. (1983b) Phototropism and polarotropism of primary chloronemata of the moss *Physcomitrella patens*: responses of the wild-type. *Planta, 158*, 357-364.

Kadota, A., Sato, Y., Wada, M. (2000) Intracellular chloroplast photorelocation in the moss *Physcomitrella patens* is mediated by phytochrome as well as by a blue-light receptor. *Planta, 210*, 932-937.

Kagawa, T. (2003) The phototropin family as photoreceptors for blue light-induced chloroplast relocation. *J Plant Res, 116*, 77-82.

Kagawa, T., Lamparter, T., Hartmann, E., Wada, M. (1997) Phytochrome-mediated branch formation in protonemata of the moss *Ceratodon purpureus*. *J Plant Res, 110*, 363-370.

Kagawa, T. and Wada, M. (2002) Blue light-induced chloroplast relocation. *Plant Cell Physiol, 43*, 367-371.

Kasahara, M., Kagawa, T., Sato, Y., Kiyosue, T., Wada, M. (2004) Phototropins mediate blue and red light-induced chloroplast movements in *Physcomitrella patens*. *Plant Physiol, 135*, 1388-1397.

Kawai, H., Kanegae, T., Christensen, S., Kiyosue, T., Sato, Y., Imaizumi, T., *et al.* (2003) Responses of ferns to red light are mediated by an unconventional photoreceptor. *Nature, 421*, 287-290.

Kern, V. D. and Sack, F. D. (1999) Irradiance-dependent regulation of gravitropism by red light in protonemata. *Planta, 1999*, 299-307.

Khorkavtsiv, O. Y. and Kardash, O. R. (2002) Gravity-dependent reactions of the moss *Pohlia nutans* protonemata. *Adv Space Res, 27*, 989-993.

Kidd, D. G. and Lagarias, J. C. (1990) Phytochrome from the green alga *Mesotaenium caldariorum*. Purification and preliminary characterization. *J Biol Chem, 12*, 7029-7035.

Klebs, B. (1893) Über den Einfluß des Lichtes auf die Fortpflanzung der Gewächse. *Biol Zentralblatt*, 641-656.

Kolukisaoglu, H. U., Braun, B., Martin, W. F., Schneider-Poetsch, H. A. W. (1993) Mosses do express conventional, distantly B-type-related phytochromes. Phytochrome of *Physcomitrella patens* (Hedw.) *FEBS Lett, 334*, 95-100.

Koprivova, A., Meyer, A. J., Schween, G., Herschbach, C., Reski, R., Kopriva, S. (2002) Functional knockout of the adenosine 5'-phosphosulfate reductase gene in *Physcomitrella patens* revives an old route of sulfate assimilation. *J Biol Chem, 277*, 32195-32201.

Kraml, M. (1994) "Light direction and polarisation." In: *Photomorphogenesis in plants*, Kendrick R.E. and Kronenberg G. H. M. (eds) Kluwer, Dordrecht, Netherlands, 417-446.

Lamparter, T. and Brücker, G. (2004) "Phytochrome in mosses." In: *New Frontiers in Bryology:Physiology, Molecular Biology and Applied Genomic*, Wood, A. J. Oliver, M. J. and Cove, D. J. (eds) Kluwer Academic Press, Dordrecht, The Netherlands, 157-176.

Lamparter, T., Brücker, G., Esch, H., Hughes, J., Meister, A., Hartmann, E. (1998a) Somatic hybridisation with aphototropic mutants of the moss *Ceratodon purpureus*: genome size, phytochrome photoreversibility, tip-cell phototropism and chlorophyll regulation. *J Plant Physiol 153*, 394-400.

Lamparter, T., Esch, H., Cove, D., Hartmann, E. (1997) Phytochrome control of phototropism and chlorophyll accumulation in the apical cells of protonemal filaments of wildtype and an aphototropic mutant of the moss *Ceratodon purpureus*. *Plant Cell Physiol, 38*, 51-58.

Lamparter, T., Esch, H., Cove, D., Hughes, J., Hartmann, E. (1996) Aphototrophic mutants of the moss *Ceratodon purpureus* with spectrally normal and with spectrally dysfunctional phytochrome. *Plant Cell Environ, 19*, 560-568.

Lamparter, T., Hughes, J., Hartmann, E. (1998b) Blue light- and genetically-reversed gravitropic response in protonemata of the moss *Ceratodon purpureus*. *Planta, 206*, 95-102.

Lamparter, T., Kagawa, T., Brücker, G., Wada, M. (2004) Positive and negative tropic curvature induced by microbeam irradiation of protonemal tip cells of the moss *Ceratodon purpureus*. *Plant Biol, 6*, 165-170.

Lamparter, T., Lutterbuese, P., Schneider-Poetsch, H. A. W., Hertel, R. (1992) A study of membrane-associated phytochrome: Hydrophobicity test and native size determination. *Photochem Photobiol, 56*, 697-707.

Lamparter, T., Podlowski, S., Mittmann, F., Schneider-Poetsch, H., Hartmann, E., Hughes, J. (1995) Phytochrome from protonemal tissue of the moss *Ceratodon purpureus*. *J Plant Physiol, 147*, 426-434.

Li, Y., Darley, C. P., Ongaro, V., Fleming, A., Schipper, O., Baldauf, S. L., *et al.* (2002) Plant expansins are a complex multigene family with an ancient evolutionary origin. *Plant Physiol, 128*, 854-864.

Lindemann, P., Braslavsky, S. E., Hartmann, E., Schaffner, K. (1989) Partial purification and initial characterization of phytochrome from the moss *Atrichum undulatum* P. Beauv. grown in the light. *Planta, 178*, 436-442.

Liscum, E. and Hangarter, R. P. (1993) Genetic evidence that the red-absorbing form of phytochrome B modulates gravitropism in *Arabidopsis thaliana*. *Plant Physiol, 103*, 15-19.

Markmann-Mulisch, U., Hadi, M. Z., Koepchen, K., Alonso, J. C., Russo, V. E., Schell, J., *et al.* (2002) The organization of *Physcomitrella patens* RAD51 genes is unique among eukaryotic organisms. *Proc Natl Acad Sci U S A, 99*, 2959-2964.

Marme, D. (1974) Binding properties of the plant photoreceptor phytochrome to membranes. *J Supramol Struct, 2*, 751-768.

Meiri, E., Levitan, A., Guo, F., Christopher, D. A., Schaefer, D., Zryd, J. P., *et al.* (2002) Characterization of three PDI-like genes in *Physcomitrella patens* and construction of knock-out mutants. *Mol Genet Genomics, 267*, 231-240.

Meske, V. and Hartmann, E. (1995) Reorganisation of microfilaments in protonemal tip cells of the moss *Ceratodon purpureus* during the phototropic response. *Protoplasma, 188*, 59-69.

Meske, V., Rupert, V., Hartmann, E. (1996) Structural basis for the red light induced repolarisation of tip growth in caulonemal cells of *Ceratodon purpureus*. *Protoplasma, 192*, 189-198.

Mittmann, F., Brücker, G., Zeidler, M., Repp, A., Abts, T., Hartmann, E., *et al.* (2004) Targeted knockout in *Physcomitrella* reveals direct actions of phytochrome in the cytoplasm. *Proc Natl Acad Sci USA, 101*, 13939-13944.

Muramoto, T., Kohchi, T., Yokota, A., Hwang, I., Goodman, H. M. (1999) The *Arabidopsis* photomorphogenic mutant hy1 is deficient in phytochrome chromophore biosynthesis as a result of a mutation in a plastid heme oxygenase. *Plant Cell, 11*, 335-348.

Murphy, J. T. and Lagarias, J. C. (1997) The phytofluors: a new class of fluorescent protein probes. *Curr Biol, 7*, 870-876.

Napier, R. M. and Smith, H. (1987) Photoreversible association of phytochrome with membranes. II Reciprocity tests and a model for the binding reaction. *Plant Cell Environ, 10*, 391-396.

Nebel, B. J. (1968) Action specta for photogrowth and phototropism in protonemata of the moss Physcomitrium turbinatum. *Planta, 81*, 287-302.

Nishiyama, T., Fujita, T., Shin-I, T., Seki, M., Nishide, H., Uchiyama, I., *et al.* (2003) Comparative genomics of *Physcomitrella patens* gametophytic transcriptome and *Arabidopsis thaliana*: implication for land plant evolution. *Proc Natl Acad Sci U S A, 100*, 8007-8012.

Olsson, T., Thelander, M., Ronne, H. (2003) A novel type of chloroplast stromal hexokinase is the major glucose phosphorylating enzyme in the moss *Physcomitrella patens*. *J Biol Chem, 278*, 44439-44447.

Pasentsis, K., Paulo, N., Algarra, P., Dittrich, P., Thümmler, F. (1998) Characterization and expression of the phytochrome gene family in the moss *Ceratodon purpureus*. *Plant J, 13*, 51-61.

Podlowski, S. (1994) Biochemische und immunologische Untersuchungen zum Photorezeptor Phytochrom aus dem Laubmoos *Ceratodon purpureus*. Diploma thesis, Freie Universität Berlin, Germany.

Porra, R. J. (1997) Recent progress in porphyrin and chlorophyll biosynthesis. *Photochem Photobiol, 65*, 492-516.

Rensing, S. A., Rombauts, S., Van de, P. Y., Reski, R. (2002) Moss transcriptome and beyond. *Trends Plant Sci, 7*, 535-538.

Rubinstein, B., Drury, K. S., Park, R. B. (1969) Evidence for bound phytochrome in oat seedlings. *Plant Physiol, 44*, 105-109.

Russell, A. J., Cove, D. J., Trewavas, A. J., Want, T. J. (1998) Blue light but not red light induces a calcium transient in the moss *Physcomitrella patens* (Hedw.). *Planta, 206*, 278-283.

Sakakibara, K., Nishiyama, T., Sumikawa, N., Kofuji, R., Murata, T., Hasebe, M. (2003) Involvement of auxin and a homeodomain-leucine zipper I gene in rhizoid development of the moss *Physcomitrella patens*. *Development, 130*, 4835-4846.

Sato, Y., Kadota, A., Wada, M. (2003) Chloroplast movement: dissection of events downstream of photo- and mechano-perception. *J Plant Res, 116*, 1-5.

Sato, Y., Wada, M., Kadota, A. (2001) Choice of tracks, microtubules and/or actin filaments for chloroplast photo-movement is differentially controlled by phytochrome and a blue light receptor. *J Cell Sci, 114*, 269-279.

Saunders, M. J. and Hepler, P. K. (1982) Calcium ionophore A23187 stimulates cytokinin-like mitosis in Funaria. *Science, 217*, 943-945.

Saunders, M. J. and Hepler, P. K. (1983) Calcium antagonists and calmodulin inhibitors block cytokinin-induced bud formation in *Funaria*. *Dev Biol, 99*, 41-49.

Sawahel, W., Onde, S., Knight, C., Cove, D. (1992) Transfer of foreign DNA into *Physcomitrella patens* protonemaltissue by using the gene gun. *Plant Mol Biol Reporter, 10*, 314-315.

Schaefer, D. G. and Zryd, J. P. (1997) Efficient gene targeting in the moss *Physcomitrella patens*. *Plant J, 11*, 1195-1206.

Schaefer, D. G., Zrÿd, J.-P., Knight, C. D., Cove, D. J. (1991) Stable transformation of the moss *Physcomitrella patens*. *Mol Gen Genet, 226*, 418-424.

Schild, A. (1981) Untersuchungen zur Sporenkeimung und Protonemaentwicklung bei dem Laubmoos *Physcomitrella patens*. PhD thesis, University Mainz, Germany.

Schipper, O., Schaefer, D., Reski, R., Flemin, A. (2002) Expansins in the bryophyte *Physcomitrella patens*. *Plant Mol Biol, 50*, 789-802.

Schulz, P. A., Hofmann, A. H., Russo, V. E., Hartmann, E., Laloue, M., von Schwartzenberg, K. (2001) Cytokinin overproducing ove mutants of *Physcomitrella patens* show increased riboside to base conversion. *Plant Physiol, 126*, 1224-1231.

Schumaker, K. S. and Gizinski, M. J. (1993) Cytokinin stimulates dihydropyridine-sensitive calcium uptake in moss protoplasts. *Proc Natl Acad Sci USA, 90*, 10937-10941.

Schumaker, K. S. and Gizinski, M. J. (1995) 1, 4-Dihydropyridine binding sites in moss plasma membranes. Properties of receptors for a calcium channel antagonist. *J Biol Chem, 270*, 23461-23467.

Schwuchow, J. M., Kern, V. D., White, N. J., Sack, F. D. (2002) Conservation of the plastid sedimentation zone in all moss genera with known gravitropic protonemata. *J Plant Growth Regul, 21*, 146-155.

Schwuchow, J. M., Kern, V. D., Kim, D., Sack, F. D. (1995) Caulonemal gravitropism and amyloplast sedimentation in the moss *Funaria*. *Can J Bot, 73*, 1029-1035.

Sineshchekov, V., Koppel, L., Hughes, J., Lamparter, T., Zeidler, M. (2000) Recombinant phytochrome of the moss *Ceratodon purpureus* (CP2): fluorescence spectroscopy and photochemistry. *J Photochem Photobiol B, 56*, 145-153.

Strepp, R., Scholz, S., Kruse, S., Speth, V., Reski, R. (1998) Plant nuclear gene knockout reveals a role in plastid division for the homolog of the bacterial cell division protein FtsZ, an ancestral tubulin. *Proc Natl Acad Sci USA, 95*, 4368-4373.

Sundqvist, C. and Björn, L. O. (1983) Light-induced linear dichroism in photoreversibly phytochrome sensor pigments - II. Chromophore rotation in immobilized phytochrome. *Photochem Photobiol, 37*, 69-75.

Thompson, J. D., GibsonT. J., Plewniak, F., Jeanmougin, F., Higgins, D. G. (1997) The ClustalX windows interface: flexible strategies for multiple sequence alignment aided by quality analysis tools. *Nucleic Acids Res, 24*, 4876-4882.

Thümmler, F., Dufner, M., Kreisl, P., Dittrich, P. (1992) Molecular cloning of a novel phytochrome gene of the moss *Ceratodon purpureus* which encodes a putative light-regulated protein kinase. *Plant Mol Biol, 20*, 1003-1017.

Thümmler, F., Herbst, R., Algarra, P., Ullrich, A. (1995) Analysis of the protein kinase activity of moss phytochrome expressed in fibroblast cell culture. *Planta, 197*, 592-596.

Tokutomi, S., Sugimoto, T., Mimuro, M. (1992) A model for the molecular structure and orientation of the chromophore in a dimeric phytochrome molecule. *Photochem Photobiol, 56*, 545-552.

Valanne, N. (1966) The germination of moss spores and their control by light. *Annals Botanici Fennici 3*, 1-40.

Voerkel, S. H. (1933) Untersuchungen über die Phototaxis der Chloroplasten. *Planta, 21*, 156-205.

von Schwartzenberg, K., Kruse, S., Reski, R., Moffatt, B., Laloue, M. (1998) Cloning and characterization of an adenosine kinase from *Physcomitrella* involved in cytokinin metabolism. *Plant J, 13*, 249-257.

Wang, T. L., Cove, D. J., Beutelmann, P., Hartmann, E. (1980) Isopentenyladenine from mutants of the moss *Physcomitrella patens*. *Phytochemistry, 19*, 1103-1105.

Yeh, K. C. and Lagarias, J. C. (1998) Eukaryotic phytochromes: Light-regulated serine/threonine protein kinases with histidine kinase ancestry. *Proc Natl Acad Sci USA, 95*, 13976-13981.

Young, J. C. and Sack, F. D. (1992) Time-lapse analysis of gravitropism in *Ceratodon* protonemata. *Am J Bot, 79*, 1348-1358.

Zank, T. K., Zahringer, U., Beckmann, C., Pohnert, G., Boland, W., Holtorf, H., *et al.* (2002) Cloning and functional characterisation of an enzyme involved in the elongation of δ-polyunsaturated fatty acids from the moss *Physcomitrella patens*. *Plant J, 31*, 255-268.

Zeidler, M., Lamparter, T., Hughes, J., Hartmann, E., Remberg, A., Braslavsky, S., *et al.* (1998) Recombinant phytochrome of the moss *Ceratodon purpureus*: heterologous expression and kinetic analysis of Pr-->Pfr conversion. *Photochem Photobiol, 68*, 857-863.

Chapter 26

CIRCADIAN REGULATION OF PHOTOMORPHO-GENESIS

Paul Devlin
Royal Holloway University of London, School of Biological Sciences, Egham, Surrey TW20 0EX, United Kingdom (e-mail: paul.devlin@rhul.ac.uk)

1. INTRODUCTION

> "Space by itself, and time by itself, are doomed to fade away into mere shadows, and only a kind of union of the two will preserve an independent reality."

From a public lecture given in 1906 by Hermann Minkowski, former college teacher in mathematics to Albert Einstein soon after Einstein announced his special theory of relativity.

In considering the environment of an organism we cannot consider the spatial environment alone – the dimension of time impinges upon all life on earth. The interaction between an organism and its spatial environment varies tremendously with time. This volume has already considered the range of responses to light throughout the life of a plant, from germination, through seedling establishment, control of plant architecture to transition to flowering. The stage in a seedling's development plays a major role in determining how it will respond to a particular light treatment. The responses involved in inducing germination are very different from those involved in triggering flowering and yet the same photoreceptors may be involved at both stages. However, environmental responses vary not only over the course of development, they vary over a much shorter timescale – that of a single day. As the earth rotates, the immediate environment of a plant will vary considerably. In particular, temperature and light intensity will gradually change over the course of the day with dramatic transitions accompanying sunrise and sunset. A response to any environmental factor that is appropriate in the early part of the day may no longer be advantageous towards evening. This is most certainly the case for plant responses to light. For example, the light-activation of genes encoding the photosynthetic machinery is of great advantage at dawn but would be of little advantage at sunset. Early investigations showed that, in fact, plants are exquisitely adapted to deal with this problem. These light-responsive genes are no longer strongly induced by light at sunset – their response varies over the course of that day to give maximum advantage to the plant. Actually, many of the plant responses that have been discussed in this volume vary in their degree of responsiveness to light depending the time of day and, in this way, plants are exquisitely adapted to the earth's rotation.

This adaptation required the development of one key feature, common to almost all organisms on earth, the ability to measure time.

2. THE CIRCADIAN CLOCK

In organisms as diverse as bacteria, fungi, plants, insects and mammals, the timing of many physiological and metabolic processes is tightly coordinated so that they occur at the optimal time of day. Our own sleep wake cycle is one such example. Our metabolism begins to gear up for activity before dawn and then to wind down again ready for sleep during the night. The benefits of an ability to predict dawn and dusk transitions are obvious – the early bird catches the worm. Such anticipation could, in theory, be achieved using an hourglass timing mechanism. The hourglass would measure time from the last day/night transition, for example. However, a simple responsive mechanism like this would not be reliable. Whilst, changes in light intensity and temperature could indicate an approaching transition and trigger a resetting of the hourglass, such variables can change dramatically over a very short timescale independent of time of day. The movement of a cloud over the sun, for example, would give a similar signal causing an incorrect setting of the hourglass. Instead, the measurement of time is achieved via a robust internal timing mechanism that is capable of continuing even in the absence of any environmental cues.

Many of us have experienced the robustness of our own internal timing mechanism at first hand. After rapidly travelling across several time-zones we become jetlagged and must adjust to new times of dawn and dusk. Our internal timing mechanism still causes us to sleep and wake as before until we have had several days to adjust to this new time-zone and our internal timing mechanism has been reset. Thus, this timing mechanism is not entirely divorced from the environment and can respond to cues that can set it to the right time.

This endogenous timing mechanism is known as the circadian clock (from the Latin for "about a day"). It has, as its basis, a beautifully simple molecular mechanism whereby levels or activity of proteins within the cell oscillate with a repeating cycle of approximately 24 hours. In effect, then, our sleep wake cycle is governed by the state of just a few proteins.

3. CIRCADIAN RHYTHMS

The first examples of circadian rhythms were observed in plants. Androsthenes, scribe to Alexander the Great, recorded that the leaves of certain trees opened during the day and closed at night. This is the first recorded example of a circadian rhythm; but it was much later, in 1729, that Jean Jacques d' Ortous de Mairan, a French astronomer, demonstrated the involvement of an endogenous timekeeper in the control of leaf movement. He showed that this leaf movement phenomenon continued with the same rhythm when the plants were kept continually in near darkness, demonstrating that this was not merely a response to the environment nor was it due to any sort of hourglass timing mechanism measuring time since the last day/night transition.

Subsequently, many circadian phenomena have been observed in all classes of organism. Circadian rhythms of gene expression have been observed in cyanobacteria (Golden *et al.*, 1997); rhythms of conidiation have been observed in fungi (Bell-Pedersen, 1998); rhythms of pupal eclosion and activity have been observed in insects (Scully and Kay, 2000); and rhythms of body temperature and alertness have been observed in mammals (Moore-Ede *et al.*, 1982).

The circadian clock allows each of these events to be synchronised to an appropriate time of day. Although the circadian clock clearly confers an advantage in allowing an organism to predict dawn and dusk, the clock is not just involved in the timing of physiological and metabolic processes associated with these transitions. Many other physiological and metabolic processes are tightly controlled by the clock to peak at a particular time during the day or, in fact, during the night. In this way, sequential processes, each dependent on previous steps in a pathway, can be timed to occur in the correct sequence. Evidence of such sequential orchestration can be most clearly seen by studying patterns of gene expression. It would appear that the timing of gene expression in several pathways is tightly orchestrated so that a gene encoding a given component of one such pathway will be upregulated at the specific time of day when that component is required. For example, in plants, genes implicated in cell elongation peak towards the end of the day so that elongation can occur at night when water pressure is greater. Then, subsequently, the genes implicated in cell wall synthesis peak towards the end of the night to stabilise the newly elongated cell walls (Harmer *et al.*, 2000). Similarly, genes involved in photosynthesis peak just after dawn whilst genes encoding starch-mobilizing enzymes peak during the subjective night (Harmer *et al.*, 2000).

4. THE CIRCADIAN CLOCK IN PLANTS

Circadian rhythms have been intensely studied in the model plant, *Arabidopsis thaliana*. *A. thaliana* shows a number of overt circadian rhythms. During de-etiolation there is a rhythm in the rate of hypocotyl elongation with maximum elongation occurring during the night when water availability is highest, ensuring maximal water pressure for cell elongation (Dowson-Day and Millar, 1999). In young seedlings there is also a clear rhythm of leaf movement, with leaves closing up to a more acute angle during the night, possibly providing protection for the meristem from colder temperatures at night (Engelmann *et al.*, 1994). However, the most frequently studied circadian rhythms in *A. thaliana* are rhythms of gene expression. In particular, the greatest insights into the workings of the plant circadian clock have been gained by studying the circadian rhythm of expression of the *CAB2* gene.

The *CAB2* gene, encodes a chlorophyll a b binding protein, part of the photosynthetic machinery of higher plants. *CAB2* transcript levels begin to rise just before dawn and peaks about 3 hours after dawn (Millar and Kay, 1991). Using a firefly luciferase (*LUC*) reporter gene fused to the *CAB2* promoter it proved possible to generate transgenic seedlings that glowed in time with *CAB2* transcription (Millar *et al.*, 1992) (Figure 1). This allowed the first large scale screening for mutants altered in the circadian rhythm of gene expression from which the *timing of cab expression 1* (*toc1*) mutant was isolated (Millar *et al.*, 1995). The *toc1* mutation results in a short period phenotype. It was demonstrated to affect all circadian rhythms tested and that this phenotype was manifest both in white light and in darkness, making it a strong candidate for a component of the central clock mechanism (Somers *et al.*, 1998b).

At about this time two other candidate central clock components were identified in *A. thaliana*. Another mutant of *A. thaliana*, *late elongated hypocotyl* (*lhy*) was demonstrated to disrupt circadian rhythms and the photoperiodic control of flowering (Schaffer *et al.*, 1998). Cloning of the *LHY* gene revealed that it encoded a MYB transcription factor (Schaffer *et al.*, 1998). Meanwhile, a search for factors binding to CIS-acting elements the *CAB2* promoter-region that are involved in circadian regulation of *CAB2* gene expression, led to the isolation of another, related, MYB transcription factor, *CIRCADIAN CLOCK-ASSOCIATED 1* (*CCA1*) (Wang *et al.*, 1997; Wang and Tobin, 1998). Both *CCA1* and *LHY* expression were found to oscillate with a circadian rhythm, whilst constitutive overexpression of either CCA1 or LHY was shown to disrupt their own and other circadian rhythms (Wang *et al.*, 1998).

Subsequent cloning of the *TOC1* gene revealed that it encoded a pseudo response regulator (Strayer *et al.*, 2000). *TOC1* expression also oscillates with a circadian rhythm but in antiphase to *CCA1* and *LHY* (Strayer *et al.*, 2000). TOC1 is a member of a small gene family of five paralogues, the *Arabidopsis PSEUDO RESPONSE REGULATOR* (*APRR*) genes (Imamura *et al.*, 1999). The others all oscillate peaking in a sequence through the day, preceding the peak of TOC1 expression (Matsushika *et al.*, 2000). In a *toc1* null allele, expression of *CCA1* and *LHY* is damped suggesting that TOC1 protein acts to positively regulate *CCA1* and *LHY* gene expression. Conversely, *CCA1* or *LHY* overexpression causes a complete loss of *TOC1* rhythmicity with *TOC1* expression fixed at a minimal level, suggesting that CCA1 and LHY proteins act to negatively regulate *TOC1* gene expression. In fact, CCA1 and LHY proteins were demonstrated to bind to the "evening element" region of the *TOC1* promoter, necessary for rhythmic expression.

This loop was proposed to form the basis of the circadian clock in plants (Figure 2) (Alabadi *et al.*, 2001). However, questions remain as to the mechanism by which TOC1 acts to regulate *CCA1* and *LHY* gene expression. TOC1 contains no recognisable DNA binding domain suggesting that it does not act directly. Furthermore, the *toc1* null mutant is still rhythmic in white light suggesting some redundancy in the role played by TOC1 (Strayer *et al.*, 2000). Likewise, a *cca1 lhy* double mutant remains rhythmic, although phase advanced, again suggesting that other factors are capable of generating a rhythm in the absence of CCA1 and LHY (Mizoguchi *et al.*, 2002; Alabadi *et al.*, 2001).

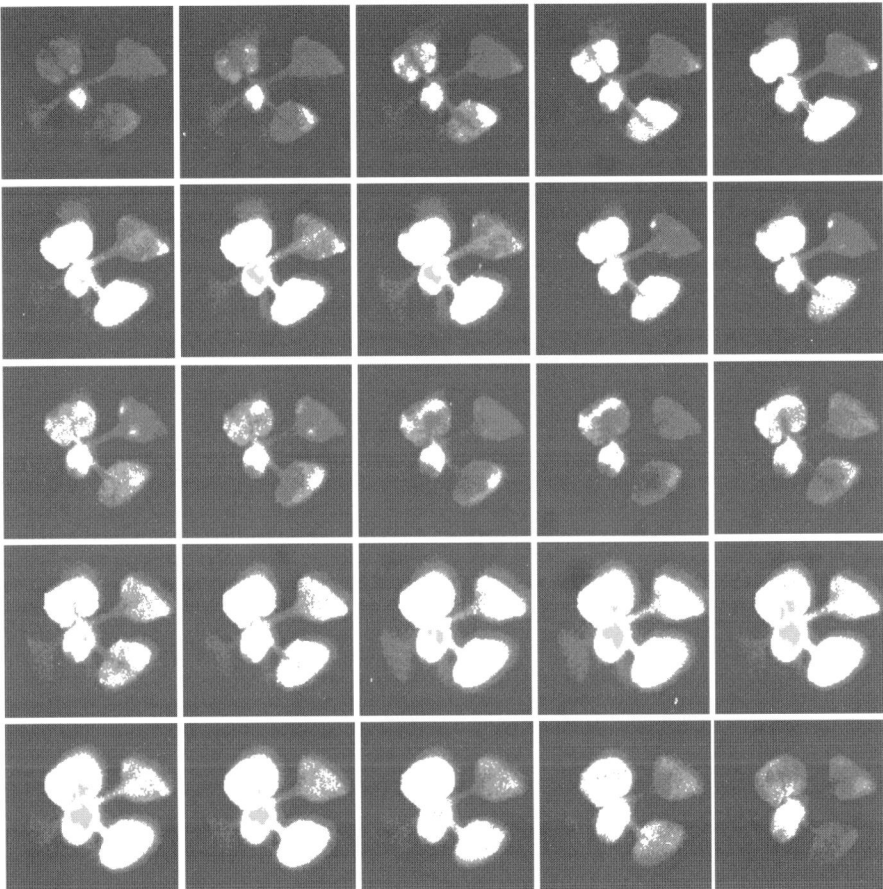

Figure 1. Time-course showing the circadian rhythm of bioluminescence in a single seedling of A. thaliana expressing the firefly luciferase gene under the control of the A. thaliana CAB2 promoter. The time-course follows the rhythm of transcription from the CAB2 promoter over 48 hours. Images are taken 2 hours apart and should be followed from left to right, one row at a time. The seedling, germinated on agar growth medium was first entrained 12 hour light,12 hour dark cycles for 6 days then transferred to constant light. The CAB2 gene encodes part of the photosynthetic machinery and appropriately shows a peak of expression during the subjective day and a trough of expression during the subjective night. Images were taken using a NightOwl cooled CCD camera Imaging System, Berthold Technologies UK Ltd.

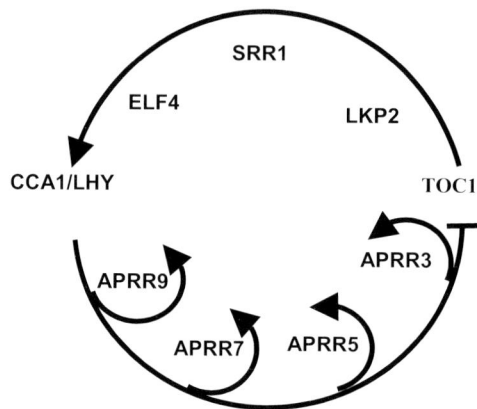

Figure 2. Schematic diagram showing the components proposed to be involved in the central circadian oscillator in *A. thaliana*. Levels of the TOC1 protein begin to rise at subjective dawn and peak around dusk. The increasing levels of TOC1 act to promote expression of the CCA1 *and* LHY *genes and levels of CCA1 and LHY proteins consequently rise during the subsequent night, peaking at dawn. CCA1 and LHY are MYB transcription factors and are capable of feeding back to suppress the expression of the* TOC1 *gene by directly binding to the* TOC1 *promoter. TOC1 levels, therefore, begin to fall as CCA1 and LHY levels rise to the point where TOC1 ceases eventually to promote CCA1 and LHY expression. As a result, levels of CCA1 and LHY then decline during as the subsequent day and this allows* TOC1 *transcription to begin again and continue the cycle. The ARABIDOPSIS PSEUDO RESPONSE REGULATORS, APRR9, APRR7, APRR5 and APRR3 are closely related to TOC1 and have also been demonstrated to be associated with the central oscillator. Each peaks in turn in the order shown during the subjective day and each plays some role in maintaining normal rhythmicity.LKP2, SRR1 and ELF4 have also been demonstrated to be involved in the central clock mechanism, though their role is poorly understood. The ELF4 protein, however, has been proposed to be involved in the promotion of* CCA1 *and* LHY *transcription.*

Some involvement of the other, related *APRR* genes in the functioning of the central oscillator has also been proposed. The transcripts encoded by *APRR9, APRR7, APRR5* and *APRR3* peak in that order in a sequence that begins at dawn and that precedes the peak of *TOC1* (Matsushika *et al.*, 2002). It has been suggested that these genes form the basis of a second interlocked circadian loop whereby each triggers the expression of the next in sequence. Mutants and overexpression analyses have been carried out for *APRR9, APRR7* and *APRR5* and all have been shown to affect circadian rhythms to some degree, though none have been shown to be essential for rhythmicity. A variety of phenotypes are observed for the different *APRR* family members. Overexpressors of *APRR9* show a short period for expression of *CCA1, LHY, TOC1* and other *APRR* genes (Matsushika *et al.*, 2002), whilst a loss of function *aprr9* mutant shows a long period for expression of these genes (Eriksson *et al.*, 2003). Conversely, a loss of function *aprr7* mutant shows no period effect but shows an early phase of expression of oscillator components

(Kaczorowski and Quail, 2003). Overexpressors of *APRR5* show no period effect but show a low amplitude of expression of oscillator components (Sato *et al.*, 2002) whilst *aprr5* mutants show a short period phenotype (Eriksson *et al.*, 2003). This range of phenotypes is difficult to reconcile with a simple loop whereby each of the *APRR* genes triggers expression of the next in sequence. Furthermore, double mutant analysis shows the *aprr5* and *aprr9* mutant phenotypes are additive, the resultant double mutant having a wild type period length. This is inconsistent with a system where both of these factors act in the same pathway and it has more recently been suggested that *APRR5* and *APRR9* affect the circadian clock by largely independent mechanisms (Eriksson *et al.*, 2003). Thus, the APRR genes clearly play a role in the clock mechanism but the precise nature of this role remains unclear (Figure 2).

Another factor identified as being closely associated with the clock mechanism is EARLY-FLOWERING 4 (ELF4). The *elf4* mutant was originally identified as being early-flowering in non-inductive photoperiods (Doyle *et al.*, 2002). Such a phenotype is typical of a defective capacity for time measurement associated with dysfunction of the circadian clock (Somers *et al.*, 1998b). Indeed, *elf4* mutants also show a very low amplitude for the circadian oscillation of *CAB2::LUC* in constant light. In addition, the *CCR2::LUC* construct; corresponding to the *COLD CIRCADIAN CLOCK-REGULATED* gene (Carpenter *et al.*, 1994); also shows a low amplitude rhythm of expression in constant darkness in *elf4*, almost to the point of arrhythmia. Such a phenotype is indicative of a key role for ELF4 in the circadian clock mechanism and such a proposed role is further supported by the fact that the *ELF4* gene is, itself, expressed with a circadian rhythm, showing a peak of expression in the evening. *ELF4* encodes a plant-specific protein of unknown biochemical activity, revealing little about its possible mechanism of action. However, a closer inspection of the *elf4* mutant phenotype reveals that the *elf4* mutant appears to go through one robust circadian cycle before this near-arrhythmicity and that this near-arrhythmicity occurs at the time of *CCA1* activation. This is perhaps suggestive that the predominant role of ELF4 may be in *CCA1* activation (Doyle *et al.*, 2002) (Figure 2).

One further factor implicated in clock function is LOV KELCH PROTEIN 2 (LKP2) (Schultz *et al.*, 2001). Seedlings of *A. thaliana* overexpressing *LKP2* show an arrhythmic phenotype in both light and dark for a number of circadian-clock regulated processes (Schultz *et al.*, 2001). The light independent, pleiotropic nature of the *LKP2* overexpression phenotype also places LKP2 close to the central oscillator (Figure 2). Notably, the LKP2 transcript, itself, is not regulated by the circadian clock (Schultz *et al.*, 2001). However, there is no requirement that all factors associated with the normal running of the clock need to be circadian clock-regulated themselves so this observation is not inconsistent.

Lastly, the SENSITIVITY TO RED LIGHT REDUCED (SRR1) protein has very-recently been proposed to influence the central clock mechanism. The *SRR1* gene encodes a nuclear and cytoplasmicly-localised protein, conserved in a range of eukaryotes, though its function in plants is not known (Staiger *et al.*, 2003). In *srr1* mutant plants in constant light, the rhythms of expression of the clock outputs, *CAB2* and *CCR2*; and the central clock components, *CCA1* and *TOC1*, showed short period length and reduced amplitude of oscillation (Staiger et al., 2003). Furthermore, in *srr1* mutant plants in constant dark, the rhythms of expression *CCR2* also showed a short period length (Staiger et al., 2003). This indicates that SRR1 is required for normal oscillator function independent of light making it another candidate clock-associated component (Figure 2). The *SRR1* transcript is not under clock control, itself, however (Staiger et al., 2003).

CCA1 and LHY are proposed to act to link the clock to the various clock-controlled outputs. CCA1 has been demonstrated to positively regulate *CAB2* gene expression by binding to an AAAAATCT site in the *CAB2* promoter (Wang et al., 1998). It is proposed that LHY also has such an effect. Hence, the clock directly regulates the level of *CAB2* transcript. At the same time, CCA1 and LHY repress *TOC1* via an evening element AAAATATCT in the *TOC1* promoter that forms a slight variant of the previously-characterised CCA1 binding site (Alabadi et al., 2001). It is proposed that both CCA1 and LHY also activate other output genes that cycle with the same phase as *CAB2* or *TOC1*, acting either as repressors or promoters depending on the nature of the CCA1/LHY binding site in the promoter regions of these genes. Indeed, the evening element has been found to be highly represented in the promoters of genes that cycle with a peak of expression in the evening, supporting this prediction (Harmer et al., 2000).

5. SETTING THE PLANT CIRCADIAN CLOCK

The internal oscillator is only of use if it is first set to the right time. Hence, the clock does not run in complete isolation but can be reset by the dawn dusk transitions that it predicts. Our own experience of jetlag is the clearest evidence of this phenomenon. On flying through a number of time-zones we find our own sleep wake cycle remains attuned to the time-zone from which we left. However, after several days in the new time-zone, our circadian clock is gradually reset to the correct time by the new set of environmental signals we are receiving. Once reset we are again able to predict dawn and dusk and to coordinate physiological processes to occur at the correct time of day. In particular, we respond to the new light signals that we are receiving. In the same way, light input plays a key part in the circadian system in plants.

In plants, where light signalling is central in numerous aspects of development an array of photoreceptors exist to give maximum information about the light environment. These photoreceptors and the responses they control have been discussed in some detail in this volume. To briefly recap, the photoreceptors of higher plants fall into three families: the phytochromes, absorbing primarily in the red and far red region of the spectrum, though to some extent in the blue too (Quail, 2002), and the cryptochromes (Lin and Shalitin, 2003) and phototropins (Briggs and Christie, 2002), absorbing primarily in the blue and UV/A region of the spectrum. Evidence also points to additional, as yet unidentified UV/B photoreceptors (Kim et al., 1998). Five phytochromes, phyA-phyE, exist in the model plant *A. thaliana*. PhyA is the most abundant phytochrome in etiolated seedlings. It is a very sensitive photoreceptor and its high abundance in etiolated seedlings allows it to act as an antenna photoreceptor. However, phyA is light labile and its abundance falls rapidly in the light. PhyA is particularly important in germination and seedling establishment but it still continues to perform a role in adult seedlings in the control of flowering time. PhyB-phyE are light stable. These phytochromes mediate responses to higher light intensities in throughout the life history of the plant (Whitelam *et al.*, 1998). The phytochromes exist in two photo-intraconvertible forms, a red-absorbing Pr form and a far red-absorbing Pfr form. Absorption of light energy by the covalently-linked tetrapyrrole chromophore triggers a conformational change from one form to the other. Pfr is generally considered the active form triggering a range of plant developmental responses (Schäfer and Bowler, 2002; see also Chapters 7, 17, 22).

Three cryptochromes, cry1-cry3, exist in *A. thaliana*. The cryptochromes are related to the photolyase DNA-repair photoreceptors. They have two attached chromophores, a pterin and a flavin allowing absorption of light energy (Cashmore *et al.*, 1999). It is thought that light activates the cryptochrome molecule by triggering an intraprotein electron transfer (Giovani *et al.*, 2003). Cry1 and cry2 are the best-characterised members of this family. Cry1 is light stable while cry2 is light labile. Like phyA, the light labile cry2 plays a major role in seedling establishment at low light intensities but also contributes to adult plant development in the control of flowering time. Like phyB-phyE, the light-stable cry1 regulates responses to higher intensities of light throughout the life of the plant (Lin *et al.*, 2003). Cry3 was only discovered very recently. It is more closely related to cyanobacterial cryptochrome sequences than to the higher plant cry1 and cry2 sequences. It is also targeted to the chloroplast leading to the suggestion that it has arisen in the nuclear genome as a result of gene transfer from the chloroplast to the nucleus, the chloroplast being regarded as having originated from an endosymbiotic ancestor of modern-day cyanobacteria (Kleine *et al.*, 2003). As yet no functions have been ascribed to cry3 (see Chapter 11).

Two phototropins, phot1 and phot2 exist in *A. thaliana*. Phototropins possess two flavin chromophores. The phototropins are primarily involved in phototropism, chloroplast migration and stomatal opening (Briggs *et al.*, 2002; see Chapter 10).

Plants employ a range of photoreceptors in setting the clock to the correct time. In *A. thaliana*, the light labile phyA photoreceptor is involved in perception of low intensity red and blue light, while the light stable phyB, phyD and phyE photoreceptors specifically mediate the resetting of the clock by high intensity red light. The cry1 photoreceptor plays a role in perceiving blue light at all intensities. However, at low intensity blue light it appears to be necessary for phyA-mediated clock resetting rather than acting as a photoreceptor in its own right. Cry2 specifically perceives low intensity blue light for clock resetting (Somers *et al.*, 1998a; Devlin and Kay, 2000). The phototropins appear to play little role in clock resetting (Somers *et al.*, 1998a). Thus the plant is able to exhibit plasticity in its recruitment of photoreceptors for resetting the clock to take maximum advantage of its light environment (Figure 3).

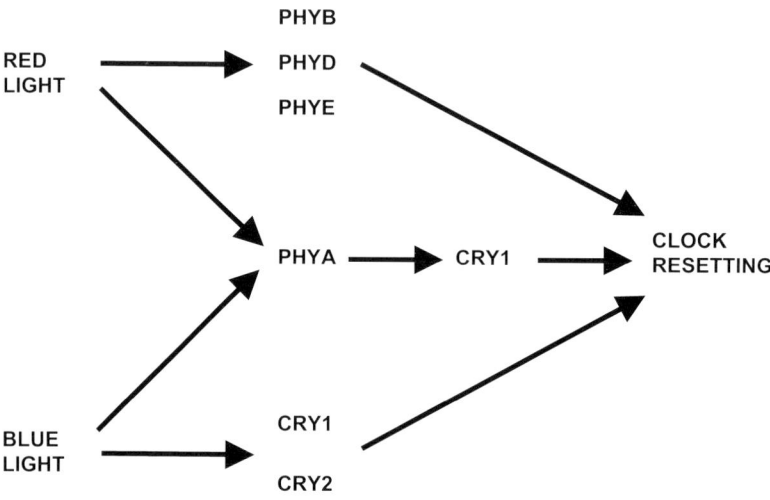

Figure 3. *Recruitment of a range of photoreceptors in clock resetting in A. thaliana. Both red and blue light act in clock resetting in A. thaliana. PhyA perceives low intensity red and blue light, whilst phyB, phyD and phyE perceive high intensity red light. Cry1 and cry2 act to perceive blue light, though cry1 is also required for proper phyA signalling.*

Resetting of the clock requires a phase shift, that is, an advance or a delay in the rhythm so that it oscillates with a new phase. A phase shift occurs as a result of a change in the level of a central clock component to a level normally associated with a different point in the circadian cycle (Figure 4). *CCA1* and *LHY* transcripts normally peak at dawn. However, it has been demonstrated that *CCA1* and *LHY* both show a strong increase in transcription in response to light (Martinez-Garcia *et al.*, 2000) and thus the onset of dawn is also accompanied by an acute spike in *CCA1* transcription (Kim *et al.*, 2003). The presence of light prior to dawn due to lengthening of the day as summer approaches would cause an earlier spike in the level of *CCA1* and *LHY* message, hence advancing the timing of the *CCA1* and *LHY* peak and causing a phase shift. For a delay in the time of dawn a late spike of *CCA1* and *LHY* transcription would occur and a phase delay would result. This light-mediated change in *CCA1* and *LHY* transcript is a good candidate for the mechanism of resetting in plants.

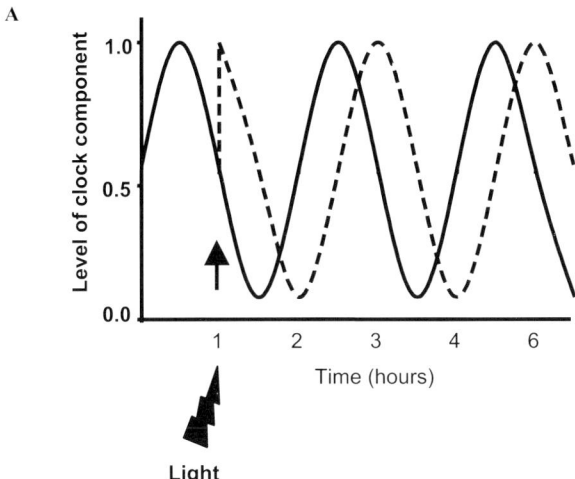

Figure 4. Resetting of the circadian clock by light involves a light induced change in one of the clock components. If light (indicated by the arrow) were to directly cause an increase in the expression of the clock component to a level normally associated with an early phase in the cycle, then the clock would be delayed. The component would subsequently be following a circadian rhythm phased a few hours later that it would have been otherwise.

The link between the photoreceptors and the circadian clock appears to be very direct. A protein named PHYTOCHROME INTERACTING FACTOR 3 (PIF3), that contains a basic helix loop helix DNA-binding domain, has been demonstrated to bind to G-box promoter sequences in the *CCA1* and *LHY* promoters (Martinez-Garcia *et al.*, 2000). These G-box elements are known to be involved in light regulation of transcription of these genes (Menkens *et al.*, 1995). In response to illumination with red light, phytochrome Pfr is transported to the plant cell nucleus where it interacts with PIF3 (Martinez-Garcia *et al.*, 2000; Kircher *et al.*, 2002; Bauer *et al.*, 2004). This interaction is thought to be the key trigger to promote transcription of the *CCA1* and *LHY* genes. More recently, an in depth study of the mode of action of PIF3 in light responses has demonstrated that PIF3 appears to act as a repressor of light responses suggesting that the action of phytochrome is to remove this repression. It should be noted, however, that some light induction of LHY protein translation has also been demonstrated in response to light pulses given around the time of dawn (Kim, *et al.*, 2003; Bauer *et al.*, 2004). The mechanism involved in the induction of LHY protein translation is not understood but this would further enhance the resetting effect of a light pulse given at this time.

In addition to the light regulation of *CCA1* and *LHY* transcription, another mechanism by which light signals may impinge upon the clock is suggested by the fact that the transcripts of both *APRR9* and *ELF4* are induced by white light (Makino *et al.*, 2001; Khanna *et al.*, 2003).

While light is the major environmental factor responsible for resetting of the circadian clock in plants, it should be mentioned that temperature fluctuations can reset the clock too. The clock can be reset by temperature changes such as those that occur at dawn and dusk transitions in the natural environment (Somers *et al.*, 1998b; Michael *et al.*, 2003). However, little is known about the way in which temperature fluctuations are perceived or about the nature of temperature resetting. It should also be mentioned at this point, though, that despite the resetting ability of temperature fluctuations, the clock is very well buffered against longer-term variations in temperature. The phenomenon known as temperature compensation has been demonstrated for all rhythms properly studied (Pittendrigh, 1954; Salisbury *et al.*, 1968; Somers *et al.*, 1998b). A temperature compensated rhythm shows no difference in period length whether it is running at a constant high temperature or a constant low temperature. Such temperature compensation is an absolute requirement for a clock to be useful in all seasons and in all climatic conditions but this behaviour is contrary to the behaviour of most biochemical reactions that will show a doubling of reaction rate for an increase in temperature of 10 degrees Celsius. Once again, however, the mechanism involved here is poorly understood.

6. DRIVEN VS ENDOGENOUS RHYTHMS

Clock resetting is not a phenomenon limited to major phase shifts such as we experience after long haul flights. It is not even limited to the changes due to the fluctuation in daylength that all organisms at higher latitudes experience as the seasons change. Conversely, clock resetting is a daily phenomenon. The circadian clock runs with a very precisely-repeating period length but it does not run with a precisely 24-hour period length. Under constant environmental conditions, the clock is said to free-run. Under such non-driven conditions, the circadian rhythm of most organisms runs with a period length of between 23 and 25 hours depending on the species or even the individual. As such the daily light, dark cycle must constantly reset the clock by a small amount at each light/dark transition to keep the clock "in synch" with the environment. Such inaccuracy is thought, perhaps, to be necessary to maintain the plasticity or flexibility that allows an organism to adapt to a constantly changing daylength. The light signals resetting or entraining the clock are often referred to as "Zeitgebers", literally time-givers. In a 24-hour light, dark cycle, Zeitgeber Time (ZT) can be measured as the time since the last dawn. In contrast, the progression of a free-running rhythm can be measured by "Circadian Time" (CT). CT, like ZT, is measured from the last dawn. However, 24 hours of CT is defined as the time to complete one full cycle of the free-running clock and therefore may be anywhere between 23 and 25 real hours depending on the species or individual.

7. GATING

If the degree of resetting in response to a light pulse is plotted against CT, the resulting graph is known as a phase response curve (Figure 5). The phase response curve is determined by examining the phase shifting effect of a standard pulse of light on individuals that have first been entrained to light, dark cycles and then transferred to constant darkness. As might be predicted, the degree of resetting of the clock in response to light is not the same at all times during the day. Clock resetting by light is due to a change in the level or activity of one or more of the central clock components. As such, the degree of response will depend upon the state of that component at the time of the light pulse. In plants, a light-triggered increase in *CCA1* or *LHY* transcript, at a time when that transcript is normally fairly low, will trigger a dramatic change in the phase of the clock. Whether we perceive that as an advance or a delay depends on whether the level of transcript was increasing or decreasing at the time. Individuals given a pulse of light just prior to subjective dawn will appear to show a phase advance. Individuals given a pulse of light just after subjective dusk will appear to show a phase delay. Conversely, a light-triggered increase in *CCA1* or *LHY* transcript, at a time when that transcript is already at its peak, will have no effect.

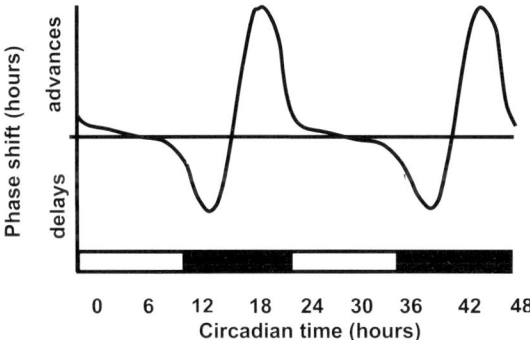

Figure 5. A stylised phase response curve for *A. thaliana* based on the work of Covington et al. (2001). The y-axis shows the phase shifting effect of light pulses given at varying time of the subjective circadian day as indicated on the x-axis. On the y-axis, phase advances are represented by a positive phase shift; phase delays are represented by a negative phase shift. On the x-axis, subjective day is represented by open boxes; subjective night is indicated by filled boxes.

However, other factors clearly also affect the shape of the phase response curve. According to the hypothesis above, we would predict that the degree of clock resetting in response to light would vary with a smooth sinusoidal rhythm but this is, in fact, not the case. Most organisms lack a response to light during a prolonged period of the subjective daytime. Such a pattern is clearly displayed by *A. thaliana* which shows very little response to pulses in the early to mid part of the subjective day (Covington *et al.*, 2001). This period of insensitivity to light for resetting during the subjective day is known as the dead-zone. From this, it is clear that the clock regulates it's own resetting. At some times the photoreceptors are able to trigger a response; at others they are not. This phenomenon is known as "gating". The gate is considered open if light is able to trigger a response and closed if it is not. If an organism receives a light pulse prior to dawn or after dusk, this acts as an indicator that its clock is "out of synch" with the environment. The perception of light during the day would not be a cause for resetting and, hence, this gate prevents unnecessary resetting (Devlin, 2002).

8. CIRCADIAN REGULATION OF PHOTOMORPHOGENESIS

In effect, gating involves the circadian regulation of photoreceptor signalling. This raised the question as to whether other photomorphogenic responses, such as those studied elsewhere in this volume, are subject to the same circadian gating. It might be predicted that the circadian clock would control the interaction of an organism with its environment. Such an adaptation would give a clear advantage. As already mentioned, a response to any environmental factor that is appropriate in the early part of the day may no longer be advantageous towards evening. Several examples have now been discovered to demonstrate that this is, indeed, the case.

8.1 Circadian regulation of light-induced changes in gene expression

Circadian regulation of light-induced changes in gene expression is, perhaps, most clearly demonstrated in plants in the way in which transcription of the *CAB* genes are induced by light. The phenomenon was first observed by Nagy *et al.* (1988). Both light and the circadian clock were demonstrated to control *Cab-1* gene expression in wheat. The authors found that during the afternoon, when light would still be exerting a positive effect, *Cab-1* expression started to decrease under circadian control. They concluded that the circadian clock was gating the light induction of *Cab-1* expression. The generation of the *CAB2::LUC* expressing plants used in screening for the *toc1* mutant (Millar *et al.*, 1995) offered the ideal opportunity to further investigate this and to test whether this regulation occurred at the level of transcription. *CAB2::LUC* expressing plants were entrained to light, dark cycles then placed into constant darkness. Groups of plants were the given light pulses at different times during the subsequent 72 hrs and the degree of light responsiveness of the *CAB2::LUC* construct was followed (Millar and Kay, 1996). A clear free-running circadian rhythm of responsiveness was observed showing peak induction of *CAB2* transcription in response to light in the early part of the subjective day, i.e. the time at which this response would be most advantageous (Millar *et al.*, 1996). Clearly the clock was suppressing light-induced transcription during the subjective afternoon and night.

The regulation of *NITRATE REDUCTASE* (*NAI2*) gene expression provides another example of gating. Pilgrim *et al.* (1993) demonstrated that *NAI2* expression is positively induced by light but also regulated by the clock. They found that *NA12* mRNA continued to cycle with a peak at subjective dawn even in continuous light. They concluded that the circadian clock was gating the light induction of *NA12* expression. However, nuclear run-on experiments demonstrated that this regulation of *NA12* expression occurred at a post-transcriptional level.

8.2 Circadian regulation of light-mediated inhibition of hypocotyl elongation

The inhibitory effect of light on hypocotyl elongation has been well documented (Koornneef *et al.*, 1980) and has become a standard assay for effects on light signal transduction pathways (Casal *et al.*, 2003). In darkness, seedlings display etiolated growth, characterised by a rapid hypocotyl elongation, an adaptation to allow seedlings to grow through a covering layer of soil into the light. On emergence into light this rapid elongation is halted. A number of phototransduction pathways converge to control hypocotyl elongation in young seedlings (Casal and Mazzella, 1998; Neff and Chory, 1998). Nevertheless, hypocotyl elongation is not a linear process in constant light. A clear rhythm of elongation growth has been observed in *A. thaliana* seedlings with maximal hypocotyl growth coinciding with subjective dusk and a growth arrest occurring around subjective dawn (Dowson-Day *et al.*, 1999). Again, the maintenance of a circadian rhythm in constant light is suggestive of a gating of light signalling by the circadian clock. Light mediated inhibition of hypocotyl elongation does not occur around subjective dawn despite the light signal being constant. It is possible that this reflects a circadian control of the factors involved directly in hypocotyl elongation. However, a study by Wildermann *et al.* (1978) in mustard that demonstrated a clear circadian rhythm of sensitivity to red light for inhibition of hypocotyl elongation, also showed no evidence of circadian regulation of the elongation itself in constant darkness. This strongly argues that the circadian clock mediates a gating of photoreceptor signalling in light-mediated inhibition of hypocotyl elongation.

8.3 Circadian regulation of light-mediated stimulation of hypocotyl hook opening

The opening of the hypocotyl hook is a classical assay for early light signalling responses. During etiolated growth when the hypocotyl is elongating rapidly the cotyledons remain closed and are hooked over so as to trail behind the leading edge. This adaptation protects the cotyledons from mechanical damage during growth through soil. On emergence into light, the hook opens. This response can be triggered by very small amounts of light leading to its characterisation as a Very Low Fluence Response (VLFR) (Yanovsky *et al.*, 1997). Horwitz and Epel (1978) showed that etiolated courgette seedlings (*Cucurbita pepo*) possessed a clear circadian rhythm of responsiveness of hook opening to pulses of red light. As the experiments were carried out with etiolated seedlings that have not seen light, dark cycles, it is impossible to ascribe the peak of responsiveness to a particular time of day. Non-the less, this demonstrates that gating not only applies to higher fluence responses in seedlings but also to at least one VLFR.

8.4 Circadian regulation of light-mediated stimulation of stomatal opening

Many plants have been demonstrated to show a circadian rhythm of stomatal opening. Stomata of most plants open during the subjective day to allow transpiration and gas exchange and close during the subjective night with a circadian rhythm that persists even in a constant environment (Meidner and Mansfield, 1968; Salome *et al.*, 2002). Rhythms in antiphase to this occur in plants possessing a Crassulacean Acid Metabolism where stomata open only at night to minimise water loss (Nimmo, 2003). Stomata are also strongly responsive to environmental signals such as light (Zeiger, 1990), humidity (Grantz, 1990) and CO_2 concentration (Morison, 1987). Both red and blue light are capable of triggering stomatal opening and phytochrome and zeaxanthin respectively have been implicated as the photoreceptors (Talbott *et al.*, 2003). Gorton *et al.* (1993) demonstrated that stomata of *Vicia faba* show a circadian rhythm of sensitivity to light in addition to the underlying circadian rhythm of stomatal conductance. Like the underlying rhythm, the circadian rhythm of sensitivity to light peaked during the middle of the subjective day. Sensitivity to both red and blue light showed the same pattern of gating (Gorton *et al.*, 1993).

8.5 Circadian regulation of sensitivity to light allows daylength perception

The ability to measure daylength allows plants to adapt to changing seasons. Flowering in many species is strongly daylength dependent, an adaptation that allows flowering within a species to be synchronised to a particular time of year. This gives an advantage in terms of ensuring the availability of pollinators or in terms of ensuring seed-set at an appropriate time of year for dispersal. A full account of the photoperiodic control of flowering can be found in the following chapter but the subject is included briefly here, as it forms another excellent example of circadian gating of photomorphogenesis.

Plants fall into three photoperiodic categories: long day plants that flower when daylength exceeds a certain critical value; short day plants that flower when daylength is shorter than a certain critical value; and day-neutral plants that will flower irrespective of photoperiod. Early work in the field of photoperiodism demonstrated that long day plants can be triggered to flower in short days if they are given a pulse of light of about 2 hours duration during the night. Similarly short day plants can be prevented from flowering in short days by a pulse of light given during the night. Such light treatments have become known as "night breaks". They demonstrated that the measurement of daylength was not achieved by an hourglass mechanism that measured the total time from dawn to dusk but by a mechanism that detected the presence or absence of light at a particular time of day. Coulter and Hamner (1964) demonstrated that this timing mechanism is controlled by a circadian clock. They grew Soybean (*Glycine max*), a short day plant, in 8-hour light 64-hour dark cycles. Such a long dark period would normally be inductive for a short day plant, however in this case, the plants were given a night break at one of a number of points during the dark period. Coulter and Hamner observed that night breaks given at about 8 hours, 32 hours or 56 hours into the dark period inhibited flowering whereas night breaks given at about 20 hours or 44 hours after transfer to darkness had no effect. This demonstrated an unmistakable circadian rhythm of sensitivity to night breaks for suppressing flowering in Soybean. The 8-hour light period was clearly sufficient to entrain the circadian rhythm such that the beginning of the 8-hour light period was perceived as dawn. The peaks of sensitivity to night breaks indicate that a 24-hour circadian cycle then continues to run during the subsequent long dark period. The peaks of sensitivity correspond to the middle of the subjective night in this 24-hour circadian cycle. The peaks of insensitivity correspond to the subjective day.

A similar circadian rhythm of sensitivity to night breaks, though in this case to promote flowering, can be demonstrated for the long day plant, *A. thaliana* (Carre, 1998). Here the availability of numerous photoperiod-insensitive mutants has allowed the mechanism of this circadian regulation of light sensitivity to be further dissected. Photoperiodic induction of flowering is dependent on the action of the photoreceptors, phyA and cry2 (Samach and Coupland, 2000). Light signals perceived by these photoreceptors are proposed to mediate a posttranslational modification of the CONSTANS (CO) protein in order to trigger flowering. Levels of the CO transcript are tightly controlled by the circadian clock with a peak about 16-20 hours after dawn. The flowering stimulus is generated when light is perceived at a time when CO is highly expressed. Such a situation could only occur in long days (Yanovsky and Kay, 2002; Suarez-Lopez *et al.*, 2001).

A summary of photomorphogenic responses regulated by the circadian clock can be found in Table 1.

9. MECHANISM OF CIRCADIAN REGULATION OF PHOTOMORPHO-GENESIS

The mechanism by which the central oscillator is able to regulate photomorphogenic responses in plants is poorly understood. Gating must be mediated by output components, regulated by the clock, and that act to regulate photoreceptor-signalling pathways. In general gating of photomorphogenic responses appears to occur around the early to middle night. For example, in constant light, maximal hypocotyl growth coincides with subjective dusk indicating that photoreceptor signalling is inhibited at this time. Likewise, the time of minimal *CAB2* or *NIA2* light responsiveness is just after subjective dusk, whilst minimal stomatal responsiveness occurs in the middle the subjective night. (It is not possible to identify the time of day that hook opening is minimally responsive to light as these experiments were carried out in etiolated seedlings that have never been entrained to light, dark cycles). It might be predicted from this that the output component responsible for gating of these photomorphogenic responses would be an inhibitor of light signalling that peaks just after dusk or a promoter of light signalling that peaks just after dawn.

Table 1. Photomorphogenic responses known to be regulated by the circadian clock in Arabidopsis thaliana and other species.

Response	Peak time of effectiveness of light	Reference
Light induction of *CAB2* gene expression in *A. thaliana*	Early part of subjective day	Millar and Kay, 1991
Light induction of *NIA2* gene expression in *A. thaliana*	Subjective dawn	Pilgrim *et al.*, 1993
Light inhibition of hypocotyl elongation in *A. thaliana*	Subjective dusk	Dowson-Day and Millar, 1999
Light stimulation of hypocotyl hook opening in *Cucurbita pepo*	N/A	Horwitz and Epel, 1978
Light stimulation of stomatal opening in *A. thaliana*	Middle of subjective day	Salome *et al.*, 2002
Night break sensitivity in *A. thaliana*	Early part of subjective night	Carre, 1998
Circadian clock resetting in *A. thaliana*	Before subjective dawn and after subjective dusk	Covington *et al.*, 2001

Two exceptions exist to this pattern of gating. One is the timing of sensitivity to night-breaks in the regulation of flowering. Here, the photoreceptors are minimally effective just after subjective dawn. However, as just described, the mechanism of gating of night-break sensitivity is fully understood and is mediated by the circadian regulation of the CO protein levels. In this response, CO acts as a promoter of light signalling. If light is incident when CO is high in the subjective evening/early night, a night-break will be effective in regulating flowering. CO acts very specifically in the flowering process and there is no evidence of a role for CO in more general regulation of photomorphogenesis.

The second exception relates to clock resetting itself. In *A. thaliana*, as in many other organisms, the clock shows a dead-zone for resetting in response to light during the subjective day (Covington *et al.*, 2001; Johnson, 1992). This indicates a gating mechanism inhibiting photoreceptor action at this time. This is possibly suggestive of the action of a distinct mechanism from that which is involved in the gating of photomorphogenic responses.

10. MUTANTS AFFECTING CIRCADIAN REGULATION OF PHOTO-MORPHOGENESIS

Two *A. thaliana* mutants have recently been identified that specifically disrupt the integration of circadian and light signalling in the control of photomorphogenic responses. The corresponding proteins are clock regulated and are, therefore, clock output components but they are not part of the central oscillator. These mutations do, however, have an effect on light input to the clock, as this is, itself, subject to circadian gating. Such an effect is manifest as a light-specific disruption of the circadian oscillator.

10.1 *early flowering 3 (elf3)*

The *elf3* mutant was identified as a daylength insensitive early flowering mutant (Zagotta *et al.*, 1992; Zagotta *et al.*, 1996). Early studies of the *elf3-1* mutant demonstrated it to be arrhythmic in light but not in darkness for the rhythm of *CAB2::LUC* expression (Hicks *et al.*, 1996). The normal rhythmicity of *elf3* mutants in darkness indicated that the central oscillator, itself, is not disrupted whilst the arrhythmicity in light suggested a light input role for ELF3.

elf3 mutants also have a long hypocotyl, long petioles and small leaf blades, all phenotypes associated with impaired light signalling (Zagotta *et al.*, 1992). Furthermore, the *elf3* mutant lacks the wild-type rhythm of hypocotyl elongation in constant light (Dowson-Day *et al.*, 1999). As this rhythm is associated with the periodic inhibition of photoreceptor action by the circadian clock, it was, therefore, first suggested at this point that *elf3* might be disrupted in the integration of circadian and light signalling in the control of photomorphogenesis.

The absence of any detectable rhythms in the *elf3* mutant in the light also raised a crucial question as to whether the *elf3* mutation caused the circadian clock to stop or whether it merely masked the normal output of the clock in response to a light signal. To address this, McWatters *et al.* (2000) tested the response of the *elf3* mutant in a "release from light" experiment. They entrained wild type and *elf3* mutant seedlings of *A. thaliana* to light, dark cycles, and then transferred them to constant light at predicted dawn. Replicate samples were then subsequently moved to constant darkness at 2-hour intervals, whereupon *CAB2::LUC* expression was followed to determine the state of the oscillator in the preceding light interval. Under these conditions, wild type seedlings retained the same phase of expression of *CAB2::LUC* irrespective of how long the seedlings were in constant light prior to transfer to darkness. The clock in wild type seedlings simply continues in phase with the entraining light, dark cycles. *elf3* seedlings behaved as wild type initially but then, beyond about 11 hours in constant light, they were no longer able to maintain their circadian rhythm. In *elf3* seedlings transferred to darkness after 11 hours of light, the phase of peak *CAB2* expression was set by that final light, dark transition rather than being in phase with the entraining light, dark cycles. Hence, the circadian clock is runs normally in *elf3* seedlings for the first 11 hours of the constant light treatment; but, beyond this subjective evening-time, the light received is most probably causing a continuous resetting signal that effectively stops the clock until the plants are subsequently released into darkness. This continuous resetting beyond the first 11 hours of constant light is presumably gated in wild type seedlings by the action of the ELF3 protein acting to suppress photoreceptor input to the clock. Effectively, the *elf3* seedlings seem to be unable to gate the light signal.

Furthermore, McWatters *et al.* (2000) demonstrated that the *elf3* mutant shows no gating of the acute induction of *CAB2::LUC* expression by light. They entrained wild type and *elf3* seedlings to 12-hour light / 12-hour dark cycles, then transferred them to constant darkness at end of a 12-hour light period. Replicate batches of seedlings were then subjected to a single 20-minute white light pulse at intervals during the subsequent dark treatment after which *CAB2::LUC* expression was followed. In wild type seedlings, the size of the acute peak of *CAB2::LUC* induction by light was clearly circadian-gated, with maximal response occurring during the subjective day and with almost no induction occurring during subjective night. In contrast, light pulses at all times during a subjective day were able to activate *CAB2::LUC* expression in *elf 3* mutant seedlings to levels equivalent to the maximal *CAB2::LUC* induction in wild type seedlings.

The lack of gating in the *elf3* mutant is therefore pleiotropic, affecting light signals that trigger both clock resetting and *CAB2* induction. In wild type seedlings, photoreceptor signalling to cause both clock resetting and *CAB2* induction is clearly moderated by the circadian clock during the night by the action of the ELF3 protein.

The subsequent cloning of the *ELF3* gene gave little clue as to how ELF3 functions. The *ELF3* gene encodes a novel plant-specific protein, 695 amino acids in length, of unknown biochemical activity (Hicks *et al.*, 2001; Liu *et al.*, 2001). The ELF3 protein was demonstrated to be nuclear localised and the protein sequence is suggestive of a possible transcriptional regulator function (Liu *et al.*, 2001). Crucially, the *ELF3* transcript is regulated by circadian clock. Analysis of *ELF3::LUC* expressing plants demonstrates that this regulation is mediated, at least in part, at the level of transcription. *ELF3* transcription shows a peak during the subjective night, exactly when the ELF3 protein has been demonstrated to act to mediate gating of light signalling.

In an attempt to further dissect the mode of action of ELF3, Liu et al, (2001) carried out a yeast two-hybrid interaction study to test for interaction between ELF3 and the phyB photoreceptor. They demonstrated that ELF3 binds PHYB apoprotein in a yeast two-hybrid assay. Hence, it is possible that the ELF3 protein directly moderates light signalling by binding to the photoreceptors themselves.

ELF3 overexpressing plants were generated and found to display a long period phenotype when grown in constant light (Covington *et al.*, 2001). Decreasing fluence rates of light cause a lengthening of period length in wild type *A. thaliana* (Somers *et al.*, 1998a) and, hence, the long period phenotype of the *ELF3* overexpressing plants fits well with the proposal that ELF3 is an inhibitor of light signalling. A closer analysis of light input to the clock in *ELF3* overexpressing plants to determine the photoreceptors affected, revealed that this long period phenotype was manifest in continuous blue light at all fluence rates but was only observed at higher fluence rates of red light (Covington *et al.*, 2001). This is consistent with ELF3 acting specifically on phyB, cry1 and cry2 signalling but not on phyA signalling. Furthermore, *ELF3* overexpression in etiolated seedlings caused a reduced acute induction of *CAB2::LUC* by light (Covington *et al.*, 2001). These two phenotypic traits combined support the finding in the *elf3* mutant and suggest that ELF3 negatively regulates light signalling both in input to the clock and in acute regulation of gene expression.

One further, elegant study by Covington *et al.* (2001) of the responses of *elf3* mutant and *ELF3* overexpressing plants looked at the phase response curves in each of these lines. Wild type *A. thaliana* shows a classical phase response curve with strong phase advances occurring just prior to dawn; strong phase delays occurring just after dusk; and a dead-zone of unresponsiveness during the subjective day (Covington *et al.*, 2001). The wild type phase response curve is further characterised by a very sudden transition from delays to advances at the breakpoint, the time during the middle of the subjective night before which pulses of light cause delays and after which pulses of light cause advances. *ELF3* overexpressing pants showed a reduced amplitude of resetting both for advances and delays. This was most pronounced in response to red light pulses. Furthermore, the breakpoint was more gradual in *ELF3* overexpressing pants. Such a topology is characteristic of phase response curves generated by subsaturating light pulses (Pittendrigh, 1981) and is, therefore, consistent with a repression of light signalling in *ELF3* overexpressing pants. Conversely, *elf3* null mutant plants display more extreme phase shifts during subjective night, sometimes becoming arrhythmic in response to a light pulse. Again, this phenotype is most pronounced in response to red light pulses (Covington *et al.*, 2001).

ELF3, therefore, acts as a suppressor of light signalling in the late evening and early night (Figure 6). The purpose of ELF3 may be to buffer transient fluctuations in light intensity occurring through the night, perhaps preventing a response to moonlight or a flash of lightning (Covington *et al.*, 2001). Circadian regulation of the *ELF3* transcription ensures that ELF3 is maximally expressed at this time. However, a number of questions remain about the *elf3* mutant phenotype. *elf3* mutants have a long hypocotyl in the light rather than the short hypocotyl that might be expected in a mutant where light signalling is derepressed. Conversely, *ELF3* overexpressing plants have a short hypocotyl (Liu *et al.*, 2001). More curiously, this short hypocotyl phenotype in *ELF3* overexpressing plants is specific to red light and is not observed in blue light (Liu *et al.*, 2001) despite the demonstration of a role for ELF3 in blue light input to the clock (Covington *et al.*, 2001). Clearly, ELF3 plays a role in the regulation of hypocotyl elongation by red light that is distinct from its role in circadian gating of photomorphogenesis. Another question is raised by the flowering time phenotype of the *elf3* mutant. The early-flowering, daylength insensitive phenotype of *elf3* is most likely also due to a disruption of the circadian clock. Analysis of *CO* transcript levels in the *elf3* mutant revealed that CO levels were high at all times tested (Suarez-Lopez *et al.*, 2001). Such an effect would be expected to trigger early flowering and is consistent with a disruption of circadian regulation. However, this, again, is not quite consistent with role for ELF3 solely in gating light signalling after about 11 hours of light. The role of ELF3 in hypocotyl elongation and in the regulation of flowering time, therefore, remains unclear.

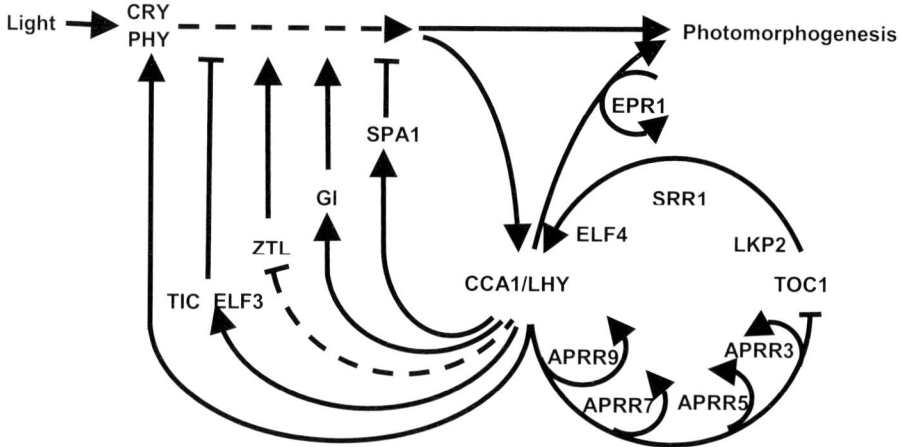

Figure 6. Schematic diagram showing the components proposed to be involved in the circadian regulation of photomorphogenesis. CCA1 and LHY are proposed to regulate output from the circadian clock. The clock regulates expression of the genes encoding the phytochromes (PHY) and the cryptochromes (CRY). The clock also regulates expression of TIC and ELF3 genes. TIC and ELF3 act to negatively regulate photoreceptor signalling both to the clock and in photomorphogenic responses during the night. ZTL, GI and SPA1 have also all been shown to be required for normal photoreceptor signalling both to the clock and in photomorphogenic responses. ZTL and GI are positively acting factors while SPA is a suppressor of light signalling. The clock controls ZTL proteins levels by regulating degradation of ZTL by the proteasome. The clock also controls expression of the genes encoding GI and SPA1. EPR1 forms a self-sustaining slave oscillator. EPR1 is a positively acting factor in photomorphogenic responses though it does not appear to play a role in light input to the clock meaning its action must be downstream of a branch-point in the light-signalling pathway.

10.2 *time for coffee (tic)*

The *tic* mutant was identified in a screen for mutants possessing an aberrant circadian rhythm of *CAB2::LUC* expression in constant light. *tic* displays a low amplitude rhythm without affecting mean luminescence (Hall *et al.*, 2003). This phenotypic trait is displayed both in darkness and in a number of light conditions. Moreover, a number of circadian regulated outputs are affected in this way: the amplitudes of *LHY*, *CCA1* and *TOC1* rhythms are all lower in *tic* suggesting that the defect lies close to the central oscillator itself. The *tic* mutation also results in a variation in the period length for various clock outputs under free running conditions. However, the nature of this effect varies according to the output studied and the conditions under which analysis is performed (for full details see Hall *et al.*;

2003). One theory put forward to explain this varied range of period length effects is that perhaps a low amplitude oscillator is more easily perturbed by subtle environmental variations that are usually buffered in a more robust system. A slight difference in temperature, for example, may have no effect on a wild type oscillator but may cause a period-changing effect in *tic*.

Crucially, however, from the point of view of this review, *tic* shows a complete lack of gating for clock resetting in response to light. As mentioned earlier, a "release from light" experiment involves entraining seedlings to 12-hour light, 12-hour dark cycles, transferring them to constant light at predicted dawn and then releasing replicate samples into constant darkness at 2-hour intervals. Plotting a graph of the timing of the first peak of *CAB2* expression, after transfer to darkness, against the preceding length of time in constant light will reveal the state of the oscillator during that preceding light treatment. In wild type seedlings, the first peak of *CAB2* expression in darkness is always in phase with the initial 12-hour light, 12-hour dark entrainment regime. In *tic* seedlings released from constant light after anything up to 19 hours of light, the peak of *CAB2* expression in darkness is always in phase with the initial 12-hour light, 12-hour dark entrainment regime. However, beyond this time in constant light, the phase of peak *CAB2* expression is set by final light/dark transition. This indicates that the oscillator in *tic* arrests after about 19 hours in constant light. Given the slightly shorter period displayed for *CAB2::LUC* expression in the tic mutant under these conditions, this equates to the time of completion one full cycle of *CAB2::LUC* expression since transfer to constant light. This point in the cycle would, therefore, coincide with subjective dawn in wild type seedlings. Consequently, the function of the TIC protein in wild type seedlings appears to be in the gating of light signals to the clock at a time around dawn. The presence of TIC allows the clock wild type seedlings to proceed through this critical time of the day (Hall *et al.*, 2003).

The apparent complete cessation of the clock in tic mutants after one circadian cycle is not quite compatible with the observation that low amplitude rhythms do persist for several days in constant light in *tic*. However, a close examination of mathematical curve fits to CAB2::LUC rhythms in individual tic seedlings in constant light reveals that these fits are statistically very poor and that many of the seedlings could be classed as arrhythmic on this basis (Hall *et al.*, 2003).

Another *tic* phenotype also suggests a role for the wild type TIC protein in the negative regulation of light signal transduction more generally. *tic* also shows a partial defect in the acute light-mediated induction of *CAB2::LUC* expression. As mentioned earlier, if wild type seedlings are entrained 12 hour light/12 hour dark cycles then transferred to constant darkness at end of a 12 hour light period, subsequent acute induction of *CAB2::LUC* expression by light pulses shows circadian rhythm of sensitivity. Maximal response is observed during the subjective day with almost no induction during subjective night. In *tic* mutant seedlings, however, light pulses at all times will strongly activate *CAB2::LUC* expression, though *tic* seedlings still show a clear rhythm of sensitivity. The peak of induction in tic seedlings occurs 4 hours earlier than that seen in wild type, though this is consistent with a shorter period length of basal *CAB2* expression observed in *tic* mutant seedlings under these conditions (Hall *et al.*, 2003).

Furthermore, *tic* mutant seedlings have a short hypocotyl phenotype, particularly pronounced in red light. It is unclear whether this represents a general role for the wild type TIC protein in the negative regulation of light signal transduction at all times of day or whether this effect is specific to light signalling at dawn as both possibilities are compatible with this observation (Hall *et al.*, 2003).

tic also shows a daylength insensitive, early flowering phenotype perhaps consistent with a shorter period circadian rhythm in light, dark cycles. This would result in an early phase of *CO* expression meaning that *CO* expression would be high in the light period even in a short day. The perception of light at a time when CO is high will trigger flowering in *A. thaliana*.

Thus, TIC gates light signals to the clock around dawn and also suppresses light signalling in the induction of CAB2 gene expression at all times (Figure 6). However, the action of TIC is not specific to light signalling as the tic mutant also shows an amplitude defect in darkness suggesting a, second, more direct effect of TIC on the central oscillator.

The combination of TIC and ELF3 would appear to allow the circadian rhythm to proceed through the whole of the night, buffered from the effects of transient fluctuations in light intensity. However, more insidious effects of both mutations become apparent in the *tic elf3* double mutant. *tic elf3* is completely arrhythmic for *CAB2::LUC* expression in both constant light and constant darkness. What's more, unlike either single mutant, the *tic elf3* double mutant shows no anticipation of dawn or dusk transitions in light, dark cycles. Even temperature entrainment failed to establish a free-running rhythm in *tic elf3*.

Non-the-less, both of these mutations act to gate light signalling during the night and their time of activity, therefore, corresponds perfectly with the observed timing of the gating of photomorphogenic responses.

11. OTHER POSSIBLE COMPONENTS INVOLVED GATING

In addition to the roles for ELF3 and TIC in circadian regulation of photomorphogenesis a number of other factors may contribute to this effect. The following examples are further candidate mechanisms or factors that may be involved in gating, though it should be stressed that these form merely hypothetical candidates and further proof is still required that they do actually contribute to gating.

11.1 Circadian regulation of photoreceptor levels

An analysis of the expression pattern of an *A. thaliana PHYB::LUC* construct in seedlings of *A. thaliana* recently revealed a surprising result (Kozma-Bognar *et al.*, 1999). The rate of transcription of from the *PHYB* promoter showed a strong circadian regulation with a peak of transcription at dawn. This was demonstrated to be true both in light, dark cycles and in continuous environmental conditions. *PHYB* mRNA levels in *A. thaliana* were shown to concur with this observed circadian regulation of transcription (Kozma-Bognar *et al.*, 1999). These authors further demonstrated that a *Nicotiana tabaccum PHYB::LUC* construct in seedlings of *N. tabaccum* revealed the same pattern of circadian regulation of *PHYB* transcription. Subsequent analysis of a *N. tabaccum* translational *NtPHYB::PHYB: LUC* gene fusion expressed in *N. tabaccum* revealed that the rate synthesis of new PHYB protein is also under circadian control. However, the authors were unable to detect any cycling in total PHYB protein levels in the same plants suggesting that somehow the cycling of new PHYB protein synthesis is buffered by a similar circadian regulation of degradation (Kozma-Bognar *et al.*, 1999). The fact that such a circadian pattern of new PHYB protein synthesis occurs is, non-the-less, intriguing and has led to much speculation as to whether this "new PHYB" protein has any unique role. If so, any effect specifically-mediated by "new phyB" would be gated by the clock.

Confirmation of the observation that *PHYB* transcript levels are circadian-regulated comes from the work of Harmer *et al.* (2000) who carried out a global analysis of gene expression patterns over 48 hours in seedlings of *A. thaliana*. This revealed that, in fact, transcript levels of a number of the photoreceptors involved in mediating developmental responses show circadian regulation. Levels of *PHYB* transcript were demonstrated to peak in the late night and early morning while levels of transcript of the blue light photoreceptors *CRY1*, *CRY2* and *PHOT1* peaked during the mid to late part of the day (Harmer *et al.*, 2000).

Circadian rhythms of transcription were subsequently revealed for the genes encoding a number of photoreceptors. The genes encoding phytochromes A, B, D and E and cryptochromes 1 and 2 all show circadian regulation of transcription. Circadian regulation of a *PHYA::LUC* transgene was demonstrated by Hall *et al.* (2001). *PHYA* transcription shows a clear circadian peak in the subjective evening. Toth *et al.* (2001) analysed the expression patterns of each of the remaining *A. thaliana* phytochromes (*PHYC* to *PHYE*) and cryptochromes (*CRY1* and *CRY2*). They also fused the promoters of these genes to the luciferase reporter gene and demonstrated that, with the exception of *PHYC*, all of these constructs displayed circadian-regulated expression. In addition, they showed that mRNA abundance of each of these genes corresponded to this pattern of regulation. *PHYD*, *PHYE* and *CRY1* showed a peak of expression in the middle part of the day, whilst *CRY2* showed a peak of expression in the evening (Toth *et al.*, 2001). Given that the plant circadian clock controls the expression of each of these photoreceptors, it is a distinct possibility that this could have a significant impact in terms of gating of photomorphogenic responses. If such a cycling of transcript levels were translated into a cycling of protein levels, then this, itself, could cause a circadian gating effect with light signalling being regulated by varying photoreceptor levels (Figure 6).

11.2 Circadian regulation of photoreceptor subcellular localisation

A key early step in phytochrome signal transduction for several responses is the light dependent nuclear entry of the phytochromes (see Chapters 9, 11). Nuclear localisation would, of course, be a prior requisite to the interaction of phyB with the

PIF3 transcription factor (Ni *et al.*, 1998; Ni *et al.*, 1999). Light dependent nuclear entry of phytochromes fused to green fluorescent protein (GFP) has been demonstrated for all of the phytochromes (phyA – phyE) (Kircher *et al.*, 1999; Yamaguchi *et al.*, 1999; Kircher *et al.*, 2002). The phytochromes are localised primarily in the cytoplasm in dark grown seedlings but upon irradiation with red light they show a translocation into the nucleus where they form speckle-like features (Kircher *et al.*, 1999; Yamaguchi *et al.*, 1999; Kircher *et al.*, 2002). Crucially, though, the nucleo-cytoplasmic partitioning and speckle formation (possibly indicative of biological activity) of the phytochromes is also regulated by a diurnal rhythm, peaking during the day (Kircher *et al.*, 2002). Furthermore, the appearance of speckles in the nuclei increased well before the light-on signal. This anticipation of dawn suggests regulation by the circadian clock (Kircher *et al.*, 2002). Thus one further possible mechanism contributing to circadian gating of photomorphogenesis could be occurring at the level of nuclear import of phytochrome and accumulation of phytochrome speckles in the nucleus.

11.3 Circadian regulation of photoreceptor signal transduction components

As well as the documented circadian regulation of photoreceptor levels, recent findings have demonstrated that the circadian clock regulates several photoreceptor signal transduction components. Such a regulation could also contribute to circadian gating of photomorphogenesis. The following proteins, therefore, form further candidates for components involved in this phenomenon.

11.3.1 GIGANTEA (GI)

The *gi* mutant was originally identified as showing delayed flowering under long days (Koornneef *et al.*, 1991; Araki and Komeda, 1993). Independently, a more recent screen for mutants displaying defects in seedling de-etiolation in red light also yielded an allele of *gi* (Huq *et al.*, 2000) The *gi* mutation results in a specific deficiency in de-etiolation responses to continuous red light suggesting that the wild type GI protein acts in signal transduction downstream of phyB. *GI* encodes a nuclear localised plant-specific protein of unknown biochemical activity (Fowler *et al.*, 1999; Huq *et al.*, 2000; Park *et al.*, 1999). Analysis of the expression patterns of the *GI* transcript revealed that it is regulated by circadian clock with a peak of expression 8-10 h after dawn (Fowler *et al.*, 1999; Park *et al.*, 1999). When *gi* mutants were tested for defects in circadian rhythmicity, it was discovered that the *gi* mutation altered amplitude and period of a number of clock outputs and clock components, specifically in continuous light. Two alleles of *gi* were tested and in both cases. Low amplitude rhythms of expression of *CAB2*, *CCA1*, *LHY* and *GI* were observed in both the *gi-1* mutant and the *gi-2* mutant. The period defect, however, varied according to the allele tested with *gi-1* causing a short period of *CAB2* expression and *gi-2* causing a long period of *CAB2* expression (Park *et al.*, 1999). The results indicate that GI is required for robust circadian rhythms in constant light, suggesting a role for GI in light input to the clock. Notably, the shortening of period length of the *CAB2::LUC* rhythm in red light caused by *gi-1* is fluence rate dependent, also supporting the argument that GI is involved in light input to the clock (Park *et al.*, 1999). However, the *gi* mutation does also have a slight effect on oscillator amplitude in darkness, indicating that this is not the only role for GI in the circadian system (Park *et al.*, 1999).

Central to this review, however, is the fact that GI forms a component of the phyB signalling pathway that is regulated by the circadian clock. GI is, therefore, a clock output that affects light signalling and forms a strong candidate for another of the components involved in the circadian gating of photomorphogenesis (Figure 6).

11.3.2 ZEITLUPE (ZTL)

ZTL was originally identified as a component of the light input pathway to the clock in *A. thaliana* (Somers *et al.*, 2000). The *ztl* mutant was discovered in the same screen that yielded the *toc1* mutant (Millar *et al.*, 1995). It displayed a long period phenotype in constant white light. However, it is the fluence rate dependency of the *ztl* mutant phenotype that places ZTL in the light input pathway as opposed to it being part of the central clock mechanism. The mutant displays a fluence rate-dependent long period phenotype for the expression patterns of numerous clock-regulated genes (Somers *et al.*, 2000). Furthermore, the fluence rate dependency of the *ztl* mutant phenotype is observed in both red and blue light, demonstrating that it is involved in both phytochrome and cryptochrome signalling to the clock. The *ZTL* gene encodes a protein possessing LOV, F-box and Kelch domains (Somers *et al.*, 2000). The LOV (Light Oxygen Voltage) domain is very similar to that of the chromophore-binding pocket of the blue light photoreceptor phot1 (Christie *et al.*, 1999), The ZTL LOV domain was very recently demonstrated to bind a flavin chromophore and to possess blue light sensing properties (Imaizumi *et al.*, 2003) leading to speculation that ZTL may act as a blue light photoreceptor in its own right (see Chapter 12). F-box domains are generally involved in targeting other proteins to E3 ubiquitin ligase complexes where these proteins are ubiquitinated, marking them for subsequent degradation (Patton *et al.*, 1998). Kelch domains, however, are found in proteins of diverse function, offering little specific clues as to the mode of action of ZTL. *ZTL* forms part of a gene family. The previously mentioned *LKP2* (Schultz *et al.*, 2001) also belongs to the *ZTL* gene family along with one further member, *FKF1*, a circadian clock output involved in determining the expression pattern of *CO* (Nelson *et al.*, 2000; Imaizumi *et al.*, 2003).

Expression of the *ZTL* gene appears to be fairly constant over the circadian day but the ZTL protein shows a clear diurnal rhythm of abundance with a peak in the evening (Kim *et al.*, 2003). Furthermore, the ZTL protein shows strong evidence of circadian regulation, initially continuing to cycle after transfer to constant environmental conditions, before damping to a high level (Kim *et al.*, 2003). Thus, the rhythmic accumulation of ZTL involves a post-translational mechanism. It was recently demonstrated that the degradation rate of ZTL protein varies throughout the circadian day, demonstrating that circadian regulation of ZTL proteolysis is at least, in part, responsible (Kim *et al.*, 2003). Furthermore, the proteolysis of ZTL was demonstrated to be proteasome dependent, implicating ZTL, itself, as substrate for ubiquitination (Kim *et al.*, 2003).

The *ztl* mutant also shows an enhanced inhibition of hypocotyl elongation, specifically in response to red light. Thus, ZTL forms another clock-regulated component involved in the phytochrome-signalling pathway. Like, GI, therefore, it may contribute to the gating of photomorphogenesis. It is particularly interesting in this respect that ZTL has been demonstrated to interact with the photoreceptors phyB and cry1 (Jarillo *et al.*, 2001). Such a close association could be indicative of a mechanism by which ZTL could moderate photoreceptor signalling (Figure 6).

11.3.3 SUPPRESSOR OF phyA 1 (SPA1)

The *spa 1* mutant was identified in a screen for extragenic mutations that suppress the morphological phenotype exhibited by a weak *phyA* mutant (*phyA-105*) (Hoecker *et al.*, 1998). *phyA-105* shows an impaired de-etiolation response to continuous far red light. The *spa1* mutant restored a wild-type responsiveness to continuous far red light in the *phyA-105* mutant background. When tested further, the *spa1* mutation was also found to enhance photoresponsiveness in a wild-type background. However, the *spa1* mutant showed no phenotype in a *phyA* null mutant background indicating that *spa1* acts specifically in the phyA light-signalling pathway. The wild type SPA1 protein, therefore, acts as a negatively regulator of phyA signalling (Hoecker *et al.*, 1998). Cloning of the *SPA1* gene revealed that it encoded a WD (tryptophan-aspartic acid)-repeat protein that also contains a coiled-coil protein-protein interaction domain and shares sequence similarity with protein kinases (Hoecker *et al.*, 1999). The SPA1 protein is localised to the nucleus (Hoecker *et al.*, 1999) where it has been demonstrated to interact with the CONSTITUTIVELY PHOTOMORPHOGENIC 1 (COP1) protein though another coiled-coil domain in COP1 (Hoecker and Quail, 2001). COP1 acts as a repressor of photomorphogenesis in plants (Deng *et al.*, 1991). It does so by exhibiting ubiquitin E3 ligase activity to regulate levels of HY5 a transcription factor that pays a key role in light signal transduction (Oyama *et al.*, 1997; Saijo *et al.*, 2003). The SPA1 protein has been demonstrated to moderate the ubiquitin E3 ligase activity exhibited by COP1 on HY5 in *in vitro* assays (Saijo *et al.*, 2003) and it seems likely that this activity of the SPA1 protein is responsible for its moderation of light signalling.

The *spa1* mutation also shows enhanced red light signalling to the circadian clock at lower fluence rates of red light, over the range of fluence rates at which phyA acts as the primary circadian photoreceptor (Devlin, unpublished data; Somers *et al.* 1998a). However, the *spa1* mutant showed an otherwise normal circadian rhythm for *CAB2::LUC* expression in constant red light. This is consistent with the action of SPA1 as a general repressor of phyA signalling. More significantly, though, the expression of the *SPA1* gene itself has been demonstrated to be regulated by the circadian clock with a peak of expression in the early morning (Harmer *et al.*, 2000). SPA1, therefore, represents another key light signal transduction component that is regulated by the circadian clock. The timing of the peak of *SPA1* expression would allow one to speculate that SPA1 action may be more intense during the early part of the subjective day and, as such SPA1 forms another candidate component involved in the circadian gating of photomorphogenesis (Figure 6).

11.3.4 *early phytochrome responsive 1 (epr1)*

The *EPR1* gene was identified by its rapid induction in response to red light in a fluorescent differential display experiment (Kuno *et al.*, 2003). EPR1 is a nuclear-localised protein, containing a single MYB DNA binding domain similar to that found in CCA1 or LHY. *EPR1* expression is regulated by both phyA and phyB. However, as well as being strongly light induced, *EPR1* expression is also regulated by the circadian clock with a peak around subjective dawn. Overexpression of either *CCA1* or *LHY* results in a loss of rhythmicity of *EPR1* expression, indicating that this rhythmicity is under the control of the central circadian oscillator. However, EPR1 also suppresses its own expression: constitutively high levels of exogenous EPR1 will suppress the rhythm of endogenous *EPR1* expression without affecting the oscillation of the central lock components CCA1 and LHY (Kuno *et al.*, 2003). This indicates that EPR1 may form a distinct "slave" circadian oscillator that is self-sustaining but, perhaps, synchronised by the central circadian oscillator. Whilst overexpression of *EPR1* does not affect overt circadian rhythmicity, it does affect photomorphogenic responses. *EPR1* overexpressors have an enhanced far red-light induced cotyledon opening and delayed flowering (Kuno *et al.*, 2003). Hence EPR1 is another positively acting component of a light-signalling pathway that is regulated by the circadian clock and thereby may also play a role in gating photomorphogenesis. However, this effect on gating does not affect light input to the clock, placing EPR1 downstream of a branch-point in the light-signalling pathway (Figure 6).

Numerous potential factors may be involved in circadian gating of photomorphogenesis in *A. thaliana*. Two factors, ELF3 and TIC, specifically function in the gating process during the early and late night, respectively, to inhibit light signalling. On the other hand, several genes encoding positively acting components of the light signal transduction pathway, including the genes encoding photoreceptors themselves, are subject to circadian regulation, with levels of these transcripts peaking at various times during the subjective day. It would appear from this that plants may employ a combination of negative and positive regulation to gate photomorphogenesis. A negative regulation is exerted during the subjective night when responses to light would not be beneficial and a positive regulation may be exerted during the subjective day. The variation in timing of the peak levels of the various positive acting factors may reflect variation in the optimum time of day at which their particular light-responsive effects are required by the plant. This variation is most apparent in the range of peak expression times for the photoreceptor genes. It is even possible that this may account for the apparent redundancy in photoreceptor function to some extent. If the variation in photoreceptor gene expression is translated into variation in protein levels then different photoreceptors may each may work in a particular temporal niche.

However, one further question remains as to the reason behind the "dead-zone" observed for clock resetting during the early to mid of the subjective day. The phase response curve for clock resetting in *A. thaliana* suggests that photoreceptor signalling is also inhibited at this time (Covington *et al.*, 2001). The phenotype of the *elf3* and *tic* mutants indicate that the primary role of ELF3 and TIC is as inhibitors of light signalling during the night. However, the phase response curve for clock resetting in response to red light pulses in *A. thaliana* suggests that ELF3 may also play a role in inhibition of light signalling during the day. Wild type seedlings of *A. thaliana* show a clear early to mid day dead-zone for clock resetting in response to both red and blue light pulses. In the *elf3* mutant and in *ELF3* overexpressing plants, a clear dead-zone is still apparent in response to blue light pulses but is absent in response to red light pulses (Covington *et al.*, 2001). Red light pulses are able to cause clock resetting at all times during the subjective day in *elf3* mutant and in *ELF3* overexpressing plants. ELF3, therefore, also appears to play a role in gating of red light signalling for clock resetting during the subjective day. However, the persistence of a dead-zone in response to blue light pulses suggests that there may be another negatively-acting factor operating that moderates clock resetting in response to blue light during the day.

One further suggestion for a mechanism involved in circadian regulation of photoreceptor signalling comes from protein-protein interaction studies. The TOC1 protein has been demonstrated to directly interact with the PIF3 transcription factor that is involved in the light signal transduction pathway. If TOC1 is able to regulate the activity of PIF3 by so doing, it is possible that this could also allow TOC1 to periodically regulate light signalling. This theory is particularly interesting in light of the fact that phyB Pfr has been shown to interact with PIF3 bound to the CCA1 promoter (Martinez-Garcia *et al.*, 2000). In addition, a number of other bHLH transcription factors, namely PIF3-LIKE 1 (PIL1), PIL2, PIL4, PIL6 and PIF4 (Huq and Quail, 2002; Yamashino *et al.*, 2003) have also been demonstrated to interact with TOC1 (Yamashino *et al.*, 2003) leading to the suggestion that these may also be involved in the mechanism of integration of light and circadian signals. Notably, PIL4 and PIL6 have been shown to be regulated by both the circadian clock and by light (Yamashino *et al.*, 2003).

12. A TWIST IN THE TALE: IS THERE JUST ONE CIRCADIAN CLOCK REGULATING PHOTOMORPHOGENESIS?

An interesting recent development has raised questions as to whether the many different clock outputs are all controlled by the same clock. Using *CAB2::LUC* and *PHYB::LUC* reporters, Hall *et al.* (2002) demonstrated that *CAB2* and *PHYB* showed different free-running periods of expression. However, the rhythmic expression of both reporters was similarly-dependent on CCA1 and LHY and similarly-affected by mutation in the light input photoreceptors. The authors suggest that this most likely represents different local factors affecting the pace of the clock in the different tissues in which these two reporters are expressed (Hall *et al.*, 2002). A more intriguing observation was made by Michael *et al.* (2003). They observed that expression patterns of the *CAB2* and *CATALASE 3* (*CAT3*) genes are regulated by distinct circadian clocks that respond differently to environmental temperature signals. When plants are grown in 12 hour-light / 12-hour dark cycles, light forms the primary environmental signal entraining the expression of *CAB2* expression, irrespective of the temperature. However, in the case of *CAT3*, temperature forms the primary entraining factor. If *CAT3::LUC* expressing seedlings are grown in 12 hour-light / 12-hour dark cycles where temperature is higher during the dark period than during the light period, *CAT3* expression cycles as if the higher temperature period were "day" and the lower temperature period were "night" (Michael *et al.*, 2003). Both *CAT3* and *CAB2* are expressed in the same tissues so these differences cannot be tissue specific. Furthermore, whilst the phase response curve for clock resetting in response to a temperature pulse is the same for both *CAB2* and *TOC1*, it is different for *CAT3*, suggesting that *CAT3* responds to a distinct oscillator (Michael *et al.*, 2003).

A third piece of evidence suggesting the existence of distinct oscillators within the same plant comes from work by Kolar *et al.* (1998). They observed that germination could synchronise a circadian rhythm of expression of the *A. thaliana CAB2::LUC* or *Triticum aestivum CAB1::LUC* transgenes in a population of *Nicotiana tabaccum* seedlings germinated in darkness. They were able to demonstrate that a 5-minute red light pulse given within 12 hours of sowing was ineffective in perturbing this oscillation. However, a 5-minute red light pulse given at 36 hours after sowing led to the induction of a new circadian oscillation without affecting the already free-running circadian rhythm. At this point the two rhythms could be observed superimposed upon each other in a single seedling. Independently of any previous light treatments a 5-minute pulse of light given at 60 hours after sowing was able to synchronise these rhythms into a single circadian oscillation. This suggests the presence of both phytochrome-coupled and phytochrome-uncoupled oscillators within the same plant for at least some stage of the life history of these seedlings. Again, it cannot be concluded whether this represents distinct oscillators within a single cell or whether this is the result of distinct oscillations running within different cells or cell types (Kolar *et al.*, 1998).

Can any single circadian output, therefore, be used to indicate the state of the "clock" within a plant? It is notable that the *tic* mutant shows slightly different period length effects depending on the output studied. This was proposed to be due to the low amplitude of the oscillator in the *tic* mutant causing it to be easily perturbed by slight changes in the environment in which measurements were taken to cause a lengthening or shortening period (Hall *et al.*, 2003). It could, however, also indicate different effects of the *tic* mutation on distinct oscillators controlling each of the outputs examined. The idea of multiple clocks raises the question as to whether there is just one circadian clock regulating photomorphogenesis or whether there are several. It has not been tested whether the same clock regulates the numerous positive-acting light-signalling components that display a circadian rhythm. A regulation of photomorphogenesis by multiple clocks is, therefore, a distinct possibility.

13. CONCLUSION: CONCERNS FOR PHOTOMORPHOGENIC STUDY

The phenomenon of circadian gating of environmental responses raises concerns for all those involved in research into photomorphogenesis or, indeed, any aspect of research into plant responses to the environment. An effect observed at one time of day may not necessarily be observed to the same extent at another time of day. Even the growth of plants in continuous light will not avoid problems due to circadian regulation of environmental signalling. Circadian rhythms can be observed in seedlings from imbibition; and imbibition itself can act as a Zeitgeber to set the clock (Zhong et al., 1998). Likewise, the very act of transferring of a group of seedlings into constant light is likely to act as a Zeitgeber, synchronising the clocks throughout the population. If maintained in constant light for some weeks, variation in period length of the rhythm between individuals in an otherwise isogenic population will eventually lead to a desynchronisation within the group. Such a strategy could, perhaps, be used to measure an average response, unaffected by time of day but this will be of little significance in the "real world". Furthermore even small daily fluctuations in temperature in "constant environment" growth chambers can entrain a population (Somers et al., 1998b) and care must be taken to determine that "constant environment" growth chambers are just that.

Further evidence of the importance of circadian rhythms in regulating photomorphogenesis comes from the observed photomorphogenic phenotypes of a number of mutants deficient in the central clock components, themselves:

Several of the APRR mutants and overexpressors demonstrate defective photomorphogenesis as well as impaired circadian function. A null allele of the *toc1* mutant shows a long hypocotyl phenotype specifically in red and far red light (Mas et al., 2003). Consistent with this, seedlings expressing an extra copy of the *TOC1* gene show an enhanced sensitivity to both red and far red for inhibition of hypocotyl elongation (Mas et al., 2003). Similarly, *aprr7* mutants display a defect in inhibition of hypocotyl elongation in response to red light (Kaczorowski et al., 2003), whilst *aprr5* and *aprr9* mutants display a slight elongated hypocotyl phenotype when grown in either red or blue light (Eriksson et al., 2003). Overexpression of either of the LHY or CCA1 proteins has also been demonstrated to result in a long hypocotyl phenotype in seedlings grown in long day photoperiods (Wang et al., 1998; Schaffer et al., 1998). Likewise, loss of ELF4 function not only results in a disruption of the normal circadian rhythm, it also specifically impairs red light-mediated de-etiolation (Khanna et al., 2003). Khanna et al. (2003) demonstrated that *elf4* mutant seedlings show a reduced inhibition of hypocotyl elongation in red light. Seedlings of *A. thaliana* overexpressing the clock-associated component, LKP2, provide another example. These seedlings also show a long hypocotyl phenotype in both red and blue light (Schultz et al., 2001). Finally, the *srr1* mutant shows reduced sensitivity specifically to red light. *srr1* mutant seedlings have long hypocotyls in red or white light whist adult *srr1* mutants have elongated petioles. *srr1* mutants also show reduced chlorophyll content in white light and a reduced promotion of elongation growth in response to end-of-day far red irradiation (Staiger et al., 2003). All of these *srr1* mutant phenotypic characteristics are reminiscent of the *phyB* mutant phenotype, suggesting that *srr1* is impaired in phyB signalling.

It remains uncertain whether the photomorphogenic phenotypes of mutants disrupted in components associated with the central clock mechanism are directly due to disruption of the clock or whether they could represent the result of disruption of other, unrelated functions of these proteins in the transduction of light signals. However, the ubiquitous association of photomorphogenic phenotypes with disruption of central clock-associated components suggests that disruption of the clock is likely to be the primary cause of impaired photomorphogenesis in these mutants. It may be, in fact, the case that such a close relationship exists between the circadian clock and photoperception in plants that disruption of one almost inevitably affects the other. Further evidence of this relationship is demonstrated by light quality dependent nature of the circadian phenotype of the *aprr* mutants. For example the circadian phenotypes described earlier for *aprr5* and *aprr9* mutants, grown in white light, are no longer apparent in red light (Eriksson *et al.*, 2003) whilst the *toc1-2* mutant, that displays a short period phenotype in other conditions, becomes arrhythmic in constant red light (Mas *et al.*, 2003).

14. EPILOGUE

"In our rhythm of earthly life we tire of light.

We are glad when the day ends, when the play ends; and ecstasy is too much pain….

Controlled by the rhythm of blood and the day and the night and the seasons.

And we must extinguish the candle, put out the light and relight it;

Forever must quench, forever relight the flame".

Excerpt from The Rock: Chorus X (final chorus) T.S. Eliot (1934)

15. FURTHER SUGGESTED READING

Hall, J. C. (2003) Genetics and molecular biology of rhythms in *Drosophila* and other insects. *Adv Genet*, *48*, 1-280.
Hayama, R. and Coupland, G. (2003) Shedding light on the circadian clock and the photoperiodic control of flowering. *Curr Opin Plant Biol*, *6*, 13-19.
Millar, A. J. (2004) Input signals to the plant circadian clock. *J Exp Bot*, *55*, 277-283.
Reppert, S. M. and Weaver, D. R. (2002) Coordination of circadian timing in mammals. *Nature*, *418*, 935-941.
Roenneberg, T. and Merrow, M. (2003) The network of time: understanding the molecular circadian system. *Curr Biol*, *13*, R198-207.
Schultz, T. F. and Kay, S. A. (2003) Circadian clocks in daily and seasonal control of development. *Science*, *301*, 326-328.
Yanovsky, M. J. and Kay, S. A. (2003) Living by the calendar: how plants know when to flower. *Nat Rev Mol Cell Biol*, *4*, 265-275.

16. REFERENCES

Alabadi, D., Oyama, T., Yanovsky, M. J., Harmon, F. G., Mas, P., Kay, S. A. (2001) Reciprocal regulation between TOC1 and LHY/CCA1 within the Arabidopsis circadian clock. *Science, 293,* 880-883.
Araki, T. and Komeda, Y. (1993) Analysis of the role of the late-flowering locus, *GI*, in the flowering of *Arabidopsis thaliana*. *Plant Journal, 3,* 231-239.
Bauer, D., Viczian, A., Kircher, S., Nobis, T., Nitschke, R., Kunkel, T., *et al.* (2004) Constitutive photomorphogenesis 1 and multiple photoreceptors control degradation of Phytochrome Interacting Factor 3, a transcription factor required for light signalling in *Arabidopsis*. Plant Cell, **16,** 1433-1445.
Bell-Pedersen, D. (1998) Keeping pace with *Neurospora* circadian rhythms. *Microbiology, 144,* 1699-1711.
Briggs, W. R. and Christie, J. M. (2002) Phototropins 1 and 2: versatile plant blue-light receptors. *Trends in Plant Sciences, 7,* 204-210.
Carpenter, C. D., Kreps, J. A., Simon, A. E. (1994) Genes encoding glycine-rich *Arabidopsis thaliana* proteins with RNA-binding motifs are influenced by cold treatment and an endogenous circadian rhythm. *Plant Physiology, 104,* 1015-1025.
Carre, I. A. (1998) Genetic Dissection of the Photoperiod-sensing Mechanism in the Long Day Plant *Arabidopsis thaliana*. In P. J. Lumsden and A. J. Millar (Eds.), *Biological rhythms and photoperiodism in plants* (pp. 257-270) Oxford: BIOS Scientific.
Casal, J. J., Luccioni, L. G., Oliverio, K. A., Boccalandro, H. E. (2003) Light, phytochrome signalling and photomorphogenesis in Arabidopsis. *Photochem PhotobiolSci, 2,* 625-636.
Casal, J. J. and Mazzella, M. A. (1998) Conditional synergism between cryptochrome 1 and phytochrome B is shown by the analysis of *phyA*, *phyB*, and *hy4* simple, double, and triple mutants in Arabidopsis. *Plant Physiology, 118,* 19-25.
Cashmore, A. R., Jarillo, J. A., Wu, Y. J., Liu, D. (1999) Cryptochromes: Blue Light Receptors for Plants and Animals. *Science, 284,* 760-765.
Christie, J. M., Salomon, M., Nozue, K., Wada, M., Briggs, W. R. (1999) LOV (light, oxygen, or voltage) domains of the blue-light photoreceptor phototropin (nph1): binding sites for the chromophore flavin mononucleotide. *Proc Natl Acad Sci U S A,* 8779-8783.
Coulter, M. W. and Hamner, K. C. (1964) Photoperiodic Flowering Response of Biloxi Soybean in 72 hour Cycles. *Plant Physiology, 39,* 848-856.
Covington, M. F., Panda, S., Liu, X. L., Strayer, C. A., Wagner, D. R., Kay, S. A. (2001) ELF3 modulates resetting of the circadian clock in Arabidopsis. *Plant Cell, 13,* 1305-1315.
Deng, X.-W., Caspar, T., Quail, P. H. (1991) *cop*1: A regulatory locus involved in light-controlled development and gene expression in *Arabidopsis*. *Genes and Development, 5,* 1172-1182.
Devlin, P. F. (2002) Signs of the time: environmental input to the circadian clock. *J Exp Bot, 53,* 1535-1550.
Devlin, P. F. and Kay, S. A. (2000) Cryptochromes are required for phytochrome signalling to the circadian clock but not for rhythmicity. *Plant Cell, 12,* 2499-2510.
Dowson-Day, M. J. and Millar, A. J. (1999) Circadian dysfunction causes aberrant hypocotyl elongation patterns in Arabidopsis. *Plant Journal, 17,* 63-71.
Doyle, M. R., Davis, S. J., Bastow, R. M., McWatters, H. G., Kozma-Bognar, L., Nagy, F. *et al.* (2002) The ELF4 gene controls circadian rhythms and flowering time in Arabidopsis thaliana. *Nature, 419,* 74-77.
Eliot, T. S. (1934) The Rock: Chorus X. In "*Choruses from 'The Rock',*" *The Complete Poems and Plays, 1909-1950* (pp. 112) New York: Harcourt Brace and World, 1952.
Engelmann, W., Simon, K., Phen, C. J. (1994) Leaf movement in *Arabidopsis thaliana*. *Z Naturforsch, 47,* 925-928.
Eriksson, M. E., Hanano, S., Southern, M. M., Hall, A., Millar, A. J. (2003) Response regulator homologues have complementary, light-dependent functions in the Arabidopsis circadian clock. *Planta, In Press.*
Fowler, S., Lee, K., Onouchi, H., Samach, A., Richardson, K., Morris, B. *et al.* (1999) GIGANTEA: a circadian clock-controlled gene that regulates photoperiodic flowering in Arabidopsis and encodes a protein with several possible membrane-spanning domains. *EMBO Journal, 18,* 4679-4688.
Giovani, B., Byrdin, M., Ahmad, M., Brettel, K. (2003) Light-induced electron transfer in a cryptochrome blue-light photoreceptor. *Nat Struct Biol, 10,* 489-490.

Golden, S. S., Ishiura, M., Johnson, C. H., Kondo, T. (1997) Cyanobacterial circadian rhythms. *Annual Review of Plant Physiology and Plant Molecular Biology, 48*, 327-354.

Gorton, H. L., Williams, W. E., Assmann, S. M. (1993) Circadian rhythms in stomatal responsiveness to red and blue light. *Plant Physiology, 103*, 399-406.

Grantz, D. A. (1990) Plant Response to Atmospheric Humidity. *Plant, Cell and Environment, 13*, 667-679.

Hall, A., Bastow, R. M., Davis, S. J., Hanano, S., McWatters, H. G., Hibberd, V. *et al.* (2003) The TIME FOR COFFEE (TIC) Gene Maintains the Amplitude and Timing of Arabidopsis Circadian Clocks. *Plant Cell, 15*, 2719-2729.

Hall, A., Kozma-Bognar, L., Bastow, R. M., Nagy, F., Millar, A. J. (2002) Distinct regulation of CAB and PHYB gene expression by similar circadian clocks. *Plant Journal, 32*, 529-537.

Hall, A., Kozma-Bognar, L., Toth, R., Nagy, F., Millar, A. J. (2001) Conditional circadian regulation of PHYTOCHROME A gene expression. *Plant Physiol, 127*, 1808-1818.

Harmer, S. L., Hogenesch, J. B., Straume, M., Chang, H. S., Han, B., Zhu, T. *et al.* (2000) Orchestrated transcription of key pathways in Arabidopsis by the circadian clock. *Science, 290*, 2110-2113.

Hicks, K. A., Albertson, T. M., Wagner, D. R. (2001) EARLY FLOWERING3 encodes a novel protein that regulates circadian clock function and flowering in Arabidopsis. *Plant Cell, 13*, 1281-1292.

Hicks, K. A., Millar, A. J., Carré, I. A., Somers, D. E., Straume, M., Meeks-Wagner, R. *et al.* (1996) Conditional circadian dysfunction of the Arabidopsis early-flowering 3 mutant. *Science, 274*, 790-792.

Hoecker, U. and Quail, P. H. (2001) The phytochrome A-specific signalling intermediate SPA1 interacts directly with COP1, a constitutive repressor of light signalling in Arabidopsis. *Journal of Biological Chemistry, 276*, 38173-38178.

Hoecker, U., Tepperman, J. M., Quail, P. H. (1999) SPA1, a WD-repeat protein specific to phytochrome A signal transduction. *Science, 284*, 496-499.

Hoecker, U., Xu, Y., Quail, P. H. (1998) SPA1: A new genetic locus involved in phytochrome A - Specific signal transduction. *Plant Cell, 10*, 19-33.

Horwitz, B. A. and Epel, B. L. (1978) Circadian changes in activity of the far-red form of phytochrome: physiological and in vivo spectrophotometric studies. *Plant Science Letters, 13*, 9-14.

Huq, E. and Quail, P. H. (2002) PIF4, a phytochrome-interacting bHLH factor, functions as a negative regulator of phytochrome B signalling in Arabidopsis. *EMBO Journal, 21*, 2441-2450.

Huq, E., Tepperman, J. M., Quail, P. H. (2000) GIGANTEA is a nuclear protein involved in phytochrome signalling in Arabidopsis. *Proc Natl Acad Sci USA, 97*, 9789-9794.

Imaizumi, T., Tran, H. G., Swartz, T. E., Briggs, W. R., Kay, S. A. (2003) FKF1 is essential for photoperiodic-specific light signalling in Arabidopsis. *Nature, 426*, 302-306.

Imamura, A., Hanaki, N., Nakamura, A., Suzuki, T., Taniguchi, M., Kiba, T. *et al.* (1999) Compilation and Characterization of *Arabidopsis thaliana* Response Regulators Implicated in His-Asp Phosphorelay Signal Transduction. *Plant and Cell Physiology, 40*, 733-742.

Jarillo, J. A., Capel, J., Tang, R. H., Yang, H. Q., Alonso, J. M., Ecker, J. R. *et al.* (2001) An Arabidopsis circadian clock component interacts with both CRY1 and phyB. *Nature, 410*, 487-490.

Johnson, C. H. (1992) "Phase Response Curves: What Can They Tell Us About Circadian Clocks?" In, *Circadian Clocks from Cell to Human*, Hiroshige T.and Honma K. (eds) Hokkaido University Press, Saporo, Japan, 209-249.

Kaczorowski, K. A. and Quail, P. H. (2003) Arabidopsis PSEUDO-RESPONSE REGULATOR7 (PRR7) Is a Signalling Intermediate in Phytochrome-Regulated Seedling Deetiolation and Phasing of the Circadian Clock. *Plant Cell, 15*, 2654-2665.

Khanna, R., Kikis, E. A., Quail, P. H. (2003) EARLY FLOWERING 4 Functions in Phytochrome B-Regulated Seedling De-Etiolation. *Plant Physiol,133*, 1530-1538.

Kim, B. C., Tennessen, D. J., Last, R. L. (1998) UV-B-induced photomorphogenesis in *Arabidopsis thaliana*. *Plant Journal, 15*, 667-674.

Kim, J. Y., Song, H. R., Taylor, B. L., Carre, I. A. (2003) Light-regulated translation mediates gated induction of the Arabidopsis clock protein LHY. *EMBO Journal, 22*, 935-944.

Kircher, S., Gil, P., Kozma-Bognar, L., Fejes, E., Speth, V., Husselstein-Muller, T. *et al.* (2002) Nucleocytoplasmic partitioning of the plant photoreceptors phytochrome A, B, C, D, and E is regulated differentially by light and exhibits a diurnal rhythm. *Plant Cell, 14*, 1541-1555.

Kircher, S., Kozma-Bognar, L., Kim, L., Adam, E., Harter, K., Schäfer, E. et al. (1999) Light Quality-Dependent Nuclear Import of the Plant Photoreceptors Phytochrome A and B. *Plant Cell, 11,* 1445-1456.

Kleine, T., Lockhart, P., Batschauer, A. (2003) An Arabidopsis protein closely related to Synechocystis cryptochrome is targeted to organelles. *Plant Journal, 35,* 93-103.

Kolar, C., Fejes, E., Adám, E., Schäfer, E., Kay, S. A., Nagy, F. (1998) Transcription of Arabidopsis and wheat Cab genes in single tobacco transgenic seedlings exhibits independent rhythms in a developmentally regulated fashion. *Plant Journal, 13,* 563-569.

Koornneef, M., Hanhart, C. J., Van der Veen, J. H. (1991) A genetic and physiological analysis of late flowering mutants in *Arabidopsis thaliana. Molecular and General Genetics, 229,* 57-66.

Koornneef, M., Rolf, E., Spruit, C. J. P. (1980) Genetic control of light-inhibited hypocotyl elongation in *Arabidopsis thaliana* (L.) Heynh. *Z Planzenphysiol, 100,* 147-160.

Kozma-Bognar, L., Hall, A., Adam, E., Thain, S. C., Nagy, F., Millar, A. J. (1999) The circadian clock controls the expression pattern of the circadian input photoreceptor, phytochrome B. *Proc Natl Acad Sci USA, 96,* 14652-14657.

Kuno, N., Moller, S. G., Shinomura, T., Xu, X., Chua, N. H., Furuya, M. (2003) The novel MYB protein EARLY-PHYTOCHROME-RESPONSIVE1 is a component of a slave circadian oscillator in Arabidopsis. *Plant Cell, 15,* 2476-2488.

Lin, C. and Shalitin, D. (2003) Cryptochrome structure and signal transduction. *Annu Rev Plant Biol., 54,* 469-496.

Liu, X. L., Covington, M. F., Fankhauser, C., Chory, J., Wagner, D. R. (2001) ELF3 encodes a circadian clock-regulated nuclear protein that functions in an Arabidopsis PHYB signal transduction pathway. *Plant Cell, 13,* 1293-1304.

Makino, S., Matsushika, A., Kojima, M., Oda, Y., and Mizuno, T. (2001) Light response of the circadian waves of the APRR1/TOC1 quintet: when does the quintet start singing rhythmically in Arabidopsis? *Plant Cell Physiol, 42,* 334-339.

Martinez-Garcia, J. F., Huq, E., Quail, P. H. (2000) Direct targeting of light signals to a promoter element-bound transcription factor. *Science, 288,* 859-863.

Mas, P., Alabadi, D., Yanovsky, M. J., Oyama, T., Kay, S. A. (2003) Dual role of TOC1 in the control of circadian and photomorphogenic responses in Arabidopsis. *Plant Cell, 15,* 223-236.

Matsushika, A., Imamura, A., Yamashino, T., Mizuno, T. (2002) Aberrant expression of the light-inducible and circadian-regulated APRR9 gene belonging to the circadian-associated APRR1/TOC1 quintet results in the phenotype of early flowering in Arabidopsis thaliana. *Plant Cell Physiol, 43,* 833-843.

Matsushika, A., Makino, S., Kojima, M., Mizuno, T. (2000) Circadian waves of expression of the APRR1/TOC1 family of pseudo-response regulators in Arabidopsis thaliana: insight into the plant circadian clock. *Plant Cell Physiol, 41,* 1002-1012.

McWatters, H. G., Bastow, R. M., Hall, A., Millar, A. J. (2000) The ELF3 zeitnehmer regulates light signalling to the circadian clock. *Nature, 408,* 716-720.

Meidner H. and Mansfield, T. A. (1968) "The Role of Rhythms in Stomatal Behaviour." In *Physiology of Stomata,* New York: McGraw-Hill, 102-169.

Menkens, A. E., Schindler, U., Cashmore, A. R. (1995) The G-box: A ubiquitous regulatory DNA element in plants bound by the GBF family of bZIP proteins. *Trends in Biochemical Sciences, 20,* 506-510.

Michael, T. P., Salome, P. A., McClung, C. R. (2003) Two Arabidopsis circadian oscillators can be distinguished by differential temperature sensitivity. *Proc Natl Acad Sci USA, 100,* 6878-6883.

Millar, A. J., Carré, I. A., Strayer, C. A., Chua, N.-H., Kay, S. A. (1995) Circadian clock mutants in *Arabidopsis* identified by luciferase imaging. *Science, 267,* 1161-1163.

Millar, A. J. and Kay, S. A. (1991) Circadian control of *cab* gene transcription and mRNA accumulation in *Arabidopsis. Plant Cell, 3,* 541-550.

Millar, A. J. and Kay, S. A. (1996) Integration of circadian and phototransduction pathways in the network controlling *CAB* gene transcription in *Arabidopsis. Proc Natl Acad Sci USA, 93,* 15491-15496.

Millar, A. J., Short, S. R., Hiratsuka, K., Chua, N.-H., Kay, S. A. (1992) Firefly luciferase as a reporter of regulated gene expression in higher plants. *Plant Molecular Biology Reporter, 10,* 324-337.

Mizoguchi, T., Wheatley, K., Hanzawa, Y., Wright, L., Mizoguchi, M., Song, H. R. et al. (2002) LHY and CCA1 are partially redundant genes required to maintain circadian rhythms in Arabidopsis. *Dev Cell, 2,* 629-641.
Moore-Ede, C. M., Sulzman, F. M., Fuller, C. A., *The Clocks That Time Us.* Cambridge, Harvard University Press, 1982.
Morison, J. I. L. (1987) "Intercellular CO_2 concentration and stomatal response to CO_2." In, *Stomatal Function,* Zeiger, E., Farquhar, G. D. and Cowan, I. R. (eds) Stanford University Press, Stanford, CA, 229-251.
Nagy, F., Kay, S. A., Chua, N.-H. (1988) A circadian clock regulates transcription of the wheat Cab-1 gene. *Genes and Development, 2,* 376-382.
Neff, M. and Chory, J. (1998) Genetic interactions between phytochrome A, phytochrome B, and cryptochrome 1 during Arabidopsis development. *Plant Physiology, 118,* 27-36.
Nelson, D. C., Lasswell, J., Rogg, L. E., Cohen, M. A., Bartel, B. (2000) FKF1, a Clock-Controlled Gene that Regulates the Transition to Flowering in Arabidopsis. *Cell, 101,* 331-340.
Ni, M., Tepperman, J. M., Quail, P. H. (1998) PIF3, a phytochrome-interacting factor necessary for normal photoinduced signal transduction, is a novel basic helix-loop-helix protein. *Cell, 95,* 657-667.
Ni, M., Tepperman, J. M., Quail, P. H. (1999) Binding of phytochrome B to its nuclear signalling partner PIF3 is reversibly induced by light. *Nature, 400,* 781-784.
Nimmo, H. G. (2003) The regulation of phosphoenolpyruvate carboxylase in CAM plants. *Trends in Plant Sciences, 5,* 75-80.
Oyama, T., Shimura, Y., Okada, K. (1997) The *Arabidopsis HY5* gene encodes a bZIP protein that regulates stimulus-induced development of root and hypocotyl. *Genes and Development, 11,* 2983-2995.
Park, D., Somers, D. E., Kim, Y., Choy, Y., Lim, H., Soh, M. et al. (1999) Control of circadian rhythms and photoperiodic flowering by the Arabidopsis GIGANTEA gene. *Science, 285,* 1579-1582.
Patton, E. E., Willems, A. R., Tyers, M. (1998) Combinatorial control in ubiquitin-dependent proteolysis: don't Skp the F-box hypothesis. *Trends Genet, 14,* 236-243.
Pilgrim, M. L., Caspar, T., Quail, P. H., McClung, C. R. (1993) Circadian and light-regulated expression of nitrate reductase in *Arabidopsis. Plant Mol Biol, 23,* 349-364.
Pittendrigh C. S. (1954) On temperature independence in the clock system controlling emergence time in Drosophila. *Proc Natl Acad Sci USA, 40,* 1018-1029.
Pittendrigh, C. S. (1981) "Circadian systems: entrainment." In *Handbook of Behavioural Neurobiology 4: Biological Rhythms,* Aschoff, J. (ed.), Plenum Press, New York, 95-124,
Quail, P. H. (2002) Phytochrome photosensory signalling networks. *Nat Rev Mol Cell Biol, 3,* 85-93.
Saijo, Y., Sullivan, J. A., Wang, H., Yang, J., Shen, Y., Rubio, V. et al. (2003) The COP1-SPA1 interaction defines a critical step in phytochrome A-mediated regulation of HY5 activity. *Genes and Development, 17,* 2642-2647.
Salisbury, F. B., Spomer, G. G., Sobral, M., Ward. R. T. (1968) Analysis of an Alpine environment. *Botanical Gazette, 129,* 16-32.
Salome, P. A., Michael, T. P., Kearns, E. V., Fett-Neto, A. G., Sharrock, R. A., McClung, C. R. (2002) The out of phase 1 mutant defines a role for PHYB in circadian phase control in Arabidopsis. *Plant Physiol, 129,* 1674-1685.
Samach, A. and Coupland, G. (2000) Time measurement and the control of flowering in plants. *Bioessays, 22,* 38-47.
Sato, E., Nakamichi, N., Yamashino, T., Mizuno, T. (2002) Aberrant expression of the Arabidopsis circadian-regulated APRR5 gene belonging to the APRR1/TOC1 quintet results in early flowering and hypersensitiveness to light in early photomorphogenesis. *Plant Cell Physiol, 43,* 1374-1385.
Schäfer, E. and Bowler, C. (2002) Phytochrome-mediated photoperception and signal transduction in higher plants. *EMBO Rep, 3,* 1042-1048.
Schaffer, R., Ramsay, N., Samach, A., Corden, S., Putterill, J., Carré, I. A. et al. (1998) The late elongated hypocotyl mutation of Arabidopsis disrupts circadian rhythms and the photoperiodic control of flowering. *Cell, 93,* 1219-1229.
Schultz, T. F., Kiyosue, T., Yanovsky, M., Wada, M., Kay, S. A. (2001) A role for LKP2 in the circadian clock of *Arabidopsis. Plant Cell, 13,* 2659-2670.
Scully, A. L. and Kay, S. A. (2000) Time Flies for *Drosophila. Cell, 100,* 297-300.
Somers, D. E., Devlin, P. F., Kay, S. A. (1998a) Phytochromes and cryptochromes in the entrainment of the arabidopsis circadian clock. *Science, 282,* 1488-1490.

Somers, D. E., Schultz, T. F., Milnamow, M., Kay, S. A. (2000) *ZEITLUPE*, a Novel Clock Associated PAS Protein from *Arabidopsis*. *Cell, 101*, 319-329.

Somers, D. E., Webb, A. A. R., Pearson, M., Kay, S. (1998b) The short-period mutant, *toc1-1*, alters circadian clock regulation of multiple outputs throughout development in *Arabidopsis thaliana*. *Development, 125*, 485-494.

Staiger, D., Allenbach, L., Salathia, N., Fiechter, V., Davis, S. J., Millar, A. J. *et al.* (2003) The Arabidopsis SRR1 gene mediates phyB signalling and is required for normal circadian clock function. *Genes and Development, 17*, 256-268.

Strayer, C., Oyama, T., Schultz, T. F., Raman, R., Somers, D. E., Mas, P. *et al.* (2000) Cloning of the Arabidopsis clock gene TOC1, an autoregulatory response regulator homolog. *Science, 289*, 768-771.

Suarez-Lopez, P., Wheatley, K., Robson, F., Onouchi, H., Valverde, F., Coupland, G. (2001) CONSTANS mediates between the circadian clock and the control of flowering in Arabidopsis. *Nature, 410*, 1116-1120.

Talbott, L. D., Shmayevich IJ, Chung Y, Hammad JW, Zeiger, E. (2003) Blue Light and Phytochrome-Mediated StomatalOpening in the npq1 and phot1 phot2Mutants of Arabidopsis. *Plant Physiology, 133*, In press.

Toth, R., Kevei, E., Hall, A., Millar, A. J., Nagy, F., Kozma-Bognar, L. (2001) Circadian clock-regulated expression of phytochrome and cryptochrome genes in Arabidopsis. *Plant Physiol, 127*, 1607-1616.

Wang, Z. Y., Kenigsbuch, D., Sun, L., Harel, E., Ong, M. S., Tobin, E. M. (1997) A Myb-related transcription factor is involved in the phytochrome regulation of an Arabidopsis Lhcb gene. *Plant Cell, 9*, 491-507.

Wang, Z. Y. and Tobin, E. M. (1998) Constitutive expression of the *CIRCADIAN CLOCK ASSOCIATED 1 (CCA1)* gene disrupts circadian rhythms and suppresses its own expression. *Cell, 93*, 1207-1217.

Whitelam, G. C., Patel, S., Devlin, P. F. (1998) Phytochromes and photomorphogenesis in *Arabidopsis*. *Philosophical Transactions of the Royal Society of London.- Series.B: Biological Sciences, 353*, 1445-1453.

Wildermann A, Drumm H, Schäfer, E., Mohr, H. (1978) Control by Light of Hypocotyl Growth in De-etiolated Mustard Seedlings. II. Sensitivity for Newly-formed Phytochrome After a Light to Dark Transition. *Planta, 141*, 217-223.

Yamaguchi, R., Nakamura, M., Mochizuki, N., Kay, S. A., Nagatani, A. (1999) Light-dependent translocation of a phytochrome B-GFP fusion protein to the nucleus in transgenic Arabidopsis. *Journal of Cell Biology, 145*, 437-445.

Yamashino, T., Matsushika, A., Fujimori, T., Sato, S., Kato, T., Tabata, S. *et al.* (2003) A Link between circadian-controlled bHLH factors and the APRR1/TOC1 quintet in Arabidopsis thaliana. *Plant Cell Physiol, 44*, 619-629.

Yanovsky, M. J., Casal, J. J., Luppi, J. P. (1997) The *VLF* loci, polymorphic between ecotypes Landsberg *erecta* and Columbia, dissect two branches of phytochrome A signal transduction that correspond to very-low-fluence and high-irradiance responses. *Plant Journal, 12*, 659-667.

Yanovsky, M. J. and Kay, S. A. (2002) Molecular basis of seasonal time measurement in Arabidopsis. *Nature, 419*, 308-312.

Zagotta, M. T., Hicks, K. A., Jacobs, C. I., Young, J. C., Hangarter, R. P., Meeks-Wagner, D. R. (1996) The Arabidopsis ELF3 gene regulates vegetative photomorphogenesis and the photoperiodic induction of flowering. *Plant Journal, 10*, 691-702.

Zagotta, M. T., Shannon, S., Jacobs, C., Meeks-Wagner, D. R. (1992) Early-flowering mutants of *Arabidopsis thaliana*. *Aust.J.Plant Physiol., 19*, 411-418.

Zeiger, E. (1990) Light Perception in Guard Cells. *Plant, Cell and Environment, 12*, 739-747.

Zhong, H. H., Painter, J. E., Salome, P. A., Straume, M., McClung, C. R. (1998) Imbibition, but Not Release from Stratification, Sets the Circadian Clock in Arabidopsis Seedlings. *Plant Cell, 10*, 2005-2018.

Chapter 27

THE MOLECULAR GENETICS OF PHOTO-PERIODIC RESPONSES: COMPARISONS BETWEEN LONG-DAY AND SHORT-DAY SPECIES

George Coupland

Max Planck Institute for Plant Breeding, Carl von Linne Weg, 10, 50829 Köln, Germany (e-mail: coupland@mpiz-koeln.mpg.de)

1. INTRODUCTION

Photoperiodism is the capacity of an organism to measure and respond to the length of the day or night. This was first described in detail in 1920, when Maryland Mammoth tobacco was shown to flower if exposed to short days, but to remain vegetative under long days (Garner and Allard, 1920). They concluded that in plants the time required for reaching and completing the flowering and fruiting stages are profoundly affected by the length of the daily exposure to sunlight." Similar responses were shown to occur widely in the plant and animal Kingdoms, and are assumed to enable organisms to adapt to their local environment by anticipating seasonal changes in temperature (Thomas and Vince-Prue, 1997). For example, genetic differences in photoperiodic responses have been described in accessions of *Xanthium strumarium* isolated from different latitudes suggesting that these play a role in determining the geographical range of this species (Ray and Alexander, 1966).

The mechanisms by which day-length responses are controlled were studied extensively at the physiological level, and in the previous edition of this book, Daphne Vince-Prue described in detail physiological analyses of day-length responses and evidence that the circadian clock acts as the time-keeper for these responses (Vince-Prue, 1994). Since then substantial progress has been made in understanding the molecular mechanisms underlying day-length responses, and much of this progress is based on the use of genetically tractable model species, such as rice and Arabidopsis, to study the effect of day length on flowering time (Hayama and Coupland, 2003; Izawa *et al.*, 2003; Mouradov *et al.*, 2002; Simpson and Dean, 2002; Yanovsky and Kay, 2003). In this Chapter, I will focus on what has been learnt over the last 10 years of the molecular processes that confer a photoperiodic response, and how this information can be related to classical physiological studies.

2. GENETIC MODEL SYSTEMS

Genetic control of day-length responses has been most extensively studied in *Arabidopsis thaliana*. This species shows a wide geographical range, and many strains (or accessions) have been collected from different locations (Alonso-Blanco and Koornneef, 2000). All of those tested behave as facultative long-day plants, flowering earlier under long than short days (Karlsson *et al.*, 1993). The critical day length, the minimum number of hours of day light required to trigger flowering, has not been systematically tested for many of these accessions, but appears to be very short, with day lengths of 4-5 hours being sufficient to trigger flowering (Laibach, 1951). In Arabidopsis, the genetic control of this response has been approached in two ways; by identifying induced mutations that alter the response to day length and by analysing the genetic differences between accessions.

The first induced mutations affecting photoperiodic control of flowering in Arabidopsis were identified and characterized by Redei (Redei, 1962; Redei *et al.*, 1974). He described the late-flowering mutants *constans* (*co*) and *gigantea* (*gi*), and showed that these flowered later than wild-type plants under long days but that under short days they flowered at a similar time to the wild-type. More extensive mutant screening extended this class of mutants to include *fha* (later renamed *cryptochrome 2* (*cry2*)), *ft*, *fwa*, and *fe* (Koornneef *et al.*, 1998; Koornneef *et al.*, 1991). The phenotype of these mutants differed from another class of late-flowering mutants, which flowered later than wild type under both long and short days, and were therefore not impaired in the response to day length. Combining mutations in double mutant plants, also separated these mutations into the same classes (Koornneef *et al.*, 1998). Double mutants carrying two mutations from the same class flowered at approximately the same time as the single mutant, whereas double mutants carrying a mutation from each class flowered much later than the corresponding single mutants. This suggested that these classes of mutations represent two genetic pathways, and that double mutants in which both pathways are impaired flower much later than double mutants in which the same pathway is impaired twice (Koornneef *et al.*, 1998). This was subsequently confirmed by the cloning and analysis of the genes in each class (Mouradov *et al.*, 2002; Simpson and Dean, 2002). The genetic pathway controlling the response to day length was called the long-day or photoperiod pathway, and is described in more detail in the following section.

Genetic loci controlling flowering time of Arabidopsis were also identified by exploiting genetic variation between accessions. This approach has been most productively used in analysing the control of flowering in response to vernalisation (extended exposure to low temperatures that mimic winter conditions), where crossing accessions that respond to vernalisation with those that do not, led to the identification of the *FLC* and *FRI* loci (Burn *et al.*, 1993; Clarke and Dean, 1994; Lee and Amasino, 1993; Napp-Zinn, 1987). Active alleles of these genes are required to repress flowering, until a vernalisation treatment induces flowering by overcoming this repression. All accessions of Arabidopsis are facultative long-day plants, precluding the direct approach of crossing accessions that show the response with those that do not. However, crossing different accessions with similar phenotypes causes a reassortment of alleles at different flowering-time loci generating phenotypes that are not shown by either parent. For example, the Columbia and Landsberg *erecta* accessions are both early flowering, but in the F2 generation derived from intercrossing these two accessions and self-fertilising the F1 plants, late-flowering plants are recovered (Jansen *et al.*, 1995). Similar analysis in crosses between the Cape Verde Islands and Landsberg *erecta* accessions identified a dominant allele of *CRY2* that causes early flowering under short-day conditions, indicating the potential in this approach for identifying genes involved in photoperiodic responses (Alonso-Blanco *et al.*, 1998; El-Assal *et al.*, 2001).

Rice is the short-day model species whose day-length responses have been analysed most systematically at the genetic level, and these analyses are being increasingly supplemented by gene cloning experiments (Izawa *et al.*, 2003). Genetic approaches in rice have relied most extensively on crossing cultivars that differ in their day-length sensitivity, and identifying the genetic differences between them (Yano *et al.*, 1997), although in some cases induced mutations have also been recovered (Izawa *et al.*, 2000). The parental accessions used in the cross between cultivars are Nipponbare, a japonica variety whose flowering responds strongly to photoperiod, and Kasalath, an indica variety with reduced sensitivity to photoperiod (Yano *et al.*, 1997). Several quantitative trait loci (QTLs) that enhance or reduce the photoperiod response have been identified segregating in the progeny of this cross and are discussed in more detail in a later section.

3. A MOLECULAR PATHWAY THAT CONTROLS FLOWERING-TIME IN RESPONSE TO DAY LENGTH IN ARABIDOPSIS BY GENERATING A LONG-DISTANCE SIGNAL FROM THE LEAF

The photoperiod pathway, which promotes early flowering of Arabidopsis under long days was originally identified based on genetic criteria and analysis of the *co, gi, ft* and *cry2* mutants (Koornneef *et al.*, 1998; Koornneef *et al.*, 1991). All of these genes have now been cloned, overexpressed in transgenic plants, and their expression patterns analysed in wild-type and mutant backgrounds (Mouradov *et al.*, 2002). In addition, more mutations that impair the pathway were identified by a range of genetic approaches (Green and Tobin, 1999; Johnson *et al.*, 1994; Nelson *et al.*, 2000; Onouchi *et al.*, 2000; Schaffer *et al.*, 1998). This section describes the function of the proteins that act in this pathway, and the hierarchy of gene action within the pathway. A summary of the interactions between the genes in this pathway is shown in Figure 1A. How the pathway is regulated by day length is elaborated on in the following section.

CO was the first gene in the pathway to be cloned (Putterill *et al.*, 1995). The predicted protein product contains at the N-terminus two adjacent regions, each 47 amino acids long, which comprise B-box zinc fingers that are also present in several proteins from animals (Robson *et al.*, 2001). At the C-terminus, it contains a domain referred to as the CCT domain, because it is present in CO, CO-like proteins of Arabidopsis and in TOC1 (TIMING OF CAB 1), an Arabidopsis protein involved in regulating circadian rhythms (Robson *et al.*, 2001; Strayer *et al.*, 2000). The seven original mutant alleles of *CO* all cause changes in residues present in the B-boxes or the CCT domain of CO, supporting the idea that these regions are critical for its function (Putterill *et al.*, 1995; Robson *et al.*, 2001).

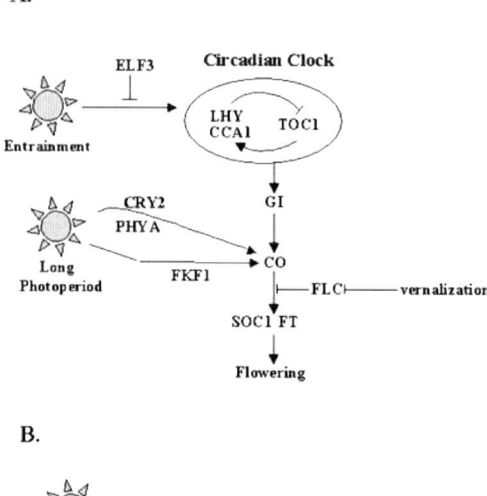

Figure 1. Hierarchy of gene action within the photoperiodic flowering pathways of Arabidopsis and rice. A. The photoperiodic response pathway of Arabidopsis. The circadian clock, which is proposed to comprise of a feed-back loop between the LHY/CCA1 and TOC1 genes, regulates the expression of an output pathway including the GI, CO, SOC1 and FT genes. These four genes represent a transcriptional cascade as shown. Light has two roles in the pathway. Exposure to light entrains the circadian clock, and is required for activation of CO function under long photoperiods. The activation of CO involves post-transcriptional regulation by the photoreceptors CRY2 and PHYA, and transcriptional regulation via FKF1. ELF3 modulates entrainment of the clock by light input. B. The photoperiodic response pathway of rice. The logic of the pathway as for A. Hd1 function is activated under long photoperiods, and represses the expression of Hd3a, thereby delaying flowering.

CO is a nuclear protein that regulates the transcription of downstream genes, although the biochemical mechanism by which this occurs is still not clear. The genes activated by CO include *SOC1* (*SUPPRESSOR OF OVEREXPRESSION OF CO 1*) and *FT* (*FLOWERING LOCUS T*), which encode a MADS box protein and a RAF-kinase-like inhibitor protein respectively (Borner *et al.*, 2000; Kardailsky *et al.*, 1999; Kobayashi *et al.*, 1999; Lee *et al.*, 2000; Samach *et al.*, 2000). Mutations in these genes delay flowering. Furthermore, overexpression of *CO* causes increased expression of *FT* and *SOC1* mRNAs, while *soc1* or *ft* mutations partially suppress the early flowering caused by *CO* overexpression (Onouchi *et al.*, 2000; Samach *et al.*, 2000). Similarly, *FT* overexpression suppresses the late flowering caused by *co* mutations (Kardailsky *et al.*, 1999; Kobayashi *et al.*, 1999). The close relationship between *FT* and *CO* function is supported by global expression analysis of wild-type and *co* or *ft* mutants (Schmid *et al.*, 2003), which showed that that these mutations affect the expression of similar classes of genes that in wild-type are induced at the apex in response to long days.

Several proteins have been identified that regulate the transcription of *CO*, and therefore encode proteins that act upstream of *CO*. The *CO* gene is circadian clock regulated (Suarez-Lopez *et al.*, 2001), and mutations that impair circadian clock function, such as *toc1* or *lhy-1*, alter the timing of *CO* expression (Suarez-Lopez *et al.*, 2001; Yanovsky and Kay, 2002). In addition, two proteins that have a strong effect on the amplitude of *CO* mRNA abundance, have been identified. These are encoded by the *FKF1* and *GI* genes, which are also both circadian clock regulated such that their mRNAs peak in the evening. GI is a large nuclear protein of unknown function (Fowler *et al.*, 1999; Park *et al.*, 1999), and mutations in the gene both delay flowering and reduce the abundance of *CO* mRNA (Suarez-Lopez *et al.*, 2001). The second protein required for increased expression of *CO* is FKF1 (Imaizumi *et al.*, 2003; Nelson *et al.*, 2000). The structure and function of this protein is described in more detail in the following section. FKF1 appears to have a day-length dependent role in promoting *CO* transcription that suggests that FKF1 activates *CO* expression specifically in response to light, but the biochemical mechanism by which it regulates CO expression is unknown.

Two further mutations that delay the regulation of flowering in response to day length and encode photoreceptors have been identified. The *fha-1* mutation, which was among the first mutations shown to delay flowering by impairing the photoperiod pathway, was subsequently shown to impair the blue-light photoreceptor Cry2 (Guo *et al.*, 1998). The abundance of this photoreceptor is regulated by photoperiod, suggesting that it is closely involved in the regulation of photoperiodic responses. Under long-day conditions Cry2 protein is present at similar levels throughout the day and night, whereas under short days it is absent during the day but accumulates at night (El-Assal *et al.*, 2001; Mockler *et al.*, 2003). This is consistent with a role in promoting flowering specifically under long days, when the protein is present throughout the time that the plant is exposed to light. Initially the *cry2* mutation was proposed to delay flowering by reducing the abundance of the *CO* mRNA, however detailed analyses of the abundance of *CO* mRNA at times throughout the day, demonstrated that this is not the case (Suarez-Lopez *et al.*, 2001). Nevertheless, *cry2* mutations reduce *FT* expression, suggesting that they might impair the capacity of CO to activate *FT* (Yanovsky and Kay, 2002). Furthermore, *cry2* mutations reduce the abundance of *FT* mRNA in *35S::CO* plants, indicating that Cry2 is likely to act on CO at the post-transcriptional level. The important role of Cry2 in promoting flowering of Arabidopsis was also demonstrated by the identification of a naturally occurring dominant *Cry2* allele in the Cape Verde Islands accession (El-Assal *et al.*, 2001). This allele promotes early flowering under short days when introduced into the Landsberg *erecta* accession. This modified form of Cry2 causes the flowering of the Landsberg *erecta* or Columbia plants to occur early under short days and to be almost insensitive to day length. This form of Cry2 contains a single amino acid change of a valine to a methionine (El-Assal *et al.*, 2001). This amino acid change appears to stabilise Cry2 protein so that it is present under short days, and this presumably then leads to early flowering under these conditions.

Cry2 appears to activate flowering by stabilising the CO protein in the nucleus (Valverde *et al.*, 2004). CO protein accumulates in the nucleus under blue light and this requires the Cry2 (and Cry1) photoreceptor, whereas in red light or darkness the protein disappears from the nucleus and is probably degraded by the proteasome after ubiquitination. In *35S::CO cry2* plants the abundance of the CO protein is reduced in plants growing under blue light or long days, and this is enhanced in *35S::CO* plants carrying both *cry1* and *cry2* mutations, suggesting there is some redundancy in the roles of these proteins (Valverde *et al.*, 2004).

The photoreceptor Phytochrome A (PhyA) is also required for the promotion of flowering in response to long days. Strikingly, like Cry2, PhyA is a light-labile photoreceptor suggesting that this may be a feature of those photoreceptors closely involved in the control of photoperiodic responses (Mockler *et al.*, 2003). The PhyA protein also shows a strong oscillation under short days, accumulating during the night, and does not oscillate under long days (Mockler *et al.*, 2003). However, unlike Cry2, PhyA does not seem to be more abundant under long than short days. Mutations in *PhyA* delay flowering under extended short days (short days of 8 hours extended with low fluence light enriched with far-red light), and reduce the accumulation of CO protein in 35S::CO plants growing under far-red light or under long days (Valverde *et al.*, 2004). Therefore, PhyA also appears to promote flowering by stabilising the CO protein, and in white light these two photoreceptors presumably have redundant roles in this function.

Classical experiments demonstrated that during the induction of flowering the day length is perceived in the leaf, and that in response to appropriate day lengths a signal is transmitted through the phloem to the shoot apex and triggers flower development. This observation was originally made using spinach plants and exposing parts of the plant to different day lengths (Knott, 1934). For example, exposing leaves to day lengths that induce flowering was sufficient to trigger floral development, whereas exposing the apex to these day lengths was not. A similar conclusion was reached using grafting experiments (Zeevaart, 1976). Strikingly, in Perilla plants grafting a single leaf that was exposed to day lengths that induce flowering onto a second plant held in non-inductive day lengths was sufficient to induce flowering of this plant (Zeevaart, 1985). Recent work has approached the problem of whether flowering-time genes identified in Arabidopsis act in the leaf to generate the transmissible signal or whether they act in the meristem to respond to this signal. The *CO* gene is expressed strongly in the phloem, but also in the meristem and protoxylem (Takada and Goto, 2003; An *et al.*, 2004). However, grafting experiments between *co* mutants and wild-type plants as well as expression of the gene from phloem specific promoters indicated that *CO* acts in the vascular tissue to regulate the expression of a long-distance signal that triggers floral development at the apex of the plant (An *et al.*, 2004; Ayre and Turgeon, 2004). Furthermore, the CO target gene *FT* is expressed specifically in the phloem, and when *CO* is expressed in the phloem from specific promoters, *FT* expression is strongly elevated in these cells (Takada and Goto, 2003; An *et al.*, 2004). Similarly, *ft* mutations strongly suppress the early-flowering phenotype caused by expression of *CO* in the phloem. This suggests therefore that the mechanism by which CO promotes flowering from the phloem involves FT, and the small size of the FT protein led to the suggestion that this protein itself may represent the transmissible signal (Takada and Goto, 2003; An *et al.*, 2004). In agreement with this, *FT* misexpression in the phloem or the meristem was sufficient to induce early flowering. However, so far no direct evidence for movement of the FT protein has been presented, and it remains possible that FT regulates the synthesis or transport of a small molecule that induces flowering.

4. AN EXTERNAL COINCIDENCE MODEL FOR THE DAY-LENGTH RESPONSE IN ARABIDOPSIS

Pittendrigh and Minis explicitly proposed an external coincidence model as the basis of the photoperiodic response (Pittendrigh and Minis, 1964), and Vince-Prue described in detail the physiological data in support of such a model (Vince-Prue, 1994). This model is based upon a photoperiod response rhythm, which is regulated by the circadian clock. The photoperiod response rhythm is proposed to be sensitive to light at a certain time of the day, and if the plant is exposed to light at this time then flowering will be promoted (in long-day plants) or repressed (in short-day plants). Light has two roles in such a model. Exposure to light entrains the circadian clock, and therefore sets the phase of the photoperiod response rhythm, thereby determining at exactly which time of day flowering will be affected by exposure to light. The second role of light is to interact with the photoperiod response rhythm to trigger flowering. Such a model is related to the gating of gene expression in response to light, as described for the Arabidopsis *CAB* gene (Millar and Kay, 1996). The expression of *CAB* is regulated by the circadian clock and by exposure to light, and the sensitivity of *CAB* expression to light shows a circadian rhythm so that if the plant is exposed to light early during the day *CAB* expression is highly induced, but exposure of plants to light during the night has little effect on *CAB* expression. Induction of *CAB* by light is said to be gated by the circadian clock so that its expression is induced only at certain times of day (Millar and Kay, 1996). A putative photoperiod response rhythm may show a similar mechanism, but with the regulated output being the activation or repression of flowering rather than the induction of CAB expression.

The circadian pattern of expression of *CO* suggested a molecular basis for such a co-incidence model (Figure 2). *CO* mRNA abundance rises late in the day, around 12 hours after dawn and stays high until the following dawn (Suarez-Lopez *et al.*, 2001). This broad peak in expression can be divided into two peaks. Under long days *CO* expression occurs when the plant is exposed to light at the end of the day, whereas under short days the peak in expression occurs exclusively in darkness. This suggested that if *CO* were regulated at the post-transcriptional level by light, then this would occur under long but not short days (Suarez-Lopez *et al.*, 2001). The hierarchy of gene action in the photoperiod pathway indicated that CO acts to activate the expression of *FT* (Samach *et al.*, 2000), and therefore *FT* expression could be used indirectly as a measure of CO activity. *FT* mRNA abundance peaks in the late evening around the time, or slightly after, CO mRNA levels rise. This diurnal peak in *FT* mRNA does not occur in a *co* mutant or in wild-type plants grown under short days. Thus the analysis of *FT* expression is consistent with the proposal that both *CO* expression and exposure to light at the end of the day are required for *FT* activation.

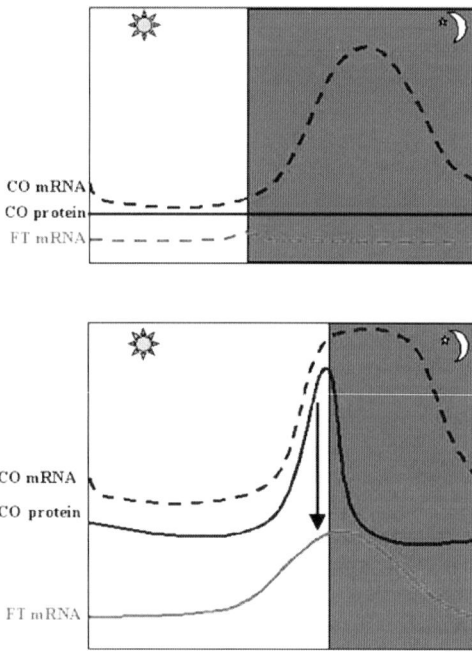

Figure 2. Coincidence model for the activation of flowering under long photoperiods in Arabidopsis. Top: Under short days circadian clock regulation of CO causes its mRNA to accumulate in the dark. CO protein does not accumulate, and FT expression is not activates. Bottom: Under long days circadian clock regulation and activation of CO transcription by light via FKF1 causes CO mRNA to accumulate in the light. The CO protein is stabilised through the activity of the PhyA and Cry2 photoreceptors. The presence of the CO protein causes activation of FT transcription, which leads to flowering.

The proposal that the co-incidence between *CO* mRNA expression and exposure to light are required for activation of *FT* and early flowering under long days was tested using mutations and physiological conditions that alter the timing of *CO* expression (Roden *et al.*, 2002; Yanovsky and Kay, 2002). The *toc1-1* mutation causes circadian rhythms to cycle faster under continuous light, and the mutant flowers earlier than wild-type under short days (Somers *et al.*, 1998). The early flowering of *toc1-1* mutants was proposed to be due to a defect of the circadian clock, because although they flowered earlier than wild-type plants under short days of 8 hours light, 16 hours dark (a 24 hour cycle), they flowered at the same time as wild-type under short days of 7 hours light, 14 hours dark (a 21 hour cycle) (Somers *et al.*, 1998). This suggests that when the total cycle time is shortened so that it is equivalent to the period of a circadian rhythm in the *toc1-1* mutant, then the early-flowering phenotype is suppressed. To test whether the effect of *toc1-1* on flowering involved *CO*, the temporal pattern of *CO* mRNA expression in the *toc1-1* mutant was tested (Yanovsky and Kay, 2002). Under short days the peak in *CO* mRNA is advanced, so that it occurs during the light phase, and *FT* is then expressed under these conditions. In addition, the early flowering of *toc1-1* mutants is delayed by *co* or *ft* mutations, demonstrating their involvement in the early-flowering phenotype (Yanovsky and Kay, 2002).

Exposure of wild-type plants to non-24 hour cycles also provided evidence that *CO* expression represents a photoperiod response rhythm. In plants grown under short days of 10 hours light, 20 hours darkness (a 30 hour cycle), *CO* expression occurred during the day, and this correlated with increased expression of *FT* and earlier flowering (Roden *et al.*, 2002; Yanovsky and Kay, 2002). These experiments greatly strengthened the previously described correlation between *CO* expression, exposure to light and *FT* activation, although they did not explain the mechanism by which light influences CO expression. Recently, this was proposed to occur both at the transcriptional and post-transcriptional levels.

The mechanism by which light regulates CO activity at the post-transcriptional level recently became clearer. The presence of CO protein was followed using specific antibodies or CO:GFP translational fusions (Valverde *et al.*, 2004).These approaches demonstrated that the stability of the CO protein is regulated by exposure of plants to light. The protein is detected in nuclei when *35S::CO* or *35S::CO:GFP* plants are grown under blue or far-red light, whereas in plants grown under darkness or red light the protein is absent, which correlates with the effectiveness of these treatments in promoting flowering since blue and far-red light are the wavelengths that promote flowering of Arabidopsis most effectively (Brown and Klein, 1971). In addition, in *35S::CO* plants grown under long days, the protein accumulates only at the end of the day, although the *CO* mRNA is present constantly. This suggests that there are post-transcriptional mechanisms that prevent accumulation of the protein early in the day and during the night. The mechanism that acts in the morning to prevent CO accumulation probably involves Phytochrome B (PhyB) since mutations in the gene encoding this photoreceptor cause the CO protein to accumulate throughout the photoperiod. However in *phyB* mutants the protein is still absent during the night, indicating that independent mechanisms prevent accumulation of the protein during the night and the morning.

The absence of the protein in the morning and in plants exposed to red light may be related, since the CO protein was detected in *35S::CO phyB* plants grown under red light, but not in *35S::CO* plants grown under these conditions. The absence of the protein in the morning and during the night is probably caused by ubiquitination followed by proteasome mediated degradation , since proteasome inhibitors caused the protein to accumulate at these times (Valverde *et al.*, 2004).The high amplitude peak in the evening requires the cryptochrome and phyA photoreceptors, suggesting that they stabilise the protein by overcoming the PhyB-mediated degradation. Stabilisation of the CO protein in the nucleus provides a molecular mechanism by which exposure to light could intersect with a circadian rhythm to trigger flowering.

In addition to stabilisation of CO protein, light also acts to boost *CO* transcription. The blue light photoreceptor FKF1 is required for the first peak in *CO* mRNA under long-day conditions, but under short days does not seem to influence *CO* transcript abundance (Imaizumi *et al.*, 2003). Mutations in *FKF1* reduce *CO* mRNA levels under long days and reduce *FT* mRNA abundance. FKF1 protein contains an F-box, indicating that it is part of an SCF complex required for the ubiquitination and degradation of target proteins, and kelch repeats, which are protein-protein interaction domains that may recruit the target proteins for degradation (Nelson *et al.*, 2000). In addition FKF1 contains a LOV domain, which are present in phototropin photoreceptors where they bind the chromophore flavinmononucleotide. This domain seems to play a similar role in FKF1 since the protein was recently shown to alter its spectral absorption when exposed to blue light (Imaizumi *et al.*, 2003). Furthermore, the transcript of *CO* was demonstrated to be more abundant when plants were exposed to day-length extensions of blue light, confirming a role for this wavelength of light in regulating *CO* transcription. This analysis of FKF1 suggests that transcriptional regulation of *CO* by light plays a role in the control of *CO* activity by day length.

Thus the flowering response to day length is controlled by a complex combination of transcriptional and post-transcriptional regulation of *CO* function (Figure 2). As originally proposed, light plays two roles. The first is to entrain the circadian clock, which then regulates the abundance of the *CO* mRNA, so that it peaks with a characteristic phase late in the day. However, CO only activates the expression of the downstream gene *FT* if the plant is exposed to light at this time. The regulation by light occurs both at the transcriptional level, because *CO* mRNA abundance is boosted by an FKF1-dependent step, and at the post-transcriptional level, because light acts to stabilise the CO protein in the nucleus. Control of CO protein stability is independent of transcription, so that these two layers of regulation appear to act separately.

5. GENETIC ANALYSIS OF THE PHOTOPERIODIC CONTROL OF FLOWERING IN RICE, A SHORT-DAY PLANT

The Nipponbare (photoperiod sensitive) and Kasalath (less sensitive to photoperiod) rice varieties were crossed to generate a population of plants used to map loci involved in photoperiod response (Yano *et al.*, 1997). A total of 14 quantitative trait loci (QTLs) affecting photoperiod response were identified and located in the rice genome using these plants. These were named *Heading date* (*Hd*) loci. To assess the roles of individual QTLs, appropriate segments of chromosomes of the Kasalath variety were introgressed into the Nipponbare background. The effects of these introgressions on the photoperiod response of Nipponbare were then assessed. This approach indicated that five QTLs, *Hd1*, *Hd2*, *Hd3*, *Hd5* and *Hd6*, confer photoperiod sensitivity. Several of these loci have now been cloned enabling a comparative analysis between Arabidopsis and rice (Izawa *et al.*, 2003) (Figure 1B).

Hd1 encodes the rice orthologue of the Arabidopsis *CO* gene, indicating that the involvement of this protein in the photoperiod response is conserved between monocotyledonous and dicotyledonous plants and between long-day and short-day response types (Yano *et al.*, 2000). However, there are important distinctions in the function of *CO* between rice and Arabidopsis. In Arabidopsis loss of *CO* function delays flowering under long-day conditions that induce flowering, whereas in rice loss of function of *Hd1* both delays flowering under inductive short-day conditions and accelerates flowering under non-inductive long-day conditions (Putterill *et al.*, 1995; Yano *et al.*, 2000). This suggests that *Hd1* has two functions in rice, both to accelerate flowering under inductive conditions and delay the response under non-inductive conditions, whilst the Arabidopsis *CO* gene only appears to regulate flowering under inductive conditions. Interestingly, there are occasional reports of *co* mutations causing earlier flowering under short days in Arabidopsis (Devlin *et al.*, 1996; Redei *et al.*, 1974), indicating a dual role for *CO* similar to that described for *Hd1*, and this may depend on other aspects of the environmental conditions, such as the light quality. There are also surprising sequence differences between *CO* and *Hd1* that suggest they might differ in the detailed mechanisms by which the proteins function. Whereas mutations in *co* that are predicted to disrupt the structure of the second B-box delay flowering, active alleles of *Hd1* encode a protein in which the second B-box would not be predicted to form due to the absence of residues crucial for zinc binding (Mouradov *et al.*, 2002; Putterill *et al.*, 1995; Yano *et al.*, 2000). These inactive forms of the second B-box also occur in CO homologues found in other monocotyledonous species, such as barley (Griffiths *et al.*, 2003).

Analysis of the *Hd3* locus, demonstrated that the function of *FT* is also conserved in rice (Kojima *et al.*, 2002). High resolution mapping of the *Hd3* locus resolved this QTL into two closely linked loci, called *Hd3a* and *Hd3b* (Monna *et al.*, 2002). The gene corresponding to the *Hd3a* locus was cloned, and shown to encode a homologue of *FT* (Kojima *et al.*, 2002). A small gene family encodes FT-like proteins in rice (Izawa *et al.*, 2002), and both *Hd3a* and another related gene, *FT-Like1*, induced very early flowering when overexpressed (Izawa *et al.*, 2002; Kojima *et al.*, 2002), confirming a role for these proteins in flowering-time control. Furthermore, expression of *Hd3a* is upregulated by *Hd1* and by exposure to short-day conditions (Kojima *et al.*, 2002). There is therefore a striking similarity in the regulation of *Hd3a* in rice and of *FT* in Arabidopsis, since the expression of both is increased by the day length conditions that induce early flowering, and this requires the function of the Hd1CO protein. Furthermore, rice also contains a homologue of *SOC1*, the other established target gene of CO that promotes flowering of Arabidopsis (Samach *et al.*, 2000). The *OsSOC1* gene promotes extreme early flowering of transgenic Arabidopsis plants when expressed from the 35S promoter, and is increased in expression at the time of floral initiation (Tadege *et al.*, 2003). However, whether *OsSOC1* is regulated by Hd1 has not yet been tested.

The *Hd6* QTL was also studied at the molecular level, and shown to correspond to the alpha subunit of casein kinase 2 (CK2) (Takahashi *et al.*, 2001). This protein kinase phosphorylates serine/threo nine residues in target proteins, and is comprised of two catalytic (alpha) subunits and two regulatory (ß) subunits (Allende and Allende, 1995). CK2 phosphorylates many substrate proteins. The Kasalath allele is predicted to encode a functional alpha subunit, whereas the Nipponbare allele contains a premature stop codon and is therefore predicted to encode a truncated, inactive protein (Takahashi *et al.*, 2001). The Kasalath allele delayed flowering under long-day conditions, suggesting that the role of this protein is to inhibit flowering under long days. Whether *Hd6* regulates flowering through the Hd1 or Hd3 proteins is unclear, but appears to interact genetically with *Hd2*, a QTL that has not so far been described at the molecular level. In Arabidopsis, overexpression of a regulatory subunit of CK2 has been shown to cause early flowering (Sugano *et al.*, 1999), and this interacts physically with CCA1 (Sugano *et al.*, 1998), a MYB-like transcription factor implicated in the circadian oscillator of Arabidopsis (Green and Tobin, 1999; Mizoguchi *et al.*, 2002). However, it is unclear whether the function of CK2 in Arabidopsis is related to the function of Hd6 in Arabidopsis. Given the likely importance of CK2 in the phosphorylation of many substrate proteins, the demonstration that the Nipponbare allele carries a null mutation in one of the catalytic domains may be surprising, particularly since no phenotypic effects could be associated with this allele apart from the effect on photoperiodic flowering (Takahashi *et al.*, 2001). Takahashi et al (2001) suggest that this may be explained by the presence of a second gene encoding this subunit and that Hd6 has a specific role in photoperiodic control of flowering, or alternatively that components of the photoperiodic response are particularly sensitive to reductions in the abundance of CK2.

The similarity in the components associated with photoperiod response in rice and Arabidopsis suggested that the hierarchy of gene action might be conserved between the two species (Figure 1). The order of gene action in the Arabidopsis pathway was established as *GI-CO-FT/SOC1*, as described above, and this set of genes represents an output pathway from the circadian clock. In rice, the *GI* gene was identified as an mRNA whose abundance is reduced in an early-flowering mutant impaired in synthesis of the phytochrome chromophore (Hayama *et al.*, 2002). The GI protein is remarkably conserved between Arabidopsis and rice, being 67% identical, and is regulated by the circadian clock so that its mRNA peaks in abundance just before dawn, as in Arabidopsis (Hayama *et al.*, 2002). However, overexpression of *GI* from a ubiquitin gene promoter delays flowering in rice under inductive short-day conditions, and causes earlier flowering under long days (Hayama *et al.*, 2003). This is in contrast to the role of *GI* in promoting flowering of Arabidopsis under inductive long days. These late-flowering transgenic rice plants have elevated levels of *Hd1* mRNA, in agreement with the role of *GI* in promoting the expression of *CO* in Arabidopsis (Hayama *et al.*, 2003; Suarez-Lopez *et al.*, 2001). Strikingly, however, these high levels of expression of *Hd1* correlate with delayed flowering of rice under inductive conditions, whereas overexpression of *CO* causes early flowering. Whether the increased levels of *Hd1* expression alter *Hd3a* expression was also examined, and *Hd3a* mRNA levels were reduced in short-day grown OsGI-ox plants (Hayama *et al.*, 2003). This suggested a model whereby the role of *Hd1* in rice is to repress *Hd3a* under long-days, thereby delaying flowering. The late flowering of the *OsGI-ox* plants would therefore be caused by increased *Hd1* expression. While this model is attractive, it probably does not explain the full complexity of the rice data. For example, it does not explain the earlier flowering of the OsGI-ox plants under non-inductive long-day conditions or the late flowering and reduced *Hd3a* expression observed in plants carrying loss of function alleles of *Hd1* (Hayama *et al.*, 2003; Kojima *et al.*, 2002).

How can the coincidence model proposed to explain the early flowering of Arabidopsis under long days be adapted for the short-day response shown by plants such as rice? The basis of the coincidence model of photoperiodic flowering in Arabidopsis is that exposure of plants to light at the end of the day causes an increase in CO transcription due to circadian clock control of CO expression and detection of light by FKF1, and then stabilisation of the CO protein in a light-dependent manner (Imaizumi *et al.*, 2003; Suarez-Lopez *et al.*, 2001; Valverde et al, 2004; Yanovsky and Kay, 2002). In short-day plants such as rice, flowering occurs under short days, when plants are not exposed to light at the times that activate CO function in Arabidopsis. The phase of expression of the *Hd1* and *OsGI* genes in rice seem very similar or identical to their Arabidopsis orthologues, excluding the possibility that shifts in the timing of expression of these genes is responsible for the change in day- length response. The current model proposed for rice therefore, suggests that the coincidence between light and *Hd1* expression under long days does control flowering, as described for Arabidopsis, but rather than accelerating flowering it inhibits flowering through the repression of *Hd3a* expression (Hayama *et al.*, 2003; Izawa *et al.*, 2002) (Figure 3). A role for phytochrome in the detection of light is suggested by the early-flowering phenotype of the *se5* mutant, which is impaired in an enzyme required for the synthesis of the phytochrome chromophore (Izawa *et al.*, 2000), and the observation that these mutants express *Hd3a* at elevated levels (Izawa *et al.*, 2002). Such a model is also consistent with the delayed flowering of *OsGI-ox* plants, and consequent increased abundance of *Hd1* mRNA and reduced level of *Hd3a* mRNA (Hayama *et al.*, 2003). However, so far this model does not incorporate an explanation for the role of *Hd1* in promoting flowering under short days. The reduced level of *Hd3a* mRNA in plants carrying loss of function alleles of *Hd1* under these conditions, suggests that *Hd1* is activated in rice under short days and that this results in increased abundance of *Hd3a* mRNA.

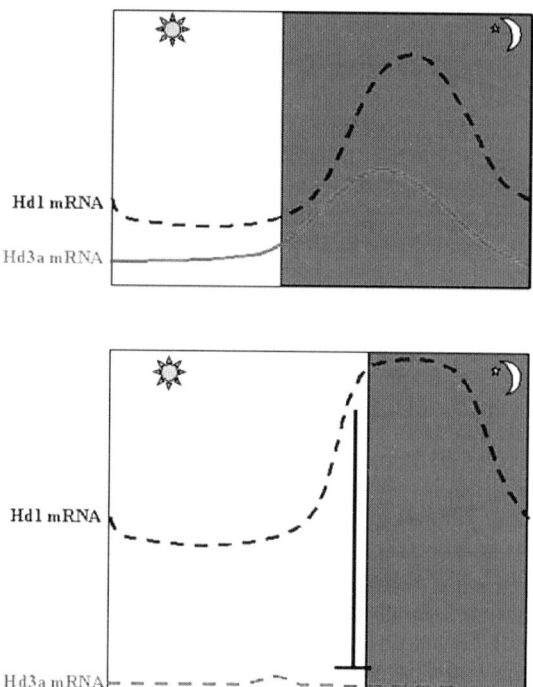

Figure 3. *Coincidence model for the activation of flowering under short photoperiods in rice. Top: Under short-day conditions, circadian clock regulation of* Hd1 *mRNA causes it to accumulate in the dark.* Hd3a *is expressed and this leads to early flowering. Bottom: Under long-day conditions circadian clock regulation of* Hd1 *mRNA causes it to accumulate in the light. This leads to activation of Hd1 function so that it represses* Hd3a *expression and delays flowering.*

6. RELATIONSHIPS BETWEEN PHOTOPERIODIC CONTROL AND OTHER ENVIRONMENTAL CUES REGULATING FLOWERING

Flowering is controlled by many environmental stimuli, as well as photoperiod, suggesting that in most locations the decision to flower will involve information from several signalling pathways. A coherent response to a complex mixture of environmental stimuli must involve prioritisation of the information from these pathways. In Arabidopsis, this concept is strengthened by the observation that the expression of genes such as *FT* and *SOC1* is regulated by several environmental stimuli, and therefore these genes are sometimes described as floral integrators, because they respond to information received from several signalling pathways (Mouradov *et al.*, 2002; Simpson and Dean, 2002).

The responses to vernalisation and photoperiod both provide cues that allow the plant to respond to the changing seasons. Both responses enable flowering in the spring, either as a response to extended exposure to low temperatures of winter (vernalisation) or to the longer day lengths experienced in spring. Therefore, the response to these two environmental parameters must be integrated. In Arabidopsis the molecular control of the vernalisation response has been extensively studied (Michaels and Amasino, 2000). The *FLC* gene plays a crucial role in this response. *FLC* encodes a MADS box transcription factor that represses flowering (Michaels and Amasino, 1999; Sheldon *et al.*, 1999). Accessions that express *FLC* at high levels are late flowering, even under long days. However, vernalisation reduces the expression of *FLC* and causes the plants to flower early under long days (Michaels and Amasino, 2001; Sheldon *et al.*, 2000). Therefore, a requirement for vernalisation is conferred by high level expression of *FLC*. The repression of flowering by FLC under long days must overcome the promotion of flowering induced by the long-day pathway and CO. One mechanism by which this occurs appears to be through direct FLC-mediated antagonism of the activation of CO target genes (Hepworth *et al.*, 2002). For example, the expression of the floral integrator genes *SOC1* and *FT* is both promoted by *CO* and repressed by *FLC*. FLC binds directly to the promoter of the *SOC1* gene *in vitro*, and mutations that disrupt the binding site increase *SOC1* expression *in vivo* (Hepworth *et al.*, 2002). This suggests that binding of FLC to the promoters of floral integrator genes blocks the photoperiodic induction of flowering by preventing the activation of the same genes by CO. This is overcome by vernalisation because this treatment leads to a reduction in the expression of *FLC*.

Flowering is also responsive to light quality (see Chapter 22). When shaded by neighbouring vegetation, many plants show extended elongation of internodes and earlier flowering (Smith and Whitelam, 1997). This response, termed shade avoidance, is induced by the lower R:FR ratios present in shade of vegetation. The earlier flowering that occurs as part of shade avoidance reduces the effect of photoperiod and in the case of Arabidopsis will induce early flowering even under short-days. The major photoreceptor involved in shade avoidance is Phytochrome B (PhyB), and *phyB* mutants show many of the features of shaded plants. These mutants retain a flowering response to photoperiod, although this is reduced compared to wild-type plants because they flower much earlier than wild-type plants under short days and only slightly earlier under long days. The mechanism by which shade avoidance regulates flowering may be closely related to the photoperiodic response. In *phyB* mutants the floral integrator *FT* is expressed at higher levels than in wild-type plants (Cerdan and Chory, 2003; Halliday *et al.*, 2003), indicating that the early flowering induced by shade avoidance probably acts through the same floral integrator as photoperiod response. In addition, the abundance of the CO protein, which acts at the end of the photoperiod response pathway, is regulated by light quality (Valverde *et al.*, 2004). The protein is absent from the nucleus in plants exposed to red light, and accumulates to high levels in far-red light. Also in the *phyB* mutant the protein is more abundant. This suggests that in the low R:FR ratios associated with shade avoidance, the CO protein is likely to accumulate and promote earlier flowering. The *co* mutation does indeed delay flowering of *phyB* mutants (Putterill *et al.*, 1995), but does not completely abolish the early-flowering phenotype suggesting that *PHYB* might control flowering through both CO-dependent and CO-independent pathways. Recently, the *pft1* mutation was also shown to delay flowering and reduce the abundance of *FT* mRNA in a *phyB* mutant background (Cerdan and Chory, 2003). PFT1, which is a nuclear protein with similarity to transcription factors, may act in a light quality pathway that promotes flowering through *FT* and in parallel to the function of CO within the photoperiod pathway (Cerdan and Chory, 2003).

7. PHOTOPERIODIC RESPONSES OTHER THAN FLOWERING

Many other plant developmental responses in addition to flowering are regulated by photoperiod (Thomas and Vince-Prue, 1997). For example, in herbaceous plants the onset of bud dormancy in autumn and the breaking of dormancy in spring are often regulated by photoperiod, and in some species, including rice, germination can be regulated by day length. Vegetative reproduction is often induced by changes in day length, and particularly tuberization in potato is triggered by exposure to short days (Ewing and Struik, 1992). These traits are less amenable to genetic analysis than the control of flowering in model species such as Arabidopsis or rice, however, the molecular control of tuberization has been studied using transgenic plants (Jackson *et al.*, 1996; Martinez-Garcia *et al.*, 2002). Potato plants expressing an antisense transgene for PhyB produced tubers under long days, whereas wild-type plants showed an absolute requirement for exposure to short days (Jackson *et al.*, 1996). This photoreceptor, therefore, plays a critical role in repressing tuberization under long days. Grafting experiments suggested that PhyB may promote the formation of an inhibitor of tuberization in the leaves, since grafting of antisense PhyB plants onto the root of a wild-type plant led to the production of tubers under long days. However, if leaves were retained on the wild-type stock then these inhibited tuberization under long days. The role of PhyB, therefore, is to promote the formation of an inhibitor of tuberization under long days (Jackson *et al.*, 1996).

As in the control of flowering in Arabidopsis, there appears to be a link between the day-length dependent regulation of tuberization by PhyB and the function of the *CO* gene. Since *CO* plays a central role in the photoperiodic response of Arabidopsis, transgenic potato plants overexpressing the gene were made to assess whether *CO* affected the ability of potato to form tubers in response to photoperiod (Martinez-Garcia *et al.*, 2002). These transgenic plants showed a significant delay in the formation of tubers under short days, indicating that the overexpression of *CO* delays tuberization. Furthermore, grafting experiments were performed in which the shoot of a plant overexpressing *CO* was grafted onto the root system of a wild-type plant. The grafted plants showed the delayed tuberization characteristic of the shoot, suggesting that *CO* acts in the shoot to alter the formation or transport of long-distance signals controlling tuber formation in the root (Martinez-Garcia *et al.*, 2002). Potato *CO*-like genes may act in wild-type potato plants to control these signals, and may play a similar role to the *CO* gene in Arabidopsis in mediating between the time-keeping circadian oscillator and an output pathway regulating a developmental event.

8. PERSPECTIVES

The 10 years since the last edition of this book have provided the first insights into the molecular mechanisms that confer photoperiodic responses, and the use of genetic model systems together with the genomic tools available has ensured that progress has been rapid. What can we expect in the next edition? The frameworks for the genetic pathways regulating the photoperiodic control of flowering that have been established in the Arabidopsis and rice model systems will provide the basis for a deeper understanding of how these are regulated by day length. This will certainly involve more sophisticated screens for mutants, the use of genomic resources to fuel reverse genetic approaches and detailed analyses of the cell types in which these genes act to control flowering time. These approaches will address obvious gaps in our understanding of the players that act in the pathway, and allow many more genes involved in these processes to be identified. However, at present we have only an elementary understanding of the biochemical function of the proteins already placed in the pathway, and no direct interactions between proteins that act in the pathway have been demonstrated. Even for COHd1 or GIOsGI, which play central roles in photoperiodic responses in Arabidopsis and rice, the biochemical mechanism by which they regulate gene expression is unknown. Deepening our understanding of these biochemical mechanisms will not only be important in demonstrating how information is transmitted through the pathway, but is likely to be essential in explaining how the function of the pathway is modified to generate the opposite response to day length in different species. Furthermore, day-length responses evolve rapidly, so that different species within the same genus can show different response types. The comparison of response types that has been initiated in Arabidopsis and rice is likely therefore to be extended to include other families of Angiosperms, thereby illustrating whether this conserved pathway can be modified in different ways during evolution to generate these response types. For example, how closely related are the molecular mechanisms that generate a short-day response in *Pharbitis nil* (a Solanaceous doctyledonous plant), which was widely used as a model to study the physiology of day-length responses, and in rice? Similarly, how related are the mechanisms that confer day-length control on other developmental transitions, such as tuberization, and those that control flowering? The deepening genetic and biochemical understanding in Arabidopsis together with the comparative analysis in other carefully chosen systems will further enhance our understanding of photoperiodism.

9. REFERENCES

Allende, J. E. and Allende, C. C. (1995) Protein Kinases .4. Protein-Kinase Ck2 - an Enzyme with Multiple Substrates and a Puzzling Regulation. *Faseb Journal,* 9, 313-323.

Alonso-Blanco, C., El-Assal, S. E.-D., Coupland, G. Koornneef, M. (1998) Analysis of natural allelic variation at flowering time loci in the Landsberg erecta and Cape Verde Islands ecotypes of Arabidopsis thaliana. *Genetics,* 149, 749-764.

Alonso-Blanco, C. and Koornneef, M. (2000) Naturally occurring variation in Arabidopsis: an underexploited resource for plant genetics. *Trends in Plant Science,* 5, 22-29.

An, H., Roussot, C., Suarez-Lopez, P., Corbesier, L., Vincent, C., Pineiro, M., *et al.* (2004) CONSTANS acts in the phloem to regulate a systemic signal that induces photoperiodic flowering of Arabidopsis. *Development, 131,* 3615-3626.

Ayre, B., and Turgeon, R. (2004) Graft transmission of a floral stimulant derived from CONSTANS. *Plant Physiol,135,* 2271-2278.

Borner, R., Kampmann, G., Chandler, J., Gleissner, R., Wisman, E., Apel, K. *et al.* (2000) A MADS domain gene involved in the transition to flowering in Arabidopsis. *Plant Journal, 24,* 591-599.

Brown, J. A. M. and Klein, W. H. (1971) Photomorphogenesis in Arabidopsis-Thaliana (L) Heynh - Threshold Intensities and Blue-Far-Red Synergism in Floral Induction. *Plant Physiology, 47,* 393-399.

Burn, J. E., Smyth, D. R., Peacock, W. J., Dennis, E. S. (1993) Genes conferring late flowering in Arabidopsis thaliana. *Genetica, 90,* 147-155.

Cerdan, P. D. and Chory, J. (2003) Regulation of flowering time by light quality. *Nature, 423,* 881-885.

Clarke, J. H. and Dean, C. (1994) Mapping FRI, a locus controlling flowering time and vernalisation response in Arabidopsis thaliana. *Molecular and General Genetics, 242,* 81-89.

Devlin, P. F., Halliday, K. J., Harberd, N. P., Whitelam, G. C. (1996) The rosette habit of Arabidopsis thaliana is dependent upon phytochrome action: Novel phytochromes control internode elongation and flowering time. *Plant Journal, 10,* 1127-1134.

El-Assal, S. E.-D., Alonso-Blanco, C., Peeters, A. J. M., Raz, V., Koornneef, M. (2001) A QTL for flowering time in *Arabidopsis* reveals a novel allele of *CRY2. Nature Genetics, 29,* 435-440.

Ewing, E. E. and Struik, P. C. (1992) Tuber formation in potato: induction, initiation and growth. *Hort Rev, 14,* 89-197.

Fowler, S., Lee, K., Onouchi, H., Samach, A., Richardson, K., Coupland, G., *et al.* (1999) GIGANTEA: A circadian clock-controlled gene that regulates photoperiodic flowering in Arabidopsis and encodes a protein with several possible membrane-spanning domains. *Embo Journal, 18,* 4679-4688.

Garner, W. W. and Allard, H. A. (1920) Effect of the relative length of day and night and other factors of the environment on growth and reproduction in plants. *J Agric Res, 18,* 553-606.

Green, R. M. and Tobin, E. M. (1999) Loss of the circadian clock-associated protein 1 in Arabidopsis results in altered clock-regulated gene expression. *Proceedings of the National Academy of Sciences of the United States of America, 96,* 4176-4179.

Griffiths, S., Dunford, R. P., Coupland, G., Laurie, D. A. (2003) The evolution of CONSTANS-like gene families in barley, rice, and Arabidopsis. *Plant Physiology, 131,* 1855-1877.

Guo, H. W., Yang, W. Y., Mockler, T. C., Lin, C. T. (1998) Regulations of flowering time by Arabidopsis photoreceptors. *Science, 279,* 1360-1363.

Halliday, K. J., Salter, M. G., Thingnaes, E., Whitelam, G. C. (2003) Phytochrome control of flowering is temperature sensitive and correlates with expression of the floral integrator FT. *Plant Journal, 33,* 875-885.

Hayama, R. and Coupland, G. (2003) Shedding light on the circadian clock and the photoperiodic control of flowering. *Curr Opin Plant Biol, 6,* 13-9.

Hayama, R., Izawa, T., Shimamoto, K. (2002) Isolation of rice genes possibly involved in the photoperiodic control of flowering by a fluorescent differential display method. *Plant and Cell Physiology, 43,* 494-504.

Hayama, R., Yokoi, S., Tamaki, S., Yano, M., Shimamoto, K. (2003) Adaptation of photoperiodic control pathways produces short-day flowering in rice. *Nature, 422,* 719-22.

Hepworth, S. R., Valverde, F., Ravenscroft, D., Mouradov, A., Coupland, G. (2002) Antagonistic regulation of flowering-time gene SOC1 by CONSTANS and FLC via separate promoter motifs. *Embo J, 21,* 4327-37.

Imaizumi, T., Tran, H. G., Swartz, T. E., Briggs, W. R., Kay, S. A. (2003) FKF1 is essential for photoperiodic-specific light signalling in Arabidopsis. *Nature, 426,* 302-306.

Izawa, T., Oikawa, T., Sugiyama, N., Tanisaka, T., Yano, M., Shimamoto, K. (2002) Phytochrome mediates the external light signal to repress FT orthologs in photoperiodic flowering of rice. *Genes and Development, 16,* 2006-2020.

Izawa, T., Oikawa, T., Tokutomi, S., Okuno, K., Shimamoto, K. (2000) Phytochromes confer the photoperiodic control of flowering in rice (a short-day plant) *Plant Journal, 22,* 391-399.

Izawa, T., Takahashi, Y., Yano, M. (2003) Comparative biology comes into bloom: genomic and genetic comparison of flowering pathways in rice and Arabidopsis. *Current Opinion in Plant Biology, 6,* 113-120.

Jackson, S. D., Heyer, A., Dietze, J., Prat, S. (1996) Phytochrome B mediates the photoperiodic control of tuber formation in potato. *Plant Journal, 9*, 159-166.

Jansen, R. C., Vanooijen, J. W., Stam, P., Lister, C., Dean, C. (1995) Genotype-by-environment interaction in genetic mapping of multiple quantitative trait loci. *Theoretical and Applied Genetics, 91*, 33-37.

Johnson, E., Bradley, M., Harberd, N. P., Whitelam, G. C. (1994) Photoresponses of light-grown phyA mutants of Arabidopsis: Phytochrome A is required for the perception of daylength extensions. *Plant Physiology, 105*, 141-149.

Kardailsky, I., Shukla, V. K., Ahn, J. H., Dagenais, N., Christensen, S. K., Nguyen, J. T., *et al.* (1999) Activation tagging of the floral inducer FT. *Science, 286*, 1962-1965.

Karlsson, B. H., Sills, G. R., Nienhuis, J. (1993) Effects of photoperiod and vernalisation on the number of leaves at flowering in 32 Arabidopsis thaliana (Brassicaceae) ecotypes. *American Journal of Botany, 80*, 646-648.

Knott, J. E. (1934) Effect of a localized photoperiod on spinach. *Proc Soc Hort Sci, 31*, 152-154.

Kobayashi, Y., Kaya, H., Goto, K., Iwabuchi, M., Araki, T. (1999) A pair of related genes with antagonistic roles in mediating flowering signals. *Science, 286*, 1960-1962.

Kojima, S., Takahashi, Y., Kobayashi, Y., Monna, L., Sasaki, T., Araki, T., *et al.* (2002) Hd3a, a rice ortholog of the Arabidopsis FT gene, promotes transition to flowering downstream of Hd1 under short-day conditions. *Plant and Cell Physiology, 43*, 1096-1105.

Koornneef, M., Alonso-Blanco, C., Vries, H. B.-D., Hanhart, C. J., Peeters, A. J. M. (1998) Genetic interactions among late-flowering mutants of Arabidopsis. *Genetics, 148*, 885-892.

Koornneef, M., Hanhart, C. J., van Der Veen, J. H. (1991) A genetic and physiological analysis of late flowering mutants in Arabidopsis thaliana. *Molecular and General Genetics, 229*, 57-66.

Laibach, F. (1951) Über sommer und winterannuelle Rasse von Arabidopsis thaliana (L.) Heynh. Ein Beitrag zur Atiologie der Bltenbildung. *Beitr Biol Pflanz, 28*, 173-210.

Lee, H., Suh, S.-S., Park, E., Cho, E., Ahn, J. H., Kim, S.-G., *et al.* (2000) The AGAMOUS-LIKE 20 MADS domain protein integrates floral inductive pathways in Arabidopsis. *Genes and Development, 14*, 2366-2376.

Lee, I. A. B. and Amasino, R. (1993) Analysis of naturally occurring late flowering in Arabidopsis thaliana. *Molecular and General Genetics, 237*, 171-176.

Martinez-Garcia, J. F., Virgos-Soler, A., Prat, S. (2002) Control of photoperiod-regulated tuberization in potato by the Arabidopsis flowering-time gene CONSTANS. *Proc Natl Acad Sci USA, 99*, 15211-15216.

Michaels, S. D. and Amasino, R. M. (1999) FLOWERING LOCUS C encodes a novel MADS domain protein that acts as a repressor of flowering. *Plant Cell, 11*, 949-956.

Michaels, S. D. and Amasino, R. M. (2000) Memories of winter: Vernalisation and the competence to flower. *Plant, Cell & Environment, 23*, 1145-1153.

Michaels, S. D. and Amasino, R. M. (2001) Loss of FLOWERING LOCUS C activity eliminates the late-flowering phenotype of FRIGIDA and autonomous pathway mutations but not responsiveness to vernalisation. *Plant Cell, 13*, 935-941.

Millar, A. and Kay, S. A. (1996) Integration of circadian and phototransduction pathways in the network controlling CAB gene transcription in Arabidopsis. *Proceedings of the National Academy of Sciences of the United States of America, 93*, 15491-15496.

Mizoguchi, T., Wheatley, K., Hanzawa, Y., Wright, L., Mizoguchi, M., Song, H. R., *et al.* (2002) LHY and CCA1 are partially redundant genes required to maintain circadian rhythms in Arabidopsis. *Dev Cell, 2*, 629-641.

Mockler, T., Yang, H. Y., Yu, X. H., Parikh, D., Cheng, Y. C., Dolan, S. *et al.* (2003) Regulation of photoperiodic flowering by Arabidopsis photoreceptors. *Proceedings of the National Academy of Sciences of the United States of America, 100*, 2140-2145.

Monna, L., Lin, H. X., Kojima, S., Sasaki, T., Yano, M. (2002) Genetic dissection of a genomic region for a quantitative trait locus, Hd3, into two loci, Hd3a and Hd3b, controlling heading date in rice. *Theoretical and Applied Genetics, 104*, 772-778.

Mouradov, A., Cremer, F., Coupland, G. (2002) Control of flowering time: interacting pathways as a basis for diversity. *Plant Cell, 14,* Suppl, S111-130.

Napp-Zinn, K. (1987) "Vernalisation. Environmental and genetic regulation." In *Manipulation of Flowering,* Atherton J.G. (ed). Butterworths, London, 123-132.

Nelson, D. C., Lasswell, J., Rogg, L. E., Cohen, M. A., Bartel, B. (2000) FKF1, a clock-controlled gene that regulates the transition to flowering in Arabidopsis. *Cell, 101*, 331-340.

Onouchi, H., Igeno, M. I., Perilleux, C., Graves, K., Coupland, G. (2000) Mutagenesis of plants overexpressing CONSTANS demonstrates novel interactions among Arabidopsis flowering-time genes. *Plant Cell, 12*, 885-900.

Park, D. H., Somers, D. E., Kim, Y. S., Choy, Y. H., Lim, H. K., Soh, M. S., et al. (1999) Control of circadian rhythms and photoperiodic flowering by the Arabidopsis GIGANTEA gene. *Science, 285*, 1579-1582.

Pittendrigh, C. S. and Minis, D. H. (1964) The entrainment of circadian oscillations by light and their role as photoperiodic clocks. *The American Naturalist, 98*, 261-322.

Putterill, J., Robson, F., Lee, K., Simon, R., Coupland, G. (1995) The CONSTANS gene of Arabidopsis promotes flowering and encodes a protein showing similarities to zinc finger transcription factors. *Cell, 80*, 847-857.

Ray, P. M. and Alexander, W. E. (1966) Photoperiodic adaptation to latitude in Xanthium strumarium. *American Journal of Botany, 53*, 806.

Redei, G. P. (1962) Supervital Mutants of Arabidopsis. *Genetics, 47*, 443-460.

Redei, G. P., Acedo, G., Gavazzi, G. (1974) Flower differentiation in *Arabidopsis*. *Stadler Genet Symposium, 6*, 135-168.

Robson, F., Costa, M. M. R., Hepworth, S., Vizir, I., Pineiro, M., Reeves, P. H., et al. (2001) Functional importance of conserved domains in the flowering-time gene CONSTANS demonstrated by analysis of mutant alleles and transgenic plants. *Plant Journal, 28*, 619-631.

Roden, L. C., Song, H. R., Jackson, S., Morris, K., Carre, I. A. (2002) Floral responses to photoperiod are correlated with the timing of rhythmic expression relative to dawn and dusk in Arabidopsis. *Proceedings of the National Academy of Sciences of the United States of America, 99*, 13313-13318.

Samach, A., Onouchi, H., Gold, S. E., Ditta, G. S., Schwarz-Sommer, Z., Yanofsky, M. F., et al. (2000) Distinct roles of CONSTANS target genes in reproductive development of Arabidopsis. *Science, 288*, 1613-1616.

Schaffer, R., Ramsay, N., Samach, A., Corden, S., Putterill, J., Carre, I. A. et al. (1998) The late elongated hypocotyl mutation of Arabidopsis disrupts circadian rhythms and the photoperiodic control of flowering. *Cell, 93*, 1219-1229.

Schmid, M., Uhlenhaut, N. H., Godard, F., Demar, M., Bressan, R., Weigel, D., et al. (2003) Dissection of floral induction pathways using global expression analysis. *Development, 130*, 6001-6012.

Sheldon, C. C., Burn, J. E., Perez, P. P., Metzger, J., Edwards, J. A., Peacock, W. J., et al. (1999) The FLF MADS box gene: A repressor of flowering in Arabidopsis regulated by vernalisation and methylation. *Plant Cell, 11*, 445-458.

Sheldon, C. C., Rouse, D. T., Finnegan, E. J., Peacock, W. J., Dennis, E. S. (2000) The molecular basis of vernalisation: The central role of FLOWERING LOCUS C (FLC) *Proceedings of the National Academy of Sciences of the United States of America, 97*, 3753-3758.

Simpson, G. G. and Dean, C. (2002) Arabidopsis, the Rosetta stone of flowering time? *Science, 296*, 285-9.

Smith, H. and Whitelam, G. C. (1997) The shade avoidance syndrome: Multiple responses mediated by multiple phytochromes. *Plant Cell and Environment, 20*, 840-844.

Somers, D. E., Webb, A. A. R., Pearson, M., Kay, S. A. (1998) The short-period mutant, toc1-1, alters circadian clock regulation of multiple outputs throughout development in Arabidopsis thaliana. *Development, 125*, 485-494.

Strayer, C., Oyama, T., Schultz, T. F., Raman, R., Somers, D. E., Mas, P., et al. (2000) Cloning of the Arabidopsis clock gene TOC1, an autoregulatory response regulator homolog. *Science, 289*, 768-771.

Suarez-Lopez, P., Wheatley, K., Robson, F., Onouchi, H., Valverde, F., Coupland, G. (2001) CONSTANS mediates between the circadian clock and the control of flowering in Arabidopsis. *Nature, 410*, 1116-1120.

Sugano, S., Andronis, C., Green, R. M., Wang, Z.-Y., Tobin, E. M. (1998) Protein kinase CK2 interacts with and phosphorylates the Arabidopsis circadian clock-associated 1 protein. *Proceedings of the National Academy of Sciences of the United States of America, 95*, 11020-11025.

Sugano, S., Andronis, C., Ong, M. S., Green, R. M., Tobin, E. M. (1999) The protein kinase CK2 is involved in regulation of circadian rhythms in Arabidopsis. *Proceedings of the National Academy of Sciences of the United States of America, 96*, 12362-12366.

Tadege, M., Sheldon, C. C., Helliwell, C. A., Upadhyaya, N. M., Dennis, E. S., Peacock, W. J. (2003) Reciprocal control of flowering time by *OsSOC1* in transgenic *Arabidopsis* and by *FLC* in transgenic rice. *Plant Biotechnology Journal, 1*, 361-369.

Takada, S., and Goto, K. (2003) TERMINAL FLOWER2, an Arabidopsis homolog of HETEROCHROMATIN PROTEIN1, counteracts the activation of *FLOWERING LOCUS T* by CONSTANS in the vascular tissues of leaves to regulate flowering time. *Plant Cell, 15*, 2856-2865.

Takahashi, Y., Shomura, A., Sasaki, T., Yano, M. (2001) Hd6, a rice quantitative trait locus involved in photoperiod sensitivity, encodes the alpha subunit of protein kinase CK2. *Proceedings of the National Academy of Sciences of the United States of America, 98*, 7922-7927.

Thomas, B. and Vince-Prue, B., *Photoperiodism in Plants, 2nd ed.* San Diego, CA: Academic Press, 1997.

Valverde, F., Mouradov, A., Soppe, W., Ravenscroft, D., Samach, A., Coupland, G. (2004) Photoreceptor regulation of CONSTANS protein and the mechanism of photoperiodic flowering. *Science, 303*, 1003-1006.

Vince-Prue, D. (1994) "The duration of light and photoperiodic responses." In: *Photomorphogenesis in Plants - 2nd Edition,* Kendrick R.E. and Kronenberg G.H.M. (eds), Kluwer, Dordrect, 447-490.

Yano, M., Harushima, Y., Nagamura, Y., Kurata, N., Minobe, Y., Sasaki, T. (1997) Identification of quantitative trait loci controlling heading date in rice using a high-density linkage map. *Theoretical and Applied Genetics, 95*, 1025-1032.

Yano, M., Katayose, Y., Ashikari, M., Yamanouchi, U., Monna, L., Fuse, T., *et al.* (2000) Hd1, a major photoperiod sensitivity quantitative trait locus in rice, is closely related to the Arabidopsis flowering time gene CONSTANS. *Plant Cell, 12*, 2473-2483.

Yanovsky, M. J. and Kay, S. A. (2002) Molecular basis of seasonal time measurement in Arabidopsis. *Nature, 419*, 308-312.

Yanovsky, M. J. and Kay, S. A. (2003) Living by the calendar: how plants know when to flower. *Nat Rev Mol Cell Biol, 4*, 265-75.

Zeevaart, J. A. D. (1976) Physiology of flower formation. *Ann Rev Plant Physiol, 27*, 321-348.

Zeevaart, J. A. D. (1985) "Perilla.": In *CRC Handbook of Flowering, Vol 5*, Halevy A. H. (ed) CRC Press, Boca Raton, Fla., 239-252.

Chapter 28

COMMERCIAL APPLICATIONS OF PHOTOMORPHOGENESIS RESEARCH

Ganga Rao Davuluri[1] and Chris Bowler [1,2]
[1] *Cell Signalling Laboratory, Stazione Zoologica Anton Dohrn, Villa Comunale, 80121 Naples, Italy*
[2] *CNRS/ENS FRE2910, Département de Biologie, Ecole Normale Supérieure, 46 rue d'Ulm, 75230 Paris Cedex 05, France (e-mail: cbowler@biologie.ens.fr)*

1. INTRODUCTION

Higher plants utilize informational light signals to modify growth and development throughout their life cycle. Research over the last decades has steadily revealed the basic cellular mechanisms that are involved (Schäfer and Bowler, 2002), and an extraordinarily sophisticated picture has emerged of how plant responses are modulated by incident light.

The principal photoreceptors controlling developmental responses to light are the phytochromes (phy) and the cryptochromes (cry). Each are encoded by small multigene families, e.g., *Arabidopsis* contains 5 phytochrome genes (*PHYA-PHYE*) and at least 3 cryptochrome genes (*CRY1-3*) (Kleine *et al.*, 2003; Møller *et al.*, 2002; Nagy and Schäfer, 2002; see Chapters 7,11). In some cases, individual members have been found to mediate specific responses (e.g., the FR-HIR and VLFR for seed germination are controlled by phyA (Shinomura *et al.*, 1996)), in other cases one specific phytochrome acts together with one specific cryptochrome (e.g., phyA and cry2 appear to be the principal photoreceptors controlling photoperiod in *Arabidopsis* (Yanovsky and Kay, 2003)), and in other examples redundancy has been found between them (e.g., several can mediate the inhibition of hypocotyl elongation during de-etiolation (Reed *et al.*, 1994)).

The fine tuning of photoreceptor activity and signal transduction is a major determinant of plant fitness in natural environments. However, productivity in modern agriculture places different constraints on plant growth that are often incompatible with survival strategies that have evolved over millions of years of natural selection to favour ecological success. This Chapter illustrates how our understanding of light signalling in plants has been utilized in an agricultural context, by focusing on the modulation of key regulators of shade avoidance and fruit quality. In addition, an example of light-based biological engineering is provided, to illustrate how commercial applications can be derived from the understanding of basic molecular processes of light perception and signaling.

2. LIGHT-MEDIATED RESPONSES IN THE NATURAL ENVIRONMENT

The finding that plants are sensitive to photoperiod, or daylength, was one of the earliest discoveries that eventually lead to the discovery of phytochrome (Sage, 1992). Daylength is an environmental cue associated with seasonal changes, and measurement of photoperiod allows the precise control of the onset of flowering in many plant species. Independently of each other, Julien Tournois in Paris, France, and Garner and Allard in Virginia, USA, reported close to 100 years ago that some plants flower faster when photoperiod is short, that others flower when photoperiod is long, and that others can flower independently of daylength (Garner and Allard, 1920; Tournois, 1912). These three groups are now known, respectively, as short-day plants, long-day plants, and day-neutral plants.

Significant advances have now been made in understanding how photoreceptors are used by plants to control photoperiod measurement, using principally *Arabidopsis* as a long-day model and rice as a short-day model (reviewed in Yanovsky and Kay, 2003). What has emerged is the importance of the phyA and cry2 photoreceptors, at least in *Arabidopsis*, as well as an amazing convergence of results that implicate the CONSTANS protein as the central player for daylength measurement in all plants (see Chapters 26, 27 for further details).

Much of the information about the precise function of individual phytochromes has been obtained through studies of null mutants in *Arabidopsis*. Such approaches have allowed a rational interpretation of the different responses to ambient light in different *Arabidopsis* ecotypes, as well as in different species.

For example, Maloof *et al.* (2001) examined hypocotyl growth inhibition in response to white light in 141 *Arabidopsis* accessions from around the northern hemisphere and found a correlation between hypocotyl length and latitude. Hierarchical clustering analysis of response patterns to different light treatments revealed that some accessions behaved similarly to known photoreceptor mutants. Notably, an accession with reduced far red light sensitivity contained a change in the *PHYA* gene which resulted in altered biochemical properties of the photoreceptor. A cryptochrome variant was also found in an *Arabidopsis* accession by similar methodologies (El-Assal *et al.*, 2001).

The central importance of phytochrome in controlling the competitive interactions between individual plants and between different species has also been shown. In many natural environments, competition for light is central for plant success and survival to reproductive maturity. Shade tolerating species acclimate to low light conditions and are optimised to growth under the shade cast by other species. On the other hand, shade avoiders aim to dominate gaps in the vegetation.

Shade avoidance has been demonstrated to be a consequence of the phytochrome-mediated perception of far red light, which is enriched in shaded environments because of absorption of shorter light wavelengths, particularly in the red region, by the chlorophyll of neighbouring leaves (see Smith, 1994 for review). Perception of far red-enriched light in shade avoiding plants results in the activation of processes aimed at modifying plant architecture that are designed to outcompete neighbouring plants by increasing stem and petiole elongation growth and by modifying leaf number and expansion (see Chapter 22).

Perception of the red:far red light ratio can be accurately measured by phytochrome because different ratios establish different concentrations of the active Pfr form (Smith and Holmes, 1977). For many plant species, the elongation growth rate of the stems is inversely related to the established concentration of Pfr (usually expressed as the proportion of Pfr to total phytochrome), and the slope of this relationship (denoted the shade avoidance response sensitivity) varies between shade avoiders and shade tolerators (Morgan and Smith, 1979). Most notably, strong shade avoiders show a steep slope, whereas weak shade avoiders display a shallow slope. Furthermore, the intrinsic elongation growth rates are often correlated with the intrinsic shade sensitivity of a particular species. In addition to stem elongation, the morphology and size of leaves and their distribution along a plant can have a major influence in the competition for space in crowded communities by generating proximity signals that can be perceived by shade sensitive species via changes in phytochrome photoequilibria (Gilbert *et al.*, 2001; see Chapter 22).

Shade avoidance responses are major determinants of fitness in crowded communities, although they can be disadvantageous because they result in resource allocation to stem growth, at the cost of leaf growth and the development of storage and reproductive structures. Consequently, although shade avoidance is of major adaptive significance to plants in natural communities, it can be a serious problem for plant productivity in modern agriculture.

3. MANIPULATION OF LIGHT RESPONSES IN AGRICULTURE

The findings described above demonstrate that plant photoreceptors play a key role in modulating plant development for adaptation to a specific environmental niche. Consequently, the flexibility in developmental programming that results from the fine-tuning of the phytochromes and cryptochromes must have been an important driving force during angiosperm evolution.

Modern agricultural practices place different constraints on plant growth that have often not been selected for during plant evolution. For example, shade avoiding species grown in dense monocultures display yield penalties to reproductive organs such as fruits and seeds, due to increased resources being allocated to elongation growth. However, by modifying light perception and signalling, plant responses can be optimised for maximizing agricultural productivity. Several examples where this has been achieved are described below.

3.1 Modulation of day length perception

Following Tournois, Garner and Allard's discovery in the early 1900's that many plants important for agriculture have critical daylength requirements for flowering and fruiting (Garner and Allard, 1920; Tournois, 1912), the optimisation of photoperiod became a standard practice. For example, new crops were grown only in areas and during seasons with the appropriate length of day, and greenhouse-grown crops were often supplemented with artificial light to extend photoperiod or were covered with black cloth to shorten it. Nowadays, the artificial manipulation of photoperiod is routinely applied to an enormous range of economically important plant species, from crop plants to flowering plants and trees. Without doubt, it has brought billions of dollars of benefits to farmers, horticulturists, and breeders.

Furthermore, utilization of knowledge derived from the physiology-based studies of phytochrome action performed largely in the first half of the 20th century have had dramatic impacts on the flower industry. For example, pulsed light is often used to substitute constant light, and a single pulse of light during the night can be used to inhibit flowering in some species. These practices exploit some of the now well known characteristics of the phytochromes; in the above two examples, that phytochrome responses have an escape time (see Chapter 2) and that phytochrome concentrations build up during the night but can be reduced significantly by a single light pulse (see Chapter 8). Again, such practices permit enormous cost savings to growers.

The molecular mechanisms controlling phytochrome behaviour are now understood to some extent (Møller et al., 2002; Nagy and Schäfer, 2002)(see also Chapters 4, 17, and 19). In principle, it has therefore become feasible to modify daylength perception by the judicious manipulation of key regulators. This could allow a high level control of plant growth and could optimise it to the particular conditions available. Many examples of manipulation of key photoperiod regulators have been reported (e.g., CONSTANS and GIGANTEA (Hayama et al., 2003; Yano et al., 2000)) although to our knowledge nothing has yet been reported in a biotechnological context, in spite of its feasibility. Overexpression of the *Arabidopsis CONSTANS* gene in potato results in a delayed tuberization phenotype, suggestive of a function for CONSTANS in the photoperiodic control of tuber formation (Martinez-Garcia et al., 2002).

3.2 Modulation of shade avoidance responses

Competition for light energy is a very important factor controlling plant architecture. As described above, when plants are grown in close proximity the shade avoidance response is activated, manifested by a dramatic increase in extension growth at the expense of leaf growth, storage organ production, and development of reproductive organs. These architectural changes are a serious problem for agriculture, because they reduce the amount of useful material that can be harvested. Concomitantly, the production of large quantities of unwanted material in a crop carries a cost in terms of wasted nutrients, which must consequently be supplied at high levels. Furthermore, it has been found that competition from weeds is directly attributable to the induction of shade avoidance responses in crop plants rather than to competition for resources (Smith, 1994).

Studies in phytochrome-deficient mutants have demonstrated convincingly that shade avoidance responses are mediated by light-stable Type II phytochromes, e.g., by phyB in *Arabidopsis* (Quail et al., 1995). This is primarily because the far red-enriched light environments in shaded conditions cause a reduction in PfrB:PrB ratios, which results in increased stem elongation and other responses associated with shade avoidance (see Chapter 22).

Harry Smith and colleagues have utilized this information to generate transgenic plants in which the shade avoidance response has effectively been disabled. This was achieved in transgenic tobacco plants overexpressing a *PHYA* cDNA from oat (Robson *et al.*, 1996). In contrast to the endogenous phyA, the transgene-derived oat phyA remained at high levels, resulting in considerably higher levels of phyA in extracts from light-grown transgenic plants compared to wild-type plants (Robson *et al.*, 1996; Smith, 1994). Presumably, the processes regulating the light-dependent degradation of endogenous phyA were not being utilized to control levels of the exogenous phyA.

Phenotypically, transgenic tobacco plants overexpressing oat *PHYA* did not show the increased stem elongation in far red-enriched white light (McCormac *et al.*, 1992a; McCormac *et al.*, 1992b), one of the characteristic features of the shade avoidance response. This remarkable result suggested that phyA must normally be removed in mature plants, in order that the Type II phytochromes (principally phyB) can activate the shade avoidance response. Furthermore, it suggested a means whereby the shade avoidance response could be effectively inactivated.

Field trials of these tobacco plants grown at different densities indeed demonstrated that the densely grown plants did not display the typical shade avoidance responses of wild-type plants (Figure 1). Furthermore, the plants elongated even more slowly in dense plantings compared with sparse plantings (Robson *et al.*, 1996), which could also be attributed to the substitution of the normal phyB responses by the transgene-derived phyA. These changes in light perception resulted in significant increases in the harvest index of leaves from the transgenic plants compared to the wild-type plants.

Figure 1. Field trial of transgenic tobacco plants overexpressing the oat *PHYA* gene. Control plants are shown on the left, PHYA-overexpressing plants are shown on the right. Disabling of the shade avoidance response is clearly apparent in the transgenic PHYA-overexpressing plants. The photograph was taken in 1994 in Rothamstead, UK, and was kindly provided by Harry Smith (University of Nottingham, UK).

The results described above demonstrate the overriding importance of phytochrome-mediated shade avoidance compared with other potential factors associated with plant performance at high densities, at least in the monoculturing situation that typifies current agricultural practices. The transgenic suppression of shade avoidance opens up possibilities for the conditional modification of architecture based upon a plant's intrinsic ability to measure plant density. This can potentially lead to improvements in crop yield, and could be used to control density-dependent dwarfing, which could reduce problems associated with lodging and nutrient utilization.

3.3 Modulation of fruit ripening

The influence of light on the development of climacteric fruits such as tomato is well known, and in the 1950s an important role for Type II phytochromes was demonstrated in the process (Piringer and Heinze, 1954). Subsequently, the importance of individual photoreceptors was revealed by the identification and study of photoreceptor mutants in single, double, and triple combinations (Weller *et al.*, 2000). For example, the loss of phyB2 in a *phyAphyB1* background results in a striking reduction in chlorophyll content of immature fruits as well as a marked increase in truss length as a result of increased distance between fruits on the inflorescence axis. The loss of cry1 in this same background also results in a reduction of chlorophyll in immature fruits but has no effect on truss architecture (Figure 2). Conversely, the RNAi-mediated suppression of HY5, a transcription factor involved in the activation of light-responsive genes (see Chapters 4, 17, and 19), resulted in reduced fruit pigmentation (Liu *et al.*, 2004).

One could therefore speculate that an increase in photoreceptor activity or light-dependent signalling in fruits would lead to enhanced pigmentation. This is of particular commercial interest because tomato fruit and derived products (e.g., ketchups, juices, soups, and sauces) are an important source of vitamins and carotenoids and are consumed by millions of people each day. Consequently, tomato cultivation and processing are billion dollar industries both in the USA and in Europe. The major carotenoid pigment found in ripe tomato fruits is lycopene, which is a potential cancer chemopreventative, particularly for prostate cancer (Kucuk, 2002; Miller *et al.*, 2002). Lycopene protects against oxidative damage by quenching photosensitisers, interacting with singlet oxygen (Krinksy, 1994) and scavenging peroxy radicals (Conn *et al.*, 1992). Although lycopene is a potent antioxidant there is evidence that it may also act in other ways to protect from cancer (Heber and Lu, 2002).

Beta-carotene is derived from lycopene and is also present in high amounts in ripe tomato fruits. It is a major provitamin A in the human diet and its deficiency can cause xerophthalmia and blindness. UNICEF has stated that improved vitamin A nutrition could prevent up to 2 million deaths annually among children aged between one and four years (Humphrey *et al.*, 1992).

A clear demonstration that the manipulation of light signalling pathways can enhance tomato fruit carotenoid levels is represented by the tomato *high pigment* (*hp*) mutants, which are characterized by dark green immature fruits that develop into mature fruits containing more than twice the normal levels of lycopene and beta-carotene (Kendrick *et al.*, 1994). Even in phy- and cry-deficient backgrounds, *hp* mutations can dramatically sensitise residual photoreceptor responses controlling fruit pigmentation, as illustrated in Figure 2. Mutations in the tomato homologues of the *DDB1* and *DET1* genes were found to be responsible the *hp-1* and *hp-2* mutant phenotypes, respectively (Liu *et al.*, 2004; Mustilli *et al.*, 1999). Previous studies in *Arabidopsis* had indicated that these genes were important negative regulators of light responses (Chory *et al.*, 1989; Pepper *et al.*, 1994; Schroeder *et al.*, 2002).

In spite of the enhanced nutritional quality of *hp* mutant fruits, they have never been commercialised because of the collateral negative effects that the mutation has on the vegetative parts of the plants, resulting in plants with bushier growth habits and reduced yields. Notwithstanding, the combination of several mutant alleles (e.g., *phy*, *cry*, *hp*) can alleviate some of these effects (for example note that in Figure 2 the elongated truss phenotype of multiple *phy* mutants is not affected by the *hp* mutation, even though *hp* is clearly epistatic to *phy* mutants in terms of fruit pigmentation phenotypes).

Figure 2. Fruit truss phenotypes from a range of tomato photomorphogenic mutants. From left to right: (1) phyAphyB1, (2) phyAphyB1hp-1, (3) phyAphyB1phyB2, (4) phyAphyB1phyB2hp-1, (5) phyAphyB1cry1, (6) phyAphyB1cry1hp-1. Note that the triple phyAphyB1phyB2 mutant displays elongated truss phenotypes whereas the phyAphyB1cry1 mutant does not. However, both show reduced fruit pigmentation. In each mutant background the hp mutant is epistatic to the photoreceptor mutant phenotypes for fruit pigmentation, but not for the defects in truss architecture. Photograph kindly provided by Ageeth van Tuinen and Dick Kendrick (Wageningen University, NL).

Negative pleiotropic effects are likely to be a general problem when attempting to manipulate such general processes as light signalling. Indeed, overexpression of oat phytochrome in tomato increased pigmentation of fruits, but also increased dwarfness (Boylan and Quail, 1989). A similar phenotype was obtained by suppression of the *DET1* gene in tomato by post-transcriptional gene silencing (PTGS) (Davuluri *et al.*, 2004). In some cases, *DET1* gene silencing was so effective that the plants actually died, analogous to null *det1* mutants in *Arabidopsis* (Pepper *et al.*, 1994). Conversely, Liu *et al.* (2004) reported that ectopic expression of another negative regulator of light responses, the E3-ubiquitin ligase COP1 (see Chapters 4, 17-19), also resulted in enhanced fruit pigmentation and exaggerated photomorphogenesis. These dramatic phenotypes are presumably a reflection of the essential functions of these proteins during plant development due to their roles in the control of photomorphogenesis.

The above examples demonstrate very clearly that manipulation of photoreceptors or of downstream signalling intermediates in transgenic plants results in highly pleiotropic phenotypes. One approach to overcome this could be to employ tissue-specific promoters. This has been attempted by us using tomato fruit-specific promoters to regulate expression of a transgene designed to suppress endogenous *DET1* expression by RNAi (Davuluri and Bowler, manuscript in preparation). A total of three promoters were used, all of which had been found previously to be expressed early during fruit development and not elsewhere in the plant (Pear *et al.*, 1989; Santino *et al.*, 1997). Furthermore, because inverted repeat constructs have been found to greatly enhance gene silencing in both plants and animals (Smith *et al.*, 2000), we employed such a strategy with the tomato *DET1* gene.

Transgenic plants containing these constructs indeed generated highly pigmented fruits (Figure 3), which matured to contain high levels of carotenoids. Importantly, the transgenic plants showed at worst only negligible yield losses and other parameters to assess tomato fruit quality did not reveal any negative collateral effects.

Figure 3. Immature fruit phenotype caused by the fruit-specific silencing of the tomato DET1 gene. The left panel shows fruit from a wild-type plant and the right panel shows fruit from a plant in which DET1 gene expression has been suppressed specifically in the fruits by RNAi.

These results therefore demonstrate that the judicial manipulation of photoreceptor signalling pathways can be used to improve characteristics of significant commercial interest. This example also demonstrates the utility of targeting key regulatory genes rather than genes encoding biosynthetic enzymes, e.g., for carotenoid biosynthesis, which had been previously reported using a bacterial phytoene synthase gene in tomato (Fraser *et al.*, 2002). In this previous example, the effects were more modest, because only one enzyme is modified, whereas loss of DET1 activity results in the upregulation of the whole biosynthetic pathway.

4. LIGHT-BASED BIOLOGICAL ENGINEERING

Regulatable transgene systems have become widely utilized in biomedical and agricultural biotechnology because they provide easily controllable approaches for the conditional regulation of gene expression (Gatz, 1997; Lewandowski, 2001). The most commonly used examples employ chemically-inducible systems, e.g., the tetracycline-based Tet repressor and the dexamethasone-regulated glucocorticoid receptor (Ohgishi and Aoyama, 2002). Photomorphogenesis research has lead to the identification of a wide range of light-regulated systems, although there are only a few examples for which biotechnological applications have been explored.

Most importantly, Peter Quail's laboratory has developed a light-regulated yeast two-hybrid assay by fusing the chromophore-binding region of phyB to the GAL4 DNA binding domain, and the Pfr-activated PIF3 transcription factor to the GAL4 activation domain (Shimizu-Sato *et al.*, 2002). In this system, GAL4-regulated promoters can be activated conditionally and precisely by a red light pulse and repressed by a far red light pulse. The system is dependent upon addition of phycocyanibilin as chromophore, which can be taken up by yeast cells. This example represents one of the few examples whereby a non-chemical physical agent (light) can precisely regulate a molecular target within eukaryotic cells. Furthermore, it allows the reversible induction and repression of such a target, thereby acting as a bistable molecular switch. By focusing a fibre optic light source on cells of interest, the system could also be used for precise spatial and temporal control.

In principle, any two-hybrid-like system can be adapted to incorporate this technology, and applications could include studies of association and disassociation rates between proteins of interest. It will therefore be interesting to see how this phyB-PIF3-based system is exploited in future years.

Another example of light-based biological engineering proposes the use of the phytochrome photoreceptor itself as a fluorescent reporter protein, similar to the widely used Green Fluorescent Protein (GFP)-based reporters. For this application, an orange fluorescent adduct of phytochrome was generated by incubation of recombinant phytochrome apoprotein with phycoerythrobilin, the linear tetrapyrrole precursor of phycoerythrin (Murphy and Lagarias, 1997). Advantages of such a reporter over current GFP-based systems are that the longer light wavelengths used for excitation are less cytotoxic, and that many cells have much lower autofluorescence at orange-red wavelengths, thus providing higher signal to noise ratios. However, this technology has not become widely utilized, perhaps because of the limited availability of the chromophore substrate or because of the difficulties with it being taken up by the cells under study.

5. CONCLUSIONS AND PERSPECTIVES

The brief summary of results presented here shows that it is possible to modulate photoreceptors and photoreceptor signalling pathways in order to obtain plants with improved characteristics of agronomic interest. Even though photoperceptory mechanisms are of importance at all stages of plant development, it is possible to manipulate them because of their inherent plasticity from one plant to another, which is a result of adaptation to different environments that has been selected for over hundreds of millions of years of evolution. However, many of the components that regulate photoperception have essential functions, meaning that their activity should be controlled judicially using, for example, tissue- and stage-specific promoters. Many such promoters are now available, as are methods for the efficient overexpression or inhibition of gene expression. Dominant-negative and dominant-positive approaches can also be employed, if there is some knowledge about function of the protein.

Furthermore, such strategies have the advantage of being based upon plant-derived sequences. This is of importance, because there is currently a great deal of public concern about the risks to human health and the environment from the cultivation and consumption of products from transgenic plants that contain genes derived from organisms other than plants. On the other hand, the exclusive use of plant genes to improve plant characteristics for agriculture is likely to be more acceptable by the general public, particularly if aimed at the generation of products that are beneficial for human health, such as increased carotenoid content in tomatoes and rice (Beyer et al., 2000). Nonetheless, given the current resistance in the public domain, particularly in Europe, to the cultivation of transgenic plants of any kind, it is difficult to envisage the commercialisation even of products containing 100% plant genes, at least before 2010.

An alternative approach is to exploit the knowledge we now have about photoperception mechanisms in plants to select for desired phenotypes by traditional breeding. One example provided here shows the possibilities of combining photoreceptor mutants to obtain novel phenotypes in the tomato (Figure 2). Such approaches could also exploit natural variation in light responses and look for specific gene combinations. Although in some cases it may not be possible to obtain a desired phenotype, e.g., the fruit-specific suppression of DET1 activity in tomato can probably only be achieved using transgenic approaches, in many cases it is likely to be possible.

6. REFERENCES

Boylan, M. T. and Quail, P. H. (1989) Oat phytochrome is biologically active in transgenic tomatoes. *Plant Cell, 1*, 765-773.
Chory, J., Peto, C., Feinbaum, R., Pratt, L., Ausubel, F. (1989) *Arabidopsis* thaliana mutant that develops as a light-grown plant in the absence of light. *Cell, 58*, 991-999.
Conn, P. F., Lambert, C., Land, E. J., Schalch, W., Truscott, T. G. (1992) Carotene-oxygen radical interactions. *Free Radic. Res. Commun., 16*, 401-408.
Davuluri, G. R., van Tuinen, A., Mustilli, A. C., Manfredonia, A., Newman, R., Burgess, D., *et al.* (2004) Manipulation of *DET1* expression in tomato results in photomorphogenic phenotypes caused by post-transcriptional gene silencing. *Plant J, 40*, 344-354.
El-Assal, S. E.-D., Alonso-Blanco, C., Peeters, A. J. M., Raz, V., Koornneef, M. (2001) A QTL for flowering time in *Arabidopsis* reveals a novel allele of *CRY2. Nature Genetics, 29*, 435-440.
Fraser, P. D., Romer, S., Shipton, C. A., Mills, P. B., Kiano, J. W., Misawa, N., *et al.* (2002) Evaluation of transgenic tomato plants expressing an additional phytoene synthase in a fruit-specific manner. *Proc Natl Acad Sci USA, 99*, 1092-1097.
Garner, W. W., Allard, H. A. (1920) Effect of the relative length of day and night and other factors of the environment on growth and reproduction in plants. *J Agric Res, 18*, 553-606.
Gatz, C. (1997) Chemical control of gene expression. *Ann Rev Plant Phys Plant Mol Biol, 48*, 89-108.
Gilbert, I. G., Jarvis, P. G., Smith, H. (2001) Proximity signal and shade avoidance differences between early and late successional trees. *Nature, 411*, 792-795.
Hayama, R., Yokoi, S., Tamaki, S., Yano, M., Shimamoto, K. (2003) Adaptation of photoperiodic control pathways produces short-day flowering in rice. *Nature, 422*, 719-722.
Heber, D. and Lu, Q. Y. (2002) Overview of mechanisms of action of lycopene. *Exp. Biol. Med., 227*, 920-923.
Humphrey, J. H., West, K. P. J., Sommer, A. (1992) Vitamin A deficiency and attributable mortality among under-5-year-olds. *Bull World Health Organ, 70*, 225-232.
Kendrick, R. E., Peters, J. L., Kerckhoffs, L. H. J., van Tuinen, A., Koornneef, M. (1994) Photomorphogenic mutants of tomato. *Biochem Soc Symp, 60*, 249-256.
Kleine, T., Lockhart, P., Batschauer, A. (2003) An *Arabidopsis* protein closely related to Synechocystis cryptochrome is targeted to organelles. *Plant J, 35*, 93-103.
Krinksky, N. I. (1994) The biological properties of carotenoids. *Pure Appl Chem, 66*, 1003-1010.
Kucuk, O. (2002) Cancer chemoprevention. *Cancer Metastasis Rev, 21*, 189-197.
Lewandowski, M. (2001) Conditional control of gene expression in the mouse. *Nature Rev Genet, 2*, 743 - 755.
Liu, Y., Roof, S., Ye, Z., Barry, C., van Tuinen, A., Vrebelov, J., *et al.* (2004). Manipulation of light signal transduction as a means of modifying fruit nutritional quality in tomato. *Proc Natl Acad Sci USA, 101*, 9897-9902.
Maloof, J. N., Borevitz, J. O., Dabi, T., Lutes, J., Nehring, R. B., Redfern, J. L., *et al.* (2001) Natural variation in light sensitivity of *Arabidopsis. Nature Genetics, 29*, 441-446.
Martinez-Garcia, J. F., Virgos-Soler, A., Prat, S. (2002) Control of photoperiod-regulated tuberization in potato by the *Arabidopsis* flowering-time gene CONSTANS. *Proc Natl Acad Sci USA., 99*, 15211-15216.
McCormac, A. C., Cherry, J. R., Hershey, H. P., Vierstra, R. D., Smith, H. (1992a) Photoresponses of transgenic tobacco plants expressing an oat phytochrome gene. *Planta, 185*, 162-170.
McCormac, A. C., Whitelam, G. C., Smith, H. (1992b) Light-grown plants of transgenic tobacco expressing an introduced oat phytochrome A gene under the control of a constitutive viral promoter exhibit persistent growth inhibition by far-red light. *Planta, 188*, 173-181.
Miller, E. C., Hadley, C. W., Schwartz, S. J., Erdman, J. W., Boileau, T. M. W., Clinton, S. K. (2002) Lycopene, tomato products, and prostate cancer prevention. Have we established causality? *Pure Appl Chem, 74*, 1435-1441.
Møller, S. G., Ingles, P. J., Whitelam, G. C. (2002) The cell biology of phytochrome signalling. *New Phytol, 154*, 553-590.

Morgan, D. C. and Smith, H. (1979) A systematic relationship between phytochrome-controlled development and species habitat for plants grown in simulated natural radiation. *Planta, 145*, 253-259.

Murphy, J. T. and Lagarias, J. C. (1997) The phytofluors: a new class of fluorescent protein probes. *Curr Biol, 7*, 870-876.

Mustilli, A. C., Fenzi, F., Ciliento, R., Alfano, F., Bowler, C. (1999) Phenotype of the tomato *high pigment-2* mutant is caused by a mutation in the tomato homolog of *DEETIOLATED1*. *Plant Cell, 11*, 145-158.

Nagy, F. and Schäfer, E. (2002) Phytochromes control photomorphogenesis by differentially regulated, interacting signaling pathways in higher plants. *Ann Rev Plant Biol, 53*, 329-355.

Ohgishi, M. and Aoyama, T. (2002) "Inducible gene expression in plants." In *Molecular Plant Biology: A Practical Approach* (Vol. 2) Gilmartin P. and Bowler C. (eds), Oxford University Press, Oxford, 97-107.

Pear, J. R., Ridge, N., Rasmussen, R., Rose, R. E., Houck, C. M. (1989) Isolation and characterization of a fruit-specific cDNA and the corresponding genomic clone from tomato. *Plant Mol Biol, 13*, 639-651.

Pepper, A., Delaney, T., Washburn, T., Poole, D., Chory, J. (1994) *DET1*, a negative regulator of light-mediated development and gene expression in *Arabidopsis*, encodes a novel nuclear-localized protein. *Cell, 78*, 109-116.

Piringer, A. A. and Heinze, P. H. (1954) Effect of light on the formation of a pigment in the tomato fruit cuticle. *Plant Physiol, 29*, 467-472.

Quail, P. H., Boylan, M. T., Parks, B. M., Short, T. W., Xu, Y., Wagner, D. (1995) Phytochromes: photosensory perception and signal transduction. *Science, 268*, 675-680.

Reed, J. W., Nagatani, A., Elich, T. D., Fagan, M., Chory, J. (1994) Phytochrome A and phytochrome B have overlapping but distinct functions in *Arabidopsis* development. *Plant Physiol, 104*, 1139-1149.

Robson, P. R. H., McCormac, A. C., Irvine, A. S., Smith, H. (1996) Genetic engineering of harvest index in tobacco through overexpression of a phytochrome gene. *Nature Biotechnol, 14*, 995-998.

Sage, L. C. *Pigment of the imagination: a history of phytochrome research*. San Diego: Academic Press, Inc., 1992.

Santino, C. G., Stanford, G. L., Conner, T. W. (1997) Developmental and transgenic analysis of two tomato fruit enhanced genes. *Plant Mol Biol, 33*, 405-416.

Schäfer, E. and Bowler, C. (2002) Phytochrome-mediated photoperception and signal transduction in higher plants. *EMBO Rep, 3*, 1042-1048.

Schroeder, D. F., Gahrtz, M., Maxwell, B. B., Cook, R. K., Kan, J. M., Alonso, J. M., *et al.* (2002). De-Etiolated 1 and Damaged DNA Binding protein 1 interact to regulate Arabidopsis photomorphogenesis. *Curr Biol, 12*, 1462-1472.

Shimizu-Sato, S., Huq, E., Tepperman, J. M., and Quail, P. H. (2002) A light-switchable gene promoter system. *Nature Biotechnol, 20*, 1041-1044.

Shinomura, T., Nagatani, A., Hanzawa, H., Kubota, M., Watanabe, M., Furuya, M. (1996) Action spectra for phytochrome A- and B-specific photoinduction of seed germination in *Arabidopsis thaliana*. *Proc Natl Acad Sci USA, 93*, 8129-8133.

Smith, H. (1994) Phytochrome transgenics: functional, ecological and biotechnological applications. *Sem Cell Biol, 5*, 315-325.

Smith, H. and Holmes, M. G. (1977) The function of phytochrome in the natural environment. III. Measurement and calculation of phytochrome equilibrium. *Photochem Photobiol, 25*, 547-550.

Smith, N. A., Singh, S. P., Wang, M. B., Stoutjesdijk, P. A., Green, A. G., Waterhouse, P. M. (2000) Total silencing by intron-spliced hairpin RNAs. *Nature, 407*, 319-320.

Tournois, J. (1912) Influence de la lumiére sur la floraison du Houblon japonais et di Chanvre. *C. R. Hebd Seances Acad Sci, 155*, 297-300.

Weller, J. L., Schreuder, M. E., Smith, H., Koornneef, M., Kendrick, R. E. (2000) Physiological interactions of phytochromes A, B1 and B2 in the control of development in tomato. *Plant J, 24*, 345-356.

Yano, M., Katayose, Y., Ashikari, M., Yamanouchi, U., Monna, L., Fuse, T., *et al.* (2000) Hd1, a major photoperiod sensitivity quantitative trait locus in rice, is closely related to the *Arabidopsis* flowering time gene *CONSTANS*. *Plant Cell, 12*, 2473-2484.

Yanovsky, M. J. and Kay, S. A. (2003) Living by the calendar: how plants know when to flower. *Nature Rev Mol Cell Biol, 4*, 265-275.

Ye, X., Al-Babili, S., Kloti, A., Zhang, J., Lucca, P., Beyer, P., Potrykus, I. (2000) Engineering the provitamin A (b-Carotene) biosynthetic pathway into (carotenoid-free) rice endosperm. *Science, 287*, 303-305.

CHAPTER 29

PHOTOMORPHOGENESIS – WHERE NOW?

Harry Smith
Division of Plant Sciences, University of Nottingham, Sutton Bonington Campus, Loughborough, LE12 5RD, United Kingdom (e-mail: Harry.Smith@Nottingham.ac.uk)
Department of Biology, University of Leicester

WHERE ARE WE GOING, DAD?

Having agreed to attempt the impossible, to write about the future, this has to be a purely personal account. I have therefore decided to take a somewhat light-hearted approach. Most of those reading this will be familiar with one of the truly exasperating experiences of life – answering questions from the child in the back seat whilst driving to the beach. Not only does the endless repetition drive one nuts, the immense difficulty of giving coherent answers to searching questions can seriously damage one's self-esteem, resulting in pathetic fobbing-off, instantly recognised as such. Readers are invited to search for the evasive answers in this chapter, as the various topics I shall cover will be introduced by a typical question from a bored ten-year old.

Looking ahead is always a risky undertaking, and nowhere is this more true than in science. Scientific advance comes from individual creativity married to opportunistic entrepreneurship, generating innovation. None of this can easily be predicted – as Joshua Lederberg said: "You rarely find the most important things by deliberately looking for them". It would be foolhardy to attempt to predict where photomorphogenesis should go. Instead, I shall try to examine where we are now, how we arrived here, and whether history offers us any guidelines for future exploration. I would be on shaky ground if I attempted to cover the cryptochromes and phototropins in the same depth as the phytochromes, so any detail to be found here will be coloured red and far-red. An article of this nature does not lend itself to exhaustive citations, and those given are consequently highly selective.

WHERE ARE WE NOW, DAD?

In science, as on a vacation trip, where one travels to depends to a large extent upon where one starts from. In the context of photomorphogenesis, the last half-Century has witnessed astonishing advances, but visualising the way ahead is no easier now than it was then. Perhaps this is true for all scientific endeavours, as opposed to technological enterprises, where the end-point is clear from the outset. 'Blue-skies' science advances unpredictably, with long, more-or-less stagnant periods interspersed by brief moments of explosive activity generated by a new idea or, more typically, a new technology. Photomorphogenesis research has grown in exactly this way. Those working feverishly to complete their PhD programme, write the next paper or struggle for the next grant may not see it as such, but the history of photomorphogenesis has been characterised by persistent confusion punctuated by flashes of insight.

Photomorphogenesis effectively started about 60 years ago with the brilliant physiological investigations of Sterling Hendricks, Harry Borthwick and their several colleagues at Beltsville. Anyone reading those papers of the 1940s, 1950s and 1960s cannot fail to be impressed by the intellectual rigour applied to the design of experiments, the interpretation of data and the economy of language, a quality I fear I myself have not learnt. Of course, it could be argued that photomorphogenesis had started years before, and the discoveries of Garner and Allard in photoperiodic control of flowering are often cited as being a starting point; indeed, Hendricks himself regarded the Garner and Allard papers as "the touchstone for studies on photomorphogenesis". One could go even further back and point to the 19th Century observations of Darwin, Caspari, Sachs and many others. We all "stand on the shoulders of giants", as Isaac Newton famously wrote to Robert Hooke, so why should the Beltsville papers be regarded as the real starting point for photomorphogenesis?

The Beltsville group unified the study of plant responses to red light by elaborating precise action spectra for several different phenomena – flowering, de-etiolation and germination. From this they knew that red light was the responsible signal. Following earlier work on lettuce seed germination they presented what is now a justly famous paper (Borthwick *et al.*, 1952) demonstrating the sequential reversibility of brief treatments of red and far-red light on germination. From this deceptively simple observation was born the hypothesis of a photoreceptor capable of existing in two forms, one active and generated by red light from the other, inactive, form itself regenerated from the active form by far-red light. The unifying power of this concept is demonstrated by its survival, unchanged, through 50 years of intensive investigation. Not only is the model persuasive, it is also impressively elegant. The reader may ask 'Why go over all this – we know it all'. First, good stories can bear endless repetition; more seriously, the incisive work of the Beltsville group gave us the basis on which all further investigations have been built. Everything before had essentially been phenomenology; from then onwards a systematic analytical approach became possible. Moreover, do all those engaged in photomorphogenesis research *really* know this story? It is depressingly common these days to find papers in otherwise reputable journals publishing data based on inadequate light sources, or with poor or non-existent quantification of the radiation parameters and, frankly, a very poor understanding of photobiological principles. The Hendricks group's papers should be required reading for all entrants to photomorphogenesis research, if only to learn how to do experiments.

One of the objectives of the Hendricks team was to isolate and purify the protein they had christened phytochrome. Spoken of as a *pigment of the imagination* (used later as the title of an excellent historical account by Linda Sage [1997]) or *a red/far-red herring*, phytochrome was regarded with a deal of humorous scepticism at the time. But when it was purified, admittedly at first in a degraded state, and shown to be blue or blue-green, depending on whether it had previously received far-red or red light, then Beltsville had the last laugh. Isolation and purification of intact phytochrome (by other groups), and the chemical characterization of the chromophore, were major achievements of the 'pre-molecular' age.

Regarding the chromophore, a quite amazing advance by the Beltsville group was Hendricks' percipience in suggesting, from little more than the similarity between the action spectra of phytochrome-mediated phenomena and the absorption spectra of the Cyanophyceæ pigments that the phytochrome chromophore would probably be a bilin. This prediction was realised in later studies and was followed by increases in understanding of the mechanisms of photoconversion, leading ultimately to a satisfying model of separate intermediary pathways in the Pr→Pfr and the Pfr→Pr directions.

The cloning of a gene encoding phytochrome A by the Quail group in the mid-1980s (Hershey *et al.*, 1984) opened the way to the next major development, as it led directly to the realisation that phytochromes are plural. This idea had been floated earlier on both physiological and biochemical grounds, but it required molecular demonstration to achieve recognition. Once the molecular biologists got their hands on the Arabidopsis *PHY* genes the (short) answer to many unresolved problems became obvious. Several phytochromes could explain the different modes of action that seemed incompatible with only one. All this is, of course, well known, but the effect at the time was truly revelatory. Apart from the trivial triumph of debunking the contorted models of the previous two decades, the molecular genetic study of the *PHY* genes has led to massive increases in our knowledge and understanding. The cloning of the phytochrome genes must therefore be seen as the next big unifying development.

Similar unifications of ideas can be seen with regard to responses to blue light. Blue light had been known to affect biological processes since the 19th Century, and indeed accurate action spectra were generated in the first half of the 20th Century for organisms in many different phyla. The spectra sparked controversy over the chemical nature of the photoreceptor(s) with much argument over the competing candidatures of carotenoid and flavin. The cloning of the cryptochrome-1 gene by the Cashmore group (Ahmad and Cashmore, 1993) resolved all this and unified not only studies on blue-light effects on plant growth but had enormous significance for research into the biological clock in both plants and animals. Phototropism in plants was similarly unified by the discovery of the phototropin family of photoreceptors, after prolonged and determined work by the Briggs group (Christie *et al.,*, 1999).

SO WHAT, DAD?

One obvious conclusion from all this is that scientific advance depends on technological innovation. Wherever one sees evidence of 'breakthrough', to use a corny old expression, one can discern the application of a new and powerful technique. In the turning-points mentioned above, this general rule is obvious; the introduction of precise and powerful light sources, the generation of action spectra and the techniques of gene cloning underpinned the advances made. The more recent spectacular progress of the last decade in understanding the mechanisms of photoreceptor action have again been made possible by new technology: reverse genetics, two-component analysis, intra-cell imaging and micro-array expression analysis are just a few.

In science as a whole, and biological sciences in particular, answers generate questions. The fascination of science is not, as the layman generally believes, the satisfaction derived from solving a puzzle and realising the solution. Even though such flashes of enlightenment can be momentarily overwhelming, what really drives a true scientist is the next big question. It may be constructive to consider, for photomorphogenesis, what answers we currently have, and what questions these answers generate. To do this at a detailed level would be tedious – even the child in the back wants a crisp answer, not an argument. So, what are the bigger 'answers' currently in our grasp?

First, the power of various forms of genetic analysis based on mutant selection has yielded an impressive array of conclusions. Plants carrying mutations in the genes encoding one or more of the five phytochromes in Arabidopsis tell us much about the independent, interactive and redundant rôles of the photoreceptors. Similar studies of mutations in phototropin and cryptochrome genes give similar pictures of independent but overlapping roles. One thing that comes out of this is that the phytochromes, whilst essential to survival, are not essential for life itself. This question – are phytochromes necessary? – has been around for decades, but at this moment there are probably several laboratories carefully nurturing sickly Arabidopsis seedlings lacking all five phytochromes. Such plants will be capable of life, but not of independent existence. Phytochrome-mediated photomorphogenesis in plants is analogous to vision in humans – not essential to life, but essential to living. Perhaps the more interesting conclusion relates to the genetic redundancy of the phytochromes. The evolutionary basis of such a 'belt and braces' system is to say the least intriguing.

Sequence comparisons have been extremely interesting and surprising conclusions have emerged regarding the early evolution of the phytochromes in photosynthetic bacteria. How did those genes get from their prokaryote hosts into the nuclear genomes of the higher plants? Their later evolution, with apparent gene duplication events at or near the points of emergence of the seed plants and the flowering plants, gives much cause for speculation regarding the evolutionary advantage of the phytochromes. How is it, for example, that the grasses are so successful as a group when they appear to possess only three, rather than the five phytochromes of dicotyledonous plants (Mathews and Sharrock, 1996)? Do monocots have less capacity for adaptation to their light environment than the dicots? Or are one or more of the dicot phytochromes not only genetically redundant but also superfluous?

Mutant selection investigations have made deep inroads into the downstream pathways by which the phytochromes, cryptochromes and phototropins transduce environmental signals into biological responses. Epistasis is a tool that has been used to great effect here, but dangers exist in a simple interpretation of epistatic interactions for signal transduction, where cross-talk may be an important consideration. Studies of epistasis worked well for the resolution of linear biochemical pathways but hold problems for signalling networks. Even so, we now have a good working knowledge of at least some of the steps in photomorphogenic signal transduction pathways. Almost weekly, it seems, papers are published with yet another mutant that affects photomorphogenesis; coordinating all the elements into a meaningful structure will be daunting. But, as micro-array studies show, photomorphogenic responses are not simple events; they comprise changes in the expression of vast numbers of genes, so we should expect to find similarly large numbers of mutants affecting the outcome. How do these pathways interact with the pathways transducing other environmental signals? Plants are exceptionally sensitive to their environment, and signals of water status, nutrient availability, gravity, temperature, pathogen attack and even wind and touch affect growth and development. We should expect some sharing of rôles between the downstream elements of the phytochromes and some of these other pathways.

More direct methods have yielded impressive results concerning the cellular actions of the phytochromes. Curiously, investigations utilising direct micro-injection of phytochromes and other components into epidermal cells seem not to have lived up to their early promise and one is left wondering about the status of the hypotheses that emanated from the results. On the other hand, visualisation of phytochromes by combination with fluorescent molecules (such as GFP) has been remarkably productive. The demonstration by the Nagy and Schäfer groups of nuclear import of phyA-GFP and phyB-GFP in the Pfr form has allowed the construction of a formidable general concept (Kircher *et al.*, 1999). Coupled with two-hybrid studies that confirm the association of Pfr with transcription factors (Ni *et al.*, 1998), we now have a more-or-less accepted opinion that phytochromes enter the nucleus and interact with the mechanisms regulating gene expression, although the devil lies in the detail, as always. The independent actions of the several phytochromes can be understood in terms of each interacting with different transduction factors, thereby controlling the expression of different genes. This is a powerful and persuasive idea, but is it enough? Can it incorporate the now near-forgotten data, which to my knowledge have not yet been gainsaid, about phytochrome action in the cytoplasm? In particular, the visually impressive evidence on the phytochrome-controlled rotation of the *Mougeotia* chloroplast that came from the Haupt laboratory in the 'pre-molecular age' seem consistent only with an ordered array of phytochrome molecules in the cytoplasm (Haupt 1960).

Micro-array analysis has provided insights into the nature and dynamics of gene expression regulation by the phytochromes. Early-onset genes tend to be those encoding further transcription factors, with the idea emerging that these factors go on to control the expression of the later-onset genes. Thus we have an answer to the long-standing fundamental question of signal amplification. Even so, the magnitude of the gene expression responses is somewhat surprising, with the expression of 10% or more of the genes in the Arabidopsis genome being regulated by phyA (Tepperman *et al.*, 2001). As mentioned earlier, this amounts to a very large change in the pattern of gene expression and satisfactorily accounts for the plethora and magnitude of physiological responses to such a simple environmental signal as R and FR light. But it raises even more interesting questions. Presumably these genes encode elements that are not all in a single linear sequence, and we should expect many branches in a tree of pathways leading to individual developmental effects. It will be a challenge to find ways to identify which gene goes into which pathway, but a generic method of doing this would be far-reaching.

Beyond the gene and beyond the cell, there is the plant. Whole-plant physiology has become somewhat of a backwater but it is as well to remember that we have arrived where we are now because of the physiological questions raised in the past. Moreover, physiology still has a rôle to play, and this can be exemplified by the research that led to the concept of the shade avoidance syndrome (Holmes and Smith 1975). Here again, technological advances were crucial. The availability of affordable spectroradiometers in the 1980s made it possible to monitor the spectral distribution of natural radiation inside and outside plant stands. The construction of controlled environment chambers in which photosynthetically active radiation could be held constant whilst the ratio of red to far-red (R:FR) could be varied over a wide range revealed that extension growth rate in many plants bore a simple linear relationship to Pfr/P (Morgan and Smith 1977). Continuous monitoring of elongation growth using position-sensitive transducers showed that elongation responded within minutes to a change in R:FR ratio – in both directions. Such findings provided the data for a new concept of the significance of the phytochromes in the field, and simultaneously posed challenges for the interpretation of the molecular and cellular data.

Some of these challenges are already being faced. The research of the Whitelam group with micro-arrays has revealed a subset of genes whose expression is rapidly and reversibly modulated by the R:FR ratio. One of these genes, PIL1, encodes a bHLHP that resembles the known phytochrome-interacting transcription factors and whose expression is regulated with lag-times sufficiently short to account for the dynamics of R:FR effects on extension growth (Salter *et al.*, 2003). The expression of the this gene is gated by the circadian clock, as is the rapid extension growth response – this characteristic may turn out to be of considerable significance for plants in the natural environment.

IS THAT ALL, DAD?

Biological science is dominated by the desire to understand how it all works. In photomorphogenesis this obsession can be seen in virtually all current work. Indeed, research groups are identified in many cases by the techniques they use and their mid-term objectives. It is important to consider whether such objectives are sufficiently ambitious. It is natural to wish to understand 'how things work'. Science is fun and should be, and some of the greatest pleasure is to be gained from the detective work of mechanistic research. There are downsides though, one being that it can become introverted; the further one delves into any mechanism the more one needs to find out at even deeper and more detailed levels. The answer to where mechanistic analysis ends is that it never does, so it is philosophically impossible to reach a conclusive understanding of 'the mechanism'.

Even if we could fully understand the mechanism, would it be enough to know 'how phytochromes work'? The reductive approach says that only by dissecting processes into increasingly independent mechanisms can we reach an understanding of the whole. There is no doubt at all that reductive analysis provides the data upon which integration can be built, but in photomorphogenesis there is as yet precious little sight of attempts to rebuild the edifice from the blitz of reductionism. The probable reason for this is that thinking of ways to generate integrative ideas *that can be tested* is exceptionally difficult, much more so than following along clear routes used by others to continue making progress in mechanistic studies. Of course, speculation is easy, and one regularly sees such fanciful blurb at the end of a mechanistic paper as 'these data may help to explain how the plant adapts to its environment'. Speculation is all part of the game, but real integration requires the construction of hypotheses that can be experimentally falsified.

In one sense, reductionism is the handmaiden of integration, putting mechanisms into an apparently subordinate position in relation to attempts to comprehend the whole. This is not to say that those attempting holistic science are any more worthy than those at the coalface of mechanisms; in contrast, it emphasises the dependence of holistic research on the advances of reductionism. More specifically, we cannot hope to understand how plants adapt, simultaneously, to variations in blue, UV, red and far-red signals unless we have a sound mechanistic understanding of the responses to individual signals. But the relationship between reductionist and holistic science does indicate that in the long term the holistic understanding is by far the more fulfilling. It follows that we need continuing in-depth investigation of the details of the actions and interactions of the several photoreceptors, not only for its own sake, but to provide a firm foundation for the broader, generalised concepts of adaptation and evolution.

There is a popular joke in English in which a drunk who has lost a sixpence in a dark place searches for it around the corner under a streetlamp; when asked why he searches under the light he replies that it is because it is too dark to see where he dropped the coin. Currently, photomorphogenesis is somewhat like that except many seem to be searching in the dark for something that is really important in the light. The use of the etiolated seedling has been of great value to our mechanistic understanding, but it is only when these endeavours move into the realm of real plants, in the real world, that we can begin the integrative interpretations that are needed.

WHY, DAD?

"Why" can only be answered by a philosopher or a cleric, and even then most of us would not believe the answer. But we can sensibly ask what advantages there are for plants in the possession of the phytochromes. To achieve a wider, as opposed to a deeper, understanding, it is almost essential to recruit expertise outside the traditional scope of plant photobiology. This is in itself valuable, as the cross-fertilisation of talents, interests and objectives is one of the best ways to get a completely new and refreshing view of problems that have seemed intransigent.

One example of this is the application of knowledge on phytochrome-mediated responses as a model for adaptive evolution. Evolutionary and population biologists, who are not themselves directly interested in photobiology, have used the phytochrome-mediated shade avoidance concept to test whether responses claimed to be adaptive do in fact enhance the fitness of the organism (Dorn et al., 2000). Using mutants, transgenic plants or plants pre-treated under non-adaptive conditions, it has been possible to demonstrate that phytochromes do confer adaptive value in crowded habitats (Schmitt et al., 1995). A start has been made on attempts to relate the range of adaptation responses to micro-evolution of the phytochrome genes (Garcia-Gil et al., 2003; Mathews et al., 2003). The shade avoidance concept has spread far and wide, and examples are known from all types of angiosperms from grasses to trees. In crops, as will be mentioned below, planting density is a strong determinant of yield, and the influence of shade avoidance as an agronomically disadvantageous process is now well-recognised.

Responses of trees to stand density have revealed hitherto unexpected aspects of the sensitivity of plants to the proximity of neighbours (Gilbert et al., 2001). In the natural environment it is important to consider not only the response of individual plants to the signals, but also the strength of the signals generated by each plant. Signal strength is a function of the quantity of chlorophyll-containing material and its distribution over the surface of the plant. There seems to be an inverse correlation between the signals generated by trees and the responsiveness of the plant to those signals. In effect, early-successional trees tend to generate small signals but respond strongly to signals from neighbours. Late-successional species generate bigger signals but respond weakly. In effect, this allows early-successional, pioneer trees to respond weakly to the small signals from conspecific neighbours but to react strongly to the large signals from mature trees – an apparent advantage when regenerating in a forest gap. As these data were from three species only, the conclusion awaits Popperian falsification! Ecological rationalisations of this type are becoming more common, but as always with ecology, variation both in the genetic nature of the organisms and the temporal characteristics of the environment, calls for caution. Ecologists love diversity and complexity, which allow them often to propose concepts that are not amenable to falsification; it also allows them to criticise hypotheses emanating from reductionism on the grounds that the world is so complicated simple ideas just will not work – let them tell Newton that!

On the other hand the variation in responses is such that there will inevitably be conditions under which shade avoidance is not adaptive. Even in *Arabidopsis thaliana*, where it is difficult even to speculate about ecological roles for shade avoidance, wide genetic variation in extension growth and floral acceleration by low R:FR ratio is evident within accessions already in the Stock Centres (Botto and Smith, 2002). We might expect to find even more variation via a focussed search for new accessions. Applying genetic variation through quantitative genetic analysis is one of the 'hottest topics' these days in the quest for ecologically important genes (Alonso-Blanco et al., 1999). The identification of quantitative trait loci (QTL) that are responsible for such variation provides clues both on signal transduction pathways and on the micro-evolutionary processes that underlie genetic adaptation to the environment (Maloof *et al.*, 2001; El-Assal *et al.*, 2001).

WHAT USE IS IT, DAD?

Some of those reading this book will have little interest in the application of photomorphogenic knowledge to the betterment of mankind, preferring to devote their energies to extending knowledge. Others will undoubtedly feel that the manipulation of plant traits by genetic engineering is potentially harmful to the natural environment. Both are arguable viewpoints, but we occupy the privileged position of being funded to pursue our intellectual curiosity on the predication that something of use will transpire. When it becomes possible to apply our discoveries towards economic aims then it behoves us at least to consider whether whatever talents we possess should be directed partly to realisable goals.

Because light signals have such manifold and often spectacular effects on plant development, there must be scope for manipulating these processes to the advantage of those growing plants as crops. Some of this has happened already through traditional plant breeding, but transgenic plant technology offers a more directed approach. There are many potential targets but as yet only the *PHY* genes themselves have been applied in this way. Crop responses to crowding are no different from those of plants in natural communities, but extension responses are not advantageous to the farmer. Deleterious effects such as lodging can be most damaging to yield and performance, and are caused by the phytochrome-mediated shade avoidance syndrome resulting in longer, thinner shoots. In principle, over expression of a *PHYA* gene will reduce extension in the field, apparently through a persistent FR high irradiance response (Robson *et al.*, 1996).

Transgenic manipulation of photoreceptor genes is a crude approach, and several observers have drawn attention to negative side effects. A more intelligent approach would be to target genes encoding downstream elements in the photoreceptor transduction pathways, hopefully resulting in specific effects on those processes that are deleterious rather the blanket modifications resulting from manipulation of the primary photoreceptors.

ARE WE NEARLY THERE YET?

There is, of course, no destination. We do not know where we are going. Science is not like a trip to the seaside, so we cannot tell whether we are nearly there yet. And if we did know where we are going would we want to go there? I doubt it – in contrast to the trip to the seaside, it is the journey that excites, not the destination. What we can try to do, though, is learn a little from where we have been that might help us plot the next part of the route in our journey to wherever we are going. That has been the objective of this article – neither I nor anyone else can predict the future of photomorphogenesis, but I hope we can all look at the past with hindsight to advantage.

In this article I have essentially drawn attention to approaches and directions that I, personally, believe will prove productive in the relatively near future. The choices are mine, and inevitably are influenced my own fascination with the exquisite sensitivity of higher plants, and to the manner of its evolution. I have used the words of some of those upon whose shoulders we stand in order to illustrate my arguments. To round up, I shall quote someone who is not a scientist, nor upon whose shoulders I would wish to stand: the US Secretary of Defence at the time of writing, Mr Donald H Rumsfeld. Whether one admires or hates his politics, his wondrously weird prose has something to teach us. He has recently stated, in a Departmental briefing:

> "As we know,
>
> There are known knowns.
>
> There are things we know we know.
>
> We also know
>
> There are known unknowns.
>
> That is to say
>
> We know there are some things
>
> We do not know.
>
> But there are also unknown unknowns,
>
> The ones we don't know
>
> We don't know."

This is as appropriate to science as to global diplomacy. The 'known knowns' are what this article has been about. It would be impossible to catalogue the 'known unknowns'. We could all list dozens of unknowns, and clearly those unknowns will become knowns, as soon as technology develops to allow their discovery. It is exactly because we know that we do not know, that soon we shall know. But it is the 'unknown unknowns' that are the key to the future of photomorphogenesis – and all science. In Donald Rumsfeld's world the 'unknown unknowns' are frightening; in science they are what drive us on.

RACE YOU TO THE BEACH, DAD!

REFERENCES

Ahmad, M. and Cashmore A. R. (1993) hy4 gene of a-thaliana encodes a protein with characteristics of a blue-light photoreceptor. *Nature, 366,* 162-166

Alonso-Blanco, C., Blankestijn-de Vries, H., Hanhart, C. J., Koornneef, M. (1999) Natural allelic variation at seed size loci in relation to other life history traits of Arabidopsis thaliana. *Proceedings of the National Academy of Sciences of the United States of America, 96,* 4710-4717

Borthwick, H. A., Hendricks, S. B., Parker, M. W., Toole, E. H. Toole, V. K. (1952) A reversible photoreaction controlling seed germination. *Proceedings of the National Academy of Sciences, USA, 38,* 662-666.

Botto J. F. and Smith, H. (2002) Differential genetic variation in adaptive strategies to a common environmental signal in Arabidopsis accessions: phytochrome-mediated shade avoidance. *Plant Cell and Environment, 25*, 53-63

Christie, J. M., Salomon, M., Nozue, K., Wada, M., Briggs, W. R. (1999) LOV (light, oxygen, or voltage) domains of the blue-light photoreceptor phototropin (nph1): Binding sites for the chromophore flavin mononucleotide. *Proceedings of the National Academy of Sciences of the United States of America, 96*, 8779-8783

Dorn, L. A., Pyle, E. H., Schmitt, J. (2000) Plasticity to light cues and resources in Arabidopsis thaliana: Testing for adaptive value and costs. *Evolution, 54*, 1982-1994.

El-Assal, S. E. D., Alonso-Blanco, C, Peeters, A. J. M., Raz, V., Koornneef M (2001) A QTL for flowering time in Arabidopsis reveals a novel allele of CRY2. *Nature Genetics, 29*, 435-440.

Garcia-Gil, M. R., Mikkonen, M., Savolainen, O. (2003) Nucleotide diversity at two phytochrome loci along a latitudinal cline in Pinus sylvestris. *Molecular Ecology, 12*, 1195-1206.

Gilbert, I. R., Jarvis, P. G., Smith, H. (2001) Proximity signal and shade avoidance differences between early and late successional trees. *Nature, 411*, 792-795.

Haupt W (1960) Die Chloroplastenbewegung bei *Mougeotia* II. Die Induktion der Schwachtlicht bewegung durch linear polarisiertes Licht. *Planta, 38*, 465-479.

Hershey, H. P., Colbert, J. T., Lissemore, J. L., Barker, R. F., Quail, P. H. (1984) Molecular-cloning of cDNA for *Avena* phytochrome. *Proceedings of the National Academy of Sciences of the United States of America, 81*, 2332-2336.

Holmes, M. G. and Smith, H. (1975) The function of phytochrome in the natural environment. *Nature, 254*, 512-514.

Kircher, S., Kozma-Bognar, L., Kim, L., Adam, E., Harter, K., Schäfer, E. et al.,(1999) Light quality-dependent nuclear import of the plant photoreceptors phytochrome A and B. *Plant Cell, 11*, 1445-1456.

Maloof, J. N., Borevitz, J. O., Dabi, T., Lutes, J., Nehring, R. B., Redfern, J. L., et al. (2001) Natural variation in light sensitivity of Arabidopsis. *Nature Genetics, 29*, 441-446.

Mathews, S., Burleigh, J. G., Donoghue, M. J. (2003) Adaptive evolution in the photosensory domain of phytochrome A in early angiosperms. *Molecular Biology and Evolution, 20*, 1087-1097.

Mathews, S. and Sharrock, R. A. (1996) The phytochrome gene family in grasses (Poaceae): A phylogeny and evidence that grasses have a subset of the loci found in dicot angiosperms. *Molecular Biology and Evolution, 13*, 1141-1150.

Morgan, D. C. and Smith, H. (1977) Linear relationship between phytochrome photoequilibrium and growth in plants under simulated natural radiation. *Nature, 262*, 210-212.

Ni, M., Tepperman, J. M., Quail, P. H. (1998) PIF3, a phytochrome-interacting factor necessary for normal photoinduced signal transduction, is a novel basic helix-loop-helix protein. *Cell, 95*, 657-667.

Robson, P. R. H., McCormac, A. C., Irvine, A. S., Smith, H. (1996) Genetic engineering of harvest index in tobacco through overexpression of a phytochrome gene. *Nature Biotechnology, 14*, 995-998

Sage, L. C., *Pigment of the Imagination: A History of Phytochrome Research.* New York: Academic Press, 1997.

Salter, M. G., Franklin, K. A., Whitelam, G. C. (2003) Gating of the rapid shade-avoidance response by the circadian clock. *Nature, 426*, 680-683.

Schmitt, J., McCormac, A. C., Smith, H. (1995) A test of the adaptive plasticity hypothesis using transgenic and mutant plants disabled in phytochrome-mediated elongation responses to neighbours. *American Naturalist, 146*, 937-953.

Tepperman, J. M., Zhu, T., Chang, H. S., Wang, X., Quail, P. H. (2001) Multiple transcription-factor genes are early targets of phytochrome A signalling. *Proceedings of the National Academy of Sciences of the United States of America, 98*, 9437-9442.

CONCLUSIONS

Our knowledge about photomorphogenesis in plants has increased enormously both in volume and depth over the last two decades. Not only several novel photoreceptors – the blue UVA photoreceptors CRY1, CRY2, PHOT1, PHOT2, the FKP family, the superchrome PHY3 in ferns – but also an ever increasing number of the elements of signal transduction have been identified. The biological importance of the light-regulated intracellular compartmentalization of photoreceptors and regulatory components for signal transduction initiated by these receptor molecules has also been recognized. Nevertheless, many fundamental questions remain unanswered. At the photoreceptor level, the UVB photoreceptor remains still elusive but thanks to the development of novel genetic methods, there is hope for its identification. The "holy grail" of photomorphogenesis research, namely elucidation of the primary function of the photoreceptors, is yet to be found. In the case of phototropines it can be speculated that the primary function is to launch a phosphorylation/ dephosphorylation cascade, but it is still an open question how this is initiated and what the substrates are. Similarily, in the case of cryptochromes it is only hypothesized that the primary event is a redox reaction, but neither the substrates nor the redox function of the molecule is documented in vivo. Bacterial phytochromes are cleary light-regulated histidine kinases and parts of a two-component system. Unfortunately, in most cases the targets of the two-component system have not been identified.

Phytochromes of higher plants are not histidine kinases and their possible kinase function in vivo is still not proven. Thus it is clearly questionable whether the kinase activity is the primary function of phytochromes. Therefore, not too surprisingly, we do not even know whether there is only one or several classes of primary functions of phytochromes exist. Is the primary function of phytochrome for membrane or membrane-associated, cytosolic responses and initiating its own light-dependent nuclear transport always the same? Phytochromes under all light conditions can be found in the cytosol and nucleus. Is the primary function of the transported receptors identical to those localized in the cytosol? How are the systems of protein degradation – COP1/ COP10/ COP9 complex and the SCF^{EID1} – coupled to the photoreceptor input?

Fortunately, due to the improvements of technology in genetics, molecular and cellular biology, an ever widening range of tools is available to address these problems. Nevertheless we can predict that the biochemical verification of the presently favoured hypothesis and sorting out the function and components of the increasing number of protein complexes involved in these processes will be a great challenge. Plants respond to biotic and abiotic stimuli, including changes in the ambient light conditions, by a concerted action of various signalling pathways. Mathematical modelling and system biology should be incorporated into photomorphogenetic research to interpret how these pathways interact hierarchically and functionally.

The authors hope that reading this book will fascinate and attract to this field many new researchers who attack these problems armed with novel ideas and techniques.

The editors
Eberhard Schäfer and Ferenc Nagy
Freiburg, Germany and Szeged, Hungary

Index

19S proteasome regulatory particle 366
1-naphthylphthalamic acid 447
26S proteasome 358, 359, 361, 362, 365, 366, 371, 372, 373, 374, 375, 376, 377, 378

ABI5 363, 375
absorption spectra 68, 80
accumulation response 174, 186, 187, 319, 525, 530
actinobacteria 66, 70
action dichroism 37
action dichroism 551
action spectra 174, 176, 177, 182, 186, 187, 195, 196
action spectroscopy 3, 13, 14, 15, 20, 23
action spectrum 2, 7, 8, 9, 13, 14, 15
acute induction 586, 587, 589
Adiantum 119, 128
Adiantum capillus-veneris 225, 235, 236, 242, 515, 523, 529, 534, 535, 536
Adiantum PHOT1 518
Adiantum PHOT2 519
Affymetrix oligoarrays 492
agriculture 629, 630, 636
Agrobacterium tumefaciens 57, 70, 73, 75, 79, 80, 81, 82, 85, 88, 89, 94, 96, 98
Allard 628, 629, 637, 642
Amborella 117, 120
amplified fragment-length polymorphisms 232
Anabaena 70, 73, 84, 88
analytical action spectroscopy 13, 15
angiosperm 100, 104, 120, 121, 122, 124, 126, 128
antagonism 423, 425
antheridia 537, 538
anthocyanin 204, 207, 212, 218, 360
antibodies 105, 107, 112, 115, 116, 127, 128
anticipation 568, 590, 592
aphototropic 549, 561, 562
apoprotein 99, 100, 101, 102, 103, 105, 106, 107, 108, 112, 113, 114, 115, 117, 118, 119, 120, 121, 122, 125, 126, 127, 128, 129
Arabidopsis 102, 103, 104, 105, 108, 109, 110, 111, 112, 113, 114, 115, 116, 121, 122, 123, 124, 125, 126, 127, 128, 129, 605, 606, 607, 608, 610, 611, 612, 613, 615, 617, 618, 620, 621, 622, 623, 624, 625
Arabidopsis PSEUDO RESPONSE REGULATOR (APRR) 570

Arabidopsis thaliana 46, 49, 50, 51, 52, 53, 54, 61, 62, 64
archegonia 537, 538
archegonium 537
ARR4 148, 149, 153, 395, 398, 399, 405
Aspergillus nidulans 71, 73, 83, 97
assays 99, 100, 101, 102, 103, 105, 106, 107, 108, 112, 113, 114, 115, 117, 118, 119, 120, 121, 122, 125, 126, 127, 128, 129
ATB2 363, 376
ATHB-2 449, 451, 472
autophosphorylation 223, 224, 228, 231, 233, 235, 240, 242, 251
auxin 396, 399, 403, 405, 445, 446, 447, 448, 449, 450, 451, 459, 468, 469, 471, 473, 540, 541, 556, 557, 562, 564
Avena sativa 100, 127, 129, 224
avoidance response 174, 186, 188, 190, 194, 542, 543, 559

bacteria, 67, 70, 92
BAS1 399, 404
Beltsville group 642
beta-carotene 633
BHF (blue light high fluence) 27
BIG/DOC1/TIR3 447
bilin 41, 44, 46, 47, 55, 56, 57, 58, 59, 60, 62, 64, 549
biliverdin 97, 549
bio-assays 99, 100, 101, 102, 103, 105, 106, 107, 108, 112, 113, 114, 115, 117, 118, 119, 120, 121, 122, 125, 126, 127, 128, 129
biotechnology 635
blue light 199, 201, 202, 203, 204, 205, 206, 207, 209, 212, 213, 214, 215, 218, 219, 220
BMAL1 203, 209, 217, 219
Bradyrhizobium ORS278 70, 73
Brassica oleracea L. 8
Brassica rapa L. 8
brassinolide 468
brassinosteroid 404, 452, 453, 454, 455, 456, 465, 466, 467, 468, 470, 472
Bunsen-Roscoe reciprocity law 7, 174, 176

Ca^{2+} 380, 384, 386, 387, 389, 391, 392, 393, 395, 399, 402, 557, 558, 559, 561
CAB2 gene 561, 570, 571, 574, 584, 590
Calothrix PCC7601 57, 70
carotenoid hypothesis 176, 178
casein kinase II 362, 368
CATALASE 3 (CAT3) 596

caulonemal cells 541, 563
caulonemata 541
CCA1 340, 341, 342, 351, 352, 354, 356
cell differentiation 541
central circadian oscillator 572, 595
Ceratodon 118, 128, 539, 540, 541, 542, 543, 544, 545, 546, 547, 548, 549, 550, 552, 553, 554, 555, 558, 559, 560, 561, 562, 563, 564, 565
Ceratodon HO-gene 549
Ceratodon purpureus 539, 540, 542, 548, 560, 561, 562, 563, 564, 565
CerpuPhy1 545, 546
CerpuPhy2 546, 547
CerpuPhy3 546, 547
cGMP 387, 404, 406
chalcone synthase (CHS) 25, 204, 207, 212
changes in stomata aperture 284
Chlamydomonas reinhardtii 500, 511
chloronemata 541, 562
chlorophyll a/b binding proteins (CAB) 25
chlorophyll precursors 499, 500, 501, 502, 505, 508, 511, 512
chlorophyll synthesis 544
chloroplast 223, 226, 227, 235, 246, 247, 249, 251, 252, 499, 500, 502, 503, 504, 505, 506, 507, 508, 509, 510, 511, 512, 513, 525, 527, 528, 535, 536
chloroplast avoidance movement 524, 531, 534
chloroplast avoidance response mutants 526, 530
chloroplast division 539, 557
chloroplast movement 543, 559, 564
chloroplast orientation 539, 551, 558, 561
chloroplast photorelocation movement 516
chromatin 370, 371
chromophore 41, 42, 43, 44, 45, 46, 47, 49, 54, 56, 57, 58, 59, 60, 61, 62, 63, 64
CIP1 364, 375
CIP7 363
CIP8 364, 374
circadian Clock 568, 601, 604, 605, 608, 609, 611, 612, 613, 614, 616, 617, 618, 622, 624
CIRCADIAN CLOCK-ASSOCIATED 1 (CCA1) 570
circadian oscillator 475, 489
circadian regulation 116, 127
circadian rhythms 203, 206, 208, 218, 219, 221
circadian Time 578
cis-regulatory elements 26
c-Jun 368, 372, 375, 376, 378
CLOCK 35, 203, 209, 217, 219
connectivity 429

CONSTANS (CO) 583, 622, 623, 624, 625, 628, 630, 638
constitutively photomorphogenic 357, 360
CONSTITUTIVELY PHOTOMORPHOGENIC 1 (COP1) 594
co-option 291
COP1 211, 212, 213, 216, 220, 221, 258, 359, 360, 361, 362, 363, 364, 368, 370, 371, 372, 373, 374, 375, 376, 377, 378
COP10 358, 360, 362, 370, 371, 377, 378
COP9 358, 360, 364, 365, 366, 369, 370, 371, 373, 374, 375, 376, 377, 378
COP9 signalosome 29
cotyledon expansion 284, 288
cry1 360, 361, 632
cry2 360, 361, 627, 628
cryptochrome 199, 201, 202, 203, 204, 206, 207, 208, 209, 210, 212, 216, 217, 218, 219, 220, 221, 360, 361, 377, 378, 556, 602
cryptochrome DASH 181
crystal structure 236, 237, 238, 241, 242
CSA 367, 372, 374
CSN 358, 360, 362, 364, 365, 366, 367, 368, 369, 371, 372
CSN5 364, 365, 366, 367, 368, 369, 372, 373, 374, 377
Cucurbito pepo 224
cullin 359, 366, 367, 371, 372, 373, 378
cyanobacteria 66, 70, 75, 77, 88, 89, 90, 92, 96
cycad 123
cyclobutane pyrimidine dimers (CPD) 281
cytokinin 447, 462, 463, 464, 466, 467, 471, 472, 473, 541, 561, 564, 565
Cytoskeleton 558

damage 279, 281, 282, 284, 287, 288, 290, 291, 292, 296, 298, 300, 301, 302, 304
dark position-deficient mutants 527
dark reversion 18, 20, 21, 131, 132, 133, 134, 143, 144, 145, 146, 147, 148, 149, 151, 152
Darwin 446, 467
day length 605, 606, 607, 609, 610, 614, 615, 619, 620, 621
daylength perception 582, 630
DDB1 358, 362, 371, 378
DDB2 367, 372, 374
deazaflavin 200
de-etiolated 357, 399
degradation 132, 133, 134, 135, 136, 138, 139, 140, 141, 142, 150, 153

Index

Deinococcus radiodurans 57, 70, 73, 77, 78, 91, 93
DELLA 444, 445, 463, 466, 467, 468
derubylation 367, 368, 369, 372
destruction 132, 133, 134, 135, 136, 137, 139, 140, 141, 142, 143, 144, 148, 149, 151, 152, 153
DET1 287, 299, 358, 360, 362, 370, 371, 373, 376, 378, 633, 634, 635, 637, 638
DET2 287
deubiquitination 366, 367, 369, 372, 377, 378
dexamethasone 635
diaphototropism 191
dichromatic irradiation 10
Dicranum scoparium 539
Dictyostelium discoideum 179, 196
dimerization 146
DNA repair 199, 217, 219, 220
DNA transfer 538
DNAse I foot printing 27

E2 ubiquitin conjugase 359, 370, 371
E3 ubiquitin-protein ligase 358, 359, 362, 363, 364, 365, 366, 367, 368, 370, 371, 372
early flowering 3 *(elf3)* 585
early phytochrome responsive 1 595
EARLY-FLOWERING 4 (ELF4) 573
ecotypes 35, 36, 39
EIF3 366, 374
ELISA 99, 100, 101, 102, 103, 105, 106, 107, 108, 112, 113, 114, 115, 117, 118, 119, 120, 121, 122, 125, 126, 127, 128, 129
escape time 630
EST 537, 539, 545
ethylene 283, 293, 299, 440, 445, 447, 449, 452, 460, 461, 462, 463, 464, 465, 466, 467, 468, 469, 470, 471, 472, 473
etiolation 1, 4, 108, 123, 124
etiolement 1
evolution 92
excision repair 282, 283, 300, 302
expression pattern 99, 100, 101, 102, 103, 105, 106, 107, 108, 109, 110, 112, 113, 114, 115, 116, 117, 118, 119, 120, 121, 122, 125, 126, 127, 128, 129
external coincidence model 611

family 99, 100, 101, 102, 103, 104, 105, 106, 107, 108, 110, 112, 114, 115, 116, 117, 118, 119, 121, 122, 123, 125

far-red High Irradiance Response (HIR) 20
F-box protein 359, 367, 372
fern photomorphogenesis 516, 535
FHY1 388, 395, 396, 402
filaments 538, 539, 541, 542, 544, 546, 547, 549, 550, 555, 556, 562, 564
FIN219 364, 374, 395, 396, 399, 403
fitness 481, 482, 629
FKF1 242, 243, 249, 250, 608, 609, 612, 614, 617, 622, 624
flavin-cysteinyl adduct 234, 236, 238, 239, 242
flavoproteins 199, 219
FLC 606, 619, 622, 624, 625
floral initiation 4, 11, 12
flowering 202, 206, 213, 218, 219, 481, 483, 484, 485, 486, 489, 490, 492, 493, 494, 496, 582, 600, 603
fluorescence correlation spectroscopy 553
fluorescence differential display (FDD) 515
fluorosulfonylbenzoyladenosine 231, 252
forward-genetic screens 336
Fremyella diplosiphon 69, 70, 73, 84, 86, 87, 90
fruit ripening 632
fruit-specific promoter 634
FT (FLOWERING LOCUS T) 609
Funaria 539, 540, 541, 543, 561, 564
functional evolution 99, 100, 101, 102, 103, 104, 105, 106, 107, 108, 110, 112, 114, 115, 116, 117, 118, 119, 121, 122, 123, 125
fungi 66, 67, 71, 83, 86, 89, 92, 94, 95
fusca 358

G proteins 286, 388, 400
GAF domain 56, 71, 72, 73, 75, 76, 77, 78, 79, 81, 83, 84, 92, 94
gametangia 537
gametophore 537, 541, 556
gametophyte 515, 516, 518, 519, 524, 525, 526, 527, 530, 531, 533, 536, 537
Garner 628, 629, 642
GATA box 27
gating 579, 580, 581, 582, 583, 584, 585, 586, 587, 589, 590, 591, 592, 593, 594, 595, 596, 598, 611
G-box 27
gene expression 499, 500, 501, 502, 503, 504, 505, 506, 508, 509, 510, 511, 512
gene-knockout 538
genome 546, 562
germination 1, 2, 4, 7, 11, 476, 479, 494, 496, 518, 532

GFP 157, 158, 159, 160, 161, 163, 164, 165, 167, 168, 635
gibberellin 439, 440, 442, 443, 466, 467, 468, 469, 470, 471, 472, 473
GIGANTEA (GI) 592, 630
gnetales 120, 123, 125
grafting 610, 620
gravitropic response 542, 551, 562
gravitropism 540, 542, 557, 560, 561, 562, 563, 564, 565
green fluorescent protein (GFP) 30
green light 172, 186, 194, 195
Grotthus-Draper's law 13
GT-1 binding protein 27
GUS reporter 157
gymnosperm 122, 123, 124, 126

haploid 537, 538
haploid insufficiency 201
Heading date (Hd) 615
Helianthus annuus 224
heliotropism 1
heme 43, 47, 49, 50, 51, 52, 59, 61, 62, 63, 64, 505, 509
heme oxygenase 47, 50, 61, 62, 63, 64, 549, 550, 560, 563
herbivorous insects 292
HIR, High Irradiance Response 141, 142, 143, 149 161
histidine kinase 97
histone 370, 373, 377
holy grail 36
homeostasis 423, 424, 433, 435
homologous recombination 538, 550, 560
Hordeum vulgare 224
hp 633
HSP70 502, 504, 511
hy mutants 34
HY5 361, 362, 363, 364, 368, 370, 371, 373, 374, 375, 376, 594, 603
HYH 363
hypocotyl elongation 569, 581, 584, 585, 587, 594, 598, 600, 602
hypocotyl growth inhibition 284, 287, 288
hypocotyl hook opening 581, 584
hypocotyls 202

immunoblot analysis 99, 100, 101, 102, 103, 105, 106, 107, 108, 112, 113, 114, 115, 116, 117, 118, 119, 120, 121, 122, 125, 126, 127, 128, 129

immunocytochemistry 99, 100, 101, 102, 103, 105, 106, 107, 108, 112, 113, 114, 115, 117, 118, 119, 120, 121, 122, 125, 126, 127, 128, 129, 156
Impatiens balsamina 172
in vitro spectroscopy 18, 20
induction of UV-protecting pigmentation 284
induction responses 7
inhibition 174, 175, 179, 180, 181, 189, 190, 192, 194, 195
inositol 1,3,4-triphosphate 5/6 kinase 368
intercellular signaling 106
interchromatin granual clusters (ICGs) 166
intracellular localization 155, 171

Jab1 367, 368, 369, 373, 374, 377
JAMM domain 367
jasmonic acid 293, 299

kineococcus radiotolerans 70, 73, 77, 83
kinetics of loss of reversibility 19

Lactuca sativa L. 2, 9, 11
LAF1 363, 364, 376
LAF6 395, 397
late elongated hypocotyl (lhy) 570
leaf and tendril curling 284
leaf expansion 171, 174, 187, 188, 189, 193, 223, 227
leaf movement 568, 600
Lepidium sativum 172
levels in Arabidopsis 99, 100, 101, 102, 103, 105, 106, 107, 108, 112, 113, 114, 115, 117, 118, 119, 120, 121, 122, 125, 126, 127, 128, 129
LHY 340, 341, 342, 351, 352, 354
light quality 476, 489, 493, 494
light-based biological engineering 635
light-dependent phosphorylation 37
light-equivalence principle 14, 15
light-responsive regulatory elements 26
LKP2 242, 243, 251
long day plant 582, 583
LOV domain 225, 228, 233, 234, 235, 236, 237, 238, 239, 240, 241, 242, 243, 244, 246, 248
LOV KELCH PROTEIN 2 (LKP2) 573
LOV-domain photocycle 238
luciferase 28, 31
lyase 57, 58, 60, 64
lycopene 632, 633, 637
Lycopersicum esculentum 224

Index

MAP kinases 294
Mesotaenium 118, 128
methenyltetrahydrofolate 199
methylation interference 27
Mg-protoporphyrin IX monomethyl ester (Mg-ProtoMe) 500
microarray 330, 335, 336, 347, 348, 352, 354
microarray technology 296, 298
microbeam 541, 550, 553, 555, 563
microfilament 558
microinjection 538, 549, 550, 560
microtubule 558
Mimosa pudica 172, 181, 191
moss transformation 538
mosses 537, 538, 539, 540, 543, 545, 547, 548, 549, 550, 551, 552, 556, 557, 559, 560, 561, 562
Mougeotia 118, 129
movement 171, 174, 186, 187, 191, 192, 193, 194, 195, 197, 227, 249, 250, 252
MPN domain 366, 367
Mulla Nasrudin tale 178
mutant 524, 525, 547, 549, 550, 552, 553, 554, 556, 562, 563
MYB21 363

NDPK2 (nucleoside diphosphate kinase 2) 340, 385, 395, 397
Neurospora crassa 35, 71, 73, 83, 179, 195
NF-κB 368
night break 582, 583, 584
nitrate reductase (NAI2) 580
nitric oxide 295, 304
nomenclature 99, 100, 101, 102, 103, 105, 106, 107, 108, 112, 113, 114, 115, 117, 118, 119, 120, 121, 122, 125, 126, 127, 128, 129
norflurazon 503, 505, 508
Nostoc 70, 73, 77
nuclear localisation 591
nuclear localisation signals (NLS) 157
Nuclear run-on experiments 26
nuclear speckles 360, 361
nucleocytoplasmic partitioning 362, 377
null-point-method 19

Oscillatoria PCC7821 70
Oxalis multiflora 172
oxidative stress 280, 293, 295, 300, 301, 302

$p27^{kip1}$ 367, 368, 369, 372, 377, 378
p53 368, 372, 373

PAS domain 56, 83, 92, 94
PAS-like domain (PLD) 76
PAT1 395, 401
PCI interaction domain 366
pelletable phytochrome 139
PERIOD 202
PFT1 620
Pharbitis nil 621
phase response curve 579, 587, 596, 597
Phaseolus 7, 8, 12
PHOR1 443, 444, 466
phosphorylation 362, 368, 369, 373, 374, 380, 381, 382, 384, 385, 397, 402, 403, 405
photoequillibrin 15
photolyase 201, 202, 203, 210, 211, 215, 216, 217, 218, 219, 220, 221, 282
photomorphogenesis 357, 358, 359, 360, 361, 362, 363, 367, 369, 370, 371, 373, 375, 376, 377, 580
photomorphogenic mutants 34
photoperiod 475, 483, 484, 490, 494, 495, 627, 628, 629, 630, 638, 642
Photoperiodism 605, 625
photoreactivation 281, 282, 283, 299
photoreceptors 515, 516, 522, 533, 534, 535
phototropic growth 284
phototropin 543, 547, 553, 555, 558, 561, 562,
phototropin phosphorylation 232
phototropism 171, 172, 173, 174, 175, 176, 177, 178, 182, 183, 184, 186, 189, 190, 192, 193, 194, 195, 196, 223, 226, 228, 229, 232, 233, 235, 244, 247, 250, 251, 252, 539, 541, 542, 547, 550, 551, 553, 558, 559, 560, 561, 562, 563
phototropism mutants 34
PHY gene evolution 99
PHY gene expression pattern 102, 103, 104, 105, 108, 109, 110, 111, 112, 113, 114, 115, 116, 121, 122, 123, 124, 125, 126, 127, 128, 129
PHY gene 99, 100, 101, 102, 103, 104, 105, 106, 107, 108, 110, 112, 114, 115, 116, 117, 118, 119, 121, 122, 123, 125, 126, 127, 128, 129
PHY mRNA 103, 104, 108, 109, 110, 112, 116, 121, 125, 126, 127
PHY promoter activities 116, 127
PHYA 99, 100, 101, 102, 103, 104, 105, 106, 107, 108, 110, 112, 114, 115, 116, 117, 118, 119, 121, 122, 123, 125
phyA 99, 100, 101, 102, 103, 105, 106, 107, 108, 112, 113, 114, 115, 117, 118, 119, 120, 121, 122, 125, 126, 127, 128, 129, 360, 361, 363, 364, 627, 628, 631

PhyA3 gene 100, 127, 129
PHYB 99, 100, 101, 102, 103, 104, 105, 106, 107, 108, 110, 112, 114, 115, 116, 117, 118, 119, 121, 122, 123, 125
phyB 99, 100, 101, 102, 103, 105, 106, 107, 108, 112, 113, 114, 115, 117, 118, 119, 120, 121, 122, 125, 126, 127, 128, 129, 363, 630, 631, 635
PHYC 99, 100, 101, 102, 103, 104, 105, 106, 107, 108, 110, 112, 114, 115, 116, 117, 118, 119, 121, 122, 123, 125
phyC 99, 100, 101, 102, 103, 105, 106, 107, 108, 112, 113, 114, 115, 117, 118, 119, 120, 121, 122, 125, 126, 127, 128, 129
phycoerythrobilin (PEB) 55
Phycoerythrobilin 553
Phycomyces 33
PHYD 99, 100, 101, 102, 103, 104, 105, 106, 107, 108, 110, 112, 114, 115, 116, 117, 118, 119, 121, 122, 123, 125
phyD 99, 100, 101, 102, 103, 105, 106, 107, 108, 112, 113, 114, 115, 117, 118, 119, 120, 121, 122, 125, 126, 127, 128, 129
PHYE 99, 100, 101, 102, 103, 104, 105, 106, 107, 108, 110, 112, 114, 115, 116, 117, 118, 119, 121, 122, 123, 125
phyE 99, 100, 101, 102, 103, 105, 106, 107, 108, 112, 113, 114, 115, 117, 118, 119, 120, 121, 122, 125, 126, 127, 128, 129
phy-interacting proteins 336, 340
phylogenetic tree 546
phylogeny 72, 99, 100, 101, 102, 103, 104, 105, 106, 107, 108, 110, 112, 114, 115, 116, 117, 118, 119, 121, 122, 123, 125
PHYN 99, 100, 101, 102, 103, 104, 105, 106, 107, 108, 110, 112, 114, 115, 116, 117, 118, 119, 121, 122, 123, 125
phyN 99, 100, 101, 102, 103, 105, 106, 107, 108, 112, 113, 114, 115, 117, 118, 119, 120, 121, 122, 125, 126, 127, 128, 129
PHYO 99, 100, 101, 102, 103, 104, 105, 106, 107, 108, 110, 112, 114, 115, 116, 117, 118, 119, 121, 122, 123, 125
phyO 99, 100, 101, 102, 103, 105, 106, 107, 108, 112, 113, 114, 115, 117, 118, 119, 120, 121, 122, 125, 126, 127, 128, 129
PHYP 99, 100, 101, 102, 103, 104, 105, 106, 107, 108, 110, 112, 114, 115, 116, 117, 118, 119, 121, 122, 123, 125
phyP 99, 100, 101, 102, 103, 105, 106, 107, 108, 112, 113, 114, 115, 117, 118, 119, 120, 121, 122, 125, 126, 127, 128, 129

PhypaPhy1 546, 551
PhypaPhy2 546, 551
PhypaPhy3 546, 551
PhypaPhy4 546, 551, 553
*Physarum polycephalu*m 81, 94, 97
Physcomitrella 118
Physcomitrella patens 516, 535, 537, 538, 539, 540, 548, 557, 560, 561, 562, 563, 564, 565
phy-specific monoclonal antibodies 105, 107, 112, 115, 116, 127, 128
phytochrome A 488, 492, 496, 497
phytochrome apoprotein 99, 100, 107, 112, 113, 118, 125, 126, 127
phytochrome apoprotein levels 116, 127
phytochrome B 489, 494, 495
phytochrome D 494
phytochrome E 494
phytochrome expression patterns 100, 127, 129
PHYTOCHROME INTERACTING FACTOR 3 (PIF3) 577
phytochrome knockout 550
phytochrome levels 102, 103, 104, 105, 108, 109, 110, 111, 112, 113, 114, 115, 116, 121, 122, 123, 124, 125, 126, 127, 128, 129
phytochrome photoequilibria 629
phytochrome signalling 379, 380, 384, 385, 386, 387, 388, 389, 392, 394, 395, 397, 399, 400, 401, 404
phytochrome synthesis 18
phytochrome 95, 96, 97, 99, 100, 101, 102, 103, 105, 106, 107, 108, 112, 113, 114, 115, 117, 118, 119, 120, 121, 122, 125, 126, 127, 128, 129, 600, 601, 602, 603, p
phytochrome-chromophore 547, 560
phytochromes 360, 376, 545
phytochromobilin 42, 48, 50, 54, 63, 64
phytochromobilin synthase 48, 54, 63, 64
PIF3 (phytochrome interacting factor 3) 157, 340, 635
PIF3-LIKE 1 (PIL1) 596
PIN1 227
Pinus 124, 126, 127, 129
Pisum sativum 224
PKS1 (phytochrome kinase substrate 1) 340, 382, 395, 397, 402, 403
PKS2 344
plant architecture 628, 630
plasma membrane 551, 552, 553, 564
plastid 499, 500, 501, 502, 503, 504, 505, 506, 507, 508, 509, 510, 511, 512, 513
plastid-to-nucleus signalling 510
polarized light 541, 542, 553
polarotropism 540, 541, 547, 551, 552, 562

Populus 121
Porphyra purpurea 53, 54
port wine 173
position 172, 174, 189, 191
post-transcriptional gene silencing 634
proteasome 358, 359, 363, 365, 366, 368, 370, 371, 373, 374, 375, 376, 378, 610, 614
protein degradation 358, 359, 360, 365, 368, 369, 371, 372, 374, 375, 376
protein kinase D 368, 377, 378
protein phosphatase 74
proteolysis 99, 100, 101, 102, 103, 105, 106, 107, 108, 112, 113, 114, 115, 117, 118, 119, 120, 121, 122, 125, 126, 127, 128, 129, 330, 332, 357, 358, 359, 363, 365, 367, 368, 373, 376
prothallus 527
protochlorophyllide (Pchlide) 501
protonema 527, 560, 561
protonemal cell 525, 528, 532
protonemata 524, 527, 528, 529, 533, 534, 536
protoplast regeneration 544
protoplasts 538
protoporphyrin IX (Proto) 501
proximity detection 480, 481, 483, 484, 485, 486, 487, 488, 489, 491, 493, 494, 495, 496, 498, 499
pseudo response regulator 570, 572
Pseudoanabaena 70
Pseudomonas syringae 70, 73, 77, 82
Psilotum 119
pyrimidine dimers 199, 200
pyrimidine-pyrimidinone dimers (6-4 photoproduct, 6-4PP) 281

R FR ratio 477, 478, 479, 480, 481, 482, 483, 484, 485, 486, 489, 490, 491, 492
rap2 (<u>r</u>ed light <u>a</u>phototropic 2) 515
Raphanus rusticanus 172
rapid inhibition of growth 223, 226
rate of photo conversion 15
RBX1 359, 372
reactive oxygen species (ROS) 280
rearrangement 223
recombination 282, 284, 302
red <u>H</u>igh <u>I</u>rradiance <u>R</u>esponse (HIR) 21
red light aphototropic mutants 524, 525
redox reaction 210, 211, 216, 217
redundancy 410, 411, 432, 433
reverse genetic 329, 336, 339
revertant screens 36
Rhizobium leguminosarium 70, 73, 77, 82

Rhodella violacea 53, 54, 64
Rhodobacter sphaeroides 70, 89
Rhodopseudomonas palustris 70, 73, 75, 77, 81, 82, 84, 88, 90
Rhodospirillum centenum 70, 73, 83, 91
ribulose-1,5-bisphosphate carboxylase/oxygenase (Rubisco) 25
rice 607
Rpn11 366, 377
RUB1 366
rubylation 367, 369

salicylic acid 293, 303
SAP 139
SCF complex 614
SCF E3 ubiquitin-protein ligase 359, 365, 367
Schiff base-type linkage 79
seed germination 2, 10, 11, 12, 99, 109, 114, 123, 126
Selaginella 118, 127
Sensitivity To Red Light Reduced (SRR1) 573
sequestered areas of phytochrome (SAP) 139
serine threonine kinase 224
shade avoidance 123, 478, 479, 481, 482, 483, 484, 485, 486, 487, 489, 491, 492, 493, 494, 496, 497, 629, 630, 631, 632, 638
shade avoidance response sensitivity 629
shade tolerance 123, 124
short day plant 582
side branch formation 551, 556, 558
side branches of protonemal filaments 539
signal transduction 329, 331, 332, 333, 334
signal transduction mutants 36
signalling 329, 330, 331, 332, 333, 334
signalling networks 465, 466, 468, 473
signalosome 358, 360, 364, 366, 369, 370, 371, 373, 374, 375, 376, 377, 378
Sinapis alba L. 8, 11
SKP1 359, 372
slave circadian oscillator 602
SOC1 (Suppressor of Overexpression of CO 1) 609
solar radiation 477
solar tracking 171, 172, 174, 191
Sorghum bicolor 224
soybean 54, 583, 600
SPA1 365, 376, 378
speckles 158, 159, 161, 162, 163, 164, 165, 166
spectral sensitivity 13, 15
spectrophotometric assay 7
spectroscopy 155, 156
sporangium 537

spore germination 518, 532, 533, 539
spores 515, 516, 523, 524, 532, 533, 534, 536
sporophyte 537, 538, 539
SPY 443
steady-state transcript levels 26
STO 363
stomata 249
stomatal opening 171, 173, 174, 184, 185, 186, 187, 189, 191, 194, 195, 196, 223, 227, 248, 250, 575, 582, 584
structure 99, 100, 101, 102, 103, 104, 105, 106, 107, 108, 109, 110, 112, 114, 115, 116, 117, 118, 119, 121, 122, 123, 125
SUB1 395, 398, 402
sunscreen 280, 285, 286, 297
Suppressor of phyA 1 (SPA1) 594
suprachiasmatic nucleus 209
Synechococcus elongatus PCC7942 70
synergism 413, 415, 416, 417, 420, 423, 431, 432, 433

Tanada effect 17
T-DNA insertional mutant 30
Tet repressor 635
tetrapyrroles 500, 510, 512
threshold fluence rate 15
time for coffee (*tic*) 588
TIMELESS 216, 218
timing of cab expression 1 (toc1) 570
tobacco 631, 638
TOC1 341, 343, 351, 352, 354, 355
tomato 103, 104, 109, 110, 121, 125, 126, 127, 632, 633, 634, 635, 637, 638
Triticum aestivum 224
truss length 632
two-component system 37
type I phytochrome 99, 100, 101, 102, 103, 105, 106, 107, 108, 112, 113, 114, 115, 117, 118, 119, 120, 121, 122, 125, 126, 127, 128, 129
type II phytochrome 99, 100, 101, 102, 103, 105, 106, 107, 108, 112, 113, 114, 115, 117, 118, 119, 120, 121, 122, 125, 126, 127, 128, 129

ubiquitin 136, 137, 138, 139, 140, 143, 151, 153, 157, 170, 358, 359, 361, 363, 364, 365, 366, 367, 368, 370, 371, 372, 373, 374, 375, 376, 377, 378
uli mutants (*U*V-B *l*ight *i*nsensitive) 288
ULI3 286, 288
UV-A (320-400 nm) 279
UV-B (280-320 nm) 279
UV-B effects 280, 289, 292, 299

UV-B tolerance 280, 281, 284, 286, 294
UV-B 299, 280, 281, 282, 283, 284, 285, 287, 288, 289, 290, 291, 292, 293, 294, 295, 296, 297, 298, 299, 300, 301, 302, 303, 304
UV-C (<280 nm) 279
UV-damaged DNA binding protein (DDB1) 287

vernalisation 606, 619, 622, 623, 624
Very Low Fluence Response (VLFR) 10, 18, 141, 142, 149, 161
Vicia faba 185, 187, 194, 195
vitamin A 632

WD-40 domain 361

yeast two-hybrid screens 336, 340, 344, 364

Zea mays 224
Zeitgeber Time (ZT) 578
ZEITLUPE (ZTL) 593
ZTL/ADO family 223, 24